Interlimb Coordination

Interlimb Coordination: Neural, Dynamical, and Cognitive Constraints

Edited by

Stephan P. Swinnen
*Laboratorium Motorische Controle
Departement Kinantropologie
Katholieke Universiteit Leuven
Leuven, Belgium*

H. Heuer
*Institut für Arbeitsphysiologie
Universität Dortmund
Dortmund, Germany*

Jean Massion
*Laboratoire de Neurosciences
  Fonctionelles
Centre National de la Recherche
  Scientifique
Marseilles, France*

P. Casaer
*Ontwikkelingsneurologische
  Onderzoekseenheid
Gasthuisberg
Departement Ontwikkelingsbiologie
Katholieke Universiteit Leuven
Leuven, Belgium*

ACADEMIC PRESS, INC.
A Division of Harcourt Brace & Company
San Diego   New York   Boston   London   Sydney   Tokyo   Toronto

This book is printed on acid-free paper. ∞

Copyright © 1994 by ACADEMIC PRESS, INC.
All Rights Reserved.
No part of this publication may be reproduced or transmitted in any form or by any means, electronic or mechanical, including photocopy, recording, or any information storage and retrieval system, without permission in writing from the publisher.

Academic Press, Inc.
525 B Street, Suite 1900, San Diego, California 92101-4495

United Kingdom Edition published by
Academic Press Limited
24–28 Oval Road, London NW1 7DX

Library of Congress Cataloging-in-Publication Data

Interlimb coordination : neural, dynamical, and cognitive constraints / edited by S. P. Swinnen ... [et al.].
    p.   cm.
   Includes bibliographical references and index.
   ISBN 0-12-679270-4
   1. Animal locomotion. 2. Motor ability. I. Swinnen, S.P. (Stephen P.)
    [DNLM: 1. Movement--physiology. 2. Psychomotor Performance. WE 103 I542 1993]
QP301.I533   1993
152.3--dc20
DNLM/DLC
for Library of Congress   93-30641
                                                                                        CIP

PRINTED IN THE UNITED STATES OF AMERICA
93 94 95 96 97 98   MM   9 8 7 6 5 4 3 2 1

Contents

Contributors xv
Preface xix
Acknowledgments xxi

1 Topics on Interlimb Coordination
Stephan P. Swinnen, Jean Massion, and H. Heuer

  I. Introduction 2
  II. Exploring the Neural Basis of Interlimb Coordination 4
  III. In Search of the Equations of Motion Underlying Interlimb Coordination: Interlimb Dynamics 14
  IV. Modulation of Coordination Patterns through Practice and Experience 19
  V. Conclusion 24
     References 25

Part I
The Neural Basis of Interlimb Coordination

2 Multisegmental Control of Axial and Limb Muscles by Single Long Descending Motor Tract Axons
Lateral versus Medial Descending Motor Systems
Yoshikazu Shinoda, Shinji Kakei, and Yuriko Sugiuchi

  I. Introduction 32
  II. Lateral versus Medial Long Descending Motor Tract Systems 33

III. Morphological Features of Single Axons in the Lateral and Medial Long Descending Motor Tract Systems 36
IV. Summary 44
References 45

3 Sensory–Motor Coordination in Crustacean Limbs during Locomotion
Daniel Cattaert, Jean-Yves Barthe, and François Clarac

I. Introduction 49
II. Review of the Central and Peripheral Mechanisms Involved in Coordination 53
III. Examples of Coordination in the Crayfish Locomotor System 62
IV. Conclusion 68
References 71

4 Interlimb Coordination during Locomotor Activities
Spinal-Intact Cats and Chronic Cats with Horizontal and Longitudinal Separation of the Spinal Cord
Masamichi Kato

I. Introduction 76
II. Normal Spinal-Intact Cat 77
III. Hemisected Chronic Cats 79
IV. Transsected Spinal Cord 84
V. Double Hemisected Spinal Cord 85
VI. Longitudinally Split Lumbar Cord 86
VII. Chronically "Isolated" Lumbar Half Spinal Cord 88
VIII. Summary 93
References 94

5 Interlimb Reflexes during Gait in Cats and Humans
Jacques Duysens and Toine Tax

I. Introduction 97
II. Crossed Reflexes in Motionless Cats 99
III. Ipsilateral Reflex Reversal in the Intact Cat 101
IV. Phase-Dependent Reversal of Crossed Reflexes 104
V. Pathways and Mechanisms Involved in Phase-Dependent Modulation of Exteroceptive Reflexes 105
VI. Fore- and Hindlimb Coordination and Reflexes during Cat Locomotion 108

VII. Flexibility and Mechanisms of Interlimb Coordination in the Cat   109
VIII. Interlimb Coordination during Human Gait   111
IX. Crossed Reflexes and Their Modulation in Humans   112
X. Conclusions   118
References   119

6 Human Locomotor Control, the Ia Autogenic Spinal Pathway, and Interlimb Modulations
J. D. Brooke, D. F. Collins, and W. E. McIlroy

I. Introduction: An Orientation to Human Bipedality and a Focus on the H (Autogenic Ia Spinal) Pathway   128
II. Review of Experimental Method   130
III. The H Pathway during Natural Movements   132
IV. Bilateral and Ipsilateral Movement Features of H Pathway Modulation   134
V. Contralateral Movement Features and H Pathway Modulation   136
VI. Supraspinal Modulation of the H Pathway   140
VII. Interlimb Spinal Path Modulation as Part of the Motoneuronal Convergence for Locomotion   141
References   143

7 Head and Body Coordination during Locomotion and Complex Movements
A. Berthoz and T. Pozzo

I. Introduction: Reference Frames   147
II. The Vestibular System: A Goal-Dependent, Euclidian, Egocentric Reference Frame   149
III. Experimental Studies of Head Stabilization during Complex Movements   151
References   163

8 Neuronal Basis of Stance Regulation
Interlimb Coordination and Antigravity Receptor Function
Volker Dietz

I. Introduction   167
II. Interlimb Coordination   170
III. Influence of Gravity   171

IV. Conclusions 176
References 176

9 Are There Unifying Structures in the Brain Responsible for Interlimb Coordination?
Mario Wiesendanger, Urs Wicki, and Eric Rouiller

I. Introduction 180
II. Cortical Areas of the Human Brain Involved in Bimanual Coordination 182
III. The Role of Sensory Guidance for Bimanual Coordination 188
IV. Some Contributions from Experiments in Subhuman Primates on the Problem of Bimanual Coordination 189
V. The Role of the SMA in Monkeys Tested in a Natural and a Highly Coordinated Bimanual Movement Sequence 193
VI. General Conclusion 201
References 203

10 The Role of the Corpus Callosum and Bilaterally Distributed Motor Pathways in the Synchronization of Bilateral Upper-Limb Responses to Lateralized Light Stimuli
Giovanni Berlucchi, Salvatore Aglioti, and Giancarlo Tassinari

I. How Does the Brain Time Motor Commands for Concurrent Actions of Different Limbs? 210
II. Lateral Reaction Time Differences in Simple Visuomotor RT 211
III. Crossed–Uncrossed Differences in Bilateral Responses to Lateralized Light Stimuli 215
IV. Response to Lateralized Flash with Movements Controlled by Bilaterally Distributed Motor Systems 220
V. Summary and Conclusions 224
References 225

11 Coordination of Cyclic Coupled Movements of Hand and Foot in Normal Subjects and on the Healthy Side of Hemiplegic Patients
Fausto Baldissera, Paolo Cavallari, and Luigi Tesio

I. Introduction and Methods 230
II. Characteristics of the "Easy" and "Difficult" Coupling 231
III. Kinesthetic Control of Coupled Movements 232

    IV. Movement Coupling on the "Healthy" Side of
        Hemiplegic Patients   236
    V. Concluding Remarks   238
        References   241

12 Bimanual Interference in a Deafferented Patient and Control Subjects
Normand Teasdale, Chantal Bard, Michelle Fleury, Jacques Paillard, Robert Forget, and Yves Lamarre

    I. Introduction   244
   II. Methods   245
  III. Results   246
   IV. Discussion   255
       References   257

13 Changes in Strength, Speed, and Reaction Time Induced by Simultaneous Bilateral Muscular Activity
T. Ohtsuki

     I. Introduction   259
    II. Maximum Muscle Strength   260
   III. Submaximal Isometric Muscle Strength   264
    IV. Speed   265
     V. Reaction Time   266
    VI. Effects of Training on Bilateral Performance   267
   VII. Exploring the Reasons Underlying the Bilateral Deficit   269
  VIII. Conclusions   272
       References   273

Part II
The Dynamics of Interlimb Coordination

14 A Low-Dimensional Nonlinear Dynamic Governing Interlimb Rhythmic Coordination
M. T. Turvey and R. C. Schmidt

   I. Introduction   278
  II. Experimental Results on 1:1 Frequency Locking   287

   III. Expectations from the Low-Dimensional Dynamic about 2:1 Frequency Locking   293
   IV. Conclusions   297
       References   298

15 Elementary Coordination Dynamics
J. A. S. Kelso

    I. Prolegomenon   301
   II. Introduction   301
  III. The Theoretical Strategy   303
  IV. Phenomena to Be Explained   304
   V. Range of Application   305
  VI. Essential Ingredients of Coordination Dynamics   307
 VII. Summary   316
      References   316

16 The Dynamical Substructure of Bimanual Coordination
Richard G. Carson, Winston D. Byblow, and David Goodman

    I. Introduction   319
   II. Informational and Mechanical Anchoring   321
  III. An Experimental Analysis of Anchoring   322
  IV. Time Scales Relations   327
   V. Asymmetries in Bimanual Coordination   329
  VI. A Preliminary Model   333
      References   335

17 From Interlimb Coordination to Trajectory Formation
Common Dynamical Principles
Gregor Schöner

    I. Introduction   339
   II. Dynamic Theory of Interlimb Coordination   342
  III. Toward a Dynamic Theory of Trajectory Formation   348
  IV. Discussion   364
      References   365

Part III
Modulation of Coordination Patterns through Practice and Experience

18 The Development of Sensorimotor Integration Underlying Posture Control in Infants during the Transition to Independent Stance
Marjorie Hines Woollacott and Heidi Sveistrup

    I. Introduction 372
   II. Posture Control of the Head and Trunk 373
  III. The Transition to Independent Stance 378
  IV. Conclusions 386
      References 388

19 The Development of Bipedal Interlimb Coordination
Jill Whitall and Jane E. Clark

    I. The Emergence of Walking 395
   II. The Role of Intralimb Coordination 402
  III. The Development of Walking 404
  IV. Later Emerging Gait Forms: Running and Galloping 405
   V. Summary 408
      References 408

20 Shifting Patterns of Interlimb Coordination in Infants' Reaching
A Case Study
Daniela Corbetta and Esther Thelen

    I. Introduction 413
   II. Part 1: The Underlying Dynamics of Reaching 417
  III. Part 2: The Transition from Bimanual to Unimanual Reaching 427
  IV. General Discussion 434
      References 436

21 Manual Strategies and Interlimb Coordination during Reaching, Grasping, and Manipulating throughout the First Year of Life
Jacqueline Fagard

  I. Strategies at the Onset of Funtional Reaching   442
 II. From Early Object Exploration to Hand-Role Differentiation in Object Manipulation   448
III. Conclusion   456
    References   458

22 The Coordination Dynamics of Learning
Theoretical Structure and Experimental Agenda
Pier-Giorgio Zanone and J. A. S. Kelso

  I. Introduction   461
 II. Theoretical Structure   465
III. Initial Tests of Predictions: Transitions and Transfer   470
 IV. Experimental Agenda   478
  V. Conclusions   485
    References   488

23 The Formation and Dissolution of "Bad Habits" during the Acquisition of Coordination Skills
Charles B. Walter and Stephan P. Swinnen

  I. Introduction   492
 II. The Nature of Task-Specific Bias (Bad Habits)   492
III. Sources of the Formation of Bad Motor Habits   496
 IV. Overcoming Systematic Coordinative Bias: Dissolving Bad Habits   502
  V. Task Differences and Individual Differences   508
 VI. Concluding Remarks   509
    References   510

24 Learning to Coordinate Redundant Biomechanical Degrees of Freedom
Karl M. Newell and P. Vernon McDonald

  I. Introduction   515
 II. Degrees of Freedom and Changes in Coordination Mode   517

III. Order to the Changes in the Coordination Mode   518
IV. Coordination Mode and the Perceptual–Motor Workspace   525
V. Search Strategies, the Perceptual–Motor Workspace, and the Acquisition of Coordination   527
VI. Concluding Comments   533
References   534

25 Constraints and Coordination in Whole-Body Actions
Wynne Ashley Lee and Aileen Marie Russo

I. Introduction   537
II. Modeling Coordination in Whole-Body Actions   539
III. Coordination during Pulls Made while Standing   546
IV. Conclusions   565
Appendix   566
References   567

26 Coordinating the Two Hands in Polyrhythmic Tapping
Jeffery J. Summers and Jeff Pressing

I. Introduction   571
II. Constraints on Bimanual Performance   572
III. The Production of Polyrhythms   575
IV. A Model of Polyrhythmic Tapping   589
V. Conclusion   590
References   591

27 Does Handedness Play a Role in the Coordination of Bimanual Movement?
Michael Peters

I. Introduction   595
II. The Role of Handedness as Inferred through Observations of Naturally Occurring Bimanual Behaviors   598
III. The Role of Handedness in Experiments That Examine Bimanual Coordination   603
IV. Summary and Conclusions   609
References   612

28 Temporal Organization of the Prehension Components in a Bimanual Task
Umberto Castiello and George E. Stelmach

I. Introduction   618
II. Discussion   626
   References   631

Index   633

Contributors

Numbers in parentheses indicate the pages on which the authors' contributions begin.

Salvatore Aglioti (209), Institute of Human Physiology, Medical School, University of Verona, 41-37134 Verona, Italy

Fausto Baldissera (229), Istituto di Fisiologia Umana II, Università di Milano, I-20133 Milan, Italy

Chantal Bard (243), Laboratoire de Performance Motrice Humaine, Université Laval, Québec, Québec, Canada G1K 7P4

Jean-Yves Barthe[1] (49), Laboratoire de Neurobiologie et Mouvements, Centre National de la Recherche Scientifique, 13402 Marseilles, France

Giovanni Berlucchi (209), Institute of Human Physiology, Medical School, University of Verona, 41-37134 Verona, Italy

A. Berthoz (147), Laboratoire de Physiologie Neurosensorielle, Centre National de la Recherche Scientifique, F-75270 Paris, France

J. D. Brooke (127), Human Neurophysiology Laboratory, School of Human Biology, University of Guelph, Guelph, Ontario, Canada N1G 2W1

Winston D. Byblow (319), Human Motor Systems Laboratory, School of Kinesiology, Simon Fraser University, Burnaby, British Columbia, Canada V5A 1S6

Richard G. Carson[2] (319), Human Motor Systems Laboratory, School of Kinesiology, Simon Fraser University, Burnaby, British Columbia, Canada V5A 1S6

Umberto Castiello[3] (617), Department of Exercise Science and Physical Education, Arizona State University, Tempe, Arizona 85287

Daniel Cattaert (49), Laboratoire de Neurobiologie et Mouvements, Centre National de la Recherche Scientifique, 13402 Marseilles, France

---

[1] *Present address:* The Nobel Institute for Neurophysiology, Karolinska Institutet, Stockholm, Sweden.

[2] *Present address:* Department of Human Movement Studies, University of Queensland, Brisbane, Queensland, Australia 4072.

[3] *Present address:* Dipartimento di Psicologia, Universita di Bologna, 40127 Bologna, Italy.

Paolo Cavallari (229), Istituto di Fisiologia Umana II, Università di Milano, I-20133 Milan, Italy

François Clarac (49), Laboratoire de Neurobiologie et Mouvements, Centre National de la Recherche Scientifique,13402 Marseilles, France

Jane E. Clark (391), Department of Kinesiology, University of Maryland, College Park, Maryland 20742

D. F. Collins[4] (127), Human Neurophysiology Laboratory, School of Human Biology, University of Guelph, Guelph, Ontario, Canada N1G 2W1

Daniela Corbetta (413), Department of Psychology, Indiana University, Bloomington, Indiana 47405

Volker Dietz (167), University Hospital Balgrist, Swiss Paraplegic Centre, CH-8008 Zurich, Switzerland

Jacques Duysens (97), Department of Medical Physics and Biophysics, Catholic University of Nijmegen, 6500 HB Nijmegen, The Netherlands

Jacqueline Fagard (441), Laboratoire de Psycho-Biologie du Développement, 75005 Paris, France

Michelle Fleury (243), Laboratoire de Performance Motrice Humaine, Université Laval, Québec, Québec, Canada G1K 7P4

Robert Forget (243), Ecole de Réadaptation, Université de Montréal, Montréal, Québec, Canada H3C 3J7

David Goodman (319), Human Motor Systems Laboratory, School of Kinesiology, Simon Fraser University, Burnaby, British Columbia, Canada V5A 1S6

H. Heuer (1), Institut für Arbeitsphysiologie, Universität Dortmund, 4600 Dortmund, Germany

Shinji Kakei (31), Department of Physiology, School of Medicine, Tokyo Medical and Dental University, Tokyo 113, Japan

Masamichi Kato (75), Department of Physiology, Hokkaido University, School of Medicine, Sapporo 060, Japan

J. A. S. Kelso (301, 461), Program in Complex Systems and Brain Sciences, Center for Complex Systems, Florida Atlantic University, Boca Raton, Florida 33431

Yves Lamarre (243), Centre de Recherche en Sciences Neurologiques, Université de Montréal, Montréal, Québec, Canada H3C 3J7

Wynne Ashley Lee (537), Programs in Physical Therapy and the Institute for Neuroscience, Northwestern University Medical School, Chicago, Illinois 60611

Jean Massion (1), Laboratoire de Neurosciences Fonctionelles, Centre National de la Recherche Scientifique, 13402 Marseilles, France

[4]*Present address:* Department of Physiology, University of Alberta, Edmonton, Alberta, Canada T6G 2E1.

P. Vernon McDonald[5] (515), Department of Kinesiology, University of Illinois at Urbana-Champaign, Urbana, Illinois 61801

W. E. McIlroy[6] (127), Human Neurophysiology Laboratory, School of Human Biology, University of Guelph, Guelph, Ontario, Canada N1G 2W1

Karl M. Newell (515), College of Health and Human Development, Pennsylvania State University, State College, Pennsylvania 16802

T. Ohtsuki (259), University of Human Movement, Department of Physical Education, Nara Women's University, Nara City, Japan 630

Jacques Paillard (243), Unité de Neurosciences Fonctionnelles, Centre National de la Recherche Scientifique, 13402 Marseilles, France

Michael Peters (595), Department of Psychology, University of Guelph, Guelph, Ontario, Canada N1G 2W1

T. Pozzo (147), Groupe d'Analyse du Mouvement, Université de Bourgoque, 21004 Dijon, France

Jeff Pressing (571), Department of Music, La Trobe University, Bundoora, Victoria 3083, Australia

Eric Rouiller (179), Institute of Physiologie, University of Fribourg, CH-1700 Fribourg, Switzerland

Aileen Marie Russo (537), Programs in Physical Therapy, Northwestern University Medical School, Chicago, Illinois 60611

R. C. Schmidt (277), Department of Psychology, Tulane University, New Orleans, Louisiana 70118, and Center for the Ecological Study of Perception and Action, University of Connecticut, Storrs, Connecticut 06218

Gregor Schöner (339), Institut für Neuroinformatik, Ruhr-Universität Bochum, D-4630 Bochum, Germany

Yoshikazu Shinoda (31), Department of Physiology, School of Medicine, Tokyo Medical and Dental University, Tokyo 113, Japan

George E. Stelmach (617), Department of Exercise Science and Physical Education, Arizona State University, Tempe, Arizona 85287

Yuriko Sugiuchi (31), Department of Physiology, School of Medicine, Tokyo Medical and Dental University, Tokyo 113, Japan

Jeffery J. Summers (571), Department of Psychology, University of Southern Queensland, Toowoomba, Queensland 4350, Australia

Heidi Sveistrup (371), Department of Exercise and Movement Science, University of Oregon, Eugene, Oregon 97403

---

[5] *Present address:* Krug Life Sciences, Houston, Texas 77058.
[6] *Present address:* Department of Physiology, University of Toronto, Toronto, Ontario, Canada M5S 1A1.

Stephan P. Swinnen (1, 491), Laboratorium Motorische Controle, Departement Kinantropologie, Katholieke Universiteit Leuven, 3001 Leuven, Belgium

Giancarlo Tassinari (209), Institute of Human Physiology, Medical School, University of Verona, I-37134 Verona, Italy

Toine Tax (97), Department of Medical Physics and Biophysics, Catholic University of Nijmegen, 6500 HB Nijmegen, The Netherlands

Normand Teasdale (243), Laboratoire de Performance Motrice Humaine, Université Laval, Québec, Québec, Canada G1K 7P4

Luigi Tesio (229), Servizio di Fisiatria, Clinica Ortopedica H. S. Raffaele, I-20132 Milan, Italy

Esther Thelen (413), Department of Psychology, Indiana University, Bloomington, Indiana 47405

M. T. Turvey (277), Center for the Ecological Study of Perception and Action, University of Connecticut, Storrs, Connecticut 06268, and Haskins Laboratories, New Haven, Connecticut 06511

Charles B. Walter (491), Department of Kinesiology, University of Illinois at Chicago, Chicago, Illinois 60680

Jill Whitall (391), Department of Kinesiology, University of Wisconsin–Madison, Madison, Wisconsin 53706

Urs Wicki (179), Institute of Physiologie, University of Fribourg, CH-1700 Fribourg, Switzerland

Mario Wiesendanger (179), Institute of Physiologie, University of Fribourg, CH-1700 Fribourg, Switzerland

Marjorie Hines Woollacott (371), Department of Exercise and Movement Science, University of Oregon, Eugene, Oregon 97403

Pier-Giorgio Zanone (461), Program in Complex Systems and Brain Sciences, Center for Complex Systems, Florida Atlantic University, Boca Raton, Florida 33431

Preface

Because of technical and conceptual developments, interest in research on human motor control, in particular the coordination of limb movements, has increased tremendously in the past ten years. This book is intended to help synthesize the rapidly expanding knowledge base on interlimb coordination. It contains a set of introductory chapters written by leading experts in their respective areas of specialization. The chapters take the reader on a journey that starts with the description of the elementary networks and pathways responsible for the production of rhythmic movements and evolves into an analysis of highly complex forms of interlimb coordination requiring tremendous amounts of practice. It is of interest to a broad spectrum of graduate students and researchers educated in the neurosciences, neurobiology, experimental and developmental psychology, kinesiology, and physical therapy. It is particularly directed at those students of motor control who wish to look beyond the borders of their own area of specialization and to expand their background.

The chapters are predominantly introductory in nature to assist readers who are not yet familiar with the various approaches to interlimb coordination currently being explored. Contributions from different subfields of science are discussed. These subfields represent different levels of analysis. Accordingly, an interdisciplinary approach is promoted. It becomes apparent that coordination emerges as a result of a complex interplay between the central nervous system, with its particular design features, and the biophysical laws of nature that pertain to the movement of levers or limb segments in a gravitational field.

This book is divided into three major parts. The allocation of chapters to these parts is not always straightforward, because some chapters fit in all three. Chapter 1 intends to bring coherence to the variety of approaches that characterize the study of interlimb coordination. In addition, it highlights the individual chapter contributions against the background of past work and elucidates their innovative features. Finally, it points to lines of convergence in the study of interlimb coordination that cross the borders of the various subfields. Chapters 2 through 13 deal with the neural basis of interlimb coordination: The first series is directed at locomotory behavior in animal and human species; the second series focuses on upper-limb control. Chapters 14 through 17 represent an introduction to the dynamical approach to interlimb coordination. Chapters 18 through 28 predominantly focus on the modulation of coordination patterns through development and learning. The enormous flexibility and adaptability that even small children display as they grow older can only humble roboticists who try to endow machines with these capabilities.

Stephan P. Swinnen

Acknowledgments

This volume grew out of a meeting entitled "The Control and Modulation of Patterns of Interlimb Coordination: A Multidisciplinary Perspective," which was held at the Catholic University of Leuven, Belgium. Generous sponsorship was provided by the Human Frontier Science Program Organization (HFSPO), which is gratefully acknowledged. In particular, I would like to thank A. Zickler, director of HFSPO fellowships and workshops, for his encouragement and support. The meeting was attended by a sampling of researchers from various areas of specialization. This illustrates that the study of the mysteries underlying interlimb coordination is not monopolized by a single subdivision of science.

The editors are grateful to the authors and co-authors of the various chapters, not only for their continuous efforts to write a text that is accessible to a broad readership, but also for their assistance in peer review. In addition, we are indebted to the following external reviewers who greatly contributed to improving the quality of the contributions: P. Crenna, S. Rossignol, K. Pearson, H. Cruse, H. J. Freund, L. Deecke, J. Armand, P. Augaut, D. Bourbonnais, J. Macpherson, B. Amblard, C. Assaiante, G. Savelsbergh, D. Sherwood, P. Beek, J. Smeets, A. Wing, S. Wallace, W. Spijkers, and T. Lee.

I am particularly indebted to Jean Massion, Herbert Heuer, and Paul Casaer for their gracious help in the organization of the workshop and in the editing of this book. They provided a great deal of expertise and assistance on the scientific and technical aspects of this endeavor. Advice and help on scientific and material matters with respect to the organization of this meeting were provided by Deborah Serrien, scientific collaborator (C. U., Leuven) and Tim Lee (McMaster University, Hamilton, Canada), who was appointed guest professor from August 1991 to July 1992 at C. U., Leuven. Their daily encouragement was invaluable for the smooth organization of this meeting. My thanks are extended to J. Pauwels, M. Buekers,

W. Helsen, I. Wuyts, C. Jardin, E. Vannecke, and P. Rowe for their support during the conference.

Finally, I owe a word of thanks to those who live in my immediate orbit. Editing a book is a major enterprise that takes time away from the family. I often experienced my family as "attracting" as well as "repelling." When I spent time with them, I felt a strong desire to close myself off in another room to work on the manuscript. But when disconnected from my family for a long time, I wanted to be with them again. I would like to thank my wife Lidy for taking over many of the family responsibilities, and my children, Ruben, Evelyne, and little Nathalie, for their patience and understanding in the face of frequent neglect. Evidently, Nathalie did not need my help to make major progress in crawling.

To all those I previously referred to, and those I most likely forgot, I express my sincere gratitude.

1

Topics on Interlimb Coordination

Stephan P. Swinnen

Laboratorium Motorische Controle
Departement Kinantropologie
Katholieke Universiteit Leuven
Leuven, Belgium

Jean Massion

Laboratoire de Neurosciences Fonctionelles
Centre National de la Recherche Scientifique
Marseille, France

H. Heuer

Institut für Arbeitsphysiologie
Universität Dortmund
Dortmund, Germany

  I. Introduction
 II. Exploring the Neural Basis of Interlimb Coordination
     A. The Neural Control of Locomotion
     B. The Neural Control of Posture
     C. Coordination of Head, Body, and Limbs
     D. The Neural Control of Goal-Directed Arm Movements
     E. Bimanual Coupling and Decoupling: Central and Peripheral Contributions
III. In Search of the Equations of Motion Underlying Interlimb Coordination: Interlimb Dynamics
 IV. Modulation of Coordination Patterns through Practice and Experience
     A. Development of Coordination Patterns
     B. Acquisition of Coordination Patterns
  V. Conclusion
     References

I. Introduction

Humans are capable of coordinating various limbs and body parts with each other, for example, the left and right hands or thumbs, the hand and foot, and the head and arm. A remarkable spatiotemporal coordination is evident in spite of the large differences in inertial characteristics of the effectors involved. This observation suggests the existence of some basic coordination principles that apply across widely different cooperative ensembles. Underlying this well-organized global behavior is the coordination of subcomponents at various levels of the movement apparatus: intrajoint, intralimb, and interlimb. At the level of the individual joint, coordination between muscles acting on a common joint is required, such as the interplay between agonists and antagonists. Within a limb, the various joints and muscles, acting on one or more of these joints, must be properly organized to function efficiently. Finally, interlimb coordination is necessary to perform the most essential animal functions like walking, swimming, and feeding.

Whereas particular expressions of interlimb coordination such as locomotion have been investigated intensively in the neurosciences during the past 30 years, the interest in coordination within the behavioral sciences is relatively recent. Due to the development of new movement registration technologies, increased computational power, and the search for new links with the neurosciences and biophysics, the way has been made free for the study of more complex motor behaviors. This is an important development, since the capability to coordinate our limbs is at the heart of everyday life.

Two scientists, who were already actively involved in interlimb coordination research more than half a century ago, can be considered pioneers in this field. First, we owe a great deal to the Russian physiologist and movement specialist Bernstein (1967) who was particularly interested in studying complex motor acts. He was mainly struck by the observation that the movement apparatus, with such a tremendous degree of multilayered complexity, can accomplish goal-directed behavior so effortlessly. This came to be known as the degrees-of-freedom problem. Second, the German behavioral physiologist Von Holst collected miles of data on the coordination of fin movements in decapitated fish (*Labrus*). He divided the observed coordination patterns into two major categories: absolute and relative coordination. Absolute coordination is characterized by the maintenance of a fixed-phase relation and by frequency synchronization of the fin movements. Relative coordination refers to a larger group of coordination patterns characterized by less stringent coupling modes, that is, the component activities are neither completely independent of one another nor linked in a fixed mutual relationship. Whereas this distinction is theoretically relevant,

Von Holst remarked that both types are often observed intermittently in fish preparations. In addition, he derived two basic principles that pertained to these coordination modes. On the one hand, he observed a tendency for each fin pattern to maintain its own frequency, referred to as the maintenance tendency (*Beharrungstendenz*). On the other hand, a tendency for one fin pattern to impose its inherent frequency on the other fin was often evident. This form of cooperation or (mutual) attraction of the fin movements was referred to as the magnet effect (*Magneteffect*). The latter effect was often associated with the superposition effect that pertains to attraction between rhythmic units in the amplitude domain.

Von Holst argued that the magnet effect and the maintenance tendency are in mutual opposition: "If the former predominates, then there is continuous agreement in frequency under absolute coordination; if the latter predominates, there is relative coordination—the frequencies of the rhythms differ, and the dependent rhythm, under the magnet effect of the dominant rhythm, exhibits periodicity whose form is determined by the reciprocal frequency relationship and whose extent is governed by the intensity of the magnet effect" (Von Holst, 1973, p. 63).

Even though these principles were extracted from research on fin movements, they currently form a major source of inspiration for the study of human coordination (Kelso, Chapter 15, this volume; Turvey & Schmidt, Chapter 14, this volume). Today, many research laboratories across the world investigate these phenomena in a variety of different contexts. Others are more concerned with the study of discrete bimanual tasks in which the limbs assume differentiated roles to accomplish goal-directed behavior (Fagard, Chapter 21, this volume; Peters, Chapter 27, this volume; Walter & Swinnen, Chapter 23, this volume).

The present book consists of a series of introductory chapters, representing various levels of research and subdivisions of science that currently address interlimb coordination, for example, the neurosciences, the behavioral sciences, kinesiology, biomechanics, and dynamics. Even though each of these fields of science is characterized by a unique approach to the study of interlimb coordination, using its own techniques to acquire knowledge, all strive for a better understanding of how the human control system manages to organize the cooperation among the limbs. Neuroscientific approaches focus on the neuronal networks and pathways underlying rhythmic and discrete coordination patterns, in particular locomotion and bimanual coordination. Some chapter contributions concentrate on the identification of the locus of the central pattern generator underlying locomotion, whereas others are mainly concerned with the reflex modulation of these patterns as a result of sensory information. Scientists advocating a dynamical approach seek to uncover the equations of motion that govern movement coordination. They attempt to identify the dynamic states at which moving animals converge when provided enough time to settle down. Finally, some scientists are mainly concerned with a better understanding of goal-directed motor behavior and the changes in coordination that occur as a result of development and learning, that is, the modulation or overcoming of preexisting/preferred coordination modes with the goal of expanding the behavioral repertoire. Those who have a strong link with cognitive psychology direct their attention to a better understanding of the nature of the central representation underlying complex coordination and the movement features it comprises.

II. Exploring the Neural Basis of Interlimb Coordination

A. The Neural Control of Locomotion

One of the most intensively studied examples of interlimb coordination in the field of neuroscience is animal and human locomotion. Since Sherrington, three areas of interest have dominated experimental studies on locomotion: (1) the role of reflexes in locomotion; (2) the capability of the spinal cord to generate intrinsic rhythms; and (3) the control of the spinal cord by higher centers. At one time or another, attention has mainly been directed at one of these mechanisms for motor control. More recent studies have concentrated on the synthesis of these mechanisms into a general framework for nervous control (Shepherd, 1988). This also typifies the chapters on locomotion in this volume. The contributions refer to the study of locomotion in invertebrates such as the crayfish (Cattaert et al., Chapter 3) and higher vertebrates, including those using a quadrupedal gait, such as the cat (Kato, Chapter 4), and a bipedal gait, such as the human (Brooke et al., Chapter 6; Duysens & Tax, Chapter 5).

Pioneering work on locomotion was conducted by Grillner (1975, 1981) and Shik and co-workers (Shik, Severin, & Orlovsky, 1966), who spent considerable efforts in demonstrating the existence of a relatively autonomous neural network, called the central pattern generator (CPG) (see also Cattaert et al., Chapter 3, this volume; Duysens & Tax, Chapter 5, this volume). This confirmed earlier ideas put forward by T. Graham Brown in 1911, who demonstrated that the rhythmic alternation between flexion and extension is not reflex in origin but is generated by neurons located in the spinal cord. CPGs have been demonstrated in most locomotory networks found in invertebrates and vertebrates. The rhythm production generally results from both membrane properties of neurons and particular network connections.

Even though these patterns of interlimb coordination can be observed in the absence of afferent information, this should not be taken to imply that afference plays a minor role in normal locomotion. Brown (1911) was well aware of this when he suggested that afferent input was probably important in grading the component movements to the specific environmental contingencies. Since locomotion is a highly automated type of motor behavior, it is not surprising that interlimb reflexes have evolved to support the coordination of the limb movements during gait and to modulate the basic patterns during unexpected perturbations (Duysens & Tax, Chapter 5, this volume).

In Chapter 4 of this volume, Kato reviews his experimenal contributions of the past 15 years, which deal with locomotor coordination after horizontal and longitudinal separation of the spinal cord in spinal intact cats and spinal lesioned cats. Lateral hemisection of the spinal cord was carried out in order to disrupt descending and ascending long tracts unilaterally. These hemisected preparations do not show any differences in step length or in step time when compared with normal control cats. However, they do demonstrate evidence for less accurate foot placement responses when walking on grid surfaces. According to Kato, this suggests that interlimb reflex pathways serve to coordinate the spatial aspects of locomotion in quadrupedal gait. On the other hand, spinal transsection or double hemisection results in a disruption of the phase relations between the fore- and hindlimbs, and this points to the importance of descending signals from brainstem locomotor centers for achieving coordination among the limbs.

Some chapters specifically deal with the reflex modulation of locomotory activities. Cattaert and co-workers have investigated two locomotor systems in the crayfish (Crustacea): (1) swimming, accomplished by four pairs of paddles; and (2) walking, by means of four to five pairs of thoracic legs. They show some nice examples of sensory–motor interactions, which are analyzed at the cellular level. A comparison of both systems reveals the existence of similarities between their central commands. But, there exist essential differences as well, pertaining to the presence or absence of contact with rigid substrates, as is the case with the legs during walking. The control of ongoing movement during walking is strongly tied with the stance phase where force coding receptors are active. These determine the switch to the swing phase in the leg where the force coding receptors reside, as well as the onset of the stance phase of the ipsilateral legs. Thus, coordination in the actual walking system results from strong interactions between central and peripheral commands that allow for maximum adaptability to environmental demands (e.g., the quality of the terrain).

Brooke and Duysens and their respective co-workers concentrate their reviews on sensori-motor interactions during human walking. Brooke and co-workers have studied one of the most rapid autogenic pathways that is represented by the H-reflex. They show that rhythmic limb movements, as generated during pedaling, cause modulation of the spinal pathway of the Ia H-reflex. For soleus H-reflexes during symmetric pedaling, they demonstrate (1) that passive movement of the legs results in movement-induced inhibition of this spinal pathway; (2) that the extent of the reflex depression is positively related to the velocity of passive movement of the two legs; and finally (3) that the relationship to velocity is maintained when the legs move actively. During asymmetric coordination of movement of one leg while the other remains passive, the inhibition is retained in the leg being moved but also in the static contralateral limb, and this is again related to the velocity of passive movement of the opposite limb. They propose that the inhibition is transmitted through the spinal pathway and that it arises from the receptors generating movement afference.

Duysens and Tax (Chapter 5) review some slower acting pathways that may be involved in the production of corrective movements during locomotory events. Whether interlimb reflexes are in operation during normal locomotion or following perturbations, they subserve the important goals of minimizing instability and securing progression. The emerging corrective movements may involve limbs other than those upon which a particular perturbation or stimulus is evoked. Two examples clearly illustrate this effect. On one hand, prolongation of the swing phase of one leg results in a longer duration of the support phase in the contralateral leg. On the other hand, a shortening of the stance phase in one leg results in a shortening of the swing phase in the contralateral leg to take up the support function earlier in time. That stimuli- or perturbation-induced inter- and intralimb responses depend on the phase of the locomotory cycle is a true example of highly efficient and goal-dependent movement organization. Many examples of such phase-dependent responses in cat and human can be found in Chapter 5 by Duysens and Tax, who conclude that there are striking similarities in patterns of interlimb coordination among these species even though we are dealing here with quadrupedal versus bipedal walkers. Differences among animals and humans are to be sought in the greater dominance of supraspinal over spinal mechanisms in the latter (Dietz, 1992).

Even though CPGs are held responsible for interlimb coupling, Grillner (1985) has also underscored the existence of individual pattern generators for each limb (see also Cattaert *et al.,* Chapter 3, this volume). Activity in one limb can be observed in the absence of activity in the other limbs. Kato supports these findings in cat preparations where all the impulses from contralateral as well as supralumbar regions are cut off. The ipsilateral hindlimb shows a walking pattern, independent of the other three legs (Chapter 4). Moreover, Grillner (1985) has suggested that these neural networks cannot only be made responsible for complex coordinations but also for the production of more specific movements. Only parts of the total unit are then activated for the volitional control of more specific ankle or knee movements. Within this perspective, the production of new movements may then require "learning to combine and sequence specific fractions of the neuronal apparatus used to control the innate movement patterns in a novel way" (Grillner, 1985, p. 148).

Whereas it has often been assumed that a CPG consists of a well-defined assemblage of neurons functionally distinguishable from others, Meyrand, Simmers, and Moulins (1991) have recently argued that CPGs may not be considered *immutable functional entities*. Instead, they show in Crustacea that neurons from different circuits can be reconfigured into a new circuit that enables a different function. This selective dismantling of preexisting networks, giving rise to the recruitment of a different functional network, provides us with an important clue toward a better understanding of the mysterious but enormous flexibility in motor coordination that can be found across the animal world. Additional evidence is required to demonstrate similar phenomena in the vertebrate nervous system, which evidently has a much larger number of neurons.

In summary, the neural substrate underlying interlimb coordination is structured in such a way that flexibility and differentiated activity is possible whenever required. Afferent information plays a crucial role in interlimb coordination and it is at the basis of the phase-dependent modulation of patterns of muscle activity.

B. The Neural Control of Posture

Even though posture looks like a rather static activity at first sight, it is a true example of interlimb coordination. A major requirement for postural equilibrium is the maintenance of the center of gravity within the base of support. This process of balance control, or, of the regaining of balance, involves coordination of many muscles. Early studies already demonstrated that the production of focal arm movements is often preceded by bilateral leg activity in order to maintain balance. For example, Belenkii Gurfinkel and Paltsev (1967) and Paltsev and Elner (1967) investigated the interactions between arm lifting and lowering and postural activity. They found that activity in the m. deltoideus (responsible for the arm action) showed an increase in activity 130–140 msec after the sound signal, whereas the biceps femoris on the ipsilateral side was activated up to 40–50 msec before onset of deltoid activity. The biceps femoris on the contralateral side showed increased electrical activity up to 30–40 msec before its contralateral counterpart. These observations suggest that even small voluntary actions, limited to a particular body part, cause a complete reorganization of muscle activity in many body parts and this occurs with the appropriate timing of the events.

In Chapter 8, Dietz describes his work on the perturbation of stance, as applied to one or both legs in the same or in opposite directions. He underscores that a combination of afferent inputs provides the necessary information to control body equilibrium. He deals with some of the interactions among the various input sources. Dietz maintains that during bilateral leg displacements, the activity induced by the respective contralateral leg is linearly summed or subtracted, depending on leg displacement in the same or in opposite directions. Thus, following unilateral displacements, a bilateral activation takes place that results in cocontraction of the nondisplaced leg. This results in a more stable base from which to compensate for the perturbation. A displacement of the legs in opposite directions causes the body's center of gravity to fall between the legs, and therefore, less invasive compensatory responses are needed. The resulting coordination is thought to be mediated by a spinal mechanism, due to the observed short latencies, which is under supraspinal control. Moreover, Dietz hypothesizes that extensor load receptors (Golgi tendon organs) signal changes with respect to the projection of the body's center of mass in reference to the feet. This possibly represents a newly discovered function of these receptors in the regulation of stance and gait.

Lee and Russo address Bernstein's degree-of-freedom problem through an attempt to characterize the constraints that underlie interlimb coordination during force pulling with the arms while standing upright (Chapter 25). They distinguish two ways to model how the system's degrees of freedom may be constrained to achieve coordination. Local constraints relate to joint kinematic and kinetic variables or muscle variables. Global constraints relate to abstract goals of the system, such as maintaining or accelerating the center of mass to a particular location over the support base. They propose an analytical model that operationally defines global coordination in the task and that is partially supported by empirical work (except for very low force pulls).

In summary, what emerges from perturbation studies during locomotion and stance is that the corrective movements thus induced are not limited to the locus of the perturbation but consist of an overall response involving the coordination of many muscles and body parts within a kinematic chain. Interlimb reflexes serve to secure the principal goals of minimizing instability and, in the case of locomotion, of enabling progression. Dietz (1992) summarizes: "Irrespective of the conditions under which stance and gait are investigated, the neuronal pattern evoked during a particular task is always directed to hold the body's center of mass over the base of support" (p. 48). For extensive reviews on the neural control of posture and movement, we refer the refer the reader to Massion (1992) and Dietz (1992).

C. Coordination of Head, Body, and Limbs

Limb movements are dependent on or covary with head position, as becomes evident in labyrinthine reflexes and symmetric and asymmetric tonic neck reflexes. These reflexes often interact with each other during movement production. For example, head dorsiflexion induces extension of the forelimbs and flexion of the hindlimbs, whereas the opposite effect occurs during ventriflexion (symmetric tonic neck reflex). The neck reflexes can be observed in the newborn but gradually disappear during the first months of life. However, some have argued that such reflexes do not fully disappear and may show up during the production of movements in sports events or other voluntary activities (Keele, 1981). Fukuda (1961) captured many skilled performers on film and found various patterns of limb coordination that are congruent with those found in the tonic neck reflexes. In addition, Hellebrandt, Houtz, Partridge, and Walters (1956) underscored the role of reflexes during the production of forceful events and demonstrated that patterns in accordance with these reflexes augment work output. It is thus reasonable to assume that some patterns of coordination are built into the organism and become subsequently integrated into voluntary activities. Coaches are aware of the important role of head position during skill acquisition. Instructions often relate to particular head movements (look at the floor during a handstand) in order to promote these built in patterns of activity. Sometimes, these patterns are conducive to the new coordination form to be acquired; at other times, they may impose persistent errors in performance that need to be suppressed (Walter & Swinnen, Chapter 23, this volume). More will be said about the two faces of preexisting coordination modes later in this chapter.

In their chapter on head an body coordination, Berthoz and Pozzo argue that the head serves as an important frame of reference during multilimb coordination (Chapter 7). Depending on the type of activity to be executed, the head is stabilized intermittently under the control of gaze. This stabilization allows the head to serve as an inertial guidance platform for the control of multilimb movement. They infer this from the strong tendency of performers to stabilize the head with respect to the sagittal as well as the frontal plane. Head stabilization may simplify the transformations necessary to set up a coherent internal representation of external space. During various types of walking, head angular displacement in the sagittal plane remains within a small range when compared with the movements of the other limbs. When trunk movements are limited, the head is locked onto the trunk. During complex balancing tasks (standing on a narrow beam or on a semicircular platform), the head is again stabilized whereas the trunk is making the compensatory movements. A remarkable stabilization of the head in the frontal plane can also be observed during downhill skiing.

D. The Neural Control of Goal-Directed Arm Movements

Through evolution, humans have evolved from a quadrupedal to a bipedal gait. This has freed the upper limbs for various manipulatory activities. Depending on the task under consideration, both hands are being used in a symmetrical manner (Ohtsuki, Chapter 13, this volume), or one hand performs the focal movement while the other serves a stabilizing function, providing a positional reference (see Peters, Chapter 27, this volume). This particular type of interlimb cooperation is currently investigated within various subdivisions of science. Neuroscientists are predominantly interested in identifying the brain structures that are uniquely involved in upper-limb control, whereas others focus on the identification and description of bimanual interactions through kinematic and kinetic analyses. On one hand, a strong tendency is observed for both limbs to be synchronized during their simultaneous performance (Kelso, Southard, & Goodman, 1979). On the other hand, bimanual activity can be differentiated to a great extent through practice and experience (Swinnen, Walter, & Shapiro, 1988a, Swinnen, Walter, Beirinckx, & Meugens, 1991a, Swinnen, Young, Walter, and Serrien, 1991b). When two tasks have to be performed that differ in their difficulty level, the left hand is usually assigned the easier and the right hand the more complex, attention-demanding task. According to Peters, this is the predominant allocation of tasks in right-handers, whereas the situation is more complicated in left-handers (Chapter 27, this volume).

In their excellent review, Wiesendanger *et al.* (Chapter 9, this volume) discuss evidence concerning the unique involvement of particular cortical areas in bimanual activities. Their review is built on the distinction between movements in which both upper limbs are strongly synchronized and movements in which asymmetrical contributions are required to accomplish a goal. The former type of activity is often characterized by simultaneous initiation of movement, which may be achieved by direct and indirect bilateral connections between motor and premotor cortical areas and the spinal cord. Each hemisphere is in principle capable of exerting this bilateral control, thereby assuring a tight coupling of the limbs. More will be said about this later in this chapter. These movements are relatively well preserved following cortical lesions outside the primary motor cortex and they do not seem to require intact commissures when they are of limited complexity (Wiesendanger et al., Chapter 9).

Many goal-directed actions involve asymmetrical contributions from both arms. This may require suppressing the strong initial tendency for bimanual coupling or synchronization (Peters, Chapter 27, this volume; Swinnen et al., 1991b). Wiesendanger and collaborators suggest that various types of cortical lesions may impair the performance and/or learning of such differentiated bimanual activities: the parietal cortex, the premotor cortex, the anterior corpus callosum, and the medial frontal cortex (including the supplementary motor area). For that reason, they conclude that it is not justified by the current evidence to hypothesize a single cortical superarea functioning as a "unifying structure." It is rather the case that widely distributed cortical association areas cooperate as an interconnected ensemble to produce goal-directed bimanual actions (Chapter 9).

Split-brains form a unique population to investigate the role of the corpus callosum in performing and learning bimanual skills and other movements involving both body sides. In Chapter 10, Berlucchi et al. review their work on reaction times for simple responses using axial, proximal, and distal muscles of the upper limbs in normal subjects and subjects with a callosal deficit. In comparison to normal subjects, they demonstrate that the time needed for interhemispheric integration of crossed responses becomes much larger in patients with a callosal deficit and this is the case for unilateral and bilateral distal responses and unilateral proximal responses. It is inferred from these findings that the intact corpus callosum allows for efficient interhemispheric communication in order to secure integration of these responses. However, when proximal bilateral responses and unilateral and bilateral axial responses are made, other pathways are hypothesized to be responsible for integration. The authors argue that a bilaterally distributed motor system secures the production of these symmetrical movements involving axial and proximal limb muscles, and this system can be called upon by each of both hemispheres. Indeed, the acallosal subjects show a strong tendency for bimanual synchronization under these circumstances. Thus, whereas a callosal contribution seems necessary for the bilateral synchronization of distal responses, it does not seem critical when proximal or axial movements are involved. In the latter case, synchronization between the two sides is made possible through the common origin of the motor commands and through the bilateral distribution of the pathways transmitting them.

This bilateral control system, to which Wiesendanger and co-workers also refer, has been demonstrated anatomically and physiologically in primates by Kuypers (1973, 1985; Brinkman & Kuypers, 1973). It is reviewed in detail by Shinoda and co-workers in Chapter 2, this volume. The descending pathways from the cerebral cortex and the brain stem to the spinal cord can be distinguished into those that connect contralaterally with the dorsolateral part of the spinal intermediate zone and those that connect bilaterally with its ventromedial parts. Cells of the dorsolateral part (the lateral descending motor tract group) seem to distribute preferentially to motoneurons of distal extremity muscles, whereas cells of the ventromedial part (the medial descending motor tract group) to motoneurons of axial and proximal limb muscles. The latter group is phylogenetically and ontogenetically older than the former group. Furthermore, the single long descending motor tract axons exert their effects on different groups of spinal neurons simultaneously through multiple axon collaterals, thereby enabling control of the excitability of multiple muscles at multisegmental levels concurrently. This wide degree of divergence that characterizes the single long descending axons may constitute the neural substrate underlying the appropriate combination of muscles into functional synergies (Shinoda et al., Chapter 2).

E. Bimanual Coupling and Decoupling: Central and Peripheral Contributions

When both arms are moved simultaneously, a strong synchronization tendency becomes evident and this is interpreted by some to indicate that the limb musculature is constrained to act as a single functional unit or coordinative structure (Kelso et al., 1979). In the previous section, we have already elaborated on the neural substrate that may subserve these synchronization effects. Viewed from the perspective of the degrees-of-freedom problem, the existence of these preferred modes of coordination is considered an optimal solution. However, if bimanual movements would always be constrained in this way, humans would largely fail to comply with daily task requirements, which are often characterized by finely differentiated patterns of limb activity (role differentiation, see Fagard, Chapter 21, this volume; Peters, Chapter 27, this volume). Luckily, the human control system is endowed with adaptability and flexibility to overcome these intrinsic coordination tendencies. A nice example of differentiated bimanual movements is provided by Castiello and Stelmach (Chapter 28, this volume). When a precision grip with one hand is performed together with a prehension task involving the whole hand, they show unique grasp-related kinematic profiles that are captured within a common temporal metric. In general, timing is proposed to be a major parameter that constrains bimanual performance (Heuer, 1991).

In a number of recent studies, one of us in collaboration with colleagues has demonstrated that learning results in a progressive dissociation of limb movements that are initially synchronized (Swinnen, Walter, Lee, & Serrien, in press; 1988, 1991b, Walter & Swinnen, 1992, Chapter 23, this volume). Dissociation of movement is not a smooth process but requires subjects to overcome the (mutual) interference that is often evident at the start of practice. Patterns of activity in one limb show up in the other limb or vice versa. One is prompted to ask what the locus of this interference is. Among other things, Cohen (1970) hypothesized that dual-task interference may result from excessive demands in monitoring the discordant proprioceptive information generated by both tasks. However, Teasdale and collaborators (Chapter 12, this volume) refute this explanation. They contend that the locus of the observed interference is to be sought at the level of movement organization and programming. Their argument is based on a comparison of the performance of normal subjects with that of a unique deafferented patient suffering from a total loss of the large sensory myelinated fibers while still possessing an intact peripheral motor system. Similar to Cohen, they have their subjects perform a continuous wrist pronation/supination movement while performing a discrete secondary task upon presentation of an auditory stimulus, for example, a wrist extension or verbal response. If Cohen's hypothesis is correct, bimanual performance should be less impaired in the deafferented patient. Instead, they show that the absence of proprioceptive information yields a more pronounced interference in performance, and the recovery from the perturbation takes longer. Thus, Teasdale and co-workers' observations suggest that dual-task interference is not a direct consequence of the monitoring of movement afference. Rather, it arises at the efferent level or during the stage of movement programming and organization. This is in agreement with observations in discrete bimanual tasks, where interference between the electrical activity of the biceps muscles of both upper limbs can already be detected before any movement has taken place (and presumably before any appreciable movement-related afferent information can be generated) (Swinnen et al., 1988, 1991b). In addition, reaction-time studies conducted by Heuer (1990) have provided evidence that intermanual interactions can already be established at the central level of control.

In summary, two important conclusions can be drawn from these observations. On one hand, interference during dual-task performance occurs in spite of the absence of proprioceptive information and can therefore be located at the level of movement planning, organization, and possibly efferent control. On the other hand, afferent information from the moving limbs is important for sustaining interlimb coordination in the face of upcoming distortions or perturbations. This is nicely demonstrated by Teasdale and co-workers, who show that the recovery from interference is less successful in the deafferented patient. More generally, we propose that interlimb coordination for various limb combinations and coupling modes depends on the monitoring of afferent information from the moving limbs. This point is also made by Baldissera and co-workers with respect to ipsilateral control of hand and feet movements (Baldissera, Cavallari, Marini, & Tassone, 1991, Baldissera et al., Chapter 11, this volume). On the basis of their findings with various limb-loading techniques, they argue that kinesthetic afferences are important for sustaining interlimb coordination, especially for anti-phase coupling. They conclude that in-phase and anti-phase coupling of the ipsilateral hand and foot are possibly controlled by means of different feedback systems and different degrees of elaboration of these systems. During in-phase coupling, the feedback system is presumably operating in a less stringent fashion, whereas during anti-phase coordination, feedback monitoring from the moving limbs is important and is argued to require constant attention. In addition, one of us has recently observed considerable phase destabilization in blindfolded subjects during the production of a homolateral coordination pattern (forearm–lower leg) as a result of passive movement generated in the contralateral side (Swinnen, Serrien, & Daelman, 1993). Presumably, passive movement generates afferent information that is discordant with the afferent information generated in the actively coordinated limbs.

III. In Search of the Equations of Motion Underlying Interlimb Coordination: Interlimb Dynamics

The dynamical approach to interlimb coordination has undergone a major development in the past 15 years (for a review, see Turvey, 1990). In the present context, dynamics is not to be understood in the strict traditional sense as the study of how objects move under the action of forces (the force–mass–acceleration approach). Dynamics is a field emerging between mathematics and the sciences. It deals with changes in systems and tries to express its existing and evolving states. In principle, it is best suited for application to cyclical tasks even though Schöner undertakes an attempt to show how a dynamical approach may address the problem of trajectory formation of a single limb (Schöner, Chapter 17, this volume). A dynamical system consists of two parts: (1) the essential information about a system or the notions of a state; and (2) a rule that describes how the state evolves with time (the dynamic) (Crutchfield, Farmer, Packard, & Shaw, 1986).

Even though the simplest of coordination tasks (like moving two fingers together) is rather complex when considering the cooperation that is required among the multiple subcomponents at various layers of the motor apparatus, the dynamical approach is directed at uncovering the basic principles or laws that characterize interlimb coordination. These principles capture the cooperative behavior of the ensemble and cannot necessarily be inferred from the individual behavior of the subcomponents. As Schöner (Chapter 17) notes, the coordination dynamics are not to be equated with the physical dynamics of the biomechanical system, even though they may contribute to those dynamics. Coordination is argued to be a consequence of evolving processes of self-organization or pattern formation, concepts that figure predominantly in Haken's synergetics (Haken, 1983; Kelso, Chapter 15, this volume).

Central to the approach is the identification of relevant macroscopic variables and their equations of motion, built around the concept of relative phase. Within the domain of interlimb coordination, relative phase can be defined as the phase difference between two oscillatory signals (in the present case, limb movements). Phase refers to the point of advancement of the signal within a cycle, that is a description of the stage that a periodic motion has reached. Relative phase is thus a useful variable for the assessment of spatiotemporal coordination. It does not provide information of each signal separately but uniquely characterizes the way two signals relate to each other. For a long time, neuroscientists interested in locomotion have used measures of phase angles to determine modes of intra- and intergirdle coordination in animals. It is only recently, however, that scientists have further explored and modeled its characteristic dynamics.

Advocates of the dynamical approach make a distinction between order parameters and control parameters. Order parameters (also called collective variables) characterize the behavioral patterns of interest. Relative phase has been proposed as a primary candidate in this respect (Kelso, Chapter 15; Schöner, Chapter 17; Turvey & Schmidt, Chapter 14). Control parameters induce changes that can be observed and characterized at the level of the order parameters even though they do not contain any specific information about the potential changes that order parameters undergo. A well-known example that aids in clarifying this distinction is the performance of cyclical wrist or finger movements under increased cycling frequency conditions (see also Carson *et al.*, Chapter 16). You can easily try this out yourself: When you start moving both fingers in front of you in the same direction (also called anti-phase or 180° out-of-phase) and you progressively increase the cycling frequency, you will suddenly experience a transition or shift toward symmetrical movements of both fingers (called in-phase), unless you intentionally oppose this rather spontaneous change. Thus, relative phase is said to capture the dynamics of this coordination system, whereas the control parameter (cycling frequency) allows these dynamics to emerge.

A number of additional points are to be made from the aforementioned example. First, although not all behavioral changes take the form of *phase transitions,* advocates of the dynamical approach argue that phase transitions provide a window into understanding behavioral patterns and, more generally, the behavior of living things. The system evolves from one level of organization to another more stable level and the transition is often preceded by instabilities. A phase transition is the simplest type of self-organization in physics. The changes from a liquid to a gas or solid are most familiar to us. Apparently, no superordinate command structure initiates or controls this transition. Currently, phase transitions are not only viewed within the small time scales evident in the aforementioned examples. Their existence is also presumed along the longer time scales of development and learning (see Section IV). Whereas the traditional experimental approach to human functioning has often concentrated on the study of stable behavioral features, the dynamical approach suggests an alternative perspective whereby instabilities are the major focus of attention. Instabilities serve to demarcate behavioral patterns. Phase transitions that arise out of these instabilities constitute a special entry point for developing a language upon which to build a deeper understanding of the behavior of living things.

Observable behavioral patterns are mapped onto attractors of the order parameter dynamics. Roughly speaking, attractors are what the behavior of a system is attracted too when allowed enough time to settle down. The attractor can be: (1) a single point; (2) a closed curve (or limit cycle), which describes a system with periodic behavior; or (3) a fractal or strange attractor for a system exhibiting chaos. From an experimental point of view, signatures of the dynamics underlying interlimb coordination can be found in the stability of coordination patterns and various stability measures can be used for that purpose. For the specific case of isofrequency finger, hand, or arm movements, these attractors refer to the in-phase and anti-phase coordination mode. The behavioral implications are a natural tendency to fall into these modes even when attempting alternative coordination patterns. Attraction toward in-phase coordination is generally stronger than anti-phase coordination (Kelso, 1984). In the specific case of arm movements in the frontal plane, cyclical movements in different directions (moving toward or away from the body midline) are produced more stably than movements in the same direction. However, a different pattern of findings can be observed for hand (lower arm) and foot (lower leg) movements in the sagittal plane (see Baldissera *et al.*, Chapter 11, this volume; Kelso & Jeka, 1992; Swinnen *et al.*, 1993). Here, movements in the same direction (both up or down) are performed with a higher degree of stability than movements in different directions (one up, one down) and this effect is not muscle specific (Baldissera, Cavallari, & Civaschi, 1982). A similar principle holds for intersegmental coordination, that is movements at the elbow and wrist joints (Kelso, Buchanan, & Wallace, 1991). This suggests that the mutual direction of limb movements is an important determinant of coordinative stability (see also Kelso & Jeka, 1992). Additional research is required to verify the generalizability of this principle to other limb combinations and planes of motion. For some of these coordination patterns, biomechanical interactions between the segments or postural disturbance effects cannot be excluded as partial accounts for differential stability. Proponents of the dynamical approach are predominantly concerned with developing equations of motion that capture the coordination dynamics, whereas neuroscientists focus at unraveling the architecture of the neural networks and pathways that give rise to these preexisting preferred coordination modes.

The aforementioned examples pertain to the specific case of interlimb coordination with frequency and phase locking. Von Holst referred to this as "absolute coordination." However, humans and other species are endowed with a great deal of adaptability and flexibility to defy these elementary coordination modes. This implies that a variety of interlimb patterns can be explored even though tendencies toward certain frequency ratios and toward in- and anti-phase coordination will remain evident ("relative coordination") Von Holst (1973/1937) gave the example of a child walking along with his father. When walking independently, both have different speed and stride amplitude preferences. When walking together, short events of phase and frequency locking will occur, alternated with periods of desynchronization and adjustments toward synchronization. In other words, there is an interplay between cooperation and competition, or in Von Holst's terms, between the magnet effect and the maintenance tendency.

Less stringent modes of coupling are more common when the effectors involved are asymmetric. The sources for asymmetry can be manifold, for example, when coordinating different effectors (the arm and leg), or when inertial differences are artificially imposed on the same effectors. In Chapter 14, Turvey and Schmidt describe an experimental task setup in which subjects oscillate hand-held pendulums that can be varied in length. These length variations affect the degree of dissociation between the pendulums' eigenfrequencies. Under these circumstances, they still show a tendency toward in-phase and anti-phase attraction even though relative phase becomes more destabilized when compared to the symmetrical conditions. The stability differences between the in-phase and anti-phase mode, as mentioned before, tend to become smaller as the difference between the pendulums' eigenfrequencies becomes larger (greater differences in their length). In addition, these phenomena interact with the overall cycling frequency at which these oscillatory movements are performed (Turvey & Schmidt, Chapter 14). As such, their work elucidates the cooperative and competitive phenomena that characterize interlimb coordination. Both are represented in the equations of motion the authors discuss in their chapter. Similar to Turvey and Schmidt, Carson and associates focus on the component oscillators to improve understanding of their combined operation (Chapter 16, this volume). They conclude that the interaction between intrinsic upper-limb asymmetries and informational and mechanical constraints may have a significant influence on the coupling dynamics.

In summary, the dynamical approach has as its main objective the development of equations of motion governing interlimb coordination and the identification of the dynamics underlying coordination. The tendency toward phase and frequency synchronization, which is ubiquitous in many species at various layers of the motor control apparatus, is thereby underscored. Whereas a strong emphasis was initially placed on the production of movement patterns with a 1:1 frequency ratio (absolute coordination), attention has recently shifted to other forms of coordination, characterized by a less stringent coupling of the limbs (relative coordination). It is now becoming evident that a great variety of coordination modes can be flexibly accomplished even though preferred coordination tendencies remain present. These are the behavioral expressions of a few elementary coordination principles. As will be discussed later, the learning of new coordination patterns can partly be understood against the backdrop of these preexisting coordination modes. Not only intrinsic or preferred coordination modes but also intended, memorized, or learned behavioral patterns are expressed in terms of the dynamics of coordination (see Zanone & Kelso, Chapter 22, this volume).

The dynamical approach brings a new challenge to the reductionist view that focuses on the study of a stystem's subcomponents. It is also committed to applying a rather universal language for describing both living and nonliving systems. Phase transitions thereby constitute a special entry point for developing a language upon which to build a better understanding of the behavior of living things. However, this approach is only in its initial stage of development and many questions remain as yet unanswered. For example, in addition to manipulating cycling frequency, there are possibly various alternative ways to probe a motor system and elucidate its archaic coordination modes. One of us has observed phase transitions during two-limb (arm–leg) coordination as a result of addition of a third limb or through passive movement of a third limb. What is perhaps common across these various manipulations is that they overload or perturb the system, causing it to regress to its most rudimentary coordination modes. The dynamical approach provides some tools to describe and characterize observed patterns and eventual transitions among patterns. The archaic movement forms that the system settles in when stressed, may also reflect the central nervous system's most easily potentiated patterns of neural wiring.

IV. Modulation of Coordination Patterns through Practice and Experience

In previous sections, we have already hinted at the flexibility and adaptability of coordination patterns. These capabilities become more prevalent as humans develop and learn. Modulation of coordination patterns is evident in the developing child, who undergoes a remarkable evolution in its basic coordination patterns during the first year of life. This pertains to the development of postural control (Woollacott & Sveistrup, Chapter 18, this volume), and quadrupedal and bipedal locomotion, with the major goal of supporting and transporting the body in a gravitational field (Whitall & Clark, Chapter 19). It is also evident, however, in the development of the upper limbs, which become specialized for various manipulative functions. (Corbetta & Thelen, Chapter 20; Fagard, Chapter 21). Even though the time scales underlying changes in behavior can be markedly different, the study of motor development and learning share many commonalities. Both involve the gradual mastering and control of the degrees of freedom inherent in the motor system. Often, preexisting or preferred coordination tendencies must be overcome. Development and learning can then be understood against the background of these preexisting patterns, which form the basic building blocks for the creation of more differentiated patterns of activity. Woollacott and Sveistrup address the degrees of freedom at the muscle synergy level, whereas the other contributors focus on the interlimb level.

A. Development of Coordination Patterns

Woollacott and Sveistrup describe the development of posture control during the first year of life, which takes place in a cephalocaudal direction (Chapter 18). They demonstrate that the calibration of input–output relationships between sensory inputs controlling posture and the neck muscles occurs before calibration of the trunk muscles. Later, during the development of pull to stand behavior, the muscles become activated in a distal to proximal sequence, that is from tibialis anterior to quadriceps and abdominal muscles. The postural synergies appear to be temporally organized in an adultlike fashion at about the onset of independent stance. Postural control is a prerequisite for the exploration of the environment and the development of locomotory behavior, discussed next.

Whitall and Clark (Chapter 19, this volume) describe the development of bipedal locomotion (walking, running, and galloping) between the first and second year of life. From the onset of walking behavior, the anti-phase (alternating) pattern is adopted, even though the phasing is more variable than in adults. Adultlike variability is evident after about three months of walking experience. Following that period, other forms of locomotion emerge, such as running and galloping. Galloping is an interesting example in that it requires the lower limbs to be out of phase about 90–120°. Thus, the preferred anti-phase mode of coordination is overcome to explore new movement forms. That small children acquire this skill relatively easy is rather striking in view of recent findings pointing to the relative difficulty of acquiring a 90° out-of-phase pattern with the upper limbs in adults (Lee, Swinnen, & Verschueren, 1993; Zanone & Kelso, 1992, Chapter 22, this volume).

Thelen and co-workers have spent considerable efforts in elucidating and describing the preferred modes of coordination in the newborn and the development of goal-directed movement patterns across the first months of life (Thelen, Kelso, & Fogel, 1987). They argue that preferred patterns of interlimb coordination are to some extent inborn, after which they undergo developmental modification and increasing modulation in the maturing child. In this respect, a primary role is assigned to the control of limb stiffness. Modulation of stiffness shifts patterns of coupling both between the joints and the limbs. Adaptive goal-directed movements require infants to have discovered optimal levels of limb stiffness.

In their chapter contribution, Corbetta and Thelen focus on the development of reaching by means of a careful and unique observation of a baby during its first year of life (Chapter 20). In addition to reaching, they have also observed and registered the broader context of spontaneous movements from which the reach emerges. The development of reaching is marked by an alternation of periods of strong bimanual coupling and decoupling. The authors propose that these shifting patterns are related to changes in the ability to control the forces imparted to the movements. In particular, the forces produced in the contralateral (nonreaching) arm become gradually under control during the transition from bimanual to unimanual reaching. In a related but more general way, Woollacott and Sveistrup argue that background stiffness levels drop with increasing experience during the development of balance control (Chapter 18). In both cases, patterns of activity become more differentiated with excess of muscle activity being reduced or eliminated.

Many bimanual skills require a differentiation of the roles assigned to each hand (see Peters, Chapter 27, this volume) and we have already elaborated on the neural mechanisms that may underlie this form of goal-directed interlimb cooperation. In Chapter 21, Fagard focuses on the development of this hand role differentiation during the first year of life. Whereas initially, hand actions are largely symmetric, there is a growing need for hand role differentiation for successfully grasping and exploring objects. Fagard shows that infants are initially successful at coordinating the hands in complementary but *successive* actions. Only by the end of the first year of life, differentiated *simultaneous* actions become evident. Thus, whereas Corbetta and Thelen have emphasized the gradual control of the contralateral nonreaching hand during reaching and grasping (Chapter 20), Fagard focuses on a next stage in development, where both hands play a complementary role in object exploration. In accouting for the development of bimanual coordination, an important role is assigned to the interaction between the maturation of neural structures and experience that generates the necessary action-produced feedback.

B. Acquisition of Coordination Patterns

Recent studies on motor learning have focused on Bernstein's premise that the initial stages of acquisition are characterized by freezing the biokinematic degrees of freedom, whereas later stages are accompanied with a release of the degrees of freedom. The work discussed by Walter and Swinnen (Chapter 23, this volume) supports this hypothesis, whereas Lee and Russo's work does not (Chapter 25). More experimental work is evidently needed to resolve this issue (see also Newell & McDonald, Chapter 24). It is conjectured that the evolution surrounding the control of the degrees of freedom across acquisition is dependent on the specific task requirements and the relation between the task to be learned and preexisting coordination modes. Freezing of the degrees of freedom is apparently evident when learning a new task is initially constrained by a bias or attraction toward preexisting modes of coordination. When these initial biases are less predominant, the way is free to explore a broader array of coordination modes, and performance is less constrained. On the other hand, as learning progresses, the degrees of freedom become gradually under control, movement invariances develop, and consistency is promoted. This, however, does not preclude the development of some degree of movement flexibility in order to cope with environmental contingencies. Movement consistency and flexibility are two distinct but complementary expressions of increased skill automation.

As mentioned before, the acquisition of new movement patterns and the associated degrees-of-freedom problem should be viewed against the background of preexisting preferred coordination modes. On one hand, preexisting modes may be coducive to the new task requirements (such as in locomotion and balance). On the other hand, the difficulties that learners encounter and the errors they make on their way to acquiring skillful behavior can often be understood as a struggle against the intrusive nature of preexisting preferred coordination modes. Here, the new pattern to be learned deviates substantially from the existing or preferred ones. Learning, then, results from the competitive and cooperative interplay between intrinsic coordination tendencies and the new extrinsic task requirements. This exemplifies the two faces of preferred or preexisting coordination modes.

In their contribution to this volume, Newell and McDonald (Chapter 24) refer to the important role of perceptual information during the acquisition of coordination and they argue that informational cues, used by the learner, may vary as a function of the task and the level of practice. Their experimental strategy is driven by the ecological approach to perception and action. They focus on the ways in which the performer explores the perceptual–motor workspace to solve a movement problem, also referred to as search strategies. Search strategies identify the path chosen by the performer through a multidimensional state space that maps perception and action in a way that is consistent with the task goal.

The approach developed in Chapter 22 by Zanone and Kelso (see also Schöner, Zanone, and Kelso, 1992) represents an attempt to describe learning within the framework of the dynamic approach. They show an example of a continuous, bimanual, finger-tapping task, where subjects learn a 90° out-of-phase pattern, located in between the previously identified stable in-phase and anti-phase coordination modes. During learning, they regularly probe the entire coordination dynamics in order to assess the changes across the whole array of relative phase modes (called the scanning procedure). They view learning as a phase transition, whereby the learner gradually overcomes the attraction exhibited by preexisting coordination modes in order to gain stability at the new mode. With increasing practice, the newly established coordination pattern displays the features of a learned attractor in the sense that nearby patterns tend to converge to the newly built attractor. The change in the coordination dynamics that is brought about by the processes of learning is ascribed to an additional layer of dynamics, be it at a slower time scale (the learning dynamics). The authors consider it important to determine the individual's existing coordination tendencies prior to exposure to the learning task. This paradigm may have the potential to provide a new élan to the study of motor learning in the near future. It helps to fill the large gap left by previous theories of motor learning concerning the nature of preexisting movement capabilities that performers bring with them when involved in the acquisition of new skills.

Whereas Zanone and Kelso discuss the production of cyclical tasks. Walter and Swinnen review their work on overcoming preferred coordination tendencies during the acquisition of discrete movements (Chapter 23, this volume). More specifically, they address the subjects' evolving capability to defy the initially strong synchronization tendency, and they refer to it as bimanual dissociation or decoupling. Individual differences in the capability to overcome these tendencies are also predominant here and the authors discuss various instructional means (information feedback, adaptive tuning) to overcome bimanual synchronization (see also Walter & Swinnen, 1992; Walter, Swinnen, & Franz, 1993).

Zanone and Kelso conceive preexisting coordination patterns mainly in terms of preferred relative phase modes. When a coordination pattern largely converges with a preexisting relative phase pattern, learning occurs rapidly. In contrast, when a new relative phase mode differs from the preexisting modes, the route to learning is more difficult because new task requirements compete with existing preferences and tendencies. However, it should be clear that the array of preexisting preferred movement patterns can be put into a much broader perspective, including movements resulting from particular patterns of neural wiring, such as the symmetric and asymmetric neck and labyrinthine reflexes, or interlimb coordination modes endowed by the neural architecture of CPGs. This point is made by Walter and Swinnen (Chapter 23, this volume) who propose many candidates for what they call systematic behavioral biases or (bad) habits.

A particular case of overcoming the limitations imposed by preferred coordination tendencies that has received considerable attention in the cognitive literature is the production of polyrhythms (Deutsch, 1983; Peters, 1985). In contrast to simple rhythms (e.g., a 2 : 1 or 3 : 1 pattern), polyrhythms require the simultaneous production of time frames that are not integer multiples of each other (e.g., a 3 : 2 or 5 : 3 pattern). Summers and Pressing (Chapter 26, this volume) address the learning of such polyrhythms; they are particularly interested in the capability of performers to override the constraints inherent in bimanual coordination through cognitive intervention. They hypothesize that successful performance is dependent on the development of an integrated representation in which both hand movements are combined into a common time base (see also Deutsch, 1983; Jagacinski, Marshburn, Klapp, & Jones, 1988; Klapp et al., 1985; Peters & Schwartz, 1989). The most common integration strategy is to insert the slow-hand responses into movements of the fast hand that serves as the general time base. Summers and Pressing are much in favor of an eclectic approach toward a better understanding of movement coordination and control within which both the dynamic and cognitive approach coexist. We fully subscribe to this viewpoint.

In summary, learning new modes of coordination can be conceived as a process of task differentiation and task integration. The learning of new coordination patterns often requires overcoming or suppressing the tendencies imposed by preexisting, preferred coordination patterns. In addition, there is also evidence for task integration, as suggested by recent polyrhythm studies. The generality of this statement needs to be further verified. In studying bimanual decoupling during the production of discrete movements, one of us has observed a variety of strategies to perform these skills. In some subjects, the major kinematic landmarks of the right and left limb movement were found to be interrelated, whereas in others they were not. The latter findings speak against task integration as a common control strategy for discrete dual-task performance (Swinnen, Walter, Lee, & Serrien, in press). Thus, interindividual differences in control strategies are evident and these may vary according to the specific nature of the task (continuous vs. discrete). The differentiation–integration hypothesis bears a link with the degrees-of-freedom problem and may provide a fruitful direction for future research on the learning of coordination patterns.

V. Conclusion

In this introductory chapter, we have provided an elementary framework and overview of the field of interlimb coordination, which is currently addressed within various subdivisions of science. These subdivisions represent different approaches to the study of interlimb coordination but they also entail different levels of description, analysis, and explanation. For that reason, neither has precedence over the other; they can be considered additive rather than mutually exclusive. The following chapters will take the reader on a journey that will provide a better understanding of some basic issues pertaining to interlimb coordination. We have deliberately chosen to go all the way from the elementary neurological substrates at the cellular level to the emergence of functional goal-directed behavior that can be observed with the naked eye. Even though the attribution of motor functions to neuroanatomical structures and pathways is not always straightforward at the current state of science, it is considered invaluable for students in motor coordination to be broadly informed about both the underlying neuroanatomical structures and the coordinative functions they bring about. This field of research is promising. Even though we are humbled by the great complexity of coordination patterns that can be found across the animal world, some basic principles appear at the horizon that may apply as equally to rhythmic fin movements in fish species as to limb movements in humans. However, what needs to be added to these principles of interlimb cooperativity is a theory of highly differentiated and flexible interlimb behavior that distinguishes humans from most other animal species.

References

Baldissera, F., Cavallari, P., & Civaschi, P. (1982). Preferential coupling between voluntary movements of ipsilateral limbs. *Neuroscience Letters,* **34,** 95–100.

Baldissera, F., Cavallari, P., Marini, G., & Tassone, G. (1991). Differential control of in-phase and anti-phase coupling of rhythmic movements of ipsilateral hand and foot. *Experimental Brain Research* **83,** 375–380.

Belenkii, Y. Y., Gurfinkel, V. S., & Paltsev, Y. I. (1967). Elements of control of voluntary movements. *Biofizika,* **12,** 135–141.

Bernstein, N. (1967). *The co-ordination and regulation of movement.* Oxford: Pergamon Press.

Brinkman, J., & Kuypers, H. G. J. M. (1973). Cerebral control of contralateral and ipsilateral arm, hand and finger movements in the split-brain rhesus monkey. *Brain,* **96,** 653–674.

Brown, T. G. (1911). The intrinsic factors in the act of progression in the mammal. *Proceedings of the Royal Society of Lond [Biol.],* **84,** 308–319.

Cohen, L. (1970). Interaction between limbs during bimanual voluntary activity. *Brain,* **93,** 259–272.

Crutchfield, J. P., Farmer, J. D., Packard, N. H., & Shaw, R. S. (1986). Chaos. *Scientific American,* **255,** 38–50.

Deutsch, D. (1983). The generation of two isochronous sequences in parallel. *Perception and Psychophysics,* **34,** 331–337.

Dietz, V. (1992). Human neuronal control of automatic functional movements: interaction between central programs and afferent input. *Physiological Reviews,* **72,** 33–69.

Fukuda, T. (1961). Studies on human dynamic postures from the viewpoint of postural reflexes. *Acta Oto-Laryngologica,* **161,** 1–52.

Grillner, S. (1975). Locomotion in vertebrates: central mechanisms and reflex interaction. *Physiological Reviews,* **55,** 247–306.

Grillner, S. (1981). Control of locomotion in bipeds, tetrapods, and fish. In V. B. Brooks (Ed.), *Handbook of physiology. Section 1: The nervous system. Motor control.* (Vol. II) (pp. 1179–1236). Bethesda: American Physiological Society.

Grillner, S. (1985). Neurobiological bases of rhythmic motor acts in vertebrates. *Science,* **228,** 143–149.

Haken, H. (1983). *Synergetics, an introduction: Non-equilibrium phase transitions and self-organization in physics, chemistry, and biology* (3rd ed.). Berlin: Springer.

Hellebrandt, F. A., Houtz, S. J., Partridge, M. J., & Walters, C. E. (1956). Tonic neck reflexes in exercises of stress in man. *American Journal of Physical Medicine*, **35**, 144–159.

Heuer, H. (1990). Rapid responses with the left or right hand: response–response compatibility effects due to intermanual interactions. In R. W. Proctor & T. G. Reeve (Eds.), *Stimulus–response compatibility* (pp. 311–342). North-Holland: Elsevier.

Heuer, H. (1991). Motor constraints in dual-task performance. In D. L. Damos (Ed.), *Multiple-task performance* (pp. 173–204). London: Taylor & Francis.

Jagacinski, R. J., Marshburn, E., Klapp, S. T., & Jones, M. R. (1988). Tests of parallel versus integrated structure in polyrhythmic tapping. *Journal of Motor Behavior*, **20**, 416–442.

Keele, S. W. (1981). Behavioral analysis of movement. In V. B. Brooks (Ed.), *Handbook of physiology. Section 1: The nervous system. Motor control* (Vol. II) (pp. 1391–1414). Bethesda: American Physiological Society.

Kelso, J. A. S. (1984). Phase transitions and critical behavior in human bimanual coordination. *American Journal of Physiology: Regulatory, Integrative, and Comparative Physiology* **15**, 1000–1004.

Kelso, J. A. S., & Jeka, J. J. (1992). Symmetry breaking dynamics of human interlimb coordination. *Journal of Experimental Psychology: Human Perception and Performance*, **18**(3), 645–668.

Kelso, J. A. S., Southard, D. L., & Goodman, D. (1979). On the coordination of two-handed movements. *Journal of Experimental Psychology: Human Perception and Performance*, **2**, 229–238.

Kelso, J. A. S., Buchanan, J. J., & Wallace, S. A. (1991). Order parameters for the neural organization of single, multijoint limb movement patterns. *Experimental Brain Research*, **85**, 432–444.

Klapp, S. T., Hill, M. D., Tyler, J. G., Martin, Z. E., Jagacinski, R. J., & Jones, M. R. (1985). On marching to two different drummers: perceptual aspects of the difficulties. *Journal of Experimental Psychology: Human Perception and Performance*, **11**, 814–827.

Kuypers, H. G. J. M. (1973). The anatomical organization of the descending pathways and their contributions to motor control especially in primates. In J. E. Desmedt (Ed.), *New developments in electromyography and clinical neurophysiology* (Vol. III) (pp. 38–68). Basel: Karger.

Kuypers, H. G.J. M. (1985). The anatomical and functional organization of the motor system. In M. Swash & C. Kennard (Eds.), *Scientific basis of clinical neurology* (pp. 3–18). Edinburgh: Churchill Livingstone.

Lee, T. D., Swinnen, S. P., & Verschueren, S. (1993). Relative phase alterations during bimanual skill acquisition. Manuscript submitted for publication.

Massion, J. (1992). Movement, posture and equilibrium: interaction and coordination. *Progress in Neurobiology*, **38**, 35–56.

Meyrand, P., Simmers, J., & Moulins, M. (1991). Construction of a pattern-generating circuit with neurons of different networks. *Nature (London)*, **351**, 60–63.

Paltsev, Y. I., & Elner, A. M. (1967). Preparatory and compensatory period during voluntary movement in patients with involvement of the brain of different localization. *Biofizika*, **12**, 142–147.

Peters, M. (1985). Constraints in the performance of bimanual tasks and their expression in unskilled and skilled subjects. *The Quarterly Journal of Experimental Psychology*, **37A**, 171–196.

Peters, M., & Schwartz, S. (1989). Coordination of the two hands and effects of attentional manipulation in the production of a bimanual 2 : 3 polyrhythm. *Australian Journal of Psychology*, **41**, 215–224.

Schöner, G. S., Zanone, P. G., & Kelso, J. A. S. (1992). Learning as change of coordination dynamics: theory and experiment. *Journal of Motor Behavior*, **24**, 29–48.

Shepherd, G. M. (1988). *Neurobiology* (2nd ed.). New York: Oxford University Press.

Shik, M. L., Severin, F. V., & Orlovsky, G. N. (1966). Control of walking and running by means of electrical stimulation of the mid-brain. *Biophysics*, **11**, 756–765.

Swinnen, S., Walter, C. B., & Shapiro, D. C. (1988). The coordination of limb movements with different kinematic patterns. *Brain and Cognition*, **8**, 326–347.

Swinnen, S. P., Walter, C. B., Beirinckx, M. B., & Meugens, P. F. (1991a). Dissociating the structural and metrical specifications of bimanual movement. *Journal of Motor Behavior*, **23**, 263–279.

Swinnen, S. P., Young, D. E., Walter, C. B., & Serrien, D. J. (1991b). Control of asymmetrical bimanual movements. *Experimental Brain Research*, **85**, 163–173.

Swinnen, S. P., Serrien, D. J., & Daelman, A. (1993). Relative phase destabilization through passive movement during the production of homolateral coordination patterns. Paper presented at the Conference on Deafferentation and the Role of Sensory Afferents in Human Motor Control, Québec.

Swinnen, S. P., Walter, C. B., Lee, T. D., & Serrien, D. J. (in press). Acquiring bimanual skills: contrasting forms of information feedback for interlimb decoupling. *Journal of Experimental Psychology: Learning, Memory, & Cognition*.

Thelen, E., Kelso, J. A. S., & Fogel, A. (1987). Self-organizing systems and infant motor development. *Developmental Review*, **7**, 39–65.

Turvey, M. T. (1990). Coordination. *American Psychologist*, **45**, 938–953.

Von Holst, E. (1973). *The behavioral physiology of animals and man: The collected papers of Erich Von Holst* (Vol. 1) (R. Martin, Trans.). London: Methuen. (Original work published 1937.)

Walter, C. B., & Swinnen, S. P. (1992). Adaptive tuning of interlimb attraction to facilitate bimanual decoupling. *Journal of Motor Behavior*, **24**, 95–104.

Walter, C. B., Swinnen, S. P., & Franz, E. A. (1993). Stability of symmetric and asymmetric discrete bimanual actions. In K. M. Newell & D. M. Corcos (Eds.), *Variability and motor control* (pp. 359–380). Champaign, IL: Human Kinetics.

Zanone, P. G., & Kelso, J. A. S. (1992). Evolution of behavioral attractors with learning: nonequilibrium phase transitions. *Journal of Experimental Psychology: Human Perception and Performance*, **18**, 403–421.

Part I

The Neural Basis of Interlimb Coordination

2

Multisegmental Control of Axial and Limb Muscles by Single Long Descending Motor Tract Axons
Lateral versus Medial Descending Motor Systems

Yoshikazu Shinoda, Shinji Kakei, and Yuriko Sugiuchi

Department of Physiology
School of Medicine
Tokyo Medical and Dental University
Tokyo, Japan

I. Introduction
II. Lateral versus Medial Long Descending Motor Tract Systems
    A. The Lateral Descending Motor Tract Group
    B. The Medial Descending Motor Tract Group
III. Morphological Features of Single Axons in the Lateral and Medial Long Descending Motor Tract Systems
    A. Morphology of Single Corticospinal Tract Axons
    B. Morphology of Single Rubrospinal Tract Axons
    C. Morphology of Single Vestibulospinal Tract Axons
IV. Summary
    References

I. Introduction

It has been tacitly assumed that the pyramidal tract consists of private lines connecting a point in the motor cortex to a single muscle, as a motoneuron innervates a single muscle. Accordingly, corticospinal tract (CST) neurons have been referred to as upper motoneurons and motoneurons as lower motoneurons. Other long descending motor tracts also have been considered to be similar to the pyramidal tract. However, the above notion of a long descending motor tract referred to as a private line is no longer tenable, since recent studies showed that axons of all major long descending motor tracts send axon collaterals to multiple spinal segments. This situation was first described by Abzug, Maeda, Peterson, and Wilson (1974), who found that 50% of lateral vestibulospinal tract (LVST) neurons, which sent axon branches in C6–Th1 segments, were also antidromically driven by stimulation of the lumbar spinal cord. A similar percentage of reticulospinal tract (RTST) neurons sent branches to the cervical gray matter as well as to the first lumbar segment (Peterson, Maunz, Pitts, & Mackel, 1975). It was also shown that 6% of CST neurons activated from the cervical gray matter (C4–C8) projected to the first lumbar segment and 24% of CST neurons activated from the cervical gray matter projected to the thoracic cord (Shinoda, Arnold, & Asanuma, 1976). Similarly, 45% and 5% of rubrospinal tract (RBST) neurons projecting to the cervical gray matter sent axon branches to the thoracic cord and below, and to the first lumbar level, respectively (Shinoda Ghez & Arnold, 1977). Furthermore, virtually all CST neurons examined in the forelimb area of the motor cortex had three to seven axon collaterals at widely separated segments of the cervical and the higher thoracic cord (Shinoda *et al.*, 1976; Shinoda, Ohgaki & Futami, 1986a). In addition, multiple axon collaterals of RBST neurons were demonstrated at different spinal segments (Shinoda *et al.*, 1977). These results indicate that single motor tract axons are not a simple private line connecting the cells of origin and motoneurons for a single muscle, but instead they may exert multiple influences on different groups of spinal interneurons and motoneurons simultaneously at widely separated segments.

The final targets of these descending motor tracts, either directly or indirectly, are motoneurons of different muscles. Somatic motoneurons, which distribute their axons through the ventral root to a given axial or limb muscle, are arranged in a longitudinal, columnlike fashion in the ventral horn (Romanes, 1951; Sprague, 1951; Burke, Strick, Kanda, Kim, & Walmskey, 1977). The longitudinal columns innervating different muscles are grouped into the medial and lateral longitudinal aggregates. The medial aggregate is made up of motoneuronal columns of vertebral muscles innervated by the dorsal rami of the ventral root and exists throughout the spinal cord; the lateral aggregate of motoneuronal columns of all other muscles innervated by the ventral rami expands in the brachial and the lumbosacral enlargements, because of the additional longitudinal columns for the muscles intrinsic to the extremities, and fuses to the medial aggregate in the upper cervical and the thoracic cord (Sugiuchi & Shinoda, 1992). As motoneurons innervating axial and limb muscles are located in different areas of the ventral horn, interneurons terminating upon motoneurons of these two groups of the muscles are also arranged in different portions of the spinal intermediate zone (laminae V–VIII) (Rexed, 1954; Kuypers, 1981). Various long descending motor tracts of supraspinal origin are known to terminate in different areas of the spinal gray matter (Brodal, 1981), so that they influence differentially the motoneurons and interneurons of these two muscle groups. When considering the functions of the long descending motor tract pathways, it should be realized that their functions are not only determined by inputs to cells of their origin, but also characterized by their terminations on their target cells in the spinal cord. The brainstem pathways terminating in the spinal intermediate zone as well as the motoneuronal cell groups probably subserve mainly motor functions. The subcorticospinal pathways have been grouped into medial and lateral systems based upon their terminal distribution in the spinal cord (Kuypers, Fleming, & Farinholt, 1962). The lateral system terminates in the dorsal and lateral parts of the intermediate zone of the spinal gray matter, whereas the medial system terminates in the ventromedial parts of the intermediate zone. Although the terminal distribution of CST fibers overlaps that of the two subcorticospinal systems, the long descending motor tract systems including the corticospinal system are subdivided into two major systems: (1) a medial system originating in the brainstem (the medial subcorticospinal system) and terminating in the mediodorsal parts of the ventral horn and the adjacent parts of the intermediate zone; and (2) a lateral system consisting of the corticospinal system and the lateral subcorticospinal system that terminate in the lateral and dorsal parts of the intermediate zone (Kuypers, 1964). The purpose of this chapter is to briefly review long descending motor tract systems controlling axial and limb muscles including motoneurons and interneurons and to describe the morphological properties of single long descending motor tract axons in the lateral and the medial descending motor systems.

II. Lateral versus Medial Long Descending Motor Tract Systems

Long descending motor tracts of supraspinal origin are classified into the lateral and medial descending groups based on anatomical and behavioral observations after lesions of these two long descending motor tract groups (Lawrence & Kuypers, 1968), although they are not completely separated. The characteristics of these two descending systems are briefly described below (Kuypers, 1981).

A. The Lateral Descending Motor Tract Group

This group is mainly composed of the CST and RBST, which run in the lateral funiculus of the spinal cord. The CST originates from the sensorimotor cortex (mainly the motor cortex) and terminates at laminae I–VII in the lower mammals and, in addition, at laminae VIII in higher mammals, and at lamina IX in the primates. Recent studies also showed that this direct corticomotoneuronal connection exists in some other mammals (Liang, Moret, Wiesendanger, & Rouiller, 1991). In the primates, the bank area of the precentral, motor cortex innervates directly the motor nuclei for distal limb muscles and the CST from the anterior portion of the motor cortex (body and limb girdle areas) and the premotor area bilaterally terminates at medial laminae VII–VIII. The RBST originating from the magnocellular portion of the red nucleus, after crossing the midline, runs in the ventrolateral part of the medulla and terminates in laminae V–VII in cats and also in lamina IX in primates. With the ascent of the phylogeny, the parvocellular portion of the red nucleus dominates, and the RBST from the magnocellular portion becomes fractional, almost negligible in the humans. The common features of the CST and RBST are summarized as follows (Phillips & Porter, 1977).

1. Both the CST and RBST mainly control distal limb muscles rather than axial and proximal muscles.
2. They exert stronger excitatory effects on flexor muscles and stronger inhibitory effects on extensor muscles.
3. In these two systems, excitatory inputs are mediated disynaptically and inhibitory inputs trisynaptically at the shortest to motoneurons, but, in addition, are mediated mono- and disynaptically in the primates, respectively.
4. The motor cortex and the red nucleus receive inputs from the interpositus and the dentate nuclei. CST neurons receive convergent inputs from the interpositus and dentate nuclei (Shinoda, Futami, & Kakei, 1992a) and RBST neurons receive input from the interpositus nucleus (Tsukahara, Toyama, & Kosaka, 1967).
5. The CST and RBST innervate interneurons spreading from laminae V–VI to lamina VII.

B. The Medial Descending Motor Tract Group

This group consists of the vestibulospinal tract (VST), RTST, tectospinal tract, interstitiospinal tract, and fastigiospinal tract. These tracts mainly run in the ventral or medial portions of the brainstem and the ventral funiculus of the spinal cord and then exert their effects bilaterally in most cases. These tracts terminate on interneurons spreading from lamina VIII to lamina VII such as long propriospinal neurons and commissural neurons and on motoneurons innervating axial and proximal limb muscles (Kuypers & Martin, 1982).

1. The medial system is phylogenetically and ontogenetically older than the lateral system.

2. The medial system characteristically steers body and integrated limb and body movements as well as movement synergisms of individual limbs involving various parts. The lesion of the medial system usually produces motor disturbance of the axial and the proximal muscles including the neck and eye, without affecting distal limb muscles.

3. The VST contains the LVST, originating from the lateral and descending vestibular nuclei, and the medial VST (MVST) from the medial and descending nuclei. The LVST, mainly receiving otolith input, descends ipsilaterally as far as the lumbar spinal cord and exerts excitatory effects on extensor motoneurons and inhibitory effects on flexor motoneurons via Ia inhibitory interneurons in postural control and body equilibrium. The MVST, receiving mainly semicircular canal input, descends bilaterally through the medial longitudinal fasciculus (MLF) and exerts monosynaptic excitatory and inhibitory effects on neck and back muscle motoneurons in vestibulocollic reflex (Wilson & Melvil-Jones, 1979).

4. The RTST is composed of the pontine RTST, originating from the pontine reticular formation that descends in the ventral funiculus ipsilaterally, and terminating in medial laminae VII and VIII, and the medullary RTST from the medulla that descends bilaterally in the ventrolateral funiculus and terminates mainly in lamina VII (partly in laminae VIII and IX). Some RTST neurons receive input from the vestibular organ. Formerly, the facilitatory and inhibitory systems in the reticular formation proposed by Magoun (Magoun & Rhines, 1946; Rhines & Magoun, 1946) were considered to control general muscle tonus of the whole body simultaneously for postural adjustment. But recent studies indicate that some RTST systems may control excitability of specific groups of muscles rather than general muscle tone, suggesting that the RTST may be involved in control of limb movements in addition to postural adjustment (Peterson, 1979).

5. Both VST and RTST neurons receive strong input from the fasti-

gial nucleus, which receives vestibular and somatosensory inputs via the cerebellar vermis. A lesion in the fastigial nucleus or the cerebellar vermis is known to produce severe truncal ataxia (Ito, 1984).

6. The tectospinal and interstitiospinal tract neurons receiving visual and vestibular inputs play an important role in eye and head coordination in orienting responses and do not project below the lower cervical cord (Fukushima, 1987; Grantyn & Berthoz, 1988).

III. Morphological Features of Single Axons in the Lateral and Medial Long Descending Motor Tract Systems

As summarized above in a generalized form, long descending motor tracts, their target motor nuclei and interneurons, may be principally segregated into the medial and lateral motor systems. Morphology of single long descending tract axons had not been documented, since it had not been possible to follow long axons to their cell bodies with anatomical techniques. Recently, it became possible to visualize the morphology of single axons in the long descending motor tracts with intraaxonal injection of horseradish peroxidase (HRP). In this section, the morphological features of single axons in the lateral and medial tract systems will be compared.

A. Morphology of Single Corticospinal Tract Axons

CST axons were penetrated in the lateral funiculus and identified by their direct responses to stimulation of the contralateral motor cortex in the cat (Futami, Shinoda, & Yokota, 1979). Then, they were injected with HRP iontophoretically. The trajectory of single stained axons was reconstructed on serial sections of the spinal cord. Figure 1A shows an example of an axon collateral that emerged from a stem axon. The primary collateral ran ventromedially and entered the gray matter. In the gray matter, fibers ramified successively, forming a deltalike branching pattern in the transverse plane, and terminating in laminae V–VII. In the horizontal plane, the fibers bifurcated successively and gradually turned their orientation in parallel to the long axis of the spinal cord. Finally, they spread over the distance of 1.3–3.0 mm in both ascending and descending directions. Virtually all CST neurons with fast and slow axonal conduction velocities had three to seven axon collaterals like this at different spinal segments (Figure 1B and 1C) (Shinoda et al., 1986a). Multiple axon collaterals were also found in CST neurons in the forelimb area of the monkey motor cortex (Shinoda, Zarzecki, & Asanuma, 1979) and single CST axons projected to different motor nuclei (Shinoda, Yokota, & Futami, 1981). Close contacts of

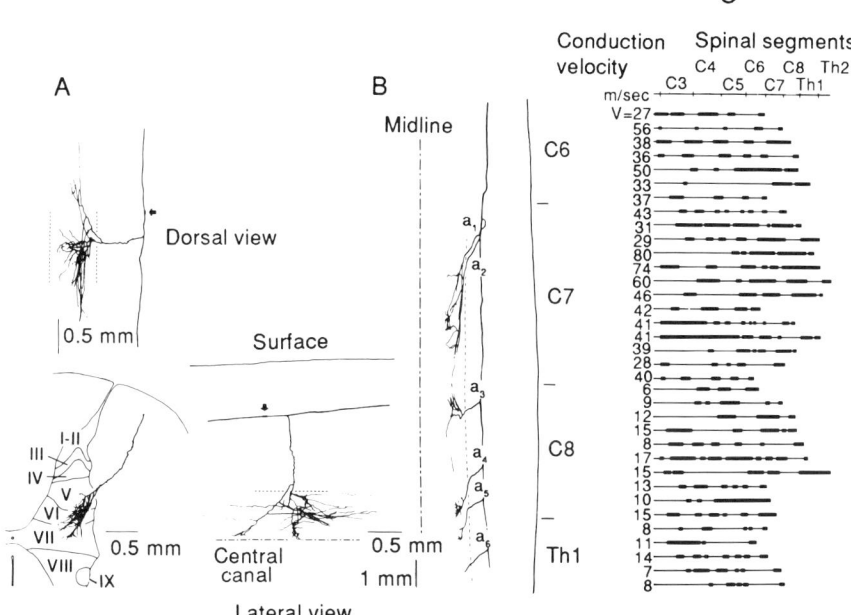

FIGURE 1 (A) Reconstruction of an axon collateral of a CST neuron at C3 in the cat. Note wide rostrocaudal extension of a single axon collateral. Arrow indicates HRP injection site. (From Futami et al., 1979.) (B) Multiple axon collaterals of an HRP-labeled CST axon in the cat spinal cord (dorsal view of the cervical enlargement). (C) Cervical axon collaterals of fast and slow CST neurons in the "forelimb area" of the cat motor cortex. Each row represents a single CST neuron whose conduction velocity is indicated on the left. Thick bars indicate locations of axon collaterals. The abscissa shows normalized spinal segments. (From Shinoda et al., 1986a.)

terminal boutons with the dendrites of some of the HRP-labeled motoneurons could be identified in multiple motor nuclei (Figure 2A).

B. Morphology of Single Rubrospinal Tract Axons

In order to confirm the existence of multiple branches, we stained RBST axons with an intracellular injection of HRP in the cat (Shinoda, Yokota, & Futami, 1982; Shinoda, Futami, Mitoma, & Yokota, 1988a). Stained stem axons could be traced both rostrally and caudally in the lateral funiculus of the cervical cord. Figure 2 shows an example of cervical distribution of axon collaterals of RBST axons. Four axons were stained over a distance of 9–22 mm. Axon collaterals emerged from the stem axons and spread in a deltalike fashion rostrocaudally and also mediolaterally. Terminals were observed mainly in lamina VI, slightly less in lamina V and lamina VII, and they were distributed not only in

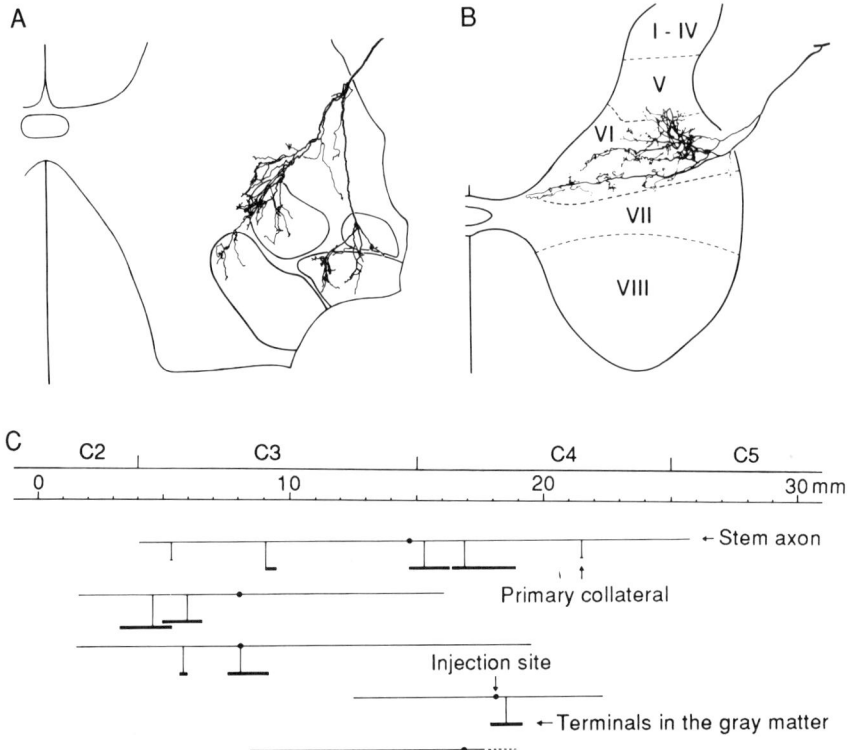

FIGURE 2 (A) Transverse reconstruction of a CST axon originating from monkey motor cortex. Ulnar nerve motoneurons (upper two motor nuclei) and radial nerve motoneurons (lower two motor nuclei) labeled with HRP (from Shinoda et al., 1981). (B) Reconstruction of an axon collateral of an RBST axon at C4 in the cat. (C) Spinal distribution of axon collaterals of single RBST axons in the cat upper cervical cord. Dots indicate injection sites and thick bars terminal distribution of single axon collaterals in a rostrocaudal direction. All five axons terminating above T3 were obtained in the same cat (from Shinoda et al., 1988a).

the lateral, but also in the medial portion of the intermediate zone (Figure 2B) (Shinoda et al., 1982). These terminal areas correspond well to the locations of propriospinal neurons found by Lundberg and his colleagues at C3—C4 of the cat (Illert, Lundberg, Padel, & Tanaka, 1975). Those propriospinal neurons receive rubral inputs and project to motoneurons in the brachial segments. Three characteristic features of branching patterns of RBST axons are shown in Figure 2: (1) RBST axons have multiple axon collaterals; (2) axon collaterals spread widely in a rostrocaudal direction (0.4–2.3 mm); (3) intervals between adjacent primary collaterals are wide, ranging from 1.6 to 6.2 mm.

Our present method allowed us to stain axons for lengths of approximately 25 mm. Therefore, we could not stain axon collaterals in both the upper and the lower cervical cord simultaneously. RBST axons in the lower cervical spinal cord were stained separately. Branching patterns of RBST axons in the lower cervical cord were similar to those described for axons in the upper cervical cord. Axon terminals were mainly distributed in both the lateral and the medial portions of the intermediate zone. Terminals were distributed in lamina VI and the ventral portion of lamina V and the lateral portion of lamina VII. Some were observed in the dorsolateral portion of lamina IX, where some terminal boutons made apparent contact with large neurons, probably motoneurons.

C. Morphology of Single Vestibulospinal Tract Axons

Single MVST or LVST axons were penetrated in the ventral funiculus of the cervical spinal cord in the cat and were identified electrophysiologically as such by their monosynaptic responses to stimulation of the vestibular nerves, and also by their direct responses to stimulation of the MLF or the LVST, respectively. Then, they were intraaxonally stained with HRP (Shinoda et al., 1986a; Shinoda, Ohgaki, Sugiuchi, & Futami, 1988b; Shinoda, Ohgaki, Sugiuchi, & Futami, 1992b).

1. Morphology of Single Lateral Vestibulospinal Tract Axons

Stem axons of LVST axons ran in the ipsilateral ventral funiculus. Over the stained distances (3.4–16.3 mm), most LVST axons terminating in the cervical cord gave off at least one axon collateral (Figure 3), although up to seven collaterals per axon were present (mean = 3.2). The collaterals arose at almost right angles from the stem axons and ran dorsally into the ventral horn. At the entrance into the gray matter, the primary collaterals ramified into a few thick branches in a delta-like or Y-shaped configuration. The branches to the dorsomedial portion of lamina VII gave off extensive thin branches to lamina VIII, including the ventromedial (VM) nucleus of lamina IX and the nucleus commissuralis. Terminal branches were thin (0.2–0.8 μm) with boutons en passant along their length and one bouton at each end. Up to six boutons en passant were strung out on the last 25–50 μm of each terminal branch. The total number of boutons per collateral ranged from 38 to 262, with a mean of 161.

According to the degeneration study of Nyberg-Hansen and Mascitti (1964), terminals of LVST axons were not present on motoneurons in the VM nucleus of the spinal enlargements, although LVST terminals were observed on motoneurons of the thoracic cord. In the present study, terminal boutons appeared to make axosomatic and axodendritic

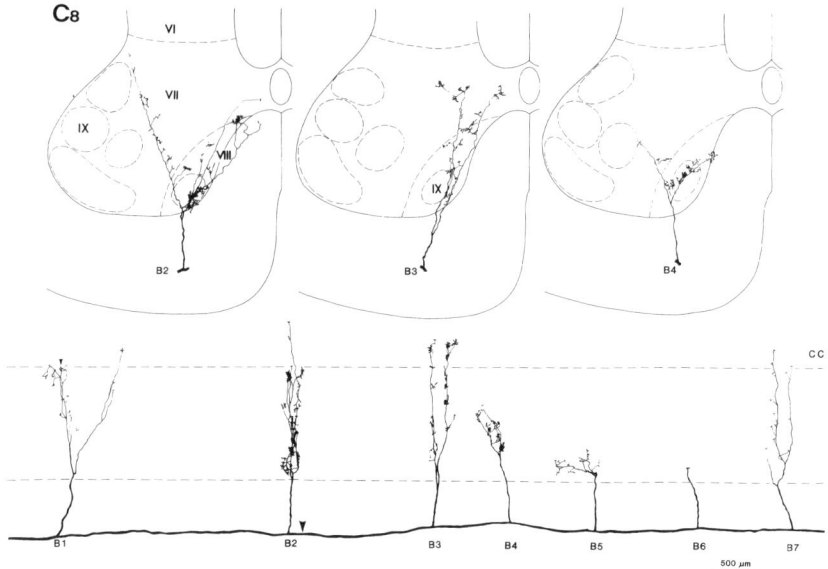

FIGURE 3 Reconstructions of axon collaterals from a single LVST axon at C8. Upper drawings are reconstructed in the transverse plane of B2–B4 axon collaterals shown in the sagittal plane in the lower drawing. The lower border of the central canal (CC) and the lower border of the ventral horn are indicated by the dashed lines. The arrow indicates the injection site. (From Shinoda et al., 1986a.)

contacts with not only small- and medium-sized neurons, but also with large neurons in the VM nucleus that are probably motoneurons to axial muscles. The commissural nucleus, a cell group close to the medial border and the base of the ventral horn, contains cells with their axons running across the midline in the anterior commissure. Some of these commissural neurons terminate on contralateral motoneurons (Harrison, Jankowska, & Zytnicki, 1986). Since a large number of terminal boutons of LVST axons were found in this nucleus, the contralateral effects following unilateral vestibular nucleus stimulation are probably mediated by way of this commissural connectivity (Hongo, Kudo, & Tanaka, 1975; Sugiuchi, Kakei, & Shinoda, 1992). Many boutons were observed in lamina VII adjacent to lamina IX in the lateral ventral horn. LVST projection to this area has not been reported before, but is important, since inhibition evoked disynaptically in flexor motoneurons from Deiters's nucleus (Grillner, Hongo, & Lund, 1970) is mediated by Ia inhibitory neurons located in this area (Jankowska & Lindstroem, 1972). In contrast to the wide spread of axon collaterals in the transverse plane of the spinal cord, the rostrocaudal extent of single axon collaterals was restricted, ranging from 230 to 1560 μm with an average of 760 μm (see the lower drawing of Figure 3). There were usually gaps free of terminal boutons between terminal fields of adjacent axon collaterals, since intercollateral intervals (mean = 1490 μm) were much longer than the rostrocaudal extent of each terminal field.

2. Morphology of Single Medial Vestibulospinal Tract Axons

MVST axons were classified into two groups, crossed and uncrossed MVST axons, which descended in the spinal cord contralateral and ipsilateral to their cell bodies, respectively. Stem axons of MVST neurons ran in the mediodorsal portion of the ventral funiculus. The branching pattern of MVST axons was very similar to that of LVST axons, but different from that of CST and RBST axons (Futami et al., 1979; Shinoda et al., 1982, 1988b). One of the most important characteristics of the branching patterns was the existence of multiple axon collaterals. One to seven axon collaterals were seen for individual MVST axons (Shinoda et al., 1988b). Both uncrossed and crossed MVST axons had many common features regarding branching pattern and terminal distribution. A typical example of an uncrossed MVST axon is illustrated in Figure 4. In this axon, nine axon collaterals arose from a stem axon at almost right angles. Primary collaterals ran laterally and entered into the ventral horn at its medial border. They divided into several thick branches immediately after the entrance to the ventral horn and spread in a deltalike fashion in the transverse plane. Three groups of branches were separable in terms of their course and destination. A typical example is depicted in Figure 5. One group of branches ran ventrolaterally into the VM nucleus and gave rise to extensive terminal arborizations there. Some of them further extended into the nucleus spinalis n. accessorii (SA). A second group of branches projected laterally to the SA nucleus or its adjacent lamina VIII. On their way, thin branchlets were given off to lamina VIII dorsal to the VM nucleus. A third group of branches ran dorsolaterally, emitting thin branchlets on their way and terminating in the medial portion of the dorsal lamina VIII and its adjacent lamina VII. These three groups of branches did not always exist and usually one or two groups were lacking. In contrast to the wide fan of terminal arborization in the transverse plane, the rostrocaudal extent of single axon collaterals was restricted (see Figure 4, right), ranging from 300 to 2100 μm with a mean of 620 μm. Since the average distance between adjacent primary collaterals (1870 μm) was much wider than the rostrocaudal extent of single axon collaterals, there were usually gaps free of terminal boutons between the terminal fields of adjacent axon collaterals. The terminal area of MVST axons occupied lamina IX, including both the VM and the SA nuclei; lamina VIII, including the commissural nucleus; and lamina VII, including the central cervical nucleus (Figure 4). Terminal boutons were most pre-

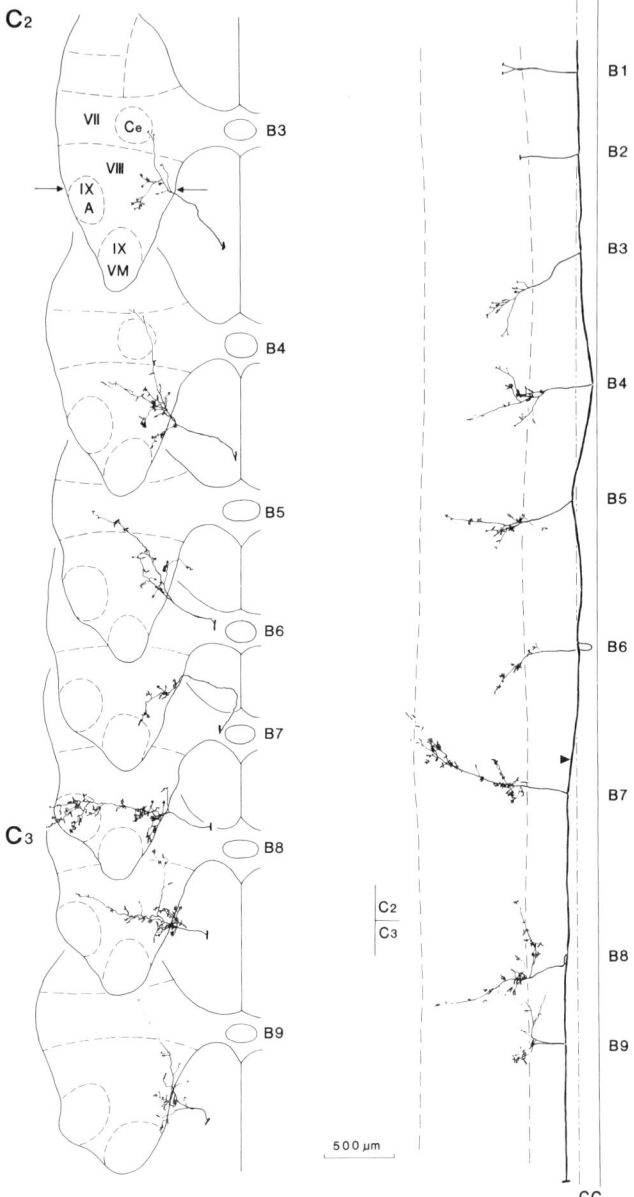

FIGURE 4 Reconstruction of an uncrossed second-order MVST axon filled with HRP in the transverse plane (*left*) and in the horizontal plane (*right*) at C2–C3 segments of the cat spinal cord. Seven axon collaterals were given off at right angles, more or less, from a stem axon to the motor nucleus in lamina VIII just dorsomedial to the ventromedial nucleus (VM) and one collateral to the nucleus spinalis n. accessorii (A). A single MVST axon innervates multiple motor nuclei of neck muscles. (From Shinoda *et al.*, 1992b.)

dominant in lamina IX, especially in the VM nucleus, where many boutons appeared to make contact with cells. Axosomatic and axodendritic contacts were observed on large, medium, and even small cells (Figure 5) (Shinoda *et al.,* 1992b). The VM nucleus contains motor nuclei of neck extensors (m. dorsi proprii) and the SA nucleus contains motor nuclei of neck flexors (Sugiuchi & Shinoda, 1992). To confirm that MVST axons indeed make contact with motoneurons in lamina IX, these motoneurons innervating different neck muscles were retrogradely labeled with HRP. Some boutons of MVST axons were observed on cell bodies or proximal dendrites of the labeled motoneurons and

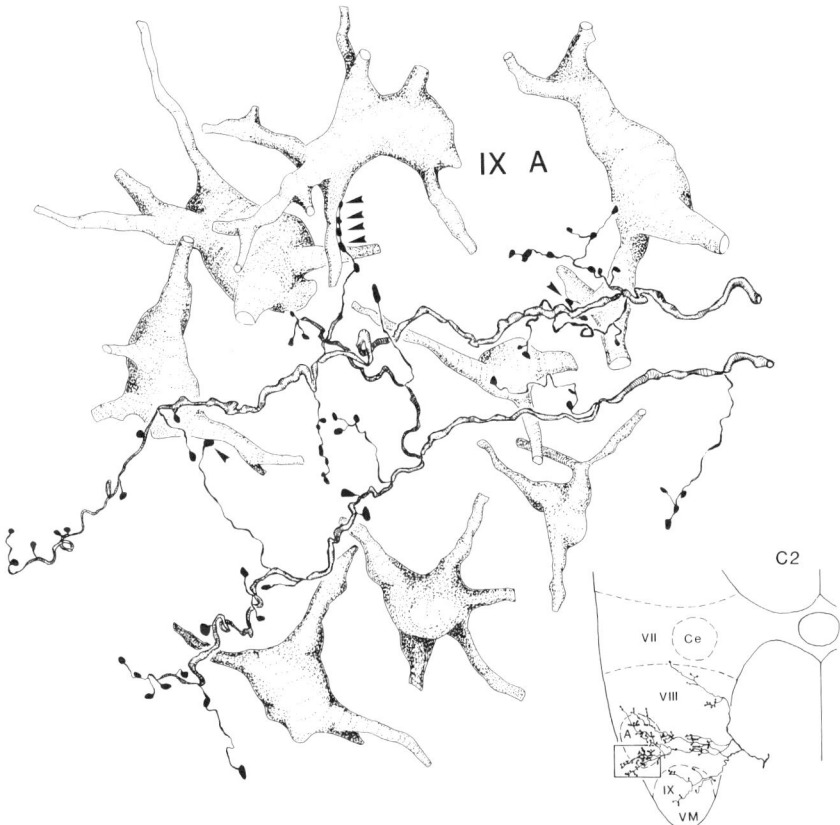

FIGURE 5 Camera lucida drawing of an axon collateral of an HRP-filled, second-order MVST axon in the cervical spinal cord of the cat. A single MVST axon innervates multiple motor nuclei of neck muscles (*bottom right*). Synaptic boutons of the MVST axon (arrowheads) make contacts on the proximal dendrites of neurons in the motor nucleus (squared area, *bottom right*). (From Shinoda *et al.,* 1992b.)

others of the same MVST axons seemed to make contact with cell bodies or proximal dendrites of unlabeled but large counterstained cells in a different portion of the VM nucleus. This finding strongly suggests that single MVST axons innervate multiple motor nuclei of different neck muscles. This possibility was further confirmed by the following finding. Three axon collaterals of the MVST axon in Figure 5 projected to both the VM and the SA nuclei. About one-third of the examined MVST axons projected to both the VM and the SA nuclei, and presumed axodendritic contacts were observed on large neurons in each nucleus. This finding gives further morphologic support for single VST axons innervating motoneurons of different neck muscles simultaneously.

IV. Summary

The present result indicates that single long descending motor tract axons are not a simple private line connecting supraspinal centers and motoneurons, but instead they may exert influences onto different groups of spinal neurons simultaneously at widely separated segments with their multiple axon collaterals. Our data provided evidence of the existence of multiple axon collaterals of single LVST and MVST axons at different spinal segments. These single axons make contacts with motoneurons innervating axial muscles. Therefore, they may control the excitability of multiple axial muscles concurrently at multisegmental levels, even at widely separate levels between the cervical and the lumbar regions in the case of LVST axons. Multiple branching of single LVST axons might function to coordinate properly widespread activation of the muscles of the neck, trunk, and limbs. MVST axons have extensive terminals in the cervical cord and innervate multiple motor nuclei. Stimulation of a semicircular canal produces head movement in the same plane as the stimulated canal. This indicates that a signal from the stimulated canal must be distributed to a proper set of neck muscles. The present data show that a single MVST axon may specify a spatial pattern of cocontraction of multiple neck muscles, which compose a functional synergy of neck muscles for a proper compensatory head movement. Obviously, a given motor control signal may be economically distributed by way of a single neuron with divergent branches to multiple target sites that participate in cocontraction of muscles to produce a purposeful movement. In a redundant system like the neck control system, where many muscles function at multiple joints to produce a particular head movement, this kind of multiple innervation of muscles by single long descending tract axons may help decrease degrees of freedom in the system. Both CST and RBST axons have multiple axon collaterals at different cervical segments. Some of their collaterals may terminate on interneurons intervening in spinal reflex pathways and control the excitability of reflex pathways, so that they may suppress or facilitate the built-in reflex systems at various spinal levels before and during voluntary movement. With some other collaterals, the same neurons may influence the excitability of ascending tract neurons transmitting internal and external feedback signals to the brain stem, the cerebellum, and the cerebral cortex. In addition to these collaterals, some other collaterals of the same single neurons may exert inhibitory as well as excitatory influences on motoneurons of different muscles. It is known that stimulation of the motor cortex causes cocontraction of many muscles and also reciprocal inhibition of antagonistic muscles. This inhibitory effect might be an important function of the motor cortex, as Sherrington (1906) pointed out that inhibition of motoneurons might be more prominent and widely represented across the cortex than excitation. Movements are brought about by proper selection of timing and spatial patterns of multiple muscle contraction, and spatial coordination of these muscles is determined by the central nervous system, but we know little about the neural mechanism of proper selection of muscles for a particular movement. The present data suggest that single long descending motor tract axons with multiple spinal collaterals may play a part for selection of appropriate combinations of muscles (functional synergies). This will be the first demonstrated example of hard-wired neural implementation of a functional synergy. The wide degree of divergence of single long descending axons in the spinal cord may provide some important clues for understanding the function of the long descending motor tract systems in control of movement.

Acknowledgments

This study was supported in part by research grants from the Human Frontier Science Program and the Japanese Ministry of Education, Science, and Culture.

References

Abzug, C., Maeda, M., Peterson, B. W., & Wilson, V. J. (1974). Cervical branching of lumbar vestibulospinal axons. *Journal of Physiology (London)*, **243,** 499–522.
Brodal, A. (1981). Neurological anatomy in relation to clinical medicine (3rd ed.) (pp. 294–393). London, New York, Toronto: Oxford University Press.
Burke, R. E., Strick, P. L., Kanda, K., Kim, C. C., & Walmskey, B. (1977). Anatomy of medial gastrocnemius and soleus motor nuclei in cat spinal cord. *Journal of Neurophysiology,* **40,** 667–680.

Fukushima, K. (1987). The interstitial nucleus of Cajal and its role in the control of movements of head and eyes. *Progress in Neurobiology,* **29,** 107–192.

Futami, T., Shinoda, Y., & Yokota, J. (1979). Spinal axon collaterals of corticospinal neurons identified by intracellular injection of horseradish peroxidase. *Brain Research,* **164,** 279–284.

Grantyn, A., & Berthoz, A. (1988). The role of the tectoreticulospinal system in the control of head movement. In B. Peterson & F. Richmond (Eds.), *The control of head movement* (pp. 224–244). New York: Oxford University Press.

Grillner, S., Hongo, T., & Lund, S. (1970). The vestibulospinal tract. Effects on alpha-motoneurones in the lumbosacral spinal cord in the cat. *Experimental Brain Research,* **10,** 94–120.

Harrison, P. J., Jankowska, E., & Zytnicki, D. (1986). Lamina VIII interneurones interposed in crossed reflex pathways in the cat. *Journal of Physiology (London),* **371,** 147–166.

Hongo, T., Kudo, N., & Tanaka, R. (1975). The vestibulospinal tract: Crossed and uncrossed effects on hindlimb motoneurones in the cat. *Experimental Brain Research,* **24,** 37–55.

Illert, M., Lundberg, A., Padel, Y., & Tanaka, R. (1975). Convergence on propriospinal neurones which may mediate disynaptic corticospinal excitation to forelimb motoneurones in the cat. *Brain Research,* **93,** 530–534.

Ito, M. (1984). *The cerebellum and neural control* (pp. 325–352, 436–451). New York: Raven Press Books.

Jankowska, E., & Lindstroem, S. (1972). Morphology of interneurones mediating Ia reciprocal inhibition of motoneurones in the spinal cord of the cat. *Journal of Physiology (London),* **266,** 805–823.

Kuypers, H. G. J. M. (1964). The descending pathways to the spinal cord, their anatomy and function. In J. C. Eccles & J. P. Schade (Eds.), *Organization of the spinal cord: Progress in Brain Research,*Vol. 11 (pp. 178–202). Amsterdam: Elsevier.

Kuypers, H. G. J. M. (1981). Anatomy of the descending pathways. In V. B. Brooks (Ed.), Handbook of physiology Section 1, Vol. II (pp. 597–666). Chicago: American Physiological Society.

Kuypers, H. G. J. M., & Martin, G. F. (1982). *Anatomy of descending pathways to the spinal cord. Progress in Brain Research,* Vol. 57. Amsterdam: Elsevier.

Kuypers, H. G. J. M., Fleming, W. R., and Farinholt, J. W. (1962). Subcorticospinal projections in the rhesus monkey. *Journal of Comparative Neurology,* **118,** 107–137.

Lawrence, D. G., & Kuypers, H. G. J. M. (1968). The functional organization of the motor system in the monkey. *Brain,* **91,** 1–15.

Liang, F., Moret, V., Wiesendanger, M., & Rouiller, E. M. (1991). Corticomotoneuronal connections in the rat: Evidence from double-labeling of motoneurons and corticospinal axon arborizations. *Journal of Comparative Neurology,* **331,** 356–366.

Magoun, H. W. & Rhines, R. (1946). An inhibitory mechanism in the bulbar reticular formation. *Journal of Neurophysiology,* **9,** 165–171.

Nyberg-Hansen, R., & Mascitti, T. A. (1964). Sites and mode of termination of fibers of the vestibulospinal tract in the cat. An experimental study with silver impregnation methods. *Journal of Comparative Neurology,* **122,** 369–388.

Peterson, B. W. (1979). Reticulo-motor pathways: their connections and possible roles in motor behavior. In H. Asanuma & V. J. Wilson (Eds.), *Integration in the nervous system* (pp. 185–200), Tokyo: Igaku Shoin.

Peterson, B. W., Maunz, R. A., Pitts, N. G., & Mackel, R. (1975). Patterns of projection and branching of reticulospinal neurons. *Experimental Brain Reearch,* **23,** 331–351.

Phillips, C. G., & Porter, R. (1977). *Corticospinal neurones. Their role in movement.* London, New York: Academic Press.

Rexed, B. (1954). A cytoarchitectonic atlas of the spinal cord in the cat. *Journal of Comparative Neurology,* **100,** 297–379.

Rhines, R. and Magoun, H. W. (1946). Brain stem facilitation of cortical motor response. *Journal of Neurophysiology* **9,** 219–229.

Romanes, G. J. (1951). The motor cell columns of the lumbo-sacral spinal cord of the cat. *Journal of Comparative Neurology,* **94,** 313–363.

Sherrington, C. (1906). *The integrative action of the nervous system.* New Haven, Connecticut. Yale University Press.

Shinoda, Y., Arnold, A., & Asanuma, H. (1976). Spinal branching of corticospinal axons in the cat. *Experimental Brain Research,* **26,** 215–234.

Shinoda, Y., Ghez, C., & Arnold, A. (1977). Spinal branching of rubrospinal axons in the cat. *Experimental Brain Research,* **30,** 203–218.

Shinoda, Y., Zarzecki, P., & Asanuma, H. (1979). Spinal branching of pyramidal tract neurons in the monkey. *Experimental Brain Research,* **34,** 59–72.

Shinoda, Y., Yokota, J., & Futami, T. (1981). Divergent projection of individual corticospinal axons to motoneurons of multiple muscles in the monkey. *Neuroscience Letters,* **23,** 7–12.

Shinoda, Y., Yokota, J., & Futami, T. (1982). Morphology of physiologically identified rubrospinal axons in the spinal cord of the cat. *Brain Research,* **242,** 321–325.

Shinoda, Y., Ohgaki, T., & Futami, T. (1986a). The morphology of single lateral vestibulospinal tract axons in the lower cervical spinal cord of the cat. *Journal of Comparative Neurology,* **249,** 226–241.

Shinoda, Y., Yamaguchi, T., & Futami, T. (1986b). Multiple axon collaterals of single corticospinal axons in the cat spinal cord. *Journal of Neurophysiology,* **55,** 425–448.

Shinoda, Y., Futami, T., Mitoma, H., & Yokota, J. (1988a). Morphology of single neurones in the cerebello-rubrospinal system. *Behavioral Brain Research,* **28,** 59–64.

Shinoda, Y., Ohgaki, T., Sugiuchi, Y., & Futami, T. (1988b). Structural basis for three-dimensional coding in the vestibulospinal reflex. Morphology of single vestibulospinal axons in the cervical cord. *Annals of New York Academy of Sciences,* **545,** 216–227.

Shinoda, Y., Futami, T., & Kakei, S. (1992a). Inputs from the cerebellar nuclei to the forelimb area of the motor cortex. *Experimental Brain Research,* **22,** 65–84.

Shinoda, Y., Ohgaki, T., Sugiuchi, Y., & Futami, T. (1992b). Morphology of single medial vestibulospinal tract axons in the upper cervical spinal cord of the cat. *Journal of Comparative Neurology,* **316,** 151–172.

Sprague, J. (1951). Motor and propriospinal cells in the thoracic and lumbar ventral horn of the rhesus monkey. *Journal of Comparative Neurology,* **95,** 103–123.

Sugiuchi, Y., & Shinoda, Y. (1992). Organization of the motor nuclei innervating epaxial muscles in the neck and back. In A. Berthoz, W. Graf, & P. P. Vial (Eds.), *The head-neck sensory motor system* (pp. 235–240). New York, Oxford: Oxford University Press.

Sugiuchi, Y., Kakei, S., & Shinoda, Y. (1992). Spinal commissural neurons mediating vestibular input to neck motoneurons in the cat upper cervical spinal cord. *Neuroscience Letters,* **145,** 221–224.

Tsukahara, N., Toyama, K., & Kosaka, K. (1967). Electrical activity of red nucleus neurones investigated with intracellular microelectrodes. *Experimental Brain Research,* **4,** 18–33.

Wilson, V. J., & Melvil-Jones, G. (1979). *Mammalian vestibular physiology.* New York: Plenum.

3
Sensory–Motor Coordination in Crustacean Limbs during Locomotion

Daniel Cattaert, Jean-Yves Barthe*, and François Clarac

Laboratoire de Neurobiologie et Mouvements
Centre National de la Recherche Scientifique
Marseille, France

I. Introduction
II. Review of the Central and Peripheral Mechanisms Involved in Coordination
    A. Central Coordination
    B. Coordination Involving Sensory Feedback
III. Examples of Coordination in the Crayfish Locomotor System
    A. Intralimb Proprioceptive Control of Output Intensity
    B. Intra- and Interleg Proprioceptive Control of Coordination
IV. Conclusion
    References

I. Introduction

The efficiency of all behavioral motor acts necessarily depends on movement coordination (Bush & Clarac, 1985). This coordination is mediated by patterns of activity of the nervous motor command controlling the timing of muscle contractions. Synchronisms, antagonisms, and lags are encoded in the nervous command, so that motor acts can be described in terms of program implementation using notions such as hierarchy and sequential activation (Humphrey & Freund, 1991). Even in the simplest metameric animals, various levels of coordination can be distinguished. Simple antagonistic operations (flexors and exten-

---

* Present address: Karolinska Institutet, The Nobel Institute for Neurophysiology, Stockholm, Sweden.

sors) are carried out at the joint level, but more complex leg movements require the activity occurring at various joints to be coordinated. In the case of the locomotor system, interleg coordination is responsible for body propulsion (Amblard, Berthoz & Clarac, 1988), but several motor systems may participate in a given task. Another way of describing motor commands consists of dissociating the static from the dynamic components, the axial muscles from the leg muscles, or the postural muscles from the dynamic muscles, for example, or those involved in body support from those mainly involved in movements (Massion, 1992).

Coordination has often been studied in the context of specific tasks. One of the tasks on which investigations of this kind have focused most intensively has been locomotion, which is a highly coordinated rhythmic activity. More than 50 years ago, Von Holst (1939) defined the relationships between the diverse elementary movements of body parts during locomotion as absolute or relative coordination. Absolute coordination occurs when two rhythmic activities are beating at the same frequency with a constant phase lock. Relative coordination refers to a looser coupling, where the two elements may be at different frequencies, but display a predominant relationship. From these considerations, the notion of an elementary oscillator emerged during the subsequent decades. Elementary oscillators were taken to be more or less closely coupled to account for the various types of coodination encountered. This approach to locomotor rhythm was widely adopted in studies such as those by Stein (1976), Wendler (1974), Barnes and Gladden (1985), and Graham (1978) on rhythm production, its variability and adaptability, the entrainment properties, and the phase-dependent responses to an input in various rhythmic motor systems. One fundamental question addressed in these studies concerned the origin of the coordination. Does it take place in the CNS itself, as the result of the central circuitry (central pattern generator, CPG) that controls lags between the various motor commands (Getting, 1989), or might it be the peripheral sensory receptors, which, depending on the external constraints, provide the precise timing needed in a behavioral task? This question has gradually faded into the background over the years because it has been demonstrated that both mechanisms are in fact involved. It is now no longer a matter of evaluating their respective contributions, but rather of determining how they interact during the execution of motor tasks.

More recently, the question of rhythm production and coordination was addressed in terms of the self-organizing processes dynamically generated in "coordinative" networks (Schöner & Kelso, 1988). This is a synergetic approach, which calls on dynamic properties of both the musculoskeletal apparatus and the nervous command (Haken, 1983).

In this chapter, we will adopt a neurobiological point of view, but we will refer to biomechanical structures of various types to demonstrate how the central nervous system is adapted to the mechanical apparatus it controls. The examples we present here were obtained on invertebrates because of their simplicity and because, in these animals, the various elements of the central nervous network are particularly accessible. These data provide good illustrations of the fundamental mechanisms involved in coordination. Analyses of this kind are more difficult in mammals because of the complexity of their structures. Nevertheless, the solutions to postural and kinetic problems adopted by arthropods seem to be analogous to those of vertebrates. Of course, their general organization, involving an external skeleton and an internal musculature (Figure 1) and a large number of segments and legs, appears quite complex at first sight. This complexity is only apparent, however. For example, in crustaceans, each joint is bicondylar, so that movements are possible in only one plane (Figure 1). The metameric structure of the body further simplifies the approach: each walking leg is more or less similar to the others, and is commanded by a central nervous network isolated in a hemiganglion. Each leg is controlled by a small number of motor neurons (<70) and interneurons (200), some of which are local and restricted to one ganglion, whereas others are intersegmental and distributed among several ganglia. In each leg there exist thousands of sensory neurons corresponding to various sensory structures that can be activated and controlled in both *in vivo* and *in vitro* experiments. Moreover, the various groups of crustaceans all have fundamentally the same musculoskeletal pattern of organization, and comparing the neurophysiology of their motor systems can be an instructive means of describing the basic organization of the central nervous networks. For example, between the pelagic swimmers and the benthic walkers, the propulsive appendages differ, although in lobsters and crayfishes both systems coexist: a thoracic walking system and an abdominal swimming system (Figure 1). In the latter case, Heitler (1983) suggested that the central commands were identical; functional differences were observed that mainly reflected adaptation to the specificity of the limbs.

Although they may possibly have a common origin, the two locomotor apparatuses of the crayfish are mechanically quite different. The walking system works by exerting forces on a hard substrate (the ground) and has to deal with problems of force division during posture and locomotion. Each movement or position of a leg in contact with the substrate produces forces that affect the body's position and its equilibrium. The situation is different in the case of the swimming system, where force is exerted on a fluid medium, and where the position of the swimming limbs does not directly affect the body equilibrium.

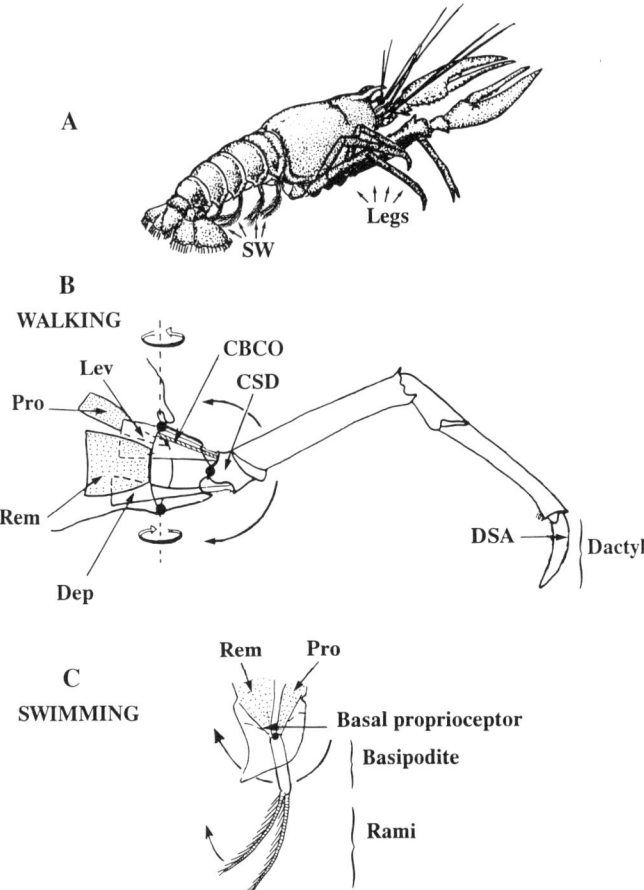

FIGURE 1 Locomotor system of the crayfish. (A) In the crayfish, locomotion consists of coordinated movements performed by four pairs of thoracic legs. The abdominal appendages (swimmerets, SW) can participate, however, in the propulsion. (B) Rear view of a leg. Only the first two joints are shown. Each joint is bicondylar and can perform movements in one plane only. The first joint makes forward and backward movements of the leg possible: These are commanded by the promotor (Pro) and remotor (Rem) muscles, respectively. The second joint is orthogonal to the first and makes upward and backward movements of the leg possible: These are commanded by the levator (Lev) and depressor (Dep) muscles, respectively. Three sensory structures are shown: Dactyl sensory afferents (DSA); cuticular stress detector (CSD); coxo-basipodite chordotonal organ (CBCO). (C) Lateral view of a swimmeret. Each swimmeret is composed of a basal segment prolonged by two flexible, contractile rami. The movements performed in the sagittal plane involve alternating contractions of the remotor muscle (Rem) for rearward powerstrokes, and the promotor muscle (Pro) for forward returnstrokes. The rami muscles contract and cause the rami to curl during the powerstroke.

The walking system of the crayfish or homarus is composed of the four most posterior pairs of appendages (the first pair is used like a chela to catch prey). Each leg movement is mainly controlled by the four basal muscles: the first joint allows for forward and backward leg movements, and is commanded by the promoter (Pro) and remotor (Rem) muscles, respectively (Figure 1B). The second joint of each leg allows for upward and downward movements, and is commanded by the levator (Lev) and depressor (Dep) muscles, respectively (Figure 1B). The animal is able to walk in all directions: forward, backward, and sideways. These various modes of locomotion are characterized by specific patterns of activities in the leg muscles. For example, during forward walking, the *swing* phase is characterized by coactivation of the levator and promotor muscles, whereas during the *stance* phase it is the depressor and remotor activities that are synchronous.

The swimmerets of the crayfish or homarus (Figure 1C) are simple abdominal appendages composed of a basal segment (basipodite) prolonged by a pair of rami. The basal movements consist of alternating backward (powerstroke) and forward (returnstroke) movements commanded by the remotor (Rem) and promotor (Pro) muscles, respectively (Figure 1C). At the end of the backward movement, the rami are actively curved to eject water backward and then exert a propelling force. They stay curved during the forward movement and then extend before and during the next backward movement.

The crayfish nervous system was one of the first isolated preparations in which rhythmic activities were studied (Ikeda & Wiersma, 1964). It is composed of a ventral chain of ganglia (one in each segment, commanding a pair of appendages). The abdominal ganglia that control the swimmerets were studied first because they elicit coordinated rhythmic activities in the four pairs of swimmeret motor nerves (Ikeda & Wiersma, 1964; Stein, 1971; Heitler, 1978). More recently, an isolated thoracic ganglia preparation has been used to analyse more complex command of the legs (Sillar & Skorupski, 1986; Sillar, Clarac & Bush, 1987; Chrachri & Clarac, 1989), as well as interactions between the two systems (Barthe, Bevengut & Clarac, 1991). Detailed studies of motor commands in similar animals (*Homarus, Palinurus, Carcinus*) have completed our knowledge of the central and peripheral mechanisms involved in coordination.

II. Review of the Central and Peripheral Mechanisms Involved in Coordination

Some coordination mechanisms exist within the central nervous system and are part of the central network; others are closely linked with sensory information. We will discuss them in this order.

A. Central Coordination

1. Existence of Central Pattern Generators

From to the literature of the past few years, it can be seen that a great number of preparations, even isolated ones such as the nervous ganglia of invertebrates or the spinal cord of lower vertebrates, are able to produce coordinated rhythmic activity. The swimmeret rhythm of Crustacea was one of the first to be described. Hughes and Wiersma (1960) demonstrated that it is possible to elicit the swimmeret bursting pattern of discharge in the first roots of ganglia 2–5 (Figure 2A, D). This has been confirmed by several authors who have studied in detail the organization of the nervous command involved (Davis, 1968; Heitler, 1978; Mulloney, Acevedo, Hall, & Sherff, 1990). The pattern of this central discharge is quite highly elaborated and contains the various parameters necessary for a totally coordinated motor command. If *in vitro* swimming is compared with that of an intact animal, the only difference concerns the variability of the response and the intensity of the discharge, which is much stronger in the intact animal. In fictive swimming, the frequency of the rhythm is more stable and more stereotyped. We have demonstrated in particular that when the nervous chain consisting of the abdominal and thoracic parts (Figure 2A, C, D) is kept intact, the swimming rhythm can last for several hours (Barthe *et al.*, 1991). The interganglionic coordination is also quite stable and similar to that observed in freely moving animals (Figure 2D). Since the output of the CNS is identical in both "isolated" and *in vivo* situations, we assume that the CNS contains motor programs that are completely preadapted to the biomechanical constraints of the limbs they command in a behavioral context.

Several other invertebrate preparations have provided similar results to those obtained on the swimmeret system, and in these animals, isolated parts of the CNS have been found to induce spontaneous rhythmic activities, such as feeding, swimming, or respiration (for review, see Getting, 1988). When isolated, various other preparations also show a pattern of motor activity as long as pharmacological activation is provided. For example, in the cat, an L-Dopa injection induces a rhythmic discharge in the lumbar root of a spinal animal (Grillner, 1981). In the neonate rat (Cazalets, Squalli-Houssaini & Clarac, 1992) and in the lamprey (Grillner, Wallén, Brodin & Lansner, 1991), excitatory amino acids (glutamate or aspartate) induce a fictive locomotor rhythm.

In the crayfish, the thoracic locomotor nervous system can be spontaneously active. This is rare, however, in totally isolated preparations. Perfusion with a solution of oxotremorine ($10^{-5}\,M$) in most cases elicits fictive locomotor activity. In a preparation of this kind, the rhythm is slow, however, in comparison with intact walking (step cycle between 10 and 40 sec). Here, the most frequently observed pattern is backward walking (Figure 2B), which involves a synchronous discharge of the

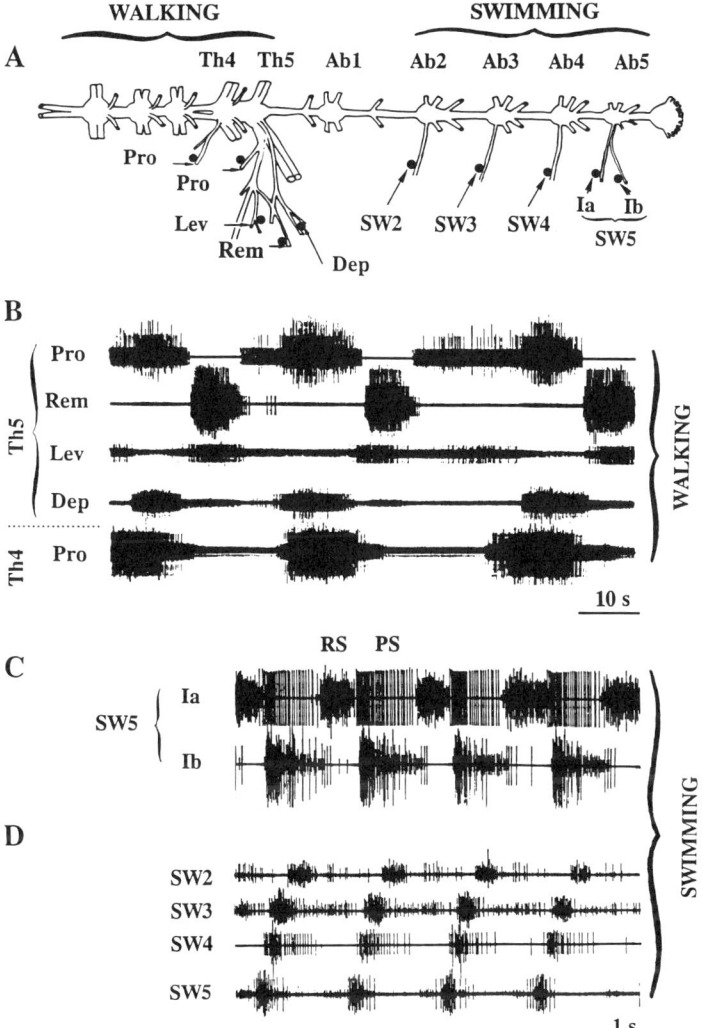

FIGURE 2 Fictive walking and swimming in the crayfish. (A) General view of the central nerve cord, composed of five thoracic ganglia, each of which is responsible for controlling the movement of one pair of legs, and five abdominal ganglia commanding four pairs of swimmerets. En passant electrodes were placed on motor nerves commanding the promotor (Pro), remotor (Rem), levator (Lev), and depressor (Dep) muscles from the fourth (Th4) and/or the fifth (Th5) thoracic ganglia. The activity of the swimmeret system was also monitored by electrodes placed on motor nerves of the second (SW2), third (SW3), fourth (SW4), and fifth (SW5) swimmerets. In some experiments, this motor nerve could be subdivided into anterior (Ia) and posterior (Ib) roots mainly controlling the return-stroke and powerstroke, respectively. (B) Fictive backward walking in isolated thoracic preparation. Note that the centrally driven promotor activities from the fourth (Th4) and the fifth (Th5) thoracic ganglia are synchronous (although they are in antiphase in intact animals during walking). (C, D) Fictive swimming in isolated abdominal preparation. (C) Alternating powerstoke (PS) and returnstroke (RS) activities recorded from the fifth swimmeret (SW5) first nerve root. (D) Coordination of swimmeret powerstroke bursts recorded from the motor nerves of the second, third, fourth, and fifth swimmerets. Note that each cycle of activity started at SW5 and propagated to more anterior ganglia.

levator and remotor nerves. A forward pattern is observed only rarely. This fictive walking activity is characterized by a clear-cut opposition between the discharge of the motor nerves innervating the antagonistic muscles of the same joint, while the interjoint coordination is often weaker than in normal walking. Moreover, the interleg walking pattern is quite different between intact and isolated preparations (Sillar et al., 1987) and is characterized by the absence of a correlation between the rhythms of contralateral legs, whereas ipsilateral legs operate practically in phase; this is the case in Figure 2B, where the discharge from the leg 4 promotor is activated at the same time as that from the leg 5 promotor.

Some isolated preparations do exist in which it is apparently impossible to characterize any rhythm at all (Bässler, 1983). This absence of patterned activities may be due either to a real lack of coordinating elements or to their being damaged, which makes any central coordination impossible. In humans, for example, this question is unsolved. Failure to recover spinal rhythmicity in paraplegic patients seems to be due to some central deficit. Indirect evidence has shown, however, that there exists a spinal CPG (Bussel, Roby-Brani, Azouri, Biraben, Yakovleff & Held, 1988).

2. Origin of the Rhythmic Coordinated Patterns

In the CNS, production of a rhythmic pattern of activity mainly results both from the membrane properties of some interneurons (INs) and motor neurons (MNs), and from the connectivity of the neurons composing the network. From the point of view of the membrane properties, some neurons can oscillate alone due to some voltage-dependent channels, which open and close alternately in a particular depolarization range (Grillner et al., 1991). Since the connectivity between neurons is the basis of the coordination, it was hypothesized that connecting INs exist at each level of complexity. Getting (1988) defined the notion of a network organized in terms of a "building block" in order to show the existence of INs that are capable of connecting momentarily together an ensemble of subnetworks able to trigger the motor activity required for a given task. A similar arrangement has been demonstrated in the stomatogastric ganglion of crustaceans, where a single IN can link together neurons belonging to different blocks to build a new CPG (Meyrand, Simmers & Moulins, 1991).

In the crayfish walking system, the intralimb coordination seems to involve the coupling of joint-related oscillators. In some isolated preparations, it is possible to observe rather independent rhythms in the nerves controlling the two basal joints: the promotor/remotor rhythm is different from the levator/depressor rhythm. These findings suggest that each joint is controlled by an oscillator. During the coordinated activities involved in forward or backward walking, some coupling mechanism must therefore exist between the joint oscillators. This central coupling is shown in Figure 3. In this experiment, an isolated thoracic nervous system displayed a backward fictive walking, that is, it was characterized by synchronized nerve activities commanding the two joints (Prom/Rem–Lev/Dep), where the promotor discharged with the depressor nerve and the remotor with the levator. This well-coordinated activity may become desynchronized if the remotor MN is hyperpolarized. Injecting hyperpolarizing current (-5 nA) into this MN prevents the rhythmic activity and subsequently perturbs the levator activity, which becomes tonic (Figure 3B). The following conclusions can be drawn from these results: (1) some MNs display active properties (the remotor MN oscillates only when depolarized, and is silent when hyperpolarized); (2) these MNs can participate in central interjoint coordination. Surprisingly, this remotor MN does not seem to influence the Pro/Rem rhythmic pattern, which indicates that an asymmetric relationship exists between this MN and the remotor MN pool. This MN is under the control of the remotor pool, but does not influence it, whereas it influences the levator pool, but is not influenced by it.

Whatever the nature of the coordinating structures involved in inter-joint coordination, they cannot be permanent, since the animal is capable of either forward or backward walking. This implies that either the strength of the connections involved or the activity of these neurons are modulated by superior INs. Evidence has been obtained that this superior control exists in various preparations. In the cat, for example, stimulation of the mesencephalic locomotor region (MLR) can evoke walking, trotting, pacing, or galloping, that is, an alternating or an "in phase" pattern (Shik & Orlovsky, 1976), depending on the intensity of the stimulus. Likewise, Steeves, Sholomenko and Webster (1986) demonstrated that in birds, mesencephalic stimulation can trigger walking at low voltages and flying at higher voltages. In the crayfish, several types of coordinating INs have been described (Paul & Mulloney, 1986). One of these INs is shown in Figure 3C (Barthe, Cattaert & Clarac, 1988). This spiking IN in the abdominal chain is able (when depolarized) to induce the swimmeret rhythm with alternating bursts in the returnstroke (Ia) and powerstroke (Ib) roots, as well as to modify the motor activity of the abdominal muscles. It facilitates extensor MNs while the flexor MN is slightly hyperpolarized. This illustrates the close relationships that can exist between limb movement and body posture (Williams & Larimer, 1981). Here, the extension of the abdomen is a prior condition for swimmeret beating as has been observed in intact animals (Murchinson & Larimer, 1990).

Up to now we have presented examples in which the central nervous system is capable of producing well-coordinated patterns of activity

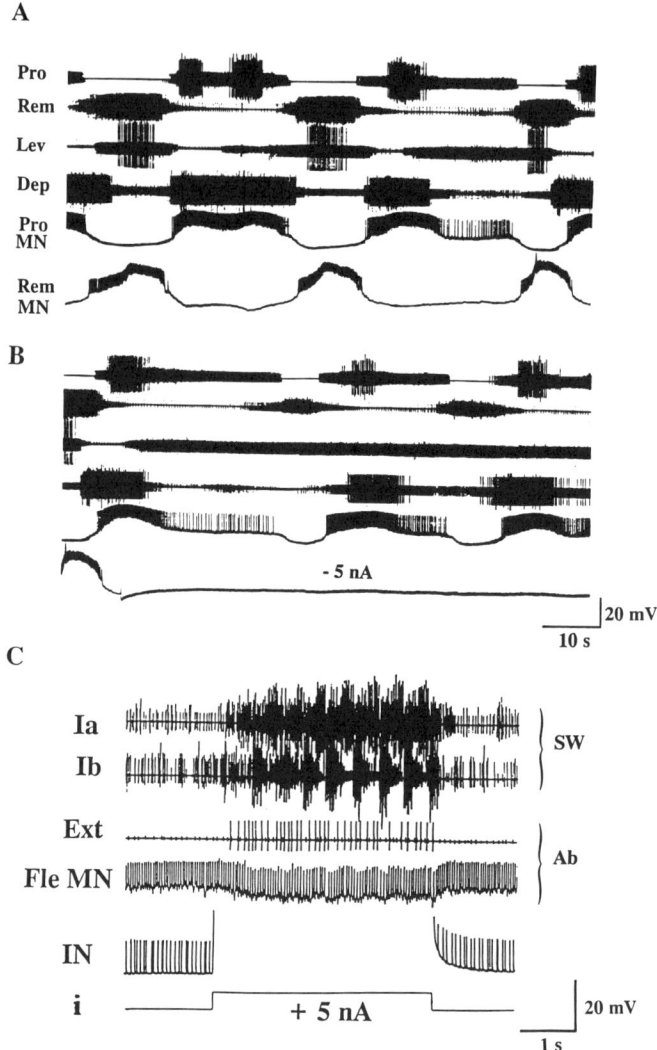

FIGURE 3 Central elements of coordination. (A, B) During backward fictive locomotion, the activities of the promotor (Pro), remotor (Rem), levator (Lev), and depressor (Dep) motor nerves were monitored using en passant electrodes, while two intracellular recordings were performed on a promotor motor neuron (Pro MN) and a remotor motor neuron (Rem MN) in the ganglion. (A) Coordination during fictive locomotion. (B) Injection of −5nA into the Rem MN perturbed the coordination: It suppressed the levator bursts, and reduced the remotor discharges recorded in the motor nerves. (C) A coordination interneuron (IN) in the abdominal nerve cord activated both swimmeret rhythm (Ia, Ib) and tonic extensor of the abdomen (Ext) when depolarized by injecting +5 nA of current. Note the inhibition of the flexor MN (Fle MN) induced by the depolarization of the coordination interneurone (IN). i, intracellular injection of current in the interneuron (IN).

that resemble those recorded in intact animals. Nevertheless, some types of coordination seem to be present only in the intact animal and disappear when proprioceptive feedback is suppressed. In these cases, another (central) type of coordination can sometimes be observed. In the crayfish, for example, the interleg coordination observed in the isolated thoracic nervous system during "fictive walking" differs considerably from that observed during real locomotion. Our results suggest the existence of INs of two types. Some INs induce an "in phase" pattern and are the only active ones when the CNS is isolated (Figure 2B). Others are responsible for the "antiphase" pattern in intact animals. Clarac and Barnes (1985) have reported similar differences in the rock lobster *in vivo*. If the legs are successively autotomized, the remaining stumps show "in phase" oscillations. This pattern was observed previously in these animals and was termed weaving (Pasztor & Clarac, 1983): here the hind legs do not contact the ground but oscillate in phase with the front legs.

B. Coordination Involving Sensory Feedback

The central network apparently possesses all the necessary equipment to accomplish coordination. One is then prompted to ask what the role of sensory information might be. To answer this question, we have to take into account the biomechanical apparatus that is controlled by the CNS, and the medium in which it operates. Since a motor command does not completely specify a movement (it also depends upon the musculoskeletal apparatus and possibly upon external perturbations), the CNS needs information about the ongoing movement to produce a suitable motor command during the performance of behavioral tasks. This information is provided by proprioceptors and exteroceptors. One can distinguish between several levels of sensory–motor interaction (MNs, INS, CPG followers, and CPG itself). The shortest loops act like monosynaptic reflexes between sensory inputs and MNs, and mainly control the intensity of MN discharge. They can also facilitate the switch between the stance and the swing phase and vice versa during the rhythmic activity.

Polysynaptic pathways involve INs that are often modulated by the CPG, and are not part of the CPG themselves. The strength of these reflexes depends on the oscillation of the INs: when they are hyperpolarized, the reflex is weaker or even suppressed. Several examples of these phase-dependent reflexes have been described (see Rossignol, Lund, & Drew, 1988; see also Duysens & Tax, Chapter 5, & Dietz, Chapter 8, this volume). For example, Sillar and Roberts (1988) have mapped reflex pathways in the tadpole, where the Rohon-Beard mechanoreceptor neurons are connected to the MNs controlling swimming through some dorsal INs that receive rhythmic inhibition from the swimming CPG. Due to this wiring, the tadpole will curve during swimming its body to avoid a lateral stimulation only when it is in the appropriate phase of oscillation, and will therefore swim away from the stimulus. Phase-dependent effects have also been described in the crayfish thoracic leg, in which the same movement of a proprioceptor, coding movements of the first joint, elicits an excitatory reflex in promotor MNs during promotor activity and an inhibitory reflex during the opposite phase (Skorupski & Sillar, 1986).

The last level of sensory control is that consisting of the CPG itself. Here, sensory inputs are able to entrain the rhythm by speeding up or slowing down the frequency. These mechanisms exist in both of the crayfish locomotor systems. In the swimmeret, the propulsive force exerted during backward movements (powerstroke) is coded by sensory structures. In *in vitro* experiments, electrical stimulation of the sensory nerve innervating these force-coding proprioceptors increases the period when delivered at the beginning of the powerstroke, and shortens the period at the beginning of the returnstroke. A rhythmic stimulation at a period slightly higher or lower than the ongoing free-running rhythm clearly entrains the swimmeret rhythm (Cattaert & Clarac, 1987). This phenomenon is quite general and is particularly important for adapting the central command to the peripheral constraints. This entrainment has been observed in all sorts of preparations, both vertebrate and invertebrate, in *in vitro* preparations, upon applying electrical stimulation to selected sensory nerves, and in *in vivo* preparations upon activating a given proprioceptor (Wendler, 1974; Rossignol et al., 1988; Pearson, Reye, & Robertson, 1983). It exerts a direct action on the CPG.

The neural mechanisms underlying the aforementioned effects can be illustrated in the walking system of Crustacea. During walking in intact animals, a stimulus delivered to the sensory nerve projecting to the contact receptors located at the tip of the leg (dactyl sensory afferents, DSA) during the stance phase reduces the ongoing period, while the subsequent ones remain unchanged (Figure 4A1). If the stimulus is delivered during the swing phase, the period is lengthened (Figure 4A2). The corresponding phase response curve (PRC.) is shown in Figure 4A3. To explain how these PRC.s are produced by the nervous system, we have analyzed the effect of DSA stimulation in an *in vitro* preparation (Figure 4B) that displayed fictive walking activities. In this example, the activity of two antagonistic MNs, a remotor and a promotor, was intracellularly recorded. Here the remotor was found to have a particular property named a bistable plateauing activity. This means that the MN possesses two states of stability: at the hyperpolar-

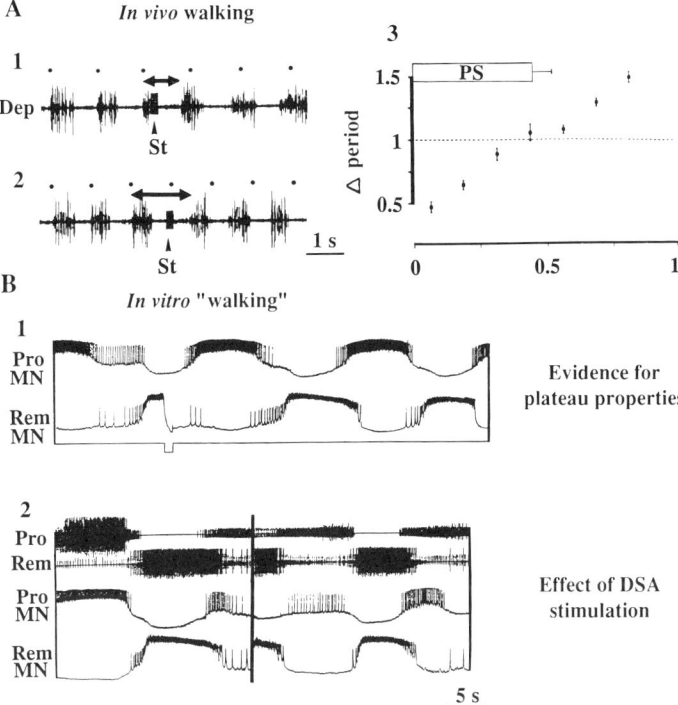

FIGURE 4 Characteristics of sensory–motor interaction during centrally generated rhythms. Resetting of the walking activity: (A) During walking activity *in vivo*, monitored by the depressor EMG, the electrical stimulation (St) of the dactyl sensory afferents (DSA) modified the timing of the ongoing locomotor cycle, which either decreased (1) or increased (2) in duration, depending on the phase of the stimulation in the walking cycle (3), see PRC. This change in the timing subsisted during the subsequent cycles (resetting). Redrawn from Liberstal *et al.*, 1987a. (B) Plateau properties of neurons and sensory–motor interactions during *in vitro* "fictive" walking. Two simultaneous intracellular recordings of the activity of a promotor (Pro MN) and a remotor (Rem MN) motor neuron during fictive locomotion. The Rem MN displayed plateau properties: A "plateau" could be interrupted by a short hyperpolarizing pulse (B1). It could be elicited by a depolarizing one (not shown) or by applying electrical stimulation to the DSA (B2).

ized level it is at rest, and at the depolarized level it discharges with a high frequency. By intracellularly injecting a pulse of current it is possible to elicit either the upper level (depolarizing pulse) or the lower (hyperpolarizing pulse) (Figure 4B1). This property is quite important, since these cells are part of the CPG. Stimulation of the DSA can play exactly the same role as the pulse of current (see Figure 4B2) and can thus induce a depolarized state, resulting in an early new step, which explains the PRC data.

Moreover, the sensory inputs interact to a large extent with the CNS, and not only produce reflexes, but cooperate with CPGs so that the actual motor command results from a permanent interplay between the CNS and the external environment. It can therefore be said that the ability of the CNS to set up rhythmic activities does not decrease the importance of the sensory information in motor pattern genesis. Moreover, some coordinating INs involved in sensory processing may be activated only if error signals are triggered by the actual movement. These INs, which enable the CNS to adapt exactly to the ongoing situation, play a major role in motor performance.

We shall now provide some precise examples of coordination in the crayfish and shall attempt to demonstrate that their principles depend on the task to be performed.

III. Examples of Coordination in the Crayfish Locomotor System

We have chosen two examples of mechanisms involved in locomotor coordination. Their levels of complexity are different and comparisons between them show up the close relationship existing between the organization of the central nervous network and the biomechanical constraints affecting the appendage they control.

A. Intralimb Proprioceptive Control of Output Intensity

The first example concerns the organization of swimmeret beating. The movements of the swimmeret and their central command have been analyzed (Cattaert & Clarac, 1987). A comparison between the EMG pattern and the movements produced confirmed that during the first part of the powerstroke, passive forces resulting from fluid resistance on the rami counteract the muscular contractions, producing the observed extension. During the powerstroke, the rami are rigid as a result of their muscular contraction, and thus exert a propulsive force against the water. At the onset of the returnstroke when the basipodite protracts because of promotor muscle contraction, the rami are bent as a result of two summed actions: the rami muscle contractions and the water resistance against the rami. The latter passive force accounts for the time lag between the movement of the proximal joint and the curling of the rami, and suggest that the difference between rami muscle EMGs and other muscle EMGs (Pro and Rem) reflects the involvement of the rami muscles in both powerstroke and returnstroke movements. This aspect of rami muscle functioning seems to be due to the fact that the innervation of these muscles is mixed; some rami muscle fibers are active in the powerstroke (PS), while others are active in the

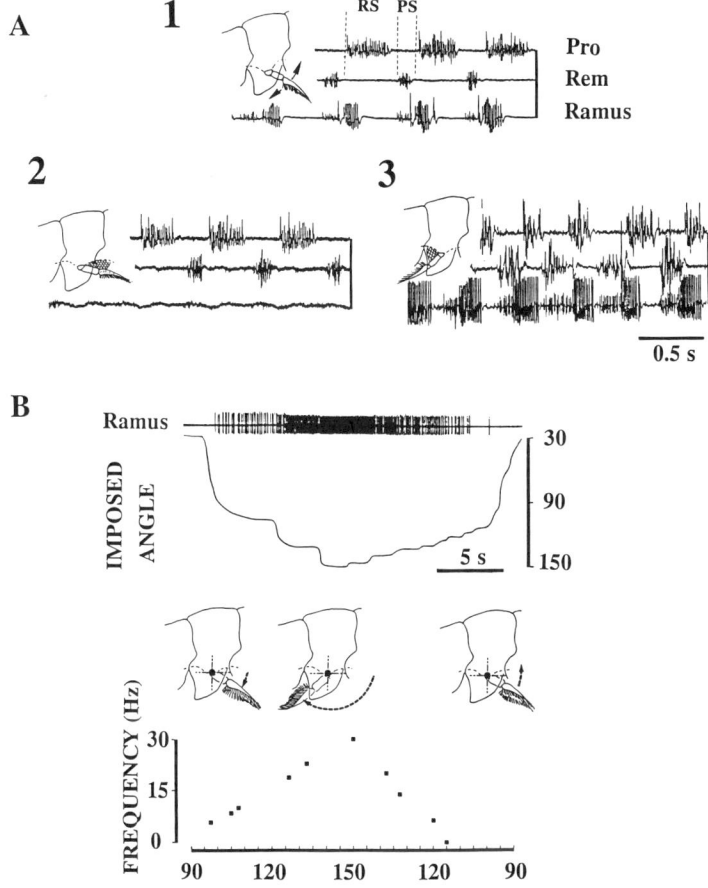

FIGURE 5 Example of simple sensory–motor coordination in the swimmeret movements. (A) In intact animals, spontaneous swimmeret beating is commanded by muscle activities (promotor, remotor, and rami curler) (1). If the basal joint is immobilized in the forward position, the alternating promotor and remotor activities are normal, but no rami curler activity occurs (2). Conversely, if the basal joint is immobilized in the backward position, the rami curler activities are considerably enhanced (3). (B) In intact animals, in the absence of swimmeret beating, the firing frequency of the rami motor neurons is correlated with the angle imposed on the basal joint.

returnstroke (RS) (Figure 5A1). It thus emerges, then, that the central command is quite stereotyped and that movements result from a combination of muscle contractions and fluid resistance. In isolated preparations, we have demonstrated (see also Figure 2) that rami MNs are activated by a central drive (Cattaert & Clarac, 1987). Here, we have analyzed the role of basal proprioceptors in this nervous command in intact animals, and we have demonstrated that rami MN activity is closely dependent on a proprioceptor, located at the base of the swimmeret.

In comparison with unrestrained swimmeret beating (Figure 5A1), the rami EMGs are completely abolished during swimming activity if the basipodite is maintained in a forward position, although the promotor and remotor activities are unchanged (Figure 5A2). If, on the contrary, the basipodite is maintained in a backward position, the rami muscle activities increase conspicuously (Figure 5A3). Moreover, in the absence of rhythmic activity (Figure 5B), imposed movements of the basipodite result in the firing of the rami MNs in backward positions, and the absence of firing in forward positions. This control is quite powerful, as indicated by the perfect correlation found to exist between the firing frequency of the rami MNs and the imposed basal angle. In fact, the basal proprioceptor that codes angle positions and movements has dual effects on the rami MNs: excitatory during retraction, and inhibitory during protraction. Finally, during real movements, the rami MNs sum the inputs from both the central command and the basal proprioceptor. This organization allows passive forces to interact with the central output through the proximal proprioceptors. We are therefore dealing here with a quite simple mechanism in which the fine distal movements are achieved on the basis of a combination of central and peripheral information.

B. Intra- and Interleg Prioprioceptive Control of Coordination

One major difference between the walking and swimming appendages centers on the presence of contact with a rigid substrate in the former case, providing both firm, resistive support for the propulsive force of the mobile segment and facilitating the postural adjustments necessary to preserve the overall body orientation in the field of gravity forces. This component involves additional regulatory mechanisms as compared to the swimming systems. Here, position-coding and especially force-coding proprioceptors, informing the CNS about the relative positions of the leg and the substrate, play a leading role. In other words, the information provided by the proprioceptors about the geometry of the leg touching the substrate is essential to ensure efficient interleg coordination, when walking on irregular ground.

In Crustacea, force-coding receptors considerably influence the interleg coordination. During walking for example, the stability of the phase relationship between the left and right legs of a given segment is enhanced when the crayfish is loaded (weight increased by 25–50%). Under these conditions, the phase histograms become unimodal, with mean phases of about 0.5 and lower standard deviations than in the case of unloaded animals (Clarac & Barnes, 1985). Among the force-coding proprioceptors of the crayfish legs, two sensory structures seem to play a major role in leg movement coordination. The first is located at the tip of each leg (the dactyl sensory afferents, DSA), and the other is situated in the proximal part of the leg (cuticular stress detectors, CSD). The CSDs are mainly active during the stance phase and reinforce the activation of muscles involved in propulsive forces. The role of the DSA, which are sensilla associated with cuticular structures in the dactyl, has been studied during locomotion in the crab (Libersat, Zill & Clarac, 1987a; Libersat, Clarac & Zill, 1987b). They are sensitive to forces exerted on the ground and code their direction and intensity. The activity of these receptors has been monitored in the crab, the rock lobster, and the crayfish. Some sensilla are phasic and code either the beginning or the end of the stance phase, while others are phasicotonic and code the intensity of the force exerted on the dactyl. Electrical stimulation of this sensory nerve results in intact animals in levation of the proper leg and depression of adjacent ones. This reflex is phase dependent, since it is more effective at the end of the stance than at the beginning.

We have stimulated the sensory nerve of DSA *in vitro,* in the crayfish, while recording from motor nerves to the levator and depressor muscles (Figure 6A). The responses were similar to those recorded in intact animals, and demonstrated that these receptors are involved in intra- and interleg reflexes. Figure 6A gives two simultaneous recordings from the levator and depressor nerves of the fourgh leg (Lev 4 and Dep 4, respectively), in an isolated preparation of the three last thoracic ganglia, with legs 4 and 5 attached. Electrical stimulation of the DSA nerve of the fourth leg resulted in the activation of the Lev 4 and the inhibition of the Dep 4 (Figure 6A, left), while electrical stimulation of the DSA nerve of the fifth leg had the opposite effect (Figure 6A, right). These effects were observed at stimulation intensities just above the threshold, and were therefore assumed not to reflect a protective reflex.

Libersat *et al.* (1987b) have investigated the role of the DSA in interleg coordination *in vivo,* during free walking in the intact crab (Figure 6B). Brief electrical stimuli delivered to the third leg DSA during walking modified the period of the ongoing cycle in a phase-dependent manner. During the stance phase of leg 3, the period was shortened, whereas it was increased during the swing phsae of leg 3 (Figure 6B, left). These stimuli affected not only leg 3 but also leg 4 (Figure 6B, right). Note that during free walking, leg 3 and leg 4 are antiphase locked (see stance phases corresponding to Dep 3 and Dep 4 activities represented by horizontal bars at 0–0.4 on left diagram and 0.4–0.8 on right diagram, respectively). In both diagrams, the horizontal scale gives the phases in the leg 3 walking cycle. This experiment demonstrates that there exists a narrow range of phase values (just at

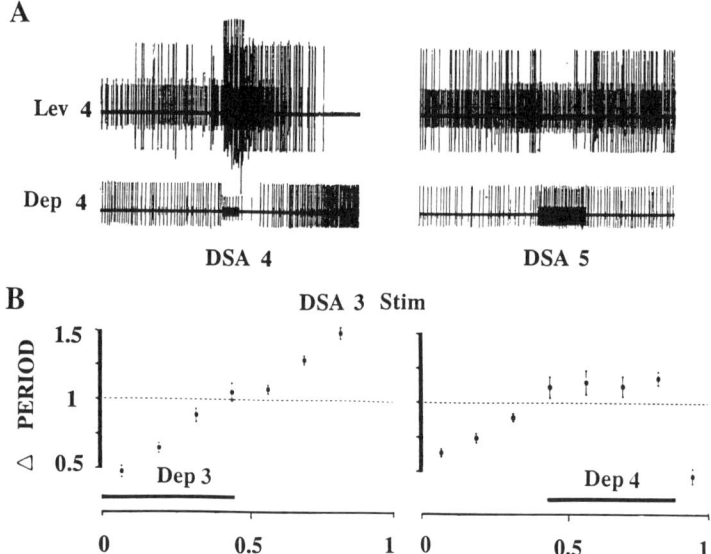

FIGURE 6 Interlimb reflexes evoked by stimulation of contact-coding receptors. (A) In *in vitro* experiments, applying electrical stimulation to the dactyl sensory afferents (DSA) coding contact of the tip of the leg with the substrate resulted in activation of the levator MNs and inhibition of the depressor MNs of that leg (*left*). Applying stimulation to the DSA of an adjacent leg resulted in the opposite effect (*right*). (B) Effect of electrical stimulation of DSA in the third leg (DSA 3 Stim) on the periods of that leg (*left*) and those of the adjacent one (*right*) during walking. As shown, when the stimulation was delivered at the end of the depressor 3 activity (horizontal bar), no change occurred in the periods of either leg 3 or leg 4. In these graphs, 0 and 1 correspond to the beginning of Dep 3 activity. Note that during walking, Dep 3 and Dep 4 are phase opposed. Redrawn from Libersat *et al.*, 1987b.

the end of Dep 3, which corresponds to the onset of Dep 4 activity), during which stimulation of DSA 3 does not change the periods of leg 3 and leg 4. Indeed, in the free-walking animal, DSA 3 are cyclically stimulated by the contact with the substrate, and these sensory cues are incorporated into the coordination process. We therefore assume that when the electrical stimuli does not modify the ongoing cycle, the timing is exactly that which occurs when natural DSA cues are integrated into the locomotor program.

This hypothesis was confirmed in an experiment in which leg 3 was artificially maintained in the levated position while the crab was walking (Figure 7). Under these experimental walking conditions, leg 2 and leg 4 were antiphase locked (phases 0.42), while leg 3 did not display any rhythmic activities, but only a tonic firing in the Dep EMGs as the result of resistance reflexes (Figure 7A2, B). This phase relationship between leg 2 and leg 4 is abnormal, since in natural walking these legs move in phase (Figure 7A1). Nevertheless, when repetitive electrical stimulation of the DSA of leg 3 was delivered at the onset of each Dep 4 activity, it induced regular bursting in the depressor muscle of leg 3 (Figure 7A3, B). More importantly, these stimuli completely reset the Dep bursting phase relationships of leg 2 and leg 3 within leg 4 (Figure

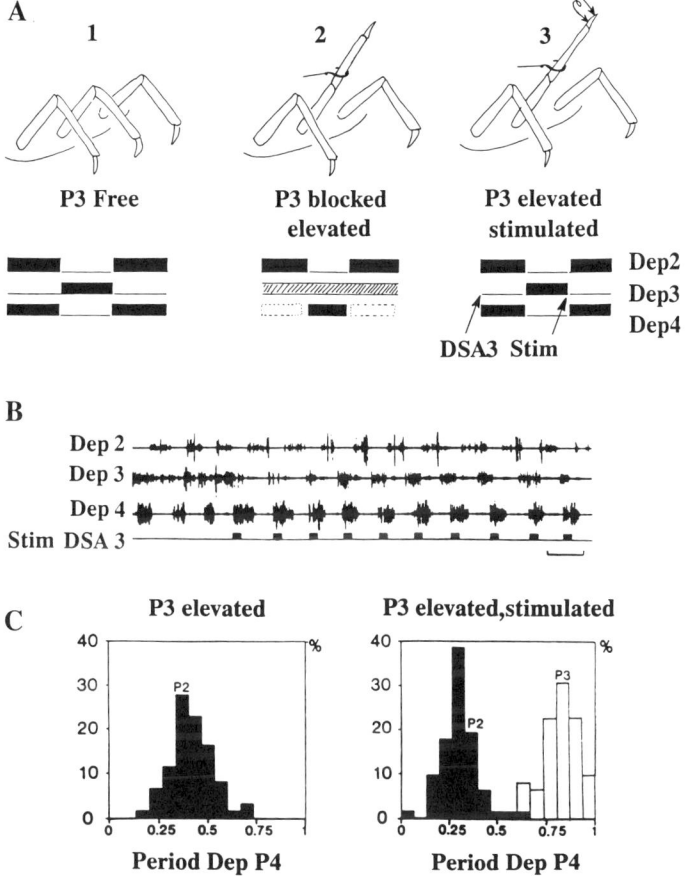

FIGURE 7 Role of contact receptors in antiphase coordination in adjacent legs. (A) Scheme of the experimental procedure. Three situations are shown: (1) normal walking (all legs free); (2) leg 3 maintained elevated; (3) leg 3 maintained elevated while DSA3 are rhythmically stimulated at the beginning of each Dep 4 burst. The resulting walking pattern is indicated below each diagram. (B) Experimental data corresponding to situations (2) and (3). During walking, the activity of three adjacent legs was monitored by recording the EMG from depressor muscles. Leg 3 was maintained elevated by experimental means, while the animal was walking on a treadmill. Under this condition, Dep 2 and Dep 4 were in phase opposition (phase histogram of Dep 2 in Dep 4 cycle around 0.4). When the DSA of the levated leg 3 was stimulated at beginning of Dep 4 activity, the bursting was restored, however, and normal coordination was established (see phase histograms on the right). Redrawn from Libersat *et al.*, 1987b.

7C, right), shifting the phase of leg 2 by 0.31 ± 0.10 and leg 3 by 0.81 ± 0.08 relative to leg 4. The value of the leg 2 phase was not significantly different from that normally observed during walking. Note also that the burst of Dep 3 activity after some walking cycles started 200–300 msec before DSA3 stimulation, and was not interrupted when stimulation was applied to DSA3, indicating that the effects of DSA3 stimulation did not constitute a protective reflex (which was evoked at stronger stimulation intensities). With the electrical stimulation used it was not possible, of course, to identify the sensory information involved. The sensilla may have been coding the loading of the leg or the unloading of the leg, since both occurred. In the latter case, the levation of the leg induced by DSA electrical stimulation would correspond to an assistance reflex reinforcing the Lev MNs of this leg while activating the Dep MNs of adjacent legs. Although this hypothesis needs to be confirmed experimentally, the interleg resetting induced by stimulation of these mechanoreceptors may have involved sensorimotor pathways controlling interleg coordination, which normally control the walking program.

Using intracellular recordings from Lev and Dep MNs in preliminary *in vitro* experiments, we demonstrated that the DSA reflexes are polysynaptic and involve central local interneurons and interganglionic coordinating interneurons. This wiring of the reflex pathways differs from the central coupling responsible for the synchronous activities of all the legs shown in Figure 3B. This central coupling is achieved through "in phase" interganglionic coordinating interneurons that may be switched off during locomotion, whereas the "out of phase" interneurons that are responsible for the alternating walking pattern are facilitated by DSA. This mechanism does not, however, involve a strict alternation between adjacent legs, but rather provides cyclical timing cues that are incorporated into the walking CPG of each leg. Depending on the strength of these effects, a whole range of situations can be observed, from the total absence of coordination to strict alternation. Although the fundamental central relationships between ganglionic oscillators may be similar in swimming and walking systems, a large part of the latter network seems to be devoted to sensory–motor integration, so that the sensory cues from the legs are predominantly involved in reshaping the interleg coordination.

IV. Conclusion

We have described two examples of sensorimotor mechanisms that control the coordination of movements. Comparing swimming and walking appendages in the crayfish reveals the existence of similarities between their central commands, although there also exist fundamental functional differences resulting from the presence of a contact with rigid substrate in the leg and its absence in the swimming appendage. Let us now compare the two types of sensory–motor organization (Figure 8).

In the swimmerets, interlimb coordination may be described as resulting from the coupling between the oscillators of each limb. The CPG is the main organizer of intralimb coordination: it successively excites the Rem, Rami, and Pro MNs in each cycle of movement according to a centrally determined pattern, via the wiring of the central network. However, the summation of the central and peripheral influences makes the rami motor output flexible, so that the rami flexion can be adapted to the actual movement of the swimmeret (Figure 8A). In the leg, the situation is quite different. Although the same central coupling exists here between oscillators of adjacent legs (Figure 8B), each joint is assumed to be commanded by a different oscillator. The oscillators of the various joints might be coupled to each other by INs that define the type of locomotion (forward, backward), since INs of this kind have been found to exist and have been described *in vitro* (Chrachri & Clarac, 1989). The difference with swimming is all the greater because here, the main control of ongoing movements seems to be linked with the force-coding receptors involved in the control of the swing phase of one leg, favoring at the same time the stance phase of the ipsilateral legs. Although in the swimming system, position receptors suffice to achieve coordination, this is not the case in the walking system, where coordination mainly depends on the contact with the substrate, which can be specified only by the force-coding receptors.

Other proprioceptors, such as the cuticular stress detector (CSD) mentioned above and proprioceptors coding joint angles (e.g., the coxo-basipodite chordotonal organ, CBCO) (Figure 1B), may participate in the control of coordination during walking. The latter proprioceptor is composed of 40 sensory cells coding the position and movements of the second joint of the leg, which are responsible for the upward and downward movements of the leg. These sensory cells that send monosynaptic negative feedback to the Lev and Dep MNs (El Manira, Cattaert & Clarac, 1991) are controlled presynaptically by the walking CPG (Cataert, El Manira & Clarac, 1992). A presynaptic control of this kind has also been observed *in vitro* in intracellularly recorded sensory terminals from CSD and DSA (A. Marchand, D. Cattaert & J. Barnes, unpublished data).

The real control of coordination in the actual crayfish walking system therefore seems to result from central and peripheral interactions that give the motor command a high level of adaptability, which is a prerequisite for efficient propulsion on irregular ground. The principles of organization described in the crayfish walking system (CPG, coupling

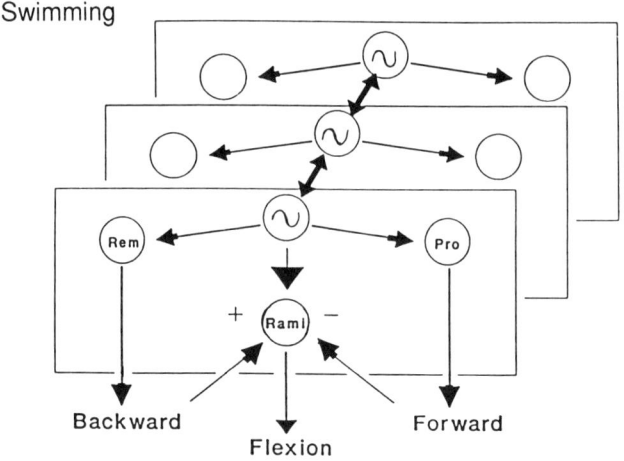

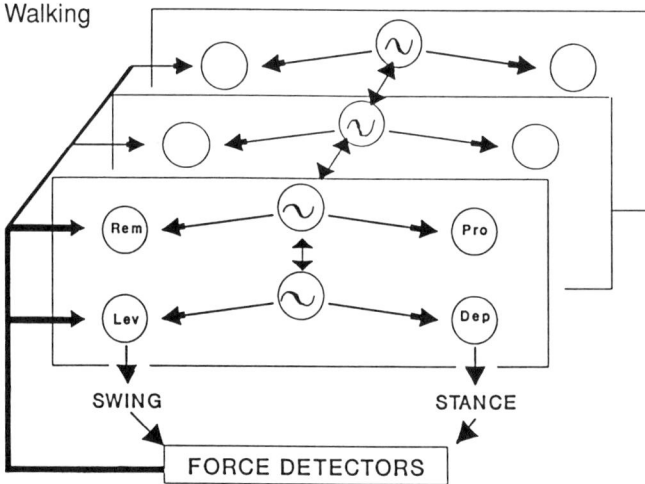

FIGURE 8 Functional diagrams of various levels of sensory–motor coordination in the crayfish. (A) Coordination in the swimming system. Interlimb coordination results from coupling INs. Interjoint coordination results both from a central command (central pattern generator, CPG) that successively triggers the remotor, the rami, and the promotor MNs during each beating cycle, and from sensory afferents from the basal angle, which are excitatory when the swimmeret is in the backward position and inhibitory when they are in the forward position. Here, the rami MNs sum the central command and the sensory information to specify the rami flexion. (B) The walking system, a more complex situation. Here, too, central interlimb coordinating INs exist, and in each leg a CPG sends a central command to various MNs in a coordinated manner. This CPG results from the coupling of oscillators responsible for each joint. For simplicity, only backward walking has been shown. Here, sensory signals coding the leg position, and especially the force exerted by the leg on the substrate, closely control the coordination. The transition time from stance to swing is largely dependent on the force receptors, which are also responsible for interlimb coordination.

of oscillators, phase-dependent effects, and presynaptic control) are not restricted to invertebrates and have been described, at least partially, in mammals. It can therefore be assumed that these principles generally underlie most "automatic" behaviors such as locomotion.

References

Amblard, B., Berthoz, A., & Clarac, F. (Eds.). (1988). *Posture and gait. Development adaptation and modulation.* Amsterdam: Elsevier Science Publishers.
Barnes, W. J. P., & Gladden, M. H. (Eds.). (1985). *Feedback and motor control in invertebrates and vertebrates.* London: Croom Helm.
Barthe, J. Y., Cattaert, D., & Clarac, F. (1988). An interneurone can induce both rhythmical and postural motor program in the abdomen of *Homarus gammarus. Proceedings of the Physiologycal Society,* **406,** 77P.
Barthe, J. Y., Bevengut, M., & Clarac, F. (1991). The swimmeret rhythm and its relationships with postural and locomotor activity in the isolated nervous system of the crayfish *Procambarus clarkii. Journal of Experimental Biology,* **157,** 207–226.
Bässler, U. (Ed.). (1983). *Neural basis of elementary behavior in stick insects.* Berlin: Springer Verlag.
Bush, B. M. H., & Clarac, F. (Eds.). (1985). Coordination in motor behaviour. *SEB seminar series* 24. Cambridge: Cambridge University Press.
Bussel, B., Roby-Brami, A., Azouri, P., Biraben, A., Yakovleff, A., and Held, J. P. (1988). Myoclonus in a patient with spinal cord transection: possible involvement of the spinal stepping generator. *Brain,* **111,** 1235–1245.
Cattaert, D., & Clarac, F. (1987). Rami motor neurons and motor control of the swimmeret system of *Homarus gammarus. Journal of Comparative Physiology,* **160A,** 55–68.
Cattaert, D., El Manira, A., & Clarac, F. (1992). Direct evidence for presynaptic inhibitory mechanisms in crayfish afferents. *Journal of Neurophysiology,* **67**(3), 610–624.
Cazalets, J. R., Sqalli-Houssaini, Y., & Clarac, F. (1992). Activation of the central pattern generators for locomotion by serotonin and excitatory amino acids in neonatal rat. *Journal of Physiology (London),* **455,** 187–204.
Chrachri, A., & Clarac, F. (1989). Synaptic connections between motor neurons and interneurons in the fourth thoracic ganglion of the crayfish, *Procambarus clarkii. Journal of Neurophysiology,* **62**(6), 1237–1250.
Chrachri, A., & Clarac, F. (1990). Fictive locomotion in the fourth thoracic ganglion of the crayfish, *Procambarus clarkii. Journal of Neuroscience,* **10**(3), 707–719.
Clarac, F., & Barnes, W. J. P. (1985). Peripheral influences on the coordination of the legs during walking in decapod crustaceans. In B. M. H. Bush & F. Clarac (Eds.), Coordination in Motor Behaviour, *SEB seminar series* 24 (pp. 249–269). Cambridge: Cambridge University Press.
Davis, W. J. (1968). The neuromuscular basis of lobster swimmeret beating. *Journal of Experimental Zoology,* **48,** 643–662.
El Manira, A., Cattaert, D., & Clarac, F. (1991). Monosynaptic connections mediate resistance reflex in crayfish (*Procambarus clarkii*) walking legs. *Journal of Comparative Physiology A,* **168,** 337–349.
Friesen, W. O. (1989). Neuronal control of leech swimming movements. In J. W. Jacklet (Ed.), *Cellular and neuronal oscillators* (pp. 269–316). New York: Marcel Dekkar.
Getting, P. A. (1988). Comparative analysis of invertebrate central pattern generators. In A. H. Cohen, S. Rossignol, & S. Grillner (Eds.) *Neural control of rhythmic movements in vertebrates* (pp. 101–128). New York: John Wiley and Sons.

Getting, P. A. (1989). Emerging principles governing the operation of neural networks. *Annual Review of Neurosciences,* **12,** 185–204.

Getting, P. A., & Dekin, M. S. (1985). Mechanisms of pattern generation underlying swimming in *Tritonia.* IV. Gating of central pattern generator. *Journal of Neurophysiology,* **53,** 466–480.

Graham, D. (1978). Unusual step patterns in the free walking grasshopper *Neoconocephalus robustus.* II. A critical test of the leg interactions underlying different models of hexapod co-ordination. *Journal of Experimental Biology,* **73,** 159–172.

Grillner, S. (1981). Motor control. In V. P. Brooks (Ed.), *Handbook of physiology* (pp. 1179–1236). Bethesda: American Physiological Society.

Grillner, S., Wallén, P., Brodin, L., & Lansner, A. (1991). Neuronal network generating locomotor behavior in Lamprey: circuitry, transmitters, membrane properties and simulation. *Annual Review of Neurosciences,* **14,** 169–199.

Haken, H. (1983). *Synergetics: An introduction* (3rd ed.). Heidelberg: Springer Verlag.

Heitler, W. J. (1978). Coupled motor neurons are part of the crayfish swimmeret central oscillator. *Nature (London),* **275,** 231–234.

Heitler, W. J. (1983). The control of rhythmic limb movements in crustacea. In A. Roberts & R. Roberts (Eds.), *Neural origin of rhythmic movements* (pp. 351–382). Cambridge: Cambridge University Press.

Hughes, G. M., & Wiersma, C. A. G. (1960). The coordination of swimmeret movements in the crayfish *Procambarus clarkii. Journal of Experimental Biology,* **37,** 657–670.

Humphrey, D. R., & Freund, H. J. (Eds.). (1991). Motor control: concepts and issues. *Dalhem Workshop Reports.* Berlin: John Wiley and Sons.

Ikeda, K., & Wiersma, C. A. G. (1964). Autogenic rhythmicity in the abdomenal ganglia of the crayfish; the control of swimmeret movements. *Comparative Biochemistry and Physiology,* **12,** 107–115.

Libersat, F., Zill, S., & Clarac, F. (1987a). Single-unit responses and reflex effects of force-sensitive mechanoreceptors of the dactyl of the crab. *Journal of Neurophysiology,* **57**(5), 1610–1617.

Libersat, F., Clarac, F., & Zill, S. (1987b). Force-sensitive mechanoreceptors of the dactyl of the crab: single-unit responses during walking and evaluation of function. *Journal of Neurophysiology,* **57**(5), 1618–1637.

Massion, J. (1992). Movement, posture and equilibrium interaction and coordination. *Progress in Neurobiology,* **38,** 35–56.

Meyrand, P., Simmers, J. & Moulins, M. (1991). Construction of a pattern-generating circuit with neurons of different networks. *Nature (London),* **351,** 60–63.

Mulloney, B., Acevedo, L. D., Chrachri, A., Hall, W. M., & Sherff, C. M. (1990). A confederation of neuronal circuits: control of swimmeret movements by a modular system of pattern generators. In K. Wiese, W. D. Krenz, J. Tautz, H. Reichert, & B. Mulloney (Eds.), *Frontiers in crustacean neurobiology* (pp. 439–441). Basel: Birkhaüser Verlag.

Murchinson, D., & Larimer, J. L. (1990). Dual motor interneurons in the abdominal ganglia of the crayfish *Procambarus clarkii:* synaptic activation of motor outputs in both the swimmeret and abdominal positioning systems by single interneurons. *Journal of Experimenal Biology,* **150,** 269–293.

Pasztor, V. M., & Clarac, F. (1983). An analysis of waving behaviour: an alternative motor programme for the thoracic appendages of decapod Crustacea. *Journal of Experimental Biology,* **102,** 59–77.

Paul, D. H., & Mulloney, B. (1986). Intersegmental coordination of swimmeret rhythms in isolated nerve cords of crayfish. *Journal of Comparative Physiology,* **158,** 215–224.

Pearson, K. G., Reye, D. N., & Robertson, R. M. (1983). Phase-dependent influences of wing stretch receptors on flight rhythm in the locust. *Journal of Neurophysiology,* **49,** 1168–1181.

Roberts, A., Soffe, S. R., & Dale, N. (1986). Spinal interneurones and swimming in frog embryos. In S. Grillner, P. S. G. Stein, D. G. Stuart, H. Forssberg, & R. M. Herman (Eds.), *Neurobiology of vertebrate locomotion* (pp. 517–534). London/New York: Macmillan.
Rossignol, S., Lund, J. P., & Drew, T. (1988). The role of sensory inputs in regulating patterns of rhythmical movements in higher vertebrates. In A. H. Cohen, S. Rossignol, & S. Grillner (Eds.), *Neural control of rhythmic movements in vertebrates* (pp. 201–283). New York: John Wiley and Sons.
Rudomin, P. (1990). Presynaptic inhibition of muscle spindle and tendon organ afferents in mammalian spinal cord. *Trends in Neuroscience*, **13**, 499–505.
Schöner, G., & Kelso, J. A. S. (1988). Dynamic pattern generation in behavioral and neural systems. *Science*, **239**, 1513–1520.
Shik, M. L., & Orlovsky, G. N. (1976). Neurophysiology of locomotor automatism. *Physiological Review*, **55**, 465–501.
Sillar, K. T., & Skorupski, P. (1986). Central input to primary afferent neurons in crayfish, *Pacifastacus leniusculus*, is correlated with motor output of thoracic ganglia. *Journal of Neurophysiology*, **55**, 678–688.
Sillar, K. T., & Roberts, A. (1988). A neuronal mechanism for sensory gating during locomotion in a vertebrate. *Nature (London)*, **331**, 262–265.
Sillar, K. T., Skorupski, P., Elson, R. C., & Bush, B. M. H. (1986). Two identified afferent neurones entrain a central locomotor rhythm generator. *Nature (London)*, **323**, 440–443.
Sillar, K. T., Clarac, F., & Bush, B. M. H. (1987). Intersegmental coordination of central oscillators for rhythmic movements of the walking legs in crayfish, *Pacifastacus leniusculus*. *Journal of Experimental Biology*, **131**, 245–264.
Skorupski, P., & Sillar, K. T. (1986). Phase-dependent reversal of reflexes mediated by the thoracocoxal muscle receptor organ in the crayfish, *Pacifastacus leniusculus*. *Journal of Neurophysiology*, **55**, 689–695.
Steeves, J. D., Sholomenko, G. N., & Webster, D. M. S. (1986). In S. Grillner, P. S. G. Stein, D. G. Stuart, H. Forssberg, & R. H. Herman (Eds.), *Neurobiology of vertebrate locomotion* (pp. 51–54). Besingstoke: MacMillan.
Stein, P. S. G. (1971). Intersegmental coordination of swimmeret motoneuron activity in crayfish. *Journal of Neurophysiology*, **34**, 310–318.
Stein, P. S. G. (1976). Mechanisms of interlimb phase control. In R. M. Herman, S. Grillner, P. S. G. Stein, & D. G. Stuart (Eds.), *Neural control of locomotion* (pp. 465–484). New York: Plenum.
Holst Von, E. (1939). Relative coordination as a phenomenon and as a method of analysis of central nervous function. *Ergeb. Physiol.*, **42**, 228–306. English translation by R. D. Martin. In *Behavioural Physiology of animals and man: The selected papers of Erich Von Holst*, Vol. 1 (1973). London: Methnen.
Wendler, G. (1974). The influence of proprioceptive feedback on locust flight coordination. *Journal of Comparative Physiology A*, **88**, 173–200.
Williams, B. J., & Larimer, J. L. (1981). Neural pathways of reflex-evoked behaviors and command systems in the abdomen of the crayfish. *Journal of Comparative Physiology A*, **143**, 27–42.

4

Interlimb Coordination during Locomotor Activities
Spinal-Intact Cats and Chronic Cats with Horizontal and Longitudinal Separation of the Spinal Cord

Masamichi Kato

Department of Physiology
Hokkaido University, School of Medicine
Sapporo, Japan

    I. Introduction
   II. Normal Spinal-Intact Cat
  III. Hemisected Chronic Cats
  IV. Transected Spinal Cord
   V. Double Hemisected Spinal Cord
  VI. Longitudinally Split Lumbar Cord
 VII. Chronically "Isolated" Lumbar Half Spinal Cord
VIII. Summary
       References

I. Introduction

Since Sherrington (1947) described interlimb reflexes of the decerebrated cats, numerous papers on interlimb coordination have been published including pure reflex experiments (e.g., Gernandt & Shimamura, 1961) and more recent studies on postural control (e.g., Gahery & Massion, 1981) and fictive motor patterns in chronic spinal cats (e.g., Pearson & Rossignol, 1991), or fictive swimming in lamprey (e.g., Matsushima & Grillner, 1992).

According to Sherrington (1947), the interlimb reflexes "produce a combined movement of remote parts." Since then, there seems to be two main opinions on the functional role of the interlimb reflexes. Henneman (1980) described in a representative textbook that . . . "A considerable amount of normal locomotor activity in four-footed animals is probably patterned in this way." The interlimb reflexes apparently contribute to diagonal postural adjustment in the quadrupeds (Massion, 1979; Gahery & Massion, 1981). The latter view is not necessarily contradictory to the former opinion, since "movements and their associated postural adjustment are coordinated" (Gahery & Massion, 1981).

The stepping generator of the hindlimbs has been reported to be located below the fourth lumbar segments from lesion experiments (Grillner & Zangger, 1979; Eidelberg, Story, Meyer, & Nystel, 1980; Deliagina, Orlovsky, & Fravlova, 1983), and activities of neurons that are believed to be parts of the generator have been studied (e.g., Edgley, Jankowska, & Shefchyk, 1988; Shefchyk, McGrea, Kriellaars, Fortier, & Jordan, 1990). The investigation of the stepping generator of the forelimbs in quadrupeds began later, and it is assumed that the forelimb stepping generator is distributed in and around the C6 and C7 segments (Yamaguchi, 1991). This assumption is based on the experiments carried out on the cat cervical cord in which transection as well as longitudinal parasagittal sections were applied at various levels (Hishinuma & Yamaguchi, 1987). Yamaguchi's group is now investigating neuronal mechanisms of the generator (Kinoshita & Yamaguchi, 1992).

In this chapter I shall review interlimb coordination during locomotor activities. The subjects were: (1) normal spinal-intact (control) cats; (2) chronic cats with lateral hemisection of the spinal cord at thoracic or cervical levels so as to disrupt descending and ascending long tracts unilaterally; (3) chronic cats with transection of the spinal cord at lower thoracic or upper lumbar levels; (4) chronic double hemisected cats whose spinal cord was serially hemisected at different levels bilaterally; (5) chronic cats with longitudinal myelotomy of the lumbar cord so as to disrupt segmental inputs from the contralateral half cord; and

(6) chronic cats whose half lumbar cord was "isolated" from both supralumbar regions and the contralateral half cord by hemisection plus longitudinal myelotomy.

Of particular interest to the author was the question if interlimb reflexes participate, or play any role, in interlimb coordination during locomotor activities.

II. Normal Spinal-Intact Cat

In neurophysiological studies on high-spinal cats, it has been reported that descending propriospinal pathways elicit excitation of mainly the reflex system controlling the physiological extensors of the hindlimbs, while ascending propriospinal pathways evoke excitation of mainly physiological flexor muscles of the forelimbs (Miller & van der Burg, 1973). Kato (1990) recorded interlimb reflex discharges from appropriate muscles of the four limbs, along with floor reaction forces (FRF) of the four limbs, in quietly standing, unanesthetized, awake, normal spinal-intact cats. The cats were encouraged to stand still on four force plates. In order to evoke segmental as well as interlimb reflexes, subnoxious electrical stimulation was applied percutaneously to the limbs. Representative examples are shown in Figure 1. A single shock stimulus applied to the foot dorsum of the right hindlimb evoked the following responses (Figure 1A). First, it activated flexion reflex in that limb, EMG discharges from tibialis anterior muscle, and a decrease of FRF of the ipsilateral hindlimb. Second, it evoked extensor reflex of the contralateral hindlimb; EMG responses of the vastus lateralis muscle were associated with an increase of FRF. Third, there were reflex extensor EMG discharges and an increase of FRF of the ipsilateral forelimb. Fourth, FRF decreased along with reflex activity in the flexor muscle of the contralateral forelimb. When the stimulus was applied to the dorsum of the forepaw of the left forelimb percutaneously, corresponding reflex discharges and FRF responses were elicited (Figure 1B). The reflex EMG discharges were, in most cases, prolonged, suggesting that the spino-bulbo-spinal reflexes of Shimamura and Livingston (1963) are elicited in addition to the segmental and interlimb reflexes. These interlimb reflex responses of EMG and FRF, both descending and ascending, were readily and stably elicited in awake, unanesthetized cats, although in high-spinal or decerebrated preparations they may be variable (Sherrington, 1947). Contralateral hindlimb nerve stimulation was always less effective to elicit forelimb muscle responses (Miller & van der Burg, 1973). Percutaneous subnoxious stimulation might have activated a kind of "homogenous" and/or "effective" population of the afferents to evoke the interlimb reflex responses

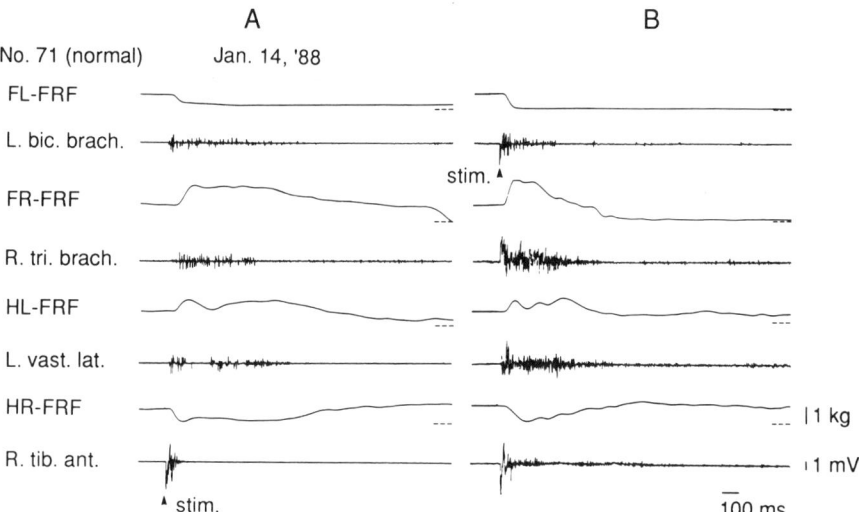

FIGURE 1 Segmental and interlimb reflexes recorded from a normal spinal-intact cat. Floor reaction forces (FRF) and EMGs were simultaneously recorded while the cat was standing on four small platforms to record floor reaction forces of each foot. FL, Left forelimb; FR, right forelimb; HL, left hindlimb; HR, right hindlimb. A single shock of electrical stimulation of subnoxious strength was applied to the right hindlimb (A), and to the left forelimb (B). Reflex EMG discharges were recorded from right tibialis anterior, left vastus lateralis, right triceps brachii, and left biceps brachii muscles, respectively. Latencies of the segmental reflex responses were about 10 msec; latencies of the interlimb responses were about 30 msec.

of awake, unanesthetized cats, contrary to the neurophysiological experiments on spinal or decerebrated preparations, where peripheral nerve bundles or dorsal roots were directly stimulated. These findings in awake cats may mean that the interlimb reflex system plays an important role in coordination among the limbs during locomotor activity. On this point, however, Stuart, Witney, Wetzel, and Goslow (1973) argued that "it is difficult for us to suggest for the present data meaningful interlimb linkages that are derived from the temporal characteristics of the two primary stride variables, strike time and contact duration."

When a cat walks straight, looking ahead, a hindlimb steps on a spot where the ipsilateral forelimb left moments earlier (see Fig. 1 in Kato, 1987). This was further investigated quantitatively by observing how the cats walk on a grid. They were encouraged to walk on a grid (80 cm × 190 cm, made of 50 mm × 50 mm steel mesh of 3 mm ø:diameter) with a food reward and affection. Normal spinal-intact cats walk on the grid with almost no error in the placement of their limbs on the grid, as representative photographs show in Figure 2A. Stick diagrams of the stepping are shown in Figure 3. Control cats made no error in forelimb placement, and in more than 97% of stepping, placed the hindlimb on the rung where the ipsilateral forelimb left moments earlier (Kato, 1992). These values correspond with those of Bregman and Goldberger (1983), although the size of their grid was larger than that of Kato (1992). This diagram also shows that the forward movement of the hindlimb is smooth and the placement is accurate. The cat does not confirm visually where the hindlimbs step during successive stepping, judging from inspection of the walking behavior and stick diagrams such as those in Figure 3 (Kato, 1992). This suggests that locomotor command signals were modified by grid information, including its size and direction, and the modified command signals affected the pattern generators of forelimbs as well as hindlimbs. In this respect, Beloozerova and Sirota (1988) investigated the role of the motor cortex (MC) and the ventrolateral nucleus of the thalamus (VL) when cats walked on the flat surface and on the ladder or flat surface with barriers. The activity of MC neurons was more strikingly modulated during locomotion on the ladder or flat surface with barriers than on the flat surface. Moreover, when the MC or the VL was destroyed bilaterally, the cats could not walk on the ladder as well as on a flat surface with barriers, although they still walked quite well on the flat surface without barriers. Their results indicate that the activity of MC neurons constitutes the cortical commands addressed to the locomotor pattern generators.

III. Hemisected Chronic Cats

Interlimb coordination, particularly that of ipsilateral fore- and hindlimbs in straight walking, was assessed by comparing step lengths and step time of both limbs (Kato, Murakami, Yasuda, & Hirayama, 1984). Hemisection was applied at lower thoracic (or upper lumbar) or high cervical level. After the hemisected cats recovered to stand and walk (on the average, 3.7 days following lower thoracic and 5.4 days after cervical hemisection), interlimb reflexes and locomotion were investigated. In chronic cats hemisected at midthoracic level, clear flexor reflex as well as crossed extensor reflex were elicited when the stimulation was applied percutaneously to the ipsilateral hindlimb. On the other hand, interlimb reflexes from both forelimbs were quite weak, and in some cats almost no response was observable. When percutaneous stimulation was applied to the ipsilateral forelimb, reflexes from three limbs except from the ipsilateral hindlimb were elicited. The step lengths and step time of the four limbs were compared with those of control cats. No difference was observed between the control and hemisected cats. Furthermore, phase relations between the fore- and hind-

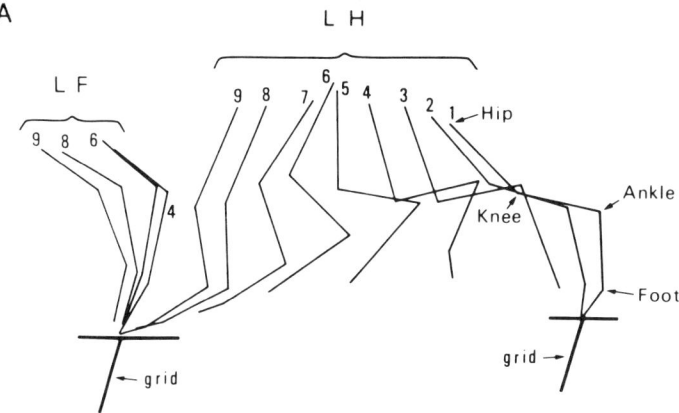

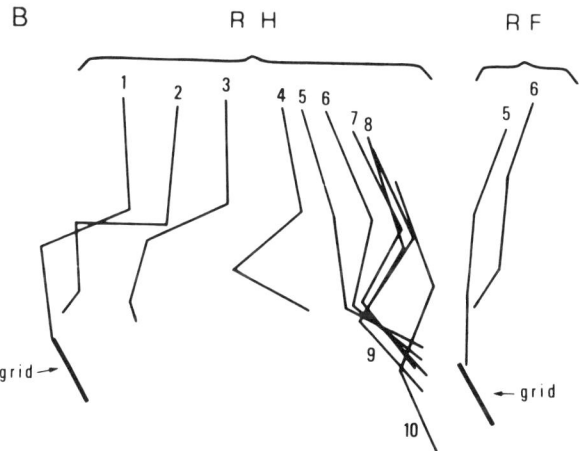

FIGURE 3 Stick diagram of walking on the grid. The diagram was constructed from movie pictures, which were taken at 24 frames/sec. L, left; R, right; F, forelimb; H, hindlimb. The stepping is numbered from the initial step of the hindlimb and the following numbers indicate frame numbers. The control cat (A) walked from right to left, and the hemisected cat (B) walked from left to right. The hemisected cat is the same cat shown in Figure 2B, but on a different day. RH of the cat failed to reach the grid and dropped through the mesh (see particularly 6–10).

FIGURE 2 Walking on the grid. (A) Normal spinal-intact cat. (B) Spinal-hemisected cat (rTh11). (A) Spinal-intact cat walks on the grid without failing to place its four limbs on the rungs of the grid. (B) This cat (No. 55) was hemisected at right Th11 on December 28, 1985. These pictures were taken on January 13, 1986, 16 days after the hemisection. For this session the cat walked 36 steps. The cat failed to place its right hindlimb on a rung, as is shown in B, in 12 out of the 36 steps (33.3%), while the other three legs stepped on the rungs. (From Kato, 1992, with permission from Elsevier.)

limb muscle activities remained unchanged after the operation. These findings seem to indicate that there existed no differences at all in stepping between the hemisected and control cats. However, inspection of the footprints of walking revealed a difference. The hemisected cats placed their hindlimbs at some distance from the forelimb footprints in many steppings (Kato, 1987, Fig. 1), indicating that the matching of spatial placement did not fully recover after the disruption of tracts in one side of the spinal cord. Combined with the results on interlimb reflexes, it may be said that interlimb reflex pathways function as spatial coordinators during locomotor activities of quadrupedal animals.

In order to collect some quantitative data on this aspect, the hemisected cats were encouraged to walk on the grid so as to investigate stepping accuracy (Kato, 1992). As illustrated in Figure 2B, the hemisected cat often failed to place the hindlimb on the rung where the ipsilateral forelimb left just moments earlier. A stick diagram of the stepping of a hemisected cat is shown in Figure 3B. This diagram illustrates how the hindlimb of the hemisected side failed to reach the rung where the ipsilateral forelimb left just moments earlier. The forward movement of the hindlimb is clumsy, when compared with the control cat. The postural stability, including distribution of floor reaction forces and lateral stability of the body, and the phase relation of locomotion recovered within one week after the hemisection. The cat walked and ran over the ground like the control cats when the test was carried out. In order to investigate how the changes developed along a time course after hemisection, foot placing on the grid was repeatedly observed. Cat No. 47 accurately placed its forelimbs on the grid during all sessions following hemisection at left Th13 (Figure 4). The left hindlimb showed great deterioration down to about a 10% success rate for about 10 days, after which rapid recovery was observed to about 70%. This percentage remained virtually unchanged for the rest of the observation until the 55th postoperative day. The right hindlimb, which is located contralateral to the hemisection, decreased it success ratio to about 70–85% for about 10 days, then recovered to the control value. Cat No. 45, which received hemisection at left C1, took a longer time to recover from the operation, and grid walking was investigated from the 36th postoperative day. Both fore- and hindlimbs of the operated side showed deteriorated placement for more than one month thereafter.

In these preparations many brainstem descending pathways as well as long propriospinal tracts in one side remained intact. Vestibulospinal as well as reticulospinal tracts exert bilateral synaptic effects on hindlimb motoneurons (Sasaki, Tanaka, & Mori, 1962; Hongo, Kudo, & Tanaka, 1975). At least some of the contralateral effects are likely to

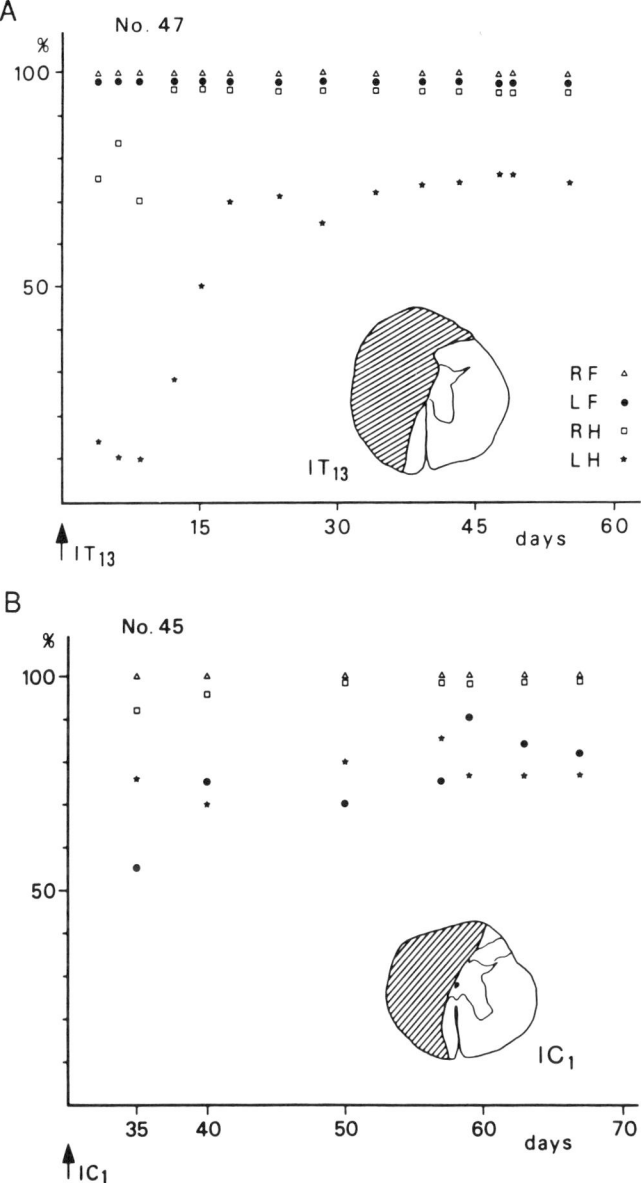

FIGURE 4. Success rate of four legs stepping on the rungs following hemisection. Cat No. 47 was hemisected at left Th13 on September 13, 1984; cat No. 45 was hemisected at left C1 on July 17, 1984. Abscissa indicates days after the hemisection and the ordinate indicates success rate of stepping on the grid. △, right forelimb; ●, left forelimb; □, right hindlimb; ★, left hindlimb. Scores were calculated from 25–50 steppings at each point. (From Kato, 1992, with permission from Elsevier.)

be direct ones; as Shinoda, Ohgaki, and Futami (1986) demonstrated, axon collaterals end in contralateral nuclei of the lumbar region. After hemisection at supralumbar level, the intact contralateral pathways may compensate the defect produced by interruption of the tracts of the hemisected side. The importance of vestibulospinal and reticulospinal tracts for locomotor recovery after spinal cord lesion was discussed by Eidelberg, Story, Walden, and Meyer (1981) and Little, Harris, and Sohlberg (1988). Moreover, Beloozerova and Sirota (1988) stressed the importance of motor cortex in adjusting stepping on the ladder, as mentioned above. Impulses flowing down either the corticospinal tract and/or via some relaying nuclei play an important role in this kind of locomotion in which the visual system is involved.

Inability to step the hindlimb over where the ipsilateral forelimb placed during successive stepping indicates that coordination of spatial placement of the fore- and hindlimbs did not recover to the control level. Normally functioning, intact, bilateral tracts are needed for this kind of precise coordination of the limbs. In cervical hemisected cats like No. 45, ipsilateral descending tracts from supraspinal centers were sectioned chronically. However, interlimb propriospinal long tracts remained intact. This suggests that descending impulses from higher centers are more important for interlimb coordination than long interlimb propriospinal tracts.

IV. Transected Spinal Cord

Grillner and Zangger (1979) reported that rhythmic alternating activity is present in the unanesthetized spinal cats treated with Nialamide and L-DOPA after spinal cord transection at L5. They concluded that the neurons responsible for such rhythmic activity were located in the lumbosacral cord. Deliagina et al. (1983) stressed the importance of the L3–L5 segments in generating rhythmic oscillation during locomotion.

Studies have shown that right and left hindlimb coordination of stepping is preserved in chronic cats after spinal cord transections at the midthoracic or lower thoracic levels (Eidelberg et al., 1980; Pearson & Rossignol, 1991), or at the rostral L4 segment (Afelt, Veber, & Maksimova, 1973), or in chronic dogs transected at Th10 (Naito & Shimizu, 1991). Kudo, Ozaki & Yamada (1991) presented data that N-methyl-D,L-aspartate (NMA) induced locomotor activity in hindlimb muscles of an isolated lumbar spinal cord–hindlimb muscles preparation of rat fetus. These experimental results indicate that neural mechanisms generating stepping cycles as well as controlling right and left hindlimb coordination exist in the lumbar spinal cord. However, Eidelberg *et al.* (1980) further mentioned that individual step cycle timing in these chronic cats became more variable and longer. These results suggest that structures in supralumbar regions play some important role for coordinating the limbs during locomotive movements. They noted that stepping rate of the forelimbs became quicker while that of the hindlimbs was prolonged after transection of the spinal cord. Similar phenomena were observed in the chronic double hemisected cats (Kato *et al.*, 1984) as described below. The mechanisms of these changes were not readily explained, but they were argued to be consequences of disruption of spinal tracts, possibly of ascending ones. Shik and Orlovsky (1976) mentioned that facilitation of a forelimb stepping by transection of the spinal cord could be considered a locomotor component of the Shiff-Sherrington phenomenon, suggesting that the phenomenon is a reflex mechanism for fore- and hindlimb coordination of locomotion in the quadrupedal animal. In this respect, Kato, Murakami, Yasuda, Hirayama, and Hikino (1983) observed increased tonus of forelimb extensors of the chronic double hemisected cats.

Pearson and Rossignol (1991) reported that in chronic spinal cats (spinalized at Th13), stimulation to the perineal region and the paw evoked fictive motor patterns such as locomotion, paw shaking, or rhythmic leg flexion, confirming the previous notion that the basic motor patterns are established in the lumbar cord.

V. Double Hemisected Spinal Cord

Kato *et al.* (1984) intended to investigate whether the interlimb reflex system plays an important role in coordination of locomotive activity of fore- and hindlimbs. As one model for the investigation they prepared chronic "double hemisected cats." In a group of cats (T–T preparations), the spinal cord was first hemisected at around Th12 and subsequently, at intervals ranging from 37 to 126 days, at the contralateral midthoracic level (around Th6). Bilateral long propriospinal as well as supraspinal descending and ascending tracts were mostly severed, and only short propriospinal tracts survived in this group. In a second group (C–T preparations) the cats received hemisection first at around C2, and then at intervals of 21–73 days at the contralateral midthoracic level. In this group, both long and short propriospinal tracts were left intact on one side, although bilateral supraspinal descending and ascending tracts and long propriospinal tracts on the midthoracic hemisected side were chronically severed. These chronic cats finally recovered to stand and walk by themselves without any support (Kato *et al.*, 1984; Kato, Murakami, Hirayama, & Hikino, 1985). It was found that in both T–T and C–T preparations, step length of the forelimbs was shortened significantly, while that of the hindlimbs was significantly lengthened. The "double hemisected" preparations resemble the chronic spinal cats (Eidelberg et al., 1980) in this respect. Furthermore, phase relations between the fore- and hindlimbs were completely lost in both T–T and C–T preparations, suggesting that the stepping generator for the forelimbs operated independently of that for the hindlimbs. The stepping pattern generator of the forelimb is located in and around C6–C7 (Yamaguchi, 1991) and that of hindlimbs is located in the lumbar spinal cord (Grillner & Zangger, 1979; Deliagina et al., 1983). Jordan and his colleagues (Jordan, 1991; Noda, Kettler, & Jordan, 1988) discussed organization of locomotor pathways descending from the mesencephalic locomotor region (MLR; Shik, Severin, & Orlovsky, 1966; Selionov & Shik, 1984) to the spinal cord via ventrolateral fasciculus (VLF) and dorsolateral fasciculus (DLF). In a C–T double hemisected cat, ipsilateral VLF and DLF to the cervical enlargement and bilateral VLF and DLF to the lumbar cord were sectioned chronically. Therefore, it is likely that command signals descending in the contralateral VLF and/or DLF influence movement of the forelimb of the hemisected side through segmental mechanisms. For the hindlimb, both descending pathways were interrupted, although the long interlimb reflex pathways in one side are intact in these preparations. In T–T double hemisected cats, bilateral interlimb long reflex pathways as well as VLF and DLF were interrupted. The close similarity of the results in T–T and C–T preparations, in spite of different degrees of impairment of propriospinal tracts, leads to the conclusion that interlimb propriospinal reflexes play little role in the coordination of locomotor movements of fore- and hindlimbs.

VI. Longitudinally Split Lumbar Cord

As described in the preceding section, the chronic animals spinalized at the lower thoracic or lumbar cord showed right and left hindlimb coordination during locomotion. The question arises whether the coordination is governed primarily by segmental mechanisms or whether some rostrally located structures play an important role in the coordination. To answer this question Kato (1988) performed longitudinal myelotomy of the lumbar cord of cats, thus preparing chronic animals with a longitudinally split lumbar cord. Following the operation, the cats could not stand up correctly on the four legs for 19 days (average of 5 cats), but eventually recovered to support the body with the four legs. After 4 more days, the chronic cats began to walk using all four legs. At that time, then the cats were tested for interlimb reflexes as well as locomotion. Figure 5 shows a representative example of interlimb reflex

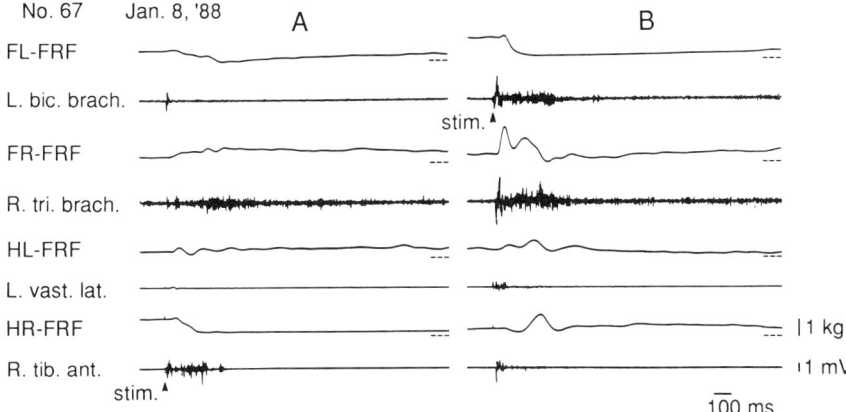

FIGURE 5 Segmental and interlimb reflex patterns from a cat with longitudinal myelotomy of the lumber cord. This cat was longitudinally myelotomized from L1 through caudal L7 on September 22, 1987. When a stimulus was applied to the right foot dorsum, ipsilateral segmental reflexes as well as reflex discharges from the forelimb muscles were elicited, while no reflex was elicited from the contralateral hindlimb. This indicates the longitudinal split of the lumbar cord was successful. Postmortem histological investigation verified the longitudinal split was successful. When a stimulus was applied to the left forepaw (B), both segmental and interlimb reflexes were elicited (see text).

responses recorded from the chronic cat. Crossed extensor reflex from the left hindlimb was not elicited by applying a single electric shock stimulus to the foot dorsum of the right hindlimb (Figure 5A), suggesting the segmental pathways were interrupted. When the stimulus was applied to the left forepaw (Figure 5B), all the interlimb reflex responses were elicited, although the responses from the hindlimbs were much weaker than those of normal animals (Figure 1). These results combined with those obtained from normal cats (Figure 1), suggest that intact lumbar segmental connections between the two sides play a formidable role in eliciting both crossed extensor reflex and interlimb reflexes, particularly between the fore- and hindlimbs.

The cats were encouraged to walk straight on the carpeted floor. Phase relations between bilateral forelimbs and hindlimbs were studied using EMG activities from appropriate muscles. A representative example obtained from the chronic lesioned cats is illustrated in Figure 6. The mean phase angles between right versus left triceps brachii muscles were from $0.96 \pm 0.05\pi$ to $1.02 \pm 0.06\pi$ in the chronic cats. These values are quite similar with those of intact control cats. The mean phase angles between bilateral vastus lateralis muscles were also quite similar with those of the control cats. These experimental data indicate that coordination of hindlimbs during stepping on flat ground

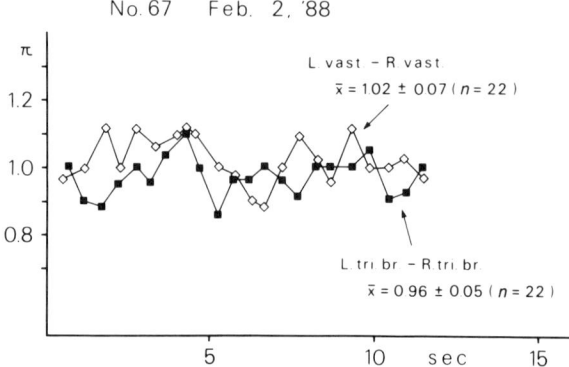

FIGURE 6 Phase relationships between bilateral forelimbs (triceps brachii) and bilateral hindlimbs (vastus lateralis) muscles plotted against time in a longitudinally myelectromized cat (No. 67). Means, standard deviations, and sample numbers are also indicated.

is well preserved after the longitudinal separation of lumbar cord. This means that structures located rostral to the lumbar cord, perhaps especially the brainstem, are dominant in controlling the coordination. As mentioned earlier, however, there exist without any doubt segmental mechanisms for the coordination. Therefore, it is a reasonable conclusion that coordination of two hindlimbs during stepping is controlled by two sources, supralumbar and segmental.

VII. Chronically "Isolated" Lumbar Half Spinal Cord

Kato (1989, 1990, 1991) investigated whether the lumbar half spinal cord, chronically isolated from both descending and contralateral impulses, could generate a stepping pattern in the ipsilateral hindlimb. Experiments were carried out on nine chronic cats whose spinal cord was hemisected at L2 or L3 and longitudinally split at the midline from the hemisected site through caudal to L7 or S1. By this operation, half the lumbar cord was "isolated" from other parts of the spinal cord. As soon as the cats recovered to stand up and walk using the four legs, interlimb reflexes and walking pattern were analyzed by recording EMGs from appropriate muscles.

Following the operation, the cat moved using only the forelimbs and the hindquarter was not supported by the hindlimbs for more than two weeks on average. During this period the ipsilateral hindlimb (the hemisected side) was extended both at the hip and knee joints and plantar-flexed at the ankle joint, while the contralateral hindlimb began to show spontaneous flexion–extension movements 3–4 days after the operation. An additional two days were needed for the cats to recover to stand up and walk with their three legs (two forelimbs and one contralateral hindlimb). The cats walked with their three legs, dragging the ipsilateral hindlimb for another two weeks. during this period, the cats stood with three legs, and the ipsilateral hindlimb was occasionally extended like a prop to support the body, allowing standing with four legs. This proplike standing was gradually prolonged, and finally recovery of locomotor activity took place in the ipsilateral hindlimb, which was innervated by the "isolated" half lumbar cord. Segmental and interlimb reflexes were investigated in these cats. Only the ipsilateral flexor reflex at a latency of about 10 msec was evoked by stimulation of the dorsum of the foot on the hemisected side (Figure 7A). On the contrary, the flexor reflex of the forelimb, the crossed extensor reflex from the contralateral forelimb, and the extensor reflex from the hindlimb of the stimulated side could be recorded by applying electrical stimulation to the forelimb contralateral to the hemisection (Figure 7B). When the stimulus was applied to the contralateral hindlimb, the flexor reflex of that limb, the extensor reflex from the forelimb of the stimulated side, and the flexor reflex from the forelimb of the hemisected side were evoked, while no appreciable reflex activity could be elicited from vastus lateralis muscle of the hemisected side. These results strongly suggest that the operation was successful.

At the initial stage of walking, that is, when the cat advanced with the three legs dragging the ipsilateral hindlimb, triceps brachii muscles of the two forelimbs and the contralateral vastus lateralis muscle of the

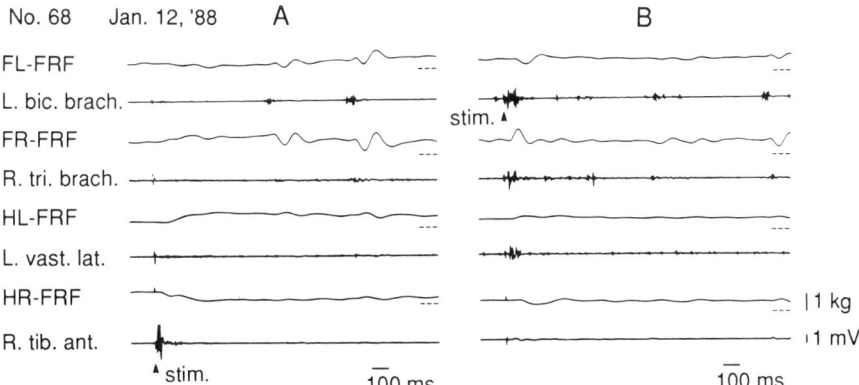

FIGURE 7 Segmental and interlimb reflexes elicited from a cat with "isolated" half lumbar cord. This cat was first hemisected at rTh13 on September 4, 1987, then longitudinally myelotomized from the hemisected site through lower L7 on October 2, 1987. Quite similar patterns were obtained by repeated observation at intervals of 1 to 3 weeks. Stimulation and recording were carried out as in Figure 1 and Figure 5.

hindlimb showed fairly regular alternate burst discharges as the cat proceeded. The ipsilateral vastus lateralis muscle showed no such burst discharges and no locomotorlike movement was observed. As further recovery took place, clear burst discharges from vastus lateralis of the isolated side could be recorded even during the extended dragging of the ipsilateral hindlimb. At the beginning, amplitudes of the burst discharges were generally lower and each burst discharge consisted of sparse spikes, indicating that only a fraction of motoneurons were active. As recovery proceeded, the ipsilateral hindlimb began to show a walking pattern, which was confirmed by inspection (Figure 8). One representative example of EMG records taken from this stage is illustrated in Figure 9. Phase relations among the muscles wee analyzed

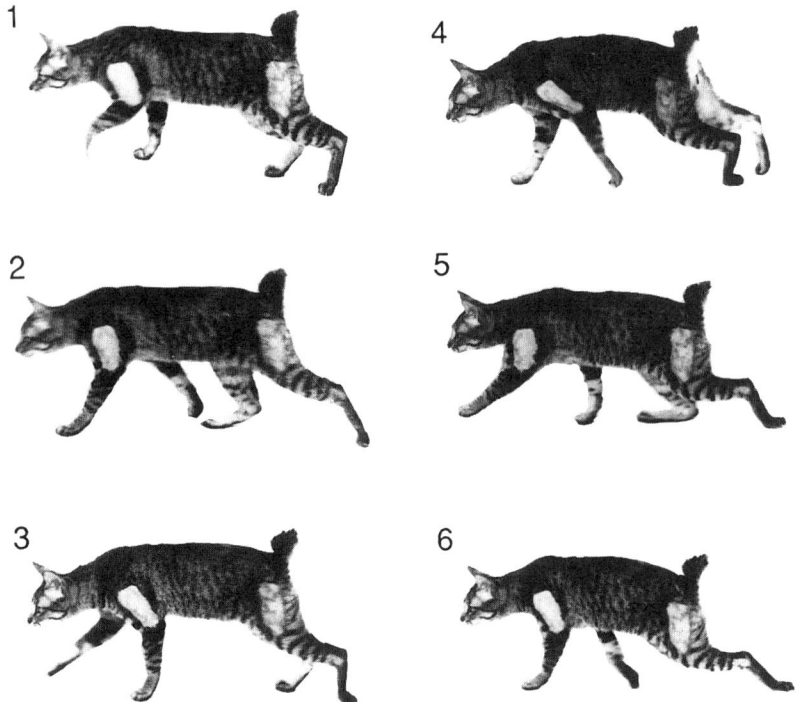

FIGURE 8 Cat No. 100. This cat was first longitudinally split from L2 through caudal L7 on May 19, 1989. On August 25, left hemisection at L2 was carried out. On the 12th day after the second operation the cat stood with its three legs for a short time. On the 24th day the cat began to walk slowly with its four legs. On the 25th day (Sept. 19) the cat walked at the speed of 30–40 cm/sec at which time these pictures were taken at 5 frames/sec. After the left leg was stretched (1) it advanced to about half the step of the right hindlimb (3), and touched down with foot dorsum (3). Then the body progressed with the right hindlimb (4, 5, 6).

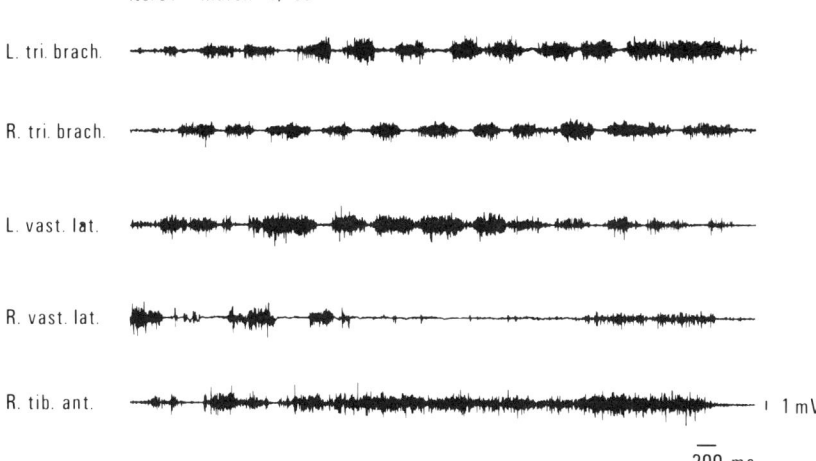

FIGURE 9 EMGs during walking of cat No. 91, 64 days after the second operation. At this stage the cat could walk with four legs, although EMGs of the right side are not so grouped.

from appropriate EMG records, one such result being illustrated in Figure 10. Phase relations between bilateral triceps were $0.97 \pm 0.13\pi$ to $1.09 \pm 0.12\pi$, indicating that the two forelimbs were stepping rhythmically and alternately. Phase relations between bilateral vastus lateralis were calculated when the measurable burst discharges were recorded. As the phase values of two sessions are plotted in Figure 10, those are highly variable step by step. Phase relations of the ipsilateral vastus lateralis versus tibialis anterior muscles of the hemisected side were $1.89 \pm 0.14\pi$ to $2.00 \pm 0.03\pi$ when tibialis anterior showed burst discharges, as compared to $1.20 \pm 0.10\pi$ to $1.47 \pm 0.17\pi$ in normal cats.

The locomotor pattern generator is assumed from transection experiment to be located below the fourth lumbar segments (Deliagina et al., 1983; Eidelberg et al., 1980; Grillner & Zangger, 1979). As described above, the ipsilateral hindlimb is extended for some time after the cat recovered standing with the two forelimbs and contralateral hindlimb, and the extended hindlimb is dragged during movement. At this time no clear locomotor activity was seen, but the ipsilateral vastus lateralis muscle showed intermittent burst discharges similar to those of a locomotor pattern. This burst activity may be interpreted as an expression of the activity of the locomotor pattern generator. In the extended hindlimb, peripheral receptors such as joint, muscle spindle, and also cutaneous receptors are likely to be stimulated, hence afferent discharges from these receptors are eventually conveyed to the isolated

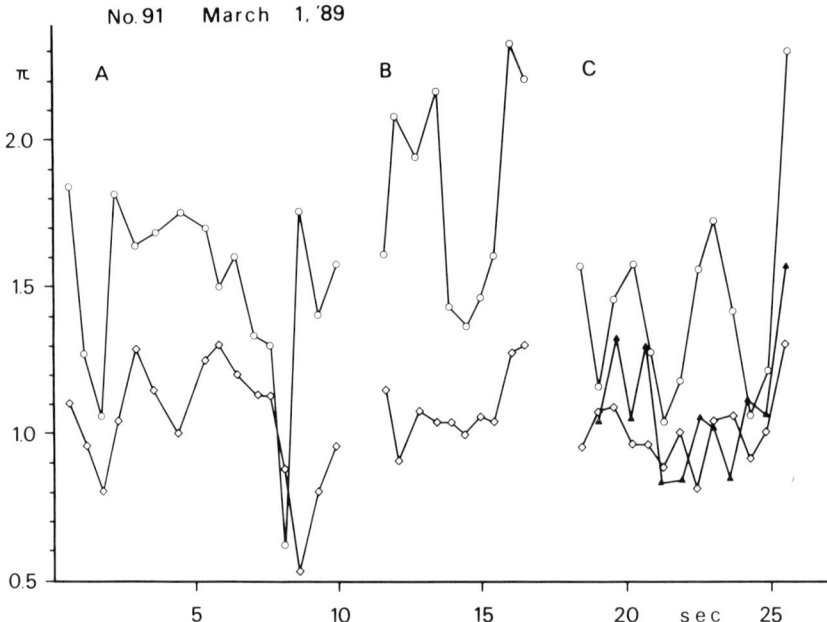

FIGURE 10 Phase relationships between bilateral triceps brachii muscles (◊), bilateral vastus lateralis (▲), and left side triceps brachii versus vastus lateralis muscles (○). This cat (No. 91) walked for about 10 steps in one session, then stopped walking either by falling to one side or by crossing the hindlimbs. After a few seconds pause the cat walked again. Abscissa indicates real time from an arbitrary time point during walking sequences, and the ordinate shows phase angles in radians, which were calculated from successive steps. Three consecutive walking sessions, interrupted twice by a few seconds pause each are shown. (A) Walking velocity 53 cm/sec. (B) Walking velocity 51.5 cm/sec. (C) Walking velocity 57.0 cm/sec. In (A) the right side hindlimb, or isolated side, was extended and dragged during walking, hence phase relationships between bilateral vastus lateralis could not be measured. Note variable values of bilateral vastus lateralis and triceps brachii versus vastus lateralis in (B) and (C) as compared with relatively stable values between bilateral triceps brachii muscles of the forelimbs. (From Kato, 1991, with permission from Japan Scientific Societies Press.)

lumbar spinal cord. Since the locomotor pattern generator receives strong innervation from supraspinal, or suprasegmental, structures (Armstrong 1986, 1988; Grillner, 1981), the isolated pattern generator receives fewer impulses as compared with normal or hemisected preparations. This reduced imput may not be strong enough to activate the pattern generating network for a period following the operation. A kind of denervation supersensitivity of the pattern generator, or an as yet unrecognized mechanism, may develop during these periods and finally gain enough excitability from the peripheral afferents to drive the generator.

VIII. Summary

Interlimb coordination during walking was investigated by recording interlimb reflexes and EMG activities from appropriate muscles in both normal spinal-intact cats and the chronically spinal-lesioned cats. The cats were encouraged to stand still on the force plates in order to record the interlimb reflexes. They were also encouraged to walk on flat carpeted floor as well as on the grid.

1. In normal unanesthetized cats, interlimb reflexes are readily elicited by percutaneous stimulation of fore- as well as hindpaw at subnoxious strength. This phenomenon suggests that interlimb reflex pathways participate in interlimb coordination during locomotor activities. Nevertheless, Stuart *et al.* (1973) were doubtful about any meaningful interlimb linkages based on their investigation of temporal characteristics of the interlimb reflexes.

Walking on the grid showed that the normal cats made no error in forelimb placement on the grid, and landed their hindlimbs on the rung where each ipsilateral forelimb left moments earlier. This suggests that grid information, including its size and direction, affected the locomotor pattern generators of both forelimbs and hindlimbs.

2. In hemisected preparations, comparison of step length as well as step time against normal control cats showed no difference. However, observation of accurate foot placement on the grid revealed differences between the hemisected and control groups: hemisected cats showed a long-lasting deficit in placing the hindlimb where the ipsilateral forelimb left moments earlier, while the normal cats adjusted quickly.

3. In spinal transected or "double hemisected" preparations, step length and step time of forelimbs are shortened, while those of the hindlimbs are lengthened, resulting in a disruption of phase relations between the fore- and hindlimbs. This could be the result of disruption of interlimb reflex pathways. However, results from our C–T preparations suggest that descending signals from brainstem locomotor centers are more important than the interlimb reflex system.

4. In chronic cats whose lumbar spinal cord was longitudinally separated into halves, coordination between bilateral hindlimbs was well preserved. This indicates that structures located on the supralumbar region of the cord are as important as the segmental mechanisms for the bilateral coordination.

5. Cats with "isolated half lumbar cord" eventually recovered to stand and walk. In these preparations, all the impulses from the contralateral as well as the supralumbar regions were cut off. Phase relationships between bilateral vastus lateralis muscles were highly variable step by step, suggesting that the stepping of the hindlimb innervated by

the "isolated half lumbar cord" was independently performed, probably elicited by the impulses from peripheral receptors such as joint, muscles, and cutaneous receptors.

References

Afelt, Z., Veber, N. B., & Maksimova, E. V. (1973). Reflex activity of chronically isolated spinal cord of the cat (in Russian). Nauka, Moscow. cited in Deliagina, T. G., Orlovsky, G. N., & Pavlova, G. A. et al., (1983). *Exp. Brain Res.*, **53,** 81–90.

Armstrong, D. M. (1986). Supraspinal contributions to the initiation and control of locomotion in the cat. *Prog. Neurobiol.*, **26,** 273–361.

Armstrong, D. M. (1988). The supraspinal control of mammalian locomotion. *J. Physiol.*, **405,** 1–37.

Beloozerova, I. N., & Sirota, M. G. (1988). Role of motor cortex in control of locomotion. In V. S. Gurfinkel, M. E. Ioffe, J. Massion, & J. P. Roll (Eds.), *Stance and motion: Facts and concepts* (pp. 163–176). New York/London: Plenum.

Bregman, B. S., & Goldberger, M. E. (1983). Infant lesion effect: II Sparing and recovery of function after spinal cord damage in newborn and adult cats. *Develop. Brain Res.*, **9,** 119–135.

Deliagina, T. G., Orlovsky, G. N., & Favlova, G. A. (1983). The capacity for generation of rhythmic oscillations is distributed in the lumbosacral spinal cord of the cat. *Exp. Brain Res.*, **53,** 81–90.

Edgley, S. A., Jankowska, E., & Shefchyk, S. (1988). Evidence that mid-lumbar neurones in reflex pathways from group II afferents are involved in locomotion in the cat. *J. Physiol.*, **403,** 57–71.

Eidelberg, E., Story, J. L., Meyer, B. L., & Nystel, J. (1980). Stepping by chronic spinal cats. *Exp. Brain Res.*, **40,** 241–246.

Eidelberg, E., Story, J. L., Walden, J. G., & Meyer, B. L. (1981). Anatomical correlates of return of locomotor function after partial spinal cord lesions in cats. *Exp. Brain Res.*, **42,** 81–88.

Gahery, Y., & Massion J. (1981). Co-ordination between posture and movement. *Trends in Neuroscience,* **4,** 199–201.

Gernandt, B. E., & Shimamura, M. (1961). Mechanisms of interlimb reflexes in cat. *J. Neurophysiol.*, **24,** 665–676.

Grillner, S. (1981). Control of locomotion in bipeds, tetrapods and fish. In V. B. Brooks (Eds.), *Handbook of physiology, Section 1, The nervous system II* (pp. 1179–1236). Bethesda, MD: American Physiology Society.

Grillner, S., & Zangger, P. (1979). On the central generation of locomotion in the low spinal cat. *Exp. Brain Res.*, **34,** 241–261.

Henneman, E. (1980). Organization of the spinal cord and its reflexes. In V. B. Mountcastle (Ed.), *Medical physiology* (14th ed.), (Vol. 1) pp. 762–786. St. Louis: C. V. Mosby Co.

Hishinuma, M., & Yamaguchi, T. (1987). Projection levels of descending pathways eliciting forelimb stepping in the cat cervical cord. *J. Physiol. Soc. Japan,* **49,** 402.

Hongo, T., Kudo, N., & Tanaka, R. (1975). The vestibulospinal tract: crossed and uncrossed effects on hindlimb motoneurons in the cat. *Exptl. Brain Res.*, **24,** 37–55.

Jordan, L. M. (1991). Brainstem and spinal cord mechanisms for the initiation of locomotion. In M. Shimamura, S. Grillner, & V. R. Edgerton (Eds.), *Neurobiological basis of human locomotion* (pp. 3–20). Tokyo: Japan Scientific Society Press.

Kato, M. (1987). Motoneuronal activity in the cat lumbar spinal cord following separation from descending or contralateral impulses. *Centrl Nerv. Syst. Trauma,* **4,** 239–248.

Kato, M. (1988). Longitudinal myelotomy of lumbar spinal cord has little effect on coordinated locomotor activities of bilateral hindlimbs of the chronic cats. *Neurosci. Lett.,* **93,** 259–263.

Kato, M. (1989). Chronically isolated lumbar half spinal cord, produced by hemisection and longitudinal myelotomy, generates locomotor activities of the ipsilateral hindlimb of the cat. *Neurosci. Lett.,* **98,** 149–153.

Kato, M. (1990). Chronically isolated lumbar half spinal cord generates locomotor activities in the ipsilateral hindlimb of the cat. *Neurosci. Res.,* **9,** 22–34.

Kato, M. (1991). Chronically isolated lumbar half spinal cord and locomotor activities of the hindlimb. In M. Shimamura, S. Grillner, & V. R. Edgerton (Eds.), *Neurobilogical basis of human locomotion* (pp. 407–410). Tokyo: Japan Scientific Society Press.

Kato, M. (1992). Walking of cats on grid: performance of locomotor task in spinal intact and hemisected cats. *Neurosci. Lett.,* **145,** 129–132.

Kato, M., Murakami, S., Yasuda, K., Hirayama, H., & Hikino, K. (1983). Increase of extensor tonus of forelimbs in chronic cats with bilateral serial hemisections of the spinal cord at different levels. *Neurosci. Lett.,* **41,** 289–293.

Kato, M., Murakami, S., Yasuda, K., & Hirayama, H. (1984). Disruption of fore- and hindlimb coordination during overground locomotion in cats with bilateral serial hemisection of the spinal cord. *Neurosci. Res.,* **2,** 27–47.

Kato, M., Murakami, S., Hirayama, H., & Hikino, K. (1985). Recovery of postural control following chronic bilateral hemisections at different spinal levels in adult cats. *Exp. Neurol.,* **90,** 350–364.

Kinoshita, M., & Yamaguchi, T. (1992). Stimulus-locked excitation of motoneurons during forelimb fictive locomotion evoked by stimulation of the lateral funiculus in decerebrated cats. *Neurosci. Res.,* Suppl. 17, S210.

Kudo, N., Ozaki, S., & Yamada, T. (1991). Ontogeny of rhythmic activity in the spinal cord of the rat. In M. Shimamura, S. Grillner, & V. R. Edgerton (Eds.), *Neurobiological basis of human locomotion* (pp. 127–136). Tokyo: Japan Scientific Society Press.

Little, J. W., Harris, R. M., & Sohlberg, R. C. (1988). Locomotor recovery following subtotal spinal cord lesions in a rat model. *Neurosci. Lett.,* **87,** 189–194.

Massion, J. (1979). Role of motor cortex in postural adjustments associated with movement. In H. Asanuma & V. J. Wilson (Eds.), *Integration in the nervous system* (pp. 239–258). Tokyo: Igaku-shoin.

Matsushima, T., & Grillner, S. (1992). Neural mechanisms of intersegmental coordination in lamprey: Local excitability changes modify the phase coupling along the spinal cord. *J. Neurophysiol.,* **67,** 373–388.

Miller, S., & van der Burg, J. (1973). The function of long propriospinal pathways in the coordination of quadrupedal stepping in the cat. In R. B. Stein, K. G. Pearson, R. S. Smith, & J. B. Redford (Eds.), *Control of posture and locomotion* (pp. 561–577). New York: Plenum.

Naito, A., & Shimizu, Y. (1991). Analysis of the stepping movements in adult spinal dogs. In M. Shimamura, S. Grillner, & V. R. Edgerton (Eds.), *Neurobiological basis of human locomotion* (pp. 395–399). Tokyo: Japan Scientific Society Press.

Noga, B. R., Kettler, J. J., & Jordan, L. M. (1988). Locomotion induced in mesencephalic cats by injections of putative transmitter substances and antagonists into the medial reticular formation and the ponto-medullary locomotive strip. *J. Neurosci.,* **8,** 2074–2086.

Pearson, K. G., & Rossignol, S. (1991). Fictive motor patterns in chronic spinal cats. *J. Neurophysiol.,* **66,** 1874–1887.

Sasaki, K., Tanaka, T., & Mori, K. (1962). Effects of stimulation of pontine and bulbar reticular formation upon spinal motoneurons of the cat. *Jpn. J. Physiol.,* **12,** 45–62.

Selionov, V. A., & Shik, M. L. (1984). Medullary locomotor strip and column in the cat. *Neuroscience,* **13,** 1267–1278.

Shefchyk, S., McGrea, D., Kriellaars, D., Fortier, P., & Jordan, L. (1990). Activity of interneurons within the L4 spinal segment of the cat during brainstem-evoked fictive locomotion. *Exp. Brain Res.,* **80,** 290–295.

Sherrington, C. (1947). *The integrative action of the nervous system.* New Haven: Yale University Press.

Shik, M. L., & Orlovsky, G. N. (1976). Neurophysiology of locomotor automatism. *Physiol. Rev.,* **56,** 465–501.

Shik, M. L., Severin, F. V., & Orlovsky, G. N. (1966). Control of walking and running by means of stimulation of the mid-brain. *Biophysics,* **11,** 756–765.

Shimamura, M., & Livingston, R. B. (1963). Longitudinal conduction systems serving spinal and brainstem coordination. *J. Neurophysiol.,* **26,** 258–272.

Shinoda, Y., Ohgaki, T., & Futami, T. (1986). The morphology of single lateral vestibulospinal tract axons in the lower cervical spinal cord of the cat. *J. Comp. Neurol.,* **249,** 226–241.

Stuart, D. G., Witney, T. P., Wetzel, M. G., & Goslow, Jr., G. E. (1973). Time constraints for inter-limb coordination in the cat during unrestrained locomotion. In R. B. Stein, K. G. Pearson, R. S. Smith, & J. B. Redford (Eds.), *Control of posture and locomotion* (pp. 537–560). New York: Plenum.

Yamaguchi, T. (1991). Cat forelimb stepping generator. In M. Shimamura, S. Grillner, & V. R. Edgerton (Eds.), *Neurobiological basis of human locomotion* (pp. 103–115). Tokyo: Japan Scientific Society Press.

5

Interlimb Reflexes during Gait in Cat and Human

Jacques Duysens and Toine Tax

Department of Medical Physics and Biophysics
Catholic University of Nijmegen
Nijmegen, The Netherlands

    I. Introduction
   II. Crossed Reflexes in Motionless Cats
  III. Ipsilateral Reflex Reversal in the Intact Cat
  IV. Phase-Dependent Reversal of Crossed Reflexes
   V. Pathways and Mechanisms Involved in Phase-Dependent Modulation of Exteroceptive Reflexes
  VI. Fore- and Hindlimb Coordination and Reflexes during Cat Locomotion
 VII. Flexibility and Mechanisms of Interlimb Coordination in the Cat
VIII. Interlimb Coordination during Human Gait
  IX. Crossed Reflexes and Their Modulation in Humans
   X. Conclusions
      References

I. Introduction

Interlimb coordination is one of the principal tasks to be accomplished during locomotion. The movements of the various limbs, used in gait, must be linked to each other to enable smooth progression and maintenance of equilibrium. Since locomotion is a highly automated motor behavior, it would not be surprising that interlimb reflexes have evolved to support the coordination of the limb movements during gait. Interlimb reflexes are also needed when an unexpected perturbation occurs in a given limb during the step cycle. Corrective movements of the perturbed limb must blend into a more generalized behavioral reaction, which may include limbs that are not directly affected by the perturbation. Whether operating during normal locomotion or following perturbations, interlimb reflexes presumably subserve common goals, such as mentioned above, namely minimizing instability and securing progression. Therefore, one might expect that there are basic similarities between some of the principles governing interlimb coordination during normal gait and some of the "rules" underlying corrective interlimb reactions. A few basic examples may suffice to illustrate this.

When for a given reason the ipsilateral swing phase is prolonged, the support of the body relies for a longer period on the contralateral leg. During alternating gait one would therefore predict that the contralateral stance is prolonged in association with prolonged ipsilateral swing phases. This is indeed what has been found in the unrestrained intact cat both for unperturbed walking (Halbertsma, 1983; Cruse & Warnecke, 1992) and for walking during which the ipsilateral swing phase was prolonged, either because of electrical stimulation (Duysens & Stein, 1978), mechanical perturbation (Matsukawa, Kamei, Minoda, & Udo, 1982; Prochazka, Sontag, & Wand, 1978), or by superposition of a paw shake response (Carter & Smith, 1986). In humans, there is a similar coupling between the duration of the ipsilateral swing and contralateral stance during locomotion with or without perturbations. When a mechanical disturbance is applied during swing, the durations of both the ipsilateral swing and the contralateral stance are prolonged by about 80 msec (Dietz, Quintern, Boos, & Berger, 1986). About the same prolongation of contralateral stance (100 msec) was obtained with electrical stimulation of the tibial nerve in the middle of the swing phase (Berger, Dietz, & Quintern, 1984).

Changes in the duration of the ipsilateral stance phase should also affect contralateral step cycle events during alternating gait. In particular, when the ipsilateral stance is shortened, the contralateral leg should take up the support function earlier than normal and one would therefore predict an earlier onset of extensor activity and a shortening of the contralateral swing phase. This is indeed observed both during unperturbed walking (Halbertsma, 1983) as well as when the ipsilateral stance phase is shortened due to a mechanical perturbation (Matsukawa et al., 1982; Udo, Kamei, Matsukawa, & Tanaka, 1982) or electrical stimulation (Forssberg, Grillner, & Rossignol, 1977, Fig. 8, top).

The aim of the present review is to summarize our current knowledge about interlimb reflexes in cat and human in relation to locomotion. The emphasis will be on the interlimb coordination of the lower limbs but some data on "intergirdle" reflexes will be discussed as well. Since little is known about crossed proprioceptive reflexes, most of the data presented here concerns reflexes following exteroceptive (skin) stimulation.

II. Crossed Reflexes in Motionless Cats

At the turn of the century, Sherrington (1906, 1910) described how exteroceptive reflexes in decerebrate cats are not limited to the flexion reflex of the stimulated leg but also involve crossed excitation of extensors and crossed suppression of flexors. This crossed extensor reflex can easily be imagined to subserve support of the animal during the withdrawal of the ipsilateral leg. However, Sherrington noted that these crossed reflexes were more labile than the ipsilateral reflexes. In fact, Graham-Brown and Sherrington (1912) described that under certain conditions (preparations recently made decapitate, exclusion of neck and labyrinthine reflexes), the same stimuli, used to evoke crossed extensor responses, induced contractions of crossed flexors and relaxation of crossed extensors. It was concluded that in such preparations "the crossed extension response is replaced by crossed flexion, and from one preparation to another varying mixtures of crossed inhibition of extensors and crossed extension may occur from one and the same stimulus" (see review in Creed, Denny-Brown, Eccles, Liddell, & Sherrington, 1932).

Special attention to crossed reflexes came at the time when they could be studied in detail in spinal cats using microelectrodes. First, Holmqvist (1961) confirmed the activation or suppression of flexor motoneurons contralateral to stimulation of high threshold afferents. Later, Jankowska, Jukes, Lund, and Lundberg (1967) injected spinal cats with DOPA and found that short-latency reflex responses were depressed and instead, late, long-lasting reflexes appeared. Long-latency flexor responses were found to be coupled to crossed extensor discharges. DOPA is thought to mimic monoaminergic transmitters, which are normally released by descending pathways during periods of locomotion. Activity in these descending paths facilitates the interneurones involved in the late discharges. The latter neurones were thought to be part of the "spinal half centers," which Graham-Brown had proposed to underlay the generation of rhythmic locomotor patterns.

In more recent years, progress has been made in the identification of the interneurones involved in crossed reflexes (reviewed by Jankowska, 1992). Using a technique for retrograde transneuronal transport, it was possible to identify neurones projecting contralaterally in Rexed lamina VIII. Many of these neurones receive excitatory input from low threshold FRA (flexor reflex afferents). Many of these neurones presumably are inhibited by ipsilateral Ib or group II afferents, indicating that ipsilateral input, such as derived from loading of the limb, may block crossed reflex actions. Reasoning along the same line, one can imagine, for example, that a crossed extensor reflex should be suppressed as long as the ipsilateral leg is loaded and weight bearing. On the other hand, it is worth noting that proprioceptive afferents may themselves yield excitatory crossed reflexes. For example, crossed reflex contractions, following presumed selective activation of group Ib afferents from soleus, have been reported (Baxendale & Rosenberg, 1977). Stimulation of the nerve to FDL (flexor digitorum longus; belonging to the muscle group FD in Figure 1) is especially effective in evoking crossed group II responses in anesthesized spinal cats (Arya, Bajwa, & Edgley, 1991). It may be noted that FDL is well suited to detect limb unloading, since stance of the cat ends with plantar flexion of the toes.

In general, it is difficult to extrapolate the results from studies on motionless animals. During movement, there is a continuous change in afferent input and this may have an effect on the reflexes. Interactions between different types of afferents could partly explain why crossed reflex actions are not stereotyped and are subject to modulatory influences (McMillan & Koebbe, 1981). One such influence is related to limb position. Grillner and Rossignol (1978a) found that changing the limb contralateral to the side of stimulation from a flexed to an extended position, induced a reversal from a crossed extensor response to a crossed flexor response in acute spinal cats injected with clonidine. In a later study, Rossignol and Gauthier (1980) used selective deafferentation at different joints to show that hip position was the most critical factor for this type of position-dependent reversal. As changes in hip position occur continuously during gait, it was speculated that these reversals were related to the modulation of reflex pathways during

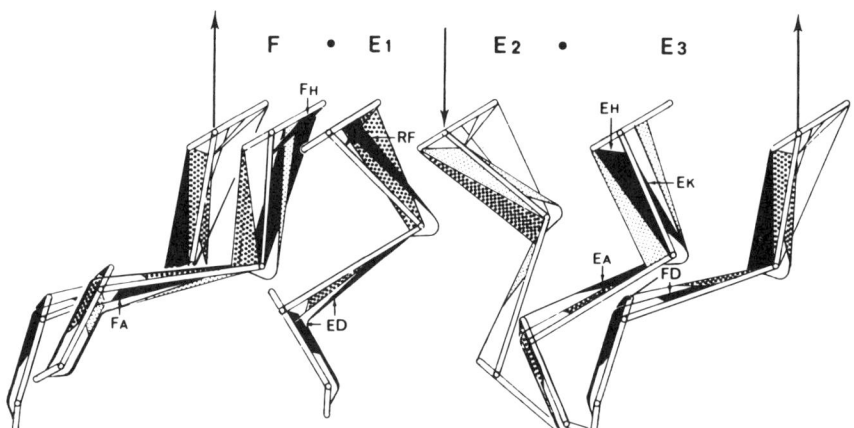

FIGURE 1 Activity of the main muscle groups of the cat hindlimb during the step cycle, divided in a flexion phase (F) and three extension phases (E1, E2, and E3). Arrows, Liftoff (upward) and touchdown (downward); FH, hip flexors; FA, ankle flexors; EH, hip extensors; EK, knee extensors; EA, ankle extensors; ED, extensors of digits; FD, flexors of digits; RF, rectus femoris. (Modified, with permission from Perret et al., 1980.)

locomotion. However, as discussed in the next section, reflex reversals can be obtained during locomotorlike activity in the absence of changes in hip position.

III. Ipsilateral Reflex Reversal in the Intact Cat

In the normal cat, the changes in burst duration induced by exteroceptive stimulation are relatively minor. Yet, such stimulation can lead to another type of phase-dependent reversal when one carefully examines the responses in detail. To understand these results, it may be helpful to first consider some theoretical situations (Figure 2). In the experiments to be described, a comparison is made between the level of spontaneous activity during gait [presumably originating from a central pattern generator (CPG) for locomotion or from some other type of motor command] and the amplitude of reflex responses, elicited by a brief electrical stimulation (Figure 2A). One normally expects that the amplitude of the responses for a stimulus of constant amplitude increases as the level of spontaneous activity increases (Figure 2B). This follows from the "automatic gain control" principle (Marsden, Merton, & Morton, 1976; Marsden, Rothwell, & Day, 1983; Matthews, 1986; Toft, Sinkjaer, Andreassen, & Larsen, 1991). The latter principle, derived from experiments on animals or humans at rest, simply states that the amplitude of reflex responses increases with increments in background contraction of the muscle involved. This is thought to be due to an increase in excitability within the motoneuron pool, for example, because of an increased central drive. If, however, during gait some instances are found when background activity is high, yet the responses are small, one has to assume the presence of an extra source of reflex modulation, either related to phase-related afferent signals or to the intervention of a central program (CPG; Figure 2).

In the intact cat walking on a treadmill, one can elicit two types of responses, one with a latency of 10 msec (P1) and another with a latency of 25 msec (P2) in a variety of hindlimb muscles using non noxious electrical stimulation of the skin (Figure 3) (Forssberg, 1979; Duysens and Loeb, 1980; Abraham, Marks, & Loeb, 1985; Loeb, Hoffer, & Marks, 1987; Pratt, Chanaud, & Loeb, 1992).

When applied at different phases of the step cycle, stimuli of constant intensity elicit P1 responses with an amplitude that is generally proportional to the level of spontaneous activity, occurring at the time of the response. This conforms to the situation described in Figure 2A and B. The modulation of the second type of responses (P2), however, conforms more to Figure 2C, since their amplitudes are not well predicted by the level of background activity during the step cycle. In flexor

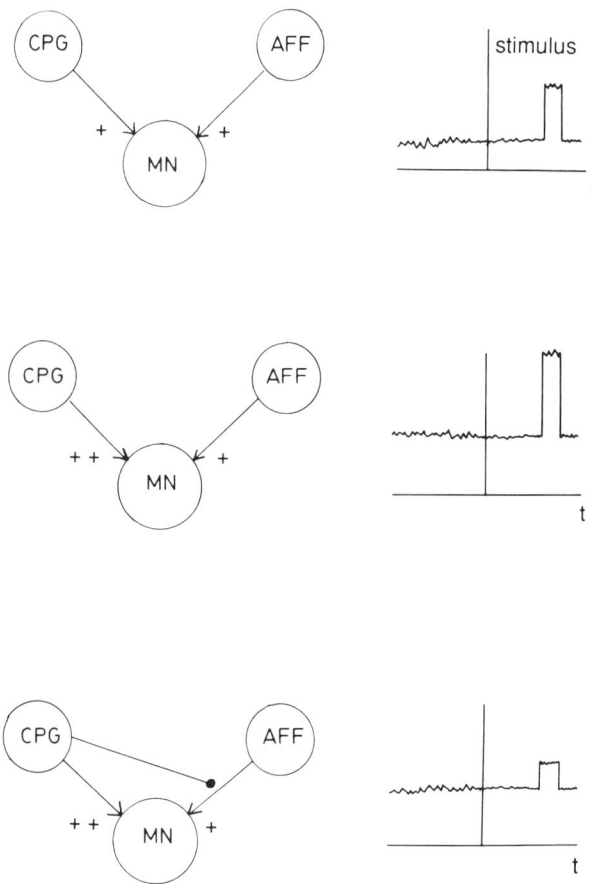

FIGURE 2 Schematic representation of three hypothetical conditions occurring during the step cycle. Relative strength of inputs to the motoneurones (indicated by pluses) explains the amplitudes of the background activity and the reflex responses, following electrical stimulation of the afferents (vertical line, "stimulus"). Mn, motoneurons; CPG, central pattern generator; Aff, afferents; t, time. For further explanation see text.

muscles, the P2 responses start to appear at end stance just prior to the spontaneous activity burst normally occurring in that period of the step cycle. The amplitude of these responses is largest in early swing and decreases near the end of swing, despite the presence of a high level of background activity. For some muscles, such as sartorius, there even was a reversal from facilitatory to suppressive responses in this period (Duysens and Loeb, 1980). During stance, the responses in flexors were absent. Instead, responses with the same latency now appeared in some extensors such as gastrocnemius medialis (GM) following stimulation

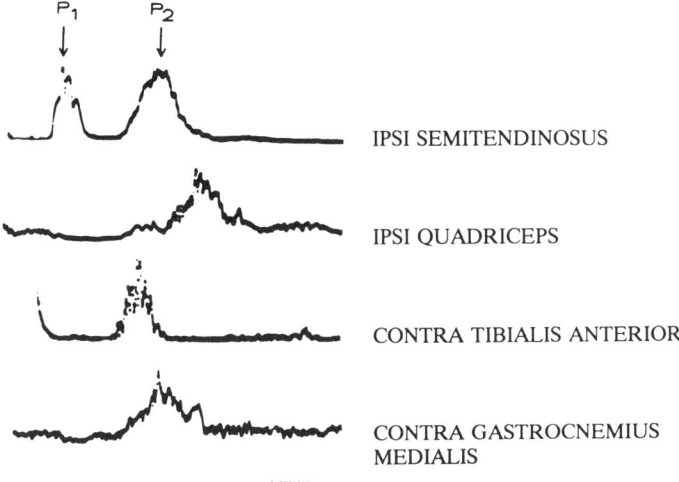

FIGURE 3 P1 and P2 responses of an intact cat following stimulation of the dorsum of the foot during walking on a treadmill. Stimulus: single pulse of 10 mA. Ipsi, ipsilateral; Contra, contralateral. Stimulus artifact: during first 10 msec in contralateral tibialis anterior and contralateral gastrocnemius medialis. Time calibration: 10 msec. The EMG responses are averages of 64 trials, taken at random periods during the step cycle. (Modified from Duysens & Loeb, 1980.)

of the sural nerves or its innervated skin (Duysens and Stein, 1978; Duysens and Loeb, 1980).

Initially, relatively little difference was observed in the responses elicited from different skin areas. However, by systematically comparing the reflex responses in a large set of muscles following stimulation of various nerves, Abraham et al. (1985) showed that the P2 activation of GM in stance was limited to sural nerve stimulation and failed to occur with stimulation of the saphenous nerve. Pratt et al. (1991) could demonstrate a high degree of specificity as well, especially for muscles showing complex activation patterns during locomotion. For example, stimulation of the saphenous nerve caused P2 responses in both semitendinosus (ST) and biceps femoris (BF), while sural nerve stimulation yielded facilitatory responses in BF but suppression in ST.

Reflexes have also been studied in the forelimbs of unrestrained walking cats (Drew and Rossignol, 1987; Rossignol and Drew, 1986; Rossignol, Lund, & Drew, 1988) using both mechanical perturbations or electrical stimulation of the superficial radial nerve. The differences between the amplitudes of the evoked responses and the level of background activity were even larger than in the hindlimb. For example, large responses were obtained in triceps during swing, when this muscle was not spontaneously active. A second important result is that the aforementioned authors found a sharp drop in the amplitude of responses in brachialis just prior to footfall. This fall in reflex gain was similar to the one found in some hindlimb flexors (Duysens and Loeb, 1980).

It should be mentioned that the results described above were mostly obtained with cutaneous stimulation. The strongest effects were seen following stimulation of distal skin sites. Nevertheless, quite similar results were obtained when muscles were stimulated through implanted electrodes. Among the muscles tested, FDL (flexor digitorum longus) was especially effective in evoking the responses (Duysens and Loeb, 1980).

IV. Phase-Dependent Reversal of Crossed Reflexes

The stimuli, which were effective in eliciting ipsilateral responses, also produced crossed responses even when stimuli were near the threshold for ipsilateral reflexes. Crossed flexor P2 responses had a lower threshold than crossed extensor P2 responses (Duysens, Loeb, & Weston, 1980). In contrast, P1 responses were never observed contralaterally. The crossed P2 reflexes, following stimulation of the plantar or lateral surface of the foot, had a latency that was quite similar to the P2 responses, observed ipsilaterally (Figure 3).

The modulation of the crossed flexor P2 responses was fully out of phase with the ipsilateral modulation and showed the same step cycle dependency as observed ipsilaterally. Crossed P2 reflexes in ankle flexors such as tibialis anterior (TA) were largest when elicited near the transition from (contralateral) stance to swing, thereby facilitating the ensuing swing phase (Figure 4; Duysens and Loeb, 1980; Duysens *et al.*, 1980). During stance, the TA responses were small, but large responses appeared in ankle extensors, such as gastrocnemius medialis (GM). In this case, the GM responses reinforced the ongoing extensor activity in this period of the step cycle.

In conclusion, the need for stability requires that the motor program, underlying locomotion, regulates the flow of activity in pathways used in interlimb reflexes to ensure that the evoked responses are appropriate for the phase of the step cycle in a given limb. Following electrical stimulation of skin or muscle, the main responses, occurring in flexor muscles both ipsi- and contralaterally, are facilitated near the transition from stance to swing. This reinforces the onset of the spontaneous activity in these muscles during this period. At end swing, however, the gain of these reflexes is often drastically reduced.

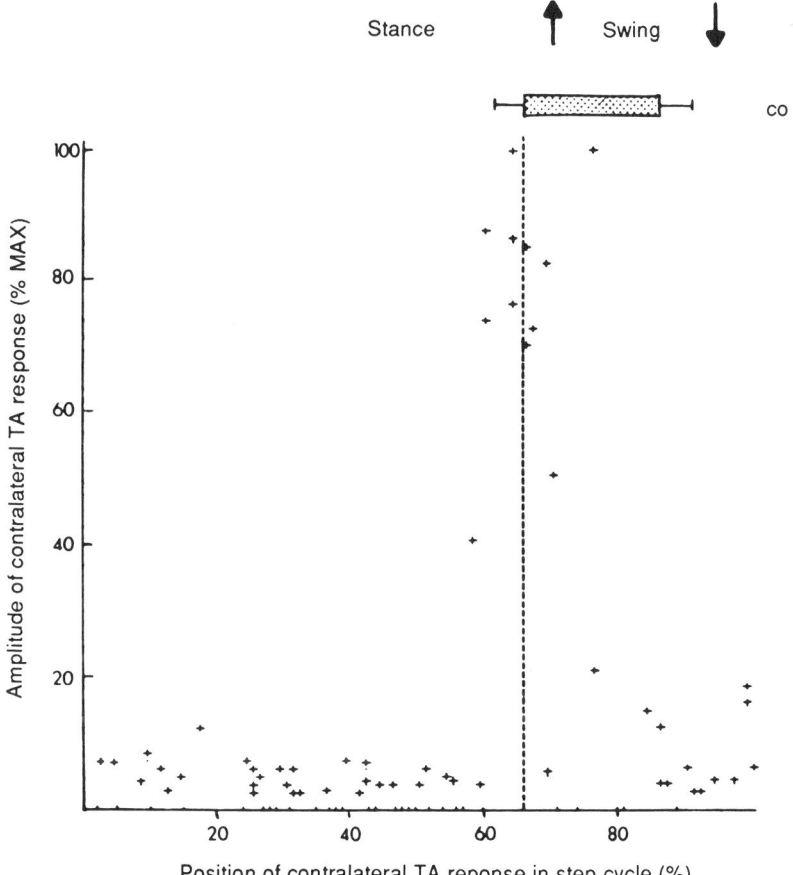

FIGURE 4 Phase-dependent modulation of contralateral tibialis anterior (coTA) P2 responses in an intact cat, walking on a treadmill. The TA response amplitude was hand-measured and normalized with respect to the largest response. The position of the responses in the step cycle was based on the onset of the responses and was expressed as a percentage of the preceding control step cycle. At top of figure, average position of the spontaneous activity period of TA in the step cycle. Note that TA responses were largest near the onset of this TA burst (vertical stippled line). Step cycle duration, 820 msec; Arrows, touchdown and liftoff. (Modified from Duysens et al., 1980.)

V. Pathways and Mechanisms Involved in Phase-Dependent Modulation of Exteroceptive Reflexes

In principle, in the intact cat the phase-dependent modulation of ipsi- and contralateral reflexes, as described above, could be due to supraspinal infuences. Cortical influences may not be essential, since phase-dependent modulation of crossed and uncrossed reflexes is present during locomotion of high decerebrate cats. Stimulation of the plantar surface of the foot in this preparation yields large extensor responses during stance, while the same stimuli evoke flexor facilitation during swing (Duysens and Pearson, 1976; Duysens, 1977). Similarly, a crossed extensor response occurs during contralateral stance, while a crossed flexor response is seen during contralateral swing (Gauthier and Rossignol, 1981; Rossignol, Gauthier, Julien, & Lund, 1981). However, the data from the latter studies cannot directly be related to those observed in the studies on the intact cat, mentioned in the previous section, since 100-msec trains were used instead of single pulses and the intensity of stimulation was higher ("strong enough to evoke an ipsilateral flexion response in all parts of the step cycle") (Rossignol *et al.*, 1981). This may explain the long latencies of the reflexes (122 msec for crossed extensor responses and 146 msec for crossed flexor responses).

Brainstem mechanisms can also be excluded as the main modulatory source underlying phase-dependent reflex gating. Forssberg and colleagues (Forssberg, Grillner, & Rossignol, 1975; Forssberg, Grillner, Rossignol, & Wallen, 1976; Forssberg *et al.*, 1977; Forssberg 1979,1981) showed that phase-dependent modulation of reflexes (with P1 and P2 latencies) is present in the chronic spinal cat, walking on a treadmill. Moreover, the modulation persists during periods of rhythmic alternating activity in nerves to flexors and extensors of the motionless (paralyzed) spinal cat, treated with injections of Nialamide and DOPA (Andersson, Forssberg, Grillner, & Lindquist; Schomburg and Behrends, 1978a,b). The latter type of preparation is commonly referred to as "fictive locomotion" (Perret, Millanvoye, & Cabelguen, 1972; Zangger, 1978; Grillner and Zanger, 1984; Jordan, 1983). The flexor excitatory postsynaptic potentials (EPSPs), evoked by cutaneous stimulation, were usually largest during the rhythmic flexor burst, while the extensor EPSPs were largest during extensor bursts. In general, Schomburg and co-workers (Schomburg and Behrends, 1978a,b; Schomberg, Roesler, & Meinck, 1977; Schomberg, Roesler, & Kenins, 1978b; Schomberg, Behrends, & Steffens, 1981) concluded that such EPSPs appeared during the active phase of the recorded motoneuron, while inhibitory postsynaptic potentials (IPSPs) were present during the inactive phase. This modulation occurred for stimulus parameters similar to those used in the intact cat (Forssberg, 1979; Duysens and Loeb, 1980). To obtain a phase-dependent reversal from a facilitatory to a suppressive response, stimulation at high intensities (sural nerve at $8 \times$ threshold) was needed. A complete phase-dependent reversal was observed from almost pure EPSPs to IPSPs in 5 out of 40 motoneurons tested (Schomburg *et al.*, 1981). Similar results on phase-dependent modulation of transmission in cutaneous pathways have been reported for the forelimb (Hishinuma & Yamaguchi, 1989).

More recent studies, however, have challenged the generalized occurrence of phase-dependent modulation during fictive locomotion. Schmidt, Meyers, Fleshman, Tokuriki, and Burke (1988) and Schmidt, Meyers, Tokuriki, and Burke (1989) found that phase-dependent modulation was present only in some motoneurons [e.g., flexor digitorum longus (FDL)] and in response to only some types of input [Saphenous and plantar nerves (SP)]. The same authors were able to show that the FDL EPSPs, evoked by the two nerves mentioned above, were differentially modulation during fictive stepping. The early components of the SP EPSPs were enhanced during the early flexion phase while those in plantar EPSPs were markedly depressed during flexion (Moschovakis, Sholomenko, & Burke, 1991). A similar picture of highly differentiated reflex pathways undergoing quite specific modulatory influences during fictive locomotion arises from the study of LaBella, Niechaj, and Rossignol (1992). The latter authors focused their attention on cutaneous reflexes in the different parts of the triceps surae. They showed that the reflexes to these muscles are generally largest during the phase of the cycle in which the nerves to these muscles are most active but subtle deviations occur and there are differences in amplitude of the responses of the various muscle parts.

A phase-dependent reversal from crossed extension to crossed flexion has been observed occasionally during fictive locomotion (spinal paralyzed cats retreated with Nialamide and L-DOPA; Rossignol et al., 1981). However, here again, long stimulus trains were used and the latencies of the responses were correspondingly much longer than those seen in the studies on the intact cats, where single shocks were used. It was noted that the reversal was much less constant than during decerebrate walking. In many cases only crossed extension or crossed flexion was obtained. Rossignol et al. (1981) attributed this reduced potency of modulation to the absence of movement-related feedback during fictive locomotion.

At least part of the modulation during fictive locomotion seems to occur presynaptically on the afferent terminals (Baev, 1978, 1980, 1981; Baev & Kostyuk, 1982; Dubuc, Cabelguen, & Rossignol, 1985; Dubuc, Cabelguen, & Rossignol, 1987; Dubuc, Cabelguen, & Rossignol, 1988; Duenas, Loeb, & Marks, 1985; Duenas & Rudomin, 1988; Gossard, Cabelguen, & Rossignol, 1989; Gossard, Cabelguen, & Rossignol, 1990; see also Cattaert et al., Chapter 3, this volume). Alternatively, the gating may occur at the level of interneurones, similar to the situation in the tadpole (Sillar & Roberts, 1988).

In summary, experiments with chronic spinal cats walking on a treadmill have convincingly shown that supraspinal structures are not essential to explain the phase-dependent modulation of reflexes observed in intact walking cats. Moreover, some evidence is available from experiments on "fictive locomotion" of spinal-paralyzed cats to demonstrate that part of this modulation persists in the absence of movement-related feedback and thus is likely to be caused by the output of a spinal locomotor CPG. However, in "fictive locomotion," the phase-dependent modulation is weaker, thereby indicating that under normal circumstances the afferent input may contribute significantly to the changes in reflex gain observed during locomotion.

VI. Fore- and Hindlimb Coordination and Reflexes during Cat Locomotion

In the cat the coordination between fore- and hindlimbs during gait has been studied by several authors (Miller, van der Burg, & van der Meché,1975; Wetzel & Stuart, 1976; English, 1979; English & Lennard, 1982; Halbertsma, 1983; Cruse & Warnecke, 1992). The main conclusions can be summarized as follows: (1) Basically, there are two types of coordinating influences, namely in-phase (walking, trotting) and out-of-phase (galloping and half-bounding); and (2) in contrast to other species (see Cruse 1990), the contralateral coupling is stronger than the ipsilateral coupling in the cat.

Little is known about the neural substrate of the coordination between fore- and hindlimbs. In a recent review, English (1989) describes the evidence for the involvement of two systems. First, long propriospinal cells, which link lumbar and thoracic portions of the spinal cord, have often been implicated (Skinner, Adams, & Remmel, 1980; Alstermark, Lundberg, Norrsell, & Sybirska, 1981; see also Kato, Chapter 4, this volume). Secondly, neurons of the ventral spinocerebellar tract may be involved, since they carry information about the timing of step cycles from more than one limb.

Reflexes could be elicited from fore- to hindlimbs or vice versa, yielding diagonal flexor and extensor patterns that were reminiscent of the patterns of muscle activations used by these animals during locomotion (see Creed *et al.*, 1932, for a review). It has been suggested initially that some reflex pathways from fore- to hindlimb could underlie some of the coupling seen during locomotion (Miller & van der Burg, 1973). However, the variability among the different interlimb patterns used is so large that it is unlikely that an association with simple neural circuitry can be made (Wetzel & Stuart, 1976; English, 1979; English & Lennard, 1982). One striking feature is the existence of a fast propriospinal inhibitory pathway from forelimb afferents to motoneurones of hindlimb FDL (Schomburg, Meinck, & Haustein, 1975). Schomburg,

Meinck, Haustein, and Roesler (1978a) have proposed that this pathway is essential to prevent plantar flexion of the toes during periods of hindlimb stance.

Phase-dependent modulation of reflex responses in hindlimb muscles, following stimulation of the forelimb, has been described in decerebrate cats by Miller, Ruit, and van der Meché (1977). The reflex responses reversed from flexor in early swing to extensor during stance. Such reversals must depend on the intervention of a spinal CPG for locomotion since phase-dependent transmission in excitatory propriospinal reflex pathways from forelimb afferents to lumbar motoneurones was present in spinal-paralyzed cats exhibiting fictive locomotion (Schomburg et al., 1977; Schomburg & Behrends, 1978a,b).

Hence, in summary, reflexes between fore- and hindlimb in the cat are present and often highly specific (FDL) but they cannot, at present, be linked to a specific form of interlimb coordination. The modulation of these reflexes during gait obeys the same rules as observed for crossed reflexes. Variations in the amplitudes of the reflexes depend primarily on the step cycle of the limb in which the reflexes are observed. As for ipsilateral and crossed reflexes, there is evidence that spinal motor centers underlie the observed phase-dependent modulation.

VII. Flexibility and Mechanisms of Interlimb Coordination in the Cat

To investigate within what limits a cat can maintain stable interlimb coordination, some authors have used locomotion on split belts, moving at different speeds, thereby imitating walking in a circle (Kulagin & Shik, 1970; Forssberg, Grillner, Halbertsma, & Rossignol, 1980; Halbertsma, 1983). Under such conditions, chronic spinal cats can maintain alternating stepping, even for two- to threefold differences in belt speed. This is achieved mainly by prolonging the flexion or first extension phases of the limb walking on the "fast" belt, and a shortening of the swing phase of the limb walking on the "slow" belt. One "rule" emerging from these studies was that overlap between different phases of opposite limbs was allowed during alternating gait except for the E1 period (first extension phase prior to touchdown). Simultaneous bilateral occurrences of E1 phases were avoided, even if it meant that the limb had to "wait" a substantial period of time at the end of the flexion phase (see also Gauthier & Rossignol, 1981, for this effect).

A second way to test the flexibility of interlimb coordination is to use backward walking. Buford, Zernicke, and Smith (1990) found that normal cats can achieve this task by using a major postural change in the pelvis. These cats then reverse the sequence of paw contacts used during slow walking. During backward walking, right hindlimb contact was followed by left forepaw, left hindpaw, and right forepaw contact. These experiments with backward walking have raised interesting questions about which units must be coordinated. Backward walking cannot be explained as the simple coupling of limb CPGs, producing a fixed synergy, since hip movement is opposite to the movements of knee and ankles. This raises the question whether the limb CPG consists of several unit CPGs, each controlling one joint, as was proposed by Grillner (1981). In this scheme, backward locomotion is primarily caused by a change from mutual excitation between hip and knee unit generators (as used during forward gait) to mutual inhibition between these unit CPGs during backward gait.

Whatever the building blocks of the neural substrate for interlimb coordination, it appears that the coupling between these blocks depends on an appropriate amount of movement-induced afferent input. Grillner and Zangger 1984) found that interlimb coordination during hindlimb walking deteriorated following deafferentation in the mesencephalic cat. Similarly, Giulani and Smith (1987) described that coupling between hindleg movements during stepping in the air was weaker following deafferentation of a hindlimb of a chronic spinal cat. They found that during the majority of locomotor bouts the bilateral stepping was characterized by irregular phasing, with the intact hindlimb stepping at a faster frequency than that of the deafferented leg.

The question whether spinal mechanisms are sufficient for interlimb coordination is discussed in the contribution of Kato to this book (Chapter 4). Evidence for spinal mechanisms involved in the coordination of movements of the different limbs during locomotion of the cat is provided by the observation that such coordinated movements persist in the high spinal cat injected with DOPA and placed on a treadmill (Miller & van der Meché, 1976). When movement feedback is absent, however, for example, during fictive locomotion, then the coordination is generally more variable than when movements are allowed. The most common type of interlimb coordination is the alternation of activity in the limbs of one girdle, but occasionally, a bilateral synchrony of flexion and extension was observed (Miller & Schomburg, 1985). The coordination of activity in fore- and hindlimbs is even more variable (Chofflon & Zangger, 1977; Viala & Vidal, 1978; Orsal, Cabelguen, & Perret, 1990). There is also some evidence for diagonal coupling between forelimbs and hindlimbs (Orsal *et al.*, 1990).

In conclusion, the results of various manipulations (split belt, backward locomotion) have shown that the coordination between movements of the various limbs during gait is quite flexible. The spinal cord is able to produce much of this coordination as well as its flexibility. In addition, movement-related afferent feedback plays an important role, as it did for the phase-dependent modulation of the responses (see above).

VIII. Interlimb Coordination during Human Gait

In humans, there is coupling between arm and leg movements during gait. Newly walking infants exhibit interlimb coordination similar to that seen in adults, although the coupling is more loose (Clark, Whitall, & Phillips, 1988). After 3 months of walking, adultlike consistency is obtained.

Arm movements are not essential for walking in humans, yet all people exhibit such movements spontaneously (Murray, Sepic, & Barnard, 1967). The most commonly observed pattern is that the arms swing forward and backward in phase with the direction of the contralateral lower limb and opposite to the directions of the ipsilateral lower limb (Craik, Herman, & Finley, 1976; Murray et al., 1967). Elftman (1939) calculated the torque of the muscles of the arms during gait and concluded that the arm swing is not a passive pendular action but results in large part from active muscle contractions. Consistent with this conclusion is the observation of EMG activity for both shoulder and arm muscles during locomotion (Hogue, 1969; Fernandez-Ballesteros, Buchtal, & Rosenfalck, 1965). Elftman (1939) suggested that the arm swing is needed to counteract pelvic rotation during stance. During running, the pattern of arm movement changes but the function remains similar. Hinrichs, Cavanagh, and Williams (1987) have suggested that arm swing during adult running is crucial for the transfer of angular momentum between upper and lower body about the vertical axis. Some subjects make asymmetrical arm movements during running. This is thought to compensate for asymmetries elsewhere in the body (Hinrichs, 1990, 1992). Little is known about the reflex origin of this coupling.

In humans, reflex responses in arm muscles have been described under stationary conditions following either ankle displacement (Kearney & Chan, 1981) or cutaneous stimulation of the foot (Kearney & Chan, 1979). In the latter experiments, it was found that electrical stimulation of the foot yielded large responses in flexors and inhibitions in extensors both in the leg and in the arm. The latency (70 msec) of the large responses in the arm was slightly shorter than the latency of the equivalent response in the leg (75 msec), which indicates, according to the authors, that a spinal-bulbo-spinal system may be involved. Similarly, Dowman (1992) recently described responses with a latency of

81 msec in biceps brachii following sural nerve stimulation (5-pulse train at 250 Hz). In view of their rapid habituation, these responses were considered to be startle responses. A similar conclusion was reached in the studies of Delwaide and Crenna (1983, 1984). They found that stimulation of cutaneous nerves of the leg resulted in facilitation first of masseter, then of arm muscles, and finally of lower limb muscles. To explain their data, they invoke the existence of a supraspinal center, which is activated by low-threshold exteroceptive afferents and which facilitates motor nuclei in a rostrocaudal sequence.

During gait, Quintern, Berger, and Dietz (1985) observed responses in arm muscles with a latency below 100 msec using mechanical perturbations (step increases in treadmill speed). These responses disappeared as the motor task became familiar. Since there was a close association between these arm responses and the simultaneously recorded cerebral potentials, Quintern *et al.* (1985) suggested a transcortical origin for the responses they observed.

Inversely, responses in leg muscles can be elicited by stimulation of the arm of normal subjects (Piesiur-Strehlow & Meinck, 1980; Meinck & Piesiur-Strehlow, 1981; Sarica & Ertekin, 1985; Gassel & Ott, 1973). Recently, the reflex interconnections of lower and upper extremity muscles have been investigated in patients with complete spinal cord lesions (Calancie, 1991). Single stimuli to lower extremity nerves induced responses in both ipsi- and contralateral upper extremity muscles. The ipsilateral responses had a latency of 52 msec while the contralateral responses had a latency of about 125 msec when single stimuli were used. Application of brief stimulus trains, however, resulted in a large reduction in the latency of the contralateral responses such that it was about equal to the ipsilateral latency. Such responses were not seen in normal controls or in quadriplegics with incomplete lesions.

In summary, as in the cat, human limb movements of all limbs, including the arms, display active contractions during gait. These contractions occur according to a limited set of interlimb coordination patterns. The role played by interlimb reflexes in this coordination is unclear. Reflexes, presumably due to propriospinal neurons, can be demonstrated in quadriplegics but they are difficult to demonstrate in normal subjects. The dominant interlimb reflexes, seen in the intact human, are most likely due to supraspinal loops.

IX. Crossed Reflexes and Their Modulation in Humans

In humans, the most important coordination occurs at the level of the legs, which under normal circumstances are the only limbs participating in weight bearing during gait. Therefore, most attention has been directed mainly to crossed responses in the legs. Crossed flexor responses were described by Kugelberg, Eklund, and Grimby (1960), following high-intensity stimulation of the sole of the foot.

As in the cat, it was found that ipsi- and contralateral reflexes were modifiable, depending on the task. For example, Lisin, Frankstein, and Rechtmann (1973) found that in hemiparetic patients, the stimulation of the sural nerve yields a flexor reflex at rest, while the same electrical stimulation elicits an extensor reflex during gait. In more recent years, the phase-dependent modulation of exteroceptive reflexes has been studied in humans, using methods similar to those used in the cat. Unfortunately, however, there were large differences in methodology, making a comparison difficult. Some groups used high-intensity noxious stimulation (Crenna & Frigo, 1984; Belanger & Patla, 1984, 1987), while others used mostly weak, nonpainful shocks (Kanda & Sato, 1983; Yang & Stein, 1990; Duysens, Trippel, Horstmann, & Dietz, 1990a; Duysens, Tax, van der Doelen, Trippel, & Dietz, 1991). The responses consist mainly of two peaks, having latencies of about 47 and 73 msec (Figure 5). As in the cat, the early peak was absent contralaterally. In view of the analogy with the cat, the two peaks will be referred to as P1 and P2.

Most studies have focused on the P2 responses, since they are larger and more consistent than the P1 responses. All studies agreed that the

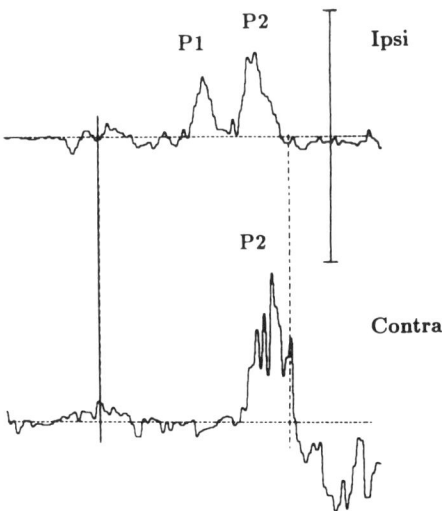

FIGURE 5 P1 and P2 responses in the ipsilateral (ipsi) and contralateral (contra) biceps femoris to sural nerve stimulation at $1.8 \times T$ (times perception threshold) during the stance phase of a human volunteer, running at 8 km/hour on a treadmill. Distance between vertical lines, 100 msec; EMG calibration, 0.5 mV. Average of 10 trials.

amplitude of the P2 responses is modulated in a phase-dependent manner during gait and that the modulation could not be simply explained by the spontaneously occurring variations in excitability of the motoneurones. In several of these studies a phase-dependent reversal was described. During early and middle swing, the P2 responses following stimulation of the posterior tibial or sural nerve were facilitatory in tibialis anterior (TA) while the same stimuli yielded TA-suppressive and GM-facilitatory responses during the E1 phase, just prior to touchdown (Figure 6; Yang & Stein, 1990; Duysens et al., 1990, 1991; Seigh-Naraghi, Herman, & D'Luzansky, 1992).

These reversals were accompanied by a movement reversal as well. During early swing, the response consisted of ankle dorsiflexion, while during late swing the responses were correlated with plantar flexion (Duysens, Tax, Trippel, & Dietz, 1992).

Similar results were obtained for the crossed responses. Such responses were obtained in a variety of muscles, both flexors and extensors (Tax, Duysens, Trippel, & Dietz, 1990; Seigh-Naraghi et al., 1992). For example, biceps femoris, which is particularly easy to recruit in ipsilateral responses (Hugon, 1973; Meinck, Küster, Benecke, & Conrad, 1985), was also quite responsive contralaterally. On the ipsilateral side, the P2 responses were often largest during stance, even when BF

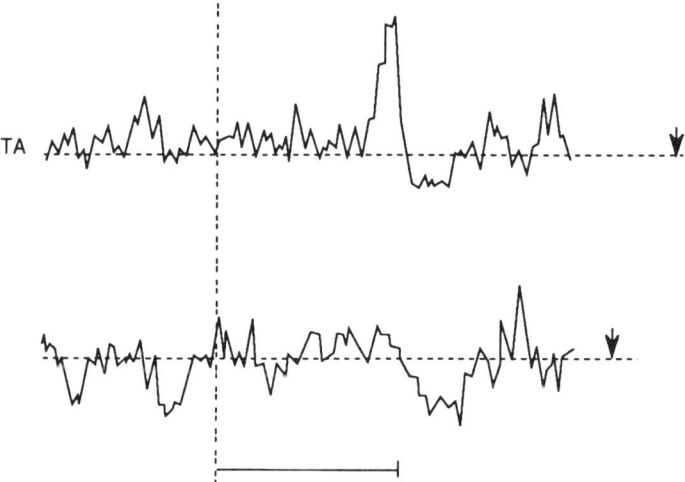

FIGURE 6 Phase-dependent reversal from a facilitatory P2 response (*top*) to a suppressive response (*bottom*) in TA (tibialis anterior) following sural nerve stimulation at 2.2 × T in a human volunteer, running on a treadmill at 8 km/hour. Downward arrow, touchdown; vertical stippled line, stimulus (20-msec train of five pulses of 1 msec). The records are averaged over 10 trials. Averaged ($n = 10$) control activity was subtracted (hence suppressive responses can be seen as deviations below the horizontal zero).

showed little spontaneous activation in that period. Similarly, the crossed P2 responses in BF were larger during contralateral stance than during contralateral swing, even when the spontaneous "control" activity was about equal in these two phases (Figure 7; Tax et al., 1990).

Both in BF and TA, there was a reduction in the amplitude of the responses near the end of the swing phase (Tax, Duysens, Gielen, Trippel, & Dietz, 1991), a feature that was also prominent ipsilaterally (Duysens et al., 1990a,b). It was concluded that the gating of the crossed responses depended on the phase of the corresponding leg in the step cycle and not on the phase of the stimulated leg. Nevertheless, there were some elements that were specific for crossed reflexes. For example, crossed extensor responses were observed during the contralateral stance phase both in GM and in SOL (soleus). However, during slow walking, instances were found when facilitatory responses appeared in the contralateral SOL, in conjunction with a suppression of GM activity (Figure 8; Duysens et al., 1991). Such opposite reflex actions were never observed ipsilaterally. Since SOL is a slow "postural" ankle extensor, while GM is faster and biarticular (knee flexor and ankle extensor), the data suggest that the crossed responses are especially aimed at maintaining stability. These crossed responses had a surprisingly low threshold, implying that the afferents involved are fast-conducting fibers, which are active during normal gait (Loeb, Bak, & Duysens, 1977).

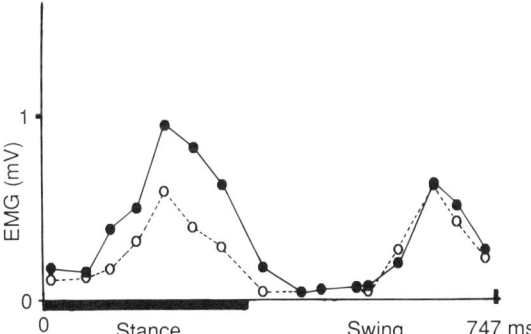

FIGURE 7 Modulation of P2 responses in the contralateral BF (biceps femoris) following sural nerve stimulation at $2 \times T$ in various phases of the step cycle of a subject running (8 km/hour) on a treadmill. For each stimulus condition the integrated area of the averaged reflex BF activity in a period of 80–110 msec after stimulation is plotted against the time of occurrence of this activity in the step cycle (●). The latter time is determined by the sum of the stimulus delay with respect to the contralateral footfall and the middle of the integration interval (95 msec). For comparison, the equivalent activity in the control cycles is shown (○). Note that the largest crossed P2 responses occurred during contralateral stance.

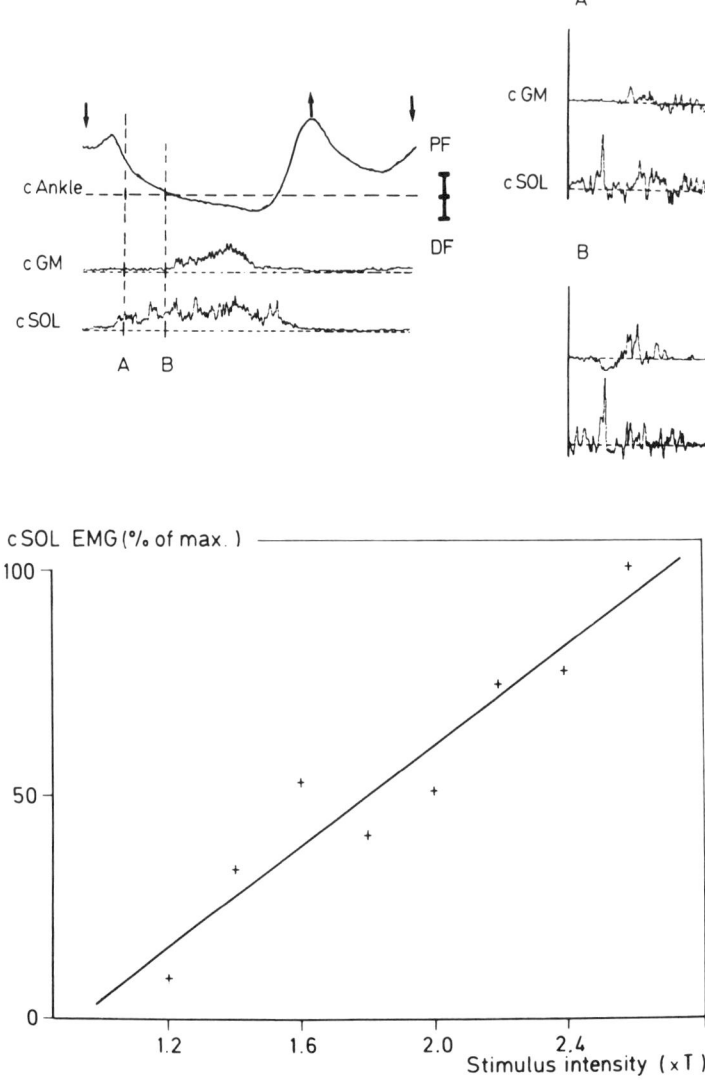

FIGURE 8 Contralateral reflex responses in gastrocnemius medialis (cGM) and soleus (cSOL) to sural nerve stimulation at 3.1 × T (*bottom*) at 2 points in the step cycle (*top* A, B: vertical stippled lines). *Top left:* control step cycle (walking at 4 km/hour, 1113-msec cycle duration). Ankle goniometer output and unsubtracted EMG activity on the side contralateral to stimulation. Arrows, onset (downward) and end (upward) of stance; PF and DF, plantar flexion and dorsiflexion; calibration, 15 degrees, 1 mV. *Bottom:* plot of the averaged amplitude of crossed P2 SOL responses (100–120 msec) as a function of stimulus intensity. At 1.4 × T the responses were already significantly above the level of the spontaneous activity. (Modified from Duysens *et al.*, 1991.)

Stimulation of the sural nerve at intensities as low as 1.4 times perception threshold was sufficient to elicit these responses (Figure 8, bottom).

The mechanisms involved in phase-dependent modulation of reflexes in humans are likely to be similar to those described for the cat. Presynaptic modulation of the afferents is thought to be the primary source of modulation of activity in the monosynaptic H-reflex pathway during human locomotory behavior (Brooke & McIlroy, 1990; Capaday & Stein, 1986), but it is presently unclear whether the same mechanism is involved in the modulation of the exteroceptive responses described above (see also Brooke et al., Chapter 6, this volume).

The crossed responses during running usually did not lead to detectable changes in joint angles or step durations when stimulus intensities were low. When strong nociceptive stimuli were given, however, a clear behavioral response was seen. The ipsilateral response then consisted of a suppression of the ongoing activity. This resulted in a plantar flexion of the ankle during swing (TA suppression) or in an ankle dorsiflexion during stance (GM suppression). Contralaterally, the dominant response was ankle dorsiflexion (Figure 9). The latter was due either to crossed flexor responses during swing or to suppression of extensor activity during stance. This suppression occurred at P2 latency and often preceded a facilitatory extensor response. The early crossed extensor suppression is unexpected in view of the commonly reported crossed extensor responses following nociceptive stimuli applied in various animal preparations (see above). During the stance phase of human gait the most common response to strong stimuli (3.4 times perception threshold in Figure 9) is a yield, leading to an unloading of the supporting limb. Apparently, such response is equivalent to a temporary arrest of the locomotory process and presumably overrules crossed extensor responses, which may otherwise accompany ipsilateral flexor reflexes.

In summary, by stimulating human skin nerves in different phases of the step cycle, one obtains responses that are similar to the P1 and P2 responses described in the cat. The phase-dependent modulation of the human P2 responses is quite similar to the modulation observed in the cat, except, for example, that in humans the suppression of responses in TA at the end of swing is more prominent. To some extent, crossed responses often follow the same modulation pattern as observed ipsilaterally and have the same time relationship with the step cycle phases, indicating that the modulation depends primarily on events related to the limb showing the responses and not to the stimulated leg. The EMG responses are accompanied by small movements ipsilaterally when stimulus is weak. In humans, strong shocks during stance cause bilateral flexion.

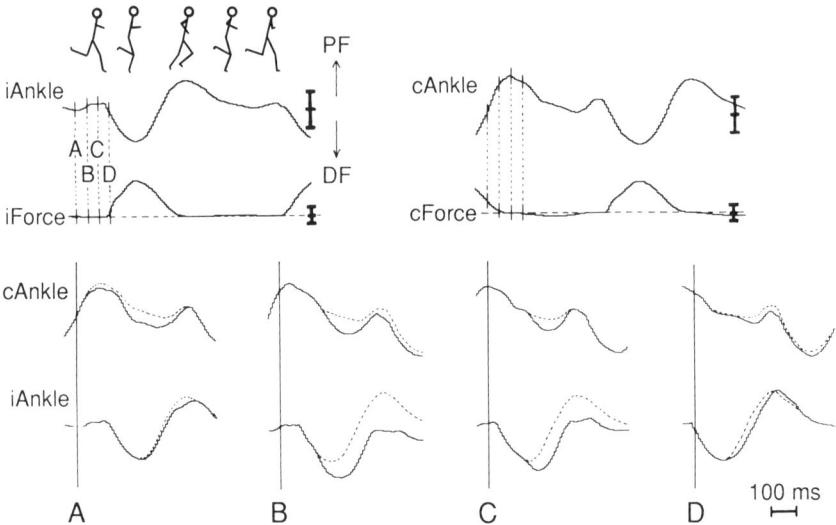

FIGURE 9 Kinesiologic changes induced by high-intensity (3.4 × T) electrical stimulation of the sural nerve at 4 phases (A, B, C, D) in the step cycle during running on a treadmill at 8 km/hour. The stimuli (20-msec train of 5 pulses of 1 msec) were given prior to ipsilateral touchdown, coinciding with the beginning of the contralateral swing (*top*). Ipsilateral and contralateral force (iForce and cForce, respectively) were measured with force plates. Ipsilaterally the stimuli induced extra dorsiflexion of the ankle (stippled line = controls). Contralaterally, the main effect was dorsiflexion as well, but with longer latency than ipsilaterally. In all ankle traces (iAnkle and cAnkle: ipsilateral and contralateral ankle angle, respectively, as measured with goniometer), plantar flexion is upward, dorsiflexion downward. Calibration: 50 N (force), 15 degrees (ankle); figures at top, step cycle starts with ipsilateral touchdown (stippled line).

X. Conclusions

Despite differences between bipedal and quadrupedal locomotion, there are striking similarities in the interlimb coordination patterns of humans and cats during gait. It is still unclear how much this coordination relies on interlimb reflexes, but it is obvious that such interlimb reflexes exist in both species and that their amplitude is modulated similarly in a phase-dependent way during gait. The primary source for this modulation is linked to the limb in which the reflexes occur and not to the stimulated limb. For cats, there is good evidence that part of the modulation is provided by the spinal motor program (CPG) for locomotion.

Acknowledgments

We gratefully acknowledge the expert secretarial assistance of A. Wanders. This work was supported by a grant to V. Dietz from the Deutsch Forschungsgemeinschaft (SFB 325), by funds from Esprit (Project 3149), and by a Nato fellowship to J. Duysens and a UOP fellowship to A.A.M. Tax.

References

Abraham, L. D., Marks, W. B., & Loeb, G. E. (1985). The distal hindlimb musculature of the cat: cutaneous reflexes during locomotion. *Journal of Neurophysiology,* **58,** 594–603.

Alstermark, B., Lundberg, A., Norrsell, U., & Sybirska, E. (1981). Integration in descending motor pathways controlling the forelimb in the cat. 9. Differential behavioural defects after spinal cord lesions interrupting defined pathways from higher centres to motoneurons. *Experimental Brain Research,* **42,** 299–318.

Andersson, O., Forssberg, H., Grillner, S., & Lindquist, M. (1978). Phasic gain control of the transmission in cutaneous reflex pathways to motoneurons during 'fictive' locomotion. *Brain Research,* **149,** 503–507.

Arya, T., Bajwa, S., & Edgley, S. A. (1991). Crossed reflex actions from group II muscle afferents in the lumbar spinal cord of the anaesthetized cat. *Journal of Physiology,* **444,** 117–131.

Baev, K. V. (1978). Periodic changes in primary afferent depolarization during fictitious locomotion in thalamic cats. *Neurophysiology,* **10,** 316–317.

Baev, K. V. (1980). Polarization of primary afferent terminals in the lumbar spinal cord during fictitious locomotion. *Neurophysiology,* **12,** 305–311.

Baev, K. V. (1981). Retuning of segmental responses to peripheral stimulation in cats during fictitious locomotion. *Neurophysiology,* **13,** 206–212.

Baev, K. V., & Kostyuk, P. G. (1982). Polarization of primary afferent terminals of lumbosacral cord elicited by the activity of spinal locomotor generator. *Neuroscience,* **7,** 1401–1409.

Barbeau, H., & Fung, J. (1992). New experimental approaches in the treatment of spastic gait disorders. In H. Forssberg, & H. Hirshfeld (Eds.), *Movement disorders in children*. Medical Sport Science (pp. 234–246). Basel: Karger.

Baxendale, R. H., & Rosenberg, J. R. (1977). Crossed reflexes evoked by selective activation of tendon organ afferent axons in the decerebrate cat. *Brain Research,* **127,** 323–326.

Belanger, M., & Patla, A. E. (1984). Corrective responses to perturbation applied during walking in humans. *Neuroscience Letters,* **49,** 291–295.

Belanger, M., & Patla, A. E. (1987). Phase-dependent compensatory responses to perturbation applied during walking in humans. *Journal of Motor Behaviour,* **19**(4), 434–453.

Berger, W., Dietz, V., & Quintern, J. (1984). Corrective reactions to stumbling in man: neuronal co-ordination of bilateral leg muscle activity during gait. *Journal of Physiology,* **357,** 109–125.

Brooke, J. D., & McIlroy, W. E. (1990). Movement modulation of a short latency reflex linking the lower leg and the knee extensor muscles in humans. *Electroencephalography and Clinical Neurophysiology,* **75,** 64–74.

Buford, J. A., Zernicke, R. F., & Smith, J. L. (1990). Adaptive control for backward quadrupedal walking I. Posture and hindlimb kinematics. *Journal of Neurophysiology* (Vol. 64) No. 3, 745.

Calancie, B. (1991). Interlimb reflexes following cervical cord injury in man. *Experimental Brain Research,* **85,** 458–469.

Capaday, C., & Stein, R. B. (1986). Amplitude modulation of the soleus H-reflex in the human during walking and standing. *Journal of Neuroscience,* **6,** 1308–1313.

Carter, M. C., & Smith, J. L. (1986). Simultaneous control of two rhythmical behaviors. I. Locomotion with paw-shake response in normal cat. *Journal of Neurophysiology,* **56,** 171–183.

Chofflon, M., & Zangger, P. (1977). Fictive locomotion in high spinal cats. *Proceeding International Congress Physiology,* **13,** 35.

Clark, J. E., Whitall, J., & Phillips, S. J. (1988). Human interlimb coordination: the first 6 months of independent walking. *Developmental Psychobiology,* **21,** 445–456.

Craik, R., Herman, R., & Finley, F. R. (1976). Human solutions for locomotion. II. Interlimb coordination. In R. M. Herman, S. Grillner, P. Stein, & D. Stuart (Eds.), *Neural control of locomotion* (pp. 647–674). New York: Plenum.

Creed, R. S., Denny-Brown, D., Eccees, J. C., Liddell, E. G. T., & Sherrington, C. S. (1932). *Reflex activity of the spinal cord.* London: Oxford University Press.

Crenna, P., & Frigo, C. (1984). Evidence of phase-dependent nociceptive reflexes during locomotion in man. *Experimental Neurology,* **85,** 336–345.

Cruse, H. (1990). What mechanisms coordinate leg movement in walking arthropods? *Trends in Neuroscience,* **13,** 15–21.

Cruse, H., & Warnecke, H. (1992). Coordination of the legs of a slow-walking cat. *Experimental Brain Research,* **89,** 147–156.

Delwaide, P. J., & Crenna, P. (1983). Exteroceptive influence on lower limb motoneurons in man: spinal and suprasinal contributions. In J. E. Desmedt (Ed.). *Motor control mechanisms in health and disease.* (Vol. 39) (pp. 797–807). New York: Raven Press.

Delwaide, P. J., & Crenna, P. (1984). Cutaneaous nerve stimulation and motoneuronal excitability II: evidence for non segmental influences. *Journal of Neurology Neurosurgery and Psychiatry,* **47,** 190–196.

Dietz, V., Quintern, J., Boos, G., & Berger, W. (1986). Obstruction of the swing phase during gait: Phase dependent bilateral leg muscle coordination. *Brain Research,* **384,** 166–169.

Dowman, R. (1992). Possible startle response contamination of the spinal nociceptive withdrawal reflex. *Pain,* **49,** 187–197.

Drew, T., & Rossignol, S. (1987). A kinematic and electromyographic study of cutaneous reflexes evoked from the forelimb of unrestrained walking cats. *Journal of Neurophysiology,* **57,** 1160–1184.

Dubuc, R., Cabelguen, J.-M., & Rossignol, S. (1985). Rhythmic antidromic discharge of single primary afferents recorded in cut dorsal root filaments during locomotion in the cat. *Brain Research,* **359,** 375–378.

Dubuc, R., Cabelguen, J.-M., & Rossignol, S. (1987). Antidromic discharges of primary afferents during locomotion. In G. N. Cantchev, B. Dimitrov, & P. Gatev (Eds.), *Motor control* (pp. 165–169). New York: Plenum.

Dubuc, R., Cabelguen, J.-M., & Rossignol, S. (1988). Rhythmic fluctuations of dorsal root potentials and antidromic discharges of primary afferents during fictive locomotion in the cat. *Journal of Neurophysiology,* **60,** 2014–2036.

Duenas, S. H., & Rudomin, P. (1988). Excitability changes of ankle extensor group Ia and Ib fibers during fictive locomotion in the cat. *Experimental Brain Research,* **70,** 15–25.

Duenas, S. H., Loeb, G. E., & Marks, W. B. (1985). Dorsal root reflex during locomotion in normal and decerebrate cats. *Society Neuroscience Abstracts,* **11,** 1028.

Duysens, J. (1977). Reflex control of locomotion as revealed by stimulation of cutaneous afferents in spontaneously walking premammillary cats. *Journal of Neurophysiology,* **40,** 737–751.

Duysens, J., & Loeb, G. E. (1980). Modulation of ipsi- and contralateral reflex responses in unrestrained walking cats. *Journal of Neurophysiology,* **44,** 1024–1037.

Duysens, J., & Pearson, K. G. (1976). The role of cutaneous afferents from the distal hindlimb in the regulation of the step cycle of thalamic cats. *Experimental Brain Research,* **24,** 245–255.

Duysens, J., & Pearson, K. G. (1980). Inhibition of flexor burst generation by loading ankle extensor muscles in walking cats. *Brain Research,* **187,** 321–332.

Duysens, J., & Stein, R. B. (1978). Reflexes induced by nerve stimulation in walking cats with implanted cuff electrodes. *Experimental Brain Research,* **32,** 213–224.

Duysens J., Loeb G. E., & Weston, B. J. (1980). Crosssed flexor reflex responses and their reversal in freely walking cats. *Brain Research,* **197,** 538–542.

Duysens, J., Trippel, M., Horstmann, G. A., & Dietz, V. (1990a). Gating and reversal of reflexes in ankle muscles during human walking. *Experimental Brain Research,* **82,** 351–358.

Duysens, J., Tax, A. A. M., Trippel, M., & Dietz, V. (1990b). Ipsilateral reflex reversal during human locomotion suggests that the biceps femoris reflex is not a flexion reflex. In T. Brandt, W. Paulus, W. Bles, M. Dieterich, S. Krafczyk, and A. Straube (Eds.) *Disorders of posture and gait* (pp. 124–127). New York: Georg Thieme Verlag.

Duysens, J., Tax, A. A. M., Doelen, van der, B., Trippel, M., & Deitz, V. (1991). Selective activation of human soleus or gastrocnemius in reflex responses during walking and running. *Experimental Brain Research,* **87,** 193–204.

Duysens, J., Tax, A. A. M., Trippel, M., & Dietz, V. (1992). Phase-dependent reversal of reflexly induced movements during human gait. *Experimental Brain Research,* **90,** 404–414.

Elftman, H. (1939). The functions of the arms in walking. *Human biology,* **11,** 529–536.

English, A. W. (1979). Interlimb coordination during stepping in the cat: an electromyographic analysis. *Journal of Neurophysiology,* **42,** 229–243.

English, A. W. (1989). Interlimb coordination during locomotion. *American Zoologist,* **29,** 255–266.

English, A. W. (1989). Interlimb coordination during locomotion. *American Zoologist,* **29,** 255–266.

English, A. W., & Lennard, P. R. (1982). Interlimb coordination during stepping in the cat: In-phase stepping and gait transitions. *Brain Research,* **245,** 353–364.

Fernandez-Ballesteros, M. L., Buchtal, F., & Rosenfalck, P. (1965). The pattern of muscular activity during the arm swing of natural walking. *Acta Physiologica Scandinavica,* **63,** 296–310.

Forssberg, H. (1979). Stumbling corrective reaction: A phase-dependent compensatory reaction during locomotion. *Journal of Neurophysiology,* **42,** 936–953.

Forssberg, H. (1981). Phasic gating of cutaneous reflexes during locomotion. In A. Taylor & A. Prochazka (Eds.), *Muscle receptors and movement* (pp. 403–412). London: Macmillan.

Forssberg, H., Grillner, S., & Rossignol, S. (1975). Phase dependent reflex reversal during walking in chronic spinal cats. *Brain Research,* **85,** 103–107.

Forssberg, H., Grillner, S., & Rossignol, S. (1977). Phasic gain control of reflexes from the dorsum of the paw during spinal locomotion. *Brain Research,* **132,** 121–139.

Forssberg, H., Grillner, S., Rossignol, S., & Wallen, P. (1976). Phasic control of reflexes during locomotion in vertebrates. In R. M. Herman, S. Grillner, P. Stein, & D. Stuart (Eds.), *Neural control of locomotion* (pp. 647–674). New York: Plenum.

Forssberg, H., Grillner, S., Halbertsma, J., & Rossignol, S. (1980). The locomotion of the low spinal cat: II. Interlimb coordination. *Acta Physiology Scandinavia,* **108,** 283–295.

Frigo, C., & Crenna, P. (1987). Neural control of locomotion: some recent advancements in the methodological approach. In J. Van Alste (Ed.), *Restoration of walking aided by Functional Electrical Stimulation* (pp. 17–27). Milan: Pro Juventute.

Gassel, M. M., & van Ott, K. H. (1973). Patterns of reflex excitability change after widespread cutaneous stimulation in man. *Journal of Neurology Neursurg. Psych.,* **36,** 282–287.

Gauthier, L., & Rossignol, S. (1981). Contralateral hindlimb responses to cutaneous stimulation during locomotion in high decerebrate cats. *Brain Research,* **207,** 303–320.

Giuliani, C. A., & Smith, J. L. (1987). Stepping behaviors in chronic spinal cats with one hindlimb deafferented. *Journal of Neuroscience,* **7,** 2537–2546.

Gossard, J. P., & Rossignol, S. (1990). Phase-dependent modulation of dorsal root potentials evoked by peripheral nerve stimulation during fictive locomotion in the cat. *Brain Research,* **537,** 1–13.

Gossard, J.-P., Cabelguen, J.-M., & Rossignol, R. (1989). Intra-axonal recordings of cutaneous primary afferents during fictive locomotion in the cat. *Journal of Neurophysiology,* **62,** 1177–1188.

Gossard, J.-P., Cabelguen, J.-M., & Rossignol, S. (1990). Phase-dependent modulation of primary afferent depolarization in single cutaneous primary afferents evoked by peripheral stimulation during fictive locomotion in the cat. *Brain Research,* **537,** 14–23.

Graham-Brown, T., & Sherrington, C. S. (1912). The rule of reflex response in the limb reflexes of the mammal and its exceptions. *Journal of Physiology, (Lond.),* **44,** 125–130.

Grillner, S. (1981). Control of locomotion in bipeds, tetrapods, and fish. In V. Brooks (Ed.), *Handbook of physiology, The nervous system II* (pp. 1179–1236). Baltimore, MD: Waverly Press.

Grillner, S., & Rossignol, S. (1978a). Contralateral reflex reversal controlled by limb position in the acute spinal cat injected with clonidine i.v. *Brain Research,* **144,** 411–414.

Grillner, S., & Rossignol, S. (1978b). On the initiation of the swing phase of locomotion in chronic spinal cats. *Brain Research,* **146,** 260–277.

Grillner, S., & Zangger, P. (1984). The effect of dorsal root transection on the efferent motor pattern in the cat's hindlimb during locomotion. *Acta Physiology Scandinavia,* **120,** 393–405.

Halbertsma, J. (1983). The stride cycle of the cat: the modelling of locomotion by computerized analysis of automatic recordings. *Acta Physiology Scandinavia supplement,* **521,** 1–75.

Hinrichs, R. N. (1990). Upper extremity function in distance running. In P. R. Cavanagh (Ed.), *Biomechanics of distance running* (pp. 107–133). Champaine, IL: Human Kinetics.

Hinrichs, R. N. (1992). Case studies of asymmetrical arm action in running. *International Journal of Sport Biomechanics,* **8,** 111–128.

Hinrichs, R. N., Cavanagh, P. R., & Williams, K. R. (1987). Upper extremity function in running. In Center of mass and propulsion considerations. *International Journal of Sport Biomechanics,* **3,** 222–241.

Hishinuma, M., & Yamaguchi, T. (1989). Modulation of reflex resonses during fictive locomotion in the forelimb of the cat. *Brain Research,* **482,** 184–189.

Hogue, R. E. (1969). Upper extremity muscular activity at different cadences and inclines during gait. *Physiology Therapeutics,* **49,** 963–972.

Holmqvist, B. (1961). Crossed spinal reflex actions evoked by volleys in somatic afferents. *Acta Physiology Scandinavia,* **52,** (Suppl. 181), 1–67.

Hugon, M. (1973). Exteroceptive reflexes to stimulation of the sural nerve in normal man. In J. E. Desmedt (Ed.), *New developments in electromyography*, Clinical Neurophysiology (Vol. 3) (pp. 713–729) Basel: Karger.

Jankowska, E. (1992). Interneuronal relay in spinal pathways from proprioceptors. *Progress in Neurobiology*, **38**, 335–378.

Jankowska, E., Jukes, M. G. M., Lund, S., & Lundberg, A. (1967). The effects of DOPA on the spinal cord. 5 Reciprocal organization of pathways transmitting excitatory action to alpha motoneurones of flexors and extensors. *Acta Physiology Scandinavia*, **70**, 369–388.

Jordan, L. M. (1983). Factors determining motoneuron rhythmicity during fictive locomotion. *Society Experimental Biology Symposia*, **37**, 423–444.

Kanda, K., & Sato, H. (1983). Reflex responses of human thigh muscles to non-noxious sural stimulation during stepping. *Brain Research*, **288**, 378–380.

Kearney, R. E., & Chan, C. W. Y. (1979). Reflex response of human arm muscles to cutaneous stimulation of the foot. *Brain Research*, **170**, 214–217.

Kearney, R. E., & Chan, C. W. Y. (1981). Interlimb reflexes evoked in human arm muscles by ankle displacement. *Electroencephalography Clinical Neurophysiology*, **52**, 65–71.

Kugelberg, E., Eklund, K., & Grimby, L. (1960). An electromyographic study of the nociceptive reflexes of the lower limb. Mechanism of the plantar responses. *Brain*, **83**, 394–410.

Kulagin, A. S., & Shik, M. L. (1970). Interaction of symmetrical limbs during controlled locomotion. *Biophysics, USSR*, **15**, 171–178.

LaBella, L. A., Niechaj, A., & Rossignol, S. (1992). Low-threshold, short-latency cutaneous reflexes during fictive locomotion in the "semi-chronic" spinal cat. *Experimental Brain Research*, **91**, 236–248.

Lisin, V. V., Frankstein, S. I.&., & Rechtmann, M. B. (1973). The influence of locomotion on flexor reflex of the hindlimb in cat and man. *Experimental Neurology*, **38**, 180–183.

Loeb, G. E., Bak, M. J., & Duysens, J. (1977). Long-term unit recording from somatosensory neurons in the spinal ganglia of the freely walking cat. *Science*, **197**, 1192–1194.

Loeb, G. E., Hoffer, J. A., & Marks, W. B. (1987). Cat hindlimb motoneurons during locomotion. IV. Participation in cutaneous reflexes. *Journal of Neurophysiology*, **57**, 563–573.

Marsden, C. D., Merton, P. A., & Morton, H. B. (1976). Servo action in the human thumb. *Journal of Physiology, (Lond.)*, **257**, 1–44.

Marsden, C. D., Rothwell, J. C., & Day, B. L. (1983). Long-latency automatic responses to muscle stretch in man: origin and function. *Advances Neurology*, **39**, 509–539.

Matsukawa, K., Kamei, H., Minoda, K., & Udo, M. (1982). Interlimb coordination in cat locomotion investigated with perturbation. In Behavioral and electromyographic study on symmetric limbs of decerebrate and awake walking cats. *Experimental Brain Research*, **46**, 425–437.

Matthews, P. B. C. (1986). Observations on the automatic compensation of reflex gain on varying the pre-existing level of motor discharge in man. *Journal of Physiology*, **374**, 73–90.

McMillan, J. A., & Koebbe, M. J. (1981). Effects of sensory inputs on the excitability of the crossed extension reflex. *Experimental Neurology*, **73**, 233–242.

Meinck, H. M., & Piesiur-Strehlow, B. (1981). Reflex evoked in leg muscles from arm afferents: a propriospinal pathway in man? *Experimental Brain Research*, **43**, 78–86.

Meinck, H. M., Küster, S., Benecke, R., & Conrad, B. (1985). The Flexor Reflex—Influence of stimulus parameters on the reflex response. *Electroencephalography Clinical Neurology*, **61**, 287–298.

Miller, S., & van der Burg, J. (1973). The function of long propriospinal pathways in the co-ordination of quadrupedal stepping in the cat. In R. B. Stein, K. B. Pearson, R. S. Smith, & J. B. Redford (Eds.), *Control of posture and locomotion* (pp. 561–577). New York: Plenum.

Miller, S., & van der Meché, F. G. A. (1976). Coordinated stepping of all four limbs in the high sinal cat. *Brain Research,* **109,** 395–398.

Miller, S., & Schomburg, E. D. (1985). Locomotor coordination in the cat. In B. M. H. Bush & F. Clarac (Eds.), *Co-ordination of motor behaviour.* Society for experimental biology seminar 24 (pp. 201–220). Cambridge: Cambridge University Press.

Miller, S., van der Burg, J., & van der Meché, F. G. A. (1975). Coordination of movements of the hindlimbs and forelimbs in different forms of locomotion in normal and decerebrate cats. *Brain Research,* **91,** 217–237.

Miller, S., Ruit, J. B., & van der Meché, A. (1977). Reversal of sign of long sinal reflexes dependent on the phase of the step cycle in the high decerebrate cat. *Brain Research,* **128,** 447–459.

Morin, C., Mazieres, L., & Pierrot-Deseilligny, E. (1982). Comparison of soleus H-reflex facilitation at the onset of soleus contractions produced voluntarily and during the stance phase of human gait. *Neuroscience Letters,* **33,** 47–53.

Moschovakis, A. K., Sholomenko, G. N., & Burke, R. E. (1991). Differential control of short latency cutaneous excitation in cat FDL motoneurons during fictive locomotion. *Experimental Brain Research,* **83,** 489–501.

Murray, M. P., Sepic, S. B., & Barnard, E. J. (1967). Patterns of sagittal rotation of the upper limbs in walking. *Physiology Therapeutics,* **47,** 272–284.

O'Donovan, M. J., Pinter, M. J., Dum, R. P., & Burke, R. E. (1982). Actions of FDL and FHL muscles in intact cats: functional dissociation between anatomical synergists. *Journal of Neurophysiology,* **47,** 1126–1143.

Orsal, D., Cabelguen, J. M., & Perret, C. (1990). Interlimb coordination during fictive locomotion in the thalamic cat. *Experimental Brain Research,* **82**(3), 536–546.

Patla, A. E., & Belanger, M. (1987). Task-dependent compensatory responses to perturbations applied during rhythmic movements in humans. *Journal of Motor Behaviour,* **19**(4), 454–475.

Perret, C., & Cabelguen, J. M. (1980). Main characteristics of the hindlimb locomotor cycle in the decorticate cat with special reference to bifunctional muscles. *Brain Res.,* **187,** 333–352.

Perret, C., Millanvoye, M., & Cabelguen, J. M. (1972). Messages spinaux ascendants pendant une locomotion fictive chez le chat curarisé. *Journal of Physiology, Paris,* **65,** 153A.

Piesiur-Strehlow, B., & Meinck, H.-M. (1980). Response patterns of human lumbosacral motoneurone pools to distant somatosensory stimuli. EEG Clin. *Neurophysiology,* **48,** 673–682.

Pratt, C. A. (1992). Specializations in the outputs of low-threshold cutaneous reflex pathways during normal locomotion. In M. Woollacott & F. Horak (Eds.), *Posture and gait. Control mechanisms* (Vol. 1) (pp. 32–35). Portland: University of Oregon Books.

Pratt, C. A., Chanaud, C. M., & Loeb, G. E. (1991). Functionally complex muscles of the cat hindlimb. IV Intramuscular distribution of movement command signals and cutaneous reflexes in broad, bifunctional thigh muscles. *Experimental Brain Research,* **85,** 281–299.

Prochazka, A., Sontag, K. H., & Wand, P. (1978). Motor reactions to perturbations of gait: proprioceptive and somesthetic involvement. *Neuroscience Letters,* **7,** 35–39.

Quintern, J., Berger, W., & Dietz, V. (1985). Compensatory reactions to gait perturbations in man: short- and long-term effects of neuronal adaptation. *Neuroscience Letters,* **62,** 371–376.

Rossignol, S., & Drew, T. (1986). Phasic modulation of reflexes during rhythmic activity. In S. Grillner, H. Forssberg, P. S. G. Stein, & D. Stuart (Eds.), *Neurobiology of vertebrate locomotion,* Int. Symp. Series 45 (pp. 517–534). London: Macmillan.

Rossignol, S., & Gauthier, L. (1980). An analysis of mechanisms controlling the reversal of crossed spinal reflexes. *Brain Research,* **182,** 31–45.

Rossignol, S., Gauthier, L., Julien, C., & Lund, J.P. (1981). State dependent responses during locomotion. In A. Taylor & A. Prochazka (Eds.), *Muscle receptors and movement* (pp. 389–402). New York: Macmillan.

Rossignol, S., Lund, J. P., & Drew, T. (1988). The role of sensory inputs in regulating patterns of rhythmical movements in higher vertebrates. A comparison between locomotion, respiration and mastication. In A. H. Cohen, S. Rossignol, & S. Grillner (Eds.), *Neural control of rhythmic movements in vertebrates* (pp. 201–283). New York: Wiley.

Sarica, Y., & Ertekin, C. (1985). Descending lumbosacral cord potentials (DLCP) evoked by stimulation of the median nerve. *Brain Research,* **325,** 299–301.

Schmidt, B. J. D., Meyers, D. E. R., Fleshman, J. W., Tokuriki, M., & Burke, R. E. (1988). Phasic modulation of short-latency cutaneous excitation in flexor digitorum longus motoneurons during fictive locomotion. *Experimental Brain Research,* **71,** 568–578.

Schmidt, B. J. D., Meyers, E. R., Tokuriki, M., & Burke, R. E. (1989). Modulation of short latency cutaneous excitation in flexor and extensor motoneurons during fictive locomotion in the cat. *Experimental Brain Research,* **77,** 57–68.

Schomburg, E. D., & Behrends, H. B. (1978a). The possibility of phase-dependent monosynaptic and polysynaptic Ia excitation to homonymous motoneurones during fictive locomotion. *Brain Research,* **143,** 533–537.

Schomburg, E. D., & Behrends, H. B. (1978b). Phasic control of the transmission in the excitatory and inhibitory reflex pathways from cutaneous afferents to α-motoneurones during fictive locomotion in cats. *Neuroscience Letters,* **8,** 277–282.

Schomburg, E. D., Meinck, H.-M., & Haustein, J. (1975). A fast propriospinal inhibitory pathway from forelimb afferents to motoneurones of hindlimb flexor digitorum longus. *Neuroscience Letters,* **1,** 311–314.

Schomburg, E. D., Roesler, J., & Meinck, H. M. (1977). Phase-dependent transmission in the excitatory propriospinal reflex pathway from forelimb afferents to lumbar motoneurones during fictive locomotion. *Neuroscience Letters,* **4,** 249–252.

Schomburg, E. D., Meinck, H. M., Haustein, J., & Roeslerr, J. (1978a). Functional organization of the spinal reflex pathways from forelimb afferents to hindlimb motoneurones in the cat. *Brain Research* **139,** 21–33.

Schomburg, E. D., Roesler, J., & Kenins, P. (1978b). On the function of the fast long spinal inhibitory pathway from forelimb afferents to flexor digitorum longus motoneurones in cats and dogs. *Neuroscience Letters,* **7,** 55–59.

Schomburg, E. D., Behrends, H. B., & Steffens, H. (1981). Changes in segmental and propriospinal reflex pathways during spinal locomotion. In A. Taylor & A. Prochazka (Eds.), *Muscle receptors and movement* (pp. 413–425). London: Macmillan.

Seigh-Naraghi, A. H., Herman, R. M., & D'Luzansky, S. C. (1992). Phase-dependent reflex modulation: effect of stimulus parameter. In M. Woollacott & F. Horak (Eds.), *Posture and gait: control mechanisms* (Vol. 1) (pp. 48–51). Portland: University of Oregon Books.

Sherrington, C. S. (1906). The integrative action of the nervous system. New Haven, CT: Yale University Press.

Sherrington, C. S. (1910). Flexion-reflex of the limb, crossed extension-reflex, and the stepping and standing. *Journal of Physiology (London),* **40,** 28–121.

Sillar, K. T., & Roberts, A. (1988). A neuronal mechanism for sensory gating during locomotion in a vertebrate. *Nature (London)* **331,** 262–265.

Skinner, R. D., Adams, R. J., & Remmel, R. S. (1980). Responses of long descending propriospinal neurons to natural and electrical types of stimuli in cat. *Brain Research,* **196,** 387–403.

Stuart, D. G., Withey, T. P., Wetzel, M. C., & Goslow, Jr., G. E. (1973). Time constraints for interlimb co-ordination in the cat during unrestrained locomotion. In R. B. Stein, K. G. Pearson, R. S. Smith, & J. B. Redford (Eds.), *Control of posture and locomotion* (pp. 537–560). New York: Plenum.

Tax, A. A. M., Duysens, J., Trippel, M., & Dietz, V. (1990). Gating of contralateral reflex responses in biceps femoris during human locomotion. In T. Brandt, W. Paulus, W. Bels, M. Dieterich, S. Krafczyk, & A. Straube (Eds.), *Disorders of posture and gait* (pp. 107–110). New York: George Thieme Verlag Stuttgart.

Tax, A. A. M., Duysens, J., Gielen, C. C. A. M., Trippel, M., & Dietz, V. (1991). A comparison of ipsilateral and contralateral cutaneous reflex responses during human running. *Society for Neuroscience* Abstracts, Vol. 17, pp. 1225.

Toft, E., Sinkjaer, T., Andreassen, S., & Larsen, K. (1991). Mechanical and electromyographic responses to stretch of the human ankle extensors. *Journal of Neurophysiology,* **65,** 1403–1410.

Udo, M., Kamei, H., Matsukawa, K., & Tanaka, K. (1982). Interlimb coordination in cat locomotion investigated with perturbation: II. Correlates in neuronal activity of Deiters' cells of decerebrate walking cats. *Experimental Brain Research,* **46,** 438–447.

Viala, D., & Vidal, C. (1978). Evidence for distinct spinal locomotion generators supplying respectively fore- and hindlimbs in the rabbit. *Brain Research,* **155,** 182–186.

Wetzel, M. C., & Stuart, D. G. (1976). Ensemble characteristics of cat locomotion and its neural control. *Progress in Neurobiology,* **7,** 1–98.

Yang, J. F., & Stein, R. B. (1990). Phase-dependent reflex reversal in human leg muscles during walking. *Journal of Neurophysiology,* **63,** 1109–1117.

Zangger, P. (1978). Fictive locomotion in curarized high spinal cats elicited by 4-aminopyridine and Dopa. *Experientia* (Basel), **34,** 904.

6

Human Locomotor Control, the Ia Autogenic Spinal Pathway, and Interlimb Modulations

J. D. Brooke, D. F. Collins*, and W. E. McIlroy*

I. Introduction: An Orientation to Human Bipedality and a Focus on the H (Autogenic Ia Spinal) Pathway
II. Review of Experimental Method
III. The H Pathway during Natural Movements
IV. Bilateral and Ipsilateral Movement Features of H Pathway Modulation
V. Contralateral Movement Features and H Pathway Modulation
VI. Supraspinal Modulation of the H Pathway
VII. Interlimb Spinal Path Modulation as Part of the Motoneuronal Convergence for Locomotion
References

I. Introduction: An Orientation to Human Bipedality and a Focus on the H (Autogenic Ia Spinal) Pathway

The purpose of this chapter is to integrate the study of the modulation of rapidly conducting spinal pathways during human locomotion

---

* *Present addresses:* D. F. Collins, Department of Physiology, University of Alberta, Edmonton, Alberta, Canada; W. E. McIlroy, Sunnybrook Health Science Centre, University of Toronto, Toronto, Ontario, Canada.

into the topic of interlimb coordination. Recent experimental results on humans permit this integration.

Coordination of the muscular activity across a limb, between limbs, and in conjunction with the axial musculature is an important control problem for animals. Activities such as walking, running, swimming, and chewing have received considerable study in vertebrates. Action potentials involved in this muscular coordination can be identified from supraspinal nuclei, from spinal pattern generators, and from primary afferent activity elicited by the movement itself (Grillner, Wallen and Brodin 1991). It is clear that such sources of control interact together. How the whole is achieved in primate (e.g., see Eidelberg, Walden and Nguyen 1981) and especially human activities is still, however, far from clear.

A number of approaches are taken to the study of the coordination of muscular activity for bipedality in humans (see Dietz, Chapter 8, this volume, and for a comprehensive review, see Dietz, 1992). The naturally occurring patterns of electromographic (EMG) and electroencephalographic (EEG) signals, together with the kinematics of the movements of joints and the dynamics of support surface reaction forces, contain useful information from which inferences can be made about neural control systems and imbedded biomechanical constraints.

Normally, in comparison to the dominant leg, the nondominant one exerts a greater ground reaction force when walking and a larger propulsive support reaction force when pedaling (Rosenrot, 1980). Two multivariate principal components account for much of the common variance in reaction forces from a number of cycles of pedaling with both legs (Hines, O'Hara Hines and Brooke 1987). The first component describes the forces exerted in the propulsive phase. The second describes the transfers of propulsive force from one leg to the other. Thus, variability in the coordination of force from one limb to the next accounts for a significant part of the total variation seen in pedal reaction forces in healthy adults. The cycle by cycle deviations about this second component significantly increase when the velocity of pedaling increases. This indicates that the transfer of propulsive force from one leg to the other is a point at which movement control is temporarily less stable.

When the legs are forced to move gradually out of phase with each other when pedaling, the EMG discharge within a leg is not greatly altered in either timing or magnitude (Boylls, Zomleffer and Zajac 1984). This stability we associate with the first principal component of propulsive pedal force. It would be wrong, however, to conclude from such results that contralateral interactions are not taking place between the legs. The second principal component, referred to above, implies that such interaction will take place. As will be seen later, these interactions actually are powerful in the modulation of at least one important spinal somatosensorimotor pathway.

When the natural movement or contractile activity of the leg is perturbed mechanically, one observes responses in EMG and kinematic variables at short, medium, and long latencies. The presentation of completely novel perturbations, that is, ones that are totally unexpected, reveals that anticipation and planning play important roles in the control of pedaling (McIlroy & Brooke, 1987) and of isolated leg movements (Brooke & McIlroy, 1990; Dietz, 1992). This planning involves a coordination of discharge between muscles that is stable within a task and varies substantially between tasks. Accordingly, the term "synergists" may mean no more than that the muscles were active at the same time in a particular task.

Analysis of the EMG responses to mechanical perturbation of the limb also identifies roles played by rapidly conducting spinal reflex pathways between somatosensory afferents and muscles. These spinal pathways have been explored further, through observing the EMG evoked in leg muscles following electrical stimulation of the afferents in peripheral mixed nerve trunks of the leg.

It is clear that there are characteristics particular to the human, in the simple spinal pathways for somatosensory afferent activity, to influence motoneuronal pools (Pierrot-Deseilligny, Morin, Bergego & Tankov 1981; Mao, Ashby, Wang and McCrea 1984; Brooke & McIlroy, 1985), in addition to those characteristics shared with other members of the mammalian family. This may reflect the adaptation of *Homo sapiens* for bipedal locomotion.

There is spinal segmental and oligosegmental (Balidessra, Hultborn and Illert 1981; Warwick & Williams, 1973) access to limb motoneurons from somatosensory receptors served by all four primary afferent groups (I, II, III, IV), through pathways that range from monosynaptic to polysynaptic (Baldissera *et al.*, 1981). In the context of coordination for human locomotion, particular attention has been paid to the most rapid autogenic pathway, represented by the H reflex (Hoffman, 1922), although recently, other, slower acting pathways have started to be examined (Yang & Stein, 1990; Brooke, Colledge, Collins and McIlroy 1991a; Duysens, Tax, van der Doelen, Trippel and Deitz 1991, Duysens, Tax, Trippel and Deitz 1992; Duysens & Tax, Chapter 5, this volume). The H reflex is observed as increased EMG activity reflexively evoked, following electrical stimulation of the large-diameter autogenic afferent fibers in a mixed-nerve trunk. It should be remembered that the leading edge of the H reflex is attributed to Ia afferents monosynaptic onto the motoneuron, but later components of the reflex may contain contributions from other fast-conducting afferents, such as cutaneous ones (Burke, Gandevia and McKeon 1984).

Part of the attraction of the H reflex pathway is the ease with which comparatively stable EMG responses can be obtained in humans. A further attraction is the abundance of the sensory receptors for Ia fibers, that is, the abundance of the primary endings of muscle spindles. The suspicion that the pathway has functional significance for the contractile locomotor activity of the muscle was recently given further support. Reflexes were elicited from stretch of the soleus muscle, a plantar flexor of the ankle. The stretch was designed to elicit discharge from muscle spindles. It was estimated that the evoked reflex activity represented 30–60% of the EMG activity seen when the soleus muscle was active and contributing to the control of walking (Yang, Stein and James 1991). It was also found that transmission in the pathway of this stretch reflex was modulated in a way similar to that seen for H reflexes during walking.

II. Review of Experimental Method

Our subjects were healthy adults, 20–50 years of age, with no reported symptoms of neurological or orthopedic diseases. They were informed, consenting, volunteers and the experiments were approved by the University committee on the ethics of using human subjects.

Surface Ag/AgCl electrodes recorded EMG activity, which was conditioned through Grass amplifiers with a band pass filter of 3–300 Hz and monitored for artifacts on a storage oscilloscope. The signal was collected over an A/D interface at 500 Hz and stored on a microcomputer. A Ag/AgCl-stimulating electrode was placed on the skin longitudinal to the fibers of the tibial nerve in the popliteal fossa, with the anode distal. Impedance at all electrode sites was <10 kΩ (Grass Impedance Meter EZM5 measuring at 30 Hz).

The delivery of the stimulus and the collection of EMGs was under computer control. Subjects' activity involved the cycle ergometer or the treadmill. On the ergometer, the spatial position for stimulation or collection was determined using pulses from an optical switch to transduce the position of the chain wheel of the pedal crank, accurate to 7° of rotation (Brooke, Hoare, & Triggs 1981). On the treadmill, position in the walking cycle was transduced through a pressure switch on the heel for ground contact.

When the stimulus current initiates action potentials in large-diameter afferents, this can lead to action potentials in the alpha motoneuronal axons, which are also of large diameter in the mixed-nerve trunk. These latter, efferent potentials can lead to muscle action potentials. This wave of activity seen in the EMG is the M wave. In our research, a small M wave was elicited. Its peak-to-peak magnitude was kept constant within ± 5% Mmax, and observed on an oscilloscope, so as to provide an indirect indication of the constancy of stimulation of the large-diameter afferent fibers. Neurographic studies have shown that a linear relationship exists between the magnitude of the afferent volley and the magnitude of both the electrically evoked efferent volley and the M wave (Abbruzzese, Ratto, & Abbruzzese 1985; Fukushima, Yamashita, & Shimada 1982).

Such biocalibration is important (Swett & Bourassa, 1981). The M wave has been used by many labs studying H reflex modulation during movements, including Morin, Katz, Maziere, and Pierrot-Deseilligny (1982) in gait; Chan and Kearney (1982), body tilting; Dietz, Quintern, and Berger (1985), gait and cerebral somatosensory potentials; Capaday and Stein (1986), walking, and (1987a), running; Crenna and Frigo (1987), stepping and walking; Hultborn, Pierrot-Deseilligny and Shindo (1987), ankle movement; Edamura, Yang, and Stein (1991), walking; Moritani, Oddson and Thorstensson (1990), hopping; Brooke, Collins, Boucher, and McIlroy (1991b), walking; Llewellyn, Yang, and Prochazka (1991), beam walking; Brooke, McIlroy, and Collins (1992), pedaling and limb posture; McIlroy, Collins, and Brooke (1992), passive movement of the limbs; Collins, McIlroy, and Brooke (1992), pedaling and contralateral leg responses; Misiaszek, Brooke, and Cheng (1992a), passive movement of single joints in humans, and Misiaszek, Brooke, and Barclay (1992b), in dogs; Wolpaw and Chen (1992), long-term adaptation in rats; and Collins, Brooke, and McIlroy (1993), pre-movement facilitation of reflexes. The latter study reported that, within the range of ± 5% Mmax at small M values, no correlation was observed between variations in magnitudes of M and H waves. Given the strong relationship between the stimulus response curves for H and M wave amplitudes, this showed that ± 5% Mmax was an acceptable band of variation that did not lead to complementary changes in reflex magnitudes.

During walking, the 95% confidence band for the mean of 20 H reflexes approximates 15% of that mean (Brooke *et al.*, 1991b). This band includes the stimulus variation allowed by the ± 5% variation in M wave magnitude. During passive movements, one can anticipate less variation. The present observations of both passive and active movement-induced changes appearing also in a nonmoving contralateral leg add further support to the validity of using M waves as biocalibration during movement. It is worth remarking that, were there presently available an additional source of biocalibration for experiments involving movement of the limbs, it would be utilized to confirm the stability inferred from the constant M wave.

For the studies of passive movement of the legs, two stationary cycle ergometers were linked in tandem. The forward ergometer was pedaled by an experimenter. The subject sat, well supported in a rigid, high-backed chair. The subject's foot was attached to the pedal of the rear ergometer, so that full pedaling movement could be achieved. The soleus surface EMG of the subject was quiet during the passive movement. We also established in pilot work that such movement did not lead to activation in any of the following muscles: soleus, tibialis anterior, vastus medialis, semitendinosus, rectus abdominus, and erector spinae. In the experiments presently being reviewed, control H reflexes were elicited from a standing condition for studies of walking and from a seated, nonmoving condition for studies of pedaling. Differences between the means for conditions were assessed by repeated-measures ANOVAs or t tests, blocked on subjects. Subsequent planned comparisons between specific means were evaluated by t test using the appropriate mean square estimate for error. *Post hoc* comparisons of means were by Scheffé's test. Statistical significance was at $p = .05$.

III. The H Pathway during Natural Movements

Recent studies of walking and pedaling clearly show that the active rhythmic limb movements involved are associated with modulation of transmission of action potentials in the spinal pathway of the H reflex. The more natural activities to study are those of walking and running (Brooke *et al.*, 1991b; Capaday & Stein, 1986; Crenna & Frigo, 1987). One of the most insightful to study is that of pedaling (Brooke *et al.*, 1992; Collins *et al.*, 1992; McIlroy *et al.*, 1992). This is because pedaling activity can easily be reduced to simpler components. These components provide additional insights, as will be seen.

Within a single cycle of movement such as bilateral pedaling, or walking, the magnitude of H reflexes systematically varies. Figure 1 shows such modulation for the soleus reflex during pedaling. (The reflex magnitudes are expressed as a percentage of the maximum M wave that can be evoked.) It can be seen that the magnitude of the H reflex is profoundly depressed in the later part of the pedal cycle (0° is top, dead center for the pedal crank). The depression is considerably offset in the first 100° of pedal rotation. This is the phase of EMG activity for the production of force to rotate the pedals (Brooke, Boyce, McIlroy, Robinson, and Wagar 1986). The soleus muscle is active at this time from approximately 10° past top, dead center to approximately 170° around the cycle for the pedal crank.

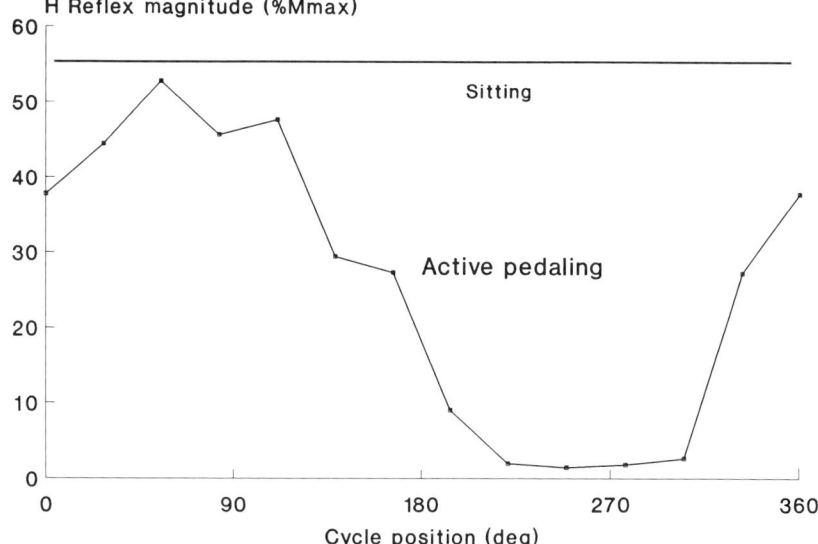

FIGURE 1 H reflex magnitudes sitting and over the cycle of active pedaling with both legs (stimulation intensity constant) in one subject. 0° = top, dead center for the pedal crank.

Compared to seated control values, this reflex is inhibited, as well as modulated. Through much of the cycle of active movement, reflex magnitudes are significantly smaller than those when the subject sits. This depression of the reflex pathway likely arises through facilitation of spinal interneurons mediating presynaptic inhibition. Studies of cat preparations (Capaday & Stein, 1989) or computer models (Capaday & Stein, 1987b) reveal that the amount of depression seen with movement-induced changes of the reflex cannot be achieved physiologically through inhibitory events that are postsynaptic for the motoneuron. This conclusion is supported when more than one reflex is passed through a motoneuron pool during walking movement. The H reflex is depressed; the non-Ia reflex is not; that is, the motoneuron pool is still in a state to be activated by reflex afferent activity (Brooke *et al.*, 1991b).

The level of contraction of the target muscle is positively related to the magnitude of the H reflex in subjects who are sitting (Verrier, 1985; Paillard, 1955; Brooke *et al.*, 1991b, 1992). However, this relationship does not hold for H reflexes in a number of leg muscles studied just after heel contact in walking (Brooke *et al.*, 1991b). Indeed, the tendency is for the reflex magnitude to decrease at that time, as the level of contraction increases in the target muscle. It appears that the inhibition of the pathway must be considerably increased, to offset the facilitatory effect associated with the increased contraction.

The human H reflex pathway to soleus is further inhibited when the pattern of gait is changed from walk to run (Capaday & Stein, 1987a). There appears to be interaction between the reflex pathway and interneurons involved in the patterning of these active movements.

It is clear that these effects on the amplitudes of H reflexes are not limited to the soleus muscle. There are now reports also on quadriceps (Dietz, Faist, and Pierrot-Deseilligny 1990a; Brooke et al., 1991b; for quadriceps stretch reflex, Dietz, Discher, Faist, and Trippel 1990b), tibialis anterior (Brooke et al., 1991b), and gastrocnemius (Moritani et al., 1990). The shape of the amplitude modulation over the cycle is, at least in part, specific to the muscle and its role in that movement cycle.

IV. Bilateral and Ipsilateral Movement Features of H Pathway Modulation

It is important to know whether the modulations of a spinal pathway, due to the task movement, occur between the limbs. That is, does the movement of one limb affect the state of the other? To appreciate the origin of the experiments performed to answer that question, and to provide prior conceptual understanding, there is a preceeding question. This question is, What is the cause of these changes in the Ia autogenic reflex pathway during these movements? Some tentative hypotheses could be as follows: (1) the brain, anticipating the movement (Prochazka, 1989), may be transmitting a particular motor plan to spinal interneurons; (2) the afferent activity from the movement may reflexively excite the brain, evoking action potentials that facilitate the spinal interneurons inhibiting the H reflex pathway; and (3) the afferent activity resulting from the movement may lead through oligosegmental spinal paths to inhibit directly the H reflex pathway. There is now information on the latter two hypotheses, particularly on the simpler question, Does the afferent activity arising from the movement cause some of the reflex modulation? This information then extends to interlimb events.

We have recently reduced pedaling to the passive movement of the legs, that is, without permitting the contraction of the subject's muscles (McIlroy et al., 1992). Soleus H reflexes were examined at various points around the pedal cycle, when the subject was sitting and when the subject's leg was passively or actively moved. Substantial inhibition of the reflex occurred when the legs of the subject were rotated passively, in a pedaling manner. This inhibition was comparatively uniform in intensity around the cycle of movement. The data showed that simply rotating the joints of the limb leads to movement modulation of the reflex, in a similar way to that seen with active movement. It appears that afferent activity elicited by the movement (movement afference) plays an important role. Such activity could be elicited from many sensory receptors, such as those for muscle stretch, joint pressure, and cutaneous pressure. Thus groups I, II, and III afferents could be involved (Baldissera et al., 1981).

The importance of the movement afference was further exposed when study was made of the effect of increasing the velocity of the passive movement (McIlroy et al., 1992). This was an indirect way of increasing the rate of firing in the movement afferents. The velocity was increased from one revolution of the pedals in 6 seconds (10 rpm) to 30, 60, and 90 rpm. Calculation of the resulting angular velocities of the knee and hip for a subject are shown in Figure 2. It was found that the increased velocity lead to increased inhibition. The inhibition started at a low velocity, in some subjects at one revolution of the pedals in 12 seconds. It was powerful by 60 rpm. At 60–90 rpm we can often observe complete inhibition of the reflex. This effect of the velocity of the movement appears to be valid to natural pedaling, as it is also seen for the recovery part of the movement cycle, when subjects actively pedaled.

The preparation of leg pedaling was further reduced to one leg passively moved (McIlroy et al., 1992). The question posed was, Is bilateral rhythmic coordination of the legs necessary to evoke the movement-

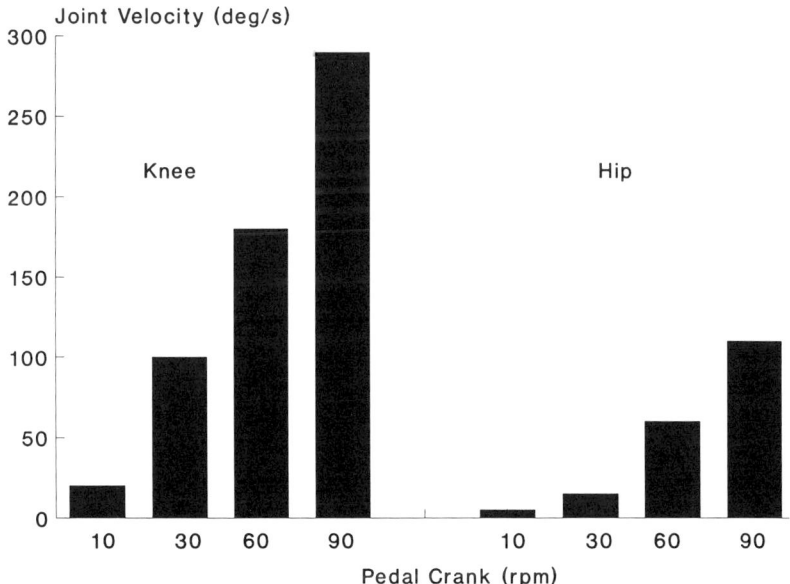

FIGURE 2 Angular velocities calculated for movement about the knee and hip in one subject, with rates of rotation of a 14-cm-long pedal crank from 10 to 90 revolutions per minute.

induced inhibition? The answer was No. The H reflex was inhibited as powerfully with one-legged as with two-legged passive movement.

During active movement at a cadence that is comfortable for the subject, the H reflex is modulated from an excited state during the stance phase of walking (the first 150° past top, dead center for pedaling) to an inhibited state during the swing phase of walking (the remaining 210° of pedal crank rotation in pedaling). The comparisons are to standing values. However, during passive movement at this comfortable rate, the reflex is profoundly inhibited over the whole cycle of pedaling movement. (Perhaps the excitatory state depends on the activation of the soleus muscle at that point in the pattern of active movement.) We therefore conclude that there is a foundation of movement-induced inhibition for the task modulations of this spinal pathway. The data reveal that the spinal reflex modulatory centers in humans are tuned by the afferent activity set up by the passive movement itself.

V. Contralateral Movement Features and H Pathway Modulation

There has been little research attention given to the effect of movement of the opposite limb on the control of this spinal path. We continued to reduce down our cycling preparation, by making observations on the reflex pathway in the stationary leg when the opposite leg is moved passively. In these experiments, the stationary leg is called the *contralateral* one. Observation of the moved leg shows that movement of the other leg is not needed to attain the depression of the short latency autogenic reflex. Nevertheless, we wanted to know if contralateral influences existed. The nonstimulated leg was passively moved at 60 rpm. The stationary, test leg was stimulated and H reflexes were observed in soleus. (The legs were 180° out of phase from each other at the time of stimulation, with the stationary leg close to its position for 270° crank rotation past top, dead center.)

The reflexes in the stationary, contralateral leg showed demonstrable and significant inhibition when the other leg was moved. The mean magnitudes were significantly lower than the seated control condition. Was this inhibition in the stationary limb related to the velocity of the passive movement of the opposite limb? That is, Was the inhibition that was occurring contralaterally also arising from the discharge of receptors stimulated by the movement? These studies of movement velocity involved the pedal crank revolution rates of 10, 30, 60, and 90 per minute, presented in random order to the subjects. The increase in rate of movement produced the increases in the angular velocities of rotation about the knee and hip joints shown previously in Figure 2. In addition to the changes in movement rate, two lengths of pedal crank were used,

14 cm and 20 cm. The latter was introduced as another way to increase the angular velocities of the joints of the limb.

It was found that the inhibition of the H reflex in the stationary leg was positively related to the velocity of passive movement of the opposite leg (Collins *et al.*, 1992). This statistically significant effect occurred in all subjects, as can be seen in Figure 3. The figure also shows that the longer crank replicated the results of the shorter one, with a small additional increment of inhibition due to the increased rate of rotation of the joints. Thus, the reflex effect of velocity of movement on the contralateral, still leg was similar to the positive relationship between H reflex inhibition and velocity of movement in the passively moving or the actively pedaling leg.

Of particular interest presently is the substantial and speed-dependent inhibition of H reflex magnitudes seen in the nonmoving leg

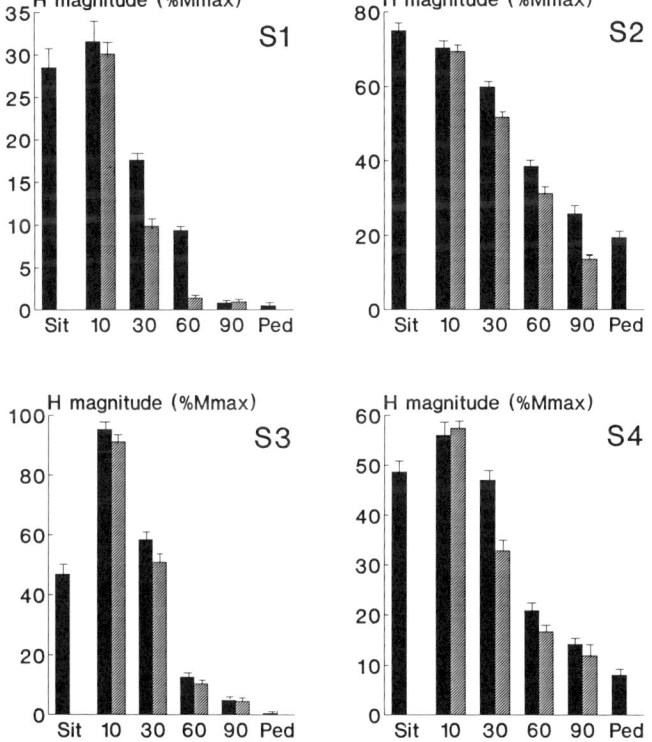

FIGURE 3 H reflex magnitudes in the contralateral leg when both legs are stationary (Sit), when the opposite leg is passively rotated at various crank rates (10–90 rpm), with short (14 cm) and long (20 cm) cranks, and when the contralateral leg is passively rotated with a crank rate of 60 rpm (Ped) (short crank length). The data for each of the four subjects are shown as S1 to S4. Vertical bars show 1 S.E.

when the other leg moves. It suggests that this inhibition is part of the neural control for the interlimb coordination that will be present during the central generation of this patterned leg movement. We observed that the H reflex inhibition from pedaling movement of one leg was similar to that for two legs. It has also been observed that the inhibition with one leg stepping is similar to that when walking (Crenna & Frigo, 1987). Such observations could easily lead to the erroneous conclusion that there was no interaction between the two limbs in the control of these spinal paths, as the movement effect from one is sufficient to depress the reflex transmission of that leg. However, the experiments reviewed above clearly show the neuronal interaction between limbs. There is quite substantial inhibition transmitted from one leg to the rapidly conducting spinal pathways of the other during leg movement in humans.

Three important questions arise: (1) Why is there apparent duplication of inhibitory effect? (2) What is the path for this inhibition, for example, is it arising via spinal oligosegmental routes from somatosensory receptors? and (3) What is the functional significance of modulating this spinal pathway in this manner?

Let us consider the first question. Active or passive movement of one leg depresses its soleus H reflex similarly to the movement of both it and the opposite leg together. Yet further inhibition is transmitted from the active or passive movement of the opposite leg. This appears to be duplication of effect. There are other instances of this apparent duplication of effect onto this spinal path. Contraction of the antagonist muscle, tibialis anterior, inhibits the soleus H reflex (Kasai & Komiyama, 1991). When tibialis anterior is contracted with the limb positioned appropriately for that muscle's activity in the pedal cycle, the soleus H reflex is depressed to values similar to those seen at that phase position with natural pedaling. Yet, if the limb is passively rotated, with no contraction from tibialis anterior, the inhibition at that point is similar to that with natural pedaling (Brooke et al., 1992). There again appears to be duplication, now between passive movement and the antagonist inhibition. It may be that such overlapping ensures that the powerful autogenic reflex is more than adequately inhibited, at points in the movement cycle when its excitation would be inappropriate in the EMG. Alternatively, these overlapping effects may have singular actions when a particular task renders one of them inoperative. For instance, when only one limb is to move in a task, it will be necessary to ensure that unanticipated reflex excitation cannot move the opposite limb and create quite inappropriate activity. The contralateral inhibitory projection we describe may help to ensure that this does not take place in a condition when the lack of activity in the contralateral antagonist removes its contribution to the inhibition.

With regard to the second question, somatosensory discharge from the movement may lead, through oligosegmental spinal paths or brain paths, to postsynaptic inhibition of motoneurons. As an alternative or additional possibility, this discharge may follow either of these paths to presynaptically inhibit the motoneurons. Further, the discharge may interact with pattern-generating mechanisms that run "silently" to elicit the inhibition pre- or postsynaptically.

The answers to this second question are not yet known. For active movements, presynaptic inhibition of H reflexes has been favored (Capaday & Stein, 1987b, 1989; Brooke et al., 1991b). However, postsynaptic effects could also contribute to that inhibition. In the present work, the discharge could be proceeding through any or all of the three routes described in the previous paragraph. We have evidence, not yet complete, for a spinal oligosegmental contribution to this inhibition. We have studied some complete quadraplegic people with severe spinal lesions from accident trauma. We still find movement-induced inhibition of the soleus H reflex pathway in this group of people. This inhibition occurs in the contralateral limb as well as the ipsilateral one. There is the possibility that some descending influence is still occurring through the incomplete lesion that characterizes even the most severely damaged of such people. However, the lesions are severe enough to stop the transmission of low-threshold somatosensory potentials to surface electrodes over the cerebral cortex. We currently feel that the most likely explanation is that the foundation for the inhibition is transmitted through a spinal pathway from the receptors activating movement afference to the H pathway.

There are crossed group I effects on muscular activity in the decerebrate cat (Baxendale & Rosenberg, 1976, 1977). Reflex conditioning in humans leads to the conclusion that there are spinal crossed group I effects (Delwaide & Pepin, 1991). It is also reported that tendon taps on one side can condition the tendon reflex of the opposite side in humans (Koceja & Kamen, 1991). Crossed effects of cutaneous stimulation are well known (e.g., Duysens et al., 1991, 1992; Duysens & Tax, Chapter 5, this volume). We clearly identify an effect on the H reflex pathway from movement in the other leg. Thus, the experimental evidence from human neurophysiological experiments suggests that crossed spinal pathways transmit inhibition to the fastest conducting reflexes, as part of the control of the contralateral leg in humans.

Further support for a spinal path contribution to the movement-induced inhibition comes from the simplicity of the relationship between increases in somatosensory discharge evoked by increasing the movement velocity, and increases in the inhibition. This clear relationship tempts one to assume that the sensorimotor transformation is passing through only a few synapses, so that the input magnitude is still reflected comparatively faithfully by the output.

The third question, that of functional significance, we address after considering the potential supraspinal modulations of this spinal pathway.

VI. Supraspinal Modulation of the H Pathway

It is reasonable to think that the brain also participates in modulating this spinal path during movement. In the final 100 msec before active movement starts, there is clear facilitation of the path on the agonist side. There is also modest facilitation on the contralateral one (Pierrot-Deseilligny & Lancert, 1973; Riedo & Ruegg, 1988). This facilitation arises from the brain's plans, after the signal to move has been given, but before any afferent discharge from the movement itself. It is complemented by inhibition of the path to the contralateral and ipsilateral antagonists (Ruegg, 1989).

It is suggested that there is presynaptic inhibition of H reflexes in the first 50 msec of a contraction. As this occurs faster than the spinal paths from sensory receptors would allow, the inhibition is proposed to be descending from the brain (Hultborn et al., 1987). At the cessation of contraction of a muscle, there remains presynaptic inhibition of the H reflex for up to five seconds. As contraction-evoked somatosensory discharge should have stopped well before this, it again is assumed that this inhibition comes from brain plans and that it is presynaptic (Schieppati, 1987).

There is indirect evidence that the "set" of the brain (Prochazka, 1989) can alter the gain in the H reflex pathway during active rhythmic movements of the two legs. When subjects walked on a narrow beam above the ground, inhibition was observed, additional to that seen with walking on the ground. The authors felt that differences in contractile and kinematic states between the two conditions of walking did not account for this change (Llewelyn et al., 1991).

It has been reported that lateral sway alters the magnitude of H reflexes in the flexor carpii radialis muscle of the forearm (Aiello, Rosati, Sau, Patraskakis, Bissakou, and Traccis 1989). There is assymetry between the change in reflex magnitude in the two arms, dependent on the direction of sway. Also, vestibular stimulation, achieved by angular acceleration of humans or static tilt of the body, can facilitate soleus H reflexes (Chan & Kearney, 1982; Scarpini, Mazzocchio, Mondelli, Nuti, and Rossi, 1991). These studies strongly implicate impulses descending from vestibular control, via brainstem nuclei and/or parietal lobe integration centers. Specific effects of these descending impulses on the coordination of the limbs have not yet been explored.

This area of the supraspinal influences on movement modulation of the human Ia autogenic path needs further investigation. This is especially the case for the neural coordination between the two limbs. It will be particularly interesting to establish whether the reflex facilitation that occurs in the phase when the muscle is contracted is simply due to motoneuronal facilitation for the contraction. As an alternative, the reflex facilitation may be a complementary input descending from the brain onto the spinal inhibitory interneurons, at the time that descending impulses also arrive to elicit the voluntary contraction. Crosscorrelation studies of the timing and magnitude of voluntary contractions to both limbs, and the timing and magnitude of modulations of the H reflex path in those limbs, may offer the opportunity to expose at least the degree of coincidence in these aspects of interlimb coordination at the activator and at the prior modulator levels.

VII. Interlimb Spinal Path Modulation As Part of the Motoneuronal Convergence for Locomotion

Between 10,000 and 15,000 synapses can occur on a single spinal motoneuron. This massive convergence sums the inputs from peripheral afferents, oligosegmental interneurons, and propriospinal neurons, and from brainstem extrapyramidal, and pyramidal neurons of the brain. [It is noteworthy that some of the spinal interneurons integrate convergence from even more sources than the motoneurons (Harrison & Jankowska, 1985).] The membrane potential resulting from summing the convergence onto the motoneuron determines in part the occurrence of an action potential. The prior state of the cellular receptors on the motoneuronal membrane also can contribute (Kiehn, 1991; Baldissera, Cavallari and Dworzak 1991).

Some of the resulting subthreshold postsynaptic potentials are comparatively weak, for example, monosynaptic pyramidal ones, and require temporal summation or background facilitation to elicit action potentials. Other postsynaptic potentials are powerful, for example, monosynaptic Ia ones. The latter neurons also diverge monosynaptically to virtually 100% of the ipsilateral autogenic motoneuronal pool, at least in the cat (Nelson & Mendell, 1978). There are also powerful projections to heteronymous motoneuronal pools in mammals (Baldissera *et al.*, 1981), including humans (Mao *et al.*, 1984; Pierrot-Deseilligny *et al.*, 1981). As a result, the Ia input has a powerful effect on the motoneuronal pool.

It is clear that movement of the limbs, even passive movement around the knee or hip joint alone, results in profound inhibition of this Ia convergence through the H reflex pathway. This appears to be the foundation upon which task modulation of that pathway is constructed by the CNS. The basic response to limb movement is inhibition of the spinal somatosensory pathway. Then, depending on the particular characteristics of the task, release of that inhibition can occur through excitation arising from control centers in the spinal cord or brain. It is appropriate that this strong Ia input is inhibited in this way. The foundation of inhibition suppresses reflex excitation that could be disruptive to the planned execution of the movement. [Recall that the stretch reflex can constitute 30–60% of the level of activity seen in the soleus EMG during human walking (Yang et al., 1991).] At the times when the muscle is excited to execute the movement, this inhibition of the reflex pathway is offset, so that the reflex is expressed close to or at the high magnitudes seen without movement.

There is coordination between the two limbs, in ensuring that this reflex inhibition occurs. The experimental evidence is now clear in humans. Moving simply one limb results in depression of the reflex pathway of the opposite limb. The control inhibition of the pathways of the opposite limb likely arises from the discharge of somatosensory receptors activated by the unilateral movement. There is human evidence suggesting that this inhibition can be delivered by spinal oligosegmental paths, to assist in the coordination of the other leg. There is also evidence, from studies of spinal-injured people, that there can occur interlimb coordination between the movement of the arms and modulation of H reflexes of the legs (Calencie, 1991). This coordination may be suppressed in the intact spinal cord of humans. Little study has taken place of upper and lower limb interaction on movement conditioning of H reflex pathways during locomotion.

Coordination of activity between the limbs in mammals is complex. It involves central pattern generating mechanisms of the spinal cord and motor control centers of the cerebrum, cerebellum, and brainstem. It appears that this coordination from control centers converges onto a spinal cord in which there is already interlimb coordination, through inhibition, at the simple level of the present segmental pathway. Thus, the state of the simplest pathway provides a secure foundation against which the interlimb coordination planned from these control centers can be executed.

In conclusion, recent human neurophysiological experiments show that interlimb movement features modulate the rapidly conducting spinal sensorimotor pathway of the H reflex during leg movement. In some tasks, the sources of modulation, within limb and between limbs, can appear to overlap. This may be "overinsurance" against disruption of motor plans or may simply be a function of the tasks studied to date. We conclude that the afferent activity surrounding movement tunes the somatosensorimotor mechanisms helping to control human locomotion. There is interlimb coordination in this control, with the afferent activity from one leg assisting in determining the contractile activity and spatial position of the opposite leg.

Acknowledgments

This work was supported by grants from NSERC (CANADA) and by scholarships to W. E. McIlroy and D. F. Collins from the University of Guelph.

References

Abbruzzese, M., Ratto, S., & Abbruzzese, G. (1985). Electroneurographic correlates of the monosynaptic reflex: experimental studies and normative data. *Journal of Neurology, Neurosurgery and Psychiatry,* **48,** 434–444.

Aiello, I., Rosati, G., Sau, G. F., Patraskakis, S., Bissakou, M., & Traccis, S. (1989). Modulation of flexor carpi radialis H reflex by lateral tilts in man. *Journal Neurological Sciences,* **93,** 191–198.

Baldissera, F., Hultborn, H., & Illert, M. (1981). Integration in spinal neuronal systems. In V. B. Brooks (Ed.), *Handbook of physiology, Sect. 1 The nervous system, Vol. II Motor control* (pp. 509–595). Bethesda MD: American Physiological Society.

Baldissera, F., Cavallari, P., & Dworzak, F. (1991). Cramps: a sign of motoneurone 'bistability' in a human patient. *Neuroscience Letters,* **133,** 303–306.

Baxendale, R. H., & Rosenberg, J. R. (1976). Crossed reflexes evoked by selective activation of muscle spindle primary endings in the decerebrate cat. *Brain Research,* **115,** 324–327.

Baxendale, R. H., & Rosenberg, J. R. (1977). Crossed reflexes evoked by selective activation of tendon organ axons in the decerebrate cat. *Brain Research,* **127,** 323–326.

Boylls, C. C., Zomlefer, M. R., & Zajac, F. E. (1984). Kinematic and EMG reactions to imposed interlimb phase alterations during bipedal cycling. *Brain Research,* **324,** 342–345.

Brooke, J. D., & McIlroy, W. E. (1985). Locomotor limb synergism through short latency afferent links. *Electroencephalography and Clinical Neurophysiology,* **60,** 39–45.

Brooke, J. D., & McIlroy, W. E. (1990). Brain plans and servo loops in determining corrective movements. In J. M. Winters and S. L.-Y. Woo (Eds.), *Multiple muscle systems: Biomechanics and movement organization* (pp. 706–715). New York: Springer Verlag.

Brooke, J. D., Hoare, J., & Triggs, R. (1981). Computerized system for measurement of force exerted within each pedal revolution during cycling. *Physiology and Behavior,* **26,** 139–143.

Brooke, J. D., Boyce, D. E., McIlroy, W. E., Robinson, T., & Wagar, D. (1986). Transfer of variability from electromyogram to propulsive force in human locomotion. *Electromyography and Clinical Neurophysiology,* **26,** 389–400.

Brooke, J. D., Colledge, M. L., Collins, D. F., & McIlroy, W. E. (1991a). Long latency responses evoked in leg muscles during sitting and pedalling. *Canadian Journal Physiology and Pharmacology,* **69,** Aiii.

Brooke, J. D., Collins, D. F., Boucher, S., & McIlroy, W. E. (1991b). Modulation of human short latency reflexes between standing and walking. *Brain Research,* **548,** 172–178.

Brooke, J. D., McIlroy, W. E., & Collins, D. F. (1992). Movement features and H-reflex modulation. I. Pedalling versus matched controls. *Brain Research,* **582,** 78–84.

Burke, D., Gandevia, S. C., & McKeon, B. (1984). Monosynaptic and oligosynaptic contributions to the human ankle jerk and H reflex. *Journal of Neurophysiology,* **52,** 435–488.

Calencie, B. (1991). Interlimb reflexes following cervical spinal cord injury in man. *Experimental Brain Research,* **85,** 458–469.

Capaday, C., & Stein, R. B. (1986). Amplitude modulation of the soleus H reflex in the human during walking and standing. *Journal of Neuroscience,* **6,** 1308–1313.

Capaday, C., & Stein, R. B. (1987a). Difference in the amplitude of the human soleus H reflex during walking and running. *Journal of Physiology,* **392,** 91–104.

Capaday, C., & Stein, R. B. (1987b). A method for simulating the reflex output of a motoneuronal pool. *Journal of Neuroscience Methods,* **21,** 91–104.

Capaday, C., & Stein, R. B. (1989). The effects of postsynaptic inhibition on the monosynaptic reflex of the cat at different levels of motoneuron pool activity. *Experimental Brain Research,* **77,** 577–584.

Chan, C. W. Y., & Kearney, R. E. (1982). Influence of static tilt on soleus motoneuron excitability in man. *Neuroscience Letters,* **33,** 333–338.

Collins, D. F., McIlroy, W. E., & Brooke, J. D. (1992). Contralateral inhibition of soleus H reflexes with different velocities of passive movement of the opposite leg. *Brain Research,* **603,** 96–101.

Collins, D. F., Brooke, J. D., & McIlroy, W. E. (1993). The independence of premovement H reflex gain and kinesthetic requirements for task performance. *Electroencephalography and Clinical Neurophysiology,* **89,** 35–40.

Crenna, P., & Frigo, C. (1987). Excitability of the soleus H reflex arc during walking and stepping in man. *Experimental Brain Research,* **66,** 49–60.

Delwaide, P. J., & Pepin, J. L. (1991). The influence of contralateral primary afferents on Ia inhibitory interneurons in humans. *Journal of Physiology,* **439,** 161–179.

Dietz, V. (1992). Human neuronal control of automatic functional movements: interaction between central programs and afferent input. *Physiological Reviews,* **72,** 33–69.

Dietz, V., Quintern, J., & Berger, W. (1985). Afferent control of human stance and gait: evidence for blocking of group I afferents during gait. *Experimental Brain Research,* **61,** 153–163.

Dietz, V., Faist, M., & Pierrot-Deseilligny, E. (1990a). Amplitude modulation of the quadriceps H-reflex in the human during the early stance phase of gait. *Experimental Brain Research,* **79,** 221–224.

Dietz, V., Discher, M., Faist, M., & Trippel, M. (1990b). Amplitude modulation of the human quadriceps tendon jerk reflex during gait. *Experimental Brain Research,* **82,** 211–213.

Duysens, J., Tax, A. A. M., van der Doelen, B., Trippel, M., & Dietz, V. (1991). Selective activation of human soleus or gastrocnemius in reflex responses during walking and running. *Experimental Brain Research,* **87,** 193–204.

Duysens, J., Tax, A. A. M., Trippel, M., & Dietz, V. (1992). Phase-dependent reversal of reflexly induced movements during human gait. *Experimental Brain Research,* **90,** 404–414.

Edamura, M., Yang, J. F., & Stein, R. B. (1991). Factors that determine the magnitude and time course of human H-reflexes in locomotion. *Journal of Neuroscience,* **11,** 420–427.

Eidelberg, E., Walden, J. G., & Nguyen, L. H. (1981). Locomotor control in macaque monkeys. *Brain,* **104,** 647–663.

Fukushima, Y., Yamashita, M., & Shimada, Y. (1982). Facilitation of H-reflex by homonymous Ia-afferent fibres in man. *Journal of Neurophysiology,* **48,** 1079–1088.

Grillner, S., Wallen, P., & Brodin, L. (1991). Neuronal network generating locomotor behaviour in lamprey: circuitry, transmitters, membrane properties and stimulation. *Annual Reviews of Neuroscience,* **14,** 169–199.

Harrison, P. J., & Jankowska, E. (1985). Sources of input to interneurones mediating group I non-reciprocal inhibition of motoneurones in the cat. *Journal of Physiology,* **361,** 379–401.

Hines, W. G. S., O'Hara Hines, R. J., & Brooke, J. D. (1987). A multivariate solution for cyclic data, applied in modelling locomotor forces. *Biological Cybernetics,* **56,** 1–9.

Hoffmann, P. (1922). *Die eigenreflexe (sehnenreflexe) menschlicher muskeln*. Berlin: Springer.
Hutborn, H., Meunier, S., Pierrot-Deseilligny, E., & Shindo, M. (1987). Changes in presynaptic inhibition of Ia fibres at the onset of voluntary contractions in man. *Journal of Physiology*, **389**, 757–772.
Kasai, T., & Komiyama, T. (1991). Antagonist inhibition during rest and precontraction. *Electroencephalography and Clinical Neurophysiology*, **81**, 427–432.
Kiehn, O. (1991). Plateau potentials and active integration in the 'final common pathway' for motor behaviour. *TINS*, **14**, 68–73.
Koceja, D. M., & Kamen, G. (1991). Interaction in human quadriceps-triceps surae motoneuron pathways. *Experimental Brain Research*, **86**, 433–439.
Llewellyn, M., Yang, J. F., & Prochazka, A. (1991). Human H-reflexes are smaller in difficult beam walking than in normal treadmill walking. *Experimental Brain Research*, **83**, 22–28.
Mao, C. C., Ashby, P., Wang, M., & McCrea, D. (1984). Synaptic connections from large muscle afferents to the motoneurons of various leg muscles in man. *Experimental Brain Research*, **56**, 341–350.
McIlroy, W. E., & Brooke, J. D. (1987). Response synergies over a single leg when it is perturbed during the complex rhythmic movement of pedalling. *Brain Research*, **407**, 317–326.
McIlroy, W. E., Collins, D. F., & Brooke, J. D. (1992). Movement features and H reflex modulation. II. Passive rotation, movement velocity and single leg movement. *Brain Research*, **582**, 85–93.
Misiaszek, J. E., Brooke, J. D., & Cheng, J. (1992a). Locomotor-like rotation of either knee or hip strongly inhibits soleus H reflexes. *Neuroscience Abstracts*, **18**, 26.
Misiaszek, J. E., Brooke, J. D., & Barclay, J. K. (1992b). Locomotor-like stifle rotation inhibits H reflexes in the dog interosseus muscle. *Physiology Canada* **23**, 3, 145.
Morin, C., Katz, R., Mazieres, L., & Pierrot-Deseilligny, E. (1982). Comparison of soleus reflex facilitation at the onset of soleus contractions produced voluntarily and during the stance phase of human gait. *Neuroscience Letters*, **33**, 47–53.
Moritani, T., Oddson, L., & Thorstensson, A. (1990). Differences in modulation of the gastrocnemius and soleus H-reflexes during hopping in man. *Acta Physiologica Scandinavia*, **138**, 575–576.
Nelson, S. G., & Mendell, L. M. (1978). Projection of right knee flexor Ia fibers to homonymous and heteronymous motoneurones. *Journal of Neurophysiology*, **41**, 778–787.
Paillard, J. (1955). *Reflexes et regulations d'origine proprioceptive chez l'homme*. Paris: Arnette.
Pierrot-Deseilligny, E., & Lancert, P. (1973). Amplitude and variability of monosynaptic reflexes prior to various voluntary movements in normal and spastic man. In J. E. Desmedt (Ed.), *New developments in electromyography and clinical neurophysiology* (pp. 539–549). Basel: Karger.
Pierrot-Deseilligny, E., Morin, G., Bergego, C., & Tankov, N. (1981). Pattern of group I fibre projections from ankle flexor and extensor muscles in man. *Experimental Brain Research*, **42**, 337–350.
Prochazka, A. (1989). Sensorimotor gain control: a basic strategy of motor systems? *Progress in Neurobiology*, **33**, 281–307.
Riedo, R., & Ruegg, D. G. (1988). Origin of the specific H reflex facilitation preceding a voluntary movement in man. *Journal of Physiology*, **397**, 371–388.
Rosenrot, P. (1980). *Asymmetry of gait and lower limb dominance*. Unpublished MSc Thesis, University of Guelph, Ontario.
Ruegg, D. G. (1989). Ia afferents of the antagonist are inhibited presynaptically before the onset of a ballistic muscle contraction in man. *Experimental Brain Research*, **74**, 663–666.

Scarpini, C., Mazzocchio, R., Mondelli, M., Nuti, D., & Rossi, A. (1991). Changes in alpha motoneuron excitability of the soleus muscle in relation to vestibular stimulation assessed by angular acceleration in man. *Journal of Otorhinolaryngology and Related Specializations,* **53,** 100–105.

Schieppati, M. (1987). The Hoffman reflex: a means of assessing spinal reflex excitability and its descending control in man. *Progress in Neurobiology,* **28,** 345–376.

Swett, J. E., & Bourassa, C. M. (1981). Electrical stimulation of peripheral nerve. In M. M. Patterson & R. P. Kesner (Eds.), *Electrical stimulation research techniques* (pp. 244–295). New York: Academic Press, Inc.

Verrier, M. C. (1985). Alterations in H reflex magnitude by variations in baseline EMG excitability. *Electroencephalography and Clinical Neurophysiology,* **60,** 492–499.

Warwick, R., & Williams, P. L. (Eds.). (1973). *Grays anatomy* (35th Ed.) (p. 815). London: Longmans.

Wolpaw, J. R., & Chen, X. Z. (1992). H-reflex conditioning in the rat: initial studies. *Neuroscience Abstracts,* **18,** 513.

Yang, J. F., & Stein, R. B. (1990). Phase-dependent reflex reversal in human leg muscles during walking. *Journal of Neurophysiology,* **63,** 1109–1117.

Yang, J. F., Stein, R. B., & James, K. B. (1991). Contribution of peripheral afferents to the activation of the soleus muscle during walking in humans. *Experimental Brain Research,* **87,** 679–687.

7

Head and Body Coordination during Locomotion and Complex Movements

A. Berthoz

Laboratoire de Physiologie Neurosensorielle
Centre National de la Recherche Scientifique
Paris, France

T. Pozzo

Groupe d'Analyse du Mouvement
Université de Bourgogne
Dijon, France

I. Introduction: Reference Frames
II. The Vestibular System: A Goal-Dependent, Euclidian, Egocentric Reference Frame
   A. The Labyrinth: A Euclidian Reference Frame under Gaze Control
   B. Other Sensory Reference Frames
   C. Nonsensory, Internal Reference Frames
III. Experimenal Studies of Head Stabilization during Complex Movements
   A. Stabilization in the Sagittal Plane
   B. Head Stabilization in the Frontal Plane
   C. Why Do Humans Stabilize Their Heads during Motor Tasks?
References

I. Introduction: Reference Frames

Multilimb coordination is often studied as if multilimb coordination could be performed without any need of a common reference frame. This may be true for extremely automatized rhythmical movements, when no precise spatial constraints are imposed on the motion. However, when movements have a spatial goal, or even in situations like locomotion or jumping, the coordination of the limbs requires one or several reference frames.

We have reviewed this question extensively elsewhere (Berthoz, 1990) and will discuss only a few points relevant to the present chapter.

By reference frames we mean a set of values to which each motor variable can be referred. Reference frames can be taken in the physical world, in the sensors themselves, or in the effector system. They do not need to be "frames"; they can be centers of rotation, for instance. The recent findings of Soechting and Flanders (1989a, b), that aiming movements of the arm are performed around a shoulder-centered reference, suggests also that the brain may choose an *ad hoc* center of rotation as a reference for various tasks, depending on the segment and the relation between the body and the environment.

Reference frames can also be virtual in the sense of being constructed internally by the brain to perform some computations in a topological space, in which, for example, relative positions or motions are the variables that are processed. Paillard (1971) has proposed the idea that, in addition to the classical "egocentric" body-centered frame of reference and "allocentric" space-centered frame of reference (in which the everchanging position of a moving body must permanently be updated), the brain can use a "geocentric" reference frame based on the direction of gravity (to which most of our spatially directed actions are to be referred). A dependence of the two systems is also suggested by Ratcliff (see chapter in Paillard, 1991), who shows that although a deficit in the allocentric system can be without consequence on the egocentric system, the opposite has never been seen.

A crucial property of reference frames is that they are hierarchically organized. A most instructive example of this hierarchy is the demonstration that, during adaptation to visual prisms in a monoarticular visuomotor pointing task, if the subject is trained to do the task only with the wrist, adaptation to the modified visual environment does not induce the capacity to perform the movement with the arm (Hay & Brouchon, 1972). More generally, if the wrist, the arm, or the shoulder are respectively used to perform the task, the adaptation follows a proximodistal hierarchy (i.e., from the shoulder to the wrist). However, if the subject can use the head, the adaptation extends to the hand. This also indicates that the head is a fundamental element as a reference in the hierarchy.

This chapter proposes a theory in which it is argued that during complex movements, the head is stabilized intermittently (during time periods that are dependent on the type of movement to be executed) under the control of gaze, and that this stabilization allows the head to serve as an inertial guidance platform for the control of multilimb movement. The head would therefore be a mobile reference frame.

Second, we believe (Droulez & Berthoz, 1986; Berthoz, 1990) that during complex coordinated multilimb movements, there must be some mechanisms of active selection of configurations of expected sensory inputs, which will be associated with each task. We can therefore expect active manipulation of reference frames, depending on the goal.

Third, we have to consider the idea that perception and movement are not always servo-controlled continuous processes, but are highly discontinuous or intermittent. Recent experimental results concerning the segmentation of hand or arm movements (Viviani & Terzuolo, 1982; Lacquaniti, Soechting & Terzuolo, 1986) are much in accord with this theory. We also must accept the idea that, as Paillard (1982) and Gurfinkel (in Paillard, 1991) have advocated, there may be several reference frames, depending on the speed of movement, the goal, the initial and final posture, and local as well as global factors related to the context of the movement.

II. The Vestibular System: A Goal-Dependent, Euclidian, Egocentric Reference Frame

One of the fundamental reasons why the head is stabilized is that it contains two of the most important sensors that allow the brain to compute the kinematic parameters of motion: the vestibular and the visual systems. The properties of the vestibular system in this context will now be reviewed, because its role in this function is not yet clearly understood.

A. The Labyrinth: A Euclidian Reference Frame under Gaze Control

The semicircular canals are involved in stabilizing reactions. They constitute a Euclidian reference frame for the measurement of the angular accelerations of the head. The reference frame of the three semicircular canals has a fundamental role as a kind of template for the geometry of the sensory systems involved in the representation of movement: it is now well established that the preferred directions of visual motion-sensitive cells in the accessory optic system are aligned with the planes of the semicircular canals.

The otolithic organs (sacculus and utriculus) are inertial detectors of linear acceleration of the head in the plane of their macula. They also detect the angle of the head with respect to gravity: when the head is tilted, the otoliths are stimulated by the component of gravity in the plane of each macula. They are "tiltmeters" and not only "accelerometers." They constitute therefore an important reference frame (see review in Berthoz & Droulez, 1982)). Can we show that they are actually used as a reference for the detection of head tilt?

Let us first consider a consequence of the capacity of the otoliths to detect the angle of the head with respect to gravity, in the elaboration and control of the posture of the head–neck system. The investigation of the spatial organization of the cervical column in various species (Vidal, Graf, & Berthoz, 1986) has shown that the cervical vertebral column at rest is kept parallel to gravitational force lines, even in total darkness. This is accomplished by a particular S-shaped geometry of the cervical vertebral column in quadrupeds. This resting position in cats, rabbits, or rodents, in general, is maintained such that their cervical column has a posture similar to birds. The relevance of this finding for our discussion on reference frame is twofold. First, animals can maintain this posture only if a sensor detects the angle of the head with respect to gravity, and the otoliths are well suited for this purpose. Second, this posture has allowed the organization of a neuronal network specialized for orienting movements in the plane of the horizontal canal, which is maintained (by this posture) perpendicular to gravity.

B. Other Sensory Reference Frames

The otoliths are, of course, not the only sensors that can contribute to this maintenance of the head in a fixed angle with respect to the gravitational vertical. Several reviews have recently dealt with this question (Berthoz, Graf, & Vidal, 1992; Howard & Templeton, 1966; Howard, 1982; Mittelstaedt, 1986). For instance, it is well known that vision also provides the so-called "visual vertical" and that the visual vertical entertains reciprocal influences on the perception of forms (Mittelstaedt, 1989; Rock, 1973). The tactile retina composed by the sole of the feet also constitutes an important array of the detection of differential pressure, which may be involved in these processes (see also a recent study by Ohlman, 1988).

However, in the absence of vision, if vision is blurred by a high-retinal slip, or during complex movements in which the ground reference is no longer available, the otoliths are indeed essential for the declaration of a subjective estimate of the vertical. But rather than analyze these rather well-documented properties, we shall discuss another point of view that is quite underestimated in most contemporary analysis of brain mechanisms relative to space, that of "internal" reference frames.

C. Nonsensory, Internal Reference Frames

It is well established that in normal gravity conditions, although the perceived gravitational vertical does deviate from the objective vertical by a few degrees (Aubert, 1967), gravitational reference is used not only for posture but also for perceptual tasks that involve orientation in space, including mental rotation (Hock & Tromley, 1978; Attneave & Olson, 1990; Corballis, Nagourney, Shetzer, & Stefanatos, 1978).

However, Mittelstaedt (1986, 1983) has proposed repeatedly that there are internal egocentric references. The "ideotropic" vector of Mittelstaedt is probably related to the trunk main axis and not to the head. We know neither the mechanism of formation nor the exact role of these putative reference frames, which are part of what has also been called the "body scheme" (Head, 1920; Gurfinkel, Debreva, & Levik, 1986).

The considerations given above should be helpful to understand the mechanisms underlying the behavioral observations that will be described below. They concern the organization of posture during complex movements.

Posture, in the past fifty years, has been studied from the wrong point of view. It has always been considered that posture is organized with the feet as a reference platform, implicitly accepting that the ground was the reference frame. This may be true in a rather limited series of cases, such as quiet stance or small amplitude perturbations. Nashner (1985) has described such a strategy of head stabilization in a particular case. However, when posture is combined with multilimb movements in which the stable platform of the feet is absent, we believe that the head is used as a reference platform.

III. Experimental Studies of Head Stabilization during Complex Movements

A. Stabilization in the Sagittal Plane

1. Locomotion

A kinematic study has revealed the capacity of human subjects to partially stabilize their head in the sagittal plane during active locomotor movements. The body movement was analyzed in three dimensions by means of an optical TV-image processor (ELITE system) (Pozzo, Berthoz, & Lefort, 1989; Berthoz & Pozzo, 1988; Pozzo, Berthoz, & Lefort, 1990; Pozzo, Bethoz, Lefort, & Vitte 1991). The computer provides several sets of data. It first constructs stick figures of the human body in motion (Figure 1A). In addition, numerical values of several parameters (in translation and rotation) are calculated.

In spite of the mechanical disparities of the tasks we have tested (natural or simulated locomotion, and more or less other dynamic tasks), head angular displacement in the sagittal plane remains within

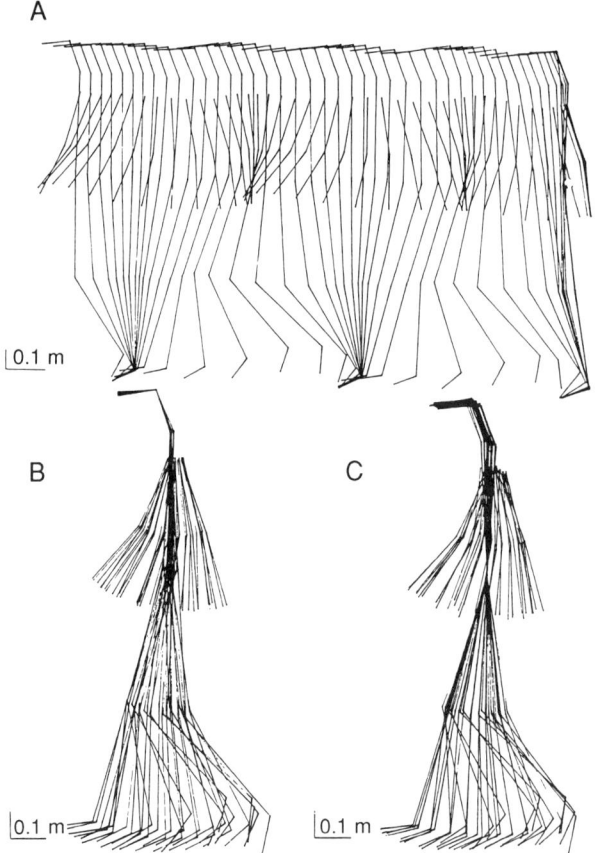

FIGURE 1 Head position during locomotion. (A) Stick figures reconstructed by computer during locomotion. Ten markers have been placed on body segments. (B) Same stick figures as in (A), but moved by translation so as to superimpose the markers on the auditory meatus. Notice the small variations of the head angular position. (C) The translation of each stick figure has been made on the tubercle of iliac crest to show the movement of the trunk with respect to the vertical.

a small range (<5°) compared with the movements of the other limbs. Our experimental analysis, which consists of superposition of the head, limb, and trunk angle relative to a single point on the body, allowed us to differentiate between, on one hand the head–trunk ensemble, which composed the standard posture and, on the other hand, the legs, which behave like actuators of the head–trunk unit.

A similar distinction was made by Hess (1943), who described the organization of postural adjustment associated with movement. He proposed two interactive mechanisms involved in voluntary movement: (1) the "ereismatic" component, that is, the support provided to the movement by the postural fixation (*the postural component*); and (2) the "teleokinetic" component, that is, the displacement of the body and the limbs oriented toward a goal (*the instrumental component*).

However, during locomotion, the trunk–head unit is, in our opinion, not only an essential mechanical support for the triggering of action but also a reference necessary for dynamic postural control. This is in part due to the fact that, during locomotion or running, there is a transient loss of foot contact and therefore of the reference that it provides by means of plantar tactile cues.

From an ontogenetic point of view, several postural strategies for locomotion have been suggested by Assaiante and Amblard (1990), which correspond to different periods of the acquisition of locomotor balance. These authors suggest that either the head or the head–trunk unit are alternatively used as stable frames of reference for dynamic locomotor equilibrium control.

Our results show that the subject minimizes the head's kinematics by locking the head on the trunk. This decreases the degrees of freedom of the head–neck system and improves head stabilization. This strategy can be used especially when trunk stabilization is the main purpose of postural equilibrium (Gurfinkel *et al.*, 1981) and when the trunk is the most highly stabilized part of the body (e.g., during walking in place and running). On the other hand, during conditions in which the trunk movements will be more pronounced because of the specific task (i.e., during hopping and probably jumping), the subject could stabilize his head in space by dissociating head and trunk movements.

2. Jumping and Running

A study made during running and jumping illustrates this statement. In Figure 2, during running in place the computer has translated the stick figures in order to superimpose the different positions of the markers located near the auditory meatus. Figure 2B gives an enhanced illustration of the head–shoulder segment angular position during several tasks: walking in place (Wip), walking (W), running (R), and hopping (H). The best stabilization is obtained during walking in

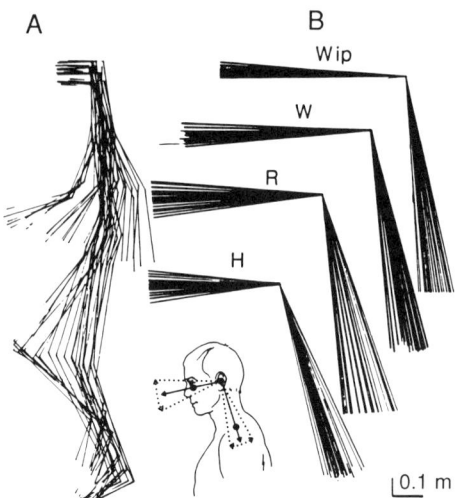

FIGURE 2 (A) Stick figures during running in place. Note the stable orientation of the Frankfort plane near the horizontal in spite of the vertical oscillations of the head. The inset at the bottom of the figure depicts the anatomical positions of the three markers placed at head level: one on the lower socket of the eye, one near the meatus of the ear, and one on the trapezius at the level of C7. (B) Enlarged views of the two links defined by the three markers during walking in place (Wip), free walking (W), running in place (R), and hopping (H).

place. These results show that the head is stabilized with respect to gravity under the condition of normal vision. The plane of the horizontal semicircular canals, which are located 25° above the canthomeatic line, remains approximately horizontal when the subject has no requirement for the direction of gaze. This angle is obviously dependent on the angle of voluntary gaze. Quantitative analysis of these movements has been reported elsewhere (Pozzo et al., 1989).

We have also demonstrated that the horizontal and vertical components of head displacement are linked by a simple phase relation in normal subjects and that this coordination is perturbed in subjects with bilateral labyrinthine lesions (Figure 3). We have proposed that this relationship can be used to assess the state of the inertial guidance mechanism. This proposal has recently been validated by its use for the study of postflight deficits in astronauts after 7-day flights (J. Bloomberg, 1992, unpublished data).

Another interesting fact, observed during jumping down from a stool, is the unbroken trajectory of head rotation in the sagittal plane during landing, in spite of the impact between foot and floor (Pozzo et al., 1989). This suggests a capacity to predict mechanical constraints resulting from physical contact. There are many examples demonstrating that, in

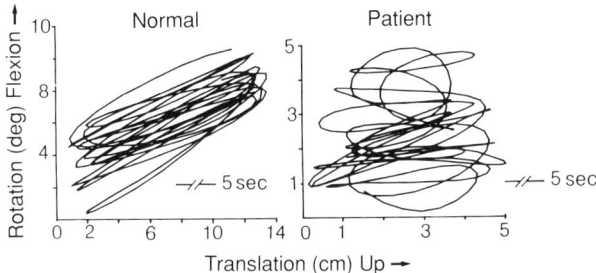

FIGURE 3 Relationship between head translation along vertical axis and head rotation in the sagittal plane. Head angular position in space, during 5 sec of recording, is plotted as a function of head position along the vertical axis, during running in the light, for a typical, normal subject and for a patient.

spite of the large body movement, the head remains stable in rotation. Surfing, or acrobatic skiing are good illustrations of mechanical and sensory independence of the head with respect to limb and trunk displacements.

3. Head Stabilization and Preselection of Sensory Inputs during Salto

The examination of the salto (which is a full backward rotation of the whole body around the center of gravity in the sagittal plane) provides a remarkable model to study the selection of sensory inputs during complex movements (Figure 4A).

Typical raw data illustrating head angular displacement and head angular velocity in the sagittal plane during a salto from one subject are shown in Figure 4B. To further evaluate the relationship between head and limb angular displacements, we have drawn on the same plot, in Figure 4B, top, the angular displacement of the neck (α), the hip (β), the trunk (γ), and the head (θ) in space. The corresponding angular velocities are displayed at the bottom of Figure 4B. The head stabilization around the horizontal axis lasts 0.25 sec, during which head angular velocity ranges from 20°/sec to 120°/sec. Peak head angular velocity during this period never exceeds 130°/sec in the 5 subjects. Such an initial plateau, which includes the initial impulse and the takeoff (phase I) is synchronous with (1) a backward rotation of the trunk with a peak angular velocity of 230°/sec (the amplitude of trunk rotation from the starting position is 60°); (2) a successive extension and flexion of the hip ($\pm 30°$); and (3) a flexion of the neck (40° of amplitude from the starting position).

Then, after a 235° amplitude head saccade (phase II), head angular displacement exhibits a second period of relative stabilization (phase III), which lasts 0.15 sec, during which head angular velocity decreases

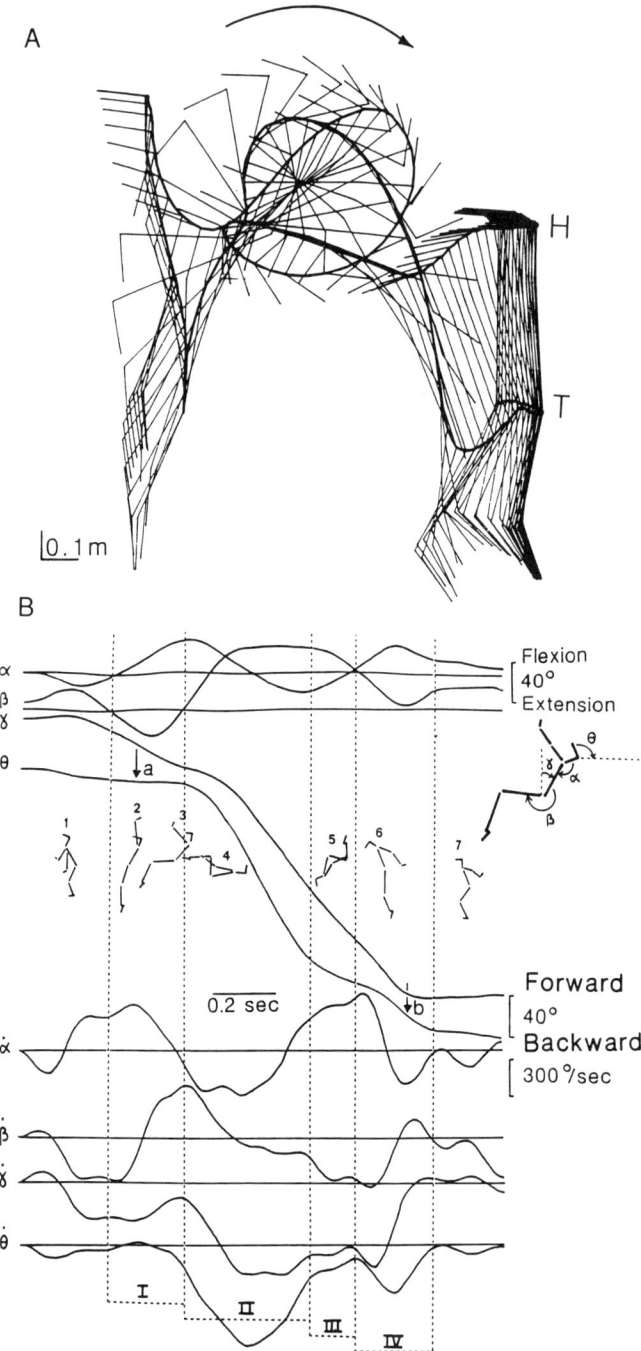

FIGURE 4 (A) Stick figures reconstructed by computer during a backward somersault. The arrow indicates the direction of the movement. Six markers have been placed on body segments. The two solid lines (H and T) connect the stick figures at the level of the

from 270°/sec to 100°/sec. Over the 5 subjects, during phase III, maximum head angular velocity ranges from 80°/sec to 120°/sec.

Head saccade (phase II) is synchronous with (1) a flexion of the hip (75° from the starting position) and the stabilization of this flexion; and (2) an extension of the neck (peak-to-peak amplitude is 60°). Peak angular velocity of this rapid head rotation in space is 750°/sec.

By contrast, the second stabilization of the head (phase III), which corresponds to the final descending part of the trajectory of the center of gravity, is associated with a flexion of the neck (60°) and an extension of the hip (70°). During this relative stabilization, the mean head angle θ, relative to the horizontal starting position, is 275° ($\pm$ 15) over the 5 subjects, that is, the Frankfurt plane (defined by a line between the lower border of the eye socket and the meatus of the ear) is roughly aligned with gravity. Last, just before the ground contact, a second head rapid rotation, in extension and with a peak angular velocity of 370°/sec, restores the initial head position. Maximum head angular velocity in the 5 subjects during the entire task ranges from 630°/sec to 800°/sec.

These results obtained during the salto suggest that the acquisition of sensory information is divided during the three main phases of the movement. During the elevation phase the head is stabilized and the brain can use a combination of visual and vestibular cues to guide the movement, control the posture, and, most important, trigger the backward rotation with the appropriate geometry and acceleration. During the backward rotation at several hundred degrees per second, by contrast, the retinal slip is too high to use the optic flow for the computation of head movement. This phase is also interesting because it occurs in a zero G condition and in addition to shifting to an inertial guidance, the subjects probably also shift to an egocentric reference frame. Frederici and Levelt (1990) have recently suggested that in microgravity, subjects have a tendency to localize the position of objects in space in a body-centered (egocentric) reference frame.

During the third phase, namely the preparation of landing, the head will be stabilized again, which allows the combined use of optic flow and

---

markers placed on the auditory meatus and on the tubercle of the iliac crest. These lines depict the trajectories of the head (H) and trunk (T) in the sagittal plane. (B) Angular displacements (*top*) and corresponding angular velocities (*bottom*) of the neck (α and α'), hip (β and β'), trunk in space (γ and γ'), and head in space (θ and θ') during a backward somersault; the figure at top right depicts the set of angles calculated. Horizontal lines indicate the starting angular position and zero angular velocity. The cartoon figures in the middle of the graph, indexed from 1–7, illustrate the temporal and spatial evolution of the movement. The four main phases of the movement (I, II, III, and IV) have been separated by vertical dotted lines.

vestibular information for the coordination of movements of legs and trunk.

B. Head Stabilization in the Frontal Plane

1. Maintenance of Dynamic Equilibrium with Bipedal Coordination on an Unstable Platform

We studied the stabilization of the head in the frontal plane during tasks requiring the maintenance of equilibrium on a beam or a moving platform.

Eight normal, human subjects and six dancers were asked to perform two different tasks: (1) maintenance of upright monopodal stance on a narrow cylindrical beam (B); (2) maintenance of upright bipedal stance on a semicircular unstable platform (rocking platform, RP). The movements of various parts of the body were measured by the ELITE system. Six markers placed on different parts of the body defining four links allowed us to compute translation and rotation parameters in three dimensions.

The two tasks were performed in light and in darkness. In the light, the experiment was first performed without upper-limb constraints and, subsequently, with the subject holding a glass full of water. In the latter situation, two visual conditions were tested: (1) without specific instructions regarding gaze; and (2) when the subject was required to fix his eyes on the glass. Each task was performed 3 times, with a duration of 5 sec each. We shall report here only the results obtained with normal subjects.

These two tasks, which require fine equilibrium in the frontal plane, have been chosen because they represent two kinds of postural tasks: (1) on the beam, the difficulty of the task is due to the small support surface. In this case, the platform provides a unipodal but stable reference frame; (2) on the rocking platform, the support surface moves in translation and in rotation. Therefore, there is no stable tactile or proprioceptive frame of reference.

Head and trunk stability in rotation were evaluated by considering the standard deviation of head and trunk angular displacements around the mean angular position, calculated over the time of one recording session.

2. Head Angular Position and Displacement

In order to compare the amplitude of head and trunk rotations, we plotted the standard deviations of head angular displacements as a function of trunk angular displacements (Figure 5). The values obtained during the two tasks, for each experimental condition and for all subjects tested, are plotted together. Several observations can be made

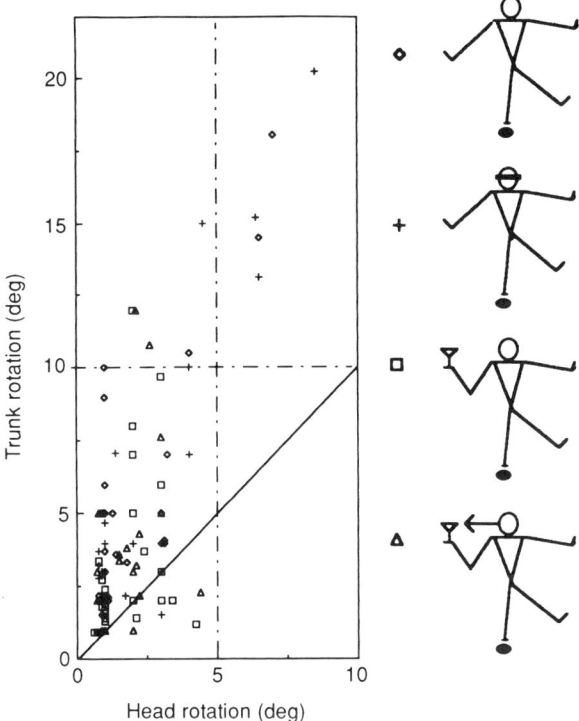

FIGURE 5 Standard deviation of head rotation is plotted as a function of the rotation of the trunk in the frontal plane when the subjects stand on the beam. We have grouped the data obtained for the 8 subjects tested during the 4 experimental conditions (*top* to *bottom*): normal vision, darkness, when the subject holds a glass full of water without visual constraints, and with the gaze fixed on the glass.

from the diagram. First, in the low-range values (below 3°), head rotation is greater than trunk rotation and each data point is situated under the diagonal line, indicating a better stabilization of the trunk. Second, above this range, all the data points are situated above the diagonal; this indicates that the head is always better stabilized than the trunk. Third, the S.D. of head rotation does not exceed 5°; this value increases for larger trunk angular displacement (above 10° in amplitude) without ever exceeding 10°. In contrast, trunk angular displacements can sometimes exceed 20°.

We have calculated the SD of head and trunk angular displacements averaged over all trials during the two tasks. In both tasks, the mean value of head angular displacement is significantly smaller than the trunk angular displacement ($p < .001$ for the beam condition and $p < .005$ for the rocking platform with a t test). The amplitudes of head and trunk angular displacements are not signficantly different during the two tasks; nevertheless, we can note a trend to better stabilize the head and the trunk when the subjects stand on the rocking platform. The relationship between head and trunk displacements is evidently dependent on the task. For example, when the subject is asked to hold a glass full of (French) wine while standing on the beam, the relationship between head and trunk rotation becomes more complex (see Figure 6, where the average trace of the plot of the evolution of the angular position of the head as a function of the trunk rotation is no longer oriented along a diagonal with a positive slop, as under the normal condition).

These results show that during complex tasks that require fine equilibrium in the frontal plane, the head is stabilized in rotation while the trunk moves underneath. They strengthen our previous studies, which

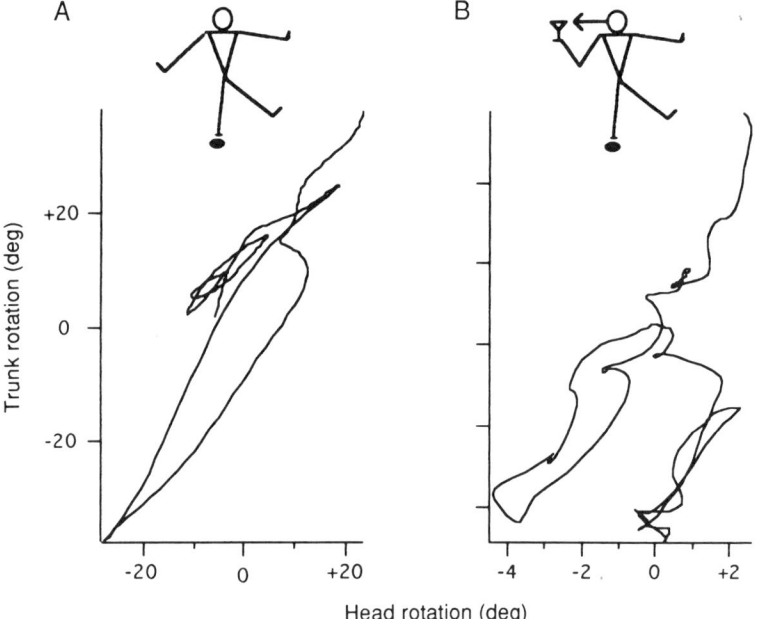

FIGURE 6 Relationship between head rotation and trunk rotation in the frontal plane. The traces correspond to the evolution of the angular position of the head in the frontal plane (abscissa) as a function of the trunk rotation in the same plane (ordinate) for a typical subject, with normal vision (A) and when the subject holds a glass full of water with the gaze fixed on the glass (B). Note that under normal conditions the relations between θ and α are almost linear. This suggests a positive correlation between the angular displacements of head and trunk: The head inclination in one direction is associated with trunk inclination to the same side. The higher density of the trace in the upper right part of the plot indicates that the rotations of head and trunk are mostly of small amplitude. In contrast, under conditions of gaze fixation on the glass the trace is more confused and the relations between θ and α are less regular.

showed a good stabilization in the sagittal plane and generalization of the behavior to the frontal plane. Accordingly, other studies have shown that the head is stabilized in the frontal plane, during a voluntary abduction of the leg (Mouchnino, Aurenty, Massion & Pedotti, 1990) or during motion on a beam (Assaiante & Amblard, 1992).

Our study also demonstrates that head stabilization is task dependent. It is the result of a strategy that improves dynamic postural control. In fact, several strategies could be described. During the two tasks, the trunk, when it is well stabilized, can be used as the stable frame of reference to estimate head and limb displacements; in contrast, for large angular displacements of the trunk, the head, which is better stabilized than the trunk, can be used as the stable frame of reference.

In addition, during the two tasks, which present different mechanical constraints, head orientation remains near the vertical, while the trunk angular position depends on the task. On the beam, the trunk, which is tilted on the same side as that of the foot support, could be used to balance the whole body in synergy with the free leg. In this case, the leg–trunk unit acts as an actuator of the equilibrium. In contrast, on the rocking platform, the subjects try to keep the mean angular position of the trunk near the vertical, while the lower limbs behave like actuators of the head–trunk unit. These differences could be explained in view of the mechanical and sensorimotor specificities of the tasks. On the beam, the head and the foot, which constitute stable frames of reference, can be used to compute the angular displacements of the actuators (the trunk and the free leg). On the rocking platform, as the head remains the only stable frame of reference, linking the trunk on the head could simplify postural control.

3. Head Stabilization Helps Bipedal Coordination during Skiing on Bumps

Skiing on bumps has now become an Olympic performance that provides us with a beautiful example of head stabilization in the frontal plane that is probably essential for multilimb coordination. It is known by experts in this sport that the motor strategy necessary during descent is not to bend down at each bump but, on the contrary, to "lift the legs up," while maintaining the head fixed in space. Films taken on the slopes with champions also show clearly that the head is remarkably stable in the frontal plane.

C. Why Do Humans Stabilize Their Heads during Motor Tasks?

What is the advantage of head stabilization? The hypothesis could be formulated as follows: during complex movements, the brain uses the head as a stabilized inertial guidance platform. This principle is used by engineers as an elegant way to control a heavy mass during 3D motion. This platform contains the sensors that will serve both to stabilize the platform and control the movements of the heavy mass. The head can then be used as a mobile reference frame, which serves as a basis for multilimb coordination.

Head rotational stabilization may also simplify the task of the brain in performing the "fusion of sensors." For instance, having only translations may simplify the visual vestibular matching, or, in other words, "fusion" between the two-dimensional detection of movement by the optic flow on the retina (Warren, Morris, & Kalish, 1988), and the 3D gravitoinertial measurement given by the vestibular organs. Such a strategy of stabilization may also be advantageous for simplifying the transformations necessary to build a coherent internal representation of space, which is also necessary for multilimb coordination.

Head stabilization could also help to solve the problem of gravitoinertial differentiation, as the otoliths cannot distinguish between a static tilt of the head (gravitational stimulus), and a linear acceleration (inertial stimulus). The stabilization of the head in rotation leaves only linear accelerations to be measured directly by the otoliths. Possible alternative mechanisms for solving this problem have been suggested by other authors (Young, 1984; Mittelstaedt, 1991). Droulez and Darlot (1989) recently formalized a model that supports the hypothesis of a low-pass filtering at the level of the vestibular nuclei to distinguish between gravitational (low-frequency) and inertial (high-frequency) signals. These two mechanisms, active positioning of the sensor by head rotational stabilization and low-pass filtering, are not mutually exclusive and could lead to a precise differentiation.

A final potential advantage of this stabilization, which is known from roboticians, is its helpfulness in updating the position of a moving vehicle when new information is available (Boissonnat, Faverson & Merlet, 1988). In the case of sensory–motor systems, a continuous updating of the internal representations of head motion is necessary because of the constant continuous changes of the forces exerted on the musculoskeletal system by gravity during displacements.

The use of the head as a reference frame for multilimb coordination is also suggested by a most probable hierarchy that exists between the head and distal musculature. Paillard and his colleagues (Hay & Brouchon, 1972) have shown that during prism adaptation, the head is at the top of the hierarchy; in other words, if one adapts the head to a prism situation, then the hand will benefit from this calibration, but if the hand is recalibrated the head will not benefit from the recalibration.

Other experiments using vibration of the neck (Biguer, Donaldson, Hein, & Jeannerod, 1988; Roll, Velay, & Roll, 1991) also point to a major role of neck proprioception for the general organization of the body scheme. Paillard and Beaubaton (1978) have suggested that movements are organized in succession from the head to more distal muscles of the limbs.

A last hypothesis concerns the temporal aspect of complex movements. As illustrated by the experiments concerning the salto, it can be suggested that when body movements cannot be measured at all by any sensor because, for instance, of the high speed of motion, the chronometry of the movement would replace the missing exteroceptive information of the visual system. Therefore, the movement is organized in successive steps, during which the head is stabilized at different angles, allowing the brain to utilize a series of different configurations of sensory inputs.

Acknowledgments

This research was supported by the French Centre National d'Etudes Spatiales (CNES) and by the Human Frontiers Sciences Program. Y. Levik and K. Dubois contributed to the experiments on head stabilization in the frontal plane.

References

Assaiante, C., & Amblard, B. (1990). Head stabilization in space while walking: effect of visual restriction in children and adults. In T. Brandt (Ed.), *Disorders of posture and gait* (pp. 229–233). Munich: Verlag.

Assaiante, C., & Amblard, B. (1992). Head trunk coordination and locomotion equilibrium in 3 to 8 year old children. In A. Berthoz, W. Graf, & P. P. Vidal (Eds)., *The head neck sensory-motor system.* New York: Oxford University Press.

Attneave, F., & Olson, R. K. (1990). Discriminability of stimuli varying in physical and retinal orientation. *Journal of Experimental Psychology,* **74,** 149–157.

Aubert, H. (1967). Über eine scheinebare Drehung von Objecten bei Neigung des Kopfes nach rechts oder links. *Virchows Archiv.,* **20,** 381–393.

Berthoz, A. (1990). Reference frames for the perception and control of movement. In J. Paillard (Ed.), *Brain and space.* New York: Oxford University Press.

Berthoz, A., & Droulez, J. (1982). Linear motion perception. In A. H. Wertheim, W. A. Wagenaar, & H. W. Leibowitz (Eds.), *Tutorials on motion perception.* (pp. 157–199). London: Plenum.

Berthoz, A., & Pozzo, T. (1988). Intermittent head stabilization during postural and locomotory tasks in humans. In B. Amblard, A. Berthoz, & F. Clarac (Eds.), *Posture and gait: Development adaptation and modulation* (pp. 189–198). Amsterdam/N.Y./Oxford: Elsevier.

Berthoz, A., Graf, W., & Vidal, P. P. (1992). *The Head-neck sensori-motor system.* New York: Oxford University Press.

Biguer, B., Donaldson, I. M. L., Hein, A., & Jeannerod, M. (1988). Neck muscle vibration modifies the representation of visual motion and direction in man. *Brain,* **111,** 1405–1424.

Boissonnat, J. D., Faverson, B., & Merlet, J. P. (1988). Technique de la robotique. *Traité des nouvelles technologies, série robotique*. Paris: Hermès.
Corballis, M. C., Nagourney, B. A., Shetzer, L. I., & Stefanatos, G. (1978). Mental rotation under head tilt: Factors influencing the location of the subjective reference frame. *Perception and Psychophysics*, **24**(3), 263–273.
Droulez, J., & Berthoz, A. (1986). Servo-controlled conservative versus topological (projective) mode of sensory motor control. In W. Bles, & Th. Brandt (Eds.), *Disorders of posture and gait* (pp. 83–97). Amsterdam: Elsevier.
Droulez, J., & Darlot, C. (1989). The geometric and dynamic implications of the coherence constraints in three dimensional sensorimotor coordinates. In M. Jeannerod (Ed.), *Attention and performance XIII* (pp. 495–526). New Jersey: Laurence Erlbaum.
Frederici, A. D., & Levelt, W. J. M. (1990). Spatial reference in weightlessness: perceptual factors and mental representations. *Perception and Psychophysics*, **43**(3), 253–266.
Gurfinkel, V. S., Lipshits, M. I., & Popov, K. E. (1981). Stabilization of the body as the main task of postural regulation. *Fiziologya Cheloveka*, **7**, 400–410. (Translated from Russian.)
Gurfinkel, V., Debreva, E. E., & Levik, Y. (1986). The role of internal model in the position perception and planning of arm movement. *Fiziologiya Chelova*, **12**(5), 769–776.
Hay, L., & Brouchon, M. (1972). Analyse de la réorganisation des coordinations visuomotrices. *Annee Psychologique*, 25–38.
Head, H. (1920). *Studies in neurology*. London: Hodder and Stoughton.
Hess, W. R. (1943). Teleokinetisches und ereimatisches krafsystem in der biomotorik. *Helv. Physiol. Pharmacol. Acta*, **1**, C62–C63.
Hock, H. S., & Tromley, C. L. (1978). Mental rotation and perceptual uprightness. *Perception and Psychophysics*, **24**, 529–533.
Howard, I. P. (1982). *Human visual orientation*. London: Wiley.
Howard, I. P., & Templeton, W. B. (1966). *Human spatial orientation*. London: John Wiley.
Lacquaniti, F., Soechting, J. & Terzuolo, C. (1986). Path constraints on point to point arm movements in three dimensional space. *Neuroscience*, **17**, 313–324.
Massion J., & Rispal-Padel, L. (1986). Thalamus: fonctions motrices. *Rev. Neurol.*, **142**, 327–336.
Mittelstaedt, H. (1983). A new solution to the problem of the subjective vertical. *Naturwissenshaften*, **70**, 272–281.
Mittelstaedt, H. (1986). The subjective vertical as a function of visual and extraretinal cues. *Acta Psychologica*, **63**, 63–85.
Mittelstaedt, H. (1989). Interactions of form and orientation. In S. R. Ellis, & M. K. Kayser (Eds.), *Spatial displays and spatial instruments*. NASA Conference Publications 10032, Moffet Field, pp. 42-1–42-14.
Mittelstaedt, H. (1991). The role of otoliths in the perception of the orientation of self and world to the vertical. *Zool. Jb. Physiol.*, **95**, 419–425.
Mouchnino, L., Aurenty, R., Massion, J., & Pedotti, A. (1990). Coordinated control of posture and equilibrium during leg movements. In T. Brandt (Ed.), *Disorders of posture and gait* (pp. 68–72). Munich: Verlag.
Nashner, L. M. (1985). Strategies for organization of human posture. In M. Igarashi, & F. O. Black (Eds.), *Vestibular and visual control of posture and locomotor equilibrium* (pp. 1–8). Basel: Karger.
Ohlman, T. (1988). *La perception de la verticale. Variabilité interindividuelle dans la dépendance à l'égard des référentiels spatiaux*. Unpublished doctoral dissertation. Paris, Université de Paris VIII.
Paillard, J. (1971). Les déterminants moteurs de l'organisation spatiale. *Cahiers de Psychologie*, **14**, 261–316.

Paillard, J. (1982). Le corps et ses langages d'Espace. In E. Jeddi (Ed.), *Le corps en psychiatrie* (pp. 53–69). Paris: Masson.
Paillard, J. (1991). *Brain and space.* New York: Oxford University Press.
Paillard, J. & Beaubaton, D. (1978). De la coordination visuo-motrice à l'organisation de la saisie manuelle. In H. Hecaen, & M. Jeannerod (Eds.), *Du contrôl moteur à l'organisation du geste* (pp. 226–261). Paris: Masson.
Pozzo, T., Berthoz, A., & Lefort, L. (1989). Head kinematics during various motor tasks in humans. In J. Allum, & M. Hulliger (Eds.), *Progress in brain research* (Vol. 30). (pp. 377–383). Amsterdam: Elsevier.
Pozzo, T., Berthoz, A., & Lefort, L. (1990). Head stabilization during various locomotor tasks in humans. I. Normal subjects. *Exp. Brain Res., 82,* 97–106.
Pozzo, T., Berthoz, A., Lefort, L., & Vitte, E. (1991). Head stabilization during various locomotory tasks in humans. II. Patients with peripheral vestibular deficits. *Exp. Brain Res., 85,* 208–217.
Rock, I. (1973). Orientation and form. New York: Academic Press, Inc.
Roll, R., Velay, J. L., & Roll, J. P. (1991). Eye and neck proprioceptive messages contribute to the spatial coding of retinal input in visually oriented activities. *Exp. Brain Res., 85,* 423–431.
Soechting, J. F., & Flanders, M. (1989a). Sensorimotor representations for pointings to targets in three-dimensional space. *J. Neurophysiol., 62,* 582–594.
Soechting, J. F., & Flanders, M. (1989b). Errors in pointing are due to approximations in sensorimotor transformations. *J. Neurophysiol., 62,* 595–608.
Vidal, P. P., Graf, W., & Berthoz, A. (1986). The orientation of the cervical vertebral column in unrestrained awake animals. I. Resting position. *Exp. Brain Res., 61,* 549–559.
Viviani, P., & Terzuolo, C. (1982). Trajectory determines movement dynamics. *Neuroscience* 7(2), 431–437.
Warren, W. H., Jr., Morris, M. W., & Kalish, M. (1988). Perception of translational heading from optical flow. *J. Exp. Psychol. Hum. Percep. Perform.* 14(4), 646–660.
Young, L. (1984). Perception of the body in space. In *Handbook of physiology. The nervous system. Sensory processes.* Vol. III, (pp. 1023–1066). Bethesda, MD: American Physiological Society.

8

Neuronal Basis of Stance Regulation
Interlimb Coordination and Antigravity Receptor Function

Volker Dietz

Swiss Paraplegic Centre
University Hospital Balgrist
Zurich, Switzerland

I. Introduction
II. Interlimb Coordination
III. Influence of Gravity
   A. Actual Body Mass and Size of EMG Response
   B. Significance of Extensor Load Receptors for Stance Regulation
IV. Conclusions
   References

I. Introduction

Regulation of bipedal stance and gait shows fundamental differences to that of quadrupedal locomotion, requiring specific neuronal mechanisms to maintain the body in an upright position. The spinal stretch reflex has been suggested as functioning so as to adapt the preprogrammed leg muscle motor patterns to the terrain encountered and to compensate for unexpected changes in ground level (Dietz, Schmidtbleicher, & Noth, 1979). While this neuronal mechanism explains quick unilateral patterns of reflex activity in leg extensor muscles, a more complex, bilateral coordination of leg muscle activation is needed to maintain the body equilibrium when gait is disturbed by an obstacle.

This is the case when a subject stands on an unstable surface: a spinal reflex mechanism rapidly compensates for a *unilateral* foot displacement by producing *bilateral* activation of tibialis anterior muscles (Dietz & Berger, 1982). Such an activation of homologous muscles of both legs is not surprising, since during normal stance upright posture was found to be regulated by an activation of homologous leg muscles (Bonnet, Gurfinkel, Lipshits, & Popov, 1976). A different mechanism might, however, be expected during gait, when homologous leg muscles are being activated reciprocally. Irrespective of the conditions under which stance and gait are investigated, the neuronal pattern evoked during a particular task is always directed to hold the body's center of mass over the base of support. All control mechanisms must, therefore, be considered and discussed in this respect. One consequence is that the selection of afferent input by the central mechanisms must correspond to the actual requirements for body stabilization. Furthermore, neuronal signals of muscle stretch or length alone are insufficient for the control of upright posture. Only a combination of afferent inputs can provide the information needed to control the body equilibrium. Some of these complex interactions between afferent inputs have been partially revealed in the last years: the function of proprioceptive reflexes and of the otoliths depends on contact forces and/or load receptors within the leg extensor muscles, which include information about the position of the body's center of gravity relative to the feet.

Two methodological approaches can be used for investigations into neuronal control of human stance and gait, namely platform rotation and translation (for review see Dietz, 1992). The EMG pattern induced by dorsiflexing rotation of the feet is different from that induced by backward platform translations, even though the triceps surae become stretched in both conditions (Figure 1). Following feet dorsiflexing *rotations* a small, short-latency response in the gastrocnemius is followed by a stronger activation of the tibialis anterior. The latter activity is required to compensate for the backward sway of the body induced by the rotation. Backward *translation* is followed only by strong, long-latency (about 70 msec) gastrocnemius activity. This EMG activity results in the maintenance of a stable, vertical position over the displaced feet. It has been suggested that the difference in the EMG pattern between the two conditions is due to a *reflex adaptation* based on the assumption that postural stabilization is the product of diminishing muscular destabilization (Nashner, 1976). This adaptation was assumed to be achieved through a selection of appropriate postural reflexes, which are initiated by ankle proprioceptors and mediated via

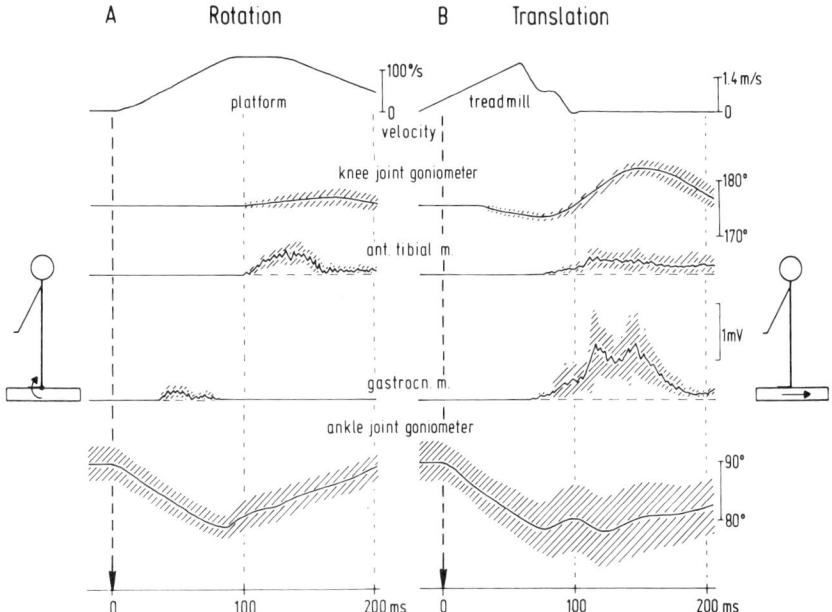

FIGURE 1 Mean (±S.D., shaded area) of rectified and averaged ($n = 30$) leg muscle EMG responses together with knee and ankle joint movements, of 7 subjects following a dorsiflexing rotation of the platform (A) and a backward translation of the treadmill (B). (From Dietz et al., 1991.)

cerebellar nuclei (Nashner, 1976; Nashner, Woollacott, & Tuma, 1979). This can be seen experimentally as a shift of the response pattern, that is, a successive "adaptation" of the reflex response, within two to four trials if the signal is "inappropriate" to a functionally directed tibialis anterior (rotational) or gastrocnemius (translational) activation.

Subsequent experiments have clearly failed to support this concept of adaptation of the response pattern and have instead shown that an immediate change in pattern occurs within the first trial, which can be seen when the type of perturbation is unexpectedly changed (Gollhofer, Horstmann, Berger, & Dietz, 1989; Hansen, Wollacott, & Debu, 1988; Nardone, Giordans, Corra, & Schieppati, 1990). This is achieved by the central (spinal) integration of multiple divergent sensory inputs.

In the following sections the afferent inputs (in addition to muscle stretch) and their interactions, which are suggested to be responsible for the generation of the appropriate EMG response pattern, will be considered. The discussion will be centered around two basic requirements for the regulation of upright stance and gait: first, the automatic interlimb coordination and, second, the influence of gravity on the EMG pattern during the control of bipedal stance.

II. Interlimb Coordination

In humans the regulation of posture and locomotion is based on the finely tuned, automatic coordination of muscle activation between the two legs. This suggests that adequate neuronal mechanisms exist to achieve task-directed coupling of bilateral leg muscle activation. Although the control of bilateral leg muscle activation and movement usually takes place at a subconscious level, voluntary interactions with these mechanisms can occur. When an obstacle is visually recognized, for example, appropriate avoidance movements of the relevant leg are voluntarily initiated. These movements are accompanied, however, by an automatic supportive activation of the other leg. This potential for independent voluntary control of the movements of a single leg, as well as the more usual automatically coordinated movements of both legs, requires adequate neuronal mechanisms to achieve a task-directed coupling of bilateral leg muscle activation.

In the spinal cat it has been shown that differences in treadmill speeds between the two sides of the body can produce some adjustments to the speed of each limb (Forssberg & Grillner, 1973; Forssberg, Grillner, & Rossignol, 1976; Grillner, 1981), indicating that the spinal cord contains networks responsible for each limb that can be connected in a variety of ways (Grillner, 1981). Similar observations were recently made for infant walking (Thelen, Ulrich, & Niles, 1986). It has been suggested that the central coordination of this pattern takes place at a spinal level.

Further experiments on interlimb coordination of leg muscle activation have demonstrated that unilateral leg displacement during stance, balancing, and gait evokes a bilateral response pattern, with a similar onset latency on both sides (Dietz & Berger, 1982, 1984). Nevertheless, the contralateral leg muscles are not activated after a unilateral displacement when they are not performing a supportive role (Dietz & Berger, 1984), that is, when they are not connected to a "postural program" (Horak & Nashner, 1986). The observations of *central* regulation of bilateral leg muscle coordination have largely been confined to the activity in the tibialis anterior. In the gastrocnemius, in contrast, *peripheral feedback* mechanisms predominate in the enhancement and modulation of EMG activity after perturbations of stance (Diener, Horak, & Nashner, 1988), during stepping (Dietz, Quintern, & Sillem, 1987; Nashner, 1980), and in running (Dietz *et al.*, 1979). In stepping, it was suggested that the functionally effective stretch reflex response evoked by gait perturbations may be incorporated into a more complex response pattern, itself generated by spinal interneuronal circuits that are closely connected with locomotor centers. This suggestion was based on the di- or triphasic EMG pattern usually induced by the perturbation, a pattern that is assumed to be, to a large extent, centrally programmed (cf. Marsden, Obeso, & Rothwell, 1983).

Further information about the organization and functional significance of interlimb coordination of leg muscle activity was obtained when perturbations were applied during stance on a treadmill with split belts (Dietz, Horstmann, & Berger, 1989a). *Unidirectional bilateral* perturbations were followed by larger EMG responses in both legs, compared to the response to a *unilateral* displacement. For a given acceleration rate, the amplitude of the response in one leg was about equal to the sum of the EMG amplitudes of both the displaced and the nondisplaced legs obtained during unilateral displacements. Opposite results were obtained when the legs were simultaneously displaced in *opposite* directions. EMG responses in both legs were significantly smaller than those obtained after unilateral displacement (Figure 2). Consequently, during *bilateral* displacements, the activity induced by the respective contralateral leg is linearly summed or subtracted, depending on whether the legs are displaced in the same or in opposite directions.

From a functional point of view, this interlimb coordination is necessary to keep the body's center of gravity over the feet. The bilateral activation seen during unilateral displacements produces rapid automatic cocontraction of the nondisplaced leg, thus providing a more stable base from which to compensate for the perturbation. A displacement of the legs in opposite directions causes the body's center of gravity to fall between the legs. Thus, a reduced level of compensatory response is needed to regain body equilibrium in this condition.

This interlimb coordination is thought to be mediated by a central (spinal) mechanism because of the short latencies of the bilateral responses (Dietz & Berger, 1982). A cerebellar contribution to interlimb coordination, via reticulospinal neurons, has been suggested both in cats (Ito, 1984) and in humans (Bonnet *et al.*, 1976).

III. Influence of Gravity

Most investigations of stance and gait have led to the conclusion that the EMG adjustments observed in a particular condition can be related to the demands of equilibrium control. Therefore, keeping the body's center of mass over the feet represents the variable controlled by the neuronal mechanisms. This suggestion is supported by the observation that, depending on the stance condition, muscle stretch does not necessarily result in a compensatory stretch reflex response, but instead results in an antagonistic muscle activation (Gollhofer *et al.*, 1989; Hansen *et al.*, 1988).

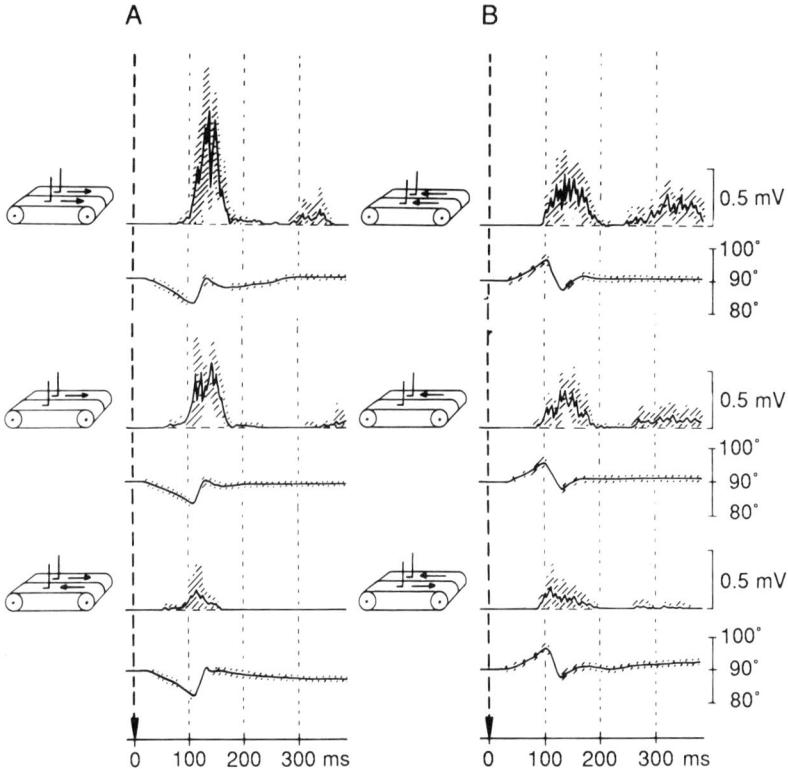

FIGURE 2 Mean values (±S.D., shaded area) of the rectified and averaged ($n=10$) gastrocnemius (A) and tibialis anterior (B) responses of the right leg, together with the movement of the right ankle joint, after backward- (A) and forward-directed (B) accelerations of the treadmill in one subject. Top to bottom: bilateral, unilateral, and opposing perturbations. Arrow indicates the onset of displacement at the ankle joint (velocity signal). (From Dietz et al., 1989a.)

How is the position of the body's center of mass relative to the feet signaled? This question has been neglected in most investigations of postural control and we are just beginning to appreciate the influence that gravity has on sensory and motor behavior. For an appropriate gain control of postural reflexes, peripheral information is required that should signal the influence of "gravity," besides inputs from muscle stretch receptors and the vestibular system. For example, input from Golgi tendon organs was suggested to contribute to the accuracy in sensing trunk position during upright stance (Jakobs, Miller, & Schultz, 1985).

A. Actual Body Mass and Size of EMG Response

The effect of weightlessness, induced by water immersion, on those receptors involved in signaling changes in the position of the body's center of mass with respect to the support surface has been studied (Dietz, Horstmann, Trippel, & Gollhofer, 1989b). The advantage of this particular technique compared with postural reactions during space flights is to leave vestibular function unaltered, while still allowing manipulation of body mass. If a gravity dependence of the compensatory EMG responses exists, then manipulation of the force between the feet and the support platform should affect the responses to destabilizing platform movements.

During immerson, postural reactions were qualitatively similar to those seen under normal conditions, as were those seen during space flights (cf. Clement, Gurfinkel, Lipshits, & Popor, 1985). There was a close relationship between actual body weight and the magnitude of the EMG responses after both backward and forward displacements (Figure 3). Out of water, however, no correlation existed between loading of the subjects and the EMG responses. This saturation of the response

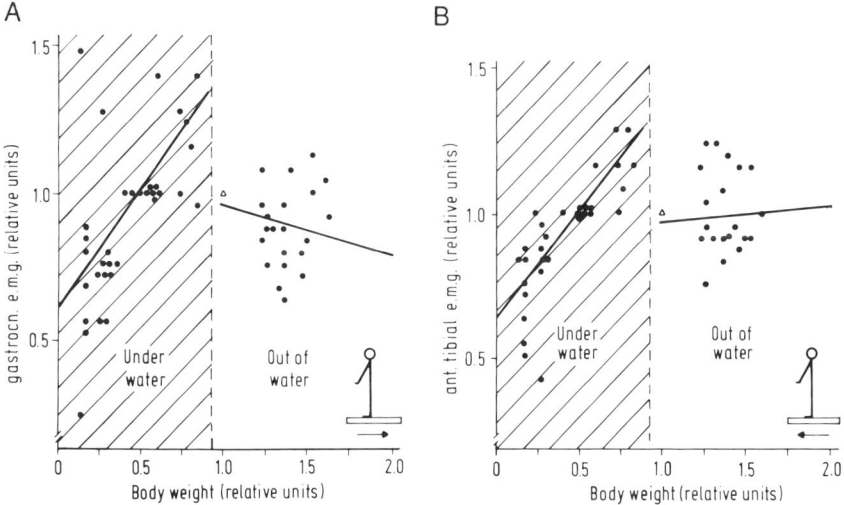

FIGURE 3 Correlative functions between body weight (abscissa) and integrated gastrocnemius (A) and tibialis anterior (B) EMG responses (ordinate) after backward- (A) and forward-directed (B) accelerations of the platform during immersion of 10 subjects. The data from all subjects were normalized with respect to the integrated EMG activity and the body weight to displacements induced unloaded out of water (Δ) and are displayed as scatter diagrams. Regression lines for both conditions, immersion and out of water, are shown. (From Dietz et al., 1989b).

out of water might represent a natural limitation of muscle activation to prevent possible injury (e.g., rupture) of the musculoskeletal system.

From results obtained in the immersion experiments, there is some evidence that the function of those reflexes known to be involved in the stabilization of human posture (e.g., muscle proprioceptive and vestibulospinal reflexes) depends on the activity of receptors within the body that indicate deviations of the body's center of mass from a certain neutral position. In addition, such receptor signals may be important for the selection of the appropriate response pattern, that is, of the centrally programmed pattern. Further studies were designed to define the type and location of the receptor responsible for signaling the position of the body's mass.

B. Significance of Extensor Load Receptors for Stance Regulation

In order to define both the type and behavior of the receptor that signals the projection of the center of the body's mass vector to the feet, rotational and "translational" dorsiflexing displacements were induced during horizontal body posture and also with different loads applied to the body (Dietz, Gollhofer, Kleiber, & Trippel, 1992). It was only during "translational" displacements that different torques were imposed by this loading, which resulted in compensatory activation of the leg extensor muscles. This feature could be studied by placing the subjects' feet 25 cm above the axis of platform rotation, thereby inducing a translational displacement component (see schematic drawings in Figure 4). The rotational impulses were followed by a small, short-latency EMG responses. The "translational" impulses were followed by a stronger gastrocnemius response of longer latency compared to the rotational response, although stretches of similar amplitude and velocity were applied. Application of a load to the body against the movable platform had a significant effect on the magnitude of the long-latency gastrocnemius response following "translational" perturbation, while no significant effect was seen on the short-latency gastrocnemius response following rotational and "translational" perturbations (Figure 4). An influence of knee joint angle on this behavior is rather unlikely, as a similar difference in the EMG responses was seen in the soleus muscle (not shown here).

A different distribution of pressure on the foot sole in the two conditions cannot account for the difference, as anesthetizing the foot does not alter the EMG pattern during upright stance (Diener, Dichgans, Guschlbauer, & Mau, 1984). Biomechanically, the major difference between both impulse modalities represents the effect on the net torque around the ankle joint. Increased loading of the body against the support area augments the torque only in the "translational" displacement

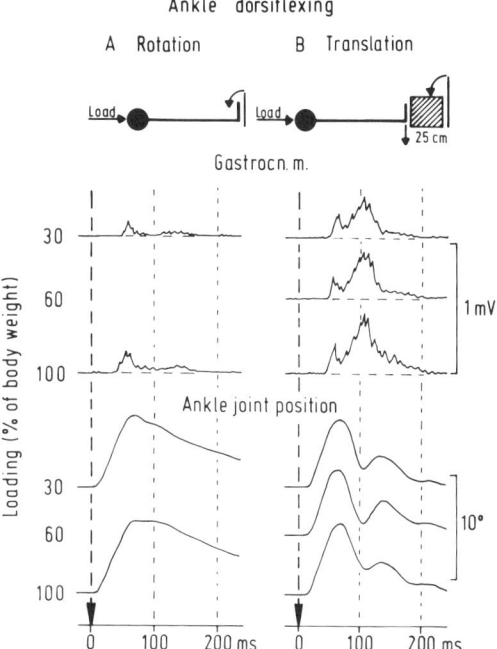

FIGURE 4 Mean of rectified and averaged ($n = 10$) gastrocnemius EMG responses, together with the ankle joint position of 10 supine subjects following dorsiflexing rotation of the platform colinear with the ankle joint (A), and with the ankle joints 25 cm above the rotational axis (B). The subjects were pressed at the shoulders against the movable platform by 30, 60, and 100% of their body weight using elastic bands (with incorporated spring balances) connecting their shoulders with the platform. The line of force created by the bands went through the ankle joints. The schematic illustrations indicate the movements induced by the two impulse modalities. (From Dietz et al., 1992.)

condition. In the rotational condition, however, the torque remains unaffected. Although ankle torque was not recorded directly, it is clear that the magnitude of torque during translational impulses is directly proportional to the amount of the loading according to the formula: $T = L \times r$ (T, torque; L, load; r, translation distance). It is, therefore, possible that load receptors in the leg extensors are mainly responsible for the different EMG patterns, which explains their antigravity function. Such a behavior does not exclude that stretch receptors themselves are also involved in the amplitude modulation of the EMG response and interact with the load receptors within spinal interneuronal circuits (cf. Fetz, Zankowska, Johannisson, & Lipski, 1979). A potentially excitatory function of load receptors has already been described for the extensor muscles of the spinal cat under specific motor conditions (Nichols, 1989; Powers & Binder, 1985a,b), and during fictive locomotion (Duysens & Pearson, 1980; Conway, Hultborn, & Kiehn, 1987). In the latter experiments it was suggested that these receptor signals arise from Golgi tendon organs and are mediated by Ib afferents to the spinal locomotor generator. In the context of stance and gait, this input obviously interacts with a spinal pattern generator and represents the load receptor mechanism responsible for the control of the body's center of mass during stance and gait.

IV. Conclusions

Two basic aspects of the neuronal control of bipedal stance and gait have been discussed, namely the interlimb coordination and the antigravity function of leg extensors. During stance and gait both legs act in a cooperative manner, as each limb affects the strength of muscle activation and the time–space behavior of the other. There are indications that interlimb coordination is mediated by spinal interneuronal circuits, which are themselves under supraspinal (e.g., cerebral and cerebellar) control. Proprioceptive reflexes involved in the maintenance of body equilibrium depend on the presence of contact forces opposing gravity. Extensor load receptors are thought to signal changes of the projection of the body's center of mass with respect to the feet. According to recent observations in the spinal cat, this afferent input probably arises from Golgi tendon organs and represents a newly discovered function of these receptors in the regulation of stance and gait. The interaction of the afferent input from these receptors with other systems involved in postural control is not yet fully understood.

References

Bonnet, M., Gurfinkel, V. S., Lipshits, M. J., Popov, K. E. (1976). Central programming of lower limb muscular activity in the standing man. *Agressologie,* **17B,** 35–42.
Clement, G., Gurfinkel, V. S., Lestienne, F., Lipshits, M. J., & Popov, K. E. (1985). Changes of posture during transient perturbations in microgravity. Aviat. *Space Environmental Medicine,* **56,** 666–671.
Conway, B. A., Hultborn, H., & Kiehn, O. (1987). Proprioceptive input resets central locomotor rhythm in the spinal cat. *Experimental Brain Research,* **68,** 643–656.
Diener, H. C., Dichgans, J., Guschlbauer, B., & Mau, H. (1984). The significance of proprioception on postural stabilization as assessed by ischemia. *Brain Research,* **296,** 103–109.
Diener, H. C., Horak, F. B., & Nashner, L. M. (1988). Influence of stimulus parameters on human postural responses. *Journal of Neurophysiology,* **59,** 1888–1905.
Dietz, V. (1992). Human neuronal control of automatic functional movements: interaction between central programs and afferent input. *Physiological Reviews,* **72,** 33–69.

Dietz, V., & Berger, W. (1982). Spinal coordination of bilateral leg muscle activity during balancing. *Experimental Brain Research*, **47,** 172–176.

Dietz, V., & Berger, W. (1984). Inter-limb coordination of posture in patients with spastic paresis: impaired function of spinal reflexes. *Brain*, **107,** 965–978.

Dietz, V., Schmidtbleicher, D., & Noth, J. (1979). Neuronal mechanisms of human locomotion. *Journal of Neurophysiology*, **42,** 1212–1222.

Dietz, V., Quintern, J., & Sillem, M. (1987). Stumbling reactions in man: significance of proprioceptive and pre-programmed mechanisms. *Journal of Physiology (London)*, **386,** 149–163.

Dietz, V., Horstmann, G. A., & Berger, W. (1989a). Interlimb coordination of leg muscle activation during perturbation of stance in humans. *Journal of Neurophysiology*, **62,** 680–693.

Dietz, V., Horstmann, G. A., Trippel, M., & Gollhofer, A. (1989b). Human postural reflexes and gravity—an under water simulation. *Neuroscience Letters*, **106,** 350–355.

Dietz, V., Trippel, M., Discher, M., & Horstmann, G. A. (1991). Compensation of human stance perturbations: selection of the appropriate electromyographic pattern. *Neuroscience Letters*, **126,** 71–74.

Dietz, V., Gollhofer, A., Kleiber, M., & Trippel, M. (1992). Regulation of bipedal stance: dependence on "load" receptors. *Experimental Brain Research*, **89,** 229–231.

Duysens, J., & Pearson, K. G. (1980). Inhibition of flexor burst generation by loading extensor muscles in walking cats. *Brain Research*, **187,** 321–332.

Fetz, E. E., Jankowska, E., Johannisson, T., & Lipski, J. (1979). Autogenetic inhibition of motoneurones by impulses in group Ia muscle spindle afferents. *Journal of Physiology (London)*, **293,** 173–195.

Forssberg, H., & Grillner, S. (1973). The locomotion of the acute spinal cat injected with clonidine i.v. *Brain Research*, **50,** 184–186.

Forssberg, H., Grillner, S., Rossignol, S. (1976). Interaction between the two hindlimbs of a spinal cat walking on a split belt (Abstract). *Proceedings of the 3rd International Conference on Motor Control*, Albena, Bulgaria, p. 9.

Gollhofer, A., Horstmann, G. A., Berger, W., & Dietz, V. (1989). Compensation of translational and rotational perturbations in human posture: stabilization of the centre of gravity. *Neuroscience Letters*, **105,** 73–78.

Grillner, S.. (1981). Control of locomotion in bipeds, tetrapods and fish. In *Handbook of physiology. The nervous system. Motor control* (Vol. II, part 2) (pp. 1179–1235). Washington D.C.: American Physiological Society.

Hansen, P. D., Woollacott, M. H., & Debu, B. (1988). Postural responses to changing task conditions. *Experimental Brain Research*, **73,** 627–636.

Horak, F. B., & Nashner, L. M. (1986). Central programming of postural movements: adaptation to altered support-surface configurations. *Journal of Neurophysiology*, **55,** 1369–1381.

Ito, M. (1984). *The cerebellum and neural control*. New York: Raven Press.

Jakobs, T., Miller, J. A. A., & Schultz, A. B. (1985). Trunk position sense in the frontal plane. *Experimental Neurology*, **90,** 129–138.

Marsden, C. D., Obeso, J. A., Rothwell, J. C. (1983). The function of the antagonist muscle during fast limb movements in man. *Journal of Physiology (London)*, **335,** 1–13.

Nardone, A., Giordans, A., Corra, T., & Schieppati, M. (1990). Responses of leg muscles in humans displaced while standing. Effects of types of perturbation and of postural set. *Brain*, **113,** 65–84.

Nashner, L. M. (1976). Adapting reflexes controlling the human posture. *Experimental Brain Research*, **26,** 59–72.

Nashner, L. M. (1980). Balance adjustments of humans perturbed while walking. *Journal of Neurophysiology*, **44,** 650–664.

Nashner, L. M., Woollacott, M., & Tuma, G. (1979). Organization of rapid response to postural and locomotor-like perturbations of standing man. *Experimental Brain Research*, **36,** 463–478.

Nichols, T. R. (1989). The organization of heterogenic reflexes among muscles crossing the ankle joint in the decerebrate cat. *Journal of Physiology (London)*, **410,** 463–477.

Powers, R. K., & Binder, M. D. (1985a). Distribution of oligosynaptic group I input to the cat medial gastrocnemius motoneuron pool. *Journal of Neurophysiology*, **53,** 497–517.

Powers, R. K., & Binder, M. D. (1985b). Determination of afferent fibres mediating oligosynaptic group I input to cat medial gastrocnemius motoneurons *Journal of Neurophysiology,* **53,** 518–529.

Thelen, E., Ulrich, D. B., & Niles, D. (1986). Bilateral coordination of stepping movements of 7-month-old human infants on a split-belt treadmill. *Society of Neuroscience Abstracts,* **12,** 881.

9

Are There Unifying Structures in the Brain Responsible for Interlimb Coordination?

Mario Wiesendanger, Urs Wicki, and Eric Rouiller

Institut de Physiologie
Université de Fribourg
Fribourg, Switzerland

I. Introduction
II. Cortical Areas of the Human Brain Involved in Bimanual Coordination
    A. The Premotor Cortex (Lateral Area 6 of Brodmann)
    B. The Supplementary Motor Area (Medial Area 6 of Brodmann)
    C. The Parietal Association Cortex
    D. The Role of Brain Commissures
    E. What Conclusions Can Be Drawn about the Specific Involvement of Cortical Areas in Coordinating Bimanual Movements?
III. The Role of Sensory Guidance for Bimanual Coordination
IV. Some Contributions from Experiments in Subhuman Primates on the Problem of Bimanual Coordination
    A. Split-Brain Monkeys
    B. Evidence for Bilateral Controls Exerted from Each Hemisphere
    C. Does the Activity of Cortical Neurones Reveal Aspects of Goal Coding?

V. The Role of the SMA in Monkeys Tested in a Natural and Highly Coordinated Bimanual Movement Sequence
  A. Background
  B. Aim of Study
  C. Demonstration of a Goal-Related Temporal Invariance of the Bimanual Movement Sequence
  D. Effects of Serial Lesions of Right and Left SMA on Bimanual Tasks
VI. General Conclusion
  References

I. Introduction

Many human skills are performed with the cooperation of both hands. The left hand is more often typically postural, while the right hand holds and manipulates objects. The left hand not only stabilizes the object acted upon by the right hand, but also provides a positional reference for the manipulative hand. In bimanual performances such as, for example, eating with a fork and knife, knitting, and dressing, the precision, both spatial and temporal, with which both hands are coordinated is most impressive; we think of such a performance as a unitary goal-directed act. Early this century in the neurological literature, the question came up as to the existence of brain structures dealing with coordination. In particular, it was proposed that for "higher units of performance" (Welford, 1968), plans of action must be constructed in the brain, well in advance of movement initiation, that concern the objective of the act rather than detailed specifications in its implementation. These ideas were born out of neurological observations of patients whose motor deficits were not caused by weakness or altered muscle tone, but were associated with the elaboration of "higher units of performance," as, for example, dressing. Since Liepmann (1920), these higher order deficits are termed "apraxia" or "dyspraxia," of which a variety has been described. It soon became clear that the brain lesions causing apraxias are in most cases localized in the association cortex. It is interesting to note that von Monakow (1914), in his critical treatise on localization of cortical functions, concluded that the pathology of apraxias was complex and could involve many different areas outside the Rolandic cortex. Thus, the idea of a hierarchically superior planning phase (*Bewegungsentwurf*) that involved widely distributed association areas was established in the first quarter of the century and, in essence, remains unchanged to the present day, as illustrated in diagrams of more recent origin (e.g., Paillard, 1982).

Quite revealing, also, is the concept of *motor equivalence* that arose from work in experimental psychology. Lashley (1930) was struck by how animals, when part of their motor apparatus had been incapacitated, immediately shifted to the use of intact parts of the motor apparatus in order to achieve the same goal. As a particularly significant example of a "higher unit of performance," he mentioned how patients with a paresis of fingers started to write with the whole arm, "or even with the pencil held in the teeth preserving the characteristics of individual chirography." Perhaps the most illuminating example of motor equivalence concerns the special case of semantic expression, normally achieved by the sound-producing orolaryngeal muscles. As shown by Poizner, Klima, and Bellugi (1987), American hand signers who suffered from a stroke that damaged their language area became "aphasic" for sign language, without any other deficit in hand use. The concept of "motor equivalence" has seen a revival in the literature of motor psychophysics because it underlines the notion that the plan of goal-directed movement is supraordinate to a lower level specification of trajectory formation and selection of muscles. The motor equivalence concept is also akin to Bernstein's ideas of variable means that lead to invariant ends or goals (Bernstein, 1967; see also Abbs & Cole, 1987). Various terms have been used to convey the notion of a high-level, goal-specifying program: "coordinative structure" (Bernstein, 1967), "generalized program" (Schmidt, 1988), "generalized schemata" (Lashley, 1951), and "coordinated central program" (Arbib, 1990).

The purpose of this chapter is to review the evidence of the existence of such high-level programs, functioning as localized "unifying structures" for bimanual movements, and in particular to restate the question in terms of localization. The fact that a lesion in Broca's speech area can produce a "hand aphasia" in hand signers begs the question of whether other cortical association areas exist that, for example, interfere specifically with bimanual coordination in purposeful acts without impairment of unimanual performance. Thus, our first approach is to review the evidence for such "coordinative structures." We shall next proceed to the question of whether electrophysiological observations reveal high-level aspects of goal coding, as one would expect to be expressed in a coordinative structure.

"Coordination" concerns muscles, joints, limbs, and the whole body and occurs at many different levels of the neuraxis. This chapter is mainly concerned with natural, goal-directed movements and in particular with cooperative actions of both hands. Accordingly, interesting observations on central oscillators considered to be important for temporal coupling in rhythmic movements of the limbs are neglected. Also omitted is a discussion of the important roles of the basal ganglia, the cerebellum, and the brainstem in coordination. Finally, it is emphasized that we are mainly concerned with lesion deficits (and hypothetical control systems) affecting specifically the cooperation of both hands, and not those deficits explicable by a weakness or dyscoordination of a single hand. As a "bimanual skill," we understand cooperative actions of both hands toward a unified end result (see also Ettlinger & Morton, 1963; Wyke, 1971).

II. Cortical Areas of the Human Brain Involved in Bimanual Coordination

A. The Premotor Cortex (Lateral Area 6 of Brodmann)

Clinical descriptions of premotor (PM) lesions have been reviewed recently (Wiesendanger, 1981; Freund, 1987) and only a few points, relevant to the question of interlimb coordination, will be addressed. In general, the effect of PM lesions that do not encroach on the precentral (primary) motor cortex (MI) are not conspicuous on clinical testing and the disturbances tend to disappear some weeks or months after the lesion. The main observation made by Foerster (1936) on the basis of excision of "area $6a\beta$" (part of PM, but possibly also including some cases with lesions on the mesial surface of area $6a\beta$) in 40 cases was that a natural movement sequence tended to be decomposed in the elemental movements, suggestive of a temporal uncoupling of the elemental movements being executed one-by-one, so that the unity of the motor act is broken down. The main conclusion on the PM deficit in the cases of Freund and Hummelsheim (1984) was that it concerns mainly proximal movement coordination. The deficit was only revealed when the patients performed movements requiring both arms for complex movements. Thus, while forward rotation of one or both arms in the shoulder were possible, "the dysfunction became most pronounced when the patient was asked to produce a windmill movement forward or backward but with both arms in alternating manners. The movement degraded and decomposed from the very beginning." Patients with unsupported arms were not able to catch a ball with both hands. The syndrome was considered to represent a "limb-kinetic apraxia that becomes apparent only during attempts at *proximal bilateral coordinated movements*" (italics added). A closer investigation, including a quantitative analysis of the temporospatial structure of the bilateral movement sequences, of such interesting (but relatively rare) cases remains to be done.

B. The Supplementary Motor Area (Medial Area 6 of Brodmann)

Since the early period of investigations of the supplementary motor area (SMA), this area was considered to function bilaterally (Penfield & Jasper, 1954; Travis, 1955). This was based on the observation that prolonged surface stimulation produced bilateral motor effects, especially of the trunk. More recent demonstrations indicating that the human SMA is strongly involved in movement preparation (Deecke & Kornhuber, 1978; Lang, Cheyne, Kristeva, Beisteiner, Lindinger, & Deecke, 1991) and in the programming of movement "subroutines" (Roland, Steinhøj, Lassen, & Larsen, 1980) renewed the interest in lesion effects (for review, see Goldberg, 1985; Roland, 1984; Wiesendanger, 1981, 1986; Freund, 1987). The clinical reports are quite varied, from minimal symptoms to dramatic deficits, which may partly be explained by the fact that, as with more lateral lesions of area 6, the deficits are transient and rapidly changing. Furthermore, lesions were mostly caused by circulatory disturbances of the anterior cerebral artery, resulting in lesioned territories not confined to the SMA (Figure 1).

A number of reports deal with bimanual deficits believed to be caused by an SMA dysfunction. Luria (1966) briefly mentions (without providing documentation) that patients with medial frontal lesions had troubles in carrying out two different things with both hands. Laplane, Talairach, Meininger, Bancaud, and Orgogozo (1977) found that "alternating serial movements of both hands remained frankly awkward," while unimanual serial movements were normal (3 cases with SMA cortectomy). Bimanual reaction times were significantly longer than unimanual reaction times in a patient with a lesion affecting both the SMA and the anterior cingulate gyrus (Meador, Watson, Bowers, & Heilman, 1986). Dick, Benecke, Rothwell, Day, and Marsden (1986) described a patient with an SMA lesion whose deficit consisted of a lack for smooth execution of movement sequences, that is, a decomposition of the performance, as seen also in cerebellar patients. This disturbance was seen in unimanual as well as in bimanual actions.

Viallet, Massion, Massarino, and Khalil (1992) found that anticipatory postural adjustments in a bimanual load-lifting task were delayed or absent when the SMA lesion was contralateral to the load-bearing hand. Similar findings with lesions of the medial frontal cortex were reported previously by Gurfinkel and Elner (1988).

Deecke and co-workers have recently proposed, on the basis of slow movement potential data as well as on lesion data, that the SMA has a *timing function*. Their main evidence (recently reviewed by Kornhuber, Deecke, Lang, Lang, & Kornhuber, 1989; see also, Lang, Obrig, Lindinger, Cheyne, & Deecke, 1990, Lang, Beisteiner, Lindinger & Deecke, 1992) is as follows: (1) the *Bereitschaftspotential* (readiness potential)

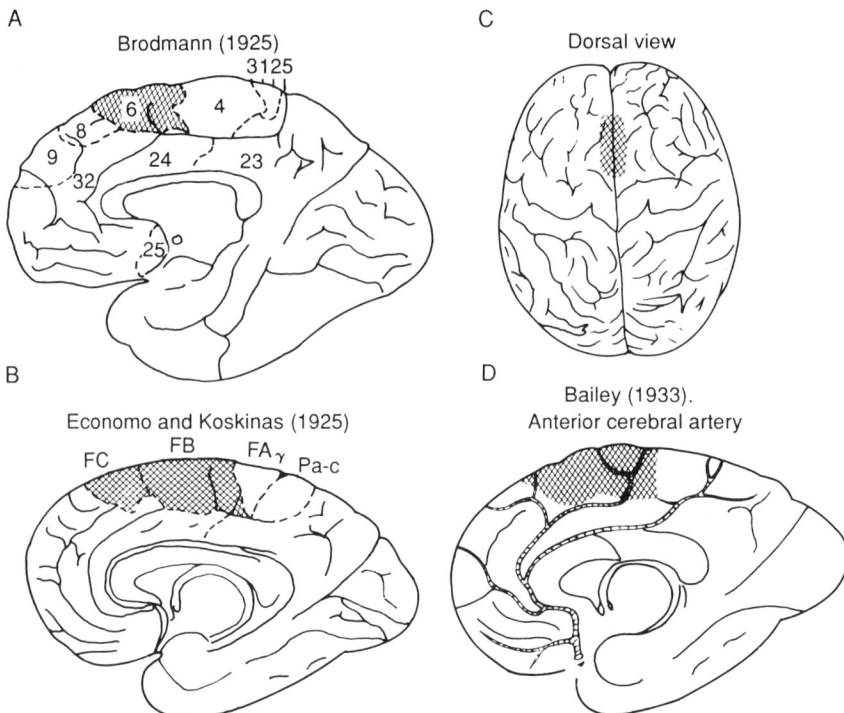

FIGURE 1 The maps (A), (B), and (C) show the relation of the SMA, defined as mesial area 6 in the terminology of Brodmann (1925) (A, C) or as FB and FC in the terminology of Economo and Koskinas (1925) (B), in relation to the distribution territory of the anterior cerebral artery (Bailey, 1933; D). In reports on human "SMA-lesions," occurring with cerebrovascular accidents of the anterior cerebral artery, the lesioned territory was often extensive, transgressing the cytoarchitectonically defined SMA, even if the larger combined areas of FB and FC were considered. From Wiesendanger (1986) with permission.

has its earliest manifestation in scalp leads overlaying the SMA; (2) in bimanual rhythmic finger movements of varying difficulties, it was shown that the most difficult task, sequential index finger movements in opposite directions, produced a large and sustained negative direct current shift in recordings over the medial frontal cortex and that the Bereitschaftspotential at the cortex was significantly larger in the sequential, as compared to the simultaneous, task. The interpretation is that "the starting signal to move the index finger of one side can only be given when control mechanisms ensure that a movement of the other side is inhibited," and that "the SMA must have information that contralateral inhibition is settled when giving the starting signal." In line with this, patients with old, unilateral lesions of the SMA had more trouble with the hand contralateral to the lesion in performing the sequential task. However, as a note of caution, it is appropriate to mention, first, that readiness potentials (Neshige et al., 1988) and EEG desynchronization (Pfurtscheller & Berghold, 1989) have been found to precede movement onset by more than one second also in the hand focus of MI, and second, that decomposition of sequential tasks was found also after cerebellar cortex lesions (Inhoff, Diener, Rafal, & Ivry, 1989), parietal cortex lesions (Hécaen, 1978), and PM cortex lesions (Freund & Hummelsheim, 1984).

More complex disturbances of bimanual performances have also been described in cases of medial frontal lesions, including the SMA. Goldberg, Mayer, and Toglia (1981) first reported on two patients with an SMA lesion who presented an "alien hand syndrome." This syndrome was first described by Brion and Jedynak (1972) in 3 patients with tumors of the corpus callosum and was interpreted as a callosal disconnection syndrome. These patients carry out seemingly natural movements with the affected arm (e.g., reach out and grasp objects) but without being motivated to do these movements and even unable to suppress the movements, as if the movement escaped their will. Interestingly, one of the 2 patients of Goldberg et al. (1981) also suffered from "intermanual conflict," whereby the affected hand interfered with the performance of the contralateral limb. Both the alien hand and intermanual conflict have been described as typical signs occurring in split-brain patients for a short period after callosal section (cf. Section II,D). Goldberg and Bloom (1990) recently published four more cases of alien hand syndrome with lesions in the medial frontal cortex. One patient with a right-sided infarction of the SMA developed mirror movements and had particular difficulty in performing nonsymmetrical movements such as threading a needle (Chan & Ross, 1988). The question is, Why do not all patients with an SMA lesion develop these most striking syndromes of the "alien hand" (AH), "intermanual conflict" (IC), and mirror movements (MM)? With a literature search covering the last 5 years, we were able to find 16 AH, 13 MM, and 7 IC cases (3 of these latter cases also had an AH sign). In 5 of 16 AH-cases, the corpus callosum seemed to have been the dominant or exclusive lesion. The other 11 AH-cases all had medial frontal pathology, typically produced by an accident of the anterior cerebral artery. Of the 7 IC cases, all had corpus callosum pathology and 2 of these had additional lesions of the medial frontal cortex. Out of 13 MM cases, only that reported by Chan and Ross (1988) had a medial frontal lesion. All other cases were of the congenital type.

It thus appears that the medial frontal cortex (a large zone including the SMA) is a critical structure for coordinating both arms for a unified cooperative and purposeful motor act. However, the exact territory that contributes to this control is not clear since most lesions are complex, comprising portions of the medial cortex extending beyond the SMA. In addition, it is also not clear whether some involvement of the corpus callosum is necessary for the expression of this high-level intermanual coordination.

C. The Parietal Association Cortex

Lesions of the parietal cortex not only lead to a perceptual dysfunction but to "major disturbances of motor functions of the hand," as revealed by objective testing of hand function (see Pause, Kunesch, Binkofski, & Freund, 1989 for a recent account). Thus, trajectory formation during active exploration was grossly abnormal (irregularities, smaller work space, paucity of movements). The grasp component of a prehension task was also abnormal, with failures to adapt the grip force exactly to the load and surface friction. It was concluded that "the essence of the motor disturbance of the hand in parietal lobe disease lies in the disturbance of the conception and generation of the movement patterns necessary to bring those receptors into action which would normally provide the information required for the identification and manipulation of objects." Severe disturbances of the grasp phase in a prehension task were also described by Jeannerod (1986, 1988). These motor disturbances seem to be restricted to the contralateral limb, but is there also evidence for a deficit in bimanual operations? No detailed and quantitative motor studies are available to answer this question. However, a large body of clinical descriptions of apractic disorders, especially with lesions in the dominant hemisphere, points to an interference with the proper execution of typical bimanual acts such as dressing or object construction. This may hold not only for the parietal, but also for other association areas, particularly when learning new bimanual tasks (Wyke, 1971). A decomposition of the synergistic relationships between the two hands is probably more the rule than the exception [see, for example, the analysis of Hécaen (1978) on 249 cases of "ideomotor apraxia" with unilateral lesions and bilateral symptomatology].

D. The Role of Brain Commissures

As an alternative to a "high-level unifying structure" that might control the motor apparatus bilaterally, as suggested in the previous sections for the premotor and posterior association cortex, the two hemispheres might be coordinated essentially via the commissures, in humans mainly via the corpus callosum. Surprisingly, split-brain patients have been less well investigated from the point of view of movement execution than from that of perception. Although it was regularly found that bimanual skills like "buttoning clothes, tying shoelaces, shuffling cards, riding a bicycle, etc. are not significantly influenced by commissurotomy," learning of new and difficult bimanual tasks was found to be almost impossible in split-brain patients (Preilowski, 1990). In an early postoperative stage, the situation is different, however. As noted by Sperry (1966), "lack of coordination in activities that require close cooperation between the hands was also an early complaint . . . ," and "voluntary control of left arm and hand was much more severely affected than the right in the early weeks after surgery." In fact, the dramatic disturbances, alluded to in the section on SMA lesions, were first described in split-brain patients: both "intermanual conflict" and the "alien hand sign" were considered typical for the callosal disconnection syndrome, expressed in the early postoperative period (cf. Bogen, 1985, for critical review). It is not clear to what extent the SMA or other medial frontal cortical regions contributes to the syndromes. However, it appears that only a minority of reported SMA cases developed this bimanual disorder. Although it is difficult to appreciate the exact extent of the lesions, it turns out that, in general, the lesions leading to the deficit were complex, concerning, for example, the entire territory of the anterior cerebral artery (Figure 1).

E. What Conclusions Can Be Drawn about the Specific Involvement of Cortical Areas in Coordinating Bimanual Movements?

There can be little doubt that the cerebral cortex, through its many intracortical communication lines and via long loops through the cerebellum and basal ganglia, has a large share in the coordination of goal-directed bimanual motor acts. A decomposition of synergistic relationships of purposeful movement sequences, including the coordination of both hands, is a typical occurrence in lesions of cortical association areas. In the previous sections, the evidence for a critical involvement of PM and parietal association areas as well as the hemispheric commissures for bimanual actions was reviewed. Relatively recent evidence points to the medial frontal cortex as an interesting and crucial structure for subjecting movements of both extremities under a unified goal. To what extent the anterior corpus callosum, the SMA, the adjacent medial prefrontal, and cingulate cortices contribute to this intermanual coordination must be further clarified. Since the number of well-investigated cases is still small, it may be premature to attribute too much emphasis on the role of any of these medial structures. It has long been known that many types of cortical lesions can affect the score of bimanual movements. Thus, bimanual disorders were described with lesions in frontal association cortex, in lateral premotor cortex, and, frequently, in parietal association cortex (e.g., Wyke, 1971; Hausmanowa-Petruscewicz, 1959; Leonard, Milner, & Jones, 1988), or in Alzheimer patients with more widespread cortical dysfunction. Unfortunately, objective spatiotemporal measures of bimanual coordination has thus far rarely been done in these patients.

The following features are common to all types of lesions affecting bimanual coordination: (1) asymmetric or out-of-phase bilateral movements are more seriously disturbed than the easier in-phase, repetitive, bilateral movements; (2) natural, everyday skills are better preserved than new or "abstract" bimanual skills; (3) bimanual disorders occurring after acute lesions tend to recover within weeks or months.

In spite of some fascinating observations of bimanual incoordination with cortical lesions, we are led to the conclusion that the hypothesis of a single cortical superarea functioning as a "unifying structure" for the proper execution of a bimanual task is, on the basis of current evidence, untenable. However, taken together, the reviewed clinical literature supports the notion, already forwarded by von Monokow (1914), that widely distributed cortical association areas, as an interconnected ensemble, cooperatively contribute to the necessary perceptual integration and its translation into a bimanual motor act. The contribution of different sets of cortical areas and subareas largely depends on the context in which a bimanual act is achieved. These ideas are not new (see, for example, Arbib, 1990), but they are different from the long-held view prevailing in clinical thinking of one-way traffic with a strict (structural) hierarchy (for entertaining reading, see Kelso & Tuller, 1981), and they are more in line with current attempts (Jordan, 1990) to understand how neural networks might implement the theories of "generalized programs" and of "schemata" (Schmidt, 1988; Arbib, 1990). Perhaps a central issue is that the postulated generalized programs ". . . are largely task-driven, reflecting the intention-related goals" (Arbib, 1990).

III. The Role of Sensory Guidance for Bimanual Coordination

As an alternative to the concept of a "coordinative structure," or as an adjunct to it, goal achievement in a bimanual task may largely depend on sensory guidance. There is indeed much recent work available, suggesting that in reach-and-grasp tasks visual and proprioceptive information is important, especially for the fine tuning of grasping (for recent treatment, see Jeannerod, 1988). Sensory inputs may not be used only for an *ad hoc* guidance of each movement in a reactive mode, but also proactively on the basis of perceptual cues. A general answer to the role of sensory guidance is not possible because the same bimanual skill may not depend on sensory information, or may require it in an obligatory fashion (e.g., when the object to be grasped is not stationary, or in an early learning phase). The problem of how much sensory input is needed for goal achievement in a given task is classically tested by removing the capability for sight from the work space and by interfering with somatosensory perturbations, such as vibration, or by altering loads. One unexplored field of research is the study of such perturbations in bimanual tasks; it would be particularly revealing to see how unilateral perturbations might affect the nonperturbed limb and how the temporal and/or spatial invariances may be disturbed if perturbations are unpredictive, and how compensation is achieved in a predictive situation.

IV. Some Contributions from Experiments in Subhuman Primates on the Problem of Bimanual Coordination

A. Split-Brain Monkeys

Ettlinger and Morton (1963) trained monkeys to perform a relatively difficult bimanual skill and measured the time course of relearning the skill to criterion following callosal section. Out of 5 animals, 4 required much fewer trials to relearn the task. On this basis, the authors concluded ". . . that the callosum is not involved in the learned coordination in the performance of the bimanual skills." However, one of the monkeys never did regain criterion of performance after many postoperative trials. Mark and Sperry (1968) noticed that only a combined lesion of both the corpus callosum and the midbrain commissure permanently affected the proper execution of a bimanual task if no vision was allowed. It was concluded that the deficit was caused by the lack of perceptual visual cues, since with vision the task was only transiently perturbed after commissurotomy. As suggested by Glickstein (1990), visual information may still be distributed to the motor centers of both hemispheres via the ponto-cerebellar-cortical route with its contralateral and ipsilateral distribution (crossed and uncrossed pontocerebellar projection). Additional coordination might result from uncrossed corticospinal projections (cf. Section IV,B and Kuypers & Brinkman, 1970; Liu & Chambers, 1964). As in human split-brain patients, some deficits in the early postoperative phase were noticed also when vision was allowed.

In elegant experiments, Brinkman and Kuypers (1973) trained split-brain monkeys (including division of the optic chiasma) to grasp for food morsels from slots of different orientation while one eye was occluded. In accord with anatomical findings of Kuypers and Brinkman (1970) that the ipsilateral corticospinal projection is directed to motoneurones of proximal limb and trunk muscles, Brinkman and Kuypers (1973) discovered that grasping, but not reaching, was grossly impaired when visual information was restricted to the hemisphere ipsilateral to the task hand. This was taken to indicate that an ipsilateral motor control is sufficient for the essentially proximal reach component of the task, but not for the fine control of distal grasping.

A severe impairment was found in split-brain monkeys when tested in a between-hand choice reaction time (Guiard & Requin, 1978). It was proposed that in this task, the monkeys have to quickly switch attention from one to the other hemisphere. This is similar to Kinsbourne's (1974) proposition that the commissures are "an assisting device, serving a stabilizing function in the interhemispheric distribution of attentional energy."

B. Evidence for Bilateral Controls Exerted from Each Hemisphere

It has long been known that motor cortical areas control trunk and proximal muscles of both sides (Liu & Chambers, 1964). However, it has been found that distal muscles are controlled exclusively from the contralateral hemisphere (Kuypers & Brinkman, 1970) and that callosal interconnection of the two hand representations in the precentral cortex is sparse (Jones, 1986). On the other hand, the entire SMA has a callosal projection to the contralateral SMA as well as to the contralateral MI (Künzle, 1978). In some recent experiments in monkeys, islands of neurones were discovered in MI that address distal hand muscles bilaterally (Aizawa, Mushiake, Inase, & Tanji, 1990). These authors speculated that "in this efferent zone between the traditional face and digit areas of MI is a site utilized for the execution of bilaterally organized hand movements." Further, neurones engaged in ipsilateral or bilateral movements have been found in the MI (Matsunami & Hamada, 1981), in the SMA (Brinkman & Porter, 1979; Tanji & Kurata, 1981; Tanji, Okano, & Sato, 1988), and in the inferolateral portion of the PM (Gentilucci, Fogassi, Luppino, Matelli, Camarda, & Rizzolatti, 1988). Cortical neurones may also address ipsilateral spinal neurons via brainstem motor centers shown to distribute their descending fibers bilaterally to interneuronal nets in the spinal cord (Kuypers, 1981). Alternatively, propriospinal neurones receiving cortical inputs (Alstermark, 1983) may distribute their actions to both sides of the spinal cord. A region around the superior precentral dimple of the monkey motor cortex was found to be particularly effective in producing ipsilateral limb movements upon repetitive electrical stimulation (Bucy & Fulton, 1933). Taken together, this anatomical and physiological evidence for a bilateral control of each motor cortical area may explain why callosal lesions and unilateral cortical lesions tend to produce only transient disturbances.

It was also argued that these bilateral controls exerted from each hemisphere are important for the execution of simultaneous actions (Kelso, Southard, & Goodman, 1979, 1983; Wing, 1982). Marteniuk et al. (1984) termed this tendency of moving simultaneously and in phase with both hands "assimilation effect." But these authors also found marked departures from synchronicity. Therefore, they consider the bilateral controls as facilitating the synchronization. They made the interesting speculation that ". . . perhaps the development and learning of bimanual skills involves the *elimination (insulation) or incorporation of neural cross talk,* depending on the task requirements" (italics added).

C. Does the Activity of Cortical Neurones Reveal Aspects of Goal Coding?

There are indeed interesting examples clearly indicating that the activity of populations of cortical neurones reflects some aspect of the task rather than muscle specifity. A well-known example is from populations of neurones in the MI coding movement direction with activity of single units broadly tuned to a preferred movement direction (Georgopoulos, 1988). The main conclusion was that the same (or similar) population of precentral neurones may be brought into play for reaching movements in various directions, however, with changed weighting of each cell's contribution, resulting in the desired appropriate net direction vector of the population. This holds perhaps mainly for neurones controlling more proximal muscles used for directing the whole arm in space. Caminiti and associates (1990) recently showed that direction-dependent population vectors were present also when the same trajectories were performed in three different work spaces (requiring a horizontal rotation of the shoulder), whereby the spatial orientation changed following the rotation of the shoulder. However, the population vector still provided a good description of movement direction, whether performed in the left, center, or right work space. These authors concluded that motor cortical neurones combine two signals, one for arm orientation in space and the other for movement direction. This observation is interesting in view of a converging result by the work of Soechting and Flanders (1992), suggesting that visual reaching in space is effected within a shoulder-centered coordination system.

It is significant that a similar population coding of movement direction was found also in the PM (Caminiti, Johnson, Galli, Ferraina, & Burnod, 1991), area 5 of the parietal association cortex (Kalaska,

Caminiti, & Georgopoulos, *et al.*, 1983), and in the cerebellum (Fortier, Kalaska, & Smith, 1989), thus revealing again the distributed nature of movement coding.

Intuitively one may consider coding aspects other than the shoulder orientation in space and movement direction, depending on the motor task. In fact, an increasing number of observations are in line with the general idea that the encoding of cortical neurones is task or goal dependent rather than muscle specific. Ingenious task requirements make it possible to dissociate actual movements from target cues. Thus, Alexander & Crutcher (1990) were able to distinguish neurones with activity changes selectively dependent on the target cue rather than on the forthcoming tracking movement. This was evident from trials during which the cursor moved in an opposite direction from the limb; in order to catch the target with the cursor, the monkey learned to move his arm in the opposite direction. During the instructed delay period, more than a third of the tested neurones exhibited target-related activity. This proportion of goal-associated activity patterns was about the same in three different structures tested in the same monkeys: MI, SMA, and putamen, revealing once more the distributed higher order coding of central neurones. Further examples of task-related associations with neuronal activity have been discovered: trajectory as well as target location for MI, SMA, dorsal PM, and ventral PM neurones (Hocherman & Wise, 1991); goal-directed hand movements as well as saccades (Mann, Thau, & Schiller, 1988); specific sequences of tracking movements for SMA neurones (Mushiake, Inase, & Tanji, 1990); reaching and/or grasping for neurones in inferior premotor cortex (Gentilucci *et al.*, 1988; Rizzolatti *et al.*, 1988); in an area rostral to the SMA (Rizzolatti, Camarda, Fogassi, Gentilucci, Luppino, & Matelli, 1988; Rizzolatti, Gentilucci, Camarda, Gallese, Luppino, Matelli, & Fogassi, 1990); and for neurones in parietal association cortex (Hyvärinen, 1982; Taira, Mine, Georgopoulos, Murata, & Sakata, 1990). Within the MI hand area, activity of pyramidal tract neurones may be associated specifically with the precision grip, but not, or much less, with the power grip (Muir & Lemon, 1983). In area 2 of somatosensory cortex (SI), neuronal activity may show a complex association with a particular natural hand manipulation (Iwamura, Tanaka, Sakamoto, & Hikosaka, 1985). These examples are not comprehensive, but they allow one to draw a number of conclusions that, together, seem to converge with some conclusions from lesion studies: brain processes involved in the preparation of a goal-directed movement, and that include numerous aspects of sensory and motor sets, are obviously distributed in many structures of the CNS. The above examples in no way exclude the possibility of a division of labor and serial processing in movement organization.

With respect to bimanual tasks, the findings of neurones in the SMA that were specifically involved in bimanual, but not only right or left hand movements, are most exciting (Tanji *et al.*, 1988). It remains to be seen whether the bimanual neurones found in a lateral PM area are also specifically engaged in a bimanual but not in a unimanual task.

V. The Role of the SMA in Monkeys Tested in a Natural and Highly Coordinated Bimanual Movement Sequence

A. Background

Previous reports indicate that the SMA might be a "hot" structure playing a role in bimanual coordination. Apart from clinical studies discussed above (cf. Section III,B), some interesting experimental results relevant for the problem of bimanual movements have been described (cf. also, Wiesendanger, 1986; Wiesendanger, Corboz, Hyland, Palmeri, Maier, Wicki, & Rouiller, 1992). Brinkman (1984), on the basis of findings that SMA neuronal activity of each hemisphere was related to right and left whole-arm movements (Brinkman & Porter, 1979), trained monkeys in a bimanual task that consisted of retrieving food morsels wedged into holes of a perspex plate. In order to be successful, the monkeys had to push the food morsel out of a hole with one finger and catch the bait with the other hand beneath the plate (a task introduced by Mark & Sperry, 1968, for studying bimanual coordination in split-brain monkeys). After a unilateral SMA lesion, the monkeys were only partly successful because "the two hands tended to behave in a similar manner instead of sharing the task between them." This clumsiness in the bimanual task persisted over a long period. When the corpus callosum was sectioned in addition to the SMA lesion, the bimanual incoordination disappeared. Also, the bimanual deficit did not occur if both SMAs were lesioned simultaneously (Brinkman, 1983). This is surprising, since, from the neurological literaure, one would expect the most dramatic effect if both the SMA and the corpus callosum were damaged (see Section II,E). Nevertheless, Porter (1990) drew "parallels between the disturbance in movement performance, and in particular the clumsiness in bimanual coordination in the monkey experiments, with modern accounts of the dysfunction experienced by patients with unilateral lesions of midline cortex involving the superior frontal gyrus and presumably damaging the SMA in humans." It was postulated that the commands about sequential actions from the SMA are available to the primary motor cortex of both sides, and that in the absence of this information, "one hand does not know what the other is doing and clumsiness in bimanual coordination will result." But, according to

Brinkman (1983), "the SMA is not essential for the performance of a bimanual coordination task."

B. Aim of Study

We have developed a bimanual task in order to learn more about the possible role of the SMA in controlling the precise temporal and spatial coordination in this task (Wiesendanger et al., 1992). Two monkeys were subjected to SMA lesions. For this comparison, we trained one of these monkeys also to perform on the manipulandum used by Brinkman (1984). The question was whether quantitative evidence can be found for the concept that the SMA serves as a "unifying" structure for bimanual coordination. The outcome of these chronic experiments is currently analyzed in detail and will be published elsewhere. A few relevant data will be described only briefly.

C. Demonstration of a Goal-Related Temporal Invariance of the Bimanual Movement Sequence

Figure 2 illustrates the pull-and-grasp synergy. All monkeys tested were highly motivated to retrieve the food morsels and performed about 100 trials daily. The entire sequence lasted about 1.5 sec. Each trial was initiated by the opening of two sliding windows through which the two arms could reach through a panel with the baited drawer now accessible to the monkey. The animal reached with the left hand for the spring-loaded drawer, pulled it and held it open, while the right hand picked up the small food morsel using a precision grip. The trial was terminated when the left hand withdrew again behind the panel. The following events were signaled by means of sensors to the computer: passage through left and right windows ("movement onset" left and right); touch drawer knob with left hand; completely open drawer; pick-in and pick-out with right hand; withdraw with right and left hands behind windows (trial end). The drawer position was provided by a linear potentiometer. In selected sessions, rectified and filtered (0.1–2 kHz) EMG of right and left biceps and triceps muscles was recorded and digitized at 500 Hz. In two monkeys, kinematic data were recorded [two dimensional (2D), 100 Hz] on an ELITE movement analysis system (Ferrigno & Pedotti, 1985) for reconstructing trajectories and time-dependent plots of position and its time derivative. All these data were obtained when visual information was accessible or in total darkness. The main result of the investigation was that, in spite of the pronounced trial-by-trial variability, in space as well as in time, the time the drawer was fully opened precisely coincided with the shaped hand reaching the food well ("pick-in") in order to remove the food morsel with the pre-

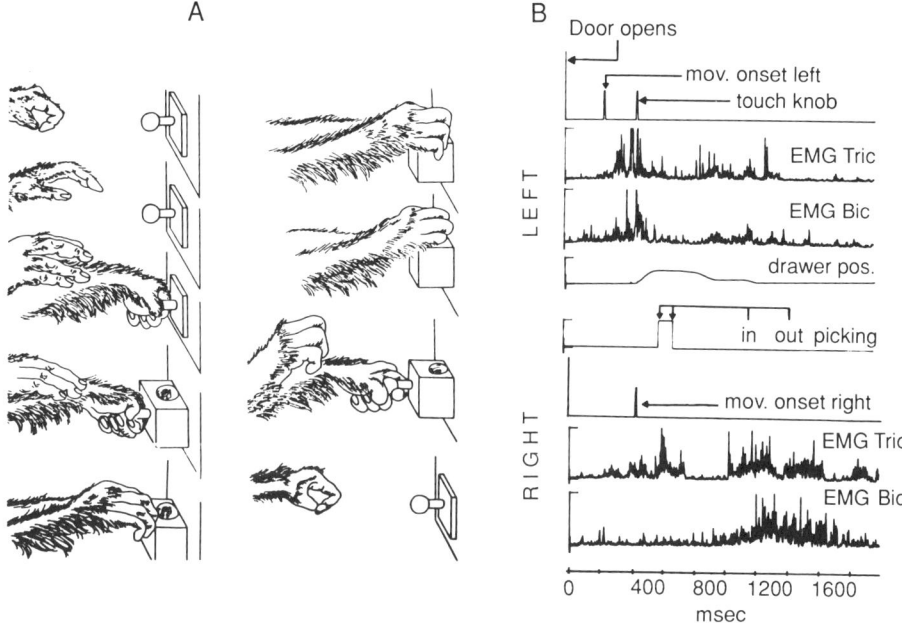

FIGURE 2 Bimanual task performed by *Macaca fascicularis*. (A) Movement sequence (*top left* to *bottom right*). (B) Left- and right-hand events, rectified EMG, and drawer position trace of a single trial. Opening of doors through which the hands can pass and reach the drawer occurs at time zero. From Wiesendanger *et al.* (1992) with permission.

cision grip. As illustrated in Figure 3, the mean timing difference between these left–right events was not only the smallest, but also showed the least variability. Similarly, of the six possible left–right timing pairs, the best correlation coefficient was also found for this same time interval.

It could be argued that this time invariance was the result of an *ad hoc* visual guidance: the monkey may have visually adjusted, for each trial, the movement speed of the right arm according to the progress of drawer opening. This was tested by comparing the synchronization in the light and in the dark condition. Apart from some increase in spatial variability (as seen in superimposed trajectories) during the first sessions when the monkeys were subjected to this new situation, the synchronization was perfectly maintained in the dark (Figure 4).

D. Effects of Serial Lesions of Right and Left SMA on Bimanual Tasks

Transient effects were observed in two lesioned animals. The present report is on the more prominent deficits noted in the second animal. Both monkeys chose to use the left hand for pulling the drawer and for

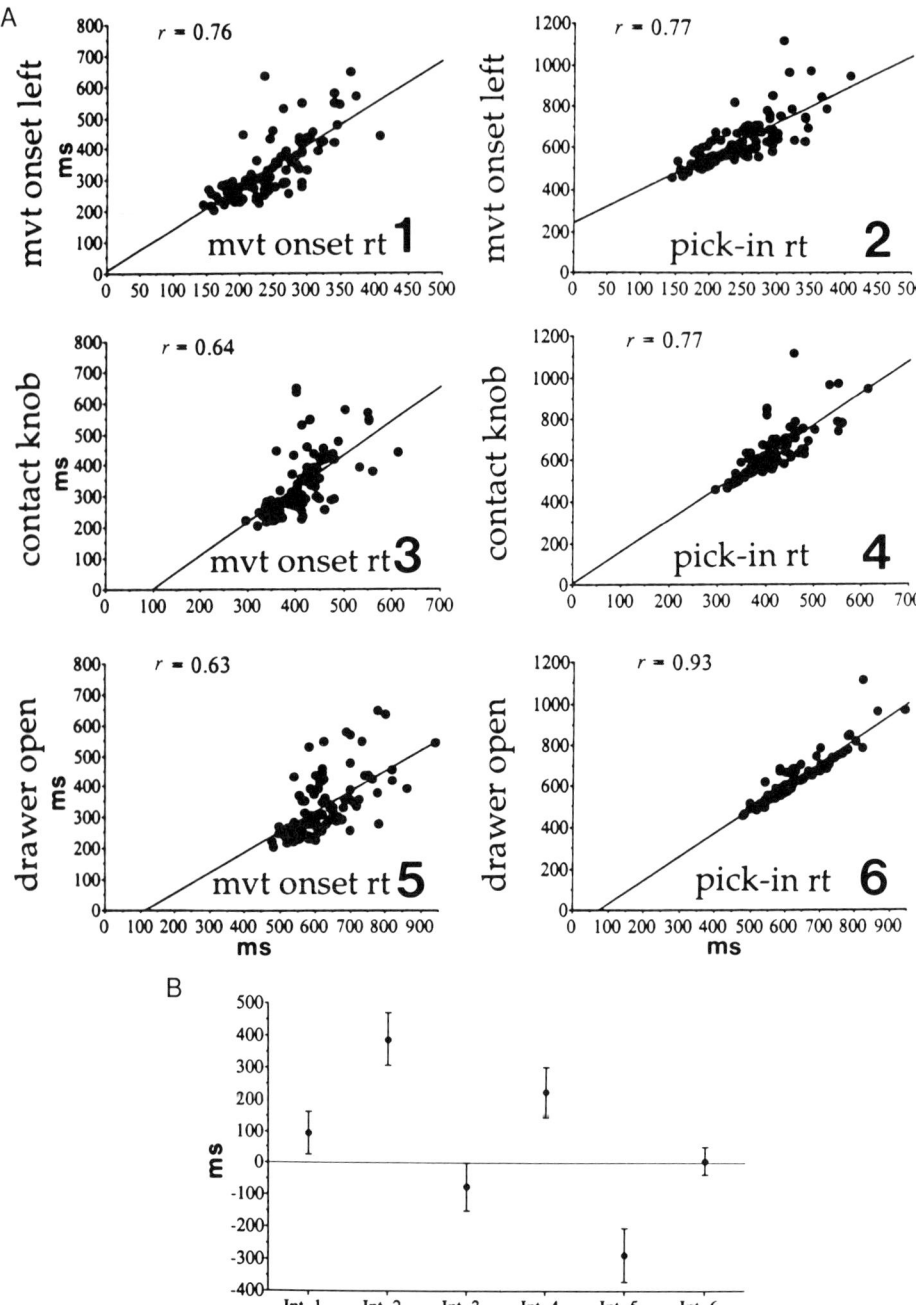

FIGURE 3 (A) Regression analysis between 6 pairs of left-hand (ordinate) and right-hand (abscissa) events with indication of correlation coefficient (r). (B) Mean and standard deviation for the 6 left-right intervals. Zero, complete synchronization; positive values, left hand is leading; negative values, right hand is leading. Note that the correla-

pushing food morsels out of the holes from the Sperry board. In both monkeys the first lesion was on the side opposite to the leading left hand.

1. First Lesion of SMA Right

In the Sperry board experiment, the monkey successfully retrieved the food from the first postoperative session, but the score of retrievals (number of successful retrievals in runs of 30 sec) was signficantly reduced by 31%, as compared to the preoperative score ($p < .01$, Mann-Whitney). In addition, recording of the trajectory of the two hands working at successive holes revealed a conspicuous neglect for the baited holes on the left part of the board. Abnormalities in bimanual retrieval were also recorded in about one-third of the trials, such as: the left index finger remaining for more than 2 seconds in the hole after successful grasping with the right hand underneath the board; perseverance of the left index finger in making pushing movements after morsels had been grasped by the right hand; attempts to push down the food with the right index (and thereby losing the food); left and right hands not matched for working on the same hole leading to loss of food. These clear-cut deficits lasted only for four postoperative sessions (five postoperative days) and recovery to preoperative levels was achieved at about the tenth postoperative day.

In the bimanual draw-and-pick task, the monkey also successfully performed from the first postoperative session without any obvious "clinical" abnormality. The trajectories, performed with and without visual feedback, also revealed no changes. However, analysis of the time structure of the bimanual synergy revealed that movement onset of the left arm was delayed, corresponding to a significant increase by 42% with respect to preoperative movement onset ($p < .001$, Mann-Whitney). Interestingly, movement onset of the right hand was also significantly prolonged by 27% ($p < .01$, Mann-Whitney). Variability of both these parameters increased. Perhaps the most interesting outcome of this timing analysis was that, in spite of the above-mentioned delays, the temporal goal-related invariance was maintained. It is tempting to speculate that the delay in movement onset of the ipsilateral right arm was a compensatory strategy, rather than a direct consequence of the right SMA lesion, in order to match again the two hands

---

tion coefficient, best synchronization, and lowest variability is for the left–right interval 6, considered to represent the goal-related temporal invariance of the bimanual task. Data are from one intact monkey performing 150 trials (5 sessions, 30 trials each). From Wiesendanger et al. (1992) with permission.

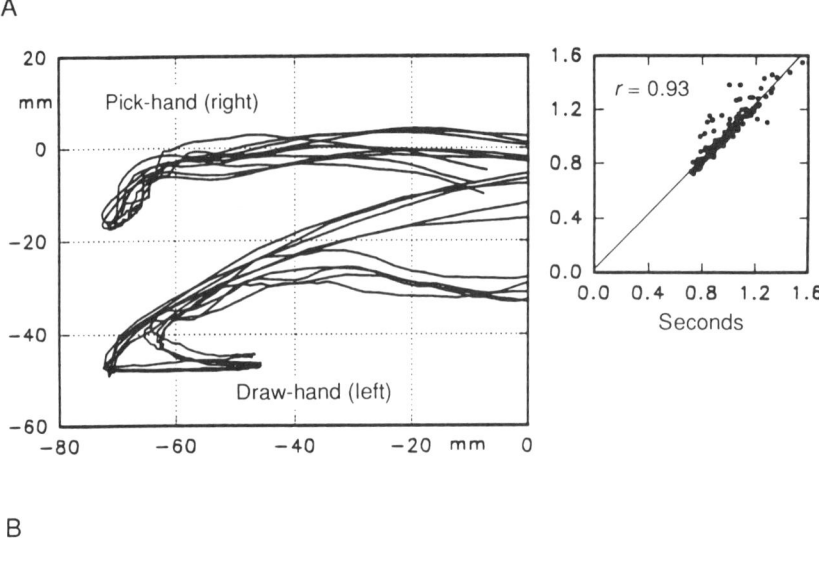

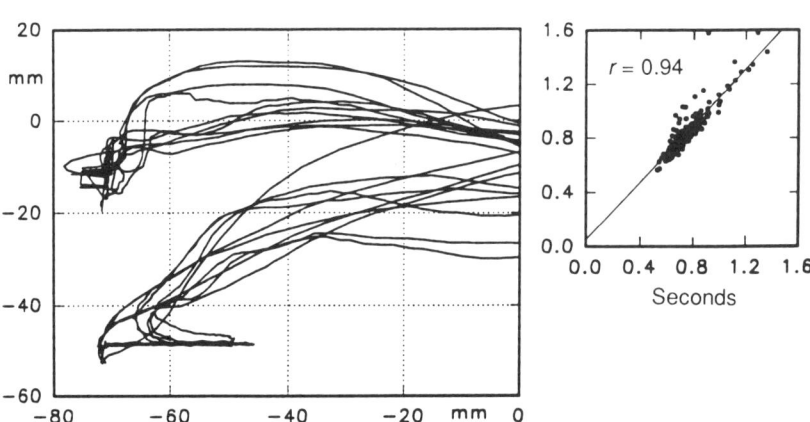

FIGURE 4 (A) Light. (*Left*) 5 superimposed trajectories in the sagittal plane of the right food-picking index finger (upper traces) and of the left index finger (lower traces) implicated in the pulling of the drawer, shown as the horizontal displacement. (*Right*) Linear regression plot between the time the drawer is fully opened by the left hand (ordinate) and the right index finger is reaching into the food well (abscissa), with indication of correlation coefficient (*r*). (B) Dark. Same as for (A), but showing performance in absolute darkness. Note the preservation of goal-related temporal invariance in spite of some increase in spatial variability.

for perfect synchronization at the goal. Whereas temporal correlation in movement onset times of the two hands was relatively weak during the control period ($r = .71$), the intermanual correlation indeed transiently improved ($r = .85$) in the period during which movement onset was delayed, and then again fell to poor correlations as the deficit in movement onset subsided ($r = .49$). The recovery was complete 2 weeks after the lesion.

2. Second Lesion of SMA Left

This lesion produced more dramatic deficits also evident on "clinical" observation: the right hand was less frequently used for grasping, which was slow, weak, and clumsy. During the first two postoperative weeks, the monkey did not perform on the Sperry board task. Only a few attempts were made with the left index finger to push down the food. Since the right hand was not ready to grasp from beneath the board, the monkey soon stopped its attempts. The first successful trials were noted in the third postoperative week. Most of the time, the right hand did not cooperate with the left and remained motionless or moved above the plate. Food morsels presented directly to the right hand were, however, seized. Improvement started during the fourth week after the lesion when the right hand became more active. The strategy was to explore with the right index from below the baited holes, as if to sense when the food was being pushed down. Although this was somewhat reminiscent of the postoperative behavior described by Brinkman (1984), it did not seem to demonstrate "mirror movements." Many morsels were lost because of unmatched grasping with the right hand. For several weeks, the performance remained unstable from session to session with great session-to-session variability in successful scores. On average, the postoperative score of successful retrievals fell significantly by 44% ($p < .001$, Mann-Whitney). The other surprising finding was that the neglect for the left part of the board, which had recovered after the first right SMA lesion, reappeared after second left SMA lesion.

In the draw-and-pick task, the effect of this second lesion was also more dramatic and longer lasting than that observed after the first lesion. However, the right arm was more actively engaged than in the Sperry board task. The grasping of the food from the drawer was clumsy and slow. Quantitative assessment of the time structure reveals that the right arm movement onset times and the duration of the right reach phase became irregular from series to series. "Bad" series with greatly prolonged values of these parameters altered with "good" series in other sessions, with mean timing values being in the preoperative range. All time values of the left arm remained unchanged. As a consequence, the time invariance for goal achievement deteriorated for "bad" runs, but was perfect in "good" runs, showing no delays in right hand performance. The timing correlation coefficients for the bimanual goal-time invariance accordingly oscillated between .85 and .97, with left–right synchronization (dt of interval 6 in Figure 3) varying between 27 ± 23 msec and 100 ± 75 msec (mean ± S.D.). This conspicuous variability between experimental series was not accompanied by any obvious motivational changes or fatigue. Performance in the dark was not different from performance in the light. When averaging over the "bad" and "good" postoperative sessions, no statistical difference was found with performance in right movement onset and reaching in preoperative and postoperative sessions, which is not surprising in view of the large session-to-session variability after the second lesion. The behavioral oscillations slowly diminished and disappeared after a postoperative period of 6 weeks.

3. Summary of Salient Features Observed after the Serial Bilateral SMA Lesions

Each of the intended SMA lesions produced a dominant deficit in the peformance of the contralateral arm and hand. It concerned movement onset of the left arm for reaching the drawer (first right lesion) and grasping with the right hand (second left lesion). Rather than the expected deterioration in the temporal coupling for bimanual goal achievement, it was found that, in spite of the delay in movement initiation following the first lesion, the bimanual synchronization at the goal was unchanged. The result after the second lesion was more complex and the deficit of the contralateral right hand was more serious. But even in this situation, a perfect synchronization for goal achievement was possible. The large oscillations in the right hand performance (movement onset and reaching) occurring between experimental series, but not within one series, remain unexplained. Possibly, these oscillations of "bad" experimental series alternating with "normal" performance with regard to the timing were due to subtle motivational and attentional factors. But even during "bad" series, deterioration in timing concerned exclusively the right arm. Interestingly, a similar day-by-day variability in bimanual performance has been recently described in a patient suffering from "intermanual conflict" as a result of a medial frontal lesion. To cite the authors Jason and Pajurkova (1992): "A striking feature of this patient's behaviour was that there were times when the left hand cooperated well and the patient could perform bimanual tasks with ease, and other times when the left hand would not cooperate even in simple tasks."

Thus, neither of the two lesions resulted in a primary deficit in the goal-oriented temporal coupling between the two hands. On the basis of this experiment (and with the proviso that the exact extent of the lesions remains to be assessed), we are forced to conclude that the temporal control of bimanual coordination in a natural task is not a major function of the supplementary motor area. On the contrary, it appears that the observed goal-related temporal invariance is an extremely robust phenomenon that may be preserved even in the presence of unilateral deficits.

VI. General Conclusion

In this chapter, we addressed the problem of whether specific cortical areas are responsible for temporal and spatial coupling of both hands when they are engaged in purposeful cooperative manipulations, as is so often the case in the natural movement repertoire of primates. One clear-cut observation is that both arms are often put into action simultaneously, as, for example, in rhythmic bilateral movements, reaching for large objects, and catching a ball. It was argued that this strategy of bilaterally synchronized initiation, also termed "assimilation effect," may be achieved by the known bilateral connections (direct and indirect) from motor and premotor cortical areas to the spinal cord. These bilateral controls, exerted by each hemisphere, would assure a tight coupling as one can observe in the easy-to-perform simultaneous and symmetrical movements. As mentioned in section IV,B, tight coupling necessary for simultaneous limb actions or eye–arm synergies is also likely to depend on lower level circuits including cerebellum, brainstem, and propriospinal networks. Typically, symmetrical limb movements are relatively well preserved in the presence of cortical lesions outside the primary motor cortex (or may even appear in exaggerated form as bilateral associated movements after cortical lesions). They are also preserved in split-brain subjects who showed an even greater attraction for synchronized movements compared to normal subjects (Tuller & Kelso, 1989).

We particularly explored the case of bimanual cooperation in skilled manipulations requiring asymmetric and complex, sequential actions of both hands for achieving a goal. It was argued that the learning of such bimanual skills may in fact involve a suppression of congenitally determined facilitation of bilateral synchronization in movement onset ("insulation" from bilateral assimilation; Marteniuk, MacKenzie, & Baba, 1984).

In multiarticulate limb movements and in speech articulation it was found that, in spite of a large trial-by-trial temporal and spatial variability of individual movement components, goal achievement was characterized by a remarkable spatial and temporal invariance. This was taken to indicate that "coordinative structures" or "generalized programs" exist that are responsible for these goal-related invariances. The concept of high-order programs is also born out by neurological observations in patients with cortical lesions leading to apraxia, characterized by a breakdown of the "higher units of performance" (Welford, 1968). We are not aware of previous reports in normal subjects and neurological patients about quantitative assessment of possible goal-related invariances in natural bimanual acts. However, it appears that various types of cortical lesions do in fact impair proper goal achivement in bimanual acts. Probably the most common lesion resulting in bimanual apraxia is that of the dominant parietal cortex. Further cortical structures include the PM, the anterior corpus callosum, and the medial frontal cortex, including the SMA. In many of the reported cases, there seems to be a truly bimanual coordination problem, but the situation is often complicated by the fact that one of the hands, used alone, may also be dyspractic. Finally, there is evidence that various types of association cortex lesions, including the temporal association cortex, the lateral premotor cortex, and diffuse cortical lesions, tend to reduce the capacity of executing, and especially of learning, complex bimanual acts.

In an animal model of a purposeful bimanual movement sequence, we present evidence for a goal-related temporal invariance. It turned out that this invariance was stable and was maintained also when visual feedback was removed. The possible role of kinesthetic feedback remains to be elucidated. In preliminary experiments, it was furthermore revealed that serial lesions of both supplementary motor areas produced only transient motor impairments, each lesion affecting chiefly the contralateral hand. The most remarkable outcome was that, even in spite of the transient disturbance in one limb, the bimanual temporal invariance for goal achievement proved to be maintained. The first lesion in both animals was on the side opposite to the left hand, which led the bimanual sequence. It is an open question whether reversal of the serial lesions would change the outcome (cf. also, Brinkman, 1984). The question also arises whether the SMA might play a more important role in the control of bimanual movements during learning. In our first lesioned monkey, the animal learned within a week to perform the Sperry board task following the first right-sided SMA lesion. On the other hand, Aizawa, Inase, Mushiake, Shima, & Tanji (1991) reported that movement-related activity recorded in the SMA gradually disappeared in an overtrained monkey after a longer period of several months. Finally, the ambiguities of involvement of neighboring structures such as the medial cortex in front of the SMA (Rizzolatti et al., 1990) and a motor field occupying the anterior cingulate cortex of area 24 (Hutchins, Martino, & Strick, 1988) in previous and possibly also in the present study make it impossible, at the present time, to arrive at a definite conclusion about the specific role of the SMA in bimanual coordination.

The experimental results, taken together with the outcome of neurological studies, indicate, however, that a number of cortical areas, including especially the dominant parietal association cortex, the medial frontal cortex, and the anterior corpus callosum, are involved in a cooperative manner in the control of goal-oriented bimanual skills that are impressive in their spatial and temporal precision.

References

Abbs, J. H., & Cole, K. J. (1987). Neural mechanisms of motor equivalence and goal achievement. In S. P. Wise (Ed.), *Higher brain functions* pp.15–43. New York: Wiley.

Aizawa, H., Mushiake, H., Inase, M., & Tanji, J. (1990). An output zone of the monkey primary motor cortex specialized for bilateral hand movement. *Exp. Brain. Res., 82,* 219–221.

Aizawa, H., Inase, M., Mushiake, H., Shima, K., & Tanji, J. (1991). Reorganization of activity in the supplementary motor area associated with motor learning and functional recovery. *Exp. Brain. Res., 84,* 668–671.

Alexander, G., & Crutcher, M. D. (1990). Neural representations of the target (goal) of visually guided arm movements in three motor areas of the monkey. *J. Neurophysiol., 64,* 164–178.

Alstermark, B. (1983). Functional role of propriospinal neurones in the control of forelimb movements. A behavioral and electrophysiological study (pp. 1–32). Thesis, University Goeteborg, Goeteborg.

Arbib, M. A. (1990). Programs, schemas, and neural networks for control of hand movements: Beyond the RS framework. In M. Jeannerod (Ed.), *Attention and Performance. Motor representation and control,* (pp. 111–138). Hillsdale, NJ: Erlbaum.

Bailey, P. (1933). In G. von Bonin (1944). Architecture of the precentral motor cortex and some adjacent areas, Fig. 36. *Precentral motor cortex* (p. 63). P. C. Bucy (ed.). Urbana: The University of Illinois Press.

Bernstein, N. A. (1967). *The co-ordination and regulation of movements.* Oxford: Pergamon Press.

Bogen, J. E. (1985). The callosal syndromes. In K. M. Heilman & E. Valenstein (Eds.), *Clinical neuropsychology* (pp 295–338). New York: Oxford University Press.

Brinkman, C. (1983). Effects of bilateral supplementary motor area lesions in the monkey. *Neurosci. Letters,* suppl 15, 523.

Brinkman, C. (1984). Supplementary motor area of the monkey's cerebral cortex: short- and long-term deficits after unilateral ablation and the effects of subsequent callosal section. *J. Neurosci. 4,* 918–929.

Brinkman, C., & Porter, R. (1979). Supplementary motor area in the monkey: activity of neurons during performance of a learned motor task. *J. Neurophysiol., 42,* 681–709.

Brinkman, J., & Kuypers, H. G. J. M. (1973). Cerebral control of contralateral and ipsilateral arm, hand and finger movements in the split-brain rhesus monkey. *Brain, 96,* 653–674.

Brion, S., & Jedynak, C. P. (1972). Troubles du transfert interhémisphérique (callosal disconnection) à propos de 3 observations de tumeurs du corps calleux. Le signe de la main étrangère. *Rev. Neurol. (Paris), 126,* 257–266.

Brodmann, K. (1925). Vergleichende Lokalisationslehre der Grosshirnrinde in ihren Prinzipien dargestellt auf Grund des Zellenbaues (2nd ed.). J. A. Barth (Ed.). Leipzig.

Bucy, P. C., & Fulton, J. F. (1933). Ipsilateral representation in the motor and premotor cortex of monkeys. *Brain, 56,* 318–342.

Caminiti, R., Johnson, P. B., Burnod, Y., Galli, C., & Ferraina, S. (1990). Shift of preferred directions of premotor cortical cells with arm movements performed across the workspace. *Exp. Brain Res., 83,* 228–232.

Caminiti, R., Johnson, P. B., Galli, C., Ferraina, S., & Burnod, Y. (1991). Making arm movements within different parts of space: The premotor and motor cortical representation of a coordinate system for reaching to visual targets. *J. Neurosci., 11,* 1182–1197.

Chan, J. L., & Ross, E. D. (1988). Left-handed mirror writing following right anterior cerebral artery infarction: evidence for nonmirror transformation of motor programs by supplementary motor area. *Neurology,* **38,** 59–63.

Deecke, L., & Kornhuber, H. H. (1978). An electrical sign of particpation of the mesial "supplementary" motor cortex in human voluntary finger movement. *Brain Res.,* **159,** 473–476.

Dick, J. P. R., Benecke, R., Rothwell, J. C., Day, B. L., & Marsden, C. D. (1986). Simple and complex movements in a patient with infarction of the right supplementary motor area. *Mov. Disord.,* **1,** 255–266.

Economo, C. von, & Koskinas, G. N. (1925). Die Cytoarchitektonik der Grosshirnrinde des erwachsenen Menschen. Berlin, Wien: Springer.

Ettlinger, G., & Morton, H. B. (1963). Callosal section: Its effect on the performance of a bimanual skill. *Science,* **139,** 485–486.

Ferrigno, G., & Pedotti, A. (1985). ELITE: a digital dedicated hardware system for movement analysis via real-time TV-signal processing. *IEEE Trans. Biomed. Eng. BME,* **32,** 943–950.

Foerster, O. (1936). Motorische Felder und Bahnen. In O. Bumke, & O. Foerster (Eds.), *Handbuch der Neurologie VI* (pp. 1–357). Berlin: Springer.

Fortier, P. A., Kalaska, J. F., & Smith, A. M. (1989). Cerebellar neuronal activity related to whole-arm reaching movements in the monkey. *J. Neurophysiol.,* **62,** 198–211.

Freund, H. J. (1987). Abnormalities of motor behavior after cortical lesions in humans. V. B. Mountcastle (Ed.), *Handbook of physiology, Section 1: Neurophysiology. Vol. 5: Higher functions of the brain, part 2* (pp. 763–810). Washington, D.C.: American Physiological Society.

Freund, H.-J., & Hummelsheim, H. (1984). Premotor cortex in man: evidence for innervation of proximal limb muscles. *Exp. Brain Res.,* **53,** 479–482.

Gentilucci, M., Fogassi, L., Luppino, G., Matelli, M., Camarda, R., & Rizzolatti, G. (1988). Functional organization of inferior area 6 in the macaque monkey: I. Somatotopy and the control of proximal movements. *Exp. Brain Res.,* **71,** 475–490.

Georgopoulos, A. P. (1988). Neural integration of movement: role of motor cortex in reaching. *FASEB Journal,* **2,** 2849–2857.

Glickstein, M. E. (1990). Brain pathways in the visual guidance of movement and the behavioral functions of the cerebellum. In C. Trevarthen (Ed.), *Brain circuits and functions of the mind. Essays in honor of Roger W. Sperry* (pp. 157–167). Cambridge: Cambridge University Press.

Goldberg, G. (1985). Supplementary motor area structure and function: review and hypotheses. *Behavioral and Brain Sciences,* **8,** 567–616.

Goldberg, G., & Bloom, K. K. (1990). The alien hand sign. Localization, lateralization and recovery. *Am. J. Phys. Med. Rehab.,* **69,** 228–238.

Goldberg, G., Mayer, N. H., & Toglia, J. U. (1981). Medial frontal cortex infarction and the alien hand sign. *Arch. Neurol.,* **38,** 683–686.

Guiard, Y., & Requin, J. (1978). Between-hand vs within-hand choice RT: A single channel of reduced capacity in the split-brain monkey. In J. Requin (Ed.), *Attention and performance, VII.* Hillsdale, NJ: Erlbaum.

Gurfinkel, V. S., & Elner, A. M. (1988). Participation of secondary motor area of the frontal lobe in organization of postural components of voluntary movements in man. *Neurophysiologia (Russian),* **20,** 7–14.

Hausmanowa-Petrusewicz, I. (1959). Interaction in simultaneous motor functions. *Arch. Neurol Psychiat. (Chic),* **81,** 173–187.

Hécaen, H. (1978). Les apraxies idéomotrices, essai de dissociation. In H. Hécaen, & M. Jeannerod (Eds.), *Du contrôle moteur à l'organisation du geste* (pp. 343–358). Paris: Masson.

Hocherman, S., & Wise, S. P. (1991). Effects of hand movement path on motor cortical activity in awake, behaving rhesus monkeys. *Exp. Brain Res.,* **83,** 285–302.

Hutchins, K. D., Martino, A. M., & Strick, P. L. (1988). Corticospinal projections from the medial wall of the hemisphere. *Exp. Brain Res.* **71**, 667–672.
Hyvärinen, J. (1982). Posterior parietal lobe of the primate brain. *Physiol. Rev.* **62**, 1060–1129.
Inhoff, A. B., Diener, H. C., Rafal, R. D., & Ivry, R. (1989). The role of cerebellar structures in the execution of serial movements. *Brain,* **112**, 565–581.
Iwamura, Y., Tanaka, M., Sakamoto, M., & Hikosaka, O, (1985). Functional surface integration, submodality convergence and tactile feature detection in area 2 of the monkey somatosensory cortex. *Exp. Brain Res.* suppl 10, 44–58.
Jason, G., & Pajurkova, E. M. (1992). Failure of metacontrol: breakdown in behavioural unity after lesion of the corpus callosum and inferomedial frontal lobes. *Cortex,* **28** (in press).
Jeannerod, M. (1986). The formation of finger grip during prehension. A cortically mediated visuomotor pattern. *Behav. Brain Res.* **19**, 99–116.
Jeannerod, M. (1988). The neural and behavioural organization of goal-directed movements. In *Oxford psychology series, No. 15* (pp. 283). Oxford: Clarendon Press.
Jones, E. G. (1986). Connectivity of the primate sensory-motor cortex. In E. G. Jones & A. Peters (Eds.), *Cerebral cortex* (pp. 113–183). New York: Plenum Press.
Jordan, M. I. (1990). Motor learning and the degrees of freedom problem. In M. Jeannerod (Ed.), *Attention and performance XIII: Motor representation and control* (pp. 796–836). Hillsdale, NJ: Erlbaum.
Kalaska, J. F., Caminiti, R., & Georgopoulos, A. P. (1983). Cortical mechanisms related to the direction of two-dimensional arm movements: relations in parietal area 5 and comparison with motor cortex. *Exp. Brain Res.,* **51**, 247–260.
Kelso, J. A. S., & Tuller, B. (1981). Toward a theory of apratic syndromes. *Brain and Language,* **12**, 224–245.
Kelso, J. A. S., Southard, D., & Goodman, D. (1979). On the nature of human interlimb coordination. *Science,* **203**, 1029–1031.
Kelso, J. A. S., Putnam, C. A., & Goodman, D. (1983). On the space-time structure of human interlimb co-ordination. *Q. J. Exp. Psychol.,* **35A**, 347–375.
Kinsbourne, M. (1974). Mechanisms of hemispheric interaction in man. In M. Kinsbourne & W. L. Smith (Eds.), *Hemispheric disconnection and cerebral function.* Springfield, IL: C. C. Thomas.,
Kornhuber, H,. H., Deecke, L., Lang, W., Lang, M., & Kornhuber, A. (1989). Will, volitional action, attention and cerebral potentials in man: Bereitschaftspotential, performance-related potentials, directed attention potentials, EEG spectrum changes. In W. A. Hershberger (Ed.), *Volition and action* (Adv. Psychol. Vol. 10) (pp. 107–168). Amsterdam: Elsevier Publ. (North Holland).
Künzle, H. (1978). Cortico-cortical efferents of primary motor and somatosensory regions of the cerebral cortex in Macaca fascicularis. *Neuroscience,* **3**, 25–39.
Kuypers, H. G. J. M. (1981). Anatomy of descending pathways. In V. B. Brooks (Ed.), *Handbook of physiology* (The nervous system) (Vo. II, part 1) (pp. 597–666). Bethesda, MD: American Physiology Society.
Kuypers, H. G. J. M., & Brinkman, J. (1970). Precentral projections to different parts of the spinal intermediate zone in the rhesus monkey. *Brain Res.,* **24**, 29–48.
Lang, W., Obrig, H., Lindinger, G., Cheyne, D., & Deecke, L. (1990). Supplementary motor area activation while tapping bimanually different rhythms in musicians. *Exp. Brain Res.,* **79**, 504–514.
Lang, W., Cheyne, D., Kristeva, R., Beisteiner, R., Lindinger, G., & Deecke, L. (1991). Three-dimensional localization of SMA activity preceding voluntary movement. A study of electric and magnetic fields in a patient with infarction of the right supplementary motor area. *Exp. Brain Res.,* **87**, 688–695.

Lang, W., Beisteiner, R., Lindinger, G., & Deecke, L. (1992). Changes of cortical activity when executing learned motor sequences. *Exp. Brain Res., 89,* 435–440.
Laplane, D., Thalairach, J., Meininger, V., Bancaud, J., & Orgogozo, J. M. (1977). Clinical consequences of corticectomies involving the supplementary motor area in man. *J. Neurol. Sci., 34,* 301–314.
Lashley, K. S. (1930). Basic neural mechanisms in behavior. *Psych. Rev. 37,* 1–24.
Lashley, K. S. (1951). The problem of serial order in behavior. In L. Jeffress (Ed.), *Cerebral mechanisms in behavior* (pp. 112–136). New York: Interscience.
Leonard, G., Milner, B., & Jones, L. (1988). Performance of unimanual and bimanual tapping tasks by patients with lesions of the frontal or the temporal lobe. *Neuropsychologia, 26,* 79–91.
Liepmann, H. (1920). Apraxia. In T. Brugsch, & A. Entenburg (Eds.), *Real-Encylopädie der gesamten Heilkunde. Ergebnisse der gesamten Medizin* (Vol. 1) (pp. 116–143). Berlin: Urban & Schwarzenberg.
Liu, C. N., & Chambers, W. W. (1964). An experimental study of the cortico-spinal system in the monkey (*Macaca mulatta*). The spinal pathway and pre-terminal distribution of degenerating fibers following discrete lesions of the pre- and postcentral gyri and bulbar pyramid. *J. Comp. Neruol. 123,* 257–284.
Luria, A. R. (1966). *Higher cortical functions in man* (pp. 215–226). New York: Basic Books, Inc.
Mann, S. E., Thau, R., & Schiller, P. H. (1988). Conditional task-related responses in monkey dorsomedial frontal cortex. *Exp. Brain Res. 69,* 460–468.
Mark, R. F., & Sperry, R. W. (1968). Bimanual coordination in monkeys. *Exp. Neurol., 21,* 92–104.
Marteniuk, R. G., MacKenzie, C. L.,, & Baba, D. M. (1984). Bimanual movement control: Information processing and interaction effects. *Q J Exp. Physiol., 36*A, 335–365.
Matsunami, K., & Hamada, I. (1981). Characteristics of the ipsilateral movement-related neuron in the motor cortex of the monkey. *Brain Res., 204,* 29–42.
Meador, K. J., Watson, R. T., Bowers, D., & Heilman, K. M. (1986). Hypometria with hemispatial and limb motor neglect. *Brain, 109,* 293–305.
Muir, R. B., & Lemon, R. N. (1983). Corticospinal neurons with a special role in precision grip. *Brain Res., 261,* 312–316.
Mushiake, H., Inase, M., & Tanji, J. (1990). Selective coding of motor sequence in the supplementary motor area of the monkey cerebral cortex. *Exp. Brain Res. 82,* 208–210.
Neshige, R., Lüders, H., & Shibaski, H. (1988). Recording of movement-related potentials from scalp and cortex in man. *Brain, 3,* 719–736.
Paillard, J. (1982). Apraxia and the neurophysiology of motor control. *Phil. Trans. R. Soc. London B, 298,* 111–134.
Pause, M., Kunesch, E., Binkofski, F., & Freund, H.-J. (1989). Sensorimotor disturbances in patients with lesions of the parietal cortex. *Brain, 112,* 1599–1625.
Penfield, W., & Jasper, H. (1954). *Epilepsy and the functional anatomy of the human brain.* Boston: Little, Brown.
Pfurtscheller, G., & Berghold, A. (1989). Patterns of cortical activation during planning of voluntary movement. *Electroencephalogr. Clin. Neurophysiol., 72,* 250–258.
Poizner, H., Klima, E. S., & Bellugi, U. (1987). *What the hands reveal about the brain.* (pp. 161–172). Cambridge, MA: MIT Press.
Porter, R. (1990). The Kugelberg lecture. Brain mechanisms of voluntary motor commands-A review. *Electroencephalogr. Clin. Neurophysiol. 76,* 282–293.
Preilowski, B. (1990). Intermanual Transfer, interhemispheric interaction and handedness in man and monkeys. In C. Trevarthen (Ed.), *Brain circuits and functions of the mind* (pp. 168–180). Cambridge: Cambridge University Press.

Rizzolatti, G., Camarda, R., Fogassi, M., Gentilucci, M., Luppino, G., & Matelli, M. (1988). Functional organization of inferior area 6 in the macaque monkey: II. Area F5 and the control of distal movements. *Exp. Brain Res.*, **71**, 491–507.

Rizzolatti, G., Gentilucci, M., Camarda, R. M., Gallese, V., Luppino, G., Matelli, M., & Fogassi, L. (1990). Neurons related to reaching-grasping arm movements in the rostral part of area 6 (area 6aβ). *Exp. Brain Res.*, **82**, 337–350.

Roland, P. E. (1984). Organization of motor control by the normal human brain. *Human Neurobiol.*, **2**, 205–216.

Roland, P. E., Steinhøj, E., Lassen, N. A., & Larsen, B. (1980). Different cortical areas in man in organization of voluntary movements in extrapersonal space. *J. Neurophysiol.*, **43**, 137–150.

Schmidt, R. A. (1988). *Motor control and learning.* Champaign, IL: Human Kinetics Publ.

Soechting, J. F., & Flanders, M. (1992). Moving in three-dimensional space: Frames of reference, vectors, and coordinate systems. *Annu. Rev. Neurosci.* **15**, 167–191.

Sperry, R. W. (1966). Brain bisection and mechanisms of consciousness. In J. C. Eccles (Ed.), *Brain and conscious experience* (pp. 298–313). Berlin: Springer Verlag.

Taira, M., Mine, S., Georgopoulos, A. P., Murata, A., & Sakata, H. (1990). Parietal cortex neurons of the monkey related to the visual guidance of hand movement. *Exp. Brain Res.*, **83**, 29–36.

Tanji, J., & Kurata, K. (1981). Contrasting neuronal activity in the ipsilateral and contralateral supplementary motor areas in relation to a movement of monkey's distal hindlimb. *Brain Res.*, **222**, 155–158.

Tanji, J., Okano, K., & Sato, K. C. (1988). Neuronal activity in cortical motor areas related to ipsilateral, contralateral, and bilateral digit movements of the monkey. *J. Neurophysiol,* **60**, 325–343.

Travis, A. M. (1955). Neurological deficiencies following supplementary motor area lesions in *Macaca mulatta. Brain,* **78**, 174–201.

Tuller, B., & Kelso, J. A. S. (1989). Environmentally specified patterns of movement coordination in normal and split-brain subjects. *Exp. Brain Res.*, **75**, 306–316.

Viallet, F., Massion, J., Massarino, R., & Khalil, R. (1992). Coordination between posture and movement in a bimanual load-lifting task: putative role of a mesial region including the supplementary motor area. *Exp. Brain Res.*, **88**, 674–684.

von Monakow, C. (1914). *Die Lokalisation im Grosshirn und der Abbau der Funktion durch kortikale Herde.* Wiesbaden: Bergmann.

Welford, A. T. (1968). *Fundamentals of skill.* London: Methuen.

Wiesendanger, M. (1981). Organization of secondary motor areas of cerebral cortex. In V. B. Brooks (Ed.), *Handbook of physiology, The nervous system II, Motor control, part II* (pp. 1121–1147). Washington D.C.: American Physiology Society.

Wiesendanger, M. (1986). Recent developments in studies of the supplementary motor area. *Rev. Physiol. Biochem. Pharmacol.*, **103**, 1–59.

Wiesendanger, M., Corboz, M., Hyland, B., Palmeri, A., Maier, V., Wicki, U., & Rouiller, E. (1992). Bimanual synergies in primates. *Exp. Brain Res.*, Series 22 (pp. 45–64). Berlin: Springer.

Wing, A. M. (1982). Timing and coordination of repetitive bimanual movements. *Q. J. Exp. Psychol.*, **34**A, 339–348.

Wyke, M. (1971). The effects of brain lesions on the learning performance of a bimanual co-ordination task. *Cortex,* **7**, 59–72.

10

The Role of the Corpus Callosum and Bilaterally Distributed Motor Pathways in the Synchronization of Bilateral Upper-Limb Responses to Lateralized Light Stimuli

Giovanni Berlucchi, Salvatore Aglioti, and Giancarlo Tassinari

Institute of Human Physiology, Medical School
University of Verona
Verona, Italy

I. How Does the Brain Time Motor Commands for Concurrent Actions of Different Limbs?
II. Lateral Reaction Time Differences in Simple Visuomotor RT
III. Crossed–Uncrossed Differences in Bilateral Responses to Lateralized Light Stimuli
IV. Responses to Lateralized Flash with Movements Controlled by Bilaterally Distributed Motor Systems
V. Summary and Conclusions
References

I. How Does the Brain Time Motor Commands for Concurrent Actions of Different Limbs?

The temporal occurrence of different muscle events involved in interlimb motor coordination depends on the timing of the motor commands from the brain as well as the anatomofunctional characteristics of the neural pathways that transmit these commands to the muscles. In one of the simplest tasks involving interlimb motor coordination, a subject is asked to contract a single muscle or muscle group in two limbs simultaneously. Although the contractions in the two limbs are highly correlated in time, as a rule they are never precisely synchronous, even though the subject reports that they are. These deviations from true synchrony do not appear to result from differences in the level of excitability of the motoneuronal pools producing the two actions, nor are they caused by noise or other random or systematic imprecisions in the system. Instead, they reveal that in different situations the brain uses different criteria for assessing the simultaneity of concomitant motor acts.

The relative nature of the judgment of the simultaneity of muscle actions is evident in experiments where subjects are instructed to perform synchronous movements of two different limbs. For example Paillard (1948) and Bard, Paillard, Lajoie, Fleury, Teasdale, Forget & Lamarre (1992) utilized two different experimental paradigms in both of which subjects attempted to achieve subjective simultaneity of muscle actions in two different limbs. In the *reactive* paradigm, subjects were instructed to extend one index finger and elevate the heel of the same side as fast as possible in response to an auditory stimulus. In the *self-paced* paradigm, the subjects performed these movements spontaneously, in the absence of any external triggers. The reaction times (RT) of the two concurrent responses in the reactive condition were indistinguishable from those of the two responses when tested alone (i.e., the RT of the finger response was systematically shorter than that of the heel response). By contrast, in the self-paced condition the elevation of the heel systematically preceded the extension of the finger. Evidently, in the reactive condition the two motor commands were released simultaneously but the difference in length between the efferent pathways caused the command for the finger response to reach its muscle effectors sooner than did the command for the heel response. By contrast, in the self-paced condition the two motor commands were shifted in time as if the brain took the different lengths of efferent and afferent pathways into account and strived to obtain a simultaneous return of reafferent signals from the two moving limbs. In accord with this interpretation, a polineuropathy patient with a somatic sensory deafferentation was found to display a precession of the short-pathway finger response over the long-pathway heel response in both reactive and self-paced conditions (Bard *et al.*, 1992). Normal subjects always claimed that the two movements were synchronous independent of the reactive or self-paced mode of performance, although the movements were in fact asynchronous in both conditions. In the reactive condition subjective synchrony was probably predicated on the effective synchronization of the two motor commands by the external trigger, while in the self-paced condition it seemed to be engendered by the expectation of synchronous peripheral consequences from two properly desynchronized motor commands.

II. Lateral Reaction Time Differences in Simple Visuomotor RT

The lengths of peripheral sensory pathways and the conduction velocities of afferent fibers have long been known to play an important role in determining the temporal aspects of sensation and perception (e.g., Piéron & Jones, 1959). The results of Paillard (1948) and Bard *et al.* (1992), illustrated above, indicate that the anatomofunctional properties of peripheral neural pathways, efferent as well as afferent, are also relevant to the timing of motor commands intended to produce synchronous movements. It would appear that the time course of sensory and motor events must similarly be affected by the organization of the underlying central neural pathways and the speed with which they transfer information. In 1912, Poffenberger applied chronometric analysis to the dissection of the central neural pathways subserving the execution of a fast manual or digital movement (such as pressing a key) in reaction to a simple visual stimulus, a light flash presented in the right or left hemifield. Because of the organization of the optic pathways, a light flash presented in such a manner is projected to the visual cortex of the opposite hemisphere. The motor reaction of each hand is under the control of the contralateral hemisphere because of the crossing of the major motor pathways. Uncrossed reactions (i.e., reactions of each hand to stimuli in the ipsilateral hemifield) can thus be integrated within a hemisphere, whereas crossed reactions (i.e., reactions of each hand to contralateral hemifield stimuli) require an interaction between the hemisphere receiving the visual stimulus and that emitting the response, probably through the corpus callosum. On this basis, Poffenberger (1912) argued that the RT of crossed responses should be longer than the RT of uncrossed responses, and that the crossed–uncrossed time difference (CUD) should correspond to the extra time needed for interhemispheric communication. He did find a CUD of a few milliseconds in the expected direction, and his finding has been repeatedly confirmed by modern studies, which have demonstrated CUDs of 2–3 msec, a difference that is accounted for by the conduction time along the largest fibers of the corpus callosum (Tassinari, Morelli & Berlucchi, 1983; Levy & Wagner, 1984; St. John, Shields & Timney, 1987; Vallar, Sterzi & Basso, 1988; Milner, Jeeves, Silver, Lines & Wilson, 1985; Clarke & Zaidel, 1989; Saron & Davidson, 1989; Aglioti, Dall'Agnola, Girelli & Marzi, 1991; Marzi, Bisiacchi & Nicoletti, 1991; Di Stefano, Sauerwein & Lassonde, 1992; for earlier studies see the review by Bashore, 1981). The direct relation of the CUD to interhemispheric transfer is supported by findings of abnormal, exceedingly long CUDs in subjects with defective interhemispheric comnication, such as subjects with callosal agenesis (Jeeves, 1969; Reynolds & Jeeves, 1974; Milner, 1982; Milner et al., 1985; Di Stefano et al., 1992; Aglioti, Berlucchi, Pallini, Rossi & Tassinari, 1993) and patients with complete callosotomy (Sergent & Myers, 1985; Clarke & Zaidel, 1989; Aglioti et al., 1993). In these patients, intact central channels for interhemispheric communication at subcortical levels suffice for ensuring the cross-midline interactions needed for performing crossed responses. Yet transfer of information by these acallosal pathways appears to be at least ten times slower than normal interhemispheric transfer via the corpus callosum (Aglioti et al., 1993).

The brief nature of the CUDs in normal control subjects is not a simple manifestation of spatial stimulus–response compatibility. Spatial stimulus–response compatibility consists of a speed advantage for a response present when stimulus and response are matched for side. Spatial compatibility effects occur in choice RT tasks, while CUDs are demonstrated by simple RT paradigms. Previous studies in normals have clearly differentiated CUDs, which depend crucially on the specific anatomic relations between visual hemifields, cerebral hemispheres, and the responding hand (e.g., Berlucchi, Crea, Di Stefano & Tassinari, 1977), from spatial compatibility effects, which arise from the match or mismatch between the code for the location of the stimulus and the code for the location of the response (e.g., Umiltà & Nicoletti, 1990). The long CUDs of subjects with a defect of the corpus callosum, whether congenital (Milner et al., 1985) or acquired (Aglioti et al., 1993), have also been convincingly distinguished from spatial compatibility effects. As an example, Figure 1 shows that the CUD of a patient with a complete section of the corpus callosum was unaffected by the spatial relations between the position of the visual stimulus and that of the responding hand. The right hand was faster than the left in responding to right-hemifield stimuli, and the left hand was faster than the right in responding to left-hemifield stimuli, regardless of whether either hand worked in the right or left hemispace. CUDs were tested on two blocks of 30 trials in which the patient pressed the right key with the right thumb and the left key with the left thumb; and in two other blocks,

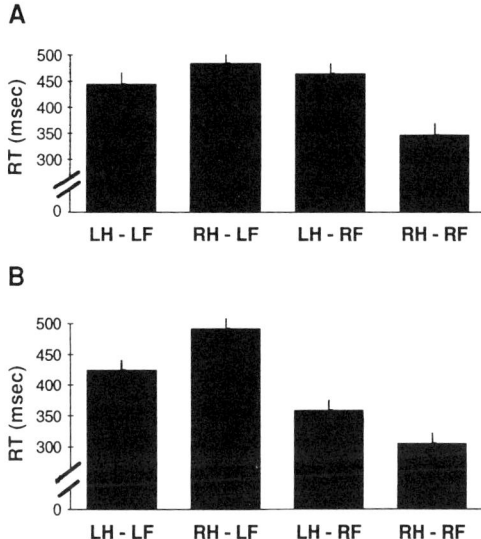

FIGURE 1 Crossed–uncrossed differences in a callosotomized subject as a function of hand position. The figure shows mean RTs (plus S.E.) of each hand (RH, right hand; LH, left hand) in each visual hemifield (LF, left hemifield; RF, right hemifield) in the normal (A) and inverted (B) hand positions. An analysis of variance having hemifield, hand, and hand position as main factors showed the hemifield/hand interaction to be highly significant ($p < .01$). The three-way interaction was completely insignificant, demonstrating that hand position did not affect the hemifield/hand interaction. The advantage of uncrossed RTs over crossed RTs was significant and similar in both hemifields and in both conditions, with overall CUDs of 78 msec in the normal hand-position condition, and 60 msec in the inverted hand-position condition. See Aglioti et al. (1993) for further details.

again of 30 trials, in which the patient held the responding arm across the midline in order to press the right key with the left thumb and the left key with the right thumb, so that uncrossed responses were performed with the hand in the hemispace contralateral to the visual stimulus, and crossed responses were performed with the hand in the hemispace ipsilateral to the visual stimulus. In each block of both conditions (the normal hand-position condition and the inverted hand-position condition), the location of the visual stimulus (a gallium phosphide green flash with a 5-msec duration and an intensity of about 1000 μcd on a background of .15 cd/m^2) varied randomly between 10° to the right and 10° to the left of fixation, with the constraint that each block included 15 right and 15 left stimuli. In each condition the trials of one block were performed with the right hand and those of the other block were performed with the left hand. The order of presentation of the four blocks was decided on a random basis according to a Latin square design.

In Figure 2, data are shown from an experiment on the same callosotomized subject as in Figure 1, in which reaction times were compared with the corresponding latencies of activation of the prime movers, that is, the muscles primarily involved in the performance of the response, as indicated by their electromyographic (EMG) activities. Behavioral reaction time can be divided into a premotor time, that is, the lag between the stimulus and the first EMG activation of the prime movers, and a motor time, or the differential between reaction time and premotor time. Only premotor time can serve as a reliable direct indicator of the temporal course of central neural processes, since motor time is mainly determined by the speed in attaining the force value necessary for overcoming the inertial load of the response device. Premotor times of crossed and uncrossed responses were assessed by recording the EMG with surface silver electrodes attached to the skin overlying the thenar muscles of the responding hand. Concurrent measures of premotor times and reaction times of crossed and uncrossed responses allow one to evaluate the relative contributions of central and peripheral factors to the crossed–uncrossed difference in reaction time. Figure 2 shows that the long CUD was by no means caused by a slower development of force on crossed as compared to uncrossed responses,

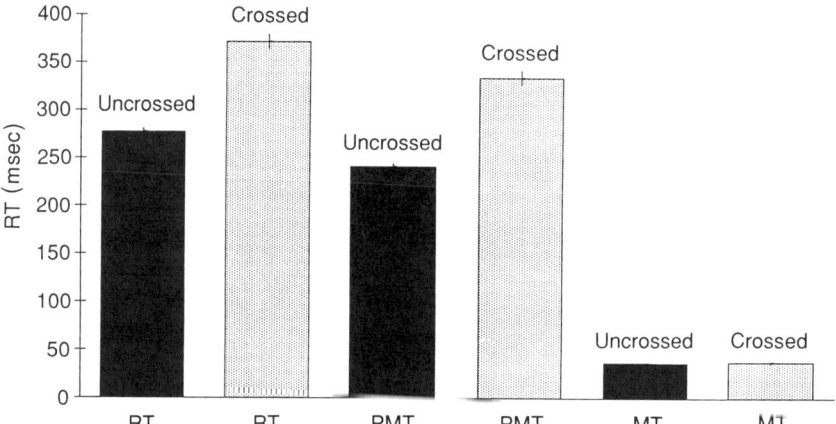

FIGURE 2 Crossed–uncrossed differences in reaction time, premotor time, and motor time. The figure shows the means (plus S.E.) of reaction times (RT), premotor times (PMT), and motor times (MT) of uncrossed responses, that is, responses made with the hand attached to the side of the body ipsilateral to the visual stimulus, and crossed responses, that is, responses made with the hand attached to the side of the body opposite the visual stimulus. Crossed–uncrossed differences in RT are entirely accounted for by crossed–uncrossed differences in PMT, since MTs are equal for the two classes of response. Crossed–uncrossed differences in RT and PMT are significant at the $p < .01$ level. Each column represents 45 data points (see Aglioti et al., 1993).

since EMG recordings from prime movers demonstrated that CUDs derived from RTs were entirely accounted for by crossed–uncrossed differences between premotor times, that is, prior to the activation of motoneurons (Aglioti et al., 1993).

III. Crossed–Uncrossed Differences in Bilateral Responses to Lateralized Light Stimuli

If the CUD is due to a difference in length between the pathways subserving crossed and uncrossed responses, then it should also be found when subjects respond bilaterally to a lateralized flash. Significant CUDs on bimanual responses to lateralized flash are indeed exhibited by normal control subjects (Jeeves, 1969; Di Stefano, Morelli, Marzi & Berlucchi, 1980) as well as by callosal agenetics (Jeeves, 1969; Reynolds & Jeeves, 1974; Milner et al., 1985). Recently we have replicated this finding also in a patient with a complete section of the corpus callosum (Aglioti et al.,1993).

Table I summarizes the results of these studies and allows a comparison between CUDs on unimanual and bimanual tasks in different groups of subjects. It is clear from the table that (1) all groups displayed positive CUDs in both unimanual and bimanual responding conditions; (2) CUDs of subjects lacking a corpus callosum were much longer than those of normals in either condition; and (3) all groups showed smaller CUDs under bimanual than under unimanual responding conditions. An additional finding was a definite tendency to synchronization of motor output in the bimanual task in all groups, as attested by the occurrence of clear-cut correlations between crossed and uncrossed RTs when these were compared on a trial-by-trial basis (Di Stefano et al., 1980; Milner, 1982; Milner et al., 1985; Aglioti et al., 1993). Figure 3 gives examples of these correlations in a callosotomized patient from the study of Aglioti et al. (1993). It is understood that given the irreducibility of the CUD in normals and acallosals alike, these strong correlations occur in spite of the absence of a true bilateral simultaneity of crossed and uncrossed responses. It should also be clear that the above crossed–uncrossed differences and correlations are quite distinct from right–left differences and correlations, which may be observed in bilateral symmetrical movements (e.g., Hongo, Nakamura, Narabayashi & Oshima, 1976). Samples of crossed and uncrossed RTs obviously include RTs for each hand and each hemifield in the appropriate combinations. Therefore, their distributions and central tendencies are unaffected by systematic differences between the right and left hands (or the right and left hemifields), as these are bound to cancel each other in the combinations.

TABLE I Crossed–Uncrossed Differences (CUDs) in Different Groups of Subjects under Unimanual and Bimanual Responding Conditions

Reference	Subject(s)	CUD (msec) Unimanual	Bimanual
Jeeves, 1969[a]	10 Normal adults	2.82	1.68
Reynolds & Jeeves, 1974[a]	1 Acallosal[b] girl	30.36	12.95
Di Stefano et al., 1980[c]	12 Normal adults	2.20	.80
Milner et al., 1985[c]	1 Acallosal[b] boy	12.60	8.00
Aglioti et al., 1993[c]	Callosotomized adult M. E.[d]	69.60	37.90
Aglioti et al., 1993[c]	Acallosal[b] adult R. B.[e]	18.05	11.70
Aglioti et al., 1993[c]	Acallosal[b] adult P. M.[e]	25.45	14.40

[a] Averaged across temporal and nasal hemiretinae in monocular stimulation.
[b] Callosal agenesis.
[c] Binocular stimulation.
[d] Male subject M. E., born in 1970, was submitted to a complete section of the corpus callosum at the Neurosurgical Institute of the Catholic University in Rome (Prof. G. F. Rossi). Callosotomy was performed in two stages (February and June 1989) in an attempt to control a form of post-traumatic epilepsy with complex partial seizures and secondary generalization which had proved totally resistent to pharmacological therapy as well as to a removal of a focus in the right prefrontal cortex. Callosotomy has resulted in a marked favorable change in both severity and frequency of the seizures. Pharmacological treatment with Phenobarbital and Phenytoin has been continued throughout the postoperative period. At the times of testing for RT (April and October 1990, June and December 1991), standard clinical examinations revealed a stationary condition with no neurological deficits except for a severe left-hand ideomotor dyspraxia on verbal command (but not on imitation), a left-hand anomia and alexia in the left hemifield. The completeness of the callosal section and the integrity of the anterior commissure have been confirmed by magnetic resonance imaging.
[e] R. B. and P. M. are two young men, aged 16 and 31 years, respectively, who have been diagnosed by MRI as congenitally lacking the corpus callosum. They are free from major neurological symptoms and appear to have normal intelligence, as indicated by their current respective performances in a technical school and a mechanical shop.

The finding of significant CUDs and crossed–uncrossed correlations in normals as well as in acallosals emphasizes the importance of the corpus callosum in bilateral motor control. Although the corpus callosum may be assumed to play a coordinating role in the synchronization of motor outputs from different hemispheres, the balance of evidence so far has instead suggested a primarily desynchronizing and differentiating callosal action in complex bimanual tasks. Thus, split-brain patients and callosal agenetics have been reported to suffer from a specific disability to suppress synchrony and symmetry of bimanual movements in dual tasks that call for differentiated actions of the two hands (Preilowski, 1972; Jeeves, Silver & Jacobson 1988; Tuller & Kelso, 1989). To the extent that strong crossed–uncrossed correlations in simple reactions to lateralized flashes occur in normal and acallosal subjects alike, at first sight the above RT findings also seem to dismiss a participation of the corpus callosum in the synchronization of concurrent symmetrical movements of the hands in a simple visuomotor task. However, owing to the different CUD magnitudes, bimanual performance comes much closer to synchrony in normals than in acallosals, justifying the assumption of a normal callosal contribution to synchronization of bimanual responses to a lateralized visual input. Since the absence of the corpus callosum leads to an increased difference in length between the pathways for crossed and uncrossed reactions, bilateral asynchrony is bound to be greater in acallosals than in normals.

In a sense, a subject performing conjoint crossed and uncrossed manual responses to a flash presented in the right or left hemifield is comparable to one making concurrent responses with the index finger and the heel of one side to a nonlateralized auditory stimulus, as in the experiments by Paillard (1948) and Bard *et al.* (1992). In both cases one of the concurrent responses is subserved by a relatively short pathway (respectively, the intrahemispheric pathway for the visuomotor response and the pathway for the finger response) and the other is subserved by a relatively long pathway (respectively, the interhemispheric pathway for the visuomotor response and the pathway for the heel response). Paillard (1948) and Bard *et al.* (1992) showed that the precession of the finger response on the heel response in the double-response task coincided with the difference in RT between the two responses when tested independently. If, for present purposes, their situation is similar to that of the bimanual visuomotor task, then the latter task should yield a CUD comparable to that found on unimanual responding. Acallosal subjects compare with normal controls in showing reduced CUDs under the bimanual relative to the unimanual responding condition. At variance with the experiments of Paillard (1948) and Bard et al. (*1992*), the difference between crossed and uncrossed RTs is smaller upon bimanual than upon unimanual responding. Theoretically, this CUD reduction may result from either an increased speed of the slower (crossed) responses, or a decreased speed of the faster (uncrossed) responses in comparison to the unimanual task. In patients with hemiparesis or hemianesthesia from lateralized brain damage, the advantage of the RT of the normal hand over the RT of the affected hand diminishes on bilateral compared to unilateral responding: the disadvantaged response is speeded up selectively in the bimanual task, as if its execution were aided in some way by the concurrent motor command from the undamaged hemisphere to the normal side (Jung & Dietz, 1975; Jeannerod, 1988, p. 82). By contrast, some of the data from our callosotomized patient suggest that the CUD reduction on bimanual compared to unimanual responses was obtained by slowing down

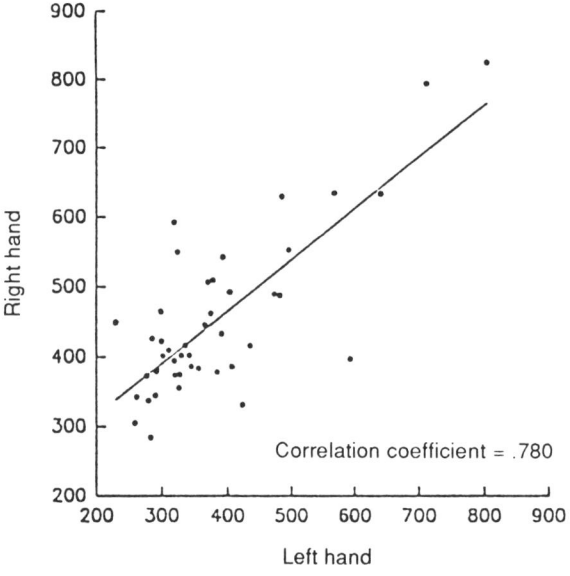

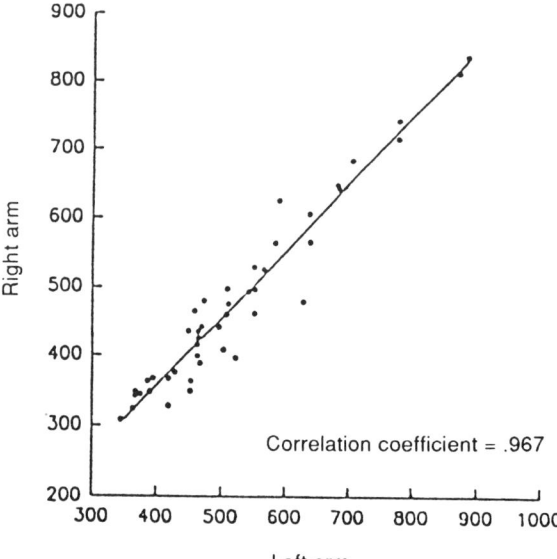

FIGURE 3 Side-to-side correlations in bilateral distal and proximal responses after callosotomy. The figure shows scattergrams and regression lines for bilateral distal (key pressing) and proximal (lever pulling) responses (in msec) upon left and right hemifield stimulation (left graphs and right graphs, respectively) in a callosotomized subject (see Aglioti et al., 1993). Each diagram is based on 45 reaction times. It is clear that in the subject, distal responses are less correlated than proximal responses. However the corre-

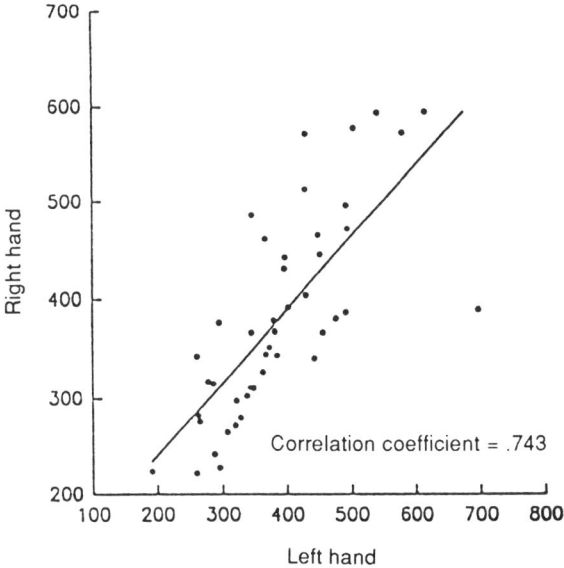

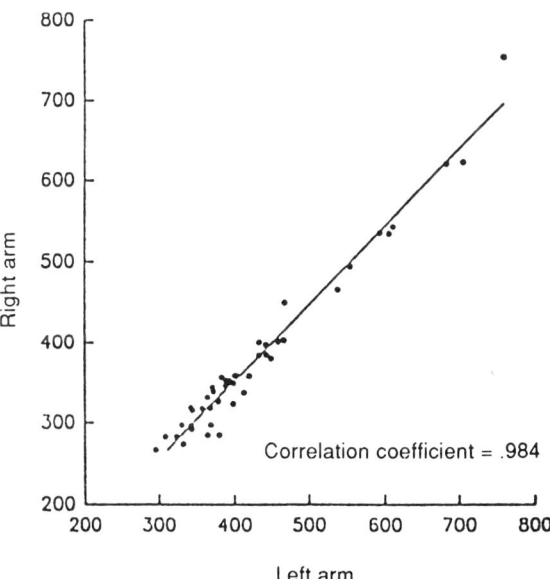

lation coefficients for the distal responses (.780 for the left hemifield and .743 for the right hemifield) fall just above the lower limit of the normal range (.730–.960) reported by Di Stefano *et al.* (1980). The subject's correlation coefficients for the proximal response (.967 for the left hemifield and .984 for the right hemifield) are fully comparable to the mean (.970) of the corresponding coefficients reported by Di Stefano *et al.* (1980) for normal controls (range .950–.980).

uncrossed responses relative to crossed responses. This finding is akin to an effect seen in normal subjects on dual tasks, where the two hands perform highly differentiated actions characterized by different degrees of difficulty. Usually the easy performance with one hand is slowed down so that its motor time comes to coincide with that of the difficult performance with the other hand (Kelso, Putnam & Goodman, 1983).

The responses studied in our simple bilateral tasks are likely to be emitted almost in a reflexlike fashion via relatively fixed and straightforward connections between visual and motor centers. Thus a perfect match between the short-pathway RT and the long-pathway RT cannot be expected on structural considerations alone, particularly where the absence of the corpus callosum increases the difference in length between the pathways. However, the speed of information transfer along these different pathways may be controlled by a superordinate center so as to effect a temporal coordination of the motor outputs of the two hemispheres. Experiments employing different paradigms from the present one have supported the notion that acallosal subjects possess mechanisms for gross synchronization and equalization of bilateral symmetrical movements (Tuller & Kelso, 1989) or for simple postural adjustments of one hand in anticipation of a movement of the other hand (Viallet, Massion, Massarino & Khalil, 1992). Alternatively, an interfering crosstalk between the motor systems controlling the two hands (e.g., Marteniuk, MacKenzie & Baba, 1984), rather than a truly coordinating action, may be responsible for the slowing down of the fastest response upon bilateral responding, and thus for the tendency to bimanual synchronization. There is no decisive argument in favor of one or the other assumption, both of which, however, concur in implying that either the coordinating or the interfering action must be able to operate between the hemispheres through both callosal and extracallosal pathways.

IV. Responses to Lateralized Flash with Movements Controlled by Bilaterally Distributed Motor Systems

We have seen that an interhemispheric transfer is necessary for responding with one hand to a stimulus in the contralateral visual field. A bilateral manual response to a lateralized flash engages motor systems in both hemispheres because moving each hand calls for a command from the appropriate hemisphere. By contrast, other upper-limb movements can be directly initiated by either hemisphere through bilaterally distributed motor pathways. Bilaterally distributed motor systems originating from each hemisphere are indeed available for the activation of axial and proximal limb muscles involved in global body movements, general postural adjustments, and integrated synergistic limb–body movements. Their existence has been demonstrated anatomically and physiologically in nonhuman primates (Kuypers, 1987; 1989) and confirmed by clinical and experimental evidence in humans (e.g., Freund, 1987; Colebatch & Gandevia, 1989; Müller, Kunesch, Binkofski & Freund, 1991; Colebatch, Deiber, Passingham, Friston & Frackowiak, 1991; Benecke, Meyer & Freund, 1991). The chief exponent of unilaterally distributed motor pathways is the crossed component of the corticospinal tract, while that of the bilaterally distributed motor systems is the cortico-reticulo-spinal tract.

Thus, it is theoretically possible for a visual input channeled into a single hemisphere to directly initiate and guide axial and proximal limb movements on both sides of the body, and CUDs may be expected to be absent when crossed as well as uncrossed visuomotor responses can be initiated by the hemisphere receiving the flash. Di Stefano *et al.* (1980) compared in normal subjects the CUD on a distal response, consisting of a keypress by a flexion of the thumb, with the CUD on a proximal response, consisting of a leverpull by a flexion of the forearm. They found indistinguishable significant CUDs for both types of response, but only when the responses were made unilaterally. A comparable result was reported by Milner, Miln & MacKenzie (1989), who found no differences in the CUD between a finger–thumb opposition response and an index lifting response presumably involving a movement of the whole hand. However, Di Stefano *et al.* (1980) described an annulment of the CUD on proximal responses (but not on distal responses) when such responses were executed bilaterally in reaction to the lateralized flash. Apparently, unilateral crossed responses to the flash, both distal and proximal, are elicited from the contralateral motor cortex and thus require an interhemispheric integration. On the contrary, bilateral proximal responses to a lateralized flash are actuated by a bilaterally distributed motor system, which ensures an approximate simultaneity of crossed and uncrossed reactions without the aid of interhemispheric integration. Quite recently we have found a similar absence of a CUD on an axial response consisting of an elevation of the shoulder, but in both unilateral and bilateral responding conditions (Tassinari, Berlucchi & Aglioti, in preparation). This finding suggests the possibility that responses of each shoulder can be effectively controlled by either the ipsilateral or contralateral hemisphere, in agreement with the pattern of motor cortex activation recently found during unilateral shoulder movements (Colebatch *et al.*, 1991).

The overall pattern of CUDs in normal subjects evidenced by the studies of Di Stefano *et al.* (1980), Milner *et al.* (1989), and Tassinari *et al.* (in preparation) allows a clear-cut distinction between crossed responses that presumably utilize interhemispheric transfer from those

TABLE II Presence or Absence of the CUD According to the Musculature and the Type of Task Involved

	Musculature involved		
	Distal	Proximal	Axial
Unilateral task	Presence	Presence	Absence
Bilateral task	Presence	Absence	Absence

that presumably do not (see Table II). The first set of responses includes unilateral and bilateral distal responses and unilateral proximal responses of the upper limb, all associated with significant CUDs, reflecting dependence on interhemispheric transfer. The second set includes bilateral proximal responses and unilateral and bilateral axial responses of the upper limb, all associated with null CUDs, reflecting independence from interhemispheric transfer. Logically, this distinction leads one to predict that impairment of interhemispheric transfer by callosal defects should alter the CUDs associated with the first set of responses, but not those associated with the second set. The CUD pattern that we found in a completely callosotomized subject fit this prediction very well (Aglioti et al., 1993). This subject showed CUDs on unilateral and bilateral distal responses and on unilateral proximal responses that were at least an order of magnitude greater than the typical 2–3 msec corresponding to CUDs of normal subjects. In the conditions that yield null CUDs in normal subjects, that is, bilateral proximal responses and both unilateral and bilateral axial responses, this subject exhibited insignificant or null CUDs (Figure 4).

Bilateral axial and proximal upper-limb responses to lateralized flashes are highly correlated. In addition, the correlation for proximal responses is substantially higher than that for distal responses, in normal subjects as well as in the acallosal subject (see Figure 3). These responses are also synchronous inasmuch as the crossed RT does not differ from the uncrossed RT, and again this is as true in the acallosal as in the normal control subjects. This difference with the systematic asynchrony of bilateral distal responses is best explained by assuming that such responses always require the concurrent activation of two motor commands, one from each hemisphere, whereas bilateral proximal and axial response can be issued after one motor command from a single hemisphere. An additional assumption is that unilateral proximal responses of the upper limb are effected through a crossed pathway, whereas the same responses, when made bilaterally, engage a different, bilaterally distributed motor system, which ensures a yoked movement of the two sides. Experimental evidence concerning facial

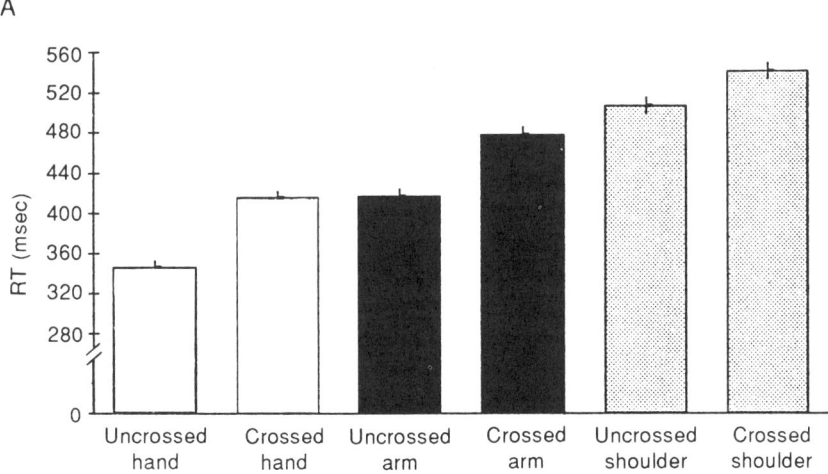

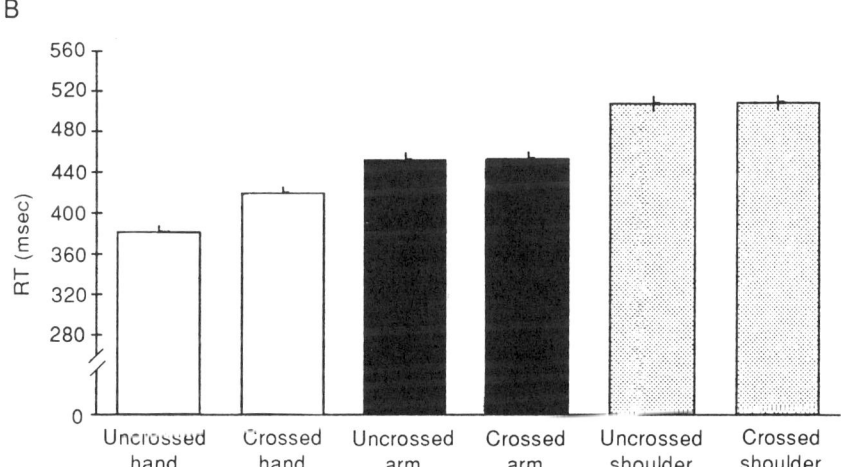

FIGURE 4 Crossed–uncrossed differences in a callosotomized subject as a function of responding effector and unilateral (A) or bilateral (B) responding condition. The figure gives mean reaction times (plus S.E.) for crossed and uncrossed distal (hand) responses, proximal (arm) responses, and axial (shoulder) responses in both the unilateral and the bilateral responding condition. Crossed and uncrossed responses are as defined in Figure 2. Uncrossed reaction times (RT) are significantly shorter ($p < .01$) than crossed RTs on unilateral and bilateral distal responses and on unilateral proximal responses; crossed–uncrossed differences on the remaining responses, including the unilateral axial (shoulder) response, are statistically insignificant. Each column represents 90 reaction times.

motility supports the hypothesis that unilateral components of a bilateral response can be produced by neural systems different, at least in part, from those mediating the same movements when made unilaterally. For example, Gazzaniga and Smylie (1990) found that in the generation of posed smiling by the left hemisphere of commissurotomy patients, the right side of the mouth began to move 90–180 msec before the left side, a facial CUD comparable in both kind and degree with those observed with hand and arm responses in our callosotomized subject. However, the same comissurotomy patients did not display any facial asymmetry during spontaneous smiling, suggesting the operation of a different bilaterally synchronizing motor system. Since the corpus callosum had been cut, bilateral synchronization was obviously achieved independent of fast interhemispheric coordination.

V. Summary and Conclusions

Speeded bilateral symmetrical responses to visual inputs restricted to one hemisphere can never be perfectly synchronous. This appears to be due to the difference in length between the neural pathways mediating the responses on the two sides. When the two responses are emitted by different hemispheres, as it occurs with distal hand movements, there is an irreducible advantage in speed for the response emitted by the hemisphere that receives the visual stimulus. However, this advantage is minimized by the corpus callosum, which allows an efficient interhemispheric communication for the fast integration of the disadvantaged response. This syncronizing callosal influence is inferred from the increased asynchrony of bilateral hand responses, which occurs in the absence of the corpus callosum, whether congenital or acquired. However, the persistence of strong temporal correlations between bilateral hand responses to lateralized visual stimuli in acallosal patients suggests a nonnegligible contribution of extracallosal mechanisms to the temporal coordination of the motor outputs from the two hemispheres. These extracallosal mechanisms for bilateral temporal coordination of hand responses are not yet understood. More information is available as to the mechanisms that ensure the bilateral synchronization of responses effected with axial and proximal arm muscles. Each hemisphere can control these responses on both sides of the body, so that the synchronization between the two sides is made possible by the common origin of the motor commands and by the bilateral distribution of the pathways transmitting them. In this case, bilateral synchronization is totally independent from the corpus callosum.

Acknowledgments

Recent studies from the authors' laboratory reported in this chapter have been aided by grants from the Ministero per l'Università e la Ricerca Scienitifica e Tecnologica and the Consiglio Nazionale delle Ricerche of Italy, and from the Human Frontier Science Programme Organization.

References

Aglioti, S., Dall'Agnola, R., Girelli, M., & Marzi, C. A. (1991). Bilateral hemispheric control of foot distal movements: evidence from normal subjects. *Cortex*, **27**, 571–581.

Aglioti, S., Berlucchi, G., Pallini, R., Rossi, G. F., & Tassinari, G. (1993). Hemispheric control of unilateral and bilateral responses to lateralized light stimuli after callosotomy and in callosal agenesis. *Experimental Brain Research* **95**, 151–165.

Bard, C., Paillard, J., Lajoie, Y., Fleury, M., Teasdale, N., Forget, R., & Lamarre, Y. (1992). Role of afferent information in the timing of motor commmands: A comparative study with a deafferented patient. *Neuropsychologia*, **30**, 201–206.

Bashore, T. R. (1981). Vocal and manual reaction time estimates of interhemispheric transmission time. *Psychological Bulletin*, **89**, 352–368.

Benecke, R., Meyer, B..-U., & Freund, H.-J. (1991). Reorgnisation of descending motor pathways in patients after hemispherectomy and severe hemispheric lesions demonstrated by magnetic brain stimulation. *Experimental Brain Research*, **83**, 419–426.

Berlucchi, G., Crea, F., Di Stefano, M., & Tassinari, G. (1977). Influence of spatial stimululs–response compatibility on reaction time of ipsilateral and contralateral hand to lateralized light stimuli. *Journal of Experimental Psychology: Human Perception & Performance*, **3**, 505–517.

Clarke, J. M., & Zaidel, E. (1989). Simple reaction times to lateralized light flashes: Varieties of interhemispheric communication routes. *Brain*, **112**, 849–870.

Colebatch, J. G., & Gandevia, S. C. (1989). The distribution of muscular weakness in upper motor neuron lesions affecting the arm. *Brain*, **112**, 749–763.

Colebatch, J. G., Deiber, M.-P., Passingham, R. E., Friston, K. J., & Frackowiak, R. S. J. (1991). Regional cerebral blood flow during voluntary arm and hand movements in human subjects. *Journal of Neurophysiology*, **65**, 1392–1401.

Di Stefano, M., Morelli, M., Marzi, C. A., & Berlucchi, G. (1980). Hemispheric control of unilateral and bilateral movements of proximal and distal parts of the arms as inferred from simple reaction time to lateralized light stimuli in man. *Experimental Brain Research*, **38**, 197–204.

Di Stefano, M., Sauerwein, H. C., & Lassonde, M. (1992). Influence of anatomical factors and spatial compatibility on the stimulus–response relationship in the absence of the corpus callosum. *Neuropsychologia*, **30**, 177–185.

Freund, H.-J. (1987). Abnormalities of motor behavior after cortical lesions in humans. In V. B. Mountcastle, F. Plum, & S. R. Geiger (Eds.), *Handbook of physiology* (Vol. V) Part 2, (pp. 763–810). Bethesda, MD: American Physiological Society.

Gazzaniga, M. S., & Smylie, C. S. (1990). Hemispheric mechanisms controlling voluntary and spontaneous facial expressions. *Journal of Cognitive Neuroscience*, **2**, 239–245.

Hongo, T., Nakamura, R., Narabayashi, H., & Oshima, T. (1976). Reaction times and their left-to-right differences in bilateral symmetrical movements. *Physiology & Behavior*, **16**, 477–482.

Jeannerod, M. (1988). *The neural and behavioural organization of goal-directed movements*. Oxford: Oxford University Press.

Jeeves, M. A. (1969). A comparison of interhemispheric transmission times in acallosals and normals. *Psychonomic Science*, 16, 245–246.
Jeeves, M. A., Silver, P. H., & Jacobson, I. (1988). Bimanual co-ordination in callosal agenesis and partial commissurotomy. *Neuropsychologia*, 26, 833–850.
Jung, R., & Dietz, V. (1975). Verzögerter Start der Willkürbewegung bei Pyramidenläsionen des Menschen. *Archiv für Psychiatrie und Nervenkrankeiten*, 221, 87–109.
Kelso, J. A. S., Putnam, C. S., & Goodman, D. (1983). On the space-time structure of human interlimb coordination. *Quarterly Journal of Experimental Psychology*, 35A, 347–375.
Kuypers, H. G. J. M. (1987). Some aspects of the organization of the output of the motor cortex. In *Ciba Foundation Symposium 132: Motor areas of the cerebral cortex* (pp. 63–820). Chichester: Wiley.
Kuypers, H. G. J. M. (1989). Motor system organization. In G. Adelman (Ed.), *Encyclopedia of neuroscience*, Suppl. 1 (pp. 107–110). Boston: Birkhäuser.
Levy, J., & Wagner, N. (1984). Handwriting posture, visuomotor integration, and lateralized reaction-time parameters. *Human Neurobiology*, 3, 157–161.
Marteniuk, R. G., MacKenzie, C. L., & Baba, D. M. (1984). Bimanual movement control: Information processing and interaction effects. *Quarterly Journal of Experimental Psychology*, 36A, 335–365.
Marzi, C. A., Bisiacchi, P., & Nicoletti, R. (1991). Is interhemispheric transfer of visuomotor information asymmetric? Evidence from a meta-analysis. *Neuropsychologia*, 29, 1163–1177.
Milner, A. D. (1982). Simple reaction times to lateralized visual stimuli in a case of callosal agenesis. *Neuropsychologia*, 20, 411–419.
Milner, A. D., Jeeves, M. A., Silver, P. H., Lines, C. R., & Wilson, J. G. (1985). Reaction times to lateralized visual stimuli in callosal agenesis: stimulus and response factors. *Neuropsychologia*, 23, 323–331.
Milner, A. D., Miln, A. B., & Mackenzie, A. M. (1989). Simple reaction times to lateralized visual stimuli using finger-thumb apposition. *Medical Science Research*, 17, 859–860.
Müller, F., Kunesch, E., Binkofski, F., & Freund, H.-J. (1991). Residual sensorimotor functions in a patient after right-sided hemispherectomy. *Neuropsychologia*, 29, 125–145.
Paillard, J. (1948). Quelques données psychophysiologiques relatives au déclenchement de la commande motrice. *Année Psychologique*, 48, 29–48.
Piéron, H., & Jones, M. H. (1959). Nervous pathways of cutaneous pain. *Science*, 129, 1547–1548.
Poffenberger, A. T. (1912). Reaction time to retinal stimulation with special reference to the time lost in conduction through nervous centers. *Archives of Psychology*, 23, 1–73.
Preilowski, B. (1972). Possible contribution of the anterior forebrain commissures to bilateral coordination. *Neuropsychologia*, 10, 266–277.
Reynolds, D. McQ., & Jeeves, M. A. (1974). Further studies of crossed and uncrossed pathway responding in callosal agenesis—Reply to Kinsbourne and Fisher. *Neuropsychologia*, 12, 287–290.
Saron, C. D., & Davidson, R. J. (1989). Visual evoked potential measures of interehemispheric transfer time in humans. *Behavioral Neuroscience*, 103, 1115–1138.
Sergent, J., & Myers, J. J. (1985). Manual, blowing, and verbal simple reactions to lateralized flashes of light in commissurotomized patients. *Perception & Psychophysics*, 37, 571–578.
St. John, R., Shields, C., & Timney, B. (1987). The reliability of estimates of interhemispheric reaction times derived from unimanual and verbal response latencies. *Human Neurobiology*, 6, 195–202.

Tassinari, G., Morelli, M., & Berlucchi, G. (1983). Interhemispheric transmission of information in manual and verbal reaction-time tasks. *Human Neurobiology,* **2,** 77–85.

Tassinari, G., Berlucchi, G., & Aglioti, S. Absence of side effects in axial responses to lateralized light stimuli (in preparation).

Tuller, B., & Kelso, J. A. S. (1989). Environmentally-specified patterns of movement coordination in normal and split-brain subjects. *Experimental Brain Research,* **75,** 306–316.

Umiltà, C. A. S., & Nicoletti, R. (1990). Spatial stimulus-response compatibility. In R. W. Proctor & T. G. Reeve (Eds.), *Stimulus–response compatibility* (pp. 89–116). Amsterdam: Elsevier.

Vallar, G., Sterzi, R., & Basso, A. (1988). Left hemisphere contribution to motor programming of aphasic speech: A reaction time experiment in aphasic patients. *Neuropsychologia,* **26,** 511–519.

Viallet, F., Massion, J., Massarino, R., & Khalil, R. (1992). Coordination between posture and movement in a bimanual load lifting task: putative role of a medial frontal region including the supplementary motor area. *Experimental Brain Research,* **88,** 674–684.

# Coordination of Cyclic Coupled Movements of Hand and Foot in Normal Subjects and on the Healthy Side of Hemiplegic Patients

Fausto Baldissera and Paolo Cavallari

Istituto di Fisiologia Umana II
Università di Milano
Milano, Italy

Luigi Tesio

Servizio di Fisiatria
Clinica Ortopedica H.S Raffaele
Milano, Italy

I. Introduction and Methods
II. Characteristics of the "Easy" and "Difficult" Coupling
III. Kinesthetic Control of Coupled Movements
    A. In-Phase (Easy) Association
    B. Anti-Phase (Difficult) Association
    C. Changes in the Timing of Muscular Activation in the Two Types of Coupling and under Different Loading Conditions
IV. Movement Coupling on the "Healthy" Side of Hemiplegic Patients
    A. Left Hemiplegia
    B. Right Hemiplegia
V. Concluding Remarks
References

1. Introduction and Methods

Several common life gestures combine movements of different limb segments. The repertoire, however, is not unlimited and a number of neural constraints make certain movements, which can be easily performed in isolation, difficult to associate. Examples are represented by the difficulty of drawing circles on the sagittal plans with the two hands moving in opposite directions (Meige, 1901), or that of beating different (nonmultiple) rhythms with two body segments (Klapp, Hill, Jagacinski, Jones, Tyler & Martin 1985).

Interest for interlimb coordination during voluntary actions has mainly been focused on bimanual synergies but, recently, attention has also been drawn to the principles governing the coordination of movements of limb segments on the same side. When examining several couples of coplanar movements of the upper and lower limb (Baldissera, Cavallari & Civaschi, 1981) it was ascertained that, as a rule, association is easily achieved when the limb segments rotate in the same direction, whereas it is difficult to move them in opposite directions. This holds true for single acts as well as for cyclic sequences, a case in which the isodirectional movements can also be described as "in-phase" and those of opposite direction as "anti-phase."

A privileged association between isodirectional, in-phase, movements was observed for several combinations of movements of the upper and lower limbs. A short list of examples includes: axial rotation of arm and leg; pronosupination of hand and foot; flexion–extension of hand and foot. In addition, moving two joints of the same limb (e.g., flexion–extension of the wrist and elbow) seems to undergo the same direction principle (cf. Kots, Krinskiy, Naydin & Shik, 1971; Kelso, Buchanan & Wallace, 1991): in this case, however, coupling is affected by mechanical influences as well.

Among the variety of possible associations, the coupling of flexion–extension movements of the hand and foot has been chosen as a model for analyzing the principles governing this type of coordination. Reasons for this choice are the absence of direct mechanical influences between the two segments and the relative simplicity, in comparison to other combinations, of instrumental monitoring and applying mechanical perturbations.

Coupled movements of the hand and foot were performed by both normal subjects, with no sign of neurological disorders (aged 20–46), and by chronic hemiplegic patients, all being right-handed.

Subjects sat on an armchair with the forearm supported in horizontal position. The hand, prone or supine, was tied to a light frame rotating coaxially with the wrist joint. In some of the experiments the frame was used to connect the hand to different perturbing loads. The mass was increased by a lead disk concentric to the rotation axis of the wrist (inertial momentum 15 g · m^2); viscous and elastic resistances were enhanced by connecting to the frame, 5 cm away from the wrist pivot, either a piston moving in an oil-filled cylinder (viscosity, 0.37 N · s/°) or two symmetric rubber threads (elasticity, 4 g/°), respectively. The foot was also lying on a platform pivoting on the ankle joint axis. Angular displacement of the frames and surface EMG from the hand and foot movers (extensor carpi radialis, ECR; flexor carpi radialis, FCR; tibialis anterior, TA; and gastrocnemius-soleus, GS) were recorded and analyzed. When not otherwise stated, movements were performed on the right side. The goniometric and electromyographic signals were digitalized for automatic analysis of the phase relations. For the movements, the cycle starting point was arbitrarily set at half-excursion, that is, when the speed was maximal, and the phase difference was then calculated between the midpoints of the hand and foot cycles. For muscular activity, the cycle duration was the delay between the onset of two successive EMG bursts and the phase was again calculated on the midpoints of the two EMG cycles.

II. Characteristics of the "Easy" and "Difficult" Coupling

When a subject with the hand prone is asked to perform coupled flexion–extension of the hand and foot in a way he feels is easy and spontaneous, he automatically associates the hand extension with the foot dorsal flexion and the hand flexion with the foot plantar flexion. Angular movements are isodirectional or, when cyclic, in-phase. This effortless coupling ("easy association") can be maintained up to a rate of 3–4 Hz. Curiously, the maximal frequency attainable by the coupled movements is somewhat higher than that reached by the slower segment (the foot) when moved alone. On its turn, the hand alone can oscillate at frequencies as high as 7–8 Hz. If during the task execution, the subject is asked to invert the phase between the two alternating movements, he cannot switch "en course" to the opposite association, but must stop and start the task *ex novo*. Once started, the anti-phase association requires continuous attentive effort and any release of voluntary control or the attempt to increase the rhythm above 1.5–2 Hz inescapably leads back to the in-phase pattern. The anti-phase coupling has therefore been termed "difficult association."

When movements are performed on the left side, the same general features are observed but the maximal frequency attainable in both the in-phase and anti-phase coupling is slightly lower (cf. Figure 5) than on the right side. The privileged coupling between hand and foot does not depend on a special pattern of muscular innervation. When movements are performed with the hand supine, the easy association in fact becomes that between the hand flexion and the foot dorsal flexion, that is, contrary to that occurring with the hand prone. Thus, the relevant factor determining the ease or difficulty of coupling appears to be the mutual direction of the two movements, quite independently of the muscles involved.

When subjects are asked to perform rhythmic oscillations of hand and foot at different frequencies, paced by a metronome, the performance duration is quite different in the easy and in the difficult association (Baldissera, Cavallari, Marini & Tassone, 1991). In the easy coupling, movements can be continued for more than 1 minute up to 2–2.5 Hz. At higher rates the exercise duration is shortened below 60 sec because of muscular pain. In the difficult association the cyclic movements can be maintained for at least one minute only up to 0.7–1.7 Hz; at higher frequencies the sequence is cut short before any muscular fatigue develops by a sudden reversal to the in-phase pattern. The duration over which the anti-phase oscillations can be maintained rapidly shortens when the frequency is increased: in the subjects so far examined, a trial could be continued for 10 seconds at frequencies not higher than 1.2–2.4 Hz. At his own limit-frequency for anti-phase coupling, each subject could keep the easy coupling for 1 minute or longer.

III. Kinesthetic Control of Coupled Movements

It is reasonable to believe that kinesthetic afferences generated by the ongoing movements are utilized for controlling the synchronism between the hand and foot oscillations. It was therefore of interest to evaluate the role of the feedback control in the two types of coupling. With this aim, the phase relations between hand and foot oscillations at different frequencies ranging between 0.6 and 3–4 Hz were studied under normal conditions and after loading the hand. Three types of loads were used, each expected to have a different effect on movements: (1) an elastic resistance, whose effect is related to the hand displacement; (2) a viscous resistance, which grows proportionally to the movement velocity; and (3) an inertial load, whose action parallels the acceleration. The two latter types of resistance should increase when the movement frequency rises, while the first one should remain constant.

Surprisingly, the easy (in-phase) movements were found to be more strongly affected by application of loads than the difficult (anti-phase) ones.

A. In-Phase (Easy) Association

With the hand free, raising the movement frequency was accompanied by an increasing lag of the hand cycle with respect to the foot cycle from about 10° at 0.6 Hz to around 30° at 3 Hz (Figure 1A). After application of the lead disk, the hand phase-delay was similar to that recorded in unloaded conditions at 0.6–0.8 Hz but it progressively grew when the frequency was increased, reaching 40–50° at around 3 Hz (Figure 1B).

The viscous load also produced a marked phase lag of the hand, comparable in size and frequency dependence to that produced by the inertial load (Figure 1C). Instead, the elastic load caused a clear-cut improvement of the phase relation, allowing the hand–foot synchronism to be maintained over the entire frequency range (Figure 1D).

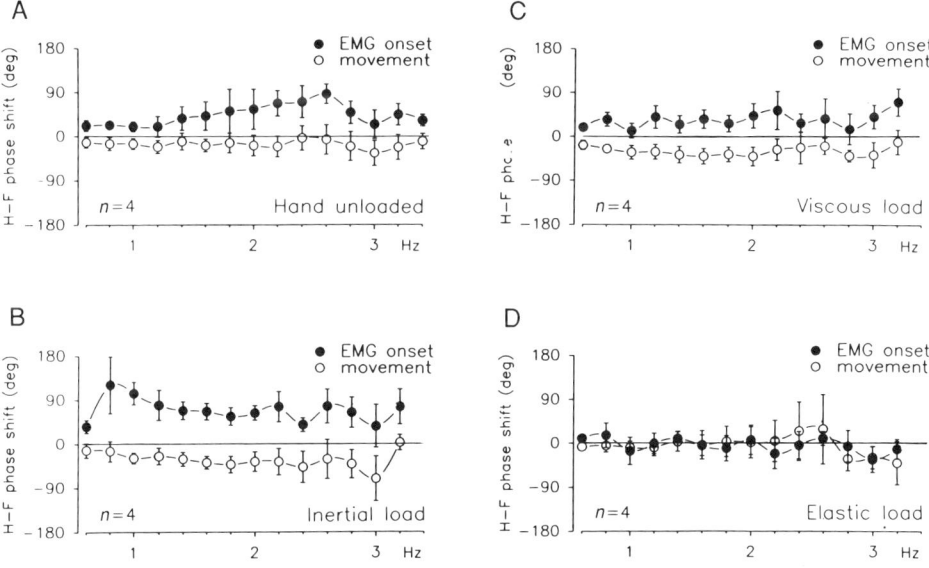

FIGURE 1 Effects of hand loading on rhythmic flexion–extension of hand and foot at different frequencies, in-phase movements. (○) The phase relations between hand and foot oscillations; (●) the phase relations between the onset of EMG activities in extensor carpi radialis and in tibialis anterior muscles. Average of 4 subjects. Vertical bars give the standard deviation of the mean. The phase angle is positive when movement or EMG of the hand leads that of the foot.

B. Anti-Phase (Difficult) Association

When performing the anti-phase movements, the task-required synchronization (180°) was strictly maintained by all subjects at all frequencies between 0.6 and 2.2 Hz, independently of the fact that the hand was unloaded or charged with any of the three previously mentioned loads (Figure 2,A–D).

C. Changes in the Timing of Muscular Activation in the Two Types of Coupling and under Different Loading Conditions

The frequency-related phase changes occurring in the easy association as a consequence of loading and, conversely, the strict maintenance of the anti-phase relation in the difficult coupling notwithstanding the loads, should rest on variations of the muscular drive. In principle, one should expect that loading the hand would induce an earlier and/or faster activation of hand movers in respect to foot movers. Since the relationships between the changes in the EMG amplitude and the force profile are not unequivocal, analysis was restricted to the changes in timing of EMG activity in the muscles moving the hand and the foot.

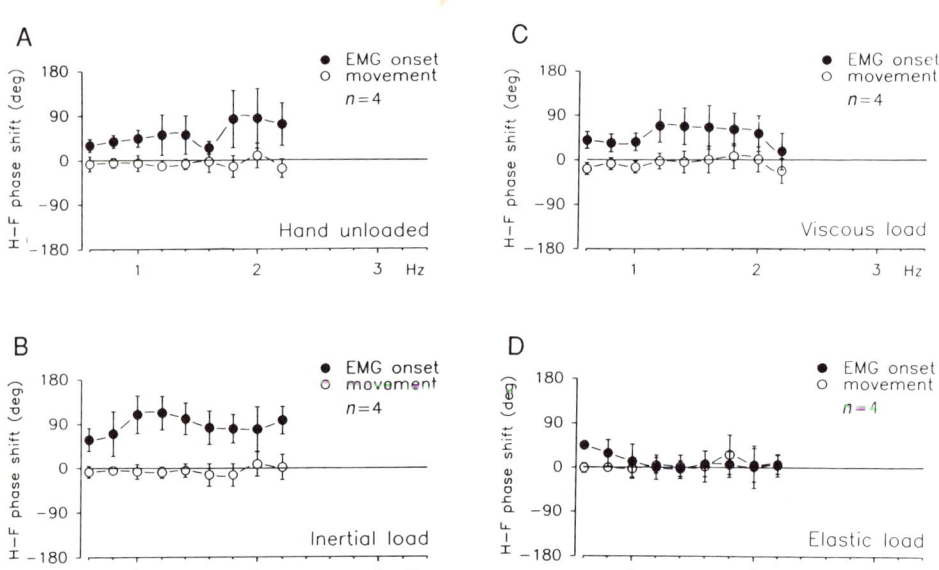

FIGURE 2 Effects of hand loading on rhythmic flexion–extension of hand and foot at different frequencies, anti-phase movements. Subjects and symbols are the same as in Figure 1.

This analysis showed that the onset of EMG bursts was earlier in the hand extensor ECR than in the foot dorsal flexor TA in all experimental situations, except under elastic loading. The time advance, however, was different under the various conditions and for the different frequencies.

In the easy association with the hand unloaded (even without the supporting frame) the EMG onset was slightly earlier in ECR than in TA at low frequencies (<1 Hz) and this time-advance progressively increased thereafter (Figure 1A). In the difficult coupling (Figure 2A), the same changes occurred, but to a greater extent.

Two points deserve a comment. First, even when the hand is free, synchronization of hand and foot requires that the forearm muscles contract before the leg muscles, the anticipation being larger as the frequency increases. This feature possibly reflects the different mechanical characteristics of the extremities, in the hand system the inertial reaction being larger, relative to the viscous and elastic ones, than in the foot. The second point regards the larger anticipation occurring (whatever the frequency) in the difficult than in the easy coupling. This feature, which is also seen under inertial and viscous loading (see below), possibly represents the main mechanism allowing the required anti-phase relations to be strictly respected in this association.

When the lead disk was applied, a clear-cut adjustment of the hand control was appreciable in the easy coupling. Already at the lowest frequency (0.6 Hz) the ECR EMG started in great advance of the TA EMG and the anticipation grew to 90–100° at 1 Hz (Figure 1B). The ECR time-advance effectively compensated for the inertial load and kept the hand phase-delay at the same value as in unloaded conditions. When the frequency rose above 1 Hz, however, the ECR phase-lead diminished, causing the progressive lag of the hand movement. In the difficult association (Figure 2B), the EMG phase differences were much greater than in the easy one, reaching 120–140° (Figure 2B), and were maintained over the whole frequency range, thus granting a more efficient phase control.

Under viscous loading the timing of ECR and TA activation during in-phase coupling was only slightly changed compared to the hand-free condition (compare Figure 1C with 1A). The paucity of the reaction to this type of loading apparently justifies the increased phase-lag of the hand movement in this condition. During anti-phase coupling, ECR precedence was larger and apparently sufficient to maintain the 180° synchronism at all frequencies (Figure 2C).

The application of the elastic resistance resulted in a synchronization of the EMG onset in the muscles moving the hand and the foot, which occurred both in the easy (Figure 1D) and in the difficult coupling (Figure 2D). The simultaneous activation of the two groups of muscles suggests that the mechanical properties of the hand and foot have become similar in this condition, possibly because of the counteraction of the hand inertia by the applied elastic resistance. Matching of actual movements with the task requirements in the various loading conditions is based on afferent signals from the moving limbs. Thus, since in all conditions the movements are more strictly coupled in the difficult (anti-phase) than in the easy (in-phase) association, it may be concluded that afferences signaling movement coupling are more proficiently utilized in the former than in the latter.

IV. Movement Coupling on the "Healthy" Side of Hemiplegic Patients

The ability to perform the associated movements, in both the in-phase and anti-phase coupling, has been tested in 6 right-handed patients (4 men and 2 women) with chronic (>6 months) hemiparesis following subarachnoid hemorrhage (2 cases) and cerebral ischemia (4 cases). None of them presented hemianopsia, hemispatial neglect, or aphasia. All could ambulate autonomously and were fully cooperative. The unaffected side was effectively employed to compensate for the paretic one in daily living tasks. Movements were performed on the unaffected side, where no clinical signs of either motor or sensory deficit were evident. The need for limiting the discomfort caused by long experimental sessions restricted examination to tests with the hand unloaded and connected to the inertial load.

A. Left Hemiplegia

When performing in-phase movements on the healthy side with the hand free, the 4 patients with left hemiplegia reached frequencies comparable to those of normal subjects, but showed a frequency-dependent lag of the hand cycle (Figure 3A) comparable to that seen in normals after inertial or viscous loading. The phase shift could be ascribed to the absence of any time advance of the ECR EMG, whose onset remained synchronous with that of the TA over the whole frequency range.

When the patients' hand was loaded with the mass, the hand phase-lag grew larger, reaching 90° at 2.2–2.4 Hz (Figure 3B). The attainment of such a large delay is probably the reason why the movement frequency could not be further increased. Compared to normal subjects in the same experimental condition, at each frequency the phase lag was about 20–30° greater (compare Figure 1B and Figure 3B). EMG analysis showed that the attempt for load compensation was scarce, consist-

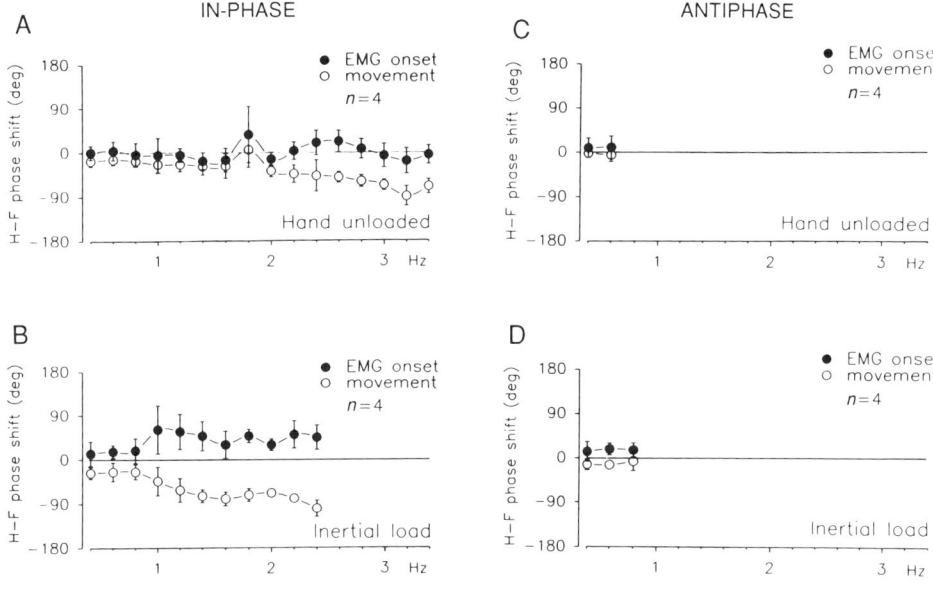

FIGURE 3 Phase relations between hand and foot movements and EMG onsets (ECR and TA muscles) in 4 patients with left hemiplegia. Symbols are the same as in Figure 1 and 2; (A, B), In-phase movements; (C, D), anti-phase movements. The hand was unloaded (A and C) or connected to an inertial load (B and D).

ing of a small time-advance of the ECR versus the TA EMGs. The impairment was even more evident when patients executed the antiphase movements, a task that none of them was able to perform at frequencies exceeding 0.8 Hz, with both the hand unloaded (Figure 3C) and mass-loaded (Figure 3D). The anticipatory shift of the ECR EMG, which connotes the difficult association in the healthy subjects, was completely absent in these patients (compare Figure 3,C–D and Figure 2, A and B).

B. Right Hemiplegia

All of the disturbances described above were even more accentuated in the two patients with right hemiplegia (Figure 4), who performed the task on the left side. In the easy association the upper frequency limit was as low as 1.4 Hz with the hand unloaded and 1 Hz after inertial loading. The difficult coupling, which was performed up to 0.8 Hz with the hand unloaded, could not be exploited at all after application of the mass. Since all subjects were right-handed, the greater impairment of

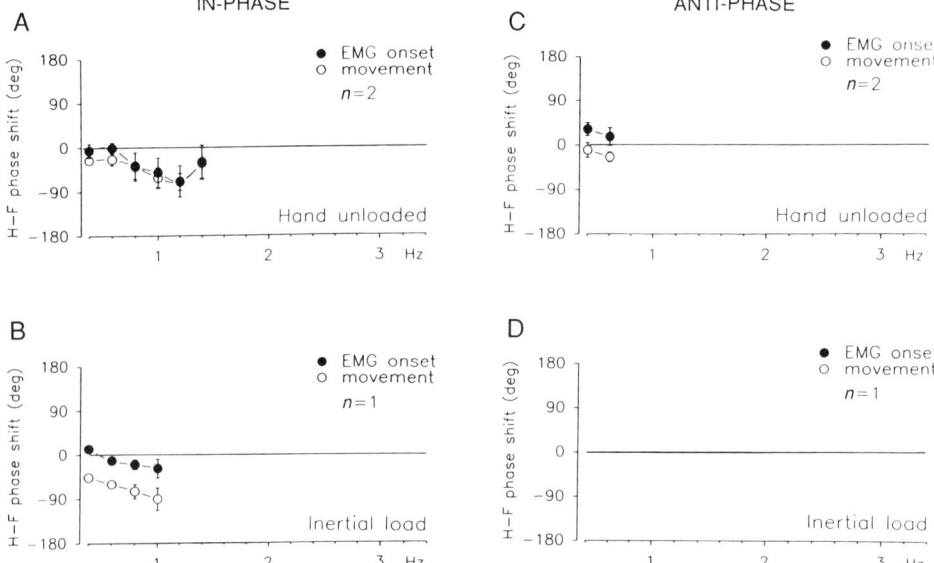

FIGURE 4 Phase relations between hand and foot movements and EMG onsets in 2 patients with right hemiplegia. Symbols are the same as in the preceding figures.

associated movements found in the right in comparison to left hemiplegics suggests that the specialization of the left hemisphere in hand control may also be extended to the coupled movements of hand and foot. This is also suggested by the higher frequency reached on the right side by normal right-handed subjects, in both the easy and the difficult associations (Figure 5).

V. Concluding Remarks

The results presented above allow some speculations about the mechanism controlling coupling of these ipsilateral limb movements.

The in-phase coupling may be imagined as the result of a parallel and synchronous innervation of the spinal circuits supporting the movements of hand and foot, perhaps assisted by intraspinal linkages like those connecting the locomotion generators for the fore- and hindlimb (Viala & Vidal, 1978), with little need of a feedback control from the periphery. However, the adjustments occurring in the EMG onset and amplitude after loading suggest that the neural oscillators that control

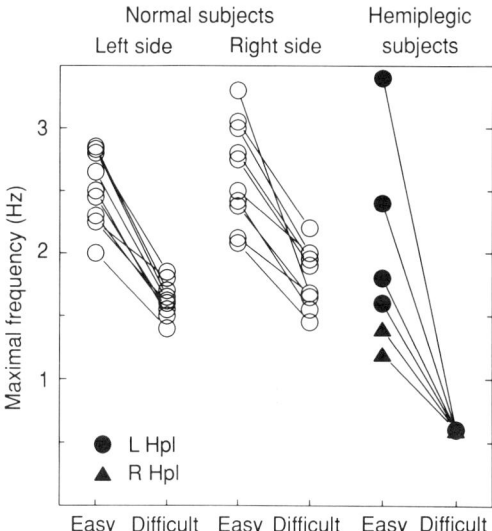

FIGURE 5 Frequency limits in the in-phase (easy) and anti-phase (difficult) coupling performed by 10 normal subjects and by 6 hemiplegic subjects on the left and right sides. Symbols refer to left hemiplegia (L Hpl, right-side movements) and right hemiplegia (R Hpl, left-side movements).

hand and foot influence each other through kinesthetic afferences. Entrainment of the rhythm of central pattern generators by afferent signals has already been described for the step generator (Andersson & Grillner, 1983; Conway, Hultborn & Kiehn, 1987) and for the respiratory cycle (Iscoe & Polosa, 1976; Kawahara, Kumagai, Nakazono & Miyamoto, 1988). Moreover, kinesthetic afferences from the upper limb have been shown to be essential for coordination of isodirectional eye and arm movements (Gautier & Mussa Ivaldi, 1988; Gautier, Vercher, Mussa Ivaldi & Marchetti, 1988). In the in-phase coupling, however, the influence exerted by the peripheral afferences is weak, since it allows the inertial load to cause a prominent lag of the hand cycle.

In this regard, it seems worth noting that during the "easy" coupling the hand delay never exceeds 90°. Actually, a 90° lag characterizes the last points of the phase plots of both normal subjects under inertial load (Figure 1B) and patients with the hand unloaded (Figure 3A and B) or loaded (Figure 4A and B), as if this delay, and not the frequency, is the limit beyond which a regular coupling is not further warranted. The existence of a limit value for the phase-lag may point to the characteristics of the control system. In most physiological movements the hand and the foot behave like second-order systems (cf. Stark, 1968), characterized by two reactive elements (elastic and viscous). In such systems, the input–output phase-relation tends to 180° when the frequency increases, the shape of the phase-frequency relation depending on the values of the mechanical parameters. A feedback control, intended to provide phase-coupling between two second-order systems with different frequency responses (e.g., hand and foot) and relying on detection of the differences between the kinematic variables of the two systems, is unable to maintain the phase within 90° if the position is monitored, but it will respect this limit if the velocity is also considered (Anderson & Moore, 1971).

Weakness of the peripheral feedback control during the in-phase coupling may actually explain why anti-phase coupling is possible, although difficult. In the latter situation one should in fact supply the properly timed muscle excitation and, at the same time, prevent the peripheral signals from entraining the in-phase coupling: it is conceivable that this inhibitory control would not prevail if the afferent feedback were too powerful.

Considering the feedback control of anti-phase coupling, experiments show that it realizes a quite strict 180° synchronization in all tested conditions. This control does not tolerate phase shifts and when they occur a phase inversion follows. To grant such an accurate coupling, EMG anticipation is always larger in anti-phase than in in-phase coupling (except when under elastic load). However, it should be noted that under inertial or viscous loading the EMG phase-advance remains constant when the frequency rises above 1 Hz. Since in this same range no phase shift takes place between movements, other mechanisms must contribute to compensate for the different mechanical properties of hand and foot. Analysis of the integrated EMG, displayed on a normalized cycle, suggests that a faster development of force may be a second compensatory mechanism. Even if the force time course cannot be directly inferred from the EMG profile, steepening of the EMG rising phase when frequency increases indicates a parallel speeding up of the force development. This would have the same mechanical effect on the rhythmic oscillation as a time-advance of a sinusoidal force.

In conclusion, the in-phase and anti-phase coupling of ipsilateral hand and foot seems to be controlled by two different feedback systems, that is, a loose one, allowing consistent phase-lags, for the in-phase coupling, and a more potent one, requiring strong attentive effort, which provides the strict anti-phase synchronization. Afferent signals seem thus to be elaborated at different levels in the two associations. During in-phase coupling afferences are likely processed by a low-gain, automatic feedback mechanism, which operates on the efferent pathways, perhaps at the spinal level. Coupling in phase-opposition appears to rely instead on conscious measurement of the phase-relation between the limbs, maybe at a certain point of the cycle. The time and attention required for this process might be the factors that limit the frequency and the duration of the anti-phase association. Hemiplegic patients are deficient both in performing anti-phase association and in compensating for mass-loading during in-phase movements (Baldissera, Cavallari & Tesio, 1990), that is, they have lost two aspects of movement control depending on kinesthetic information. Since they did not exhibit any sensory loss on the unaffected side, damage should affect the central mechanisms that utilize the kinesthetic inflow to coordinate movements association. Thus, the hand–foot incoordination that follows unilateral brain injuury of either side shows that, although these mechanisms exert a unilateral action, their proper function requires the integrity of both hemispheres.

A number of motor disturbances affecting the "healthy" side of hemiplegic patients have been recently described. These include changes in the biceps and triceps stretch reflexes (Thilmann, Fellows, & Garms, 1990), muscle weakness in upper and lower limbs (Adams, Gandevia & Skuse, 1990; Colebatch & Gandevia, 1989), prolonged reaction times, and alterations of ballistic and tracking movements of the arm (Jones, Donaldson, & Parkin, 1989). These disturbances, however, are mild and unlikely to be detected in routine clinical examination. By contrast, the inability to perform anti-phase movements can easily be ascertained at the bedside without need of any instrumental aid. This defect, which is recorded regardless of the overall severity of the clinical picture, may be taken as a sensitive and reliable marker of organic neural damage whenever hysteria or malingering is suspected to underlie the motor deficit.

References

Adams, R. W., Gandevia, S. C., & Skuse, N. F. (1990). The distribution of muscle weakness in upper motor neuron lesions affecting the lower limb. *Brain,* **113,** 1459–1476.
Anderson, B. D. O., & Moore, Y. B. (1971). *Linear optimal control.* New York: Prentice-Hall.
Andersson, O., & Grillner, S. (1983). Peripheral control of the cat's step cycle. *Acta. Physiol. Scand.,* **118,** 229–239.
Baldissera, F., Cavallari, P., & Civaschi, P. (1982). Preferential coupling between voluntary movements of ipsilateral limbs. *Neurosci. Lett.,* **34,** 95–100.
Baldissera, F., Cavallari, P., & Tesio, L. (1990). Coupled movements of hand and foot reveal a motor impairment on the healthy side of hemiplegic patients. *Neurosci. Lett.,* **39** (Suppl.), S14.
Baldissera, F., Cavallari, P., Marini, G., & Tassone, G. (1991). Differential control of in-phase and anti-phase coupling of rhythmic movements of ipsilateral hand and foot. *Exp. Brain Res.,* **83,** 375–380.
Colebatch, J. G., & Gandevia, S. C. (1989). The distribution of muscular weakness in upper motor neuron lesions affecting the arm. *Brain,* **112,** 749–763.

Conway, B. A., Hultborn, H., & Kiehn, O. (1987). Proprioceptive input resets central locomotor rhythm in the spinal cat. *Exp. Brain Res.*, **68,** 643–656.

Gautier, G. M., & Mussa Ivaldi, F. (1988). Oculo-manual tracking of visual targets in monkey: role of the arm afferent information in the control of arm and eye movements. *Exp. Brain Res.*, **73,** 138–154.

Gautier, G. M., Vercher, J.-L., Mussa Ivaldi, F., & Marchetti, E. (1988). Oculo-manual tracking of visual targets: control learning, coordination control and coordination model. *Exp. Brain Res.*, **73,** 127–137.

Iscoe, S., & Polosa, C. (1976). Synchronization of respiratory frequency by somatic afferent stimulation. *J. Appl. Physiol.*, **40,** 138–148.

Jones, R. D., Donaldson, I. M., & Parkin, P. J. (1989). Impairment and recovery of ipsilateral sensory-motor function following unilateral cerebral infarction. *Brain*, **112,** 113–132.

Kawahara, K., Kumagai, S., Nakazono, Y., & Miyamoto, Y. (1988). Analysis of entrainment of respiratory rhythm by somatic afferent stimulation in cats using phase response curves. *Biol. Cybern.*, **58,** 235–242.

Kelso, J. A., Buchanan, J. J., & Wallace, S. A. (1991). Order parameters for the neural organization of single, multijoint limb movement patterns. *Exp. Brain Res.*, **85,** 432–444.

Klapp, S. T., Hill, M. D., Jagacinski, R. J., Jones, M. R., Tyler, J. G., & Martin, Z. E. (1985). On marching to two different drummers: perceptual aspects of the difficulties. *J. Exp. Psych. Hum. Perc. Per.*, **11,** 814–827.

Kots, Y. M., Krinskiy, V. I., Naydin, V. L., & Shik, M. L. (1971). The control of movements of the joints and kinesthetic afferentation. In I. M. Gell'fand, V. S. Gurfinkel, S. V. Fomin, & M. L. Tsetlin (Eds.), *Models of structural functional organization of certain biological systems* (pp. 371–381). Cambridge: MIT Press.

Meige, H. (1901). Les mouvements en mirroir: leur applications pratiques et thérapeutiques. *Rev. Neurol.*, **19,** 780.

Stark, L. S. (1968). *Neurological control systems. Studies in bioengineering.* New York: Plenum Press.

Thilmann, A. F., Fellows, S. J., & Garms, E. (1990). Pathological stretch reflexes on the "good" side of hemiparetic patients. *J. Neurol. Neurosurg. Psychiatry*, **53,** 208–214.

Viala, D., & Vidal, C. (1978). Evidence for distinct spinal locomotion generators supplying respectively fore- and hind-limbs in the rabbit. *Brain Res.*, **155,** 182–186.

12

Bimanual Interference in a Deafferented Patient and Control Subjects

Normand Teasdale, Chantal Bard, and Michelle Fleury

Laboratoire de Performance Motrice Humaine
Université Laval
Québec, Canada

Jacques Paillard

Unité de Neurosciences Fonctionnelles
Centre National de la Recherche Scientifique
Marseille, France

Robert Forget

École de Réadaptation
Université de Montréal
Montréal, Canada

Yves Lamarre

Centre de Recherche en Sciences Neurologiques
Université de Montréal
Montréal, Canada

   I. Introduction
  II. Methods
      A. Subjects
      B. Apparatus
      C. Procedure
      D. Data Analysis
 III. Results
      A. Pronations/Supinations in the Absence of a Secondary Task
      B. Pronations/Supinations When the Secondary Task Is a Wrist Extension
      C. Pronations/Supinations When the Secondary Task Is a Verbal Response
      D. Spatial Reference without Vision
 IV. Discussion
    References

I. Introduction

Performance decrements are often observed when two tasks are initiated and performed simultaneously. For example, when bimanual aiming movements are made to targets with different index of difficulty, Fitt's Law is violated (Fowler, Duck, Mosher, & Mathieson, 1991; Kelso, Southard, & Goodman, 1979; Kelso, Putnam, & Goodman, 1983). Several hypotheses have been proposed to explain this result. On one hand, Kelso and colleagues suggested that limbs are constrained to act together as a unit to overcome the degrees-of-freedom problem (Bernstein, 1967). On the other hand, Marteniuk, MacKenzie and Baba (1984) argued that separate streams of commands are issued to each limb, yielding a neural crosstalk (as opposed to a coordinative structure, and presumably a single stream, governing both limbs in Kelso et al., 1979). Swinnen and Walter (1991; Swinnen, Walter & Shapiro, 1988; Swinnen, Young, Walter, & Serrien, 1991) have also argued that neural crosstalk plays a dominant role in dual-task performance.

When a repetitive movement is produced and a second movement is initiated with the contralateral limb, an interference between the two limbs is also observed (Cohen, 1970, 1973; Cohn, 1951). Cohen (1970) suggested that, when the second movement is initiated, movements of both limbs must be monitored and that this monitoring (proprioceptive information) imposes an excessive demand on the central signal processing mechanism. Processing resources hypotheses (Wickens, 1984) can also account for bimanual interference. Within such a framework, tasks are assumed to demand resources (these resources are limited in their availability) and structures. When the joint resource demand exceeds the available supply, or when there is competition for the sharing of specific information-processing mechanisms or structures, efficiency drops.

If the monitoring of proprioceptive information is a determinant for the emergent interactions observed when two tasks have to be produced simultaneously, a deafferented patient should exhibit less interference than normal subjects. The purpose of this chapter is to test this hypothesis by examining how a patient showing a total loss of all the large sensory myelinated fibers but intact peripheral motor system can produce two tasks at once. More specifically, we have examined how a deafferented patient performed a continuous wrist pronation/supination task with and without vision of the limb when she was requested to perform a discrete secondary task (wrist extension from the contralateral limb or a verbal response) upon the presentation of an auditory stimulus.

II. Methods

A. Subjects

A deafferented patient (42 years old) showing a total loss of the senses of touch, vibration, pressure, and kinesthesia, as well as absent tendon reflex in the four limbs was tested. Motor nerve conduction velocities of the muscles of the arm are normal (Cooke, Brown, Forget, & Lamarre, 1985). A more detailed clinical description of the patient can be found in Forget and Lamarre (1987). Four neurologically normal subjects (age 22 to 34) also participated.

B. Apparatus

The apparatus consisted of a handle connected perpendicularly to a shaft. The handle rotated freely along the axis of the forearm with low inertia and friction. A linear potentiometer was attached at the end of the rotating shaft. A removable pointer, secured onto the shaft, was used to align the handle on two targets (.5 cm wide) located on a black, semicircular path (30° pronation and 30° supination). Rotation of the handle caused simultaneous movement of the pointer and the potentiometer, allowing the recording of angular displacement. A shield could be positioned over the hand to prevent vision of the forearm, wrist, handle, and pointer. Vision of the targets, however, was still available. A simple normally closed telegraphlike response key (3 cm × 5 cm) was also used. Raising the key sent a digital impulse that was recorded by a computer. Subjects also wore headphones through which a 1-kHz tone (250-msec duration) could be sent. An analog microphone secured on the headphones, 10 cm from the mouth, was used to detect the verbal response's onset. The signals from the potentiometer, response key, and microphone were all digitized on-line (12 bits A/D conversion, 100-Hz sampling rate).

C. Procedure

For the patient, data acquisition was conducted on two consecutive days. Thirty-five trials were collected on Day 1 and 20 trials on Day 2. For control subjects, 20 trials were collected. Each trial consisted of performing continuous 60° pronation and supination movements with the preferred hand for 20 sec. With the contralateral hand on the response key, the subjects were also required to produce a wrist extension, or a verbal response ("top") as fast as possible upon the presentation of an auditory stimulus. The auditory signal always occurred

TABLE I Average Reaction Time to the Auditory Stimulus

	Wrist response with vision	Wrist response without vision	Vocal response with vision	Vocal response without vision
Controls	351 (11)[a]	351 (18)	441 (41)	413 (43)
Patient	372	367	459	474

[a] Values in parentheses are the between-subject standard deviations.

more than 8 sec after the onset of the pronation/supination movements. The average motor and verbal reaction time obtained for the patient and control subjects is presented in Table I. For half of the trials, vision of the pronating/supinating hand (and of the pointer) was precluded.

D. Data Analysis

The position-time traces were filtered with a Butterworth second-order, 8-Hz, low-pass, cutoff frequency with dual-pass to remove phase shift. This filtering procedure results in a fourth-order, zero-phase shift filter with a cutoff frequency of 6.4 Hz (Oppenheim & Willsky, 1983; Winter, 1979). The displacement signals were then differentiated numerically with a central finite difference technique to obtain velocity-time curves.

III. Results

A. Pronations/Supinations in the Absence of a Secondary Task

Without a secondary task, the initial performance of the patient was variable and characterized by several discontinuities in the velocity profiles. Across trials, the frequency of these discontinuities decreased. Figure 1 illustrates this observation; the velocity-time profiles (initial 8 sec before the presentation of an auditory stimulus) for the first and last two trials with vision on Day 1 are presented. Curves for a representative control subject are also presented. In contrast to the patient, the control subjects showed "smoother" velocity-time profiles on the first trials. The deafferented patient's behavior is reminiscent of Pew's (1966) hypothesis of a shift in the control process (presumably from a high-level to a lower level control). This shift in control process is thought to free the attentional mechanisms for use on higher order aspects of the task or for doing other simultaneous tasks. However, the shift in the control process (and presumably more optimal performance) did not permit the patient to cope efficiently with the dual task.

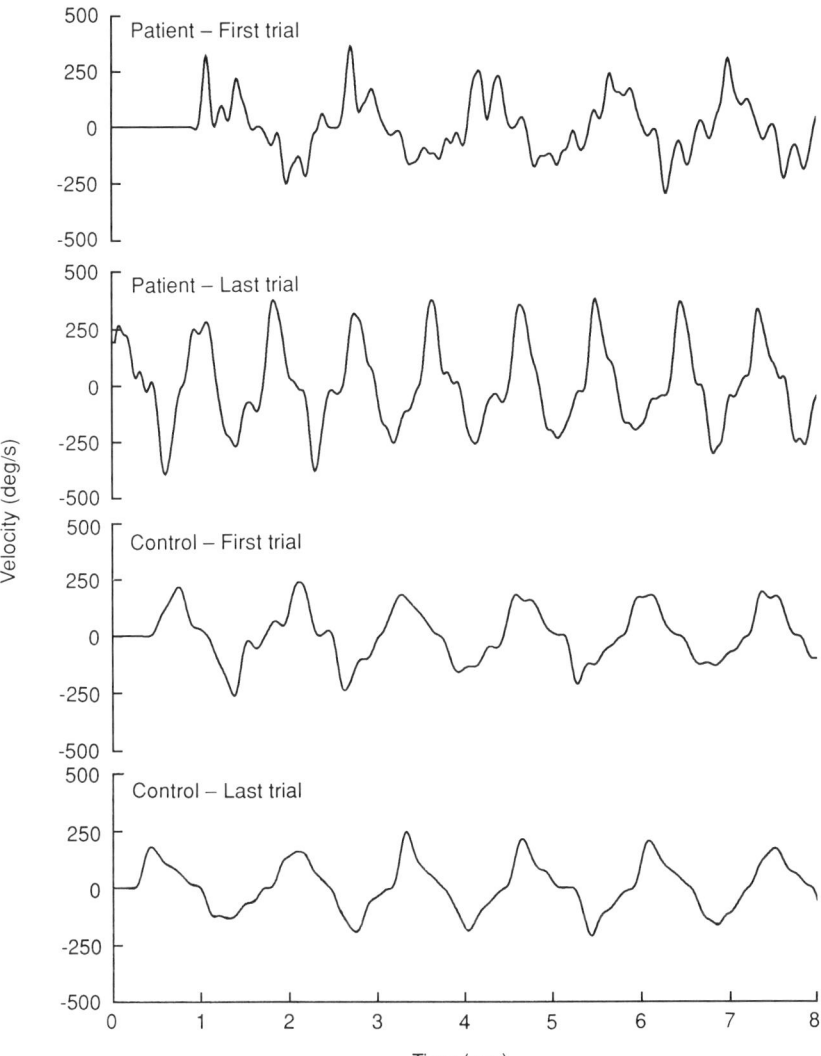

FIGURE 1 Velocity-time profiles for the first and last trials in the absence of a secondary task. Vision of the pronating/supinating limb was available. The data for the patient are from the first day of testing.

B. Pronations/Supinations When the Secondary Task Is a Wrist Extension

The most striking observation of this experiment, and contrary to our original hypothesis, was that the pronation/supination movements of the patient were altered in nearly all conditions. These observations clearly suggest that, in bimanual movements, the interference is of central origin and does not result from an inability to monitor afferent signal arising from the movements. Figure 2 presents the average movement amplitude and the within-subject variability for a pronation/supination cycle before and following the presentation of the auditory stimulus when vision was available. Dashed lines represent the data for the deafferented patient. Following the presentation of the

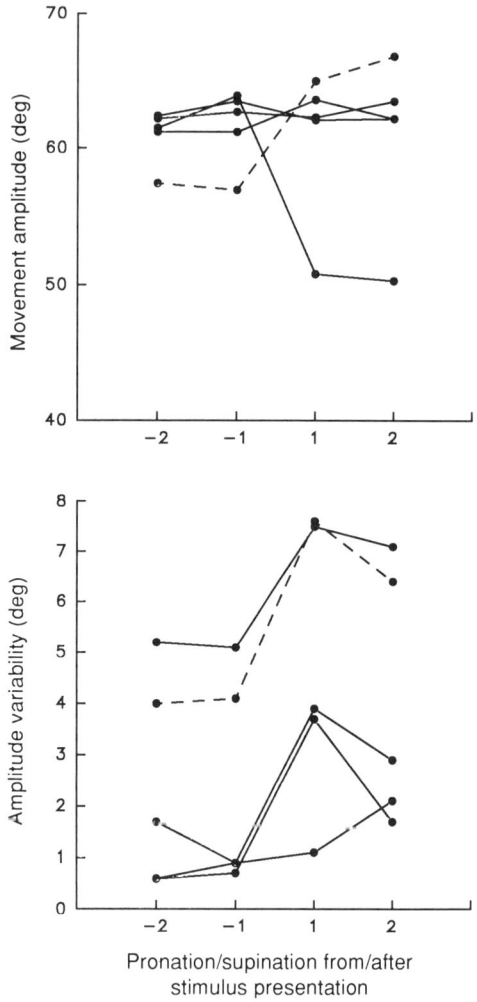

FIGURE 2 Movement amplitude and amplitude variability (degree) for the pronation/supination movements prior to and following the presentation of the stimulus when the secondary task was a hand extension and when vision was available. The dashed lines represent the data for the deafferented patient.

stimulus, the patient and one control subject showed large changes in movement amplitude (increased amplitude for the patient and decreased amplitude for the control subject). The bottom portion of the figure shows that across trials, the stimulus presentation yielded an increased amplitude variability in all subjects. For the patient, the effects of the perturbation were most often represented by an increased amplitude and an increased velocity. Figure 3 presents phase-plane trajectories for a trial showing a nearly constant amplitude but an increased velocity following the stimulus presentation, and a trial showing both an increased amplitude and an increased velocity. The top portions of the figure present the behavior before the presentation of the stimulus, whereas the bottom portions present the behavior following the stimulus presentation (for each figure, the dotted line is the cycle immediately following the stimulus presentation). For all trials, the

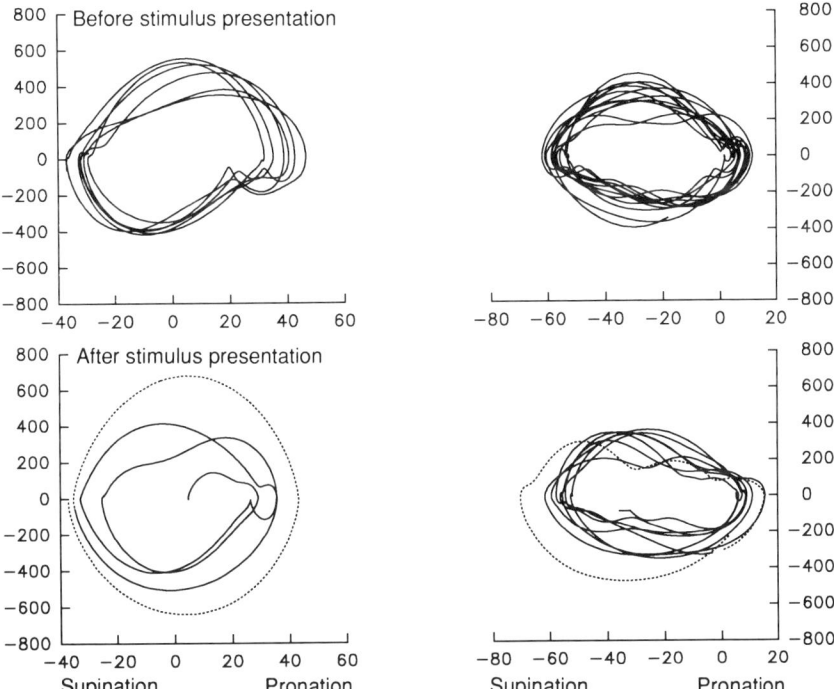

FIGURE 3 Phase-plane trajectories (velocity in °/sec vs. position in °) from the patient for trials with vision of the pronating/supinating limb when the secondary task was a wrist extension. The top portions of the figure present the behavior before the stimulus was given; the bottom portions present the behavior following the presentation of the stimulus. The dotted lines illustrate the pronation/supination cycle immediately following the presentation of the stimulus.

patient returned to her previous movement pattern after the production of the wrist extension.

Figure 4 illustrates that, without vision, the patient produced movements that were nearly twice as large as those produced with vision. The secondary task also yielded more perturbation than that for control subjects (on average, a decreased amplitude of 15° vs. 5° for the control subjects). Interestingly, the patient also returned to her original move-

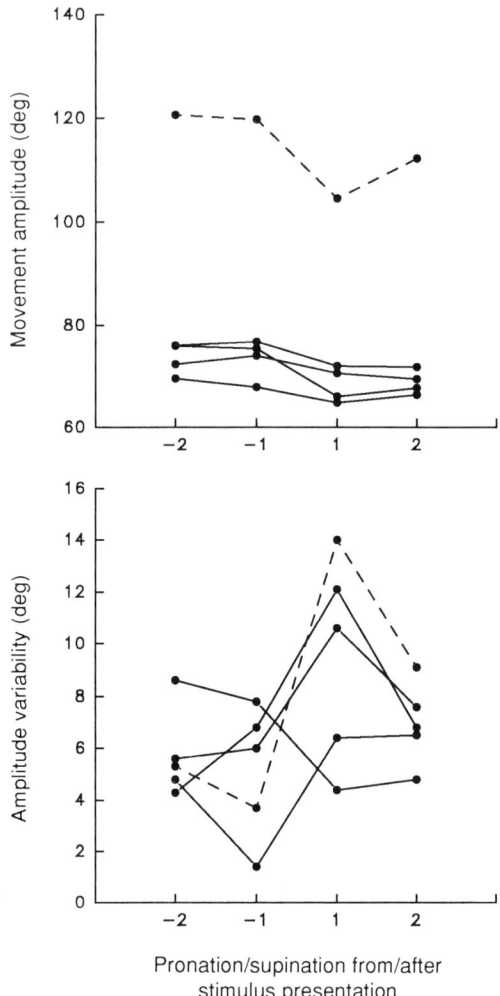

FIGURE 4 Movement amplitude and amplitude variability (degree) for the pronation/supination movements prior to and following the presentation of the stimulus when the secondary task was a hand extension and when vision of the pronating/supinating limb was absent. The dashed lines represent the data for the deafferented patient.

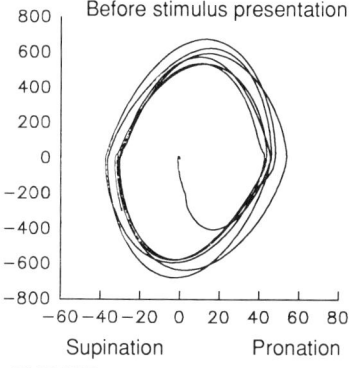 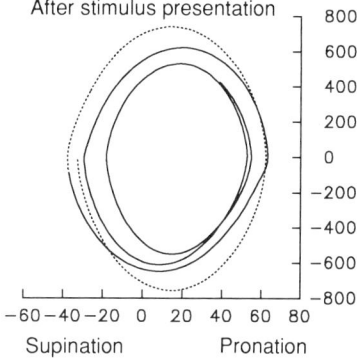

FIGURE 5 Phase-plane trajectories (velocity in °/sec vs. position in °) from the patient for a trial without vision of the pronating/supinating limb when the secondary task was a wrist extension. The left portion of the figure presents the behavior before the stimulus was given; the right portion presents the behavior following the presentation of the stimulus. The dotted line illustrates the pronation/supination cycle immediately following the presentation of the stimulus.

ment pattern after producing the wrist extension, suggesting the presence of a mechanism (presumably of central origin) for monitoring the limb's performance. A representative phase-plane trajectory illustrating this behavior is presented in Figure 5.

On some trials, a near complete stop in the pronating/supinating hand was observed simultaneously with the wrist extension. One such trial (without vision of the pronating/supinating limb) for the deafferented patient is presented in Figure 6. The position-time profile shows the severe disruption and the near complete abortion of the movement when the wrist extension was initiated. The patient recovered and reinitiated her movements following the perturbation. When asked about her behavior, the patient reported not knowing about the perturbation.

In contrast, the performance of control subjects generally showed more stable behavior; on several trials, the perturbation was hardly detectable. Figure 7 shows representative phase-plane trajectories for a control subject with and without vision of the pronating/supinating hand. For both trials, the production of the wrist extension yielded an increased velocity in the pronating/supinating hand.

C. Pronations/Supinations When the Secondary Task Is a Verbal Response

The production of a verbal response yielded results similar to that obtained with the wrist extension task. Figures 8 and 9 present move-

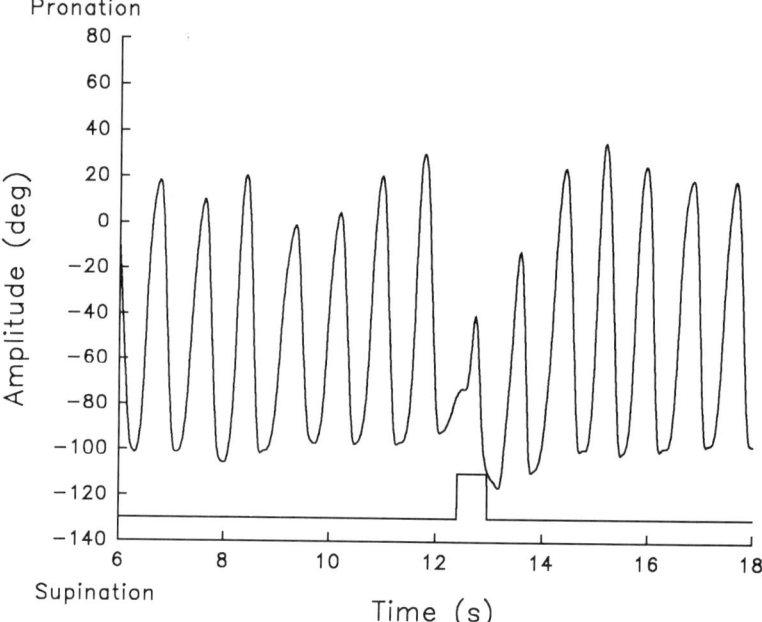

FIGURE 6 Patient's position-time profile of a trial showing a near complete abortion of the pronation/supination movement when a wrist extension was produced. The trial was performed without vision of the pronating/supinating limb. The data are shown starting from the 6th second only to graphically highlight the perturbation. The bottom trace indicates the onset of the wrist extension.

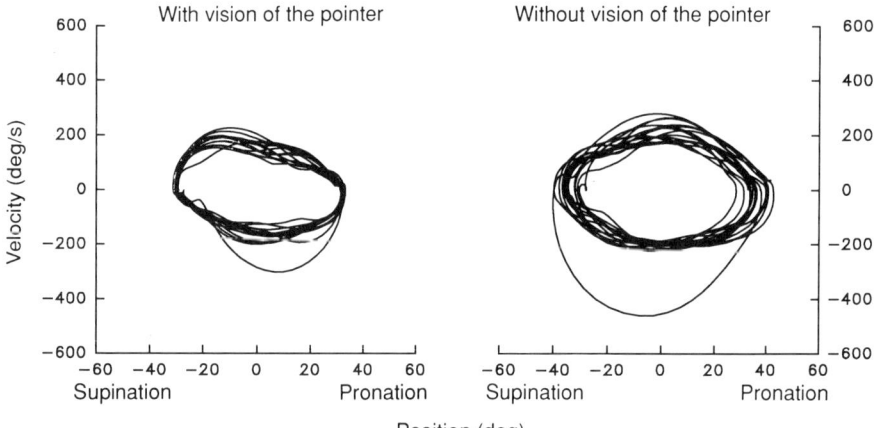

FIGURE 7 Phase-plane trajectories (velocity in degree/sec vs. position in degree) with and without vision of the pronating/supinating limb when the wrist extension was the secondary task. The data are from a control subject.

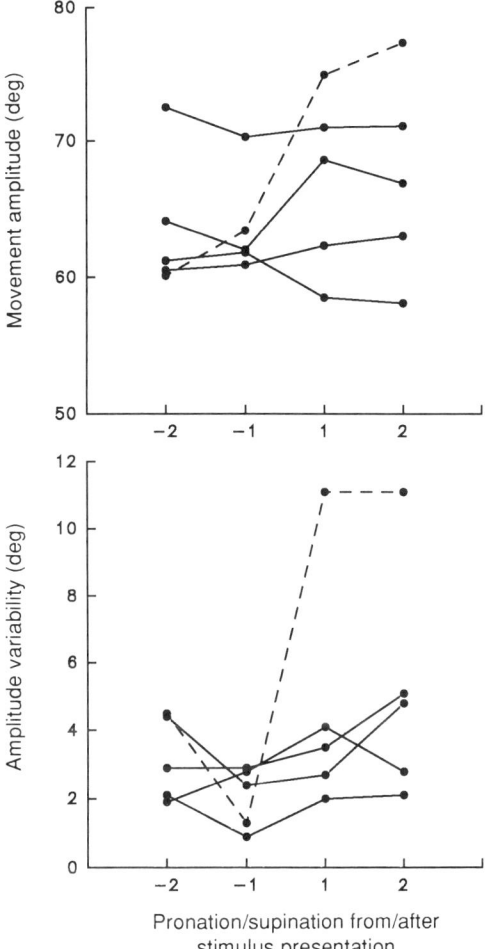

FIGURE 8 Movement amplitude and amplitude variability (degree) for the pronation/supination movements prior to and following the presentation of the stimulus when the secondary task was a verbal response and when vision was available. The dashed lines represent the data for the deafferented patient.

ment amplitude and amplitude variability for the pronations/supinations before and following the stimulus presentation with and without vision, respectively. With vision, and as for the wrist extension condition, the patient generally exhibited an increased amplitude and a large increased variability following the presentation of the stimulus. Without vision, the patient produced movements that, on average, were nearly twice as large as those produced with vision. The near absence of perturbation on the average data of the patient is explained by the fact

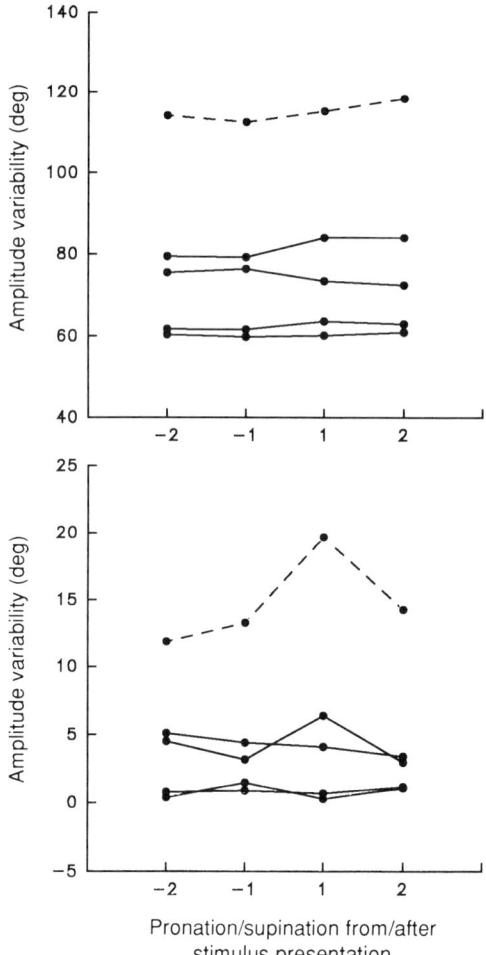

FIGURE 9 Movement amplitude and amplitude variability (degree) for the pronation/supination movements prior to and following the presentation of the stimulus when the secondary task was a verbal response and when vision of the pronating/supinating limb was absent. The dashed lines represent the data for the deafferented patient.

that the production of the verbal response yielded a decreased amplitude on half of the trials and an increased amplitude on the other half of the trials.

D. Spatial Reference without Vision

Without vision, the deafferented patient rapidly lost her spatial frame of reference. This result is illustrated in Figure 10 for Day 1 and Day 2. The amplitudes of the pronation and supination movements for

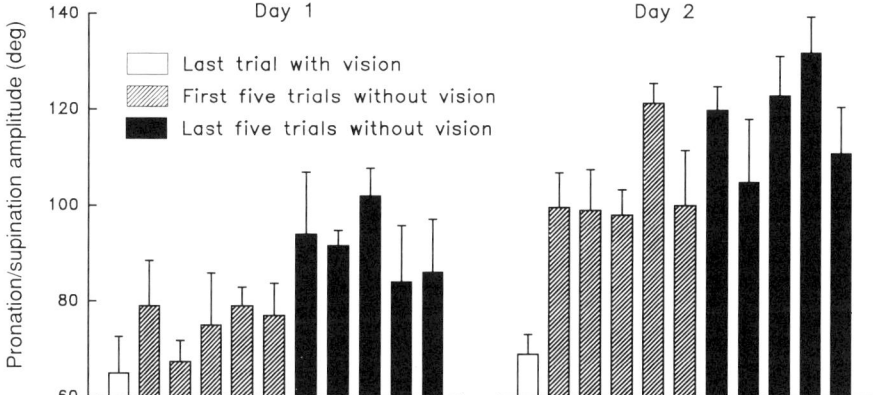

FIGURE 10 Amplitude of pronation/supination movements for the last trial with vision and the first and last 5 trials without vision for the deafferented patient on Day 1 and Day 2. The amplitudes were calculated for the movements occurring before any stimulus was presented.

the last trial with vision are compared with trials without vision. Amplitudes were computed on movements occurring before the stimulus presentation (i.e., on the first 8 sec of each trial). Clearly, an increased amplitude can be observed on the first trial without vision. In spite of this initial loss of calibration, the patient kept a relatively stable behavior throughout the trial, suggesting access to a central monitoring mechanism. This result replicates recent findings obtained with the same patient (Larue, Fleury, Bard, & Teasdale, 1992). On Day 1, there is a systematic increase across trials. On Day 2, a more rapid increase is observed and, by the end of the session, the patient was producing movements that were nearly twice as large as those requested. For normal subjects, no loss of calibration (i.e., movement amplitudes) was observed without vision of the limb.

IV. Discussion

The present experiment clearly shows that the deafferented patient was unable to perform a sequence of voluntary pronating/supinating movements with one limb when another response (whether manual with the contralateral limb or verbal) had to be carried out simultaneously. These results are at variance with Cohen's (1970) and our initial hypothesis that the monitoring of proprioceptive information could be a determinant for the interference. Briefly, Cohen had suggested that bimanual interference could result from the operation of a central regulatory mechanism responsible for reducing the competition of afferent signals arising from the limbs. More specifically, he had proposed that, when a contralateral movement is initiated during the performance of alternating movements, the movements of both limbs have to be monitored, and that this monitoring imposes an excessive demand on the central signal processing mechanism. At least for the deafferented patient, the perturbed behavior does not support the suggestion that proprioceptive information is a determinant for the interference.

Overall, the absence of proprioceptive information, rather than reducing or inhibiting the interference, yielded a more pronounced interference. The patient exhibited larger amplitude changes and greater across-trial variability than control subjects and often needed more than two cycles for recovering from the interference induced by the secondary task (whether manual or verbal). Without vision, the patient, although unaware of the interference, could sometimes recover (although partially) from the interference. In contrast, normal subjects (with or without vision) generally needed less than a pronation/supination cycle to recover and produced alternating movements that were much more stable than that of the patient (see Figure 7). Thus, it can be proposed that proprioceptive information, rather than enhancing the interference, served to rapidly correct the perturbed movement. This suggestion rejoins that of Baldissera, Cavallari, Marini, and Tassone (1991), who proposed that the performance level attained in a task requiring coupling of anti-phase upper- and lower-limb movements was limited by the efficient utilization of afferent signals. They suggested that the time required for perception of the limb dynamics and the attentive effort required might be the critical factor for efficient performance. In the present experiment, afferent information could have permitted control subjects to rapidly detect the interference and recover from it. On the other hand, the absence of afferent information could require a continuous attentive effort (e.g., Kahneman, 1973) to modulate and monitor the movement performance that in normal conditions are processed automatically by feedback mechanisms.

The interference observed in our subjects most likely occurred at the program level and could be the result of separate streams of motor commands yielding neural crosstalk (e.g., Marteniuk et al., 1984; Swinnen et al., 1991). In support of this suggestion, the patient commented to us that it had taken her several years of practice to accomplish simple, dual-tasks performance such as speaking and producing a simultaneous hand gesture, and that dual-tasks performance still requires a lot of attention. For example, when asked to draw circles with both hands, she reported needing "to think about it" before initiating the movements. After the initiation, she had to shift her attention from one hand to the other, rapidly leading to a temporal desynchronization of both hands.

The loss of spatial reference observed in the no-vision conditions replicates recent findings obtained with the same patient (Teasdale, Forget, Bard, Paillard, Fleury, & Lamarre, 1993). Briefly, we have observed that, without vision, the patient produced all required handwriting shapes (morphocinetic components). It was the spatial location of the shapes (topocinetic components) that was greatly degraded (e.g., letters were superimposed, bars on t's and dots on i's were misplaced), suggesting that proprioceptive information is the basis for spatial calibration.

Acknowledgments

Our special appreciation is first directed to the patient. Thanks to Benoit Genest and Gilles Bouchard for programming and technical expertise. This project was supported by various NSERC and FCAR grants. Robert Forget is a "chercheur-boursier" of the Fonds de la Recherche en Santé du Québec. Yves Lamarre is a member of the MRC Groupe de Recherche en Sciences Neurologiques at Université de Montréal.

References

Baldissera, F., Cavallari, P., Marini, G., & Tassone, G. (1991). Differential control of in-phase and anti-phase coupling of rhythmic movements of ipsilateral hand and foot. *Experimental Brain Research,* **83,** 375–380.

Bernstein, N. (1967). *The coordination and regulation of movements.* Oxford: Pergamon Press.

Cohen, L. (1970). Interaction between limbs during bimanual voluntary activity. *Brain,* **93,** 259–272.

Cohen, L. (1973). Synchronous bimanual movements performed by homologous and non-homologous muscles. *Perceptual and Motor Skills,* **32,** 639–644.

Cohn, R. (1951). Interaction in bilaterally simultaneous voluntary motor function. *Archives of Neurology and Psychiatry,* **65,** 472–476.

Cooke, J. D., Brown, S., Forget, R., & Lamarre, Y. (1985). Initial agonist burst duration changes with movement amplitude in a deafferented patient. *Experimental Brain Research,* **60,** 184–187.

Forget, R., & Lamarre, Y. (1987). Rapid elbow flexion in the absence of proprioceptive and cutaneous feedback. *Human Neurobiology,* **6,** 27–37.

Fowler, B., Duck, T., Mosher, M., & Mathieson, B. (1991). The coordination of bimanual aiming movements: Evidence for progressive desynchronization. *Quarterly Journal of Experimental Psychology,* **43**A, 205–221.

Kahneman, D. (1973). *Attention and effort.* Englewood Cliffs, NJ: Prentice-Hall.

Kelso, J. A. S., Southard, D. L., & Goodman, D. (1979). On the nature of human interlimb coordination. *Science,* **203,** 1029–1031.

Kelso, J. A. S., Putnam, A., & Goodman, D. (1983). On the space-time structure of human interlimb coordination. *Quarterly Journal of Psychology,* **35**A, 347–375.

LaRue, J., Fleury, M., Bard, C., & Teasdale, N. (1992). Rôle des afférences proprioceptives et cutanées dans le contrôle de mouvements de pointage en amplitude. In M. Laurent, J. F. Marini, R. Pfister, & P. Therme (Eds.), *Recherches en A.P.S. 3.* (pp. 101–110). Paris: Actio/Université Aix-Marseille II (UFR STAPS).

Marteniuk, R. G., MacKenzie, C. L., & Baba, D. M. (1984). Bimanual movement control: Information processing and interaction effects. *Quarterly Journal of Experimental Psychology,* **36A,** 335–365.

Oppenheim, A. V., & Wilsky, A. S. (1983). *Signals and systems.* Englewood Cliffs, NJ: Prentice-Hall.

Pew, R. W. (1966). Acquisition of hierarchical control over the temporal organization of a skill. *Journal of Experimental Psychology,* **71,** 764–771.

Swinnen, S. P., & Walter, C. B. (1991). Toward a movement dynamics perspective on dual-task performance. *Human Factors,* **33,** 367–387.

Swinnen, S. P., Walter, C. B., & Shapiro, D. C. (1988). The coordination of limb movements with different kinematic patterns. *Brain and Cognition,* **8,** 326–347.

Swinnen, S. P., Young, D., Walter, C. B., & Serrien, D. (1991). Control of bilateral asymmetrical movements. *Experimental Brain Research,* **85,** 163–173.

Teasdale, N., Forget, R., Bard, C., Paillard, J., Fleury, M., & Lamarre, Y. (1993). The role of proprioceptive information for the production of isometric forces and for handwriting tasks. *Acta Psychologica,* **82,** 1–13.

Wickens, C. D. (1984). Processing resources in attention. In R. Parasuraman & D. R. Davies (Eds.), *Varieties of attention* (pp. 63–102). Orlando, FL: Academic Press.

Winter, D. A. (1979). *Biomechanics of human movement.* New York: Academic Press.

13

Changes in Strength, Speed, and Reaction Time Induced by Simultaneous Bilateral Muscular Activity

T. Ohtsuki

Laboratory of Human Movement
Department of Physical Education
Nara Women's University
Nara City, Japan

  I. Introduction
 II. Maximum Muscle Strength
     A. Decrease in Arm Strength during Simultaneous Bilateral Exertion
     B. Decrease in Grip Strength by Simultaneous Bilateral Exertion
     C. Lower Limb Strength
III. Submaximal Isometric Muscle Strength
 IV. Speed
  V. Reaction Time
 VI. Effects of Training on Bilateral Performance
VII. Exploring the Reasons Underlying the Bilateral Deficit
     A. Division of Attention
     B. Reciprocal Inhibition
     C. Interhemispheric Inhibition
VIII. Conclusions
     References

I. Introduction

On many occasions in daily life and sports, simultaneous bilateral functioning of the body is required to perform normal movements. To hold, push, or pull a large and heavy object, to pass a basketball, or to wash our face, we often contract finger flexors, elbow extensors, or elbow flexors of both arms simultaneously to apply appropriate force. Some studies report that simultaneous bilateral use of two limbs causes deterioration of performance of each limb. In the present chapter, I shall discuss the deficit phenomena observed when humans perform simple simultaneous bilateral motor tasks, and then elaborate on the possible causes of the phenomena.

II. Maximum Muscle Strength

A. Decrease in Arm Strength during Simultaneous Bilateral Exertion

The first researchers who reported about maximum strength developed by the two arms simultaneously were Henry and Smith (1961). They found that maximum isometric strength exerted by the dominant arm during simultaneous bilateral exertion decreased by 3% in comparison with unilateral maximum strength. Later, Kroll (1965) also reported similar data with isometric wrist extension. Recently, I also have conducted experiments that produced comparable results. This chapter will start with a description of my own research on arm strength (Ohtsuki, 1983). Ten normal, healthy, right-handed adult subjects were instructed to slowly and steadily pull a chain connected to the wrist by flexing or extending the elbow joint kept at a right angle to the table, applying maximum isometric strength for 5 seconds. The subjects were not permitted to jerk the chain. In the bilateral condition, the subject contracted the corresponding muscles of both arms simultaneously.

Figure 1 shows a typical example of the recorded traces. For both arms, simultaneous bilateral contraction reduced the maximum amplitude of strength developed at the wrist in comparison with the unilateral contraction. Overall mean values of strength (Figure 2) of bilateral conditions were consistently smaller than unilateral conditions (118 and 110 N for right flexion; 113 and 102 N for left flexion; 79 and 59 N for right extension; and 70 and 56 N for left extension). In all cases, the unilateral–bilateral difference was statistically significant.

The percentage of bilateral strength to unilateral was calculated as 100B/U, where B and U are the mean value of the three observations of the bilateral and unilateral conditions in one session. Means over sessions and subjects were 93.7 and 92.4% for right and left flexion, and 75.4 and 81.2% for right and left extension. In other words, decrease ratio was 6.3, 7.6, 24.6, and 18.8%, respectively.

The electrical activity of the agonist muscles (T. B. in Figure 1) was also reduced as seen in the raw and integrated electromyograms. Surface EMG from the elbow flexor and extensor muscles was integrated

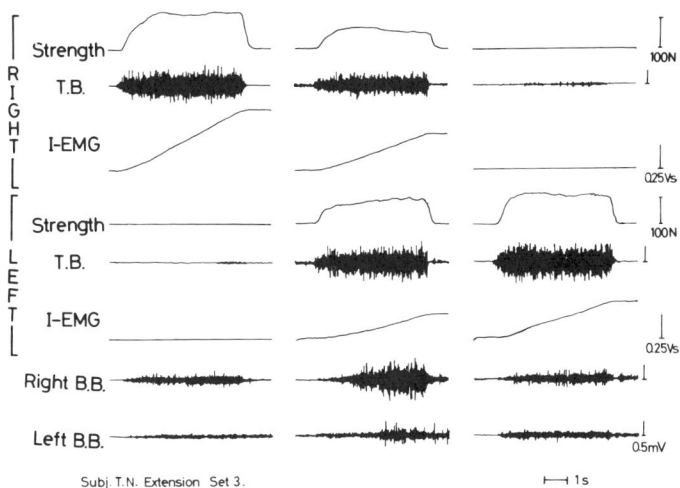

FIGURE 1 An example of the recorded traces of elbow extension. Traces 1 and 4, the output of the force transducers; 2 and 5, surface EMG of agonist muscle (triceps brachii); 3 and 6, integrated EMG of agonist; 7 and 8, surface EMG of antagonist (biceps brachii). Calibration of EMG refers to all. (From Ohtsuki, 1983.)

during the time period when the strength value was 90% of its peak or more, and divided by that time period to obtain integrated EMG per unit time. These EMG values were transformed to percentages in the same manner as strength and then were correlated with percent strength values by pooling subjects. Highly significant positive correlations were obtained ($r = .560$ and $r = .542$ for right and left flexion, respectively; $n = 30$, $p < .05$, for both; $r = .726$ and $r = .750$ for right and left extension, respectively; $n = 30, p < .001$, for both). This means that the strength decrease in simultaneous bilateral exertion was due to the decrease in the activity of the agonist muscle itself and not from the mere mechanical counterpull of the antagonist muscle activity.

B. Decrease in Grip Strength by Simultaneous Bilateral Exertion

Similar observations have also been made for maximum voluntary isometric grip strength (Ohtsuki, 1981a), using sixteen male and eight female normal, healthy, right-handed adults. Figure 3 shows typical examples of the records obtained. During bilateral force production, the strength and the surface EMG from the finger flexors clearly decreased. For both hands of males and females, mean simultaneous bilateral grip strength was smaller than unilateral strength (for unilateral and bilateral, respectively, 414 and 356 N for male right; 401 and 384 N for male left; 308 and 278 N for female right; and 277 and 252 N for female left).

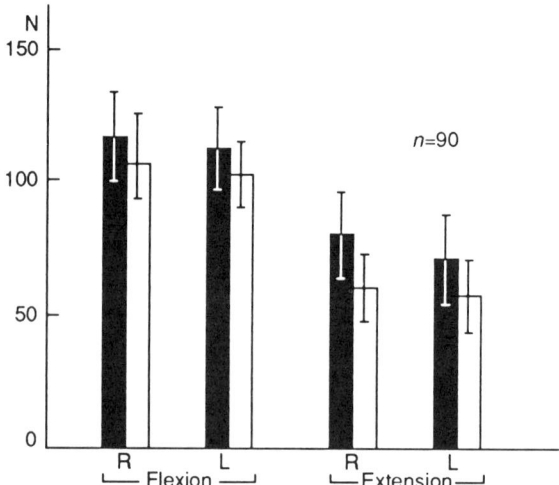

FIGURE 2 Overall mean ± SD values of maximum isometric strength. R, Right; L, left. Unilateral, ■; bilateral, □. (From Ohtsuki, 1983.)

Decrease ratios calculated in the same manner as arm strength (see Section II,A) were 13.6 and 4.9% for male right and left hands, and 9.5 and 8.8% for female right and left hands, respectively.

Surface EMG from the finger flexor muscles in the forearm was integrated in the same manner as arm strength (see Section II,A). Observed values of strength of each subject were transformed to a percentage of the largest value of each subject. Values of integrated EMG were transformed in the same way as strength except that the values obtained with the largest strength were employed as 100%. Correlation coefficients between these two variables calculated by pooling subjects were highly significant for male and female and for right and left ($r = .657$ and $r = .593$, $n = 240$, $p < .001$ for male right and left; $r = .771$ and $r = .596$, $n = 128$, $p < .001$ for female right and left).

C. Lower Limb Strength

The pattern of data found for the lower limbs is not as simple as that for the upper limbs. Currently, there seems to be several different findings. The study of bilateral simultaneous strength exertion by the lower limbs begins with Asmussen and Heebøll-Nielsen (1961). Their method of measuring leg extension strength is now widely accepted by many researchers studying simultaneous bilateral leg extension. The subject is kept seated on a chair mounted in a horizontal iron frame with the feet placed on a foot plate attached to a steel bar mounted on

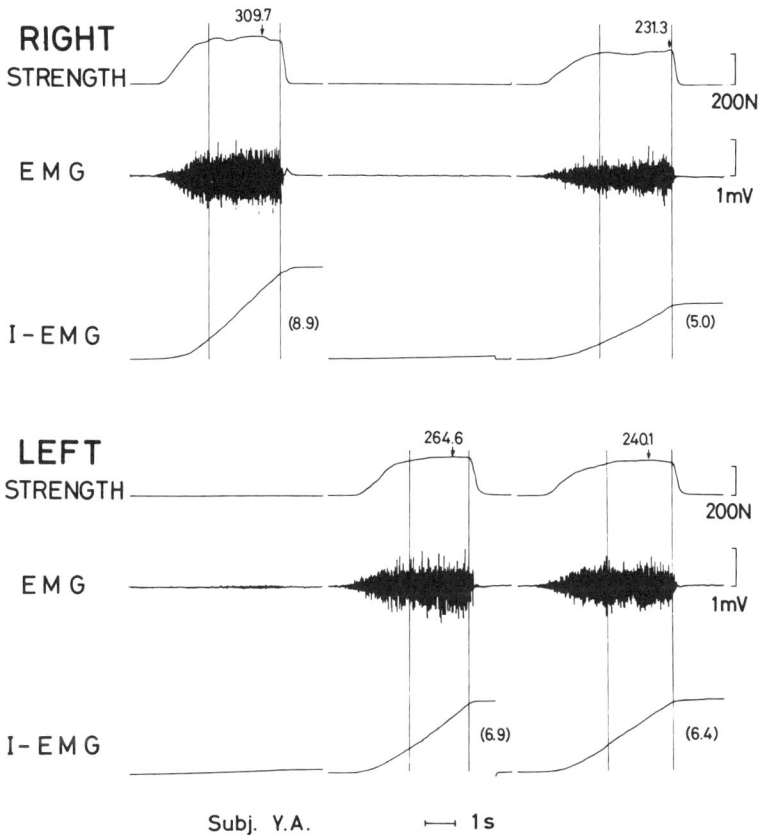

FIGURE 3 An example of recorded traces. Numerals represent measured values of maximum grip strength indicated with small arrows. Vertical lines show time interval during which 90% or more of maximum strength is maintained. Numerals in parentheses are integrated EMG per second. (From Ohtsuki, 1981a.)

the frame so that the trochanter major and malleolus lateralis are horizontally aligned. The angle of ankle, knee, and hip joints can be changed by adjusting the position of the chair, bar, and plate; however, usually the angles of ankle and knee are kept at 90°. The subject, pressing his or her back against the back of the chair, tries to straighten the lower limb by pushing the foot plate with either one or both feet to exert the isometric maximum voluntary leg extension strength. This type of exercise needs simultaneous force exertion by hip, knee, and ankle joints.

Asmussen and Heebøll-Nielsen (1961) showed that the isometric two-leg extension force was smaller compared to the sum of the isometric extension strength of each leg. Using the same methods, Secher *et al.* (1978) found that maximum two-leg extension strength was 25% smaller than the sum of the maximum one-leg extension strength of both legs. Later, Secher *et al.* (1988) reported a mean decrease of 20% (knee angle 90°) and 19% (knee angle 150°), using untrained subjects. In the same way Vandervoort *et al.* (1984) found a 9% decrease with a knee angle of 90°. Schantz *et al.* (1989) reported 10, 12, 13 and 19% decreases in four groups of physical education students, and 11 and 14% decreases in two groups of untrained subjects with a knee angle of 90°.

However, when the strength exertion is confined to the knee joint only, contradictory results have been reported by different authors. Howard and Enoka (1991) found an approximately 8–10% decrease with a knee angle of 109° in untrained subjects. They kept their subjects supine on the bench so that the knees could be flexed downward at the edge of the bench, and the ankles attached to force transducers. In contrast, Schantz *et al.* (1989), keeping subjects in an upright chair made them extend the knee isometrically at a knee angle of 90° and found a 4% increase in bilateral extension. These subjects were the same as those who showed a 10% decrease in the bilateral leg extension (see above), a disagreement still unresolved.

Thus, to my knowledge, for leg extension, which needs co-contraction of the muscles acting on the hip, knee, and ankle joints, simultaneous bilateral exertion without exception causes the decrease of maximum voluntary strength in normal subjects who have no special strength training. For knee extension, however, the effect of bilateral exertion is still unclear.

III. Submaximal Isometric Muscle Strength

The bilateral deficit in voluntary strength has also been observed for submaximal isometric hand grip, elbow flexion, and extension strength. Seki and Ohtsuki (1990) had subjects exert forces of 25, 50, and 75% of maximal strength of each arm (or hand), based on their subjective judgment without any knowledge of the results. As exemplified in Figure 4, when the subject was required to exert the same proportion of maximum strength of each limb by both limbs simultaneously, muscle strength of both sides was smaller than that of the same proportion of strength exerted by either hand alone. The decrease ratio ranged between 7 and 22% of the unilateral strength, which is in the same range as the decrease ratio of maximum strength reported in Section II. Thus, when the same subjective level of isometric muscle strength is exerted by both upper limbs, bilateral simultaneous strength always decreases irrespective of whether the intended magnitude is maximal or submaximal.

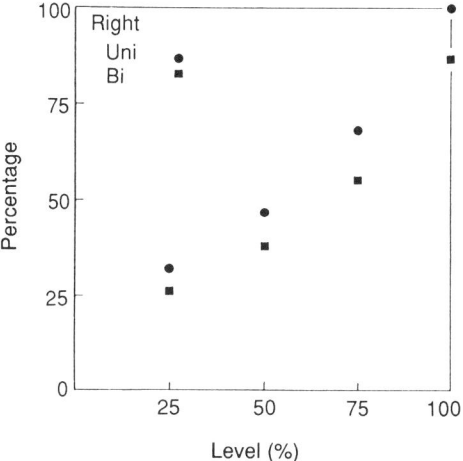

FIGURE 4 Overall means of unilateral and bilateral elbow flexion strength of maximal and submaximal exertion (right arm). Uni, unilateral; Bi, bilateral. (Modified from Seki & Ohtsuki, 1990.)

IV. Speed

The speed of movement execution also reduces when both limbs are moved simultaneously. Wyke (1969, 1971) reported that the speed of hand tapping (tapping frequency per unit time) was reduced when both hands were used simultaneously. Corcos (1984) showed an increase in movement time during a bilateral reaction time task in comparison with unilateral. We had subjects flex the elbow from 140 to 60° for flexion, or extend from 60 to 140° as fast as possible, resisting a gravitational load of 40% of the isometric maximum strength of each arm measured at 90° elbow angle, by using the same apparatus and subjects as described previously (Ohtsuki, 1983; see Section II,A). As shown in Figure 5, for both left and right flexion and extension, a consistent decrease was observed in the velocity of the wrist, which was calculated from the time spent between 70 and 120°. The decrease ratio was 11.6 ± 4.2% (mean ± SD) and 11.4 ± 6.1% for right and left flexion, and 11.9 ± 5.0% and 10.2 ± 5.1% for right and left extension, respectively.

These results clearly show that the speed of the limb decreased during bilateral exercise. It seems reasonable that the speed of muscle contraction depends on the magnitude of the electrical activity of the agonist muscle, as confirmed by several authors (Marsden *et al.*, 1983; Benecke *et al.*, 1985; Mustard & Lee, 1987).

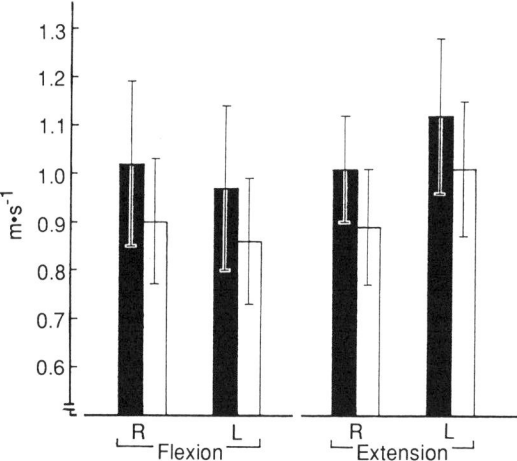

FIGURE 5 Overall mean ± SD values of speed of muscle contraction measured as a tangential velocity of the wrist in elbow flexion and extension. Unilateral, ■; bilateral, □.

V. Reaction Time

When simultaneous bilateral activation of the muscles is used in a voluntary reaction time task, the latency of initiation of the response from the onset of the external stimulus is lengthened relative to unilateral activation.

Kerr et al. (1963) had their subjects perform a keypressing task either by a single hand or by both hands in response to visual stimuli. Although their experiment was not designed for direct comparison between unilateral and bilateral reaction times, their data clearly indicated that the average keypressing latencies for simultaneous bilateral responses in various conditions ranged between 341 and 350 msec for the dominant hand and between 354 and 356 msec for the nondominant hand. These times were longer than those for unilateral responses (324–338 msec and 333–343 msec for the dominant and nondominant hands, respectively). The relative increase in latency from unilateral to bilateral responses was 3.6 to 6.3%.

Jeeves (1969) found for various conditions that simple reaction time to a single visual stimulus (latency of key release) was longer in subjects who were ordered to respond by both hands simultaneously (means, 256–273 msec) than in subjects who were ordered to respond by a single hand (219–238 msec). Using the same methods, Jeeves and Dixon (1970) confirmed that simple reaction time to a visual stimulus was longer for simultaneous bilateral responses (means, 253–273 msec) than for single-handed responses (217–239 msec). The mean unilateral–bilateral difference was 34–37 msec, corresponding to a 14–16% increase.

Di Stefano et al. (1980), using keypressing and lever pulling to a single visual stimulus, reported a slight but statistically significant delay of movement onset for simultaneous bilateral responses (key, 200.2 msec; lever, 227.7 msec) relative to unilateral (key, 198.4 msec; lever, 223.2 msec).

I also observed longer reaction times during simultaneous bilateral knee extension in response to auditory stimuli (Ohtsuki, 1981b). The mean latency of EMG recorded from the knee extensor muscle (m. rectus femoris) was 139 msec for the unilateral and 155 msec for the bilateral response. The difference was 16 msec, which was an 11.6% increase relative to the unilateral condition. The latency of knee extension was 219 msec for the unilateral and 231 msec for the bilateral response. The difference was 12 msec (a 5.6% increase).

VI. Effects of Training on Bilateral Performance

Several studies report a reduction of the bilateral deficit with specific bilateral training. For isometric leg extension strength measured by the method described in Section II,C, Secher (1975) reported that the Olympic or World gold and bronze medalist oarsmen showed 820 kilopond (kp) [8036 newtons (N)] of the sum of the right and left single maximum isometric leg extension strength and 850 kp (8330N) of simultaneous bilateral leg extension strength, while lower level club oarsmen showed 740 kp (7252N) and 680 kp (6664N) for the corresponding strengths, respectively, at a 150° knee angle. Thus, the well-trained oarsmen showed an increase (3.7%), while the less-trained oarsmen still showed a decrease (8.1%) comparable to nontrained men.

However, using the same apparatus and a different knee angle (90°), Secher et al. (1988) found a 24% decrease in trained bicyclists ($n = 8$) and an 18% decrease in trained weight lifters ($n = 38$). Schantz et al. (1989) also reported a bilateral decrease in professional ballet dancers (13%), volleyball players (12%), and the heavy-resistance trained subjects (8%). These decrease values were not significantly different from the untrained normal subjects.

The "well-trained" subjects in these studies had much experience in their special activities; however, except for oarsmen and weight lifters, they had no specific training in bilateral leg extension. For example, bicycling needs alternating extension of two legs but not simultaneous bilateral extension. Weight lifting requires simultaneous bilateral leg extension, but the direction of force application is vertical, while it is horizontal in rowing.

However, when the activated joint was confined to the knee, Howard and Enoka (1991) found no significant bilateral deficit in weight lifters ($n = 6$) and cyclists ($n = 6$). This result is contradictory to those for leg extension (Secher et al., 1988; see above), which also involves knee extension. The discrepancies between these studies has yet to be resolved.

Recently, I tested trained pianists with more than 10 years experience on bilateral simultaneous hand tapping. Two pairs of copper plates were mounted on a horizontal table and subjects tapped a pair of lateral and medial copper plates alternately with each hand (see Wyke, 1969, for methodological detail). As shown in Figure 6, for bilateral symmetrical tapping requiring activity in the contralateral homologous muscles (BS), normal subjects showed a great decrease of tapping frequency in comparison with unilateral tapping, whereas the pianists showed little changes. When both hands moved in anti-phase with each other using the contralateral antagonistic muscles (BP), normal subjects showed further decreases in the slower hand compared to the in-phase task, whereas the pianists did not show any changes. On the other hand, for the maximum isometric flexion strength of the index finger or little finger (Figure 7), both pianists and normal subjects

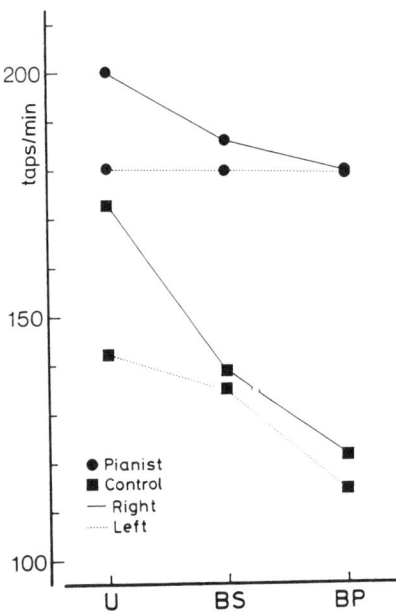

FIGURE 6 Overall mean tapping speed of pianists and normal subjects. U, unilateral; BS, bilateral symmetrical (in-phase); BP, bilateral parallel (anti-phase) tapping.

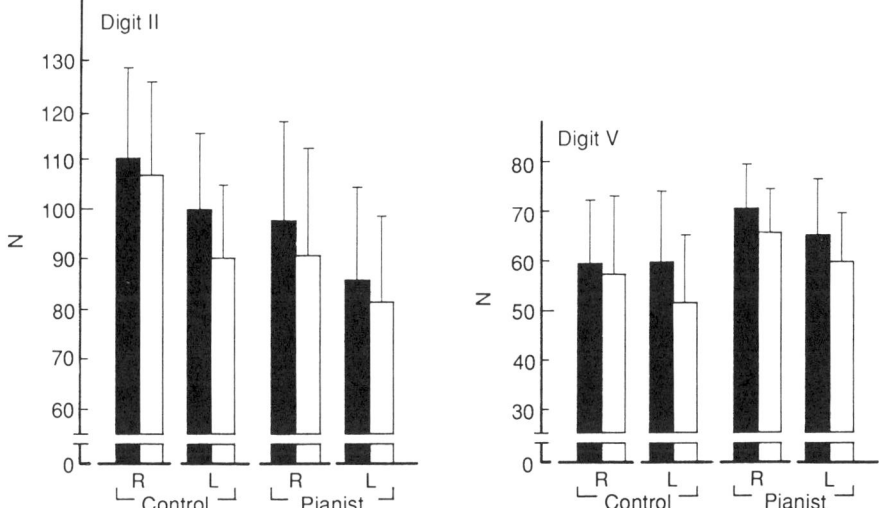

FIGURE 7 Overall mean and SD values of maximum strength of finger flexion. Unilateral, ■; bilateral, □.

showed a decrease of the same degree. This suggests that the magnitude of bilateral deficit is reduced in a task that closely resembles actual piano playing, in which parallel anti-phase movements of both arms are predominant, but is unaffected in the untrained tasks.

In summary, the findings mentioned above suggest that the decrease in strength diminishes only in experimental settings that resemble the well-practiced task, and that the bilateral deficit is a robust innate physiological phenomenon in ordinary humans and may be reduced to null (or even to reversal) by specific bilateral training.

VII. Exploring the Reasons Underlying the Bilateral Deficit

A. Division of Attention

One of the functions of attention is to focus processing resources on the attended process. Translating this notion to the task of producing force, we can consider attention to facilitate the processing of the optimal agonist–antagonist relations necessary to produce an appropriate force. Thus, any diminution of attentional resources would result in a reduction of force.

This idea was introduced at the end of the nineteenth century. Welch (1898) found that the maximum grip strength of one hand decreased while subjects performed a task that needed careful observation (e.g., pressing a rubber tube with the other hand in time with the arrival of a pendulum at a particular point). She stated that, "in order to obtain the maximum static contraction curve, the constant undivided attention is necessary," and assumed that the strength exerted without accompanying tasks is "the expression of the attention in the case that the whole available attention is given to muscular work." Her assumption indicates that attention is a quantity with a certain limit and the division of attention is a cause of reduction of maximum strength.

If a bilateral simultaneous task is performed as a combination of two independent tasks by each limb, and if a limited amount of attention is distributed to each task, attention allocated to each limb would likely decrease and, as a result, reduction of force and speed (integral of force divided by mass) occurs. Increased task difficulty by doing two tasks bilaterally could delay the initiation of task movements.

However, there are arguments against the division of attention. For example, Howard and Enoka (1991) conjectured that if the division of attention to different body parts were the main cause of the bilateral deficit, then not only the simultaneous use of contralateral muscles but also muscles anatomically distant from each other should cause the deficit. Therefore, they had their subjects exert maximum isometric strength of the left arm and the right leg simultaneously. However, they found no deficit in the leg or the arm, and concluded that the division of attention is not primarily responsible for the bilateral force deficit. Therefore, at present, division of attention as an account for bilateral deficit is still controversial.

B. Reciprocal Inhibition

One possible physiological mechanism responsible for the bilateral deficit in isometric strength may be a reflexive inhibition through the "double" reciprocal innervation in the spinal cord first reported as "crossed extension-reflex" (Sherrington, 1906; see also Henneman 1980; Rothwell, 1987) for flexion reflex afferents (FRA); the afferent impulses from muscle spindle or skin receptors of the agonist muscles inhibit the motoneurons innervating ipsilateral antagonists and contralateral agonists, and excite those innervating contralateral antagonists. Enhanced EMGs of antagonists under the bilateral condition (B. B. in Figure 1) testify that this mechanism is working during bilateral exertion.

If this "double" reciprocal innervation is an exclusive reason for the bilateral deficit, simultaneous contraction of contralateral antagonists (reciprocal contraction) must increase the force developed by each limb. I tested this by making subjects flex the right elbow and extend the left elbow simultaneously (Ohtsuki, 1983). However, as Figure 8 shows, no

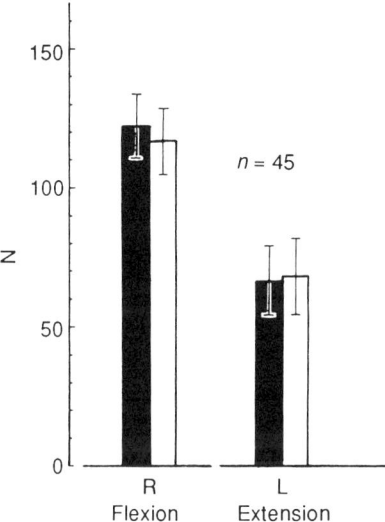

FIGURE 8 Overall means and SD values of maximum strength of bilateral simultaneous reciprocal exertion. Unilateral, ■; bilateral, □.

such clear increase was observed. Therefore, at least reciprocal innervation cannot fully explain the decrease of maximum voluntary strength in simultaneous bilateral exertion, though it can partly contribute to the bilateral deficit (for detailed discussion, see Ohtsuki, 1983).

C. Interhemispheric Inhibition

It is well known that the distal musculature of one side of the body is mainly controlled by the contralateral cerebral hemisphere. Thus, bilateral distal movement requires simultaneous activation of both hemispheres. The two hemispheres are connected by commissural nerve fibers such as the corpus callosum and the anterior commissure.

Gazzaniga and Sperry (1966) asked split-brain patients, whose interhemispheric fibers were sectioned, to perform two different choice reaction time tasks with each hand simultaneously, and found that reaction times for this double choice reaction task was the same as the single hand task, while normal subjects and patients before commissurotomy showed a much longer reaction time for the double task than for the single task. They concluded that the split-brain patients lacked the mutual interference through interhemispheric commissures that is present in normal subjects to keep the two hemispheres working in harmony. This study suggests that an increase in reaction time under the bilateral tasks reported in this review is due to interhemispheric interference with the information processing and motor programming.

Wyke (1971) found that the speed of bilateral synchronous hand tapping slowed down in normal subjects, whereas no difference occurred in patients with lesions on the cerebral cortex. Recently, Ferbert et al. (1992) observed an interhemispheric inhibition of the motor cortical neurons controlling a finger muscle (first dorsal interosseous); a transcranial magnetic stimulus applied on the motor cortex of one cerebral hemisphere inhibited the contralateral corresponding motor cortical neurons and reduced the EMG of the muscle controlled by these neurons. Reduction of EMG would be closely related to a decrease of the force and speed produced by that muscle.

These findings suggest that, for bilateral simultaneous movements, interhemispheric inhibition mediated through interhemispheric fibers plays some role in reducing motor performance, as indicated by such measurements as movement initiation, force, and speed.

VIII. Conclusions

This chapter presents studies that report the deterioration of motor performance induced by bilateral simultaneous exercise (i.e., bilateral deficit) and discusses possible causes of the phenomenon. In summary, there may be at least three possible mechanisms subserving the bilateral deficit. However, it seems that each does not hold true for all types of exercise. It is possible that these three mechanisms are working together with different degrees of contribution in different tasks. How these mechanisms are interrelated and whether there are other mechanisms are our future questions. In addition, there were some cases where the bilateral deficit did not occur. Investigation of whether some unknown mechanisms are hidden behind these seemingly contradictory results would also give us better understanding of this phenomenon.

Furthermore, in this chapter, I focused on the same level of maximal and submaximal exercise performed bilaterally and simultaneously. A different story might possibly be written for bilateral tasks performed by more than one body part at different relative intensities.

From a practical viewpoint, it is interesting that when one tries to activate multiple units of the body, the performance capacity of each unit declines. It may have a functional significance in decreasing fatigue by limiting the energy consumption of simultaneously working muscles to achieve optimal performance.

References

Asmussen, E., & Heebøll-Nielsen, K. (1961). Isometric muscle strength of adult men and women. *Comm. Danish Nat. Ass. Infant Paral.*, **11**, 3–43.

Benecke, R., Meinck, H.-M., & Conrad, B. (1985). Rapid goal-directed elbow flexion movements: limitations of the speed control system due to neural constraints. *Exp. Brain Res.*, **59**, 470–477.

Corcos, M. D. (1984). Two-handed movement control. *Res. Quart. Exerc. Sport.*, **55**, 117–122.

Di Stefano, M., Morelli, M., Marzi, C. A., & Berlucchi, G. (1980). Hemispheric control of unilateral and bilateral movements of proximal and distal parts of the arm as inferred from simple reaction time to lateralized light stimuli in man. *Exp. Brain Res.*, **38**, 197–204.

Ferbert, A., Priori, A., Rothwell, J. C., Day, B. L., Colebatch, J. G., & Marsden, C. D. (1992). Interhemispheric inhibition of the human motor cortex. *J. Physiol.*, **453**, 525–546.

Gazzaniga, M. S., & Sperry, R. W. (1966). Simultaneous double discrimination response following brain bisection. *Psychon. Sci.*, **4**, 261–262.

Henneman, E. (1980). Organization of the spinal cord and its reflexes. In: Mountcastle, V. B. (ed) Medical Physiology, 14th ed. Chapter 28 (p. 779). The C. V. Mosby Company: St. Louis.

Henry, F. M., & Smith, L. E. (1961). Simultaneous vs. separate bilateral muscular contractions in relation to neural theory and neuromotor specificity. *Res. Quart.*, **32**, 42–46.

Howard, J. D., & Enoka, R. M. (1991). Maximum bilateral contractions are modified by neurally mediated interlimb effects. *J. Appl. Physiol.*, **70**, 306–316.

Jeeves, M. A. (1969). A comparison of interhemispheric transmission times in acallosals and normals. *Psychon. Sci.*, **16**, 245–246.

Jeeves, M. A., & Dixon, N. F. (1970). Hemisphere differences in response rates to visual stimuli. *Psychon. Sci.*, **20**, 249–251.

Kerr, M., Mingay, R., & Elithorn, A. (1963). Cerebral dominance in reaction time responses. *Brit. J. Psychol.*, **54**, 325–336.

Kroll, W. (1965). Isometric cross-transfer effects under conditions of central facilitation. *J. Exp. Psychol.*, **20**, 297–300.

Marsden, C. D., Obeso, J. A., & Rothwell, J. C. (1983). The function of the antagonist muscle during fast limb movements in man. *J. Physiol.*, **335**, 1–13.

Mustard, B. E., & Lee, R. G. (1987). Relationship between EMG patterns and kinematic properties for flexion movements at the human wrist. *Exp. Brain Res.*, **66**, 247–256.

Ohtsuki, T. (1981a). Decrease in grip strength induced by simultaneous bilateral exertion with reference to finger strength. *Ergonomics*, **24**, 37–48.

Ohtsuki, T. (1981b). Increase in simple reaction time of knee extension induced by simultaneous bilateral performance. *Percept. Motor Skills*, **53**, 27–30.

Ohtsuki, T. (1983). Decrease in human voluntary isometric arm strength induced by simultaneous bilateral exertion. *Behav. Brain Res.* **7**, 165–178.

Rothwell, J. C. (1987). Control of human voluntary movement. pp. 114–115. An Aspen Publication: Rockville.

Schantz, P. G., Moritani, T., Karlson, E., Johansson, E., & Lundh, A. (1989). Maximal voluntary force of bilateral and unilateral leg extension. *Acta Physiol. Scand.*, **136**, 185–192.

Secher, N. H. (1975). Isometric rowing strength of experienced and inexperienced oarsmen. *Med. Sci. Sports*, **7**, 280–283.

Secher, N. H., Rørsgaard, S., & Secher, O. (1978). Contralateral influence on recruitment of curarized muscle fibres during maximal voluntary extention of the legs. *Acta Physiol. Scand.,* **103,** 456–462.

Secher, N. H., Rube, N., & Elers, J. (1988). Strength of two- and one-leg extension in man. *Acta Physiol. Scand.,* **134,** 333–339.

Seki, T., & Ohtsuki, T. (1990). Influence of simultaneous bilateral exertion on muscle strength during voluntary submaximal isometric contraction. *Ergonomics,* **33,** 1131–1142.

Sherrington, C. (1906). The Integrative action of the nervous system. Yale University Press: New Haven.

Vandervoort, A. A., Sale, D. G., & Moroz, J. (1984). Comparison of motor unit activation during unilateral and bilateral leg extention. *J. Appl. Physiol.: Respirat. Environ. Exerc. Physiol.,* **56,** 46–51.

Welch, J. C. (1898). On the measurement of mental activity through muscular activity and the determination of a constant of attention. *Am. J. Physiol.,* **1,** 283–306.

Wyke, M. (1969). Influence of direction on the rapidity of bilateral arm movements. *Neuropsychologia,* **7,** 189–194.

Wyke, M. (1971). The effects of brain lesions on the performance of bilateral arm movements. *Neuropsychologia,* **9,** 33–42.

Part II

The Dynamics of Interlimb Coordination

14

A Low-Dimensional Nonlinear Dynamic Governing Interlimb Rhythmic Coordination

M. T. Turvey

Center for the Ecological Study of Perception and Action
University of Connecticut
Storrs, Connecticut
and Haskins Laboratories
New Haven, Connecticut

R. C. Schmidt

Department of Psychology
Tulane University
New Orleans, Louisiana
and Center for the Ecological Study of Perception and Action
University of Connecticut
Storrs, Connecticut

I. Introduction
    A. The Bistability of 1 : 1 Interlimb Frequency Locking
    B. Equating Von Holst's Maintenance Tendency and Magnet Effect with $\Delta\omega$ and $K\sin\phi$, Respectively
    C. A Methodology for Investigating Interlimb Coordination
    D. Developing a Low-Dimensional, Nonlinear Dynamic for a 2-Oscillator Coupled System
    E. Defining the Relevant Experimentally Manipulable and Measurable Observables
II. Experimental Results on 1 : 1 Frequency Locking
    A. Time Series of Interlimb Coordination: The Basic Data Pattern
    B. Magnet Effect (Strength of Nonlinear Coupling) and Maintenance Tendencies (Contrasting Eigenfrequencies) Interact in Determining Average Phase

C. The In-Phase and Anti-Phase Equilibrium Points Exhibit a Symmetry over Transformations in the Competition and Cooperation Parameters
  D. The Equilibrium Points Engendered by $\Delta\omega$ and ω_c Differ in Stability
  E. A "Changing Potential" Perspective on the Interactive Effects of Competition ($\Delta\omega$) and Cooperation (ω_c qua $K\sin\phi$)
III. Expectations from the Low-Dimensional Dynamic about 2 : 1 Frequency Locking
  Contrasting Dependencies of ϕ and ξ on the Competitive Processes (Component Eigenfrequencies Difference) and Cooperative Processes (Nonlinear Coupling Functions)
IV. Conclusions
  References

I. Introduction

The standard circle map studied originally by Arnold (1965) has been used to provide a general theory of phase locking in 2-oscillator systems via nonlinear coupling (e.g., Baker & Gollub, 1990; Jackson, 1989), and its potential application to coordination patterns in biological movement systems has been duly recognized (Glass & Mackey, 1988; Kelso, DeGuzman, & Holroyd, 1990a; Schmidt, Beek, Treffner & Turvey, 1991a). This discrete map is commonly given in the form

$$\theta_{n+1} = \theta_n + \Omega - (K/2\pi)\sin(2\pi\theta_n) \quad mod\ 1 \tag{1}$$

where θ is a point on the unit circle (specified by the angle of the unit vector in radians), Ω may be interpreted as the ratio of the two uncoupled frequencies, and K is the coefficient of a 2π periodic (that is, nonlinear) coupling function. Given the nth point, Equation (1) yields the $(n + 1)$th point in an iterative sequence.

Equation (1) is a difference equation. Such equations often arise when it becomes necessary to resort to numerical methods to obtain approximate solutions to differential equations that do not have "closed form" solutions. In obtaining a numerical solution, a discrete set of points is selected, such as the temporal points $t_0, t_1, \ldots, t_n$, with the simplifying requirement that the distance between any two consecutive points, the step size h, be constant, for example, $h = t_{n+1} - t_n$. Approximate values of the solution $x(t)$ at these equally spaced points are then calculated. Suppose the differential equation is $\dot{x} = f(x)$ and $x(t_0)$ is given, then the challenge is to find an algorithm to approximate $x(t_1)$. The algorithm can then be used to determine any required approximation $x(t_n)$, designated by x_n, knowing $x(t_{n-1})$. Euler's algorithm is the simplest (Hale & Kocak, 1991). It takes $t_n = nh$ and replaces $\dot{x}(t)$ by the difference quotient $(x_{n+1} - x_n)/h$. The differential equation $\dot{x} = f(x)$ thus becomes the difference equation: $x_{n+1} = x_n + hf(x_n)$. How well the difference equation approximates the dynamics of the differential equation depends, among other things, on the step size h.

Equation (1) is the difference equation obtainable by Euler's algorithm for the differential equation $\theta = a - b\sin(2\pi\theta)$ where a and b are real parameters. In arriving at Equation (1), $\Omega = ah$ and $K = 2\pi bh$, with the end points 0 and 1 of the unit interval defined by the modular 1 operation (Hale & Kocak, 1991). Our purpose in the present chapter is to show that a particular differential equation of the preceding type— one representable, therefore, by Equation (1) or an analogous equation—is of considerable significance for understanding the stable states of interlimb coordination. The equation in question is from Haken (1983, Section 8) and Rand, Cohen, and Holmes (1988). It was developed separately to address, in Haken's case, the general issue of bifurcation to frequency locking, and in Rand et al.'s case, the specific conditions of intra- and intersegmental coordination in *Lamprey*. The equation has the form

$$\dot{\phi} = \Delta\omega - K\sin\phi \qquad (2)$$

where ϕ is the phase difference $(\theta_1 - \theta_2)$ between two oscillators (with θ_i being the phase of an individual oscillator), $\Delta\omega$ is the uncoupled eigenfrequency difference $(\omega_1 - \omega_2)$, and K is the strength of the nonlinear coupling. If $\Delta\omega = 0$, and if K is positive in sign, then a graph in which the right-hand side of Equation (2) is plotted against ϕ exhibits a negatively sloped zero-crossing (defining a stable point) at each of the intercept values of $\phi = 0$ and $\phi = 2\pi$, and a positively sloped zero-crossing (defining an unstable point) at $\phi = \pi$. These conditions characterize in-phase coordination. Conversely, if the sign of K is negative, then there is a negatively sloped zero-crossing at the intercept value of π and a positively sloped zero-crossing at each of 0 and 2π. These conditions characterize anti-phase coordination. If $\Delta\omega \neq 0$, then the stable points will deviate from 0 and π to a degree depending on the magnitude of $|\Delta\omega|$.

The development of Equation (2) for purposes of analyzing animal rhythmic movements (Rand et al., 1988) expresses a general modeling strategy, which makes no assumptions about the "state space" (the neurobiology) containing the dynamics of individual biological oscillators and their couplings, referring only to the phenomenologically observed oscillations. Complicated (high-dimensional) dynamics are reduced to simpler (low-dimensional) dynamics by assuming that the interoscillator coupling depends on $\phi = (\theta_1 - \theta_2)$ and not on θ_1 and θ_2. ϕ is a coordination or order parameter (e.g., Haken, 1977, 1983; Haken, Kelso, & Bunz, 1985; Kelso, Schöner, Scholz, & Haken, 1987), a collective variable that captures the spatiotemporal organization of the component subsystems (here, rhythmic movement units), and which changes more slowly than the variables characterizing the states of the component subsystems (e.g., velocity, amplitude). A key idea behind Equation (2), therefore, is that even though the coupling influences may occur at varied times in a rhythmic appendage's cycle, and even though their effects could depend on more than ($\theta_1 - \theta_2$), stable interappendage coordinations may be predictable solely from a consideration of the behavior of the cycle-averaged order parameter (e.g., Haken et al., 1985; Kopell, 1988a,b).

The stable points of Equation (2) can be conveniently conceptualized as the minima of the potential (V) function:

$$V(\phi) = -\Delta\omega\phi - K\cos(\phi). \qquad (3)$$

The derivative of Equation (3) with respect to ϕ gives Equation (2). As identified in synergetics (Haken, 1977), equations for order parameters are often interpretable as the differentiation of a potential function. In analogy with classical mechanical systems, Equation (3) defines a potential "well" with the behavior of the 2-oscillator system described by identifying ϕ with the coordinate of a particle that moves in an overdamped fashion in this well (Haken et al., 1985). Equation (2), however, is deterministic, meaning that the system can change its position only if the potential defined by Equation (3) changes. In order to permit motion within a well fixed by the parameters $\Delta\omega$ and K, and to address, thereby, the fluctuations evident in a given instance of 1:1 interlimb frequency locking, a stochastic force needs to be added (Schöner, Haken, & Kelso, 1986). A good candidate is ζ_t, a Gaussian white noise process with characteristics $\langle\zeta_t\rangle = 0$ and $\langle\zeta_t\zeta_{t'}\rangle = \delta(t - t')$, and strength $Q > 0$ (Kelso et al., 1987; Schmidt, Treffner, Shaw, & Turvey, 1992; Schöner et al., 1986). Behind this choice of stochastic force is the assumption that the degrees of freedom (e.g., underlying subsystems), acting as noise on the interlimb or intersegmental coordination, operate on a time scale that is considerably faster than the time scale of the order parameter (see Haken, 1983, Section 6). Conceptualizing the noise as a random sequence of brief kicks of equal probability in different directions means that, on the average (represented by the brackets $\langle\rangle$), the stochastic force will be zero (Haken, 1977, Section 6). Forming the product of the stochastic force at time t with the stochastic force at another time t' and taking the average over the times of the kicks and their directions yields a correlation function equal to the Dirac δ-function (see Haken, 1977, Section 6, for details). Consequently, Equation (2) can be elaborated to include both deterministic and random influences on the behavior of the order parameter:

$$\dot\phi = \Delta\omega - K\sin\phi + \sqrt{Q}\zeta_t. \qquad (4)$$

As expressed by Kugler and Turvey (1987), interlimb coordination can be imaged as a virtual single macroscopic particle undergoing Brownian motion due to the random action of many microscopic particles.

A. The Bistability of 1 : 1 Interlimb Frequency Locking

To reiterate, the particular mode of 1 : 1 frequency coordination—in-phase versus anti-phase—depends on the sign of the coupling coefficient K. For the interlimb order parameter dynamics expressed by Equation (4), it is assumed that only one of the two coordinative modes exists as an attractor for those dynamics during any given bout of interlimb rhythmic coordination. An alternative assumption can be made, however: Both attractors are present, one usually explicitly and one usually implicitly, during any given bout of 1 : 1 frequency locking between two limbs. This alternative is underscored by research on spontaneous transitions in human interlimb rhythmic patterns. The experimental setting is one in which a person is required to oscillate the two index fingers (or two hands) at the coupled frequency ω_c, where ω_c is varied by a metronome that the person tracks (e.g., Kelso, 1984; Kelso, Scholz, & Schöner, 1986; Scholz, Kelso, & Schöner, 1987). Results show that there are only two steady states: in-phase and anti-phase. With increasing ω_c, anti-phase switches rapidly to in-phase. In-phase, however, does not switch to anti-phase, and the anti-phase–to–in-phase transition is not reversed by a reduction in ω_c. Further, ϕ exhibits increases in relaxation time (the time taken to return to its value prior to a brief perturbation) and fluctuations (measured, for example, by its standard deviation) as the transition point is approached. The same basic pattern of results is found when two limbs are connected optically between two people rather than anatomically within a person (Schmidt, Carello, & Turvey, 1990).

The fundamental aspects of the above behavioral transitions have been successfully modeled, and their subtleties successfully predicted, by the following order parameter equation expressed in successively developed forms by Haken et al. (1985), Schöner et al. (1986), and Kelso, DelColle, and Schöner (1990b):

$$\dot{\phi} = \Delta\omega - a\sin(\phi) - 2b\sin(2\phi) + \sqrt{Q}\,\zeta_t \quad (5)$$

where a and b are coefficients such that b/a decreases as coupling frequency ω_c increases and coupling strength (stability) decreases. As with Equation (4), the plotting of the right-hand side of Equation (5) (minus the noise) against ϕ leads to the identification of stationary values (the zero-crossings). The two sine terms in Equation (5) are obtained from differentiating the potential function

$$V(\phi) = -\Delta\omega\phi - a\cos(\phi) - b\cos(2\phi) \qquad (6)$$

which provides an "energy landscape" characterized, when $\Delta\omega = 0$, by a global minimum at $\phi = 0$ and local minima at $\phi = \pm\pi$. If $b = 0$, then during a bout of interlimb 1:1 frequency locking, there is only one minimum (0 or π, depending on the sign of a) as in Equation (3); if $a = 0$, then there are three equal minima $(-\pi, 0, +\pi)$. In the superposition of the two cosine functions of Equation (6) the minima are distinguished, with stability greatest for $\phi = 0$ (meaning that the potential is least at $\phi = 0$) when $\Delta\omega = 0$, as suggested by the experiments summarized above. With respect to the dissolution of the anti-phase interlimb pattern in those experiments, this landscape is modulated by ω_c, such that at a critical value (when $a = 4b$) the local minima are annihilated (see Haken et al., 1985).

In the present chapter we shall show, through a summary of a number of important results, that Equations (4) and (5) not only facilitate the experimental study of interlimb coordination but promise to capture its fundamental, universal character. In so doing we shall underscore the general significance and importance for the theory of interlimb coordination of exploring the nonlinear dynamics implied by Equation (1) and its important extension, the phase-attractive circle map, developed to accommodate the bistability expressed by Equation (5) (deGuzman & Kelso, 1991). For simplicity and for expositional purposes we restrict our focus to Equation (4). The conclusions to be drawn, however, apply with minor qualifications to the interlimb coordination dynamics expressed by Equation (5).

B. Equating von Holst's Maintenance Tendency and Magnet Effect with $\Delta\omega$ and $K\sin\phi$, Respectively

It is important to see how the dynamics expressed in Equation (4) dovetail with long-term intuitions on the nature of interappendage rhythmic coordination. Investigating the rhythmic fin movements of *Labrus*, a fish that swims with its longitudinal axis immobile, von Holst (1937/1973) observed two archetypal scenarios. In one, the fins maintained a fixed phase relation and oscillated at the same frequency; von Holst referred to this as *absolute coordination*. In the other observed scenario, there was an absence of phase and frequency locking; von Holst referred to this as *relative coordination*. Over a period of observation, examples of both scenarios would be seen intermittently, and whenever one dominated, signs of the other would still be present. An interfin relation that was on the average a strong case of absolute coordination would exhibit momentary deviations from the strict mode-locked state. Similarly, an interfin relation that was on the average a strong case of relative coordination would exhibit momentary advances toward phase and frequency locking. On the basis of such observations, von Holst concluded that even when absolute coordination was achieved, competition among rhythmic units (each fin proceeding at its own pace) remained, and even when relative coordination was occurring, a tendency to cooperate (each fin proceeding at the pace of the others) was still in evidence. He referred to the competitive aspect as the "maintenance tendency" and the cooperative aspect as the "magnet effect."

In Equation (4) there is a term referring to the contrast or competition between the preferred dynamics of the component oscillatory systems, namely, $\Delta\omega$. Also present in Equation (4) is a term referring to the coming together or coupling of the component oscillators, and the degree to which they will tend to do so, namely $K\sin\phi$. In short, the dynamics expressed in Equation (4) comport with von Holst's insights on interappendage coordination. For von Holst, an assembled rhythmic coordination reflects the individual tendencies of the components to continue to do what they prefer to do (viz., $\Delta\omega$) and their collective tendency to act as a single unit (viz., $K\sin\phi$).

C. A Methodology for Investigating Interlimb Coordination

The research program initiated by von Holst (1937/1973) suggests three major requirements for an experimental method directed at the dynamics of interappendage rhythmic coordinations: (1) the eigenvalues of the individual rhythmic movement units should be manipulable and easily quantified; (2) the interlimb system should be easily prepared in one of the two basic patterns of in-phase and anti-phase; and (3) the focus of measurement and dynamical modeling should be on the interactions of phase. Requirement 3 is satisfied by Equation (4); requirements 1 and 2 are satisfied by an experimental procedure developed by Kugler and Turvey (1987). That procedure is depicted in Figure 1. A person is shown seated and holding a pendulum in each hand. The pendulums can vary physically in shaft length and/or the mass of the attached mass bob. Each of the two pendulums is swung parallel to the sagittal plane about an axis in the wrist (with other joints essentially immobile). The eigenfrequency of an individual "wrist-pendulum system" can be approximated by the eigenfrequency of the equivalent simple gravitational pendulum, $\omega = (g/L)^{1/2}$, where L is the simple pendulum length and g is the constant acceleration due to gravity. The quantity L is calculable from the magnitudes of shaft length, added mass, and hand mass, through the standard methods for representing any arbitrary rigid body oscillating about a fixed point as a simple pendulum (den Hartog, 1950; Kugler & Turvey, 1987).

It is evident from the preceding that if the pendulums oscillated in each hand differ in physical dimensions (length, mass), then their eigenfrequencies (their maintenance tendencies) will not correspond. The

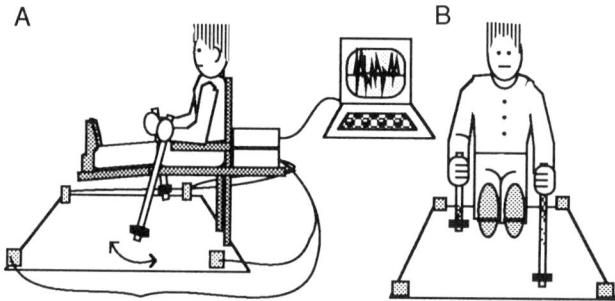

FIGURE 1 Side (A) and front (B) views of the experimental arrangement in which the subject oscillates two hand-held pendulums. The schematic depicts the four microphones of an ultrasonic 3-phase digitizer arranged at the corners of a horizontal, square grid beneath the subject. An ultrasonic emitter is attached to the lower end of each pendulum. The time-varying positions of the emitters are recorded by computer.

component rhythmic units will be in frequency competition. It is also evident from the preceding that preparing the interlimb coordination in either in-phase or anti-phase is a relatively simple matter of instructing the subjects to maintain one or the other basic pattern in the course of a trial.

D. Developing a Low-Dimensional, Nonlinear Dynamic for a 2-Oscillator Coupled System

Let us now return to the strictly deterministic part of Equation (4), that is, Equation (2), and identify the steps in its development after Rand et al. (1988). The modeling assumptions behind Equation (2) focus on (1) the characterization of the individual biological oscillator, and (2) the characterization of two coupled biological oscillators. It is assumed at the outset that an individual oscillator is on a steady-state limit cycle at all times. A limit cycle is a closed orbit (which is commonly noncircular) in the phase space of velocity and displacement, to which the system returns following a perturbation. It is further assumed that the oscillator's state can be specified by a single variable $\theta(t)$ representing the phase of the limit cycle, where θ passes from 0 rad to 2π rad in one cycle, and where θ can be made proportional to the fraction of the cycle period that has elapsed. Consequently,

$$\dot{\theta} = \omega \tag{7}$$

where ω is the eigenfrequency of the oscillator. By integration, and recognizing that successive oscillations are not distinguished, Equation (7) gives

$$\theta(t) = \omega t + \theta(0) \, (mod \, 2\pi). \tag{8}$$

It is assumed that, for sufficiently weak coupling, oscillators characterized by Equation (8) continue on limit cycle orbits and that the state space for the dynamics of the 2-oscillator coupled system is a 2-dimensional torus, a donut-shaped geometric object represented mathematically by T^2. The motion of the two coupled oscillators is visualized as a particle moving on the surface of the donut. If the particle motion is the position vector $r(t)$, then $r = R + p$, with $R(t)$ and $p(t)$ being the motions around the donut's major and minor axes, respectively, at frequencies of ω_1 and ω_2, respectively. A first pass, therefore, on the behavior of an individual oscillator under coupling yields

$$\dot{\theta}_i = \omega_i + H_{ij}(\theta_i, \theta_j). \tag{9}$$

The next step is to determine the coupling function H_{ij}. Because θ_i is the single-state variable of the individual oscillator, the coupling function H_{ij} can depend only on θ_1 and θ_2 and cannot depend, by assumption, on factors such as the size of the limit cycle or the wave form of the oscillation. Further, in order for the motion expressed by Equation (9) to be defined uniquely on T^2, the effect of the one oscillator on the same phase point of the other oscillator must always be the same. H_{ij} must be 2π periodic. A particular simple form of H_{ij}—known as *diffusive* coupling (Kopell, 1988a; Murray, 1990; Rand et al., 1988)—is defined by assuming that the coupling vanishes when $\theta_1 = \theta_2$. It allows that

$$H_{12}(\theta_1, \theta_2) = H_{12}(\theta_2 - \theta_1) = H_{12}(\phi). \tag{10}$$

Given Equation (10) and H_{ij} as 2π periodic, the requisite coupling function can be approximated by taking the first terms of the Fourier series. For simplicity, the coupling function can be considered as a pure sine function with coefficient k_{12} by taking the coefficient on the cosine term to be equal to zero (Rand et al., 1988). Hence,

$$H_{12}(\theta_1, \theta_2) = k_{12} \sin(\theta_2 - \theta_1). \tag{11}$$

As a consequence, the motion equations of the two oscillators take the form

$$\begin{aligned} \dot{\theta}_1 &= \omega_1 + k_{12} \sin(\theta_2 - \theta_1) \\ \dot{\theta}_2 &= \omega_2 + k_{21} \sin(\theta_1 - \theta_2). \end{aligned} \tag{12}$$

Subtracting the motion equation for oscillator 2 from that of oscillator 1, using the identity $\sin(\theta_1 - \theta_2) = -\sin(\theta_2 - \theta_1)$, provides us with the low-dimensional dynamic in ϕ captured by Equation (2), where $\Delta\omega = (\omega_1 - \omega_2)$ and $K = (k_{12} + k_{21})$.

As intimated above, a general expression for the varied equilibria of Equation (2) can be determined under the assumption that, at equilibrium, the time derivative of ϕ equals 0. Consequently, the equilibrium values of ϕ are given by

$$\phi = \arcsin(\Delta\omega/K). \qquad (13)$$

Obviously, given the nature of the arcsine function, if $\Delta\omega > K$, then there are no stabilities, meaning no frequency locking (Haken, 1983; Rand et al., 1988). With respect to von Holst's observation of relative coordination, an irregular waxing and waning of coupling strength in the medulla-transected *Labrus* could engender an irregular shifting back and forth from $\Delta\omega/K < 1$, with a resulting irregular pattern of frequency wandering and frequency locking.

E. Defining the Relevant Experimentally Manipulable and Measurable Observables

In the form given, Equation (13) is not fully open to investigation. At issue is how to bring K under experimental control. Fortunately, the observations on phase transitions in interlimb coordination and their theoretical explanation in the form of Equation (5) provide resolution. With increasing frequency ω_c of the 2-oscillator anti-phase coupling, coupling strength weakens (Haken et al., 1985). According to Equation (5), the phase transition occurs because the potential well surrounding $\phi = \pm\pi$ becomes increasingly more shallow and eventually disappears. Loss of stability with increasing ω_c causes the transformation from the anti-phase organization. But what of the potential well surrounding $\phi = 0$? It also changes with increasing ω_c in the direction of decreasing stability. The magnitude of V increases and the surrounding gradients become less steep (see Haken et al., 1985), meaning that the fluctuations due to $\sqrt{Q}\zeta$ are realized as ever larger standard deviations in the behavior of $\phi = 0$. The important implication of these observations for present purposes is that K can be systematically manipulated by manipulating ω_c, with the understanding that K varies inversely with ω_c. Equation (13) can be approximated, therefore, by

$$\phi \cong \arcsin[(\Delta\omega)(\omega_c)] \qquad (14)$$

with each of the right-hand quantities manipulable and with the right-hand quantity measurable directly from the 3-space motions of the two wrist-pendular rhythmic units. The average value of ϕ over a bout (an experimental trial) of interlimb coordination is ϕ_{ave} for an intended value of ϕ_ψ (with ψ symbolizing "intended"). From the discussion above, $\Delta\omega$ is a control parameter that governs the degree of competition between the oscillators. During a trial in the wrist-pendulum paradigm, with the subject exhibiting coupled oscillations of a given pair of hand-held pendulums, the frequency competition between the rhythmic subsystems will be constant. The common or coupled frequency ω_c of the rhythmic units is another control parameter. It likewise can be held constant during a trial (by means of a pacing metronome).

It is intuitively apparent that, for different coupled wrist-pendulum systems, a given ω_c cannot mean (in dynamical terms) the same thing. The physical significance of any given ω_c must be defined relative to a time quantity characteristic of the coupled system. When two pendular rhythmic units are coupled, the system thus formed possesses, like the subsystems from which it is formed, an eigenfrequency. Kugler and Turvey (1987) presented arguments and data (see also Turvey, Schmidt, Rosenblum, & Kugler, 1988) for characterizing a coupling of left and right wrist-pendulum systems as a single virtual system with an equivalent simple pendulum length. If the two wrist-pendulum systems were coupled such that θ_1 was always, at every instant, identically equal to θ_2, or to $(\theta_2 + \pi)$, then the two oscillators could be considered as rigidly connected. The simple pendulum equivalent $L_{equivalent}$ of a compound pendulum so composed (that is, of two pendulums connected by a rigid bar) is given by

$$L_{equivalent} = (m_1 l_1^2 + m_2 l_2^2)/(m_1 l_1 + m_2 l_2) \qquad (15)$$

where m_i and l_i refer to the mass and the equivalent simple pendulum length, respectively, of an individual (compound) pendulum system. Through Equation (15), two coupled pendulums of lengths l_1 and l_2 can be interpreted as a virtual (v) pendulum of length L_v with an eigenfrequency $\omega_v = (g/L_v)^{1/2}$. Just as the eigenfrequencies $\omega_1 = (g/l_1)^{1/2}$ and $\omega_2 = (g/l_2)^{1/2}$ of the uncoupled wrist-pendulum systems characterize their respective maintenance tendencies, so it is that ω_v characterizes the maintenance tendency of the coupled wrist-pendulum system. The importance of the latter is that ω_c is most appropriately expressed in units of ω_v—the control parameter is ω_c/ω_v. Accordingly, Equation (14) becomes

$$\phi \cong \arcsin[(\Delta\omega)(\omega_c/\omega_v)]. \qquad (16)$$

II. Experimental Results on 1:1 Frequency Locking

A. Time Series of Interlimb Coordination: The Basic Data Pattern

Figure 2 shows the continuous phase relation between two wrist-pendulum systems in 1:1 frequency locking under $\phi_\psi = \pi$ for a typical subject (see Schmidt *et al.*, 1991a). Three conditions are depicted, one with the two pendulums nearly identical (left/right ratio Ω of uncoupled eigenfrequencies is nearly unity, $\Omega = 1.03$), one with the eigenfrequency of the left pendulum lower than that of the right pendulum

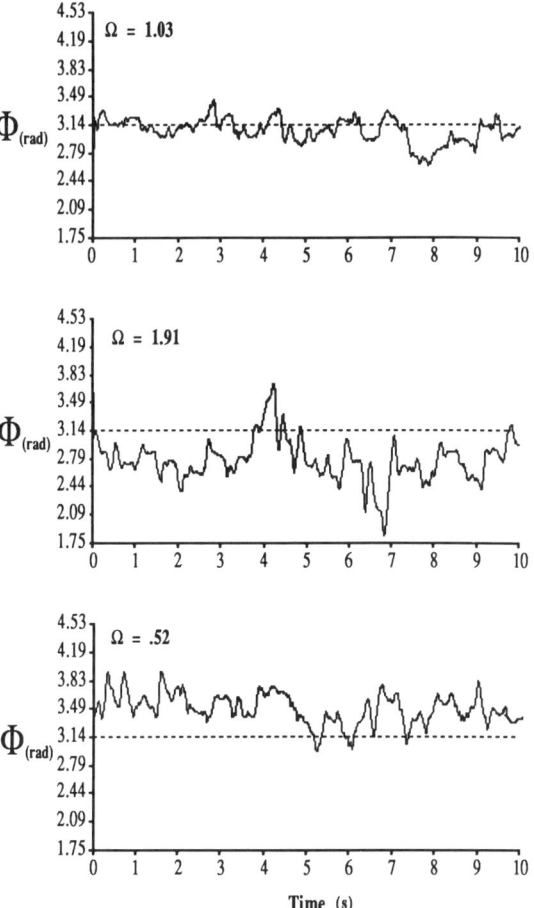

FIGURE 2 Examples of φ time series as a function of Ω.

(Ω = 0.52), and one with the eigenfrequency of the left pendulum higher than that of the right pendulum (Ω = 1.91). As can be seen, there is considerable moment-to-moment variation in phase, the average deviation from the intended phase relation is larger for Ω = 0.52 and Ω = 1.91 than for Ω = 1.03, and the direction of the deviation from anti-phase (180° or π) depends on which rhythmic unit, right or left, had the higher eigenfrequency (compare Ω = 0.52 and Ω = 1.91).

The actual phase relation between the limbs evident in Figure 2 would seem to be one of phase entrainment. Phase locking refers to a situation in which the phases of the two component oscillators are related linearly, that is, $\theta_2(t) = \theta_1(t) +$ a constant. In contrast, phase entrainment refers to a situation in which θ_2 completes one full cycle in the same amount of time as θ_1 completes one full cycle in the absence of a fixed magnitude of $\theta_2 - \theta_1$. Phase entrainment and mode attraction rather than strict mode locking appear to be the rule rather than the exception in biological movement systems (Kelso et al., 1990a; Schmidt, Shaw, & Turvey, 1993; Turvey Schmidt, & Beek, 1993). While the phase relation between limbs may be thought of conveniently as a static property characterizing an interlimb coordination, it is in fact a mean or average of a continuously fluctuating process. As remarked above, a reasonable surmise is that the presence of fluctuations is a consequence of the complexity of the biological movement system—multiple components functioning at multiple space and time scales smaller and faster, respectively, than those of ϕ (Schmidt et al., 1991a; Schöner et al., 1986). In Equations (4) and (5), the basis for these fluctuations is identified with the stochastic force $\sqrt{Q}\zeta$.

The coordination patterns evident in Figure 2 can be seen in another behavioral context in which one person swings a pendulum in the right hand, one person swings a pendulum in the left hand, and between the two of them they achieve anti-phase 1 : 1 frequency locking by watching the motions of each other's pendulum (Schmidt & Turvey, in press). The similarities suggest that the description given to interlimb coordination within a person, and the principles used to account for it, may be applicable to interlimb coordination between two persons (Schmidt et al., 1990). The dynamics at work in fashioning interappendage rhythmic coordinations seem to generalize over different experimental settings involving different numbers of microcomponents (e.g., the nervous system of one person vs. the nervous systems of two people) and different perceptual bases for the coordinations (haptic perceptual system vs. visual perceptual system).

B. Magnet Effect (Strength of Nonlinear Coupling) and Maintenance Tendencies (Contrasting Eigenfrequencies) Interact in Determining Average Phase

Von Holst (1939/1973) established that interappendage coordination arose from competitive and cooperative processes. As noted above, the competitive processes may be quantified by $\Delta\omega$ and the cooperative processes may be quantified by $K\sin\phi$. Two recent papers, Schmidt et al. (1993) and Sternad, Turvey, and Schmidt (1992), report experiments that provide direct evaluations of the way in which these two quantities interact in producing stable interlimb coordinations. We focus on one experiment for brevity. Sternad et al. (1992, Experiment 3) compared $\omega_c/\omega_v < 1$, $\omega_c/\omega_v = 1$, and $\omega_c/\omega_v > 1$ under anti-phase coordination, $\phi_\psi = \pi$. According to Equation (2) in the form of Equation (16), the dependency of ϕ_{ave} on $\Delta\omega$ should be steepest for $\omega_c/\omega_v > 1$ and shallow-

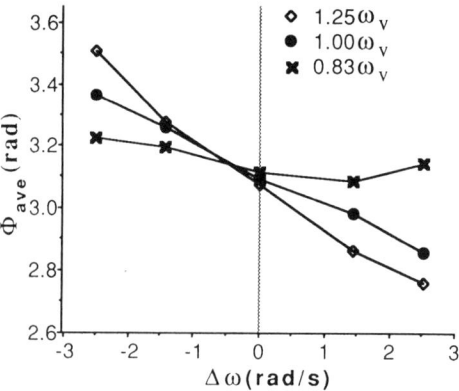

FIGURE 3 Average phase difference ϕ_{ave} as a function of the difference in uncoupled frequencies $\Delta\omega$ and the relative coupled frequency ω_c/ω_v. Each data point represents the mean of five subjects.

est for $\omega_c/\omega_v < 1$. For $\Delta\omega = 0$, ϕ_{ave} should equal π for each of $\omega_c/\omega_v < 1$, $\omega_c/\omega_v = 1$, and $\omega_c/\omega_v > 1$. In the experiment, the two inequalities were satisfied with values of 0.83 and 1.25. The magnitude of ϕ_{ave} as a function of $\Delta\omega$ and ω_c/ω_v is presented in Figure 3. Inspection suggests that, in accordance with Equation (16), the rate of change of ϕ_{ave} as a function of $\Delta\omega$ was greater for $\omega_c/\omega_v > 1$ than for $\omega_c/\omega_v = 1$, and greater for $\omega_c/\omega_v = 1$ than for $\omega_c/\omega_v < 1$. It is also evident that for $\Delta\omega = 0$, $\phi_{ave} = \pi$ regardless of ω_c/ω_v. An ANOVA with ϕ_{ave} as the dependent variable yielded a significant main effect of $\Delta\omega$ ($p < .0001$), no main effect of ω_c/ω_v, ($p > .05$), and a significant interaction between $\Delta\omega$ and ω_c/ω_v, ($p < .0001$). An ANOVA on the data for $\Delta\omega = 0$ found no effect of ω_c/ω_v, $F < 1$. Evaluation of the contrast between ω_c/ω_v and ω_c through simple linear regressions of $|\phi_{ave}-\pi|$ on ω_c/ω_v and ω_c revealed that ω_c/ω_v accounted for significantly more of the variance in ϕ_{ave} than ω_c.

C. The In-Phase and Anti-Phase Equilibrium Points Exhibit a Symmetry over Transformations in the Competition and Cooperation Parameters

Equation (13), and its experimentally testable forms Equations (14) and (16), suggest that a given magnitude of $\Delta\omega \neq 0$ induces $\phi_{ave} \neq 0$ but in the vicinity of 0 (when K is positive) and $\phi_{ave} \neq \pi$ but in the vicinity of π (when K is negative), with the magnitudes of these deviations from 0 and π identical. This means that for any given $\Delta\omega \neq 0$, the stable points under a sign transformation of a given K are always separated by π. There is a symmetry of the "in-phase" equilibrium points and "anti-phase" equilibrium points under the transformation

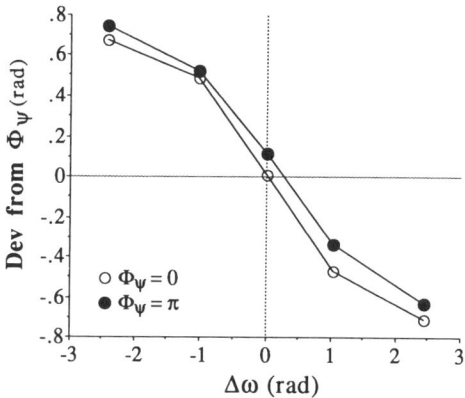

FIGURE 4 ϕ_{ave} deviation from $\phi_\psi = 0$ and ϕ_{ave} deviation from $\phi_\psi = \pi$ as a function of $\Delta\omega$. Data points are the mean values of five subjects for each of three values of $\Delta\omega$.

$\Delta\omega$. Figure 4, based on the results of Schmidt et al. (1993), shows that the deviation of ϕ_{ave} from ϕ_ψ as a function of $\Delta\omega$ is identical for $\phi_\psi = \pi$ and $\phi_\psi = 0$ (see also Sternad et al., 1992; Turvey, Rosenblum, Schmidt, & Kugler, 1986).

Equation (13) further suggests that for a given $\Delta\omega$, a given magnitude of K results in a ϕ_{ave} different from $\phi_\psi = 0$ when K is positive by the same amount that the resultant ϕ_{ave} is different from $\phi_\psi = \pi$ when K is negative. This expectation can be examined explicitly through Equation (14), which exploits the proportionality $K \propto \omega_c^{-1}$. Figure 5, also based on the results of Schmidt et al. (1993), shows that the deviation of ϕ_{ave} from ϕ_ψ as a function of ω_c is identical for $\phi_\psi = \pi$ and $\phi_\psi = 0$ (see also Sternad et al., 1992).

D. The Equilibrium Points Engendered by $\Delta\omega$ and ω_c Differ in Stability

The total power σ_ϕ^2 of the spectral density analysis of the ϕ time series is a summary measure of the fluctuations of interlimb coordination as that coordination is captured by ϕ. Figure 6 shows the dependence of σ_ϕ^2 on $\Delta\omega$ for both $\phi_\psi = 0$ and $\phi_\psi = \pi$. As can be seen, σ_ϕ^2 is larger for $\phi_\psi = \pi$. This overall stability difference between the in-phase and anti-phase interlimb coordination patterns comports with other observations, namely, the hysteresis of the spontaneous transition in coordination induced by ω_c (with increasing ω_c, a transition in ϕ_{ave} occurs in the direction from π to 0 but not in the direction from 0 to π) (Haken et al., 1985; Kelso, 1984; Schmidt et al., 1990), and the larger "clock variance" (Wing & Kristofferson, 1973) under $\phi_\psi = \pi$ (Turvey et al., 1986). In terms of the potential function identified by Equation (3),

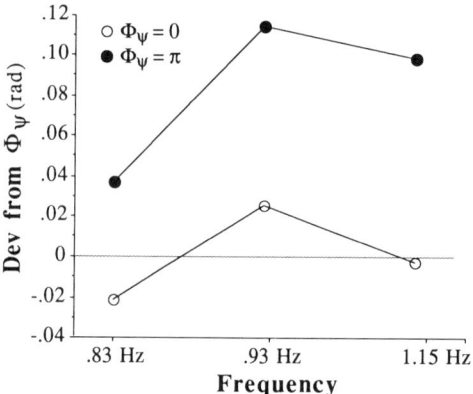

FIGURE 5 ϕ_{ave} deviation from $\phi_\psi = 0$ and ϕ_{ave} deviation from $\phi_\psi = \pi$ as a function of ω_c. Data points are the mean values of five subjects for each of three values of ω_c.

the weaker stability of $\phi_\psi = \pi$ relative to $\phi_\psi = 0$ for a given $\Delta\omega$ would have to be interpreted as a general tendency for coupling strength in anti-phase to be less than that for in-phase. In terms of the potential function identified by Equation (6), the observed stability difference follows straightforwardly from the potential difference favoring $\phi_\psi = 0$.

It can also be seen in Figure 6 that σ_ϕ^2 increases as the departure of $\Delta\omega$ from 0 increases, with the rate of growth higher for $\phi_\psi = 0$ than for $\phi_\psi = \pi$. There is an interaction between $\Delta\omega$ and ϕ_ψ in determining σ_ϕ^2 (Schmidt et al., 1993), with the result that stability differences between the two coordination modes tend to wash out at larger $\Delta\omega$.

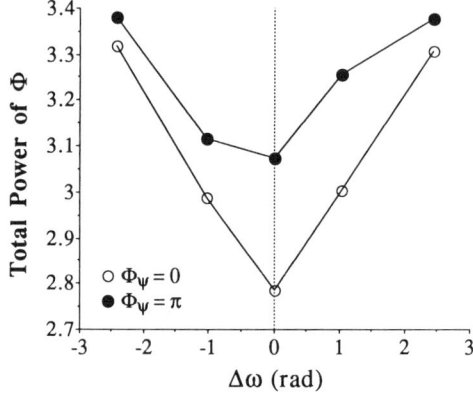

FIGURE 6 Total power of ϕ as a function of $\Delta\omega$ for $\phi_\psi = 0$ and $\phi_\psi = \pi$. Data points are the mean values at each $\Delta\omega$ of five subjects.

Significantly, there is no interaction between ω_c and ϕ_ψ in determining σ_ϕ^2 (Schmidt et al., 1993). For both coordination modes, the increase in ω_c tends to increase σ_ϕ^2 and reduce the differences due to $\Delta\omega$. The interaction between ω_c and $\Delta\omega$ in determining σ_ϕ^2 is depicted in Figure 7A. Decomposing σ_ϕ^2 into that part that is associated with the peaks in the power spectrum (Figure 7B) and that part that remains when the peak power is removed (Figure 7C) is further revealing of the interplay among ϕ_ψ, $\Delta\omega$, and ω_c. Only ϕ_ψ affects the residual and only $\Delta\omega$ and ω_c affect the power associated with the peaks. The implication is that the kinds of fluctuations induced by $\Delta\omega$ and ω_c are harmonically organized fluctuations or rhythms in the behavior of ϕ, whereas the kinds of fluctuations induced by ϕ_ψ are stochastic.

E. A "Changing Potential" Perspective on the Interactive Effects of Competition ($\Delta\omega$) and Cooperation (ω_c qua $K\sin\Phi$)

In sum, the manipulations of $\Delta\omega$ and ω_c affect the mean state and variance of interlimb coordination, as quantified through relative phase, in the systematic fashion predicted by the low-dimensional dynamic of Equation (4). When $\Delta\omega = 0$, the equilibrium points lie at 0 and π. Departures of the equilibrium points from 0 and π are induced by both $|\Delta\omega| > 0$ and ω_c. As suggested by Equation (3), we may identify a potential well with each of the equilibrium points 0 and π, such that each well has its minimum at its respective equilibrium value with the 0 well more steep sided than the π well. This image allows for the conclusion that the potential wells governing 1:1 interlimb frequency locking are both displaced and rendered more shallow by the interaction of $|\Delta\omega| > 0$ and ω_c. Where the displacement of the potential wells from 0 and π are the same for a given $|\Delta\omega| > 0$ and ω_c, the reduction in the slope of the well is not; the slopes of the wells displaced from 0 become shallower at a faster rate than those displaced from π. With increasing shallowness, the same stochastic force $\sqrt{Q}\zeta_t$ can spawn increasingly larger fluctuations of the order parameter (see Figure 8).

III. Expectations from the Low-Dimensional Dynamic about 2:1 Frequency Locking

Let us now expand the deterministic part of Equation (4), that is, Equation (2), so that we might bring 2:1 interlimb coordinations into the picture. Following Rand et al. (1988), an additional coupling term can be introduced into Equation (12) to yield

$$\dot{\theta}_R = \omega_R + k_{RL}\sin(\dot{\theta}_L - \dot{\theta}_R) + p_{RL}\sin(2\dot{\theta}_L - \dot{\theta}_R)$$
$$\dot{\theta}_L = \omega_L + k_{LR}\sin(\dot{\theta}_R - \dot{\theta}_L) + p_{RL}\sin(\dot{\theta}_R - 2\dot{\theta}_L). \qquad (17)$$

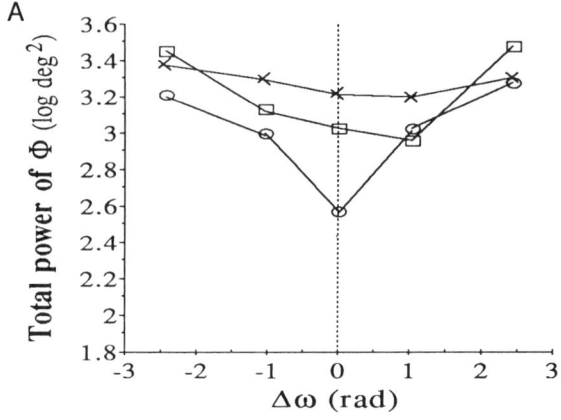
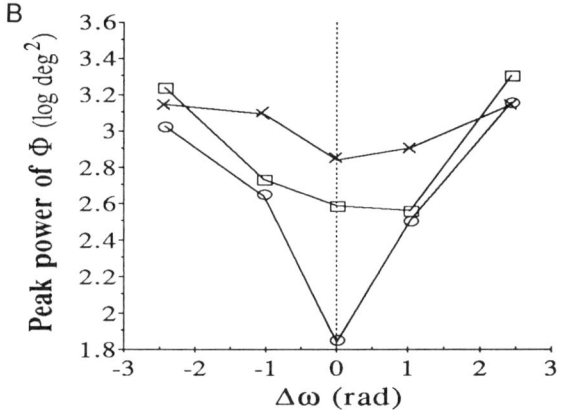
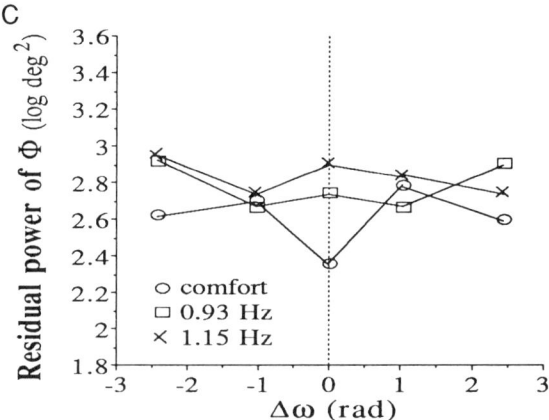

FIGURE 7 The spectral power of ϕ as a function of $\Delta\omega$ for three ω_c values. (A), (B), and (C) display the total power, the peak power, and the residual power, respectively. Data points are the mean values of five subjects.

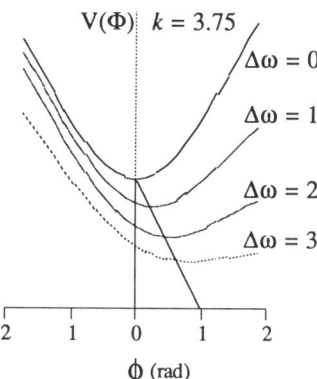

FIGURE 8 The changing potential function $V(\phi)$ of $\dot{\phi}_\psi = 0$ as a function of $\Delta\omega$ given a constant K. Note that (1) the stable point at the bottom of the well moves away from $\phi = 0$, and (2) the sides of the well become more shallow as the competition parameter ($\Delta\omega$) increases.

In the system given by Equation (17), θ_R and θ_L are the phases of the right (R) and left (L) oscillators, respectively; ω_R and ω_L are the uncoupled frequencies (eigenfrequencies) of the R and L oscillators, respectively; k_{RL}, k_{LR}, and p_{RL}, p_{LR}, are the coupling coefficients that determine 1:1 and 2:1 mode locking, respectively; and the sine functions identify that the two kinds of coupling are 2π periodic. When R:L = 2:1, the phase rate of change of R is twice that of L. Parallel equations to Equations (2) and (13) for this 2:1 phase locking can be obtained by defining

$$\zeta(t) = \theta_R(t) - 2\theta_L(t) \tag{18}$$

and setting k_{RL} and k_{LR} (simply K in discussions above) to zero in Equation (17) (Rand et al., 1988). Thus, on subtracting twice $\dot{\theta}_L$ from $\dot{\theta}_R$ (see Equation [17]) we obtain the low-dimensional dynamic in ζ

$$\dot{\zeta} = (\omega_R - 2\omega_L) - P\sin\zeta \tag{19}$$

where $P = (p_{RL} + 2p_{LR})$. As in the 1:1 case, a general expression for the varied equilibria of Equation (19) can be determined under the assumption that, at equilibrium, the time derivative of ζ equals 0. Consequently, the equilibrium values of ζ are given by

$$\zeta = \arcsin[(\omega_R - 2\omega_L)/P] \tag{20}$$

Contrasting Dependencies of Φ and ξ on the Competitive Processes (Component Eigenfrequencies Difference) and Cooperative Processes (Nonlinear Coupling Functions)

Nontrivial predictions follow from the comparison of Equations (13) and (20). To identify one of them, consider a pairing of R and L pendulums in the Kugler and Turvey (1987) procedure satisfying the conditions $\omega_R = \omega_L$. Let the two pendulums, as a coupled system, oscillate with frequency ω_c (which refers, in the 2:1 case, to the frequency of the faster rhythm). Assume that the coupling strength P identified in Equation (20) is like the coupling strength K in that it varies inversely (but not necessarily to the same degree) with ω_c.

As already observed for 1:1 coupling, pursuant to Equation (13) in the form of (14) or (15), when the pendulum pair satisfies $\omega_R = \omega_L$, ϕ should not vary with ω_c. This is because when $(\omega_R - \omega_L) = 0$, variations in $K (\propto \omega_c^{-1})$ will be immaterial. However, from Equation (20) it follows that for 2:1 coupling, when the pendulum pair satisfies $\omega_R = \omega_L$, ζ should vary with ω_c; $(\omega_R - 2\omega_L) \neq 0$ and variations in $P (\propto \omega_c^{-1})$ will act multiplicatively on $(\omega_R - 2\omega_L) \neq 0$ in determining ζ. Figure 9 shows the results for a typical subject in a coordination situation for which $\omega_R = \omega_L$, there are four magnitudes of ω_c (each presented 20 times and determined by a metronome governing the pace of R), and the intended coupling is 2:1. Figure 9 should be compared to Figure 3. As is evident from the comparison, when the coupling is 2:1, ω_c affects the mean state ζ of the interlimb coordination; when the coupling is 1:1, ω_c does not affect the mean state ϕ of the interlimb coordination. The expectations from the comparison of Equations (13) and (20) are upheld. As an aside, it is important to note that Equation (20) does not offer an understanding of the direction of change in ζ accompanying variations in ω_c; it merely identifies that a dependency should be observed.

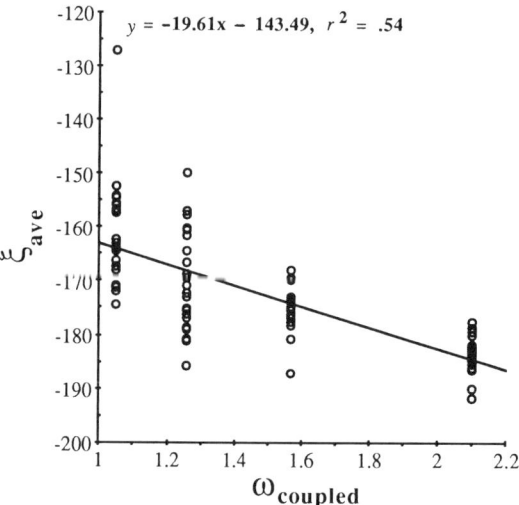

FIGURE 9 ζ_{ave} (in degrees) as a function of ω_c (rad/sec) for the condition $\omega_R = \omega_L$. Data are from one subject; 20 observations at each ω_c.

IV. Conclusions

One major conclusion to be drawn from the theory and research summarized above is that the predictive success of Equation (4) reinforces the claim that Equation (1) relates closely to the dynamics of real biological systems and deserves, therefore, careful study and elaboration (e.g., deGuzman & Kelso, 1991; Kelso, et al., 1990a; Schmidt et al., 1991a). Despite the great complexity of the neural subsystems involved, 1:1 and 2:1 coordinations between rhythmically moving limbs are potentially understandable through low-dimensional dynamics.

A second major conclusion is that von Holst's arguments about the processes formative of interappendage coordination, and the character of the interappendage coordinated state, are correct in their essentials. As suggested elsewhere (e.g., Kugler & Turvey, 1987; Turvey et al., 1986; Turvey et al., 1993), an interlimb coordination, like physical cooperatives in general, is three tiered: The lower level comprises the left and right rhythmic units with their inherent dynamical preferences; the upper level comprises the intention (goal, plan, schema); and the middle level is the interlimb coordination. All three levels—the (lower) atomistic, the (middle) coordinational, and the (upper) intentional—are characterized dynamically, each governed by one or more attractors. The atomistic level dynamics are expressed in terms of the eigenfrequencies, ω_{left} and ω_{right}, and in terms of the actual frequencies, ω_c. (Fluctuations in the periodic timing of a rhythmic pendular unit are least at the eigenfrequency and substantially larger at frequencies other than the eigenfrequency [Rosenblum & Turvey, 1988].) With respect to the intentional level dynamics, one conjecture is that they are expressable through the collective variables used to characterize the cooperative or coordination level dynamics (Schöner & Kelso, 1988; Scholz & Kelso, 1990). In the case of 1:1 frequency locking, the intentional state would be expressed as ϕ_ψ with substates of 0 and π. In the case of 2:1 frequency locking, the intentional state would be expressed as ζ_ψ with as yet unknown substates. At the intentional level these phase relations are attractors of equal strength—they are describable by potential wells that are of the same depth (identical minimum potential value) and slope (identical gradient descents to minimum value) (Schöner & Kelso, 1988). In terms of Equation (4), the intentional state ϕ_ψ of 1:1 coordination would translate as the specification of $K\sin\phi$, with the intentional substates $\phi_\psi = 0$ and $\phi_\psi = \pi$ translating as specifications of the sign of K. Similarly, in terms of Equation (19), the intentional state ζ_ψ of 2:1 coordination would translate as the specification of $P\sin\zeta$.

With respect to the cooperative or coordination level, we reiterate that the dynamics at this level are expressed through the mean state

ϕ_{ave}. The analyses presented in the present chapter have pinpointed the dependence of ϕ_{ave} on the atomistic and intentional levels; the collective variable expressing the spatial and temporal order of the middle level (interlimb coordination) was shown to be determined by the interplay of the level above (e.g., $\phi_\zeta = \pi$) and the level below (e.g., $\Delta\omega$). A major implication of von Holst's (1937/1973) perspective reinforced by the results summarized above, is that an interlimb rhythmic coordination is a pattern-selection process. Competition and cooperation among different attractors, both within a level and between levels, select the interlimb system's order parameter dynamics (Turvey et al., 1993).

Acknowledgment

The reported research and preparation of the chapter were made possible by grants from the National Science Foundation (BNS 88-11510, BNS 91-09880).

References

Arnold, V. I. (1965). Small denominators. I. Mapping of the circumference onto itself. *American Mathematical Society Translations,* **46,** 213–284.
Baker, G. L., & Gollub, J. P. (1990). *Chaotic dynamics: An introduction.* Cambridge: Cambridge University Press.
deGuzman, G. C., & Kelso, J. A. S. (1991). Multifrequency behavioral patterns and the phase attractive circle map. *Biological Cybernetics,* **64,** 485–495.
den Hartog, J. P. (1950). *Mechanics.* New York: Dover.
Glass, L., & Mackey, M. C. (1988). *From clocks to chaos: The rhythms of life.* Princeton, NJ: Princeton University Press.
Haken, H. (1977). *Synergetics.* Berlin: Springer Verlag.
Haken, H. (1983). *Advanced synergetics.* Berlin: Springer Verlag.
Haken, H., Kelso, J. A. S., & Bunz, H. (1985). A theoretical model of phase transitions in human hand movements. *Biological Cybernetics,* **51,** 347–356.
Hale, J., & Kocak, H. (1991). *Dynamics and bifurcations.* Berlin: Springer Verlag.
Jackson, E. A. (1989). *Perspectives of nonlinear dynamics.* Cambridge: Cambridge University Press.
Kelso, J. A. S. (1984). Phase transitions and critical behavior in human bimanual coordination. *American Journal of Physiology: Regulatory, Integrative, and Comparative Physiology,* **15,** R1000–R1004.
Kelso, J. A. S., Scholz, J. P., & Schöner, G. (1986). Nonequilibrium phase transitions in coordinated biological motion: Critical fluctuations. *Physics Letters,* **118,** 279–284.
Kelso, J. A. S., Schöner, G., Scholz, J. P., & Haken, H. (1987). Phase-locked modes, phase transitoins, and component oscillators in biological motion. *Physica Scripta,* **35,** 79–87.
Kelso, J. A. S., deGuzman, G. C., & Holroyd, T. (1990a). The self organized phase attractive dynamics of coordination. In A. Babloyantz (Ed.), *Self organization, emerging properties and learning* (pp. 41–62). New York: Plenum.

Kelso, J. A. S., DelColle, J. D., & Schöner, G. (1990b). Action-perception as a pattern formation process. In M. Jeannerod (Ed.), *Attention and performance* XIII (pp. 139–169). Hillsdale, NJ: Erlbaum.

Kopell, N. (1988a). Toward a theory of modelling central pattern generators. In A. H. Cohen, S. Rossignol, & S. Grillner (Eds.), *Neural control of rhythmic movements in vertebrates* (pp. 369–413). New York: Wiley.

Kopell, N. (1988b). Chains of oscillators and the effects of multiple couplings. In J. A. S. Kelso, A. J. Mandell, & M. F. Schlesinger (Eds.), *Dynamic patterns in complex systems* (pp. 156–161). Singapore: World Scientific.

Kugler, P. N., & Turvey, M. T. (1987). *Information, natural law and the self-assembly of rhythmic movement*. Hillsdale, NJ: Erlbaum.

Murray, J. D. (1990). *Mathematical biology*. Berlin: Springer Verlag.

Rand, R. H., Cohen, A. H., & Holmes, P. J. (1988). Systems of coupled oscillators as models of central pattern generators. In A. H. Cohen, S. Rossignol, & S. Grillner (Eds.), *Neural control of rhythmic movements in vertebrates* (pp. 333–367). New York: Wiley.

Rosenblum, L. D., & Turvey, M. T. (1988). Maintenance tendency in coordinated rhythmic movements: relative fluctuations and phase. *Neuroscience, 27,* 289–300.

Schmidt, R. C., & Turvey, M. T. (in press). Phase-entrainment dynamics of visually coupled rhythmic movements. *Biological Cybernetics*.

Schmidt, R. C., Carello, C., & Turvey, M. T. (1990). Phase transitions and critical fluctuations in the visual coordination of rhythmic movements between people. *Journal of Experimental Psychology: Human Perception and Performance, 16,* 227–247.

Schmidt, R. C., Beek, P. J., Treffner, P. J., & Turvey, M. T. (1991a). Dynamical substructure of coordinated rhythmic movements. *Journal of Experimental Psychology: Human Perception and Performance, 17,* 635–651.

Schmidt, R. C., Treffner, P. J., Shaw, B. K., & Turvey, M. T. (1991b). Dynamical aspects of learning an interlimb rhythmic movement pattern. *Journal of Motor Behavior, 24,* 67–83.

Schmidt, R. C., Shaw, B. K., & Turvey, M. T. (1993). Coupling dynamics in interlimb coordination. *Journal of Experimental Psychology: Human Perception and Performance, 19,* 397–415.

Scholz, J. P., Kelso, J. A. S., & Schöner, G. (1987). Nonequilibrium phase transitions in coordinated biological motion: Critical slowing down and switching time. *Physics Letters, 123,* 390–394.

Scholz, J. P., & Kelso, J. A. S. (1990). Intentional switching between patterns of bimanual coordination depends on the intrinsic dynamics of the patterns. *Journal of Motor Behavior, 22,* 98–124.

Schöner, G., & Kelso, J. A. S. (1988). A dynamic pattern theory of behavioral change. *Journal of Theoretical Biology, 135,* 501–524.

Schöner, G., Haken, H., & Kelso, J. A. S. (1986). A stochastic theory of phase transitions in human hand movement. *Biological Cybernetics, 53,* 247–257.

Sternad, D., Turvey, M. T., & Schmidt, R. C. (1992). Average phase difference theory and 1:1 phase entrainment in interlimb coordination. *Biological Cybernetics, 67,* 223–231.

Turvey, M. T., Rosenblum, L. D., Schmidt, R. C., & Kugler, P. N. (1986). Fluctuations and phase symmetry in coordinated rhythmic movements. *Journal of Experimental Psychology: Human Perception and Performance, 12,* 564–583.

Turvey, M. T., Schmidt, R. C., Rosenblum, L. D., & Kugler, P. N. (1988). On the time allometry of coordinated rhythmic movements. *Journal of Theoretical Biology, 130,* 285–325.

Turvey, M. T., Schmidt, R. C., & Beek, P. J. (1993). Fluctuations in interlimb rhythmic coordinations. In K. Newell and D. Corcos (Eds.), *Variability in motor control* (pp. 381–411). Champaign, IL: Human Kinetics.

von Holst, E. (1937/1973). On the nature and order of the central nervous system. In R. Martin (Ed. and Trans.), *The collected papers of Erich von Holst: Vol. 1. The behavioral physiology of animal and man*. Coral Gables, FL: University of Miami Press.

Wing, A. M., & Kristofferson, A. B. (1973). Response delays and the timing of discrete motor responses. *Perception & Psychophysics,* **14,** 5–12.

15

Elementary Coordination Dynamics

J. A. S. Kelso

Program in Complex Systems and Brain Sciences
Center for Complex Systems
Florida Atlantic University
Boca Raton, Florida

I. Prolegomenon
II. Introduction
III. The Theoretical Strategy
IV. Phenomena to Be Explained
V. Range of Application
VI. Essential Ingredients of Coordination Dynamics
VII. Summary
References

I. Prolegomenon

The basic coordination law for coupled symmetric components accommodates the facts of absolute but not relative coordination. When the symmetry of this law is broken, *both* types of coordination emerge in a unified way, corresponding to different régimes of the *same* coordination dynamics. A principled solution to the so-called degrees of freedom problem in complex, multivariable systems is hypothesized to lie in the identification of collective variables whose low-dimensional dynamics exhibits a restricted class of archetypal bifurcations.

II. Introduction

In this chapter, I wish to discuss the elementary structure of coordination. I shall limit discussion to the coordination between two functional components, even though the same considerations are known to extend to coordinative interactions among multiple components. By elementary, I mean a simple formulation (but not so simple that the essence of the problem is lost) that in turn may provide a foundation for understanding other issues (learning, trajectory formation, intentional changes in coordination, and so forth). By structure, I mean a mathematical law or principle very much in the style of physics. Of course, in the context of coordination in complex living systems, appropriate observables at a given level of description are not usually provided by Newtonian mechanics, but have to be found. The structure presented here possesses novel features of considerable interest, but one of its main attractions is that it encapsulates experimental results in physiology and behavior that have been known for a long time. In fact, some of the basic phenomena to be understood have been discovered and rediscovered in a wide variety of biological systems over the last half century. It is beyond the scope of this chapter to review this large body of literature[1] or to provide a detailed mathematical exposition, which can be found elsewhere. Rather, it is my intent to choose some examples that not only are amenable to rudimentary presentation but that capture something essential about the problem. I take this to be a necessary evil of science, namely, that to get a handle on a behaviorally complex phenomenon, such as coordination, it is necessary to strip it down to its essential features. Already this poses a difficult problem because it means that we have to choose appropriate windows, that is, experimental model systems and appropriate theoretical concepts and tools with which to investigate them.

There are many definitions of coordination (see Turvey, 1990, for review), but the one that I prefer defines coordination as the state of *being* or *becoming* coordinate. In this definition, coordination may be interpreted as an *a posteriori* consequence of evolving processes of *self-organization* or *pattern formation*. Imagine a living system composed of elements that did not interact with each other or with their surround. Such an organism would possess neither structure (in the anatomical sense) nor function. Principles of self-organization, expressible in the language of dynamical systems, may be seen to apply at neural, behavioral, and cognitive levels of description; they serve to bridge evolutionary, developmental, and learning processes. The topic of *interlimb coordination* is just one means of studying how components are put together in such a way that some recognizable function results.

---

[1] For reviews of related work, however, the reader is referred to the contributions in this volume, for example: Chapter 11, Baldissera *et al.;* Chapter 16, Carson *et al.;* Chapter 20, Corbetta and Thelen; Chapter 24, Newell and McDonald; Chapter 17, Schöner; Chapter 26, Summers; Chapter 14, Turvey and Schmidt; Chapter 23, Walter and Swinnen; and Chapter 19, Whitall and Clark. See also Beek, 1989; Vereijken, 1991; Wallace *et al.,* 1990; Wimmers *et al.,* 1992.

The essential questions concern the form that this basic interaction takes, how it occurs, and why it is the way it is (see Section V for one answer).

For coordination to be functional, it must adapt and change over time according to external or internal conditions. Thus, any coordinative structure must be dynamical (e.g., Saltzman & Kelso, 1987). Our elementary structure will express the system's existing and evolving patterns in terms of the dynamics of macroscopic observables. I assume that the innumerable microscopic degrees of freedom contributing to coordinated states may, for the present, be ignored and that the main goal, *on any level of description,* is to identify relevant macroscopic variables and their equations of motion. It should be kept in mind, however, that what is "macro" at one level may be "micro" at another. In brief, *elementary coordination dynamics* uses concepts of self-organization and pattern formation (Haken, 1977, 1983) as part of a theoretically motivated experimental strategy, and the tools and language of coupled nonlinear dynamics to express lawfully (in continuous or discrete form) how coordination patterns form and change.

III. The Theoretical Strategy

How do we find relevant coordination variables on a given scale of observation? Many observables on various levels, neural, muscular, kinematic, or kinetic may, in principle, contribute to a description of coordination. Even on a single level, a difficult problem is to determine, from the broad range of possibilities, which variables are essential and which are not. This problem is exacerbated if experimental model systems are studied only in the linear range of the system's operation where many things can be measured (but not all are relevant) and change is smooth. Our focus is in and around phase transitions or bifurcations (these terms will be clarified shortly) where *qualitative* change occurs. Bifurcations provide a special entry point because: (1) they allow a clear distinction between patterns of coordinated behavior, enabling the identification of the dimension on which pattern change occurs, a so-called *collective variable.* The implication is that this collective variable is relevant to the system in the linear range of its operation as well; (2) they open a path into modeling the collective variable dynamics; (3) well-defined observables are available to evaluate predictions about the dynamics (e.g., stability, loss of stability) near critical points; (4) instabilities provide a generic mechanism for flexibly switching among multiple coordinative states, that is, for entering and exiting coherent patterns. It is important to emphasize that phase transitions (and pattern formation, in general) may be instantiated in a multitude of ways on many different levels (e.g., neural, energetic, physical-maturational, hormonal) but the generic mechanism of instability is universal to all of them.

IV. Phenomena to Be Explained

What phenomena do we want our elementary coordination dynamics to explain? A central criterion for any law of coordination is reproducibility of the phenomena in question. The study of rhythmical forms of coordination thus offers an obvious entry point. Rhythmical behavior is ubiquitous in biological systems, from the simplest to the most complex. Oscillations are as prevalent in the inanimate world as they are in living organisms. Notions of rhythm and oscillation, as many have pointed out, are essential to the study of the dynamics of nonmonotonic evolution, regardless of whether the motion is regular or irregular. Although rhythmical behaviors may be quite complicated, we have the deep impression that the principles underlying them possess a beautiful simplicity. Rhythms are known to confer positive functional advantages for the organism. Among these are temporal and spatial organization, prediction of events, efficiency, and precision of control. Rhythmic oscillations may be viewed as archetypes of time-dependent behavior, but there is no reason to confine analysis to strictly periodic events. As shown elsewhere, it is possible to extend the basic coordination dynamics to discrete movements, as well as to accommodate posturally stable states (see Schöner, Chapter 17, this volume, for review and theoretical development).

Table I lists a sample of the coordination phenomena predicted and (to be) explained by our elementary coordination dynamics. Any law of coordination worth its salt must be able to accommodate the two classes of coordination that von Holst called *absolute* and *relative* coordination. The former, almost idealized, type of coordination, involves constant phase and frequency relationships between two or more interacting components (Property 1. in Table I). The latter refers to situations in which occasional phase slippage or drift between coordinating components occurs even though an intrinsic attraction to certain preferred phase relations (the so-called magnet effect) still remains (Property 10. in Table I). Figure 1 shows typical time series and corresponding phase distributions of these two types of coordination. Note that in relative coordination every possible phase relation between interacting components occurs, even though, on the average, a single phase relation characteristic of absolute coordination predominates.

Our aim is to show that *all* the features associated with absolute and relative coordination fall out of one coherent theory of coordination dynamics. Specifically, states of absolute coordination and relative co-

TABLE I Elementary Coordination Dynamics: Phenomena Predicted and Explained

1. Small set of phase (e.g., $\phi = 0$; $\phi = \pi$) and frequency (e.g., $1:1, 2:1, 3:2, \ldots$) relations among interacting components, so-called *coordination modes*.
   [COROLLARIES. *Temporal stability; "invariance" across parameter changes*]
2. Multi- (e.g., bi-) stability: coexistence of several coordination modes for the same parameter value
3. Hysteresis: coordination mode observed depends on *direction* of parameter change
4. Switching among coordination modes at critical values of a parameter
5. Loss of stability en route to switching
6. If system is symmetric, no systematic drift of phase-locked modes before switching (pitchfork bifurcation). [COROLLARY. *No preferred transition pathways*]
7. If system is asymmetric, *directed* drift of phase-locked modes before switching (saddle-node or tangent bifurcation). [COROLLARY. *Preferred transition pathways*]
8. Signatures of instability before switching, for example, critical slowing down; fluctuation enhancement
9. Characteristic time scales for coordination, for example, switching time distribution
10. Small set of *tendencies* to synchronize phase and frequency relations, that is, no strict mode locking
11. Metastability; intermittency
12. Characteristic dwell time distribution near metastable states
13. Extra steps or cycles between nearly mode-locked states
14. Phase slippage and wandering; desynchronization or loss of entrainment

ordination will be seen to emerge as two sides of the same coin corresponding to *different* parameter regimes of the *same* underlying law of coordination, an essential feature of which is *broken symmetry*.

V. Range of Application

Basic laws of coordination are not structure specific or dependent on physicochemical processes per se. Although patterns of coordination are always realized and instantiated by physical structure, the laws themselves are abstract and mathematical. It is wrong, however, to think of the coordination dynamics as simply mathematical curiosities. On the contrary, they have been shown to express spatiotemporal relationships between: (1) components of an organism (e.g., Kelso, 1984; Baldissera, Cavallari, Marini, & Tassone, 1991); (2) organisms themselves (e.g., Schmidt, Carello, & Turvey, 1990); and (3) organisms and their environment (e.g., Wimmers, Beek, & van Wieringen, 1992; see Table II). Moreover, the theory has been extended, for example, to the case of learning (see Zanone & Kelso, Chapter 22, this volume) and to

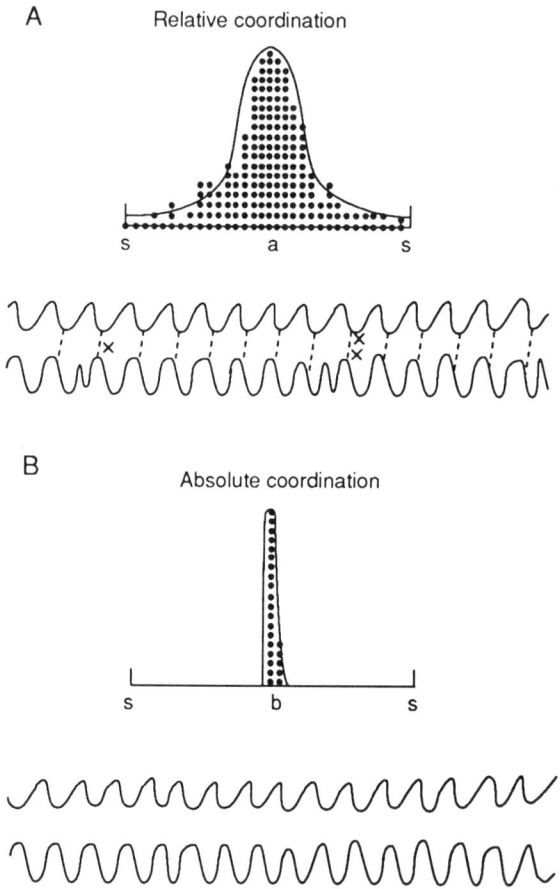

FIGURE 1 (A) Relative coordination. The distribution of possible phase relations between two signals (*top*) and the corresponding time series from which the phase relation is extracted (*bottom*). The distance from s to s on the abscissa spans the phase interval 0 to 2π rad. The signals come from pectoral (upper time series) and dorsal (lower time series) fin movements of a fish; (B) Absolute coordination. Distribution of phase relations and corresponding time series as in (A). (Adapted from von Holst, 1973.)

the coordination of multiple, anatomically different components (Kelso & Jeka, 1992; Schöner, Jiang, & Kelso, 1990). In all these situations, equations of motion are formulated for observable variables and often result in predictions that can be (and have been) experimentally tested. In such theoretical modeling, variability plays an essential *constructive* role, providing, on one hand, a source of flexibility, and on the other, a means to probe the stability of coordinative states.

TABLE II Range of Application of Elementary Coordination Dynamics[a]

Between components of an organism
  Bimanual coordination
  Interlimb coordination
  Within-limb multijoint coordination
  Speech–manual coordination
Between an organism and the environment
  Visually specified coordination patterns
  Patterns of auditory–motor coordination
Between organisms themselves
  Visual coupling between people
  Interactional synchrony?
  Mother–infant interactions?

[a] Elaboration of the basic coordination dynamics is possible (e.g., to development, Whitall and Clark, Chapter 19, this volume; Corbetta and Thelen, Chapter 20, this volume) and in a number of cases has been carried out (e.g., for trajectory formation, Schöner, Chapter 17, this volume; learning, Zanone and Kelso, Chapter 22, this volume; and multicomponent coordination, Kelso and Jeka, 1992; Schöner *et al.*, 1990). A huge range of neural and physiological rhythms (cardiovascular, respiratory, locomotor, etc.) is also accessible (see, e.g., Köpchen, 1991).

VI. Essential Ingredients of Coordination Dynamics

The aim is to convey the essential ingredients of the theory by pictures and words, and, for present purposes, a single equation of motion that may be expressed in continuous or discrete form. As noted in Section III, in the case of biological coordination, the choice of collective variables must be based on empirical insights: collective variables define stable and reproducible relationships among interacting components and are, in general, function or task specific. Examples of collective variables are relative phases (ϕ), which capture patterns of coordination in the nervous system and behavior (see Table I). Although the same coordination phenomena may exist at different levels, the first step is to find dynamic laws *within* a given level. The main idea is that understanding at any level of organization requires knowledge of: (1) the parameters acting on the system; (2) the interacting subsystems or components themselves; and (3) the patterns or modes of coordination that emerge from component interactions. This is not a rigid organizational scheme. A cooperativity at one level of organization may act as a parametric boundary condition for the level below it. Conversely, the former may act as component processes for the level above. Elsewhere, it has been possible to show that coordinated behavioral patterns arise as self-organized stable states due to nonlinear coupling among the components (e.g., Haken, Kelso, & Bunz, 1985). We do not pursue this aspect of "vertical integration" further here.

The theoretical strategy is to map empirically stable patterns of coordination onto attractors of the collective variable dynamics. Generally, we assume sufficiently high order dynamics to accommodate observed patterns and expand the vector field of the dynamics accordingly.[2] Here, for simplicity, we restrict ourselves to the interaction among *two* components, as shown in Figure 1, although, as mentioned earlier the same considerations apply, for example, to the treatment of multiple components.

Figure 2 plots the vector field of the relative phase dynamics ϕ (x-axis) versus $\dot\phi$ (y-axis) for different parameter values of Equation 1, introduced originally to model action-perception as a pattern formation process (Kelso, DelColle & Schöner, 1990):

$$\dot\phi = \delta\omega - a\sin\phi - 2b\sin2\phi + \sqrt{Q\xi_t} \qquad (1)$$

where the parameters a and b are coupling parameters related to task requirements and $\delta\omega$ represents intrinsic frequency differences between the *uncoupled* components. The coupling ratio is proportional to the experimental driving frequency and expresses the relative importance of the intrinsic phase-attractive states at $\phi = 0$ and $\phi = \pi$ (see below). Because fluctuations are conceptually important (e.g., as tests of stability of attractive states and in effecting transitions) a stochastic term, ξ_t of strength Q is needed. From the stochastic dynamics of relative phase (Eq. 1) it is easy to quantify the coupling strength among interacting components from experimental measures of phase fluctuations and local relaxation times (Schöner, Haken, & Kelso, 1986; for one application, see Scholz, Kelso, & Schöner, 1987).

The only difference between the two plots shown in Figure 2 is that the parameter $\delta\omega$ is zero in Figure 2A and a constant nonzero value in Figure 2B. This simple fact has consequences for the dynamics, and hence for predicting and explaining the various phenomena listed in Table I. To understand the figures (which, remember, represent the flow of ϕ, not the potential, $V(\phi)$ as in Haken et al. 1985), note that the system defined by Equation 1 contains stationary patterns or *fixed points* of ϕ where the time derivative, $d\phi/dt$ or $\dot\phi$ is zero and crosses the ϕ-axis. When the slope of $\dot\phi$ is negative at the abscissa the fixed points are *stable* and *attracting;* when the slope is positive the fixed points are *unstable* and *repelling*. In Figure 2, the arrows indicate the direction of

---

[2] For additional background information on dynamical systems concepts, see, for example, Bergé et al. (1984) for a more physically based approach, or Guckenheimer and Holmes (1983) for a more mathematical treatment. A recent tutorial by Kelso et al. (1992) is aimed more at a biological and psychological audience.

flow, for example, initially in both (A) and (B) there are two stable fixed points of ϕ separated by an unstable fixed point. As one travels from bottom to top in these figures, decreasing the ratio b/a, note that in Figure 2A, one of the stable fixed points disappears eventually. In Figure 2B first one, then eventually both stable fixed points disappear.

The essential conceptual ingredients of our elementary coordination dynamics (Eq. 1) may be gleaned from Figure 2. Now, let us make them explicit and see to what extent they relate to the features listed in Table I (which we refer to, from now on, in parentheses):

• The *collective variable* or *order parameter* (Haken, 1983) is the relative phase, ϕ, in this case, corresponding to stable phase- and frequency-locked states. In Figure 2, the components are initially coupled at the same frequency (1:1) either in-phase (or close to in-phase) or anti-phase. Note that these states remain stable or "invariant" over a range of parameter values (1. and its COROLLARIES).

• The *control parameter* is the ratio b/a, corresponding in Equation (1) to the experimental movement frequency. Generally speaking, there is nothing that requires the control parameter to always be frequency (likewise, relative phase for the collective variable). Buchanan and Kelso (1993), for example, show that the spatial orientation of the forearm is a control parameter in single-limb, multijoint movements (see Walter & Swinnen, Chapter 23, this volume, for other potential candidates). The role of control parameters in the theory is to move the system through collective states. In Figure 2A, for ratio values greater than 1, both $\phi = 0$ and $\phi = \pi$ are stable fixed points. The strength of attraction to these coordinative modes decreases as frequency increases (and the coupling ratio, b/a decreases), until one (the less stable pattern near $\phi = \pi$) and ultimately both (in Figure 2B) cease to be stationary solutions.

• *Multistability*. Which of the two coordination modes is observed initially is determined only by the initial conditions, that is, in which basin of attraction the system is either prepared or finds itself. Nevertheless, *the two modes coexist*, necessitating nonlinear models. This is what I meant in my opening remarks by an elementary coordination dynamics that is simple, but not too simple. If we drop the second order term in Equation 1 we give up both the multistability and bifurcation properties of the system (2. and 4. in Table I). Such a step ignores many experimental findings and is too high a price to pay for simplicity. Elsewhere (e.g., Kelso & Schöner, 1987; Schöner & Kelso, 1988) we have discussed the important connection between multistability of the coordination dynamics and *multifunctionality*, which is now considered an essential aspect of real neural networks and biological systems in general.

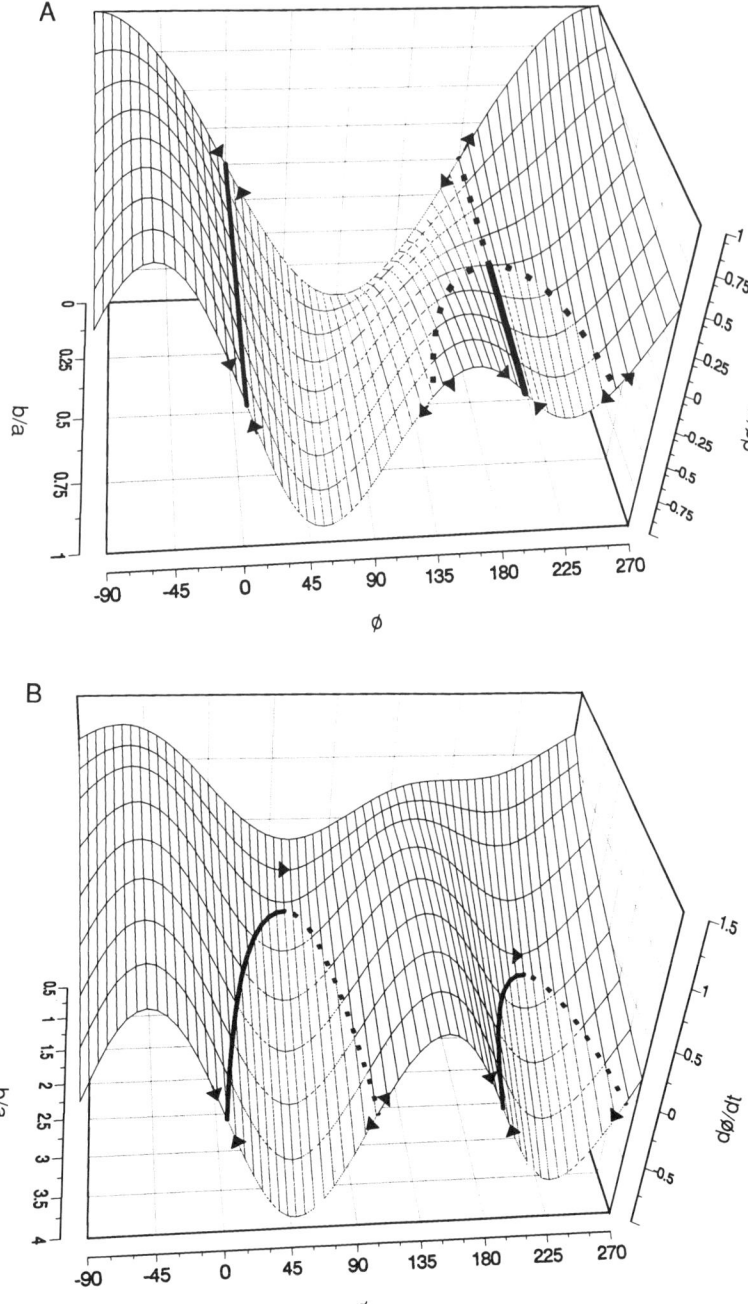

FIGURE 2 (A) The symmetric coordination law (Eq. 1) with the parameter, $\delta\omega = 0$. Arrows indicate the direction of flow. Thick solid and dashed lines correspond to attractive and repelling fixed points of the dynamics (sinks and sources, respectively). Note the inverse pitchfork bifurcation of the collective variable, ϕ as the coupling ratio, b/a, is

- *Symmetry and Broken Symmetry.* It is important to distinguish between the symmetry of the specific situation that the law (Eq. 1) is trying to explain and the symmetry of the law itself. Symmetry serves two purposes. On one hand, it identifies the basic patterns that are to be captured theoretically; on the other, it imposes restrictions on the dynamics themselves, that is, only certain solutions are stationary. All patterns that remain invariant under the same transformation belong to the same symmetry group. Yet, not all solutions of the coordination dynamics are equally stable. For example, in Figure 2A, the slope of the function around the stable fixed point at $\phi = 0$ is greater than the slope near $\phi = \pi$, even though the system (Eq. 1, with $\delta\omega$ equal to zero) is perfectly symmetric, that is to say, invariant under the operation $\phi \rightarrow -\phi$. This is another way of saying that when we look at the relative phase between two components, it does not matter whether one component leads or lags the other. The system is identical under left–right reflection and a shift of $\phi = \pi$. *Deep consequences for coordinative phenomena lie in this fact.* Indeed, the *only* phase relations possible under left–right exchange and a phase shift of π are in-phase and anti-phase! (So, one answer to the question of why we often see in-phase and anti-phase relations in biological systems is because of a fundamental, and quite abstract symmetry.)

Of course, nature thrives on broken symmetry for its diversity, and the richness of coordination is no exception. The sources of such spontaneous symmetry breaking are manifold. Coordination often occurs between different structural components, for example, or between the same components with different functional requirements. More specifically, when we introduce a term, $\delta\omega \neq 0$, in Equation (1), the $\phi \rightarrow -\phi$ reflection symmetry is broken. Such is the case when different anatomical components must be coordinated, each possessing their own characteristic spatial and temporal properties. Or, as in some of the examples listed in Table II, task requirements dictate that some action component must be coordinated in a particular fashion with an environmental event. In our theory, *any* situation that introduces or amplifies intrinsic differences between the interacting elements may break the symmetry of the dynamics (see, e.g., Turvey & Schmidt, Chapter 14, this volume, for possible examples). Such examples enhance the importance of the symmetric coordination law proposed originally by Haken

---

varied. (B) The broken symmetry coordination law (Eq. 1) with the parameter $\delta\omega \neq 0$. When the symmetry of the coordination dynamics is broken, inverse saddle-node bifurcations are seen. Eventually no stationary solutions exist (see text for definitions and discussion).

et al. (1985) and illustrated in Figure 2A. The reason is that the manifold consequences of *broken* symmetry in the coordination dynamics, the violation of the $\phi \rightarrow -\phi$ symmetry constraint, become accessible to understanding (see the following).

- *The Bifurcation Structure.* Changes in coordination may be smooth or abrupt. Regardless, loss of stability is at the origin of coordinative change (5.). In the symmetric dynamics (Figure 2A), the bifurcation is of *pitchfork type* (6.). The stable coordination state near $\phi = \pi$ (solid line) is surrounded by two unstable fixed points (dashed lines). At a certain critical point, the stable state at $\phi = \pi$ is annihilated and the only remaining stable mode is at $\phi = 0$. Technically speaking, this kind of bifurcation is called a subcritical or inverse pitchfork.

In the asymmetric system (Figure 2B), the bifurcation is of *saddle-node* type ("saddle" referring to repelling direction, "node" to attracting direction). Stable (solid lines) and unstable (dashed lines) fixed points coalesce onto a tangent, leaving first a stable fixed point near $\phi = 0$ and then, by exactly the same process, no stable fixed point (coordination mode) whatsoever. Now we begin to see the richer dynamics as a result of the broken symmetry law. Whereas the stable fixed points in the symmetric case do not change their value as the parameter is changed, systematic drift toward the (unobservable) unstable fixed point is predicted in the asymmetric case (7.). Moreover, because of the greater slope of the function surrounding the fixed point near $\phi = \pi$, the system has a preferred transition direction, that is, from $\phi \approx \pi$ to $\phi \approx 2\pi$ rather than to $\phi \approx 0$. Experiments by Jeka and Kelso (submitted; see also Kelso & Jeka, 1992) have tested and confirmed these predictions by manipulating inertial differences between the arms and legs in a coordination task.

In addition to drifting fixed points as indicators of upcoming tangent or saddle-node bifurcations, theoretical predictions associated with nonequilibrium phase transitions are a feature of the basic coordination dynamics (Eq. 1). *Critical slowing down* is predicted by the fact that the negative slope around stable fixed points progressively decreases with parameter change, especially evident in the symmetric case near $\phi = \pi$ (Figure 2A). This means that the time to return to the stable state (τ_{rel}) following a small perturbation becomes longer and longer as the system approaches the critical point. By the same argument, the influence of fluctuations that act to continuously kick the system away from attractive fixed points increases near the upcoming instability (8.). Predicted features of the transient switching process itself, such as switching time and its distribution (9.) have also been confirmed in experiments on spontaneous and intentional switching. Further, we note that *hysteresis* is a predicted feature of the coordination dynamics (Eq. 1). It is quite obvious, for example, that once the system has switched to the in-phase state, it will stay there even though the direction of parameter range is reversed (3.). Of course, in real systems with noise there is always a finite probability of switching between stable states.

Before leaving the topic of bifurcations let us draw a further generalization. As we have argued (Section III), a great strength of an approach that focuses on bifurcations and instabilities is that complex multivariable systems may reduce to a simplified, low-dimensional (order parameter) description near points of qualitative change. The essence of this qualitative behavior is governed by a small set of evolution equations called *normal forms*. Our elementary coordination dynamics (Eq. 1) exhibits two of these normal forms, the pitchfork and saddle node. A quite astonishing result, that we make here in passing, is that *all* the simplest (called *codimension*-1) bifurcations have been found in studies of movement coordination. Codimension is the term given to the minimum number of parameters that have to be varied to observe a certain type of bifurcation. In the top diagram of Figure 3, we illustrate this notion: even though the parameter space is three-dimensional, by moving along any arbitrary line, D (of dimension 1), that cuts the surface, Ω, we encounter a bifurcation (change of stability). Were it necessary to move along a surface (which has two dimensions) to find bifurcations, they would be referred to as of codimension-2 type, and so on.

The four codimension-1 bifurcations corresponding to the basic normal forms are also shown in Figure 3. The pitchfork and saddle-node bifurcations require no further discussion since they are characteristic of symmetric and asymmetric coordination systems. The *transcritical* bifurcation (two stable solutions coexist with exchange of stability at the bifurcation point) has been seen in studies of multijoint coordination where one *stable* coordination pattern switches to another *stable* pattern at a critical value of spatial orientation (Buchanan & Kelso, 1993).

The *Hopf* bifurcation is the complex equivalent of the pitchfork. Here a stable fixed point exists for $\mu < 0$, but bifurcates into a circle on the x,y plane of radius $\sqrt{\mu}$. An example of the Hopf bifurcation is when rhythmic motion, for example, of a single finger, in the horizontal (abduction–adduction) plane switches to circular and eventually flexion–extension motion in the vertical plane as movement rate is increased (cf. Kelso & Scholz, 1985). The Hopf mechanism, we hypothesize, may be quite crucial to understanding how biological systems *spontaneously recruit and annihilate degrees of freedom,* in the present example, spatial components of the motion.

The remarkable finding that movement coordination exhibits these archetypal signatures in which a single collective variable or order parameter and a single control parameter are sufficient to describe the

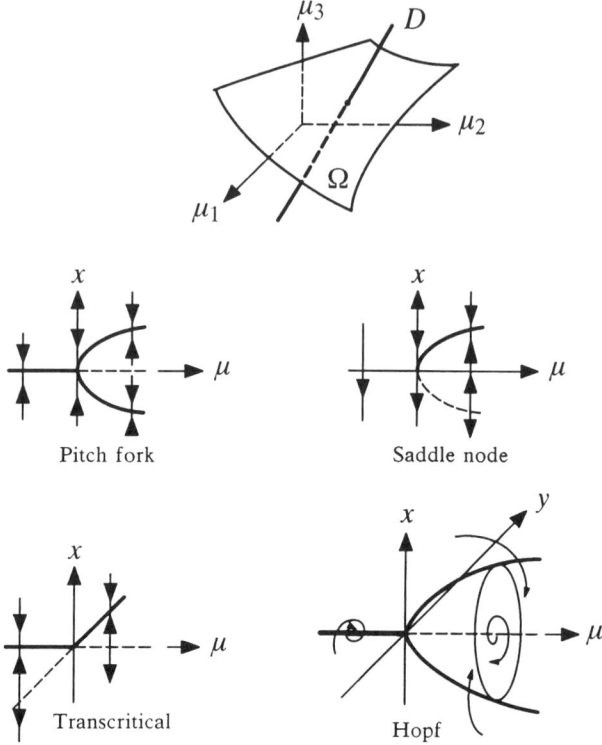

FIGURE 3 (*Top*) Codimension-1 bifurcation in a three-dimensional parameter space. The locus of points in parameter space is defined on a surface, Ω. A parameter varied along the line, D, cuts the surface Ω and induces a codimension-1 bifurcation. (*Bottom*) The four most frequently encountered codimension-1 bifurcations of fixed points. The collective variable is denoted as x and the control parameter as μ. Thick solid lines and thin dashed lines correspond to stable and unstable steady states. Arrows on vertical lines indicate the direction of vector field flow. All four of these archetypes have been observed in research on coordination (see text for discussion).

system's behavior supports the notion that this is one way the motor system handles complexity. We cannot resist hypothesizing that in these archetypal forms (e.g., Figure 3) lies a solution to the so-called degrees-of-freedom problem. Note that in all these cases it has been necessary to identify the order parameters and the control parameters and to relate them specifically to the normal form dynamics.

• *Loss of Entrainment.* In Figure 2B it is easy to see that as the function is flattened by decreasing the control parameter, stationary solutions of Equation 1 eventually disappear. Instead, "running" solutions (ϕ ever increasing or decreasing) or, if the 2π periodicity convention is applied to ϕ, "wrapping" solutions occur. In this régime (exemplified by the solitary arrow at the top of Figure 2B) there is no longer any phase or frequency locking, a condition called *loss of entrainment* or *desynchronization* (14.). Such desynchronization is not present in the symmetric version of the coordination dynamics (Eq. 1 and Figure 2A), but is again a consequence of symmetry breaking. Note that desynchronization does not always mean irregular behavior: its magnitude depends on how close the system is to its critical points (which depends on all three parameters, $\delta\omega$, a, and b). This simply reflects the fact that in Equation 1, the frequency difference between components and the basic frequency of coordination are control parameters, affecting the onset of the running solution.

 • *Intermittency*. When the saddle nodes vanish, indicating loss of entrainment, the coordination system may stay near the previously stable, fixed point. Note that in Figure 2B there is still *attraction* to certain phase relations, even though the relative phase itself is unstable ($\dot{\phi} > 0$). Thus "wrapping" solutions can possess a fine structure, spending more time around relative phase values at which $\dot{\phi}$ is minimal (the remnant or ghost of the attractive fixed point). Such behavior is especially significant because it shows that although there is no strict mode locking, there is partial coordination interrupted, occasionally, by phase wandering (10., 11.). This form of partial coordination was termed *relative coordination* by von Holst (1973), who attributed the phase attraction to a central control principle, the so-called magnet or M-effect (see Figure 1).

Relative coordination has a theoretical interpretation in terms of *intermittent dynamics* (Kelso & DeGuzman, 1991): In Figure 2B, as the curve is lifted up from the x-axis, a phantom fixed point appears. Motion hovers around this point most of the time (giving rise to a characteristic dwell-time distribution, 12., see DeGuzman & Kelso, 1992), but occasionally escapes away along the repelling direction. What happens after the escape depends on the attractor layout around the saddle node. In our case, due to the 2π periodicity of ϕ, the relative phase slips and is then reinjected (13. and Figure 1). This phenomenon is called (type 1) intermittency and represents one of the generic processes found in low-dimensional dynamical systems near tangent or saddle-node bifurcations (Pomeau & Manneville, 1980). A statistical distribution of the relative phase in this intermittent regime of the coordination dynamics contains all possible phase values, but a concentration around "preferred" phase relations (the previously stable fixed point[s]) exactly like that shown in Figure 1 for relative coordination. Our identification of relative coordination with intermittent behavior of periodic flows is consistent with the idea that biological systems tend to live near boundaries separating regular and irregular behavior, surviving best, as it were, in the margins of instability. Relative coordination qua intermittency allows for low-energy, flexible switchings among metastable coordinative states.

VII. Summary

The elementary coordination dynamics (Eq. 1) contain (1) no coordination; (2) absolute coordination (when two or more components oscillate at the same frequency and maintain a fixed phase relation); and (3) relative coordination (the tendency toward phase attraction even when the component frequencies are not the same). These spatiotemporal forms of organization have an explanation, namely, they are patterns that emerge in different parameter régimes of the identified coordination dynamics. Transitions between régimes add rich behavioral complexity and are an expression of the principle of self-organization. The existence of a small set of bifurcation structures argues strongly for cooperative effects that simplify the control problem in multidegree-of-freedom systems. This theory accommodates simple kinds of coordination between (1) components of an organism; (2) organisms themselves; and (3) organisms and their environment (Table II). It provides a foundation for additional theoretical and experimental developments. Attesting to the veracity of the elementary coordination laws (Figure 2, A and B) is that most, if not all, the effects predicted by them have been found in experiments (Table I). We note again that the collective dynamics specified by Equation 1 can also be derived from the component level where the various coordination patterns arise as self-organized states because of nonlinear coupling among the components involved.

Acknowledgments

The research described here is supported by NIMH (Neurosciences Research Branch) Grant MH 42900, contract N00014-88-J-1191 from the U.S. Office of Naval Research and NSF Grant DBS-9213995. I thank Pier Zanone and William McLean for their aid in preparing Figures 2 and 3, and three anonymous reviewers for their helpful comments.

References

Baldissera, F., Cavallari, P., Marini, G., & Tassone, G. (1991). Differential control of in-phase and anti-phase coupling of rhythmic movements of ipsilateral hand and foot. *Exp. Brain Res.*, **83,** 375–380.

Beek, P. J. (1989). Timing and phase-locking in cascade juggling. *Ecological Psychology,* **1,** 55–96.
Bergé, P., Pomeau, Y., & Vidal, C. (1984). *Order within chaos.* Paris: Hermann.
Buchanan, J. J., & Kelso, J. A. S. (1993). Posturally induced transitions in rhythmic multijoint limb movements. *Experimental Brain Research,* **94,** 131–142.
DeGuzman, G., & Kelso, J. A. S. (1992). The flexible dynamics of biological coordination: Living in the niche between order and disorder. In A. B. Baskin & J. E. Mittenthal (eds.), *Principles of organization in organisms, in Proc. Vol. XII.* (pp. 11–34) Santa Fe, NM: Addison-Wesley.
Guckenheimer, J., & Holmes, P. (1983). *Nonlinear oscillations, dynamical systems, and bifurcations of vector fields.* NY: Springer-Verlag.
Haken, H. (1983). *Synergetics, an introduction: Non-equilibrium phase transitions and self-organization in physics, chemistry and biology* (3rd edition). Berlin: Springer.
Haken, H., Kelso, J. A. S., & Bunz, H. (1985). A theoretical model of phase transitions in human hand movements. *Biological Cybernetics,* **51,** 347–356.
Jeka, J. J., & Kelso, J. A. S. (submitted). Manipulating symmetry in the coordination dynamics of human movement.
Kelso, J. A. S. (1984). Phase transitions and critical behavior in human bimanual coordination. *American Journal of Physiology: Regulatory, Integrative and Comparative Physiology,* **15,** R1000–R1004.
Kelso, J. A. S., & DeGuzman, G. C. (1991). An intermittency mechanism for coherent and flexible brain and behavioral function. In J. Requin & G. E. Stelmach (eds.), *Tutorials in motor neuroscience* (pp. 305–310). Dordrecht: Kluwer.
Kelso, J. A. S., & Jeka, J. J. (1992). Symmetry breaking dynamics of human multilimb coordination. *Journal of Experimental Psychology: Human Perception and Performance,* **18**(3), 645–668.
Kelso, J. A. S., & Scholz, J. P. (1985). Cooperative phenomena in biological motion. In H. Haken (Ed.), *Complex systems: Operational approaches in neurobiology, physical systems and computers* (pp. 124–149). Berlin: Springer.
Kelso, J. A. S., & Schöner, G. S. (1987). Toward a physical (synergetic) theory of biological coordination. *Springer Proceedings in Physics,* **19,** 224–237.
Kelso, J. A. S., DelColle, J. D., & Schöner, G. (1990). Action-perception as a pattern formation process. In M. Jeannerod (Ed.), *Attention and performance XIII* (pp. 139–169). Hillsdale, NJ: Erlbaum.
Kelso, J. A. S., Ding, M., & Schöner, G. (1992). Dynamic pattern formation: A primer. In A. B. Baskin & J. E. Mittenthal (Eds.), *Principles of organization in organisms, in Proc. Vol. XII.* (pp. 397–440) Santa Fe, NM: Addison-Wesley.
Köpchen, H. P. (1991). Physiology of rhythms and control systems: An integrated approach. In H. Haken & H. P. Köpchen (Eds.), *Rhythms in physiological systems* (pp. 3–20). Berlin: Springer-Verlag.
Pomeau, Y., & Manneville, P. (1980). Intermittent transition to turbulence in dissipative dynamical systems. *Communications in Mathematics and Physics,* **74,** 189–197.
Saltzman, E. L., & Kelso, J. A. S. (1987). Skilled action: A task dynamic approach. *Psychological Review,* **94,** 84–106.
Schmidt, R. C., Carello, C., & Turvey, M. T. (1990). Phase transitions and critical fluctuations in the visual coordination of rhythmic movements between people. *Journal of Experimental Psychology: Human Perception and Performance,* **16**(2), 227–247.
Scholz, J. P., Kelso, J. A. S., & Schöner, G. S. (1987). Non-equilibrium phase transitions in coordinated biological motion: Critical slowing down and switching time. *Physics Letters,* **A,** 390–394.
Schöner, G., & Kelso, J. A. S. (1988). Dynamic pattern generation in behavioral and neural systems. *Science,* **239,** 1513–1520.

Schoner, G. S., Haken, H., & Kelso, J. A. S. (1986). A stochastic theory of phase transitions in human hand movement. *Biological Cybernetics,* **53,** 442–452.

Schoner, G., Jiang, W. Y., & Kelso, J. A. S. (1990). A synergetic theory of quadrupedal gaits and gait transitions. *Journal of Theoretical Biology,* **142**(3), 359–393.

Turvey, M. T. (1990). Coordination. *Am. Psychologist,* **45,** 938–953.

Vereijken, B. (1991). *The dynamics of skill acquisition.* Ph.D. Thesis, Vrije Universiteit.

von Holst, E. (1973). Relative coordination as a phenomenon and as a method of analysis of central nervous function. In R. Martin (Ed.), *The collected papers of Erich von Holst* (pp. 33–135). Coral Gables, FL: University of Miami.

Wallace, S. A., Weeks, D. L., & Kelso, J. A. S. (1990). Temporal constraints in reaching and grasping behavior. *Human Movement Science,* **9,** 69–93.

Wimmers, R. H., Beek, P. J., & van Wieringen, P. C. W. (1992). Phase transitions in rhythmic tracking movement: A case of unilateral coupling. *Human Movement Science,* **11,** 217–226.

16

The Dynamical Substructure of Bimanual Coordination

Richard G. Carson,* Winston D. Byblow, and David Goodman

Human Motor Systems Laboratory
School of Kinesiology
Simon Fraser University
Burnaby, British Columbia, Canada

    I. Introduction
   II. Information and Mechanical Anchoring
  III. An Experimental Analysis of Anchoring
  IV. Time Scales Relations
   V. Asymmetries in Bimanual Coordination
  VI. A Preliminary Model
      References

I. Introduction

Recently, considerable attention has been focused on the application of the principles of dynamics to the study of coordination (e.g., Jeka & Kelso, 1989; Schöner & Kelso, 1988a; Turvey, 1990). Generally, these approaches have stressed the self-organizing, autonomous nature of coordinative systems comprising multiple degrees of freedom representing neural, muscular, and metabolic components (Kelso, Schöner, Scholz, & Haken, 1987). These systems are conceived of as being "softly

---

* *Present address:* Human Movement Studies, University of Queensland, Brisbane, Queensland, Australia 4072.

assembled" in a functionally specific manner as a means to perform behavioral tasks. A coordinative system thus assembled "lives" in the low-dimensional space of order parameters or collective variables (Jeka & Kelso, 1989). Laws of coordination derived in terms of the order parameter dynamics thus represent a reduction of the degrees of freedom, which describe the system from the potential to the essential.

In undertaking dynamical analyses, a general problem is to obtain an appropriate level of description and identify the task-relevant degrees of freedom therein (Kay, 1988). One means of delineation derives from the examination of situations in which there is a qualitative change of system behavior, or "phase transition." As qualitative change by definition permits one pattern of behavior to be distinguished from another, examination of the pre- and post-transition behavior allows identification of the essential dimensions of the patterns or "order parameters" (Jeka & Kelso, 1989). The study of phase transitions also allows one to examine the action of control parameters. These are the parameters responsible for inducing changes in the topology of the system's dynamics as a new pattern is achieved.

It may also be the case that the boundary conditions encountered in a natural or an experimental context preclude the occurrence of abrupt transitions. However, there exist forms of dynamical analyses whose currencies are not phase transitions. Analytic techniques based on the paradigm introduced by Kugler and Turvey (1987) have been employed with some success in exploration of the dynamical basis of bimanual coordination (e.g., Bingham, Schmidt, Turvey, & Rosenblum, 1991; Kugler, Turvey, Schmidt, & Rosenblum, 1990; Rosenblum & Turvey, 1988; Schmidt, Beek, Treffner, & Turvey, 1991; Schmidt, Treffner, Shaw, & Turvey, 1992). In these cases the dynamics of order parameters including relative phase are monitored in a region of state space in which phase transitions are atypical.

The seminal application of the phase transition methodology to the study of human movement regulation was conducted by Kelso (1981, 1984). In these tasks, subjects performed rhythmic voluntary oscillations of the hands (Kelso, 1984) or of the fingers (Kelso, 1981). The frequency of oscillation was increased, usually through the use of a pacing metronome. In these circumstances, only two phase-locked modes could be stably and reliably reproduced. These were the in-phase mode, in which there was simultaneous contraction of homologous muscle groups, and the anti-phase mode, in which homologous muscle groups contracted in an alternating fashion. When the system was initially prepared in the anti-phase mode, an involuntary shift to the in-phase mode was observed as the cycling frequency was increased. When, however, the system was initially prepared in the in-phase mode no switching was observed. Relative phase (the latency of one finger with respect to the cycle of the other finger) was identified as the appropriate order parameter, as it characterized all observed coordinative patterns or stationary states.

II. Informational and Mechanical Anchoring

The coupling of perception and action may itself be characterized as a pattern formation process (e.g., Kelso, Delcolle, & Schöner, 1990). This strategy relies on the identification of collective variables that correspond to perception–action patterns, which are in turn mapped onto attractors of the collective variable dynamics. The conception of information is highly specific. Perceptual information is viewed as relevant only in terms of the behavior it modifies. More explicitly, "information is viewed as *meaningful* and *specific* to the extent that it contributes to the collective dynamics attracting the system toward an (environmentally-specified, memorized, intended) behavioral pattern" (Schöner & Kelso, 1988a). Studies dealing with the role of information in the pattern formation process have concentrated primarily on its contribution to the dynamics at the level of the collective variable relative phase (e.g., Scholz & Kelso, 1990; Schöner & Kelso, 1988b; Zanone & Kelso, 1991). However, a central tenet of the dynamical approach is that organizational principles apply across levels of description (e.g., Kelso & Schöner, 1987). As such, the collective variable dynamics may be derived by coupling, generally in a nonlinear fashion, the dynamics of the individual oscillatory components. In the bimanual coordination paradigm this strategy has been applied with some success by modeling the component joints as limit cycle oscillators (Kay, Kelso, Saltzman, & Schöner, 1987). It has also been extended to encapsulate the contribution of environmental information at this level of observation (Schöner & Kelso, 1988c).

The dynamics of the softly assembled structures, whether observed at the level of the collective variable or that of the component oscillators, are constrained by the morphology/biomechanics of the muscle joint complexes (Beek, 1989). There appear to exist mechanical "anchor points" within the limit cycles of component oscillators. These are typically exhibited at points of maximum angular excursion. The swing cycles of rhythmically oscillating hand-held pendulums (Kugler & Turvey, 1987) comprise a "hardening spring" in the abduction phase of the wrist motion and a "softening spring" in the adduction phase (Beek, 1989). Phase plots representing rhythmic flexion–extension of the index fingers (e.g., Kay, Saltzman, & Kelso, 1991, Fig 1b, p. 185) reveal a greater degree of mechanical anchoring at the flexion phase of the movement cycle. Given a finer grained level of analysis, mechanical anchoring could itself be conceived of as a process mediated by the flow of haptic information accompanying mechanical events in the musculature (Kugler & Turvey, 1987.)[1]

Characterization of the dynamics of the component oscillators may also elucidate features that are not adequately captured at the level of the collective variable. Beek (1989) has suggested that the presence of discontinuities occupying the same relative temporal position on each cycle indicates the occurrence of discrete information-based, "externally" applied, forcing pulses. If, in modeling the system, there is closure of description (the temporary restriction of analysis to a single level of observation) at the level of the component oscillators, such external forcing is problematic, as it represents a violation of autonomy. However, it is possible that when descriptive boundaries are subsequently expanded, autonomy reasserts itself within the context of the whole. As Beek (1989, p. 185) notes, the strategy of descriptive closure "opens a window into understanding how information can enter into the kinetic description of action. Once this has been achieved, the nonautonomy can be discarded." On an a priori basis, it is impossible to determine whether discontinuities reflect forcing pulses or information-based anchors of the limit cycle, arising from the localization of information to discrete regions of phase space for each component oscillator (cf. Kelso & Jeka, 1992).

III. An Experimental Analysis of Anchoring

In most studies of coordinated rhythmic movement, pacing information is mediated via auditory or visual metronomes. The contribution to the dynamics of these informational anchors is likely to depend on the relative temporal location of the metronome cue relative to the movement cycle. In extant bimanual paradigms this relationship is not generally controlled (cf. Kelso et al., 1990). In a recent study conducted in our laboratory (Byblow, Carson, & Goodman, in press) we sought to determine how the interaction of anchoring contingent on the information-mediated pacing of the movement and that arising from the mechanical properties of the muscle–joint system mapped onto patterns of behavior at the level of the collective variable relative phase. The experimental system we chose to employ was essentially that of Kelso (1984) in which subjects were required to cycle the hands at the wrists in two modes of coordination, in-phase and anti-phase. As has become conventional, scaling of movement frequencies was induced by

---

[1] We thank an anonymous reviewer for this insight.

an auditory metronome. Unlike previous experiments, however, the position of the metronome pulse relative to the movement cycle of each hand was explicitly controlled through instructions to subjects.

A number of preferred frequency trials were first performed to establish whether, in the absence of informational constraints, the degree of mechanical anchoring was equivalent in each portion of the movement cycle. The data obtained were consistent with previous reports in confirming that the points of maximal angular excursion were distinguished by their relative stability. As Figure 1 illustrates, points of maximal supination were characterized by greater endpoint (amplitude) variability than were points of maximal pronation. It was suggested that in this particular coordinative system, constraints imposed by the morphology/biomechanics of the muscle joint complexes were such that movement reversals at maximum pronation were subserved largely by passive mechanical properties, whereas reversals at maximum supination required active mediation.

In the externally paced trials, subjects were required to produce one full cycle of their movement for each beat of the metronome while maintaining the prescribed mode of coordination. The frequency of the metronome was increased from an initial frequency of 1.25 Hz to a frequency of 3.00 Hz in steps of 0.25 Hz. In each mode of coordination, subjects were prepared in one of two conditions (Figure 2). In the in-phase mode, subjects were required to synchronize each beat of the metronome with either maximum pronation (in-phase pronation) or maximum supination (in-phase supination). In the anti-phase mode, subjects were required to synchronize each beat of the metronome with either maximum excursion to the left (anti-phase left) or maximum excursion to the right (anti-phase right).

Subjects were instructed to maintain the pattern in which they were prepared as accurately as possible, but were also told that should the pattern change they were not to intervene. That is, they were not to actively resist pattern change but were to establish the most comfortable pattern compatible with the prevailing frequency. A number of "basic phenomena" were observed, most notably spontaneous transitions from the anti-phase to in-phase mode as the frequency of oscillation was increased. Transitions in the reverse direction were not present.

Of particular interest in this study was the manner in which informational constraints were superimposed on mechanical anchors. In the in-phase pronated condition, the metronome cue was, for both hands, coincident with the more stable mechanical anchor (maximal excursion at pronation). Whereas the less stable mechanical anchor (maximal excursion at supination) generally occurred midway between metronome cues. In the in-phase supinated condition, the metronome was

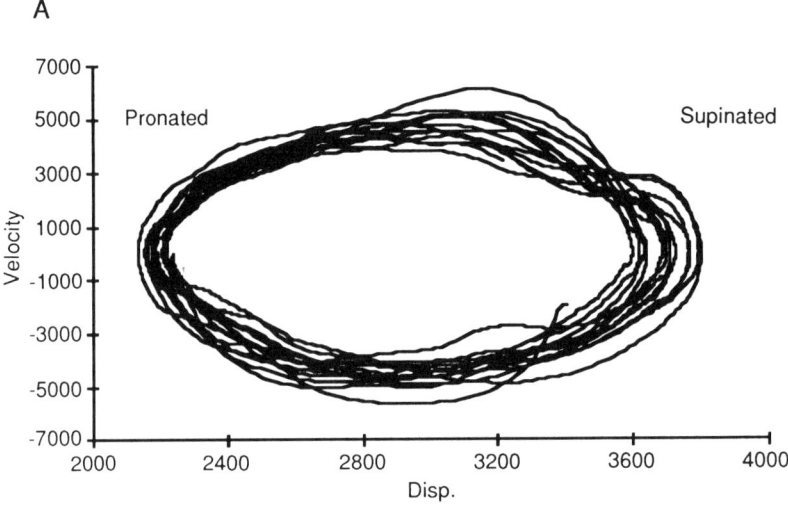

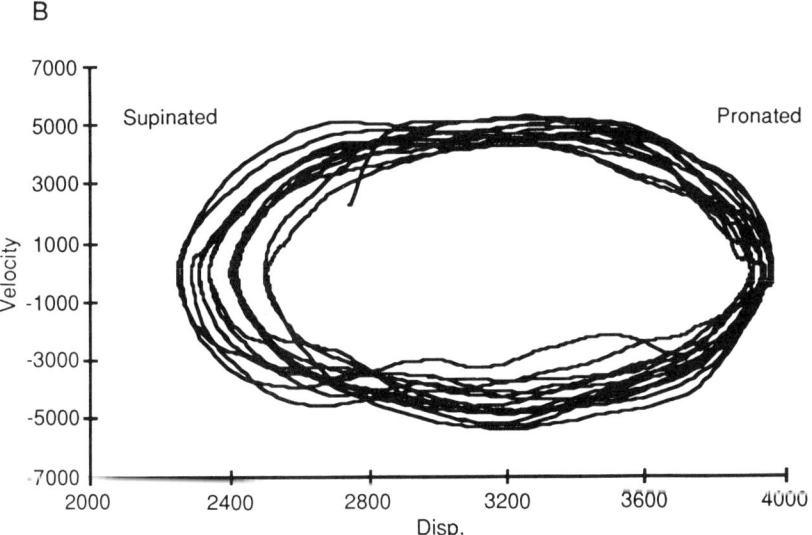

FIGURE 1 Representative phase plots from a single preferred frequency trial. (A) The motion of the left hand; (B) the motion of the right hand. Angular displacements and velocities are given in arbitrary units.

paired with the less stable of the mechanical anchors, with the more stable mechanical anchor occurring midway between the metronome cues.

The supposition that the most stable combination would be that in which the informational anchor and more stable mechanical anchor (in-phase pronated—peak pronation) were coincident was supported by examination of phase plots (Figure 3) and by analysis of the variability of maximal angular excursions. The least stable combination was, as anticipated, that in which the less stable mechanical anchor was not paired with the metronome cue (in-phase pronated—peak supination). Both in-phase supinated combinations were intermediate in terms of the variability of maximal excursions. These data suggest that the

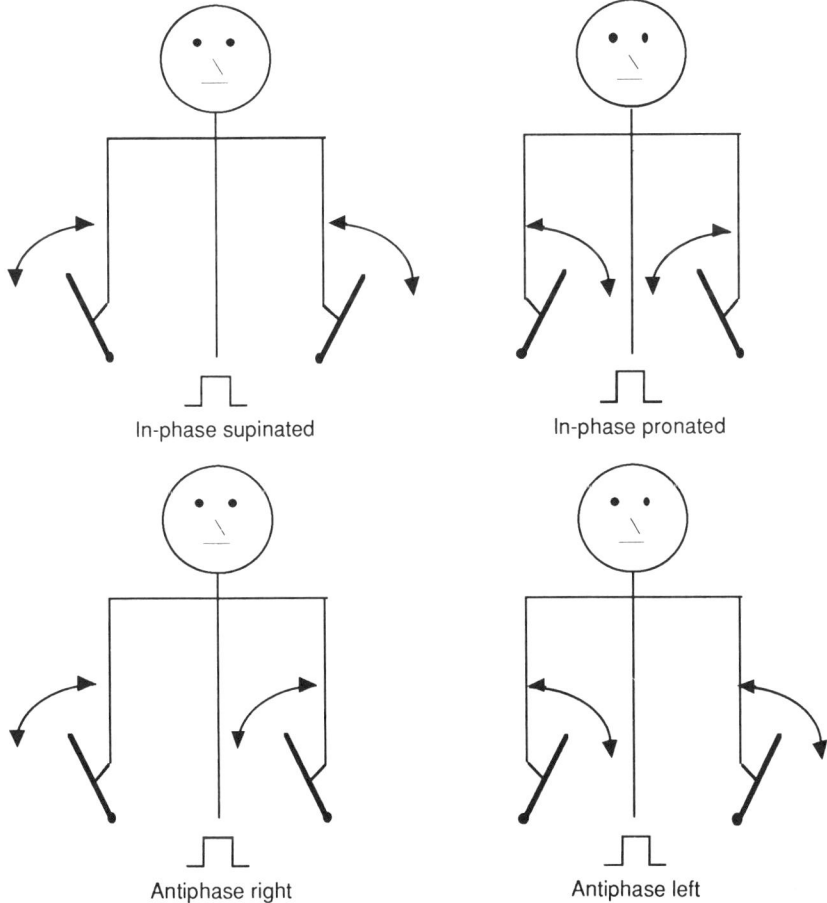

FIGURE 2 Schematic representation of the four experimental preparations.

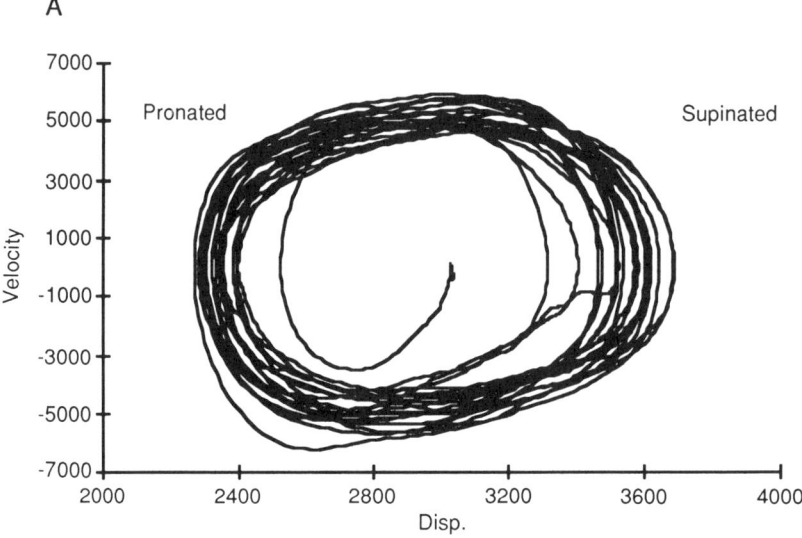

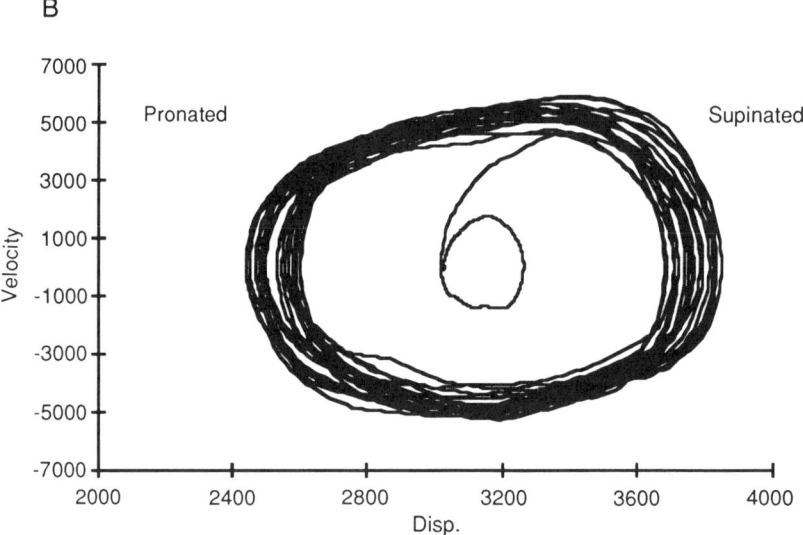

FIGURE 3 Representative phase plots. (A) A single in-phase pronated trial; (B) a single in-phase supinated trial. Both plots represent the motion of the left hand.

information provided by the metronome cue contributes to the dynamics at the level of the component oscillators. They also indicate that information-based modifications of the dynamics are superimposed in a consistent fashion on constraints imposed by the biomechanics of the muscle joint complexes.

It was noted that following transitions, the in-phase pronated mode of coordination was considerably more prevalent than the in-phase supinated mode (Figure 4). The in-phase pronated mode comprises both the least stable and most stable combinations of mechanical and informational anchors. Although our analyses revealed no overall advantage for the in-phase pronated mode in terms of stability, the presence of the most stable combination was clearly of considerable significance.

IV. Time Scales Relations

In modeling the bimanual system, considerable emphasis has been placed on analysis of system time scales (e.g., Scholz, Kelso, & Schöner, 1987; Schöner, Haken, & Kelso, 1986; Schöner & Kelso, 1988d). In particular, the interpretation of observed states as (local) attractor states is consistent if the following relation is fulfilled

$$\tau_{rel} \ll \tau_{obs} \ll \tau_{equ} \qquad (1)$$

where τ_{rel} is an index of the time required for the system to relax onto an attractor in its immediate vicinity, τ_{obs} is the typical time period over

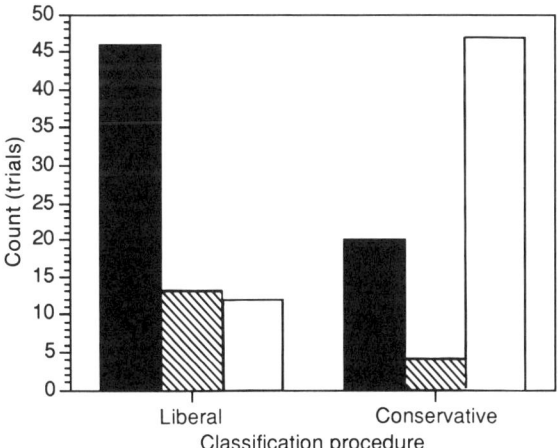

FIGURE 4 Frequency histogram illustrating the dominant patterns of coordination following transitions. Note the differences in distribution between the in-phase pronated (black columns) and in-phase supinated (striped columns). Ambiguous classifications are represented as white columns.

which the system is observed and over which ensemble averages are calculated, and τ_{equ} is the equilibration time (Schöner & Kelso, 1988d).[2] It is anticipated that as switching occurs the time scales relation in Equation (1) is violated. The phenomena of critical fluctuations and critical slowing down are predicted as the transition region is approached and τ_{rel} tends toward infinity. In the pretransition region τ_{rel} is thus predicted to exceed τ_{obs}.

As Equation (1) is violated, an additional time scale becomes important. This is the time scale of parameter change τ_{par}. The nature of the phase transition is crucially determined by the time scale of parameter change with respect to other system time scales. Consider cases in which

$$\tau_{rel} \ll \tau_{par} \ll \tau_{equ}. \qquad (2)$$

In these circumstances the system remains at a specific value of the control parameter appreciably longer than the time required to return to a locally stable state (Jeka & Kelso, 1989). The system may, for conditions in which Equation (2) holds, remain in a particular state even as fluctuations increase. Therefore, the system changes state only as the initial state becomes unstable. As the system changes state, the feature of critical fluctuations is anticipated. Consider on the other hand the case in which

$$\tau_{rel} \ll \tau_{equ} \ll \tau_{par}. \qquad (3)$$

In Equation (3) the time scale of parameter change is substantially greater than the time required for the system to equilibrate. In these circumstances no enhancement of fluctuations is anticipated as the system moves to the lowest potential minimum prior to the old state becoming unstable. The delineation of critical fluctuations is dependent on the relationship of τ_{rel} to τ_{par} in the region of the transition (Scholz et al., 1987).

In our recent study, while spontaneous transitions from the antiphase to the in-phase mode were observed, analysis of variability of relative phase prior to the transition failed to reveal the presence of critical fluctuations. This was found to be the case both when τ_{obs} was the duration of the scaled frequency plateaus and when τ_{obs} was the duration of the individual movement cycle. In the absence of indices of local relaxation time it was not possible to conduct a full analysis of time scales relations. It is conceivable that the use of smaller values of τ_{par}

---

[2] The equilibration or global relaxation time is "the time it takes the system to achieve the stationary probability distribution from a typical initial distribution" (Schöner & Kelso, 1988d, p. 1516). In the bimanual case, the equilibration time is determined largely by the time required to traverse from one basin of attraction (stationary state) to the other.

would have served to reveal the signature features of transitional behavior (cf. Jeka, 1992).

However, it is also possible that the failure to observe critical fluctuations may have been directly attributable to the imposition of additional task constraints.[3] Unlike previous studies, subjects were required not only to maintain the metronome frequency but were also directed to synchronize with the metronome at a particular point in the movement cycle. The dynamic coupling of the limb and periodic environmentally specified information has been treated in some detail for the unimanual case (Kelso *et al.*, 1990). In the current experiment, wandering of the relative phase relation between the limbs and the metronome, one of the predicted features of broken symmetry, was observed on occasion at higher frequencies of oscillation. Clearly, further work is required to establish how the broken symmetry that characterizes limb–metronome couplings impinges on the order parameter dynamics for the limb–limb coupling.

Fluctuations in the control parameter are thought to generate covarying fluctuations in the order parameter (Haken & Wunderlin, 1990). We noted that when movements were prepared in the in-phase mode of coordination, the variability of continuous frequency (phase velocity) decreased as the pacing frequency was increased. In addition, the variability of continuous relative phase covaried with the variability of continuous frequency, consistent with the contention that noise in the order parameter may be induced by noise in the control parameter. Equivalent compression of fluctuations with increasing frequency in the anti-phase mode would be anticipated to offset fluctuations arising from deformation of the potential function as the anti-phase mode becomes unstable.

V. Asymmetries in Bimanual Coordination

The essential features of the switching phenomena evident in bimanual coordination were modeled by Haken, Kelso, and Bunz (1985) using the basic tools of the synergetic approach (e.g., Haken, 1983). Symmetry of the component oscillators was a fundamental assumption under which the system was modeled. That is, the behavior of the system would not change if the labeling of the left and right hands was reversed. However, symmetry is unlikely to prove a ubiquitous feature of biological systems and recent advances have been made in delineating the effects of broken symmetry on coordination dynamics (e.g.,

---

[3] We thank John Jeka and John Buchanan for bringing this possibility to our attention.

Jeka, 1992; Kelso, DeGuzman, & Holroyd, 1991; Sternad, Turvey, & Schmidt, 1992). Differences in the characteristic or eigenfrequencies of the component oscillators have been encapsulated, in an extension of the Haken *et al.* (1985) model, through the inclusion of a symmetry-breaking term (Kelso *et al.*, 1990). Specific predictions of the augmented model include fixed point drift, isodirectional state transitions, and phase wandering.

There is considerable evidence to document the case for asymmetries in both unimanual and bimanual activity (see Carson, 1989; Peters, 1989, for a brief discussion). Indeed, even in tasks requiring absolute coordination, the right hand is characterized by a lesser degree of variability in frequency (Kay *et al.*, 1987) and in juggling, appears to "dominate" the left hand (Beek, 1989). While in the bimanual case intrinsic frequency differences between the limbs have generally been considered negligible, when the limbs are uncoupled these differences are consistently reproduced (R. G. Carson, unpublished data). In our recent study, the presence of asymmetries was revealed by larger coefficients of variation of discrete frequencies, and a greater degree of movement endpoint (amplitude) variability for movements made by the left hand. Perhaps most striking was the observation that when movements were initially prepared in the anti-phase left condition, transitions to the in-phase mode occurred earlier than when trials were prepared in the anti-phase right condition (Figure 5). These data suggested to us that the interaction between intrinsic asymmetries and informational and mechanical constraints may have a significant influence on the coupling dynamics at the point of transition.

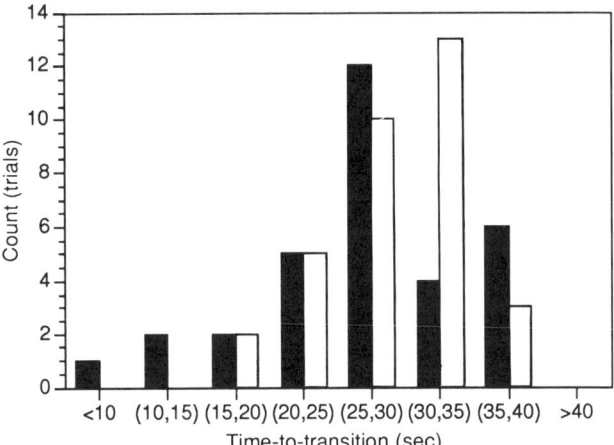

FIGURE 5 Frequency histogram illustrating times to the onset of transition for each anti-phase mode. [Anti-phase left (black columns) and anti-phase right (white columns)].

During phase transitions from the anti-phase to in-phase mode of coordination, in which the relative phase relation between the limbs changes by an integer multiple of π, it is axiomatic that one limb loses or gains half a cycle with respect to the other limb. In circumstances in which movements are paced by a discrete signal such as a metronome, this relative change may also be expressed as a composite of the respective limb–metronome relative phase relations. In the bimanual case, possible relations range from a "mutual transition," in which each limb is subject to an (opposite sign) alteration of the limb–metronome relative phase relation of $\pi/2$, to instances in which a single limb undergoes an alteration of π. Any phase transition can conveniently be categorized in these terms.

In our study, it was generally observed to be the case that a single limb deviated from the initial limb–metronome preparation during transition. Furthermore, when initially prepared in the anti-phase left condition, it was *always* a deviation of the left limb that precipitated the transition (Figure 6), with the result that the in-phase pronation pattern was exhibited following the transition. Thus, the left limb moved from a pattern in which the least stable mechanical anchor (supination) was coincident with the metronome to a pattern in which the most stable mechanical anchor (pronation) was coincident with the metronome. In contrast, when prepared in the anti-phase right condition, transitions precipitated by both the left and the right limb were observed (Figure 6). As a consequence, both the in-phase pronated and in-phase supinated patterns were exhibited following transition.

An explanation for these patterns of behavior may be sought in terms of the interaction between the "forces" that accrue from an asymmetry

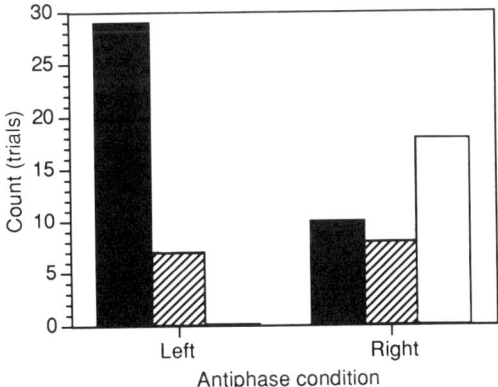

FIGURE 6 Frequency histogram illustrating the contribution of the individual limbs to phase transitions. Note the differences between the left limb (black columns) and right limb (white columns). Mutual contributions are indicated by the striped column.

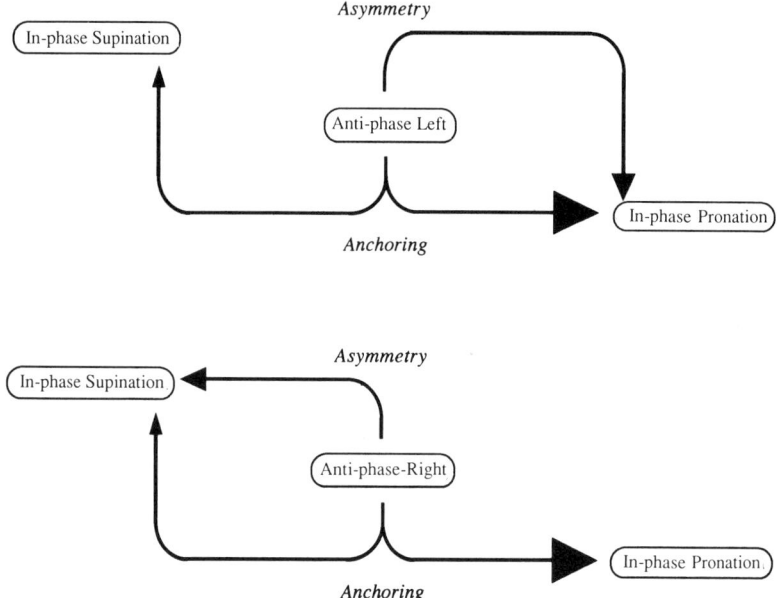

FIGURE 7 Putative transition pathways of the anti-phase left and anti-phase right conditions. Forces of anchoring (bidirectional) and asymmetry (unidirectional) bias the system toward one of two potential patterns of in-phase coordination. The size of the arrowheads approximates the inferred relative magnitudes of the forces observed experimentally.

in the strength of the coupling between the left and right limbs, and those that are contingent upon the presence of mechanical and informational anchors (Figure 7). This account is predicated upon the assumption that the magnitude of the coupling influence of the right limb on the left limb is greater than that in the opposite direction in the region of the transition. In a sense this may be viewed as an expression of the magnet effect (von Holst, 1937/1973). However, it is not conceived of as arising from differences in constituent eigenfrequencies (see Section VI).

When prepared in the anti-phase left condition, the attraction to the most stable combination of mechanical and informational anchors for the left limb (at in-phase pronation) and the force arising from the attraction to the right limb originating from intrinsic asymmetries are coincident and presumably additive. As stability is lost in the region of the transition, these forces are such as to induce with overwhelming probability a transition in which the left limb precipitates a half-cycle change in the limb–limb pattern through a change of π in its limb–metronome relation. In these circumstances the right limb maintains its preexisting (stable) limb–metronome relation in which points of full pronation are coincident with the metronome. In contrast, when prepared in the anti-phase right condition, while the forces arising from the attraction to the most stable combination of mechanical and informational anchors for the right limb (at in-phase pronation) would tend to induce a transition precipitated by the right hand, the greater attraction of the right limb on the left limb is such as to promote a transition precipitated by the left hand. Thus, the forces arising from the anchoring influences and those arising from intrinsic asymmetries are to some extent in opposition. Transitions from the anti-phase right condition, which are mediated by a change of π in the limb–metronome relation of the right hand, are conceived of as those in which the forces accruing from informational and mechanical anchoring predominate over those arising from intrinsic asymmetries. On the other hand, transitions precipitated by a change of π in the limb–metronome relation of the left hand are those in which asymmetries in the coupling strength are transcendent. Our data indicated that transitions to the in-phase supination pattern, in which the left hand mediated the transition, were observed in approximately one-third of the trials prepared in the anti-phase right condition, while transitions in which the right hand mediated the transition were observed in one-half of trials. Differences observed in the time to transition between the two anti-phase conditions (Figure 5) are also consistent with this account. In the anti-phase left condition, the shorter times to transition support a view in which the forces accruing from anchoring and asymmetries are regarded as additive. The significantly longer times to transition onsets observed when movements were prepared in the anti-phase right condition comports with the presence of counteracting forces.

VI. A Preliminary Model

Recently we (Carson, in press) have given some consideration to the manner in which the proposed asymmetries in coupling strengths might arise. The account relies heavily on Average Phase Difference (APD) Theory (Kopell, 1988; Rand, Cohen, & Holmes, 1988). This is a general modeling scheme in which no assumptions are made regarding the dynamical substructure or "state space" of the individual oscillatory components. Notwithstanding these limitations, this phenomenological approach provides a convenient means through which the bimanual system may be studied.

In the nomenclature adopted by Rand *et al.* (1988), the characteristic frequencies of the left and right oscillators are given by ω_l and ω_r, respectively. It is assumed that the magnitude of the coupling between them is a function of their phase difference, such that

$$\frac{d\theta_1}{dt} = \omega_l + k_{lr}\sin(\theta_r - \theta_l) \tag{4}$$

and

$$\frac{d\theta_r}{dt} = \omega_r + k_{rl}\sin(\theta_l - \theta_r) \tag{5}$$

where k_{rl} is the coupling coefficient representing the influence of the oscillator corresponding to the left hand on the oscillator corresponding to the right hand. The reverse applies in the case of the coupling coefficient k_{lr}. The collective variable relative phase $\phi(t)$ is defined as

$$\phi(t) = \theta_l(t) - \theta_r(t). \tag{6}$$

Subtracting Equation (5) from Equation (4) we obtain

$$\frac{d\varphi}{dt} = (\omega_l - \omega_r) - (k_{lr} + k_{rl})\sin\varphi. \tag{7}$$

In Equation (7) stationary states (1 : 1 phase locking) exist as special solutions in which $\dot\phi = 0$ and in which relative phase is constant over time. Thus

$$\varphi = \arcsin\frac{\omega_l - \omega_r}{k_{lr} + k_{rl}} \tag{8}$$

(Rand et al., 1988).

Phase and frequency locking are predicted to occur if the difference in the characteristic frequencies of the component oscillators (ω_l and ω_r) is sufficiently small relative to the net coupling between them. Given two oscillators of fixed characteristic frequencies, a decrease in the net coupling strength will, at a critical value, engender a transition from phase-locked motion to phase drift (Rand et al., 1988).

In modeling the interlimb coordination of rhythmic movements, it has generally been assumed that the coupling strengths (k_{lr} and k_{rl}) are inversely proportional to the frequency of oscillation ω_v. Sternad et al. (1992) have suggested that relevant quantities should not be expressed simply in terms of the frequency of oscillation but rather as a ratio of the actual and characteristic frequencies.

We may therefore introduce the following expressions

$$k_{rl} = G_l\frac{\omega_l}{\omega_v} \tag{9}$$

$$k_{lr} = G_r\frac{\omega_r}{\omega_v} \tag{10}$$

where G_l and G_r are constants of proportionality, expressing for each hand the degree to which the respective coupling strengths scale in magnitude with the frequency of oscillation.

Considerable emphasis has been placed on the use of preferred frequency as an index of the characteristic frequency (e.g., Schmidt et al., 1991; Turvey, Schmidt, & Rosenblum, 1989) or eigenfrequency (e.g., Jeka, 1992) of component oscillators. The limited amount of data available pertaining directly to this issue provides little evidence to suggest that the hands differ in terms of preferred frequency (Kay et al., 1987). In addition, fixed-point drift or phase wandering are not usually observed in bimanual coordination. On initial inspection, these observations appear to support the assumption of symmetry (Haken et al., 1985). However, our data concerning transition pathways reveal that asymmetries are clearly present. The finding of preferred transition pathways between stationary states fulfills at least one of the predictions of symmetry-breaking dynamics. It also suggests that there may exist asymmetries that are not fully expressed in terms of unimanual preferred frequencies.

In Equation (8) the key quantity determining stability is the ratio of the difference in the component eigenfrequencies to the coupling strength. As the frequency of oscillation ω_v is increased and the transition region is approached, $|k_{lr} + k_{rl}|$ decreases to a critical value relative to $|\omega_l - \omega_r|$. In order for the transition to proceed predominantly in a single direction, that of left to right at transition, it is necessary that $k_{lr} > k_{rl}$, is larger at the transition frequency. This would occur in circumstances in which G_l is smaller than G_r. Although there is no evidence directly in support of this proposition, it is known that maximal rates of response are consistently higher for the right hand than for those of the left hand (e.g., Peters, 1977, 1987, 1990), while preferred frequencies appear to be equivalent. In our recent study (Byblow et al., in press) we also observed that preferred frequencies were sensitive to task exposure, suggesting that G_l and G_r are not stationary over time. It remains to be determined whether G_l and G_r are distinguished from each other on a consistent basis. Appraisal of the maximum sustainable frequencies of oscillation for single limbs on these types of task may help establish whether this is the case.

References

Beek, P. J. (1989). *Juggling dynamics*. Amsterdam: Free University Press.

Bingham, G. P., Schmidt, R. C., Turvey, M. T., & Rosenblum, L. D. (1991). Task dynamics and resource dynamics in the assembly of a coordinated rhythmic activity. *Journal of Experimental Psychology: Human Perception and Performance, 17*, 359–381.

Byblow, W. D., Carson, R. G., & Goodman, D. (in press). Expressions of asymmetries and anchoring in bimanual coordination. *Human Movement Science*.

Carson, R. G. (1989). Manual asymmetries: In defense of a multifactorial account. *Journal of Motor Behavior*, **21**, 157–162.

Carson, R. G. (in press). Manual asymmetries: Old problems and new directions. *Human Movement Science*.

Haken, H. (1983). *Synergetics, an introduction: Non-equalibrium phase transitions and self-organization in physics, chemistry, and biology*. Berlin: Springer-Verlag. 3rd edition.

Haken, H., & Wunderlin, A. (1990). Synergetics and its paradigm of self-organization in biological systems. In H. T. A. Whiting & P. C. W. van Wieringen (Eds.), *The natural-physical approach to movement control*. Amsterdam: VU University Press.

Haken, H., Kelso, J. A. S., & Bunz, H. (1985). A theoretical model of phase transitions in human hand movements. *Biological Cybernetics*, **39**, 139–156.

Jeka, J. J. (1992). *Asymmetric dynamics of human limb coordination*. Unpublished doctoral dissertation, Florida Atlantic University, Boca Raton, Florida.

Jeka, J. J., & Kelso, J. A. S. (1989). The dynamic pattern approach to coordinated behavior: A tutorial review. In S. A. Wallace (Ed.), *Perspectives on the coordination of movement* (pp. 3–45). North Holland: Amsterdam.

Kay, B. A. (1988). The dimensionality of movement trajectories and the degrees of freedom problem. A tutorial. *Human Movement Science*, **7**, 343–364.

Kay, B. A., Kelso, J. A. S., Saltzman, E. L., & Schöner, G. (1987). The space–time behavior of single and bimanual movements: Data and model. *Journal of Experimental Psychology: Human Perception and Performance*, **13**, 178–192.

Kay, B. A., Saltzman, E. L., & Kelso, J. A. S. (1991). Steady-state and perturbed rhythmical movements: A dynamical analysis. *Journal of Experimental Psychology: Human Perception and Performance*, **17**, 183–197.

Kelso, J. A. S. (1981). On the oscillatory basis of movement. *Bulletin of the Psychonomic Society*, **18**, 63.

Kelso, J. A. S. (1984). Phase transitions and critical behavior in human bimanual coordination. *American Journal of Physiology*, **240**, R1000–R1004.

Kelso, J. A. S., & Jeka, J. J. (1992). Symmetry-breaking dynamics of human limb coordination. *Journal of Experimental Psychology: Human Perception and Performance*, **18**, 645–668.

Kelso, J. A. S., & Schöner, G. (1987). Toward a physical (synergetic) theory of biological coordination. In R. Graham & A. Wunderlin (Eds.), *Lasers and synergetics (Springer Proceedings in Physics, Vol. 19)*. New York: Springer-Verlag.

Kelso, J. A. S., Schöner, G., Scholz, J. P., & Haken, H. (1987). Phase-locked modes, phase transitions and component oscillators in biological motion. *Physica Scripta*, **35**, 79–87.

Kelso, J. A. S., Delcolle, J. D., & Schöner, G. S. (1990). Action–perception as a pattern formation process. In M. Jeannerod (Ed.), *Attention and performance XIII* (pp. 139–169). Hillsdale, NJ: Erlbaum.

Kelso, J. A. S., DeGuzman, G. C., & Holroyd, T. (1991). The self-organized phase attractive dynamics of coordination. In A. Babloyantz (Ed.), *Self-organization, emerging properties, and learning* (pp. 41–62). New York: Plenum.

Kopell, N. (1988). Toward a theory of modelling central pattern generators. In A. H. Cohen, S. Rossignol, & S. Grillner (Eds.), *Neural control of rhythmic movements in vertebrates* (pp. 369–413). New York: Wiley.

Kugler, P. N., & Turvey, M. T. (1987). *Information, natural law, and the self assembly of rhythmic movement*. Hillsdale, NJ: Erlbaum.

Kugler, P. N., Turvey, M. T., Schmidt, R. C., & Rosenblum, D. (1990). Investigating a nonconservative invariant of motion in coordinated rhythmic movements. *Ecological Psychology*, **2**, 151–189.

Peters, M. (1977). Simultaneous performance of two motor activities: the factor of timing. *Neuropsychologia*, **15**, 461–465.

Peters, M. (1987). A nontrivial motor performance difference between right handers and left handers: Attention as intervening variable in the expression of handedness. *Canadian Journal of Psychology,* **41,** 91–99.

Peters, M. (1989). Do feedback processing, output variability and spatial complexity account for manual asymmetries? *Journal of Motor Behavior,* **21,** 151–155.

Peters, M. (1990). Subclassification of non-pathological left-handers poses problems for theories of handedness. *Neuropsychologia,* **28,** 279–289.

Rand, R. H., Cohen, A. H., & Holmes, P. J. (1988). Systems of coupled oscillators as models of central pattern generators. In A. H. Cohen, S. Rossignol, & S. Grillner (Eds.), *Neural control of rhythmic movements in vertebrates* (pp. 333–367). New York: Wiley.

Rosenblum, L. D., & Turvey, M. T. (1988). Maintenance tendency in co-ordinated rhythmic movements: relative fluctuations and phase. *Neuroscience,* **27,** 289–300.

Schmidt, R. C., Beek, P. J., Treffner, P. J., & Turvey, M. T. (1991). Dynamical substructure of coordinated rhythmic movements. *Journal of Experimental Psychology: Human Perception and Performance,* **17,** 635–651.

Schmidt, R. C., Treffner, P. J., Shaw, B. K., & Turvey, M. T. (1992). Dynamical aspects of learning an interlimb rhythmic movement pattern. *Journal of Motor Behavior,* **24,** 67–83.

Scholz, J. P., & Kelso, J. A. S. (1989). A quantitative approach to understanding the formation and change of coordinated movement patterns. *Journal of Motor Behavior,* **21,** 122–144.

Scholz, J. P., & Kelso, J. A. S. (1990). Intentional switching between patterns of bimanual coordination is dependent on the intrinsic dynamics of the patterns. *Journal of Motor Behavior,* **22,** 98–124.

Scholz, J. P., Kelso, J. A. S., & Schöner, G. (1987). Non-equilibrium phase transitions in coordinated biological motion: Critical slowing down and switching time. *Physics Letters A,* **123,** 390–394.

Schöner, G, & Kelso, J. A. S. (1988a). Dynamic patterns of biological coordination: Theoretical strategy and new results. In J. A. S. Kelso, A. J. Mandell, & M. F. Shlesinger (Eds.), *Dynamic patterns in complex systems* (pp. 77–102). Singapore: World Scientific.

Schöner, G., & Kelso, J. A. S. (1988b). A synergetic theory of environmentally-specified and learned patterns of movement coordination. I. Relative phase dynamics. *Biological Cybernetics,* **58,** 71–80.

Schöner, G., & Kelso, J. A. S. (1988c). A synergetic theory of environmentally-specified and learned patterns of movement coordination. II. Component oscillator dynamics. *Biological Cybernetics,* **58,** 81–89.

Schöner, G., & Kelso, J. A. S. (1988d). Dynamic pattern generation in behavioral and neural systems. *Science,* **239,** 1513–1520.

Sternad, D., Turvey, M. T., & Schmidt, R. C. (1992). Average phase difference theory and 1:1 phase entrainment in interlimb coordination. *Biological Cybernetics,* **67,** 223–231.

Turvey, M. T. (1990). Coordination. *American Psychologist,* **45,** 938–953.

Turvey, M. T., Schmidt, R. C., & Rosenblum, L. D. (1989). 'Clock' and 'motor' components in absolute coordination of rhythmic movements. *Neuroscience,* **33,** 1–10.

von Holst, E. (1937/1973). On the nature and order of the central nervous system. In R. Martin (Ed. and Trans.), *The collected papers of Erich von Holst: Vol. 1. The behavioral physiology of animal and man.* Coral Gables, Florida: University of Miami Press.

Zanone, P. G., & Kelso, J. A. S. (1991). Relative timing from the perspective of dynamic pattern theory: Stability and instability. In J. Fagard & P. H. Wolff (Eds.), *The development of timing control and temporal organization in coordinated action* (pp. 69–92). Amsterdam: North Holland.

17

From Interlimb Coordination to Trajectory Formation:
Common Dynamical Principles

Gregor Schöner

Institut für Neuroinformatik
Ruhr-Universität Bochum
Bochum, Germany

I. Introduction
II. Dynamic Theory of Interlimb Coordination
III. Toward a Dynamic Theory of Trajectory Formation
    A. Levels
    B. Coordinate Systems
    C. Variables
    D. Dynamics
    E. Results: One-Dimensional
    F. Results: Two-Dimensional
IV. Discussion
    References

I. Introduction

Over the last few years an approach to interlimb coordination has emerged that posits: (1) coordination patterns can be characterized by task-related collective variables; (2) the emergence and change of coordination patterns is governed by dynamics, that is, equations of motion of these collective variables; and (3) both intrinsic and behavioral constraints, the latter arising, for example, from the perceived environment, through learning, or intentionally, act as parts of the coordination dynamics. The approach is variedly referred to as dynamic approach, dynamic pattern theory, or, here, dynamic theory of coordination. The ideas are embedded in the historical context of concepts such as synergism (Bernstein, 1947/1967) and relative coordination (von Holst, 1939/1973) stressing collective and dynamic properties, respectively, as well as the more recent concept of coordinative structures (e.g., Kelso, Holt, Kugler, & Turvey, 1980; Kugler, Kelso, & Turvey, 1980). The starting point was observations of spontaneous change of coordination pattern in rhythmic movement (Kelso, 1984) and analogies drawn with concepts from physical theories of pattern formation (Haken, 1983; Haken, Kelso, Bunz, 1985). In its more recent formulations (Schöner & Kelso, 1988a,b; Schöner, Zanone, & Kelso, 1992) the approach consists of a theoretical strategy for arriving at mathematical dynamical models and a related set of experimental measures and paradigms. Central to the theoretical language is the concept of temporal stability, that is, the capability of coordination systems to retain coordinated patterns in the face of fluctuations and external perturbations. Identifying stability as an essential property of coordination systems leads to a number of general model-independent predictions, and their successful experimental test has lent support to the basic premises of the dynamic theory of coordination. Moreover, stability can also be used as a diagnostic tool through which specific assumptions about particular coordination systems can be tested.

It is important to realize that the coordination dynamics are not identical to the physical dynamics of the biomechanical system (although the latter may contribute to the coordination dynamics). This is obvious for the case of the coordination of limbs that are not mechanically coupled, for instance, the index fingers of the left and right hand. The dynamics of coordination as described, for example, by an equation of motion of a relative phase is not determined by the biomechanics of the individual limbs and certainly reflects in large part the coordination activity of the central nervous system. Moreover, the coordination dynamics may also describe properties that lie outside the scope of the classical approach of control theory to biological motion. This becomes clearest when we consider multifunctional (or multistable) systems in which the nervous system may stabilize a number of different relationships among moving limbs under the same conditions, and switching among these patterns is possible and is governed by the coordination dynamics as well. A more subtle point of distinction from these classical theories (that will be taken up at several points in this chapter) is the abstract, task-related nature of the variables in which such dynamics can be identified (see also Saltzman & Kelso, 1987, for such an emphasis). The logical mirror image of multifunctionality makes this clear: different structures may be governed by the same type of coordination dynamics if put into the same task context.

It has become clear that the methods of analysis and the concepts of dynamic theory can be extended to the problem of coordinating movement with the perceived environment, that it, to the formation of action–perception patterns (e.g., Kelso, Delcolle, & Schöner, 1990; Schöner, 1991; Dijkstra, Schöner, & Gielen, 1993). It was found that the temporal relationship between movement and a temporally structured environment can be measured through collective variables, and that this relationship is characterized by temporal stability. Hence an action–perception dynamics can be postulated. This work gives a first indication about the potential generality of the concepts of the dynamic approach.

In this chapter we raise the question to what extent the principles formalized in the dynamic theory of coordination are specific to interlimb coordination and to what extent these principles can be usefully extended to trajectory formation in single and multiple limb movement. Because the dynamic theory aims at analysis in functional terms we must inquire whether from its perspective the distinction between interlimb and intralimb coordination, which is given in anatomical terms, is a relevant one at all. It will emerge that pure interlimb coordination can be regarded as a limit case of a more general form of coordination, the limit being defined through weak coupling at a behavioral level dealing with physical forces.

More concretely, a number of observations suggest that the problem of trajectory formation may have to be analyzed in a behavior-based and dynamic fashion. Recent evidence from the neurophysiology of movement indicates that cortical neurons may be tuned to effector-independent movement parameters such as direction of movement in body-centered coordinates, and are not always rigidly linked to the skeletomuscular system (for review, see, e.g., Georgopoulos, 1991; Kalaska & Crammond, 1992). Neurophysiologists also tend to recognize more strongly today that the cortical control of motion is effected via multiple closed sensorimotor loops rather than by unidirectional information processing (for a review, see Evarts, 1984). More peripherally, a dominant theme of recent work in central pattern generators is their extreme flexibility in the sense of both strong tunability of the produced pattern by sensory feedback and multifunctionality (e.g., Delcomyn, 1991). This form of flexibility is not limited to the coordination among limbs but may affect the form of motion of individual limbs as well. The concepts of dynamics to deal with such flexibility should likewise not be limited to the neural basis of interlimb coordination. At the behavioral level, classical results on effector independence of movement skills (reviewed, e.g., in Wright, 1990) as well as recent results on the psychophysics and kinematics of pointing movement (reviewed, e.g., in Soechting & Flanders, 1991) indicate that trajectory formation can also be described in terms of relatively abstract task-related variables rather than in effector variables. At the muscle–joint level, the equilibrium point hypothesis and subsequent theoretical developments (for reviews, see Feldman, 1986; Bizzi & Mussa-Ivaldi, 1990; Flash, 1990) have introduced aspects of dynamics to the description of trajectory formation. Saltzman and Kelso (1987) have attempted to generalize the mass-spring dynamics from the description of posture at the muscle–joint level to the description of movement at a task level, considering also nonlinear dynamics and limit cycles. There is clearly a need for clarifying the relationship between the mass-spring type of dynamics and the coordination dynamics.

The plan of this chapter is as follows: In Section II we briefly review the dynamic theory of interlimb coordination and then expand on the special case of coordination of discrete movement. Predictions of the theory on remote compensatory movements when coordination patterns are perturbed are discussed in some detail and the limitations of the theory as applied to the individual components are highlighted. In Section III these limitations are lifted by addressing explicitly the level at which mechanical loads act on the movement system. We formulate in general terms a framework for providing a dynamic theory of trajectory formation, which is implemented in a model of two-dimensional endeffector motion and compared to recent experiments by Latash and collegues. We discuss the effect of the additional load level on smoothness of trajectories, the behavior under perturbation, the relationship to the model of Wing and Kristofferson (1973), as well as the properties of movement paths in two dimensions, such as the two-thirds law and the stability of component movement.

II. Dynamic Theory of Interlimb Coordination

The three language elements of the dynamic theory of interlimb coordination are (1) collective variables (2) coordination dynamics, and (3) constraints parameterized as intrinsic dynamics (constraints not specific to coordination) or behavioral information (constraints specific to coordination). The term "collective variables" is used here in two different meanings, each of which fixes a level of resolution at which a system is studied. On one hand, variables may explicitly characterize the relationship among the movements of different limbs. In the case of rhythmic movement, the relative phases among moving limbs are candidate variables of this type. In goal-directed movement covariance of trajectories of different limbs indicates that reduced descriptions by a common factor are possible. Formally, such factors may be identified by statistical analysis, although this is not of much practical value. On the other hand, variables may characterize the movements of the individual limbs. These variables are collective only with respect to the multiple mechanical, muscular, and nervous components of each limb, but not with respect to the coordination problem at hand. Coordination comes in because the dynamics of these variables for different limbs are coupled.

The coordination dynamics may consist of various contributions. We distinguish two types of contributions: (1) Intrinsic dynamics account for coordination patterns that are observed in the absence of any specific influences of the coordination system. Such spontaneously available or intrinsic patterns define attractors of the intrinsic dynamics. Behavioral constraints that do not specify a particular coordination pattern (e.g., frequency of rhythmic movement does not specify relative phase) may act as parameters of the intrinsic dynamics. The parameter ranges over which various patterns are stable may be limited and at their boundaries instabilities lead to observable predictions (see below). Mathematically, normal form theory (for introduction, see, e.g., Guckenheimer & Holmes, 1983) can be used to find the simplest dynamics that includes all observed intrinsic patterns as well as their instabilities. (2) Behavioral information defines contributions to the coordination dynamics that capture specific constraints (Schöner & Kelso, 1988c). These can be characterized in terms of the same type of variables used to describe coordination patterns, that is, as specified patterns. The corresponding contributions to the coordination dynamics are defined such as to attract toward the specified patterns. We refer to these types of contributions as behavioral information.

The experimental measures derived from these theoretical language elements revolve around the concept of temporal stability. Direct measures of stability are (Schöner, Haken, & Kelso, 1986): (1) (Local) relaxation time, τ_{rel}, is the time for the coordination system to return to its stable pattern following an external perturbation; (2) the spontaneous fluctuations of coordination patterns in time can be measured through the variability SD of the collective variables in time under stationary conditions; (3) the correlation time is the time scale over which correlations of the pattern variables decay and is measured either directly from correlation functions or as the inverse spectral width of the corresponding power spectrum (see, e.g., Gardiner, 1983, for explanation). Theoretically, the correlation time is identical to the relaxation time, although experimentally the first is estimated in the absence, the second in the presence of external perturbations. Related measures include: (4) the switching time of pattern change, estimated as the length of the transient switching process. This is a measure of the difference of stability of the two states involved in the pattern change (Schöner & Kelso, 1988b). Paradigmatically, the variation of external parameters, either specific or nonspecific, that lead the system through state changes can be used to probe the coordination dynamics. Essentially, pattern stability is indicative of how close a pattern is becoming unavailable as external parameters are changed. Also, systematic dependence of pattern and pattern stability as specific behavioral requirements are varied allows one to evaluate or scan the coordination dynamics (Schöner & Kelso, 1988c; Schöner et al., 1992): (1) The influence of instrinsic coordination tendencies leads to systematic deviations from required patterns; (2) Patterns closer to the instrinsically stable patterns are more stable than patterns in conflict with the intrinsic dynamics.

Both the theoretical strategy and the experimental measures and paradigms have been extensively applied to the study of the coordination of rhythmic movement. Using relative phase as a simple collective variable, the observation of instabilities as unspecific parameters are changed has provided the most convincing evidence to date for the validity of the dynamic approach: in a series of experiments it has been shown that the various stability measures covary as predicted and indicate the predicted loss of stability as a point of abrupt pattern change is approached (Kelso, Scholz, & Schöner, 1986b; Scholz, Kelso, & Schöner, 1987; Scholz & Kelso, 1989). Also, the role of behavioral information to modify patterns in either continuous or abrupt fashion has been elucidated for rhythmic movement (Schöner & Kelso, 1988c; Tuller & Kelso, 1989). In theoretical models for locomotory gait patterns the role of symmetry has been highlighted (Schöner, Jiang, & Kelso, 1990): symmetries may lead to invariance under change of parameter. Such invariance must be distinguished from stability, that is, persistence in time under external or random perturbation. The nervous system may keep patterns invariant to comply with symmetries. On the other hand, the nervous system may be capable of stabilizing a wealth of patterns that are not invariant under change of parameters. Such patterns may change continuously with or without instability even if under other conditions abrupt changes occur. Abrupt changes, however, are always associated with instablities.

To see that the concepts of dynamics are not limited to the understanding of the coordination of rhythmic movement consider discrete, goal-directed movement of two limbs. Experimentally, Kelso, Southard, & Goodman (1979) have studied fast movement of the two hands from initial positions of two visually indicated targets. Target distance and width were varied such that the movements of the individual hands, when performed separately, varied in movement time as accounted for by Fitts's law (Fitts, 1954). Coordination was then observed when the two hands initiated movement simultaneously in the sense that movements were close to synchronous even when distance and target size for the two limbs were sufficiently disparate to imply movement times of the individual hands differing by a factor of two.

A dynamical model of this system can be formulated (Schöner, 1990) by choosing the endeffector coordinates, x_i, and their velocities, $\dot{x}_i$ (i = left or right) as collective variables. To capture the behavioral patterns of single-limb posture, goal-directed and rhythmic movement, the dynamics of these variables must be defined to contain in separate parameter regimes (1) a stable fixed point modeling posture, (2) two stable fixed points modeling initial and target posture of a goal-directed movement, and (3) a limit cycle modeling rhythmic movement. To ease the mathematical analysis, an exactly solvable mathematical model was used, although more realistic trajectories can be obtained by systematic fit of the vector field. In the case of discrete movement, movement initiation was modeled by casting the intention to move into behavioral information, that is, a force stabilizing the movement state (a piece of a limit cycle). Fitts's law can be built into such a model by adequate parameterization. Coordination is then described through the coupling of two such component dynamics. The coupling function can be identified by requiring that in the oscillatory movement regime the dynamics of interlimb coordination (stable in-phase and anti-phase; Haken et al., 1985) are recovered. The model is then capable of predicting synchronization of two-limb, goal-directed movement even when the individual movements have disparate movement times. Moreover, the breakdown of this synchrony as movement parameters become even more disparate (Kelso, Putnam, & Goodman, 1983; Marteniuk, MacKenzie, & Baba, 1984) is also predicted.

In such a dynamical account synchronization of the two goal-directed movements is due to the formation of a stable pattern of coordination, even if the time window over which this pattern exists is limited by the duration of the movement. This view has important testable consequences because the stability property implies that the nervous system restores the coordination pattern if this pattern is perturbed. For instance, if one limb is restrained by external perturbations, compensatory reactions also in the other limb are predicted. These compensatory reactions are directed such as to restore the zero-lag relative timing relationship between the limbs.

While direct experimental tests of this prediction in this particular model system have not yet been performed, there is some evidence for compensatory reactions of the type postulated here. For instance, Smeets, Erkelens, and Denier van der Gon (1990) perturbed goal-directed arm movement by adding or reducing the inertial mass unexpectedly. Although this mechanical change affected primarily the elbow, compensatory reactions were observed in the shoulder EMG within approximately 40 msec of the point where the change in inertial mass induces a measurable change in velocity. Arguments raised by Smeets et al. (1990) exclude a simple servo-control on joint position or velocity as a mechanism for these reactions. Analogous observation had been made earlier for articulatory movements during speech production (Kelso, Tuller, Vatikiotis, & Fowler, 1984; Gracco & Abbs, 1988). The relative timing of articulator movement is specific to the target gesture and highly reproducible. When one articulater, for example, the lower lip or the jaw, is perturbed so that the relative timing of articulators is affected, then compensatory movements occur within 40 msec in remote articulators with only weak mechanical coupling, for example, the upper lip. These compensatory movements tend to restore the adequate relative timing to the intended gesture. These experiments highlight dramatically the functional, task-specific character of the coordination dynamics: remote compensatory movements occur only if the gestures specify the relative position of the articulators in question (such as bilabial closing for the final /b/ in /bæb/). When no such relationship is specific (e.g., no relationship between upper and lower lip required for the final /z/ in /bæz/), remote compensatory reactions are not observed (Kelso et al., 1984). In the theoretical language this means that what is specified by the central nervous system to achieve a particular behavioral goal (here, a particular gesture) is the coordination dynamics including the stability rather than the particular movement patterns!

We note that these results on remote compensatory reactions in speech articulation were the primary motivation behind the theoretical work of Saltzman and Kelso (1987; for the speech articulation problem, see Kelso, Saltzman, & Tuller, 1986a; see also Saltzman & Munhall, 1989) on task dynamics. The task-dynamic account of the remote compensatory reactions differs, however, from the one given here. Here, we argued that these reactions can be understood in terms of the fundamental concept of temporal stability of patterns of coordination. They are strict analogues for the case of discrete movement of the relaxation process to coordination pattern following a phasic perturbation during ongoing rhythmic movement or during posture. Hence, they are predicted whenever, under circumstance of continuous coordination, relaxation is observable. In task dynamics, the reactions arise from a weakened form of equifinality: only degrees of freedom relevant for the task have a stable end configuration, while degrees of freedom with little influence on the task fulfillment are marginally stable. The remote compensatory reactions arise because mechanical degrees of freedom and task degrees of freedom are not identical: the mechanical perturbation shifts a marginally stable task degree of freedom. This shift is distributed over a number of mechanical degrees of freedom, including remote ones.

The stability of the coordination pattern formed during a goal-directed, multiarticulator movement can be observed also in the absence of external perturbations by examining variability of kinematic parameters from trial to trial. In the theory, stochastic forces must be included to account for pattern fluctuations and behavioral change (cf. Schöner & Kelso, 1988b) and these random perturbations also probe the stability of the underlying dynamics. Less stable patterns formed between weakly coupled components yield more to stochastic perturbations, leading to larger variability in time or from trial to trial than stable patterns formed between coupled components. For instance, the model of discrete movement coordination predicts that variability of relative timing is less than variability of absolute timing of each individual component.[1] This observation was made experimentally by Gracco and Abbs (1986), analyzing the timing variability of upper and lower lip and jaw movements during speech production. The strong covariance of the timing parameters of the individual components indicates that these articulators are dynamically coupled during the utterance studied. This again shows how stability can be used diagnostically to identify what the central nervous system plans about speech articulatory movements: that is, what it keeps dynamically stable (not necessarily invariant across different utterances). In interlimb coordination such stability is generated by coupling of component dynamics and leads to an interlimb form of motor equivalence: multiple absolute movements of the individual articulators may generate the same stable pattern of coordination. Below, we shall argue that this linkage of motor equivalence to stability and, hence, dynamics is valid beyond interlimb coordination.

We have introduced dynamic descriptions for the individual limb movements in order to understand coordination through coupling. How valid is the dynamical picture provided for the movements of individual limbs? We point out one limitation of the description offered so far (cf. Schöner, 1990, for details). We have assumed that external perturbations act directly on the modeled level of coordination dynamics. As a consequence, the system has no additional level of time keeping. In rhythmic movement, for instance, perturbations lead in the model to resetting of the oscillator phase. This has indeed been observed experimentally (Yamanishi, Kawato, & Suzuki, 1979; Kay, Saltzman, & Kelso, 1991). In discrete movement, perturbations shift the trajectory planned by the nervous system. For instance, if we perturb by abruptly moving a limb to its final target position, the movement dynamics may simply lead to the system resting at the target position. By contrast, in experiments on monkeys moving with deafferented limbs to visually instructed targets, the animals reacted to the same type of perturbation by moving back to a "planned" trajectory and reaching the target again

---

[1] More precisely, the phase sum is unstable while the phase difference is stable.

later by active movement. Curiously, this effect can also be obtained in the model if the active movement state is stabilized by intention while the perturbation acts. In that case the system starts to move back to the initial position similar to the experimental result. However, what is clearly needed here is to deal more rationally with the way mechanical perturbations act on the movement dynamics. Sensory feedback to the central nervous system must play a role in this process, and it is not surprising, of course, that deafferented animals show most strongly a level of movement planning that is little affected by external perturbations (see also Saltzman & Kelso, 1987, for a similar concern). A second motivation to deal with the meaning of external mechanical loads for the nervous system coordination dynamics is to try to link the theoretical work here with the insights about the dynamical properties of the muscle–joint system as epitomized by the mass-spring model (Feldman, 1966, 1986).

III. Toward a Dynamic Theory of Trajectory Formation

Typical approaches toward trajectory formation are based on a structural analysis of the peripheral effector system and its control, and on information-processing analysis of the central aspects of motion planning. Structural levels of description are the biomechanics of effectors, the muscle–joint system, and spinal, cerebellar, and cortical levels of motor control. In terms of information processing, the problems of transformations among coordinate systems (retinal, head-centered, body-centered, effector-centered, object-centered) and motor programming have received special attention. Recently, there is growing awareness that the structural levels of analysis are much less clearly demarcated and in the functional state intimately linked into elementary motor–sensory loops (Evarts, 1984). We shall argue below that levels of motor planning and coordinate transforms are similarly closely interwoven.

A. Levels

Here we propose to follow through on these insights and perform a strictly behavioral (or functional) analysis of trajectory formation. For the present purposes we distinguish three levels of such a behavioral analysis:

1. Load level. The level at which effectors interact with external mechanical forces (which we may take to include biomechanical forces such as those arising from gravity, inertia, and coriolis effects) (see Hasan, 1991, for review). Structurally, this level relates to biome-

chanics and muscle–joint synergisms as described by the mass-spring model, but also sensory feedback on all levels, including transcortical loops insofar as such sensory feedback is involved in the behavioral response to mechanical loads. The formal definition of this level is given by choice of (collective) variable: endeffector position, $\mathbf{x}$ and velocity, $\dot{\phi}$ This means that we restrict ourselves, for the present purposes to describing the effect of loads that act on the endeffector. These variables are collective with respect to the coordination of joints, for instance. The dynamical description at this level makes use of the mass-spring model and is parameterized by equilibrium points, and inertial, stiffness, and viscosity matrices.

2. Timed trajectories. The level at which temporal order inherent to posture, and discrete and rhythmic movement is generated. Again, structurally, contributions across all levels are thinkable. Our formal definition is to consider virtual trajectories of equilibrium points (e.g., Flash, 1990; Latash & Gottlieb, 1991; Latash, 1992). These variables act as parameters on the load level. Essentially, the dynamics considered earlier (Schöner, 1990; see above) to model posture and discrete and rhythmical movement are the pattern dynamics at this level. Interlimb coordination (see above), as well as the generation of temporal coherence between action and perception (Schöner, 1991), take place at this level in the form of dynamical coupling among different components and coupling to behavioral information, respectively. The dynamics are parameterized by spatial movement parameters such as target positions, amplitudes, and path parameters, as well as by global temporal parameters such as movement rate or frequency.

3. Setting movement goals. The level at which movement goals such as target positions, amplitudes, and curve form are determined. Structurally, this involves, in particular, the visual system and its neurophysiological substrate, but may include as well all other sensory systems and, through reafference, motor systems. The variables at this level are the parameters of the previous timing level. In this chapter we shall not elaborate on possible dynamics at this level and consider only stationary states. For example, a movement target will be assumed to be given (not be behaviorally determined such as in a mental rotation task, for instance). Therefore, we do not at the moment further parameterize this level.

For the present purposes we restrict our attention to one- and two-dimensional endeffector motion. By treating the goal-setting level only as providing constants, we are left with a two-level dynamical system. For a somewhat related mathematical model of locomotory gait movements see Taga, Yamaguchi, and Shimizu (1991). The so-called network coupling task dynamic model in Saltzman and Kelso (1987) is formally also quite similar. Note, however, that in these cases the definition of levels is given in structural terms rather than through behavioral operations.

B. Coordinate Systems

To concretely formulate dynamical models we must make choices about coordinate systems. Note that in the structural analysis coordinate systems play an important role because associations are made between particular structures (e.g., cortical areas) and particular coordinate systems (see, e.g., Paillard, 1991; Wise & Desimone, 1988; Kalaska & Crammond, 1992). By contrast, in the present approach the choice of variables is the important step, while the coordinate systems in which these variables are expressed are a *matter of convenience* only (but may be guided by neurophysiological insight where available). Without mapping coordinate systems onto structures and thus making them observable, coordinate systems do not have any operational meaning of their own.

In the following, we use the notational convention that vectors have upper indices to designate which coordinate system their coefficients refer to, and lower indices to identify different vectors. For vectors we use lower case, boldface letters and for matrices, upper case, boldface or Greek letters. We introduce three coordinate systems (see Figure 1):

1. Body-centered coordinates (bcc): These are world coordinates, not linked to a sensor surface, for instance, independent of eye and head position. The shoulders serve as reference points. Tasks are presented in this reference frame. Neurophysiologically, these coordinates may be reflected in the body map as observable, for instance, in parietal cortex. Notation: $r^b = (x^b, y^b)$.

2. Load coordinate system (lc): The lc system is centered in a posture position of the endeffector as defined, for instance, by the initial condition of a goal-directed movement. The lc system is oriented such that one axis points toward the body center. The idea is that in this system the stiffness tensor of the muscle–joint system is diagonal for most points in work space and varies smoothly and weakly as a function of the position in work space (Mussa-Ivaldi, Hogan, & Bizzi, 1985). Notation: $\mathbf{r}^l = (x^l, y^l)$, parameterized by the "initial" posture position, $\mathbf{r}^b_{init}$ in bcc. In polar coordinates:

$$\alpha_{init} = \arctan\left(\frac{y^b_{init}}{x^b_{init}}\right) \qquad r_{init} = |\mathbf{r}^b_{init}|. \tag{1}$$

Transformation bcc ↔ lc

$$\mathbf{r}^l = \mathbf{T}(\alpha_{init})(\mathbf{r}^b - \mathbf{r}^b_{init}) \tag{2}$$

$$\mathbf{r}^b = \mathbf{r}^b_{init} + \mathbf{T}^T(\alpha_{init})\mathbf{r}^l \tag{3}$$

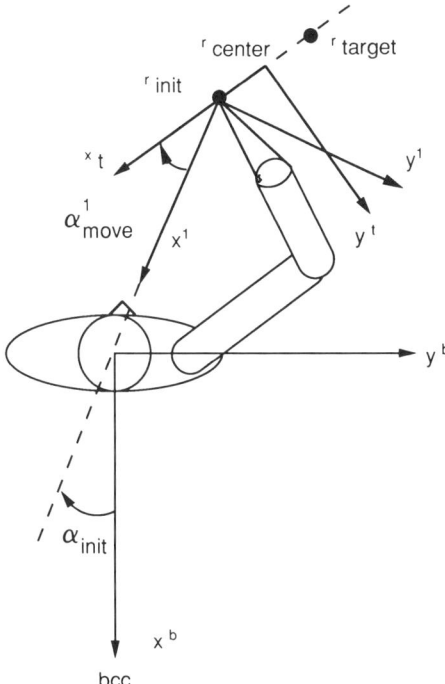

FIGURE 1 The coordinate systems used to formulate dynamical models: (1) the body-centered coordinate system (bcc) is fixed to the shoulder; (2) the load coordinate system (lc) is centered at a posture position of the endeffector, r_{init}, and its x-axis is oriented toward the body center; (3) the task coordinate system (tc) is centered at the midpoint of a discrete or rhythmic movement task and its x-axis is oriented toward the initial posture position of the endeffector.

where

$$\mathbf{T}(\alpha) = \begin{pmatrix} \cos(\alpha) & \sin(\alpha) \\ -\sin(\alpha) & \cos(\alpha) \end{pmatrix} \quad (4)$$

and T indicates the transpose.

3. Task coordinate system (tc): We define the tc system from the point of view of the lc system for the three tasks posture, discrete movement, and rhythmic movement.

(a) In posture tasks, tc is identical to lc.

(b) In a discrete movement task the lc system is centered at the initial posture position, and the target position as viewed from the origin of the lc system defines the (negative) x-axis of the tc system. To express symmetry between target and initial posture the y-axis is shifted such that the origin lies halfway between initial and target postures. A neurophysiological motivation is that movement direction is represented relative to the position of the shoulder (Caminiti, Johnson, & Urbano 1990) in motor cortex. This is captured here approximately by the angle α_{move} between the lc and tc coordinate systems. For the parameterization note:

$$\alpha_{\text{move}} = -\alpha_{\text{init}} + \arctan\left(\frac{y^b_{\text{target}} - y^b_{\text{init}}}{x^b_{\text{target}} - x^b_{\text{init}}}\right) \quad (5)$$

$$\mathbf{r}^b_{\text{center}} = (\mathbf{r}^b_{\text{target}} + \mathbf{r}^b_{\text{init}})/2. \quad (6)$$

(c) In rhythmic movement tasks, the tc is again centered around a midpoint reflecting some form of symmetry. The axes are defined by the principle axes of the curve—here, ellipse—so that in tc the curve can be parameterized in a simple manner.
Notation: $\mathbf{r}^t = (x^t, y^t)$.
Transformation lc ↔ tc:

$$\mathbf{r}^t = \mathbf{T}(\alpha_{\text{move}})(\mathbf{r}^l - \mathbf{r}^l_{\text{center}}) \quad (7)$$

$$\mathbf{r}^l = \mathbf{r}^l_{\text{center}} + \mathbf{T}^T(\alpha_{\text{move}})\mathbf{r}^t \quad (8)$$

where $\mathbf{T}(\alpha)$ is given by Equation (4).
Transformation tc ↔ bcc:

$$\mathbf{r}^b = \mathbf{T}^T(\alpha_{\text{init}})\mathbf{T}^T(\alpha_{\text{move}})\mathbf{r}^t + \mathbf{r}^b_{\text{center}} \quad (9)$$

$$\mathbf{r}^t = \mathbf{T}(\alpha_{\text{move}})\mathbf{T}(\alpha_{\text{init}})(\mathbf{r}^b - \mathbf{r}^b_{\text{center}}). \quad (10)$$

C. Variables

In terms of these coordinate systems we formalize the choice of collective variables:

1. Load level: physical endeffector position and velocity, $(\mathbf{r}, \dot{\mathbf{r}})$ (no lower index)
2. Timing level: virtual trajectory, $(\mathbf{r}_p, \dot{\mathbf{r}}_p)$ (lower index p for "planned")
3. Goal-setting level: we consider the variables at this level as time-invariant parameters here and define these separately for three tasks:
   (a) posture: posture position, $\mathbf{r}_{\text{init}}$.
   (b) goal-directed movement: initial posture position, $\mathbf{r}_{\text{init}}$, and target posture position, $\mathbf{r}_{\text{target}}$, from which movement direction, α_{move} (cf. Eq. [5]) and amplitude, $A = |\mathbf{r}_{\text{final}} - \mathbf{r}_{\text{init}}|$, can be determined. We consider movement time to be determined by an amplitude–movement time relationship modeling Fitts's law.
   (c) rhythmic movement: here we consider the drawing of ellipses parameterized by their center point, $\mathbf{r}_{\text{init}} = \mathbf{r}_{\text{center}}$ (at which

both lc and tc systems are centered), the orientation of the principle axis, α_{ellipse} relative to the x-axis of the bcc system ($\alpha_{\text{move}} = \alpha_{\text{ellipse}} - \alpha_{\text{init}}$), and the two half-axes, (A_x, A_y).

D. Dynamics

1. The physical endeffector dynamics: We assume that the effective dynamics are of the mass-spring type. This means that we extrapolate the known mass-spring properties of individual muscle–joint systems (Feldman, 1966, 1986) to a more abstract form of a mass-spring model, which is defined in endeffector spatial variables, not in muscular or joint variables. This is consistent with both the observations of Mussa-Ivaldi *et al.* (1985) that parameters of such a spatial mass-spring model are well defined and vary slowly in work space, and recent reports identifying spinal neurons that code for spatial equilibrium points in the multijoint frog hindleg (Bizzi, Mussa-Ivaldi, & Giszter, 1991). The mathematical model is given in the lc system in which the parameters take a particularly simple form. Deviating from the typical models, we assume not only that the virtual trajectory defines an instantaneous equilibrium point, $\mathbf{r}_p$, but also an instantaneous movement velocity, $\dot{\mathbf{r}}_p$ by coupling $\dot{\mathbf{r}}_p$ into the viscous term:

$$\mathbf{M} \cdot \ddot{\mathbf{r}}^l = -\mu^l \cdot (\dot{\mathbf{r}}^l - \dot{\mathbf{r}}_p^l) - \mathbf{K}^l \cdot (\mathbf{r}^l - \mathbf{r}_p^l) + nonlinearities\ (\mathbf{r}^l, \dot{\mathbf{r}}^l) + \mathbf{q}^l \xi^l(t) + \mathbf{f}_{\text{extern}}^l. \quad (11)$$

In this formulation the timing level variables act as behavioral information onto the load level. The extra velocity coupling leads to a near in-phase relationship between virtual and real trajectory (see below).

In this equation the inertial tensor $\mathbf{M}^l$, the viscosity tensor, μ^l, and the stiffness tensor $\mathbf{K}^l$ are all 2×2 matrices in the lc system. We interpret Mussa-Ivaldi *et al.*'s (1985) result by approximating the stiffness tensor as diagonal: $K_{12}^l = K_{21}^l = 0$. In the simulations below we have taken the inertial tensors to be diagonal, although Mussa-Ivaldi *et al.* (1985) showed that the main axes of the inertial tensor are rotated by between 20 and 40° from those of the stiffness tensor. In the absence of any experimental information of the viscosity tensor, we have for simplicity also set that diagonal. Note that experimental estimates of these parameter tensors show that the parameters values do depend on the position of the equilibrium posture in work space. Here, we neglect this dependence, which is a good approximation for relatively small movement amplitudes because the tensors change slowly in work space. Finally, we neglect nonlinearities even though these are known to occur. The external forces

$$\mathbf{f}_{\text{extern}}^l = \mathbf{T}(\alpha_{\text{init}}) \mathbf{f}_{\text{extern}}^b \quad (12)$$

are the physical forces. Stochastic forces have been included as Gaussian white noise sources, independent for each component with variances $\mathbf{q}^l = (q_1^l, q_2^l)$.

2. The timing dynamics: For each component we define four contributions to the dynamics: (a) intrinsic dynamics poses for different parameter regions a single fixed point, two fixed points, and a limit cycle, and capture the basic tasks posture, discrete movement, and rhythmic movement; (b) behavioral information stabilizes limit cycle motion over a limited time interval and describes the intentional initiation of movement; (c) coupling among the different spatial components stabilizes the phase relationship of $\pi/2$, capturing the capability to perform smooth two-dimensional curves; (d) the load level is assumed to couple into the timing dynamics, generating a tendency to synchronize real and virtual trajectory. Mathematically, the first two parts are chosen as in earlier work (Schöner, 1990).

$$\ddot{\mathbf{r}}_p^t = \begin{pmatrix} f_{\text{intr},x}(x_p^t, \dot{x}_p^t) \\ f_{\text{intr},y}(y_p^t, \dot{y}_p^t) \end{pmatrix} + c_{\text{int}} \chi[t_0, t_0 + \Delta t](t) \mathbf{r}_p^t + \begin{pmatrix} c_{xy} y_p^t \\ c_{yx} x_p^t \end{pmatrix} + \begin{pmatrix} c_x x^t \\ c_y y^t \end{pmatrix} + \mathbf{q}_p^t \zeta_p^t(t). \tag{13}$$

Herein the intrinsic contribution ($i = x, y$)

$$f_{\text{intr},i}(x, \dot{x}) = -(a_i^2 + \omega_i^2)x + 2a_i \dot{x} - 4b_i x^2 \dot{x} + 2a_i b_i x^3 - b_i^2 x^5 \tag{14}$$

defines an exactly solvable nonlinear dynamical system (Gonzalez & Piro, 1987). Its phase diagram is sketched in Figure 2. The three model parameters are mapped onto task parameters as follows (cf. Schöner, 1990):

- ω_i^2 determines rate of movement, for instance, as cycle time $T = 2\pi/\omega_i$ for the limit cycle. Note that this parameter can both be positive and negative (corresponding to purely imaginary ω_i).
- a_i determines stability, for instance, as relaxation time $\tau_{\text{rel}} = 1/(2a_i)$ for the limit cycle.
- b allows control of the spatial extent of movement, for instance, as amplitude $A_i = 2\sqrt{2a/b}$.

The strength, c_{int} of the intention to move matters little, as does the time interval Δt over which this signal acts. (The χ-function is one in the time interval $[t_0, t_0 + \Delta t]$ and zero elsewhere.) The coupling functions are parameterized by c_{xy} and c_{yx} and are only effective in the case of rhythmic movement, because in all other cases the amplitude of the y component in the tc system is zero. The coupling to the load level is parameterized by (c_x, c_y). Finally, $\mathbf{q}_p^t$ parameterizes the noise term. The

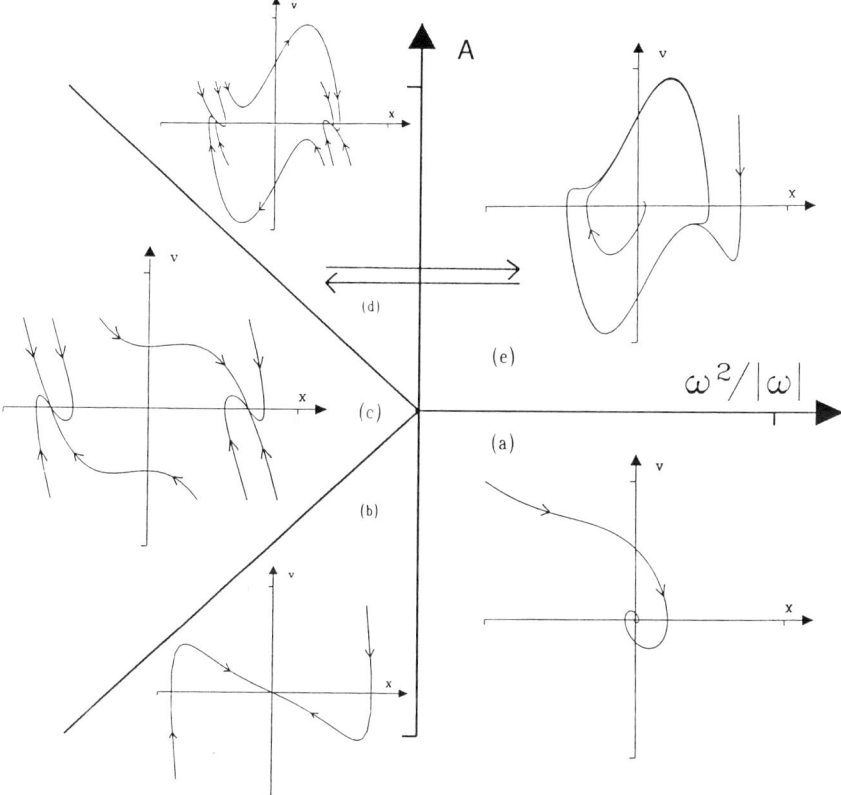

FIGURE 2 The phase diagram of the nonlinear dynamical system Equation (14) used to model the timing level of trajectory generation contains three regions (or five subregions): (1) In the lower right (regions [a] and [b]) a single, globally stable fixed point exists; (2) in the upper left (regions [c] and [d]) two stable fixed points exist; (3) in the upper right (region [e]) a globally stable limit cycle exists. The three regions are used to model posture, discrete movement, and rhythmic movement, respectively. The phase diagram depends only on the two model parameters, a and ω, shown here as long as the third parameter, b, is positive. The arrows indicate how the system is shifted from one region to the other by behavioral information modeling the intention to move.

Gaussian white noise sources in all equations are assumed independent.[2]

---

[2] We note that the dynamics of the network coupling task dynamic model of Saltzman and Kelso (1987) are formally similar although the two levels are defined differently. Their effort is to lift the mass–spring model from muscle–joint systems to a higher task level and also consider limit cycle dynamics at that level in an *ad hoc* fashion. Note, however, that this strategy involves inconsistent treatment of discrete and rhythmic movement: While in rhythmic movement, the trajectory is an attractor; in discrete movement only the endpoint is assumed stable. The initial posture is treated as an initial condition, which is therefore forgotten once the system has started to move. This is clearly not realistic in a behavioral analysis.

The concrete mathematical form of the dynamical models at the two levels was chosen such that model parameters can be mapped in a relatively direct way onto observable parameters in experiments that manipulate the adequate behavioral constraint. For the timing level this is apparent from the brief discussion of the parameters of the intrinsic dynamics, which can be determined on the basis of measured movement times, relaxation times, and amplitudes (cf. Schöner, 1990, for examples). For the load level, the measured stiffness and inertial tensors can, in principle, be directly used (less information is available on viscosity, but see Latash & Gottlieb, 1991, for estimates). The strategy in this chapter has been to choose a fixed set of rather generic parameter values that reproduce roughly the correct orders of magnitudes for most observables. No effort has been made to optimize the fit of the parameters. Instead, the emphasis is on effects of a qualitative nature that arise irrespective of the exact functional form of the models and the exact set of parameter values chosen. The results displayed below are all based on the following choices: Timing dynamics ($i = x,y$); for discrete movement: $a_i = 10$ Hz, $\omega_i^2 = -3$ Hz^2, $b_i = 1$ Hz/cm^2, $c_{\text{int}} = 20$ Hz, $c_{xy} = c_{yx} = -10$ Hz^2, $c_x = c_y = -10$ Hz (except Figure 4). These choices lead to movement times of about 600–800 msec with realistic stability properties (cf. Schöner, 1990). For rhythmic movement: same except $\omega_i^2 = 100$ Hz^2 leading to movement frequencies of about 1.6 Hz. Load level dynamics: $M_{1\,1} = M_{2\,2} = 1$, $K^l_{1\,1} = K^l_{2\,2} = 49$ Hz^2, $\mu^l_{1\,1} = \mu^l_{2\,2} = 14$ Hz. Here we have set the units of force to unity, set stiffness and viscosity to the critical damping ratio, and chosen a mechanical eigenfrequency $\nu = \sqrt{K/M}/(2\pi) \approx 1.1$ Hz, close to the movement frequencies generated at the timing level (cf. Latash, 1992, for experimental support). The noise levels for both dynamics were $q = 0.01$ Hz.

E. Results: One-Dimensional

We first consider again the case of one-dimensional movements and discuss how treating the load level increases the range of validity of the theory over that of Schöner (1990). Figure 3 shows virtual and real trajectory for a discrete and a rhythmic movement task. First, the shape of the trajectory becomes smoother and more sinusoidal by the load level, that is, the load level acts essentially as a low-pass filter. Structurally, this is in part achieved by the passive viscoelastic properties of the muscle–joint system (see also Feldman, Adamovich, & Ostry, Flanagan, 1990, for a discussion of smoothness in relation to the mass-spring model). For discrete movement the shape may approximate the experimentally observed Gaussian form. Note that the trajectory shape

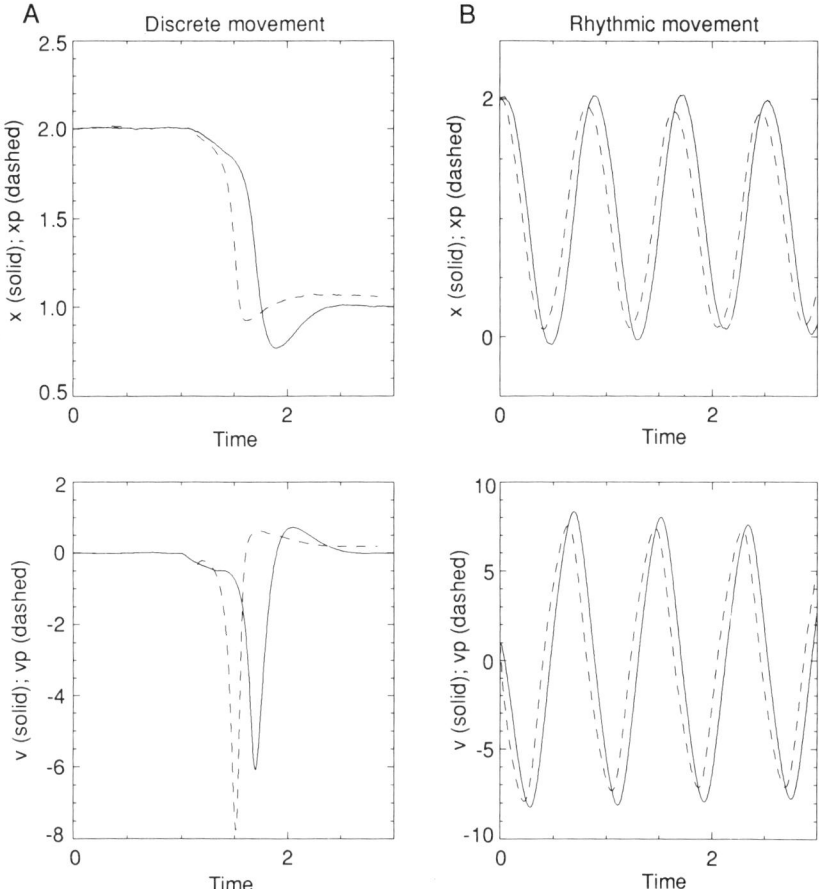

FIGURE 3 Discrete (A) and rhythmic (B) movement trajectories as generated by the 2-layer dynamical model are shown. In each case the load level real endeffector position (*top*) and velocity (*bottom*) are shown as solid lines and the corresponding data for the timing level virtual trajectories are shown as dashed lines. Note (1) that the load level smooths the virtual trajectories, and (2) that the two layers are approximately in-phase with a slight phase lead of the virtual trajectory. Overall, the trajectory shape is not necessarily realistic, because the timing level dynamics was not chosen to fit experimental data, but to ease analysis.

at the level of the virtual trajectory is more or less arbitrary here because the dynamics were chosen purely for analytical convenience. More realistic descriptions can be achieved by fitting dynamics at that level to experimental estimates of virtual trajectories (Latash & Gottlieb, 1991; Latash, 1992).

The second observation is that the two layers are almost in-phase. This is due to our assumption of both elastic and viscous coupling among the two layers and thus accomodates recent experimental results by Latash (1992). Latash reconstructed virtual trajectories by applying ramp torques and regressing torque and instantaneous arm position. He found that virtual trajectories are either in-phase or anti-phase to the real trajectory.[3]

What happens now in the two-layer model under mechanical perturbation? In the limit of weak (back-)coupling of the load level into the timing dynamics ($c_i \ll a_i$) or of weak perturbation, the mechanical perturbation does not affect strongly the trajectory timing level. Hence, in discrete movement, return to the virtual trajectory is observed after a perturbation takes the system to its eventual target position (see Figure 4A). This captures the results of the monkey experiments with deafferented limbs (Bizzi, Accornero, Chapple, & Hogan, 1982). The approximation of weak back-coupling is appropriate here because deafferentiation is liable to reduce the coupling between the two behavioral levels. In the case of rhythmic movement, no phase resetting occurs, as illustrated in Figure 4C. Such lack of phase resetting was observed by Kay et al. (1991) for weak perturbation and during some phases of the movement. In the other limit with sufficient strength of back-coupling, the timing layer is affected by the perturbations and the timing variable is shifted in its phase. This is shown in (A) and (B) of Figure 5. For discrete movement this corresponds to the more typical behavior observed readily in humans, for instance. In the case of rhythmic movement, Kay et al. (1991) reported phase resetting by external torques as averaged over all phases of the movement.[4]

Feldman (1980) has reported experiments in which he unloaded a limb during rhythmic movement while the subject was under the instruction not to interfere with the subsequent shift of the trajectory.

---

[3] The type of coupling suggested here would lead to a somewhat different estimate of the virtual trajectory, requiring regression against velocity as well. In that case the antiphase pattern may disappear.

[4] Kay et al. (1991) report that an average phase advance results, which cannot be explained by a direct insertion of extern forces into a coordination dynamics. Such an average advance reflects an asymmetry of the coupling function between the two layers such that both speeding up the endeffector and slowing down the endeffector advances the timing layer. The linear coupling considered here does not reproduce this result, but an additional nonlinear coupling, for example, of the form $(\dot{x}^t)^2$, does.

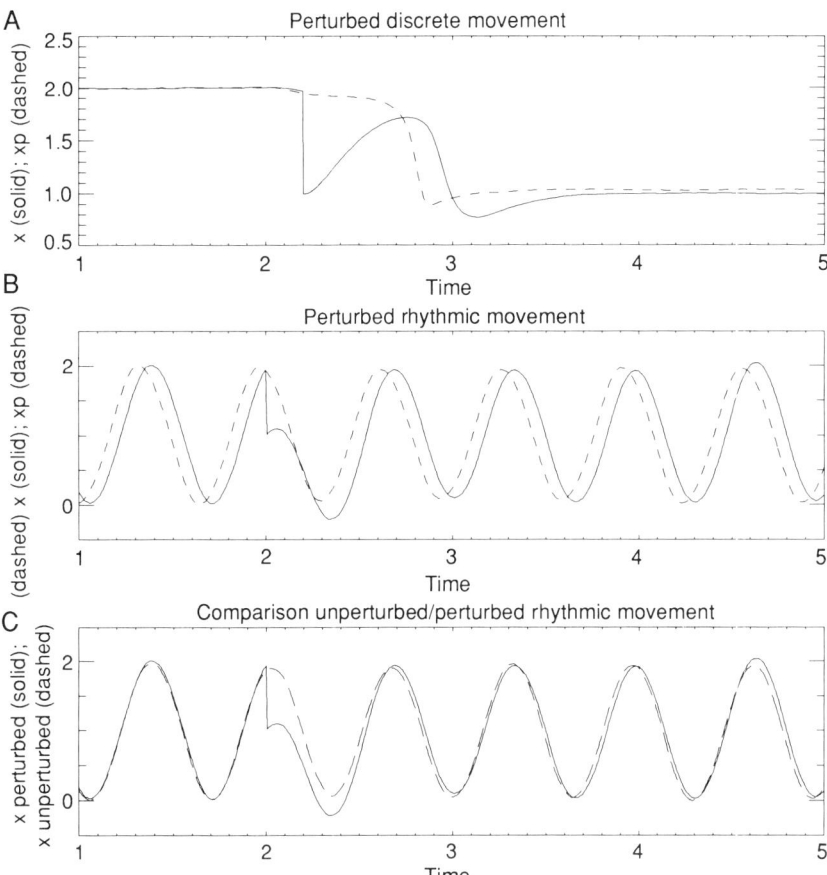

FIGURE 4 The effect of perturbations on discrete (A) and rhythmic (B, C) movement. Here, the back-coupling of the load level to the timing level is assumed weak ($c_i = -1$). For discrete movement, the virtual trajectory (dashed line) is little affected by the perturbation, which advances the real trajectory toward its target state. As a result, the endeffector returns to the virtual trajectory, thus moving away from its target state only to return later. In rhythmic movement the virtual trajectory is likewise only weakly affected (B, dashed line). As a result, there is little phase resetting, as shown in (C), where the perturbed real trajectory (solid line) is plotted jointly with a trajectory from an unperturbed simulation (dashed line). The perturbed trajectory returns to its old phase after the perturbation because of the influence of the weakly perturbed virtual trajectory.

The model correctly describes this observation (Figure 6). Only the unloading itself may act as a perturbation shifting the phase of the overall motion, but the phase between the two layers remains stably at in-phase. This is consistent with Feldman's (1980) observation that there was no phase shift between the loaded and the unloaded segment.

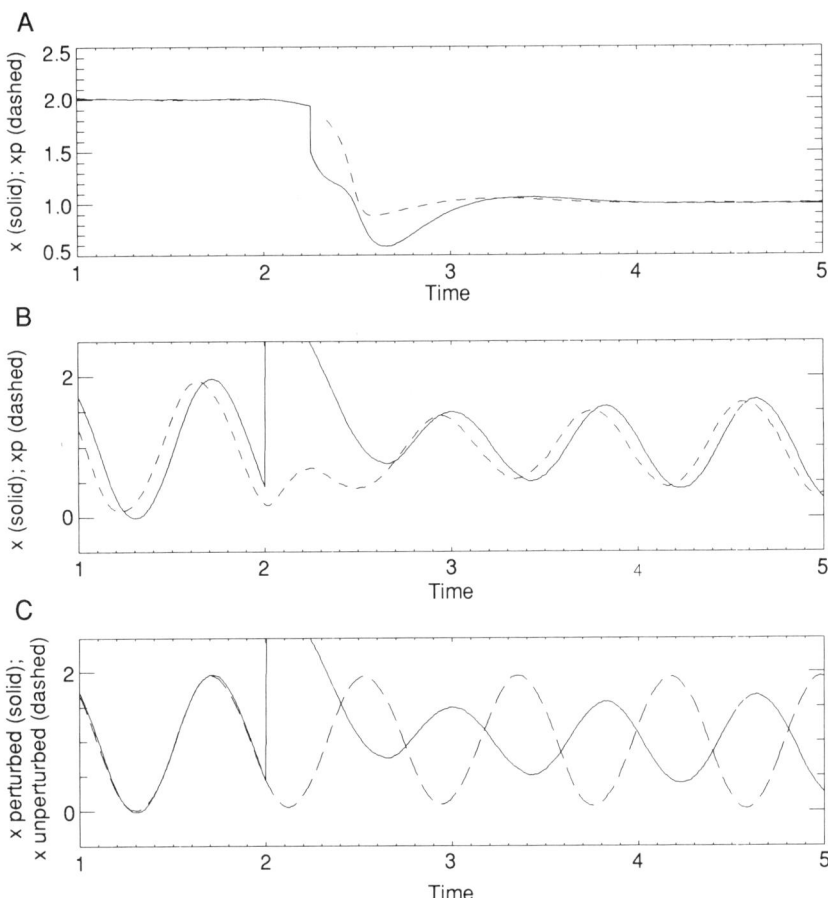

FIGURE 5 The effect of perturbations on discrete (A) and rhythmic (B, C) movement is shown in the limit case of strong back-coupling of the load level to the timing level ($c_i = -10$), or, equivalently, of strong perturbation. In (A) note that the virtual trajectory (dashed line) is advanced considerably in its phase because of the mechanical perturbation shifting the real endeffector position toward its target state (cf. Figure 4). Likewise, for rhythmic movement the virtual trajectory is strongly perturbed by the mechanical perturbation applied at the load level (B, dashed line). As a result, the phase is reset: In (C) the phase relation between an unperturbed endeffector trajectory (dashed line) and the perturbed trajectory (solid line) remains nonzero after the perturbation.

The analysis of these perturbation experiments illustrates that a stable pattern is formed between the two levels of trajectory formation. In analogy to our earlier remarks on the stable coupling among limbs, this stability can likewise be observed also in the absence of external perturbations. Fluctuations in relative timing between the two layers are smaller than the fluctuations in absolute timing of any layer. An

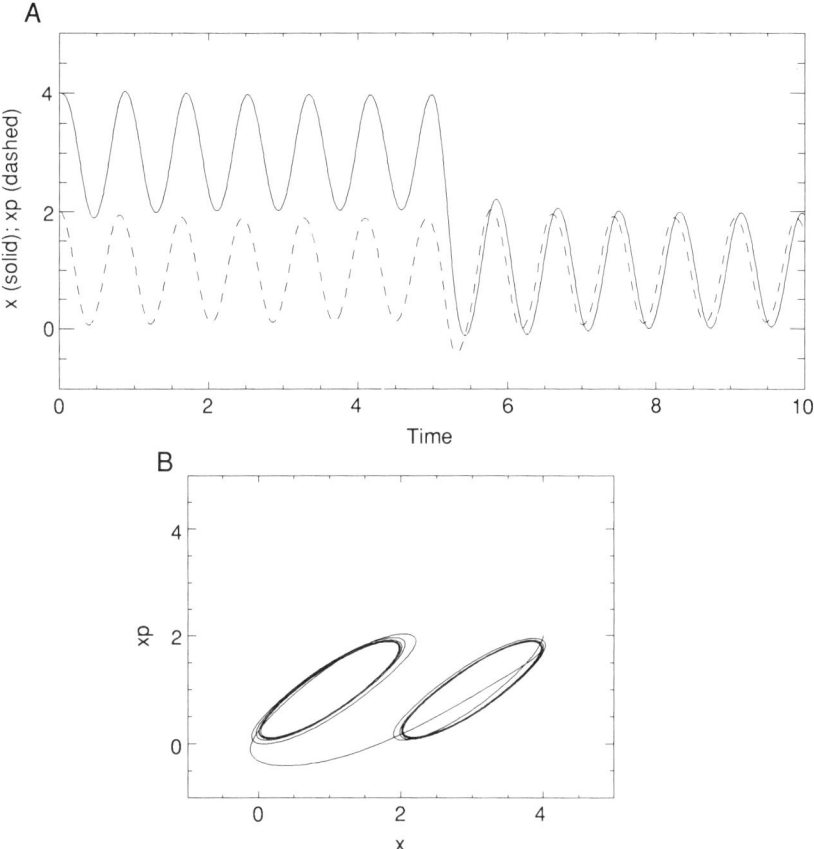

FIGURE 6 Unloading of a constant mechanical load during rhythmic movement leads to a shift of the center of oscillation. The relative phase between the two levels is little affected, however. This can be seen both from the constant relative timing in (A) (real trajectory, solid line; virtual trajectory, dashed line) as well as in the Lissajous plot of the virtual versus the real trajectory in (B). The position of the ellipse shifts, but its excentricity—reflecting relative phase between the two layers—remains constant.

observable consequence of this phenomenon relates to a model by Wing and Kristofferson (1973). These authors assumed that two sources of variability of cycle time in rhythmic movement can be distinguished: fluctuations of an underlying clock and fluctuations of the motor system executing clock commands. This model predicts anticorrelation between subsequent cycle times (lag-one cycle time autocorrelation) that lies in the range between -0.5 and 0.0 if the correlation function is normalized by cycle time variance. In the dynamic model presented here it would be tempting to view the two dynamic layers as implementations of the two putative underlying sources of error. The stochastic perturbations applied to each level would generate the two types of variabilities. In fact, one can demonstrate that the real endeffector motion possesses the observed anticorrelation of subsequent cycle times. This is true over the entire available parameter range (with the anticorrelation converging to 0.0 as we reduce the noise strength on the load level). As a note of caution we point out, however, that even a single-layer oscillatory model can generate the observed anticorrelation! To test this we simulated the timing layer in isolation and considered its output to be the observed trajectory. If the relaxation time of the limit cycle is of the same order of magnitude as the cycle time then the relaxation processes generate—on average—anticorrelation. To understand this intuitively, consider a cycle that deviates from the limit cycle outwardly in the phase plane, so that the cycle is slower than the exact limit cycle. On the next cycle the system will have—on average—relaxed toward the limit cycle and will therefore have again a smaller cycle time. The reverse is true for a cycle fluctuating inward. As a result, subsequent cycle times are anticorrelated. As it appears, the condition relaxation time approximately equal to the cycle time is typically fulfilled in voluntary rhythmic movement (see, e.g., Kay et al., 1991). This is also reflected, for instance, by the critical damping observed when the parameters of mass-spring models are fitted (Mussa-Ivaldi et al., 1985). Hence, the observation of Wing and Kristofferson (1973) does not as such justify dissecting the system into the two putative processes of clock and motor system. In fact, in their derivation of how the two sources of timing variability can be identified from cycle time variability and lag-one autocorrelation of cycle time, these authors have neglected the lag-one autocorrelation of the clock rhythm itself (as well as of that of the motor error contributions). In the dynamic model, the timing level has nonzero lag-one autocorrelation, and under the conditions just discussed, this clock contribution alone can account for the observed negative lag-one autocorrelation of cycle time. The contribution from the clock dynamics disappears in the limit where relaxation is much faster than a cycle time (as may happen, for instance, for slow oscillatory motion). More generally, however, the observed anticorrelation is a dynamic property of the entire, fully coupled system. This issue clearly deserves further experimentation.

F. Results: Two-Dimensional

When we consider two spatial components, the statements about smoothing of the trajectory also affect the relationship between trajectory, $\mathbf{r}(t)$, and curve or path (the relation between the two components, x and y, induced by the trajectory). First, by construction, goal-directed movements generate straightline movement paths in the model. Rhythmic movements drawing out an ellipse generate roughly sinusoidal trajectories with a constant and stable relative phase between the two components (Figure 7) (the stability of the relative phase between spatial components being the point rather than the sinusoidal form of each individual component). This implies that these trajectories fulfill the

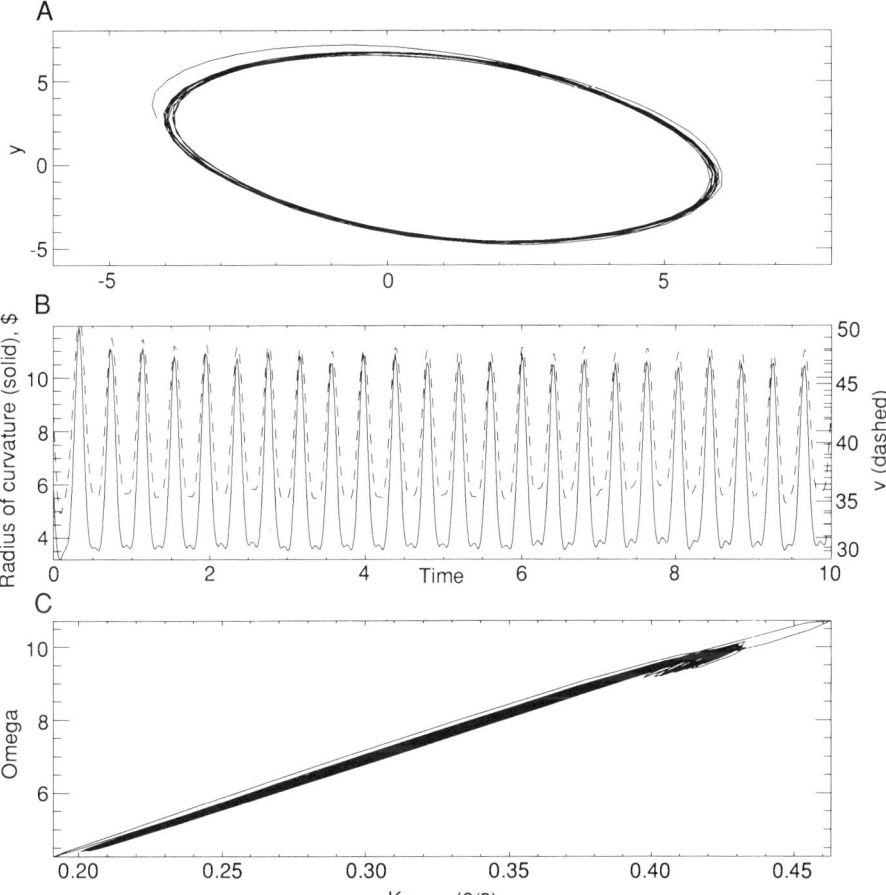

FIGURE 7 Two-dimensional trajectories are performed with stable relative timing between the spatial components. Hence, the curve in the plane approximates a definite shape (ellipse) that closes within one cycle (A). This constant and stable relative phase leads to the linkage between timing and geometry first noted in human motion by Viviani and collegues (for a review, see Viviani, 1990; Lacquaniti, 1989). In (B) this is shown by plotting the radius of curvature as a function of time (solid line) on top of the tangential velocity (dashed line). Note the close covariation. In (C), the two-thirds law is illustrated by plotting the angular velocity against the curvature raised to the 2/3 power. The linear function that emerges shows that the model trajectories do conform to the two-thirds law.

"2/3" scaling law in which curvature and angular velocity covary (Lacquaniti, Terzuolo, & Viviani, 1983; for a review, see Lacquaniti, 1989; Viviani, 1990). Again, the temporal relationship of the two spatial components is due to the formation of a stable pattern, so that (1) perturbations are compensated to restore the relative timing of these components, and (2) fluctuations of the two time series covary. Note that—in contrast to experiment—this is not the case in models that impose the relationship between different spatial components through initial conditions (e.g., Hollerbach, 1981).

IV. Discussion

In this chapter we have reviewed the dynamic approach to interlimb coordination and have sketched how a dynamic approach may address the problem of trajectory formation also of a single limb. The common principles that emerge are summarized as follows: (1) Relevant components of a movement system are determined by task constraints, not necessarily in terms of structure. Hence, interlimb coordination is in principle not different from the coordination among different task components in trajectory formation. (2) Even more abstract, the principles of pattern formation through dynamic coupling apply equally to interlimb coordination, to the coordination among multiple levels of movement behavior, and to the coordination of action and perception. In each case coordination is expressed as relative temporal order. (3) What the nervous system "plans" about a particular movement behavior may best be determined in terms of what it keeps stable against external or stochastic perturbations. Motor equivalence then means that relationships that are not stable patterns can take on a range of values, while the task-relevant relationships that are stable patterns return to their fixed value following external perturbation.

The brief sketch given in this chapter of the beginnings of a dynamic theory of trajectory formation stressed that the definition of levels of description should be strictly operationalized in terms of behavioral constraints. Only in this case is there a reasonable chance to identify separately different parts of what ultimately is a fully coupled dynamical structure. Essentially, the operational definitions refer to limit cases in which coupling among levels and among degrees of freedom within levels is weak. For instance, the behavioral constraint of using one limb versus using two limbs can be used to separately identify, as sketched, dynamic principles of trajectory formation in individual limbs, and to study dynamic coupling among such modules as those in the dynamic models of interlimb coordination. This segregation is successful only where the dynamic coupling between limbs is weak in relation to the dynamics of intralimb coordination. In other cases, no operational distinction between intra- and interlimb coordination is possible. Another limit case, where coupling occurs only at one level of behavioral constraint, the timing level, with negligible coupling at the load level as well as at the movement goal-setting level, could be defined as pure interlimb coordination. This is the limit case studied previously within the framework of the dynamic approach (see Schöner & Kelso, 1988a, for review).

An apparent difference between interlimb coordination and trajectory formation remains: we can readily list many examples of multistable interlimb coordination systems, while it is more difficult to find similar flexibility in trajectory formation. In the laboratory this may be due to our choice of experimental paradigms; in interlimb coordination we often look for spontaneously emerging coordinations without specifying from the outside a particular pattern. In trajectory formation the behavioral goal is usually fixed from the outside and is unique. In nature, multistability occurs as a form of flexibility. For instance, horses do not use their different gaits (only) for their enjoyment, but each gait has its own functional characteristics (e.g., energetically, see Hoyt & Taylor, 1981). The fact that these gaits overlap in the parameters' ranges enables the system to maintain stable performance in the face of fluctuating demands. Imagine if there were no bistable range of trot and gallop: at particular speeds, horses would locomote unstably, oscillating stochastically between the different gaits. It is quite possible that this form of flexibility can also be discovered for trajectory formation (consider, for instance, the different forms of swiping movements performed by spinal frogs; Berkenblit, Feldman, & Fukson, 1986).

Acknowledgments

Support through BMFT (Bonn) (01IN101A6), MWF (Düsseldorf), and DFG (Bonn) (Di/Sch 334/5-1) is acknowledged. The initial impetus for this study came from discussions with Anatol Feldman. Discussions with Scott Kelso and Mark Latash were helpful. Axel Steinhage provided programming support. Two referees made useful suggestions that improved the manuscript.

References

Berkenblit, M. B., Feldman, A. G., & Fukson, O. I. (1986). Adaptability of innate motor patterns and motor control mechanisms. *Behavioral and Brain Sciences*, **9**, 585–638.
Bernstein, N. A. (1947/1967) *The coordination and regulation of movements*. London: Pergamon Press.

Bizzi, E., & Mussa-Ivaldi, F. A. (1990). Muscle properties and the control of arm movement. In D. N. Osheron, S. M. Kosslyn, & J. M. Hollerbach (Eds.), *Visual cognition and action* (pp. 213–242). Cambridge, MA: The MIT Press.

Bizzi, E., Accornero, N., Chapple, N., & Hogan, N. (1982). Arm trajectory formation in monkeys. *Experimental Brain Research, 46,* 139–143.

Bizzi, E., Mussa-Ivaldi, F. A., & Giszter, S. (1991). Computations underlying the execution of movement: A biological perspective. *Science, 253,* 287–291.

Caminiti, R., Johnson, P. B., & Urbano, A. (1990). Making arm movements within different parts of space: dynamic aspects in the primate motor cortex. *Journal of Neuroscience, 10,* 2039–2058.

Delcomyn, F. (1991). Perturbation of the motor system in freely walking cockroaches. i. rear leg amputation and the timing of motor activity in leg muscles. *Journal of Experimental Biology, 156,* 483–502.

Dijkstra, T., Schöner, G., & Gielen, C. (1993). Temporal stability of the action-perception cycle for postural control in a moving visual environment. *Experimental Brain Research* (in press).

Evarts, E. V. (1984). Hierarchies and emergent features in motor control. In G. M. Edelman, W. E. Gall, & W. M. Cowan (Eds.), *Dynamical aspects of neocortical function* (pp. 557–579). New York: Wiley.

Feldman, A. G. (1966). Functional tuning of the nervous system during control of movement or maintenence of a steady posture. iii. mechanographic analysis of the execution by man of the simplest motor acts. *Biofizika, 11,* 667–675.

Feldman, A. G. (1980). Superposition of motor programs. i. rhythmic forearm movements in man. *Neuroscience, 5,* 81–90.

Feldman, A. G. (1986). Once more on the equilibrium point hypothesis (λ-model) for motor control. *Journal of Motor Behavior, 18,* 15–54.

Feldman, A. G., Adamovich, S. V., Ostry, D. J., & Flanagan, J. R. (1990). The origin of electromyograms—explanations based on the equilibrium point hypothesis. In J. M. Winters & S. L.-Y. Woo (Eds.), *Multiple muscle systems* (pp. 195–213). Berlin: Springer-Verlag.

Fitts, P. M. (1954). The information capacity of the human motor system in controlling the amplitude of movement. *Journal of Experimental Psychology, 47,* 381–391.

Flash, T. (1990). The organization of human arm trajectory control. In J. M. Winters & S. L.-Y. Woo (Eds.), *Multiple muscle systems* (pp. 282–301). Berlin: Springer-Verlag.

Gardiner, C. W. (1983). *Handbook of stochastic methods for physics, chemistry and the natural sciences.* Berlin: Springer-Verlag.

Georgopoulos, A. P. (1991). Higher order motor control. *Annual Reviews of Neuroscience, 14,* 361–377.

Gonzalez, D. L., & Piro, O. (1987). Global bifurcations and phase portrait of an analytically solvable nonlinear oscillator: Relaxation oscillations and saddle-node collisions. *Physical Review,* A36, 4402–4410.

Gracco, V. L., & Abbs, J. H. (1986). Variant and invariant characteristics of speech movements. *Experimental Brain Research, 65,* 156–166.

Gracco, V. L., & Abbs, J. H. (1988). Central patterning of speech movements. *Experimental Brain Research, 71,* 515–526.

Guckenheimer, J., & Holmes, P. (1983). *Nonlinear oscillations, dynamical systems, and bifurcations of vector fields.* New York: Springer-Verlag.

Haken, H. (1983). *Synergetics—An introduction* (3rd edition). Berlin: Springer-Verlag.

Haken, H., Kelso, J. A. S., & Bunz, H. (1985). A theoretical model of phase transitions in human hand movements. *Biological Cybernetics, 51,* 347–356.

Hasan, Z. (1991). Biomechanics and the study of multijoint movements. In D. R. Humphrey & H.-J. Freund (Eds.), *Motor control: Concepts and issues* (pp. 75–84). New York: Wiley.

Hollerbach, J. M. (1981). An oscillatory theory of handwriting. *Biological Cybernetics*, **39,** 139–156.
Hoyt, D. F., Taylor, C. R. (1981). Gait and energetics of locomotion in horses. *Nature* **292,** 239–240.
Kalaska, J. F., & Crammond, D. J. (1992). Cerebral cortical mechanisms of reaching movements. *Science*, **255,** 1517–1523.
Kay, B. A., Saltzman, E., & Kelso, J. A. S. (1991). Steady-state and perturbed rhythmical movements: A dynamical analysis. *JEP:HHP*, **17,** 183–197.
Kelso, J. A. S. (1984). Phase transitions and critical behavior in human bimanual coordination. *Am. J. Physiol: Reg. Integ. Comp.*, **15,** R1000–R1004.
Kelso, J. A. S., Southard, D. L., & Goodman, D. (1979). On the nature of human interlimb coordination. *Science*, **203,** 1029–1031.
Kelso, J. A. S., Holt, K. G., Kugler, P. N., & Turvey, M. T. (1980). On the concept of coordinative structures as dissipative structures. ii. empirical lines of convergence. In G. E. Stelmach & J. Requing (Eds.), *Turorials in motor behavior* (pp. 49–70). Amsterdam: North-Holland.
Kelso, J. A. S., Putnam, C. A., & Goodman, D. (1983). On the space-time structure of human interlimb coordination. *Quarterly Journal of Experimental Psychology*, **35A,** 347–375.
Kelso, J. A. S., Tuller, B. T., Vatikiotis-Bateson, E., & Fowler, C. A. (1984). Functionally specific articulatory cooperation following jaw perturbations during speech: Evidence for coordinative structures. *J. Exp. Psychol: Hum. Perc. Perf.*, **10,** 812–832.
Kelso, J. A. S., Saltzman, E. L., & Tuller, B. (1986a). The dynamical perspective on speech production: data and theory. *Journal of Phonetics*, **14,** 29–59.
Kelso, J. A. S., Scholz, J. P., & Schöner, G. (1986b). Nonequilibrium phase transitions in coordinated biological motion: critical fluctuations. *Physics Letters*, **A118,** 279–284.
Kelso, J. A. S., Delcolle, J. D., & Schöner, G. (1990). Action-perception as a pattern formation process. In M. Jeannerod (Ed.), *Attention and performance* (pp. 139–169). Hillsdale, NJ: Erlbaum.
Kugler, P. N., Kelso, J. A. S., & Turvey, M. T. (1980). On the concept of coordinative structures as dissipative structures: I. theoretical lines of convergence. In G. E. Stelmach & J. Requing (Eds.), *Tutorials in motor behavior* (pp. 1–47). Amsterdam: North-Holland.
Lacquaniti, F. (1989). Central representations of human limb movement as revealed by studies of drawing and handwriting. *Trends in Neuroscience*, **12,** 287–292.
Lacquaniti, F., Terzuolo, C., & Viviani, P. (1983). The law relating the kinematic and figural aspects of drawing movements. *Acta Psychologica*, **54,** 115–130.
Latash, M. L. (1992). Virtual trajectories, joint stiffness, and changes in the limb natural frequency during single-joint oscillatory movements. *Neuroscience*, **49,** 209–220.
Latash, M. L., & Gottlieb, G. L. (1991). Reconstruction of shifting elbow joint compliant characteristics during fast and slow movements. *Neuroscience*, **43,** 697–712.
Marteniuk, R. G., MacKenzie, C. L., & Baba, D. M. (1984). Bimanual movement control: information processing and interaction effects. *Quarterly Journal of Experimental Psychology*, **36A,** 335–365.
Mussa-Ivaldi, F. A., Hogan, N., & Bizzi, E. (1985). Neural, mechanical and geometric factors subserving arm posture in humans. *Journal of Neuroscience*, **5,** 2732–2743.
Paillard, J. (1991). Motor and representational framing of space. In J. Paillard (Ed.), *Brain and space* (pp. 163–182). Oxford: Oxford University Press.
Saltzman, E., & Kelso, J. A. S. (1987). Skilled actions: A task dynamic approach. *Psychological Review*, **94,** 84–106.
Saltzman, E. L., & Munhall, K. G. (1989). A dynamical approach to gestural patterning in speech production. *Ecological Psychology*, **1,** 333–382.

Scholz, J. P., & Kelso, J. A. S. (1989). A quantitative approach to understanding the formation and change of coordinated movement patterns. *Journal of Motor Behavior,* **21,** 122–144.

Scholz, J. P., Kelso, J. A. S., & Schöner, G. (1987). Nonequilibrium phase transitions in coordinated biological motion: critical slowing down and switching time. *Physics Letters,* **A123,** 390–394.

Schöner, G. (1990). A dynamic theory of coordination of discrete movement. *Biological Cybernetics,* **63,** 257–270.

Schöner, G. (1991). Dynamic theory of action-perception patterns: the "moving room" paradigm. *Biological Cybernetics,* **64,** 455–462.

Schöner, G., & Kelso, J. A. S. (1988a). Dynamic pattern generation in behavioral and neural systems. *Science,* **239,** 1513–1520.

Schöner, G., & Kelso, J. A. S. (1988b). A dynamic theory of behavioral change. *Journal of Theoretical Biology,* **135,** 501–524.

Schöner, G., & Kelso, J. A. S. (1988c). Synergetic theory of environmentally-specified and learned patterns of movement coordination. *Biological Cybernetics,* **58,** 71–89.

Schöner, G., Haken, H., & Kelso, J. A. S. (1986). A stochastic theory of phase transitions in human hand movement. *Biological Cybernetics,* **53,** 247–257.

Schöner, G., Jiang, W. J., & Kelso, J. A. S. (1990). A synergetic theory of quadrupedal gaits and gait transitions. *Journal of Theoretical Biology,* **142,** 359–391.

Schöner, G., Zanone, P. G., & Kelso, J. A. S. (1992). Learning as change of coordination dynamics: Theory and experiment. *Journal of Motor Behavior,* **24,** 29–48.

Smeets, J. B. J., Erkelens, C. J., & Denier van der Gon, J. J. (1990). Adjustments of fast goal-directed movements in response to an unexpected inertial load. *Experimental Brain Research,* **81,** 303–312.

Soechting, J. F., & Flanders, M. (1991). Deducing central algorithms of arm movement control from kinematics. In D. R. Humphrey & H.-J. Freund (Eds.), *Motor control: Concepts and issues* (pp. 293–306). New York: Wiley.

Taga, G., Yamaguchi, Y., Shimizu, H. (1991). Self-organized control of bipedal locomotion by neural oscillators in unpredictable environment. *Biological Cybernetics,* **65,** 147–159.

Tuller, B. T., & Kelso, J. A. S. (1989). Environmentally-specified patterns of movement coordination in normal and split brain subjects. *Experimental Brain Research,* **75,** 306–316.

Viviani, P. (1990). Common factors in the control of free and constrained movements. In M. Jeannerod (Ed.), *Attention and performance* (Vol. XIII). Hillsdale, NJ: Erlbaum.

von Holst, E. (1939/1973). Die relative Koordination als Phänomen und als Methode zentralnervöser Funktionsanalyse. *Ergebnisse der Physiologie,* **42,** 228–306. Englisch translation in R. Martin (Ed.), *The behavioral physiology of animals and man. (Collected works of Erich von Holst).* Coral Gables, FL: University of Miami Press.

Wing, A. M., & Kristofferson, A. B. (1973). Response delays and the timing of discrete motor responses. *Perception and Psychophysics,* **14,** 4–12.

Wise, S. P., & Desimone, R. (1988). Behavioral neurophysiology: Insights into seeing and grasping. *Science,* **242,** 736–740.

Wright, C. E. (1990). Generalized motor programs: Reexamining claims of effector independence in writing. In M. Jeannerod (Ed.), *Attention and performance XIII* (pp. 294–320). Hillsdale, NJ: Erlbaum.

Yamanishi, J., Kawato, M., & Suzuki, R. (1979). Studies on human finger tapping neural networks by phase transition curves. *Biological Cybernetics,* **33,** 199–208.

Part III

Modulation of Coordination Patterns through Practice and Experience

18

The Development of Sensorimotor Integration Underlying Posture Control in Infants during the Transition to Independent Stance

Marjorie Hines Woollacott and Heidi Sveistrup

Department of Exercise and Movement Science
University of Oregon
Eugene, Oregon

- I. Introduction
- II. Posture Control of the Head and Trunk
    - A. Development of Sensorimotor Input–Output Relationships in Head Control
    - B. Development of Intersensory Integration Relationships for Head Control
    - C. Development of Head–Trunk Control
    - D. Higher Level Adaptive Processes Involved in Head and Trunk Control
    - E. Contributions of the Development of Muscle Strength to the Emergence of Independent Sitting
- III. The Transition to Independent Stance
    - A. Development of Intralimb Coordination of the Legs
    - B. Development of Visual/Motor Coordination
    - C. Contributions of the Development of Muscle Strength to the Emergence of Independent Stance
    - D. Development of Higher Level Adaptive Processes
- IV. Conclusions
    - References

I. Introduction

In the 1960s the perception–action theory of motor control was put forth by Gibson (1966) and later elaborated to extend to the control of many different motor subsystems such as posture control (Lee & Lishman, 1975; Reed, 1982) and eye–hand coordination (Hofsten, 1986). According to this theory, the developing organism is not simply reflexively driven to perform actions, but instead is a perceiving and active participant in the environment. According to this theory, it is the active exploration of the environment by the developing organism that shapes the sensory experiences and therefore the ultimate behavior of the adult.

Concurrent with the emergence of the perception–action or "ecological" theory of motor control, "systems theory" was also being developed by Bernstein (1967) as a way of taking into account the contributions of multiple subsystems of the body (neural and musculoskeletal) as well as environmental variables, to the ultimate outcome of an action. Bernstein stated that motor coordination is produced by a self-regulating "system" of mechanisms that ensures the control of the motor apparatus and permits its rich and complex flexibility (Bernstein, 1967).

As research on motor development has expanded, these theories have received strong support among scientists working in the area of the development of posture control. Research has shown that multiple sensory systems contribute to postural control, including the visual, vestibular, and somatosensory systems (Nashner & Woollacott, 1979). It has been hypothesized that, during development, these sensory inputs create perceptions or internal representations of the body and its relationship to the environment, which could be called a body schema. These internal representations may actually consist of specific input–output relationships, which govern the ways in which perceptual information is transformed into functional postural actions.

In support of these theoretical approaches, it has recently been shown that different sensory experiences can shape motor cortex representations of specific muscles (Humphrey, Qiu, Clavel, & O'Donoghue, 1990; Cohen, Brasil-Neto, Pascual-Leone, Wall, Jabir, & Hallett, 1991). This implies that the creation of the input–output relationships, which are the basis of the perception–action system, is highly dependent on experience.

In this chapter we review the literature on the development of postural control abilities in infants through the transition to independent stance and locomotion. We use the systems approach in discussing the literature, examining studies on contributions of sensory, motor, higher level adaptive, and musculoskeletal systems to the emergence of postural control for the head, for sitting, and for stance. In reviewing the literature we discuss the way in which each study adds to our information on the way possible input–output relationships are formed in the postural control system.

II. Posture Control of the Head and Trunk

A. Development of Sensorimotor Input–Output Relationships in Head Control

In 1946, Gesell summarized what he called the general sequences of early behavioral growth in the first few years of life (Gesell, 1946). He noted that the general direction of behavioral development was from head to foot, and from proximal to distal within segments. Thus, he described development as beginning with control of the eye muscles within the first 16 weeks (first quarter) of postnatal development, followed by control of the head and arms from 16 to 28 weeks (second quarter), the trunk and hands from 28 to 40 weeks (third quarter), and the legs and feet (stance) from 40 to 52 weeks (fourth quarter). This would be followed by walking and running in the second year. He called this the law of developmental direction. He described development as being a process that results in a progressive spiral kind of reincorporation of sequential forms of behavior, where one sees the reintegration of control processes within higher, more complex behaviors as the child matures.

Since 1946, many researchers have found exceptions to Gesell's general developmental rules. For example, it has been shown that infants show early control of the legs in kicking and supported walking behaviors (Thelen, Ulrich, & Jensen, 1989; Forssberg, (1985). However, in the realm of postural control, it does appear as if development occurs in a cephalocaudal direction. In fact, much of Gesell's descriptions of motor development appear as if they are more appropriate to the development of the postural system than to other systems of the body. Though Gesell's original behavioral description is of interest, it does not give us information on (1) the way in which body segments are progressively coordinated together as units, as the infant progressively gains control over additional segments; (2) developmental changes underlying the infant's integration of perceptual information to control posture; and (3) whether any input–output relationships can be described for the transformation of perceptual information into postural responses.

In the 1980s a number of investigators began to examine the early development of postural control of the head and neck (Jouen, 1990; Prechtl, 1984). Prechtl has described the newborn infant's antigravity function as being very poor. This could be due either to lack of muscle tone or it could alternatively be due to lack of maturity of the motor processes controlling posture of the head and neck at this age. Prechtl used both electromyographic recordings and video recordings to determine whether antigravity movements were absent only because of muscle weakness, or whether there was actually an absence of patterned motor activity itself. He recorded from 24 full-term infants, who were 5 days old, the left (L) and right (R) sternocleido muscles and the L and R posterior neck muscles, among other body muscles. When recording from the infant while lying supine or prone, he noted epochs of activity in all the muscles, with activity in the neck muscles associated with occasional active head turns. However, there was no evidence of activity patterns or postures that would counteract the force of gravity on any maintained basis.

In order to see if infants would show clear antigravity responses to specific stimuli, he placed a second group of infants (4 days–25 weeks) on a rocking table, which could rotate in the transverse axis $\pm 15°$. He noted that newborns and infants up to 8–10 weeks did not respond either to head-downward or head-upward tilts. However, at 8–10 weeks, with the onset of spontaneous head control, the infants showed clear EMG patterns, along with head retroflexion and extension of the arms when the head was tilted downward. From about 3 months of age, this response became consistent. Thus, Prechtl's work, using a rocking table to elicit postural adjustments, suggests that the emergence of coordinated postural responses in neck muscles occurs at about 2 months of age. Since use of a rocking table may stimulate vestibular, visual, and neck proprioceptive afferents, this study does not give us specific information about the ability of individual sensory systems to drive postural responses in the neck, but it allows us to see the time of the first emergence of coordinated postural behavior.

Though the postural control of the neck does not appear to be coordinated until 8–10 weeks of age according to the above study, an interesting study by Jouen (1988) indicates that the visual system in isolation may have the capacity to drive postural muscles in a primitive manner in the neonate. It has previously been shown that newly walking children, and adults in certain environmental conditions, are highly reliant on visual cues in the control of posture.

In order to determine if the newborn infant was also responsive to visual flow, Jouen built an apparatus that would stimulate the optical flow created by body movement. This apparatus consisted of sequentially activated lights, which were placed in the lateral visual fields, either in the forward or backward direction. The infants were placed in a baby seat, with pressure-sensitive airbags placed behind the head. He studied 3-day-old infants, (3 groups of 12), and found that about 80% of the infants demonstrated a sensitivity to the optical flow. At low visual accelerations (3 cm/sec) 67% of the infants responded with postural reactions in the same direction as the optical flow (33% of the infants responded with movements in the opposite direction). This is the same behavior that had been previously observed in newly standing infants (Lee & Aronson, 1974), with the exception of the response probability. Lee and Aronson saw optically induced responses 82% of the time and found only 2 opposite-direction responses in 92 trials.

Since both newborns and new walkers respond at least 80% of the time to optical flow, they may be considered to be equally sensitive to this basic information. However, since the newborns respond in the correct direction in only 67% of the trials, the directional specificity of the newborn's neuromuscular connections may not be calibrated. Thus it appears that the input–output relations between visual flow and postural responses in the newborn's neck muscles to correct for head sway have not yet been well established.

B. Development of Intersensory Integration Relationships for Head Control

In a study to determine the development of visual–vestibular interactions Jouen (1984) examined two groups of infants, the first group ranging in age from 28 to 135 days and the second group 84 to 210 days. The infants were tilted in an infant seat 25° to the left or the right from vertical, at 25°/sec; the seat was placed in a room that could also tilt in the same fashion. The infant's head movements were measured with a helmet fitted with potentiometers.

Jouen thus could measure vestibular responses (the infant tilting in a dark environment) with vestibular–visual interactions (the infant tilting with various room tilts). He noted that the visual tuning of the vestibular reaction was not the same for the two groups ($p < .01$). For the younger infants the visual stimulation generally gave a better compensatory behavior than an empty field, but in the older infants the visual tuning of the reaction was differentiated according to stimulus orientation (head displacement increased for horizontal stripes and ipsilateral room-tilt conditions and decreased for vertical stripes and contralateral room-tilt conditions). He noted that there was an initial vestibular reaction to the tilt during the first second after the body tilt, then a visual reaction to the tilt in seconds 1–6. He also noted that the amplitude of the head tilt was smaller for the older infants, indicating that the compensatory reaction is better in this group ($p < .01$).

It is thus possible that this progressive fine-tuning of the visual–vestibular response is due to the laying down of a perceptual map in the nervous system of visual and vestibular space. Thus, at this point we may see rules for intersensory integration for postural control of the neck being established. What do we mean by "rules" in this context? We hypothesize that the brain is calibrating complementary input–output relations between visual–motor and vestibular–motor systems to aid each other in their redundant motor functions involving head control.

C. Development of Head–Trunk Control

As children begin to develop trunk control they must now learn to coordinate two body segments together in the control of posture. This will involve extending both the sensorimotor input–output relationships and the intersensory integration relationships used in postural control to the new set of muscles controlling the trunk. One might hypothesize that once these rules have been established for the neck muscles, they would be easily extended by the infant to the trunk musculature as well. However, it is not clear that this is the case. A recent study by Bertenthal and Bai (1988) examined the development of postural responses to optical flow in infants from 5 to 9 months of age. They noted that 5-month-olds showed no evidence of compensating to optical flow information. Seven-month-olds showed evidence of postural compensations as long as the visual stimulation was global, while 9-month-olds were sensitive to visual flow in only the peripheral portions of the visual field.

In the above study of the development of trunk posture control Bertenthal and Bai (1988) assessed the control of only the magnitude of the force associated with postural compensation. They have more recently studied the development of the control of timing of these movements as well. They noted that the timing of the responses of infants to room movements was the same over the age range of 7–13 months (Bertenthal, 1990), though response amplitudes changed over that period, thus indicating that timing control may develop before force or amplitude control.

Woollacott, Debu, and Mowatt (1987), in a cross-sectional study, and H. Sveistrup and M. H. Woollacott (unpublished observations) in a longitudinal study, have examined the progressive development of neck and trunk control in infants from 2 months to 8 months of age. In this paradigm, infants (seated in an infant seat) were placed on a platform that moved (2 cm) in the anterior or posterior direction (duration of movement, 250 msec) to cause a movement of the center of mass and require a stabilizing response in the neck and/or trunk muscles to bring the center of mass back to its original position. Surface electrodes measured the responses of the neck flexors (NF), neck extensors (NE), abdominals (A), and trunk extensors (TE) on the left side of the body during platform perturbations. Two 2-month-olds tested did not show consistent directionally appropriate responses to the platform perturbations. Responses were elicited in some of the trials, but they were not directionally specific. For posterior platform movements, causing forward sway, a 3 1/2-month-old showed directionally specific responses in the neck extensor muscle, compensating for the sway, in 40% of the trials, and a 4-month-old, in 60% of the trials (in the remaining trials there was either an inappropriate response or no response). Of the two 5-month-olds who were beginning to sit independently, trunk muscle responses were activated appropriately 40% of the time in one of the infants, while the second infant showed no response in trunk muscles, though responses were appropriate in the neck. By 8 months of age, the responses showed a mature response pattern of NE and TE activation for forward sway, and NF and A activation for backward sway.

The absence of vision did not cause any disruption in response organization. Therefore, results indicate that the somatosensory and vestibular systems are capable of driving the responses in isolation from vision.

Thus, as in the previous study, data indicate that neck muscle and trunk muscle responses develop gradually over time as infants develop neck and trunk control. First, there appears to be a calibration of input–output relationships between sensory inputs controlling posture and the neck flexor and extensor muscles themselves; this is later followed by calibration of the trunk musculature, as the child begins to sit independently. From these studies it is impossible to say whether it is nervous system maturation or experience that allows neck and trunk muscle responses to emerge, since maturation and the refinement of synergies through experience are both gradual, and they appear to be occurring concomitantly.

D. Higher Level Adaptive Processes Involved in Head and Trunk Control

It has been shown in previous research (Nashner, et. al 1982) that adults can show context-dependent reweighting of their use of sensory inputs relevant to posture control (e.g., they can switch from the use of somatosensory to visual cues when somatosensory cues are no longer sway related). It has been proposed that this reweighting is established by a higher level process acting on the sensory activation of lower level synergies of postural muscles.

If higher level adaptive responses are present in the infant in the independent sitting stage, infants should show attenuation of the responses to the visual flow caused by the moving room, since these responses are inappropriate. No studies have mentioned such an attenuation at this stage in development.

E. Contributions of the Development of Muscle Strength to the Emergence of Independent Sitting

Though there is no doubt a clear progression in the development of neck muscle strength and then trunk muscle strength in infants first learning to control their head and then their trunk, this is a difficult parameter to measure in young children, since one cannot ask them to make voluntary muscle contractions. Thus, there are no data in the literature indicating the developmental changes in the strength of these muscles as independent head and trunk control emerge.

III. The Transition to Independent Stance

A. Development of Intralimb Coordination of the Legs

When normal young adults are asked to stand on a platform that is subsequently moved forward or backward (a situation similar to standing on a bus that starts to move), they show stereotyped responses in muscles of the legs and hip that have the following characteristics. The responses are organized in a distal-to-proximal response sequence, with the stretched muscle that is closest to the base of support being activated first, followed by the muscles of the upper leg and hip (Nashner & Woollacott, 1979). Muscles in both the left and right legs are activated similarly. Thus, if the platform is moved forward, the response is activated in the tibialis anterior (TA), quadriceps (Q), and A muscles for both the left and right legs, at approximately 90, 110, and 130 msec, respectively. This strategy is called the ankle strategy, because sway is primarily about the ankle joint of the two legs. Similarly organized responses are seen in the gastrocnemius (G), hamstrings (H), and TE muscles of the two legs for platform movements in the opposite direction. Thus, responses are directionally specific and temporally organized in a distal-to-proximal sequence for both legs.

If the platform movement is of high velocity or of large amplitude, the person sways closer to his or her limits of stability and tends to use a different strategy in which sway is primarily about the hip joint, called the hip strategy. In this strategy a different set of muscles is activated, to cause flexion or extension at the hips. Nashner and McCollum (1985) have hypothesized that there are a discrete number of balance strategies that are used by young adults, including the ankle, hip, and suspensory strategies (in this strategy, there is a flexion/extension movement at the hip/knee/ankle). They describe these strategies as seen in Figure 1, with the suspensory strategy represented by flexion–extension movements along the vertical axis of the figure, and the ankle and hip strategies represented by ankle or hip movements, along the

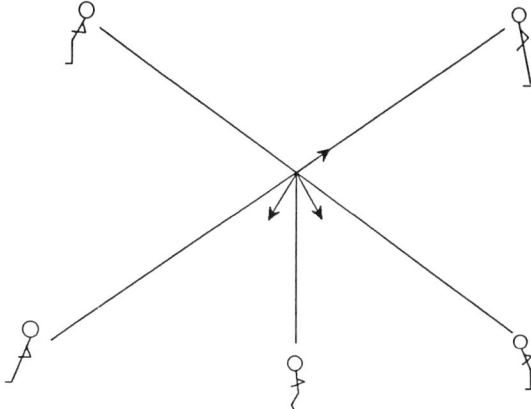

FIGURE 1 Possible axes for movements in postural position space. The three axes shown represent movements involved in pure ankle rotation (what could be called an "ankle strategy" for balance), pure hip rotation ("hip strategy"), and a suspensory or flexion movement mixing ankle, knee, and hip rotations ("suspensory strategy," involving lowering the center of mass toward the floor). (Adapted from Nashner & McCollum, 1985, Fig. 2.)

two horizontal axes of the figure. They describe the movements about these axes mathematically as vectors representing postural movement in position space and conclude that the positions that a body will move through in controlling posture will be limited by biomechanical constraints. They thus predict that these constraints will also limit the number of postural response patterns seen in the compensation for balance loss.

Previously (Sveistrup, Woollacott, Shumway-Cook, & McCollum, 1990), we have hypothesized that, during the process of development, infants will practice movements in this specific position space because of the body's biomechanical constraints. In order to test this hypothesis, Woollacott and Sveistrup (1992) and Sveistrup and Woollacott (1993) have longitudinally studied the emergence of postural response synergies underlying the intralimb coordination of posture control during the transition to independent stance. We analyzed the changing characteristics of postural responses in the muscles of the legs and trunk in 7 infants from 2 months of age (presitting) through the pull-to-stand stage of behavior, independent stance, and walking (18 months).

Figure 2 shows the responses of one of the infants to one of the forward platform displacements for each session from the time she was 16–49 weeks old. In the first sessions she could not sit or stand independently, and thus was supported at the hips by her mother during the experimental sessions. In the figure you can see that the infant showed

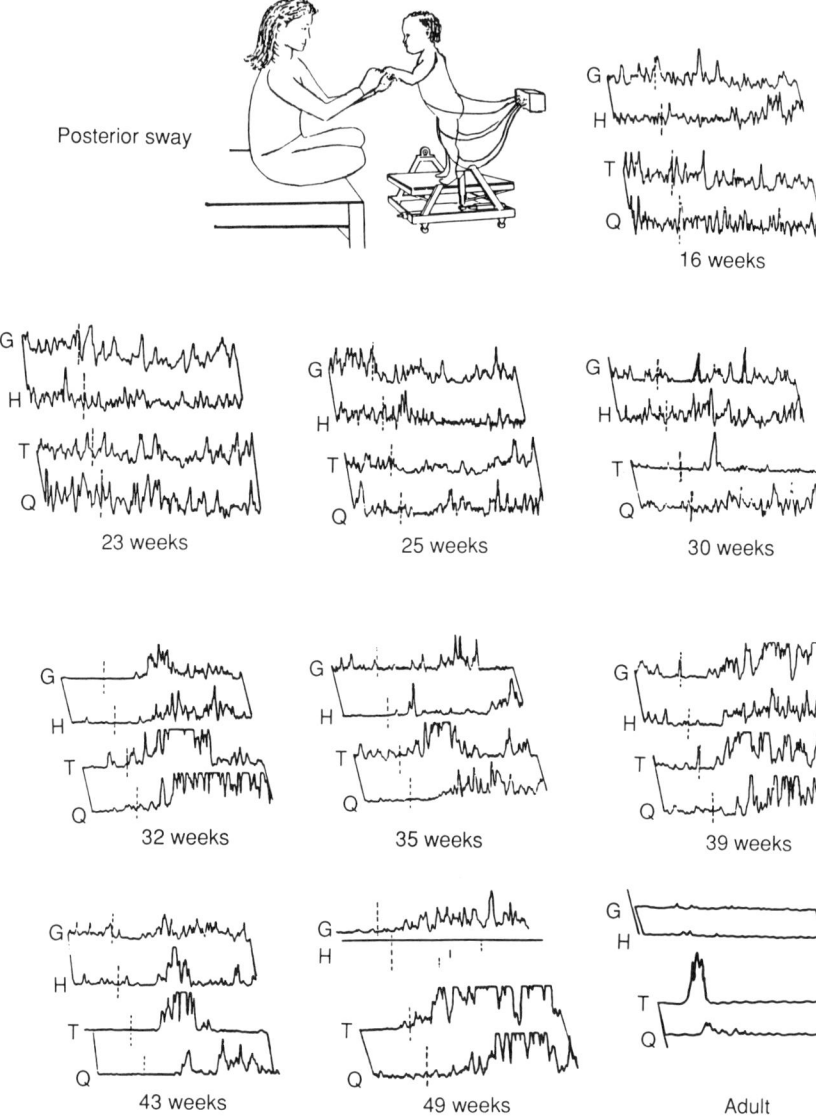

FIGURE 2 Examples of postural muscle response patterns to forward platform perturbations causing backward sway in a child from 16 weeks through 49 weeks of age. (Hamstrings data at 49 weeks are missing.) Each trial represented is the first trial from the testing session on that day. Abbreviations are explained in the text.

high levels of background tonic activity in all muscles, and there were no organized responses to platform perturbations. During this period, the infant showed a high variability of foot placement on the floor, with the feet often being placed at unusual angles to one another, or one foot being placed in partial overlap with the other, as if the infant was not even aware of somatosensory information coming from the soles of the feet and the ankles (see Figure 3). At this age, the infant possessed neither intralimb coordination nor interlimb coordination for postural control.

With the onset of pull-to-stand behavior at about 30 weeks of age, the infant showed a lowering of tonic background activity coupled with the first responses in the stretched TA muscle. Thus, at this time, directional specificity begins to emerge in the development of stance postural control.

Over the next weeks of pull-to-stand behavior, the TA and Q muscles began to be activated together in a consistent distal-to-proximal sequence, showing the emergence of intralimb coordination in posture control. Finally, in late pull-to-stand behavior, the abdominal muscles were consistently added to the synergy. Figure 4A shows the probability of seeing a response in the TA, Q, and A muscles following an anterior platform movement. One sees that the TA muscle first shows consistent responses, followed by the Q and the A muscles. This same trend was observed for posterior platform perturbations and the emergence of the G-H-TE synergy. When muscle responses from all 7 infants (categorized according to level of behavioral development) were averaged, similar results were seen (Figure 4B), with the TA muscle first showing consistent responses, followed by Q and A muscles.

An additional process in the development of stance control involves the ability to use alternate response synergies under different conditions. As mentioned above, an adult may use an ankle strategy to compensate for small, slow platform perturbations. However, the adult shifts to a hip strategy when platform perturbations are larger or faster, and the center of mass is moved closer to the limits of stability.

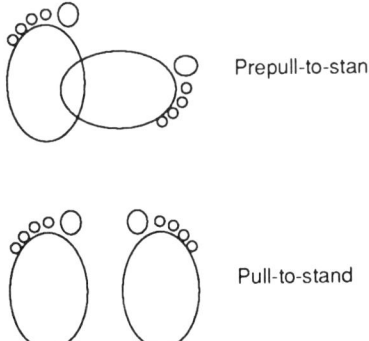

FIGURE 3 Types of foot placement seen in pre-pull-to-stand behavior (feet at a variety of angles) versus pull-to-stand behavior (feet facing forward).

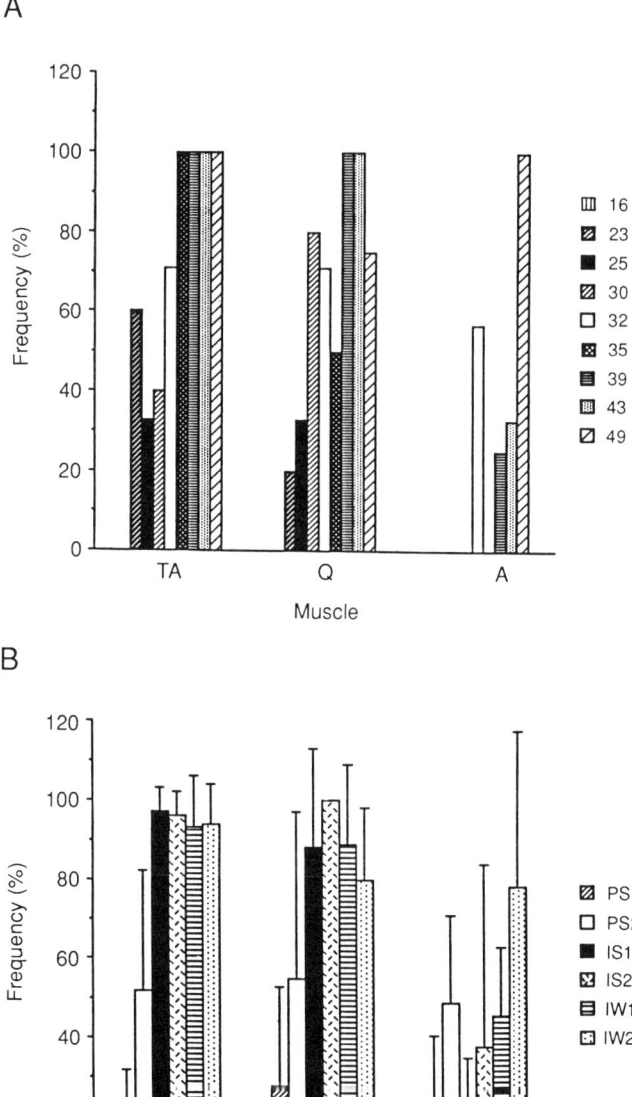

FIGURE 4 The probability of seeing a response in each of the agonist muscles for forward platform perturbations in the synergy over the time period from 16 to 49 weeks of testing. (A) Data from one child; (B) mean values from 4 children. PS1 & PS2, early & late pull to stand; IS1 & IS2, early and late independent stance; IW1 & IW2, early and late independent walking.

At the onset of independent walking, we observed an unusual change in the structure of the postural response patterns in two of the children. During one visit to the laboratory, during forward platform translations, each child showed a consistent inhibition of the Q muscle whenever the TA was activated. This coincided with a fall on the trials in which this was observed. On the next laboratory visit, the child's strategies were again normal. We hypothesize that the child was attempting different strategies for balance control. It is possible that, in this case, the child was using the suspensory strategy, in which the center of gravity is lowered in order to recover balance. This strategy would involve activation of TA and inhibition of Q.

B. Development of Visual/Motor Coordination

Since the above study indicated that postural response synergies activated by support surface perturbations did not emerge until the appearance of pull-to-stand behavior, it could be hypothesized that visual stimuli related to postural control would not be able to access these synergies until this period in time as well. In order to test this hypothesis, Sveistrup, Foster, and Woollacott (1992) explored the emergence of visually activated postural responses in infants during the development of independent stance. In this cross-sectional study (Sveistrup *et al.*, 1992), a total of 39 infants and children were divided into the following behavioral categories: (1) independent sitting (5–8 months); (2) pull-to-stand (8–10 months); (3) new walkers (11–14 months); (4) experienced walkers 1 (2–3 years); (5) experienced walkers 2 (4–6 years); (6) experienced walkers 3 (7–10 years). These children were compared to 5 young adults (20–29 years). Visual flow activating postural responses was created by using a "moving room" similar to that used in the experiments of Lee and Aronson (1974) described earlier in this paper. Total amplitude of room movement was 60 cm, with a movement velocity of .5 m/sec. The child's sway was recorded through a one-way mirror with a video camera mounted outside the room. EMGs were recorded as described above. Infants who were unable to stand independently were supported by their parents about the hip.

In this paradigm (Lee & Aronson, 1974; Butterworth & Hicks, 1977), the room moves, for example, forward, causing an illusory perception of backward sway in the subject, and a subsequent forward sway response, which destabilizes the subject. Results of the study (Sveistrup *et al.*, 1992) indicated that children as young as 5 months of age showed sway (averages of 45% of their stability limits) in response to room movements (see Figure 5A), with sway amplitudes increasing in the pull-to-stand infants (50% of stability limits) and peaking in the inde-

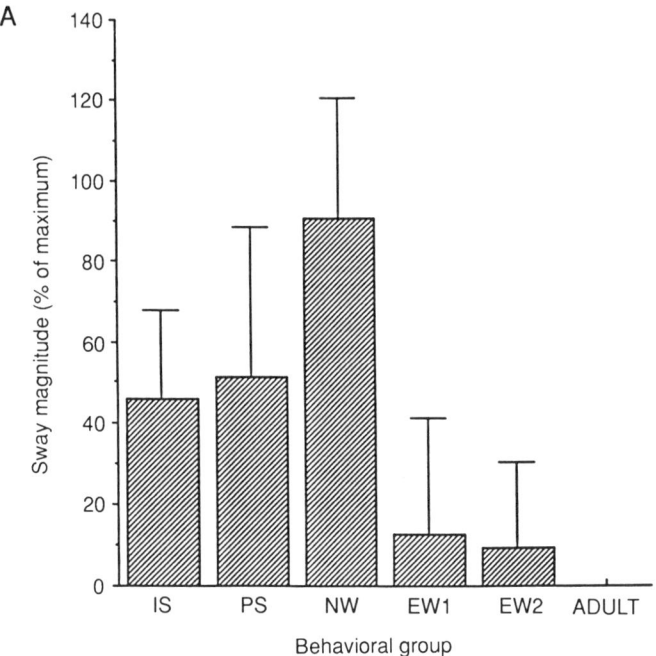

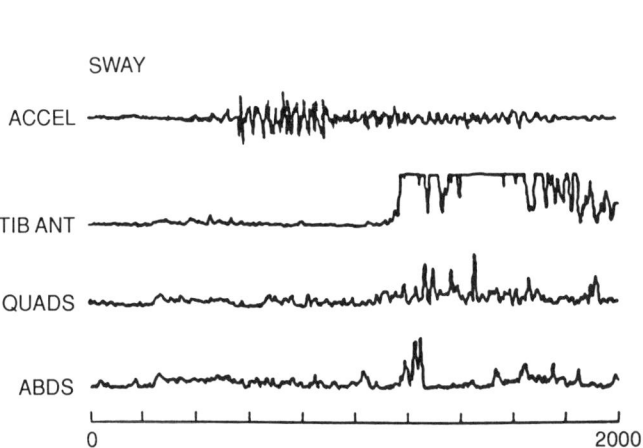

FIGURE 5 (A) Histograms showing the magnitude of sway for children in developmental stages from independent sitting (IS), through pull to stand (PS), new walking (NW), and experienced walking (EW). These are compared to adult values. Note that sway magnitudes increase through independent walking, then decrease. (B) Example of muscle response patterns of a 6-month-old independent sitter to room movements causing forward sway. An accelerometer (accel) detected the onset of room movement.

pendent walkers (90% of stability limits). Responses then dropped to low levels of sway in experienced walkers (13%). Correlated with these sway responses were clear patterns of responses in the muscles that would serve to pull the child in the direction of the visual stimulus. Thus, in one independent sitter (see Figure 5B) a response of TA-Q-A was seen in a trial in which the room moved in the anterior direction, away from the infant. Though responses were available to the infant in pre-pull-to-stand stages of behavior, the responses at all behavioral levels were more variable than those seen in response to platform perturbations. Thus, it can be concluded that perturbations to the visual system during supported stance will elicit postural responses at an earlier time than support-surface perturbations.

C. Contributions of the Development of Muscle Strength to the Emergence of Independent Stance

In observing the movement of the hip, knee, and ankle joints of the children when they were quietly standing on the platform, we noted a large amount of background "wobble" or movement of the joints over the first few months of dependent stance, which disappeared with further development. This could be due either to (1) a low level of postural stiffness (possibly including muscles and tendons) at early ages, possibly due to low muscle strength, which was correlated with an inability to stand independently, or (2) what might be called postural "babbling," a time in which the infant was exploring postural space, in order to calibrate a "sensorimotor map" to be used at the onset of independent stance. In the second case one would not expect a low level of background stiffness during this period, since the wobble or "babble" would be due to an active exploration of space. In order to explore these two hypotheses we measured the slope of the torque trace during the initial 50 msec of the platform perturbation, before the onset of any active muscular compensation, and determined if there were changes in these slopes with development. In one child the slopes of the torque trace (for forward platform movements, causing backward body sway) almost doubled between 10 and 12 months of age. This corresponds to a drop in stiffness as the child learned to stand independently. Thus, this does not support the first hypothesis, which predicts that the background movement was wobble due to lack of stiffness or of muscle strength, and gives some support for the second hypothesis, that the background movement was a form of postural "babbling," or exploration of the postural sensorimotor space.

It is also possible that the children are using the high muscle stiffness levels as a means of reducing large oscillations in sway because they have not yet refined their kinematic strategies for postural control.

These data also show that with experience in balance control, the background stiffness levels drop, a characteristic change that is found in the learning of other motor tasks as well.

In order to further explore changes in the infants' ability to support their weight during the development of independent stance behavior, we monitored their weight under two conditions: (1) sitting on a scale; and (2) standing on the same scale, while supported minimally by the parent, who was standing on the floor. We noted that the infants supported at least 98% of their full weight well before the emergence of independent stance behavior.

D. Development of Higher Level Adaptive Processes

In order to determine if higher level adaptive processes were available to the infant during pull-to-stand, independent stance, and early walking, we monitored the ability of the infants to attenuate postural responses to the visual flow created by the moving room. None of the infants in any of these behavioral categories was able to adapt their inappropriate postural responses to low levels, over a period of 5 trials. Thus, we have concluded that higher level adaptive processes related to postural control have not yet matured by the emergence of independent walking.

IV. Conclusions

In summary, this chapter describes research data from a number of laboratories, which document the gradual development of posture control in the cephalocaudal direction. Research by Jouen (1990) and Prechtl (1984) documents the development of head posture control in the first few months of life, with organized antigravity and visually induced responses in neck muscles appearing at about 2 months of age, and continuing to show refinement in the following months.

The beginning of postural control of the trunk occurs at about 3-4 months of age, with responses maturing by 7-9 months of age (Woollacott et al., 1987; Bertenthal & Bai, 1989).

Experiments on the transition to independent stance (Woollacott & Sveistrup, 1992; Sveistrup & Woollacott, 1993) indicate that stance balance control also develops gradually. Data suggest that a sequential process occurs in the development of muscle response synergies. During this process, infants calibrate the sensory inputs contributing to stance control and the relevant muscles that would act to stabilize stance.

For example, infants who were tested before the onset of pull-to-stand behavior did not show either intralimb or interlimb coordination of muscle response organization in response to threats to balance. All of the relevant postural muscles showed equal response probability during the infant's attempts at stability; the responses were thus not yet directionally specific.

During early pull-to-stand behavior a new process occurs. The infants begin to segregate the muscles within the leg by direction. For anterior or posterior sway, any combination of directionally appropriate muscles may be activated. Thus, the sensory stimuli may be providing directional signals that allow this process to occur.

In late pull-to-stand behavior, the infants begin to organize the muscles within the leg into appropriate synergies, with the ankle muscles being consistently activated first, followed by the upper-leg and hip muscles.

At about the onset of independent stance, intralimb coordination has become well organized, with the synergies being consistently activated as a unit, and temporally organized in an adultlike fashion. At this point fine-tuning of the synergies occurs, with their latencies and amplitudes beginning to undergo adjustments to allow control of the intended postural movement.

Finally, once the ankle synergy is established there is an expansion of the repertoire of synergies, to include other synergies such as the suspensory and/or hip synergies.

Data also indicate that the proprioceptive topography of stance is mapped later in development than the visual topography, and this allows the visual activation of postural muscles to appear earlier in development than that by proprioceptive inputs. This is possible because the visual map of postural space is calibrated early, with the respective onset of head control and trunk control within the first 2–6 months of postnatal development.

Muscle strength does not seem to be a rate-limiting factor in the emergence of independent stance, since infants show high postural (muscles, tendons, etc.) stiffness in early pull-to-stand behavior, and they support their full weight before the appearance of independent stance.

Research evidence indicates that higher level adaptive abilities have not emerged in infants by the time they have begun to show independent walking behavior, since infants are not able to attenuate inappropriate postural responses to visual flow created by the moving room.

Acknowledgments

This research was supported by a grant from the National Science Foundation #BNS-9190897 (NSF) (to M. H. Woollacott). H. Sveistrup was the recipient of a postgraduate scholarship from the Fonds pour la Formation de Chercheurs et L'Aide à la Recherche, Quebec (FCAR).

References

Bernstein, N. (1967). *Coordination and regulation of movement.* New York: Pergamon Press.
Bertenthal, B. I. (1990). Application of biomechanical principles to the study of perception and action. In H. Bloch & B. I. Bertenthal, (Eds.), *Sensory-Motor Organizations and Development in Infancy and Early Childhood* (pp. 243–260). Dordrecht: Kluwer.
Bertenthal, B. I. & Bai, D. L. (1988). Infants' sensitivity to optical flow for controlling posture. In *Visual-vestibular integration in early development: technical and clinical perspectives,* eds. C. Butler & K. Jaffe, pp. 43–61. Washington, D.C.: RESNA.
Bertenthal, B. I., Bai, D. L. (1989). Infants' sensitivity to optical flow for controlling posture. Developmental Psychology **25,** 936–945.
Bullinger, A., & Jouen, F. (1983). La Sensibilité du champ de détection périphérique aux variations posturales chez le bebe. *Archives de Psychologie,* **51,** 41–48.
Butterworth, G., & Hicks, L. (1977). Visual proprioception and postural stability in infancy. A developmental study. *Perception,* **6,** 255–262.
Butterworth, G., & Pope, M. (1983). Origine et fonction de la proprioception visuelle chez l'enfant. In S. de Schonen (Ed.), *Le dèveloppement dans la premiére annèe* (pp. 107–128). Paris: Presses Universitaires de France.
Cohen, L. G., Brasil-Neto, J., Pascual-Leone, A., Wall, R. T., Jabir, F. K., & Hallett, M. (1991). Rapid reversible reorganization in maps of outputs of human motor cortex following transient restrictive deafferentation. *Neuroscience Abstracts,* **17,** 1112.
Forssberg, H. (1985). Ontogeny of human locomotor control I. Infant stepping, supported locomotion and transition to independent locomotion. *Experimental Brain Research,* **57,** 480–493.
Foster, E. (1991). *The effect of visual flow on infants during critical periods of development: a cross-sectional study.* Honors Thesis. University of Oregon.
Gesell, A. (1946). The ontogenesis of infant behavior. In L. Carmichael (Ed.) *Manual of child psychology.* New York: Wiley.
Gibson, J. J. (1966). *The senses considered as perceptual systems.* Boston: Houghton-Mifflin.
Hofsten, von, C. (1986). The emergence of manual skills. In M. Wade & H. Whiting (Eds.), *Motor development in children: Aspects of coordination and control* (pp. 167–186). Boston: Martinus Nijhoff Publishers.
Humphrey, D. R., Qiu, X. O., Clavel, P., & O'Donoghue, D. L. (1990). Changes in forelimb motor representation in rodent cortex induced by passive movements. *Neuroscience Abstracts,* **16,** 422.
Jouen, F. (1984). Visual-vestibular interactions in infancy, *Infant behavior and development,* **7,** 135–145.
Jouen, F. (1988). Visual-proprioceptive control of posture in newborn infants. In *Posture and Gait: Development, Adaptation and Modulation,* B. Amblard, A. Berthoz, & F. Clarac, (Eds.), pp. 13–22. Amsterdam: Elsevier.
Jouen, F. (1990). Early visual-vestibular interactions and postural development. In H. Bloch & B. I. Bertenthal (Eds.), *Sensory-motor organizations and development in infancy and early childhood* (pp. 199–215). Dordrecht: Kluwer.
Lee, D. N., & Lishman, R. (1975). Visual proprioceptive control of stance. *Journal of Human Movement Studies,* **1,** 87–95.

Lee, D. N. & Aronson, E. (1974). Visual proprioceptive control of standing in human infants. *Perception & Psychophysics,* **15,** 529–532.

Nashner, L., & McCollum, G. (1985). The organization of human postural movements: A formal basis and experimental synthesis. *Behavioral and Brain Sciences,* **9,** 135–172.

Nashner, L., & Woollacott, M. (1979). The organization of rapid postural adjustments of standing humans: an experimental-conceptual model. In R. E. Talbott & D. R. Humphrey (Eds.), *Posture and movement* (pp. 243–257). New York: Raven Press.

Nashner, L., Black, F. O., & Wall, C. (1982). Adaptation to altered support and visual conditions during stance: patients with vestibular deficits. *Journal of Neuroscience,* **2,** 536–544.

Pope, M. J. (1984). *Visual proprioception in infant postural development.* Unpublished Ph.D. Thesis, University of Southampton.

Prechtl, H. F. R. (1984). Continuity and changes in early human development. In H. F. R. Prechtl (Ed.). *Continuity of neural functions from prenatal to postnatal life* (pp. 1–15). Oxford: C. D. M. Blackwell.

Reed, E. (1982). An outline of a theory of actions systems. *Journal of Motor Behavior,* **14,** 98–134.

Sveistrup, H., & Woollacott, M. H. (1993). Systems contributing to emergence and maturation of stability in postural development. In G. J. P. Savelsbergh (Ed.) *The development of coordination in infancy.* pp. 319–336. Amsterdam: Elsevier.

Sveistrup, H., Woollacott, M., Shumway-Cook, A., & McCollum, G. (1990). A longitudinal study on the transition to independent stance in children. *Neuroscience Abstracts,* **16,** 893.

Sveistrup H., Foster, E., & Woollacott, M. H. (1992). Changes in the effect of visual flow on postural control across the lifespan. In M. H. Woollacott & F. B. Horak (Eds.), *Posture and gait: Control mechanisms* (pp. 224–227). Eugene, OR: University of Oregon Books.

Thelen, E., Ulrich, B., & Jensen, J. (1989). The developmental origins of locomotion. In M. Woollacott & A. Shumway-Cook (Eds.), *Development of posture and gait across the lifespan* (pp. 25–47). Columbia, SC: University of South Carolina Press.

Woollacott, M., & Sveistrup, H. (1992). Changes in the sequencing and timing of muscle response coordination associated with developmental transitions in balance abilities. *Human Movement Science,* **11,** 23–36.

Woollacott, M., Debu, B., & Mowatt, M. (1987). Neuromuscular control of posture in the infant and child: is vision dominant? *Journal of Motor Behavior,* **19,** 167–186.

19

The Development of Bipedal Interlimb Coordination

Jill Whitall

Department of Kinesiology
University of Wisconsin
Madison, Wisconsin

Jane E. Clark

Department of Kinesiology
University of Maryland
College Park, Maryland

I. The Emergence of Walking
II. The Role of Intralimb Coordination
III. The Development of Walking
IV. Later Emerging Gait Forms: Running and Galloping
V. Summary
   References

Humans are unique primates. We have four limbs, yet, with few exceptions, our major forms of locomotion are bipedal. While the upright, two-legged gait frees our upper limbs for other tasks (such as carrying objects or throwing implements), it also puts unique biomechanical constraints on the neuromusculoskeletal system. For example, balance in the bipedal posture is considerably more tenuous than it is in the quadrupedal form. Producing propulsive forces with two limbs rather than four also offers a distinct biomechanical challenge. Indeed, the demands of bipedal gait are so daunting that it takes the human infant about a year before the first independent walking steps are taken. With those first steps comes a struggle for the infant as she seeks a workable solution to remaining upright while progressing forward.

How infants come to find the interlimb coordination that affords upright independent locomotion is the focus of the present chapter.

The developmental story of upright bipedal locomotion begins *in utero*, where early kicking motions signal a fetus capable of flexing and extending the legs (Prechtl, 1986). At birth, the neonate demonstrates a well-organized "walking" reflex, which, though not functionally suitable for independent upright locomotion, nonetheless reveals a well-organized coordinative relationship between the legs. Throughout the first year of life, orderly, well-documented changes occur in the infant's efforts to travel in her environment. However, it is not until the latter part of the first year that the infant achieves an independent bipedal stance that eventually gives way to the first attempts at upright forward locomotion. It is at this point that we pick up our story about the development of the interlimb coordination of upright locomotion.

For decades, explanations of how independent locomotion develops have been situated in a maturational perspective. Changing behaviors were viewed as the result of maturing neuromuscular structures and connections (McGraw, 1940). However, as Thelen (cf. 1983, 1989; Thelen & Ulrich, 1991) has repeatedly argued, explanations of how locomotion develops cannot be found solely in maturing neuroanatomical structures, but rather require consideration of the context within which the behavior occurs. Understanding how behaviors develop is sought not in prescribed biological programs, but in self-organizing processes of nonlinear cooperative systems. Using the principles of nonlinear dynamics, Thelen (Thelen, Kelso, & Fogel, 1987) and Kelso (Kelso & Tuller, 1984; Kelso & Schöner, 1988) and their colleagues have attempted to describe the control and coordination of movement and its development. This approach, variously called the dynamical approach, dynamic pattern theory, synergetics, or, as we favor, dynamical systems theory, is a theory that seeks to describe and ultimately understand how change may occur.

Clearly, any theory that seeks principles for understanding the emergence and evolution of new forms has appeal to developmentalists such as ourselves. Within this approach, the development of a new behavior, such as independent upright bipedal locomotion, is viewed as a potential transition from one dynamical state to another. In other words, crawling can be seen as a behavioral state that gives way to a new form, walking. Our quest, then, is to understand the dynamical principles that govern behaviors when they are stable as well as those principles that underlie the transition to new forms.

To examine stability and transition in upright bipedal interlimb coordination, we first sought a generic dynamical model of the behavioral system. A dynamical system can be described by two elements:

(1) a *state* that represents the system at a given instant, and (2) a *dynamic* or *rule* by which the state evolves over time (Crutchfield, Farmer, Packard, & Shaw, 1987; Rosen, 1970). Mapped in this way, dynamic systems tend toward one of three qualitatively different trajectories: point (equilibrium), periodic (limit cycle), or chaotic (mixing) (Thompson & Stewart, 1986). Inspection of the leg action during locomotion suggests the periodic cyclic motion characteristic of the limit-cycle attractor. With each walking cycle, the leg oscillates to and fro, dissipating energy during the first half of stance and injecting energy in the latter half of stance. Indeed, if we examine the phase portrait of the shank (lower leg) or the thigh (upper leg) action during a walking cycle, we see a limit-cycle attractor (Figure 1). That is, we observe trajectories that form an "attractive" closed periodic orbit (See Clark, Truly, & Phillips, in press). In this way, we can classify the system's behavior into the generic category of limit-cycle attractors; an essential first step in qualitative dynamics (Kay, Kelso, & Saltzman, 1991).[1]

If a walking leg can be modeled dynamically as a limit-cycle attractor, then understanding the coordination between the two legs would suggest a model in which limit-cycle systems are *coupled* together into a behavioral unit. A similar approach was first proposed by Kelso and his colleagues in modeling the interlimb coordination between the fingers (Kelso, Holt, Rubin, & Kugler, 1981). A system of coupled nonlinear limit-cycle oscillators also has been employed to model central pattern generation in lamprey locomotion (cf. Kopell, 1988; Kopell & Ermentrout, 1988; Rand, Cohen, & Holmes, 1988; Williams, Sigvardt, Kopell, Ermentrout, & Remler, 1990). More recently, Turvey and colleagues have extended Rand *et al.*'s modeling to human interlimb coordination (see Schmidt, Treffner, Shaw, & Turvey, 1992; Turvey & Schmidt, Chapter 14, this volume). In each of these studies, the properties of specific theoretical coupled limit-cycle systems were validated against experimental data from various behavioral systems. The dynamical approach, then, offers a means by which specific properties derived from the dynamics of a coupled limit-cycle system are compared with

---

[1] It should be noted, however, that the form of the attractor is dependent on the level of description employed. For example, if we were to model the leg as a whole by taking the time of footstrike as the level of description we could then plot the interheelstrike interval (stride time) at time x versus the previous stride time $(x - 1)$. In this case we would find a fixed-point attractor, given steady-state conditions (cf. Garfinkel *et al.*, 1992). In fact, this level of description is not useful for modeling a single leg in infants, since few strides are obtained at the onset of walking. Further, in rhythmical tasks, a more appropriate qualitative dynamical description appears to be a limit cycle (Kay *et al.*, 1991). In addition, although the thigh and shank phase portraits appear to exhibit limit-cycle behavior, a plot of the knee is actually a torus. The major point here is that alternative examples of mapping the attractor for a particular system illustrate the significance of selecting a level of description.

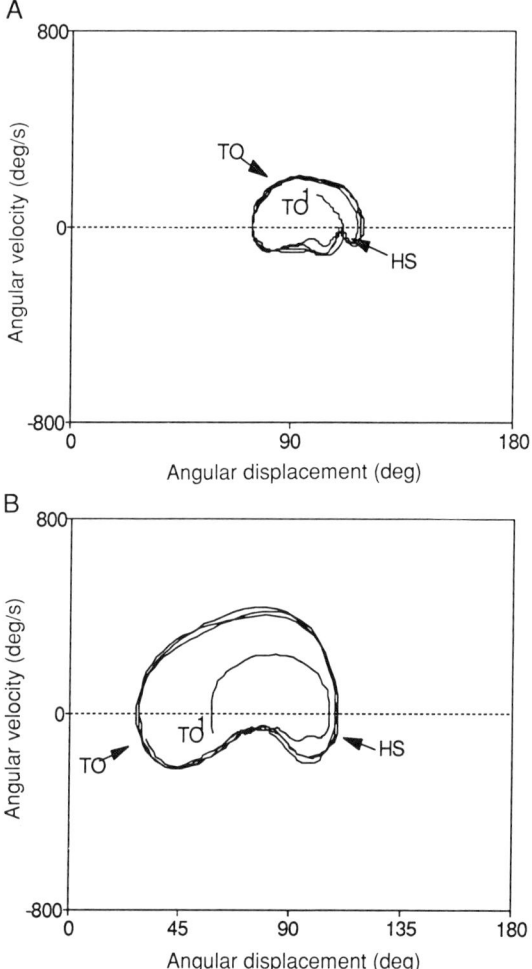

FIGURE 1 The phase portraits of the thigh (A) and shank (B) motion during four walking cycles in an adult. Trajectories move clockwise and represent the cycles from the first toe-off (TO^1) at gait initiation to toe-off of the last cycle. The regions of the phase portraits where toe-off (TO) and heelstrike (HS) exist for all but the first gait cycle also are designated.

the observed behavioral system. This interplay between the model and the empirical behavior offers the researcher a rich arena for exploring the nature of interlimb coordination.

To date, our own work has evolved within a conceptual framework based on principles of dynamical systems as we build empirical knowledge about the development of bipedal interlimb coordination. Rather than testing a specific mathematical model of coupled oscillators against our empirical data, we have sought the existence of three general properties of coupled nonlinear limit-cycle oscillators, namely phase locking, entrainment, and structural stability. In addition, we have explored the qualitative dynamics found in phase portraits of the limb segments. To demonstrate how we have used these approaches, we first present our work on interlimb coordination at the emergence of the first upright, independent gait form, walking. We then describe our work on two of the later emerging forms of bipedal locomotion, running and galloping.

I. The Emergence of Walking

In our first study on the development of interlimb coordination in independent walking, we defined the relationship between the two limbs in terms of their temporal and spatial (amplitude or distance) phasing between footfalls (Clark, Whitall, & Phillips, 1988). Phasing was defined as the proportion of a leg cycle when footstrike occurred in the contralateral leg. Newly walking infants, infants who had been walking for 0.5, 1, 3, and 6 months and adults were filmed while traversing a runway. The data revealed that at the onset of walking, infants adopted the same alternating coordination pattern (both temporal and amplitude measures) as the adults, a 50% (or 180°) phasing relationship between the legs at footfall.[2] That is, at the onset of independent locomotion, the infant discovers the solution to the problem of upright forward progression while maintaining the upright posture, namely, an alternating symmetric gait (Raibert, 1986). The constraints on a biped with symmetric morphology would appear to promote the alternating phasing pattern over other potential gaits such as an asymmetric or symmetric in-phase gait. Indeed, right from the onset of walking, the infant appears to demonstrate a preference for a coordination that is phase locked at footfalls—a behavior that can arise from a system of coupled nonlinear limit-cycle oscillators (Rand et al., 1988).[3] However, while the infant has achieved the same average 50% phase locking as the adult, she has significantly greater variability of phasing

---

[2] Quantifying the phase relationship by percentage is an alternative method (originating in the animal literature) to the more common use of degrees. Conversion is straightforward, with 50% equivalent to 180° out-of-phase.

[3] Strictly speaking, if we consider the interlimb coordination over the entire stride cycle (i.e., from footfall to footfall rather than at footfalls only) the term phase entrainment would be more appropriate than phase locking. Phase locking implies a constant phase difference throughout the cycle, whereas phase entrainment requires only a consistent 1:1 frequency between the limbs (Rand et al., 1988). If the phasing of the legs is followed throughout the stride, each leg has a time period (swing) where it is ahead of the other leg (Clark et al., 1990). Again, this is an example of how choosing different levels of description can change and enhance our understanding of the behavior in question.

relationships for both the temporal and distance measures. It was not until the infants had been walking three months that they demonstrated adultlike low variability. We interpret these results as evidence of a less stable interlimb coordination at the onset of walking, but one that becomes more stable with walking experience (See Figure 2). These fluctuations of phasing around the footfalls in walking can be seen as a period in which the system settles into a new behavioral state.

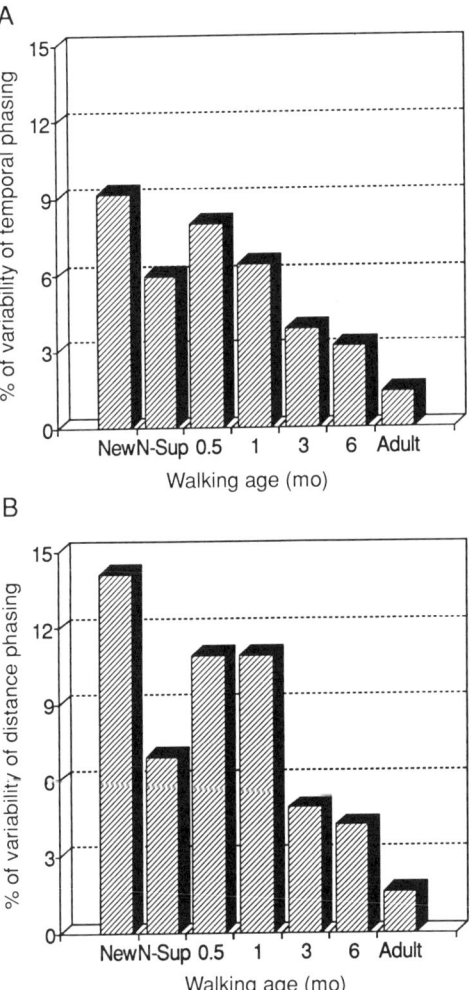

FIGURE 2 Mean variability of temporal (A) and distance (B) phasing for seven walking groups: New independent walking (New); newly walking with support (N-Sup); walking 0.5 months (0.5); 1 month (1); 3 months (3); 6 months (6); and adults. (Redrawn with permission from Clark *et al.*, 1988.)

Our results mirror those found in the development of interlimb coordination in other vertebrates such as rat pups (Bekoff & Trainer, 1979) and kittens (Bradley & Smith, 1988). Like humans, these animals are not able to locomote immediately. In addition, their early attempts are characterized by EMG and kinematic patterns of interlimb coordination that are remarkably adultlike yet also show an increased variability of temporal phasing compared to later recordings.

Returning to humans, then, if the interlimb system is unstable after transitioning from quadrupedal to bipedal gait, how does the system attain the coordinative stability observed after three months of walking? One property of coupled nonlinear limit-cycle ensembles is entrainment. Entrainment is the tendency of one oscillator to match the frequency (or the subharmonic) of the other oscillator. Hypothetically, if the legs were uncoupled they would be independent oscillators, unaffected by each other. If coupled, however, they will interact and influence each other, that is, they will entrain. Less stable systems, it could be argued, are not well entrained. If we consider, for a moment, a general mathematical model of coupled oscillators (Cohen, 1988; Koppell, 1988; Rand *et al.*, 1988), changes in phase-lags between oscillators are produced by changes in either relative frequencies of the component oscillators (eigenfrequencies) or changes in the coupling strength between them. Since there is no anthropometric evidence for the lower limbs to have different relative eigenfrequencies in young children, it seems plausible that it is the coupling strength that is weaker in the newly walking infants, resulting in a less well entrained coordination. This does not address the question, however, of why the infants' legs do not appear as strongly coupled/well entrained when compared to adults.

One possibility is that infants are not able to walk a sufficient number of steps to entrain the two legs. While that might be the case, inspection of the adults' interlimb phasing reveals that the limbs are tightly entrained within one stride. By an adult's second step, the interlimb phasing is near 50%. This is in contrast to the infant who, as a new walker, fluctuates, sometimes rather dramatically (as evidenced by phasings of 35 and 65%), but rarely achieves the exact 50% phasing at any time during the walking trial (see Figure 3). The young walkers do not walk the same number of steps as the older walkers, and they rarely seem to settle into a consistent phasing until after three months of walking. After this time, like the adults, they do find the 50% phasing relationship within the first 2–3 steps.

If entrainment occurs within the first steps in adults and 3-month walkers, why does it not occur in the younger walker? A clue to this question can be found by examining what happened to the newly walking infants' interlimb coordination when we provided postural support for them. While the mean phasing remained the same (50%), the tempo-

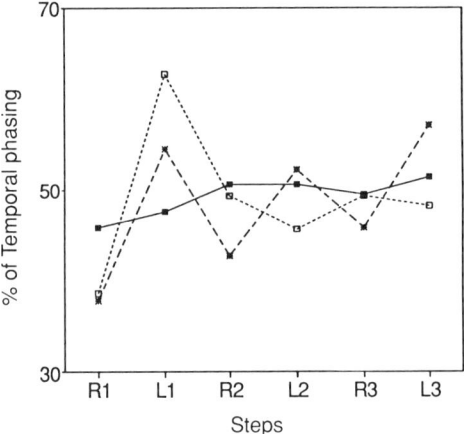

FIGURE 3 Temporal phasing across individual steps for a 2-week walker, the same infant after 3 months of walking, and an adult. R1, first right step; L1, first left step; and so forth.

ral and distance variability were significantly decreased. Indeed, the new walker's variability was about that of an infant that had been walking for a month (refer to Figure 2). This finding suggests that the ability to control posture is at least one of the contributing subsystems involved in refining the entrainment of the lower limbs. Or, to put it another way, the coupling of the legs' oscillations is affected by the postural requirements of the whole body. It is rather obvious that balance is important because the infants will still occasionally fall during the first weeks of walking.

Again, similar results have been found in animal studies where manual support of kittens (Bradley & Smith, 1988), mechanical support of baby mice (Fentress, 1978), and buoyancy support of rat pups (Bekoff & Trainer, 1979) have all accelerated the onset and development of coordinated interlimb behavior. Perhaps, as Bradley and Smith (1988) suggest, the prolonged postnatal development of locomotion reflects the time required to establish adaptive mechanisms such as postural control and agility, rather than the basic patterns (coordinative structures) for locomotion. Others have argued similarly. For example, in a recent paper, Bril and Brenière (1992) suggest that gaining greater postural control in the first few months of walking is related to the infants' ability to change their progression velocity. Clearly, the infant is struggling to find the coordinative organization that permits forward progression without losing the upright bipedal posture. The tendency for the two limbs to entrain may well be masked by the daunting postural demands of independent locomotion.

Another property of coupled nonlinear limit-cycle oscillators is that of structural stability. Structural stability is the tendency of the system to return to its original behavior after small perturbations or changes in initial conditions (Pavlidis, 1973). To explore this property in the development of lower limb coordination, we undertook a second study. This time, we followed the infants longitudinally for six months after they had begun walking (Whitall, Block, & Clark, 1992). To perturb the system, we added a small mass (5% body weight) to one leg on some trials. This enabled us to assess the stability of coupling between the two lower limbs. In case the mass had any effect on walking per se, a third condition was utilized, where the same mass was strapped to the infant's back. The two weight conditions were counterbalanced in order while the no-weight conditions occurred at the beginning and end of the sessions.

Until one month of independent walking, none of the infants walked more than a couple of steps with the weight placed on one ankle, and one infant would not take a single step. Apparently, the addition of 5% body weight was too much for the newly walking infants to manage. From one to six months of walking, however, both the mean and variability temporal phasing demonstrated the same general result as the earlier experiment, with no significant differences between the three conditions. In contrast, the distance phasing illustrated a significant effect on the coordination at first, both in the mean phase relationship and in the variability (stability) of this relationship (see Figure 4). Not until six months of walking did the infants appear to maintain equivalent phasing between the two legs. It would appear, therefore, that the legs had a structurally stable form of coupling in their temporal relationship before their spatial relationship. The reason for this discrepancy is less clear. One possibility is that the infants' muscular strength grew disproportional to the body weight so that the extra torque needed to place the weighted leg halfway between the unweighted leg's stride was not possible until later despite the uninterrupted temporal phasing relationship. In terms of the general dynamical model, adding a mass to one limb will alter the inertial properties of that limb and therefore change the uncoupled eigenfrequency of the oscillator/leg. However, as von Holst (1939/1973) noted long ago, in biological systems there is a tendency for individual rhythmic units to act together in absolute or relative coordination regardless of the individual uncoupled frequencies of the component units.

For a more complete assessment of the effect of adding mass to the leg, it is necessary to look at the changes in absolute frequency and amplitude of each leg in the weighted conditions versus the no-weight conditions. Do the limbs change their "no-weight" frequency and amplitude in order to entrain together in the weighted condition, or is the

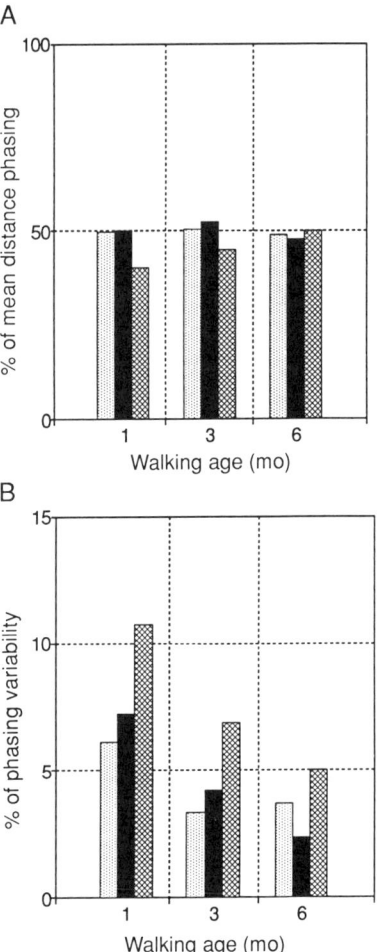

FIGURE 4   Mean (A) and variability (B) of distance phasing across age and condition. (Reprinted with permission from Woollacott and Horak, 1992.)

"no-weight" condition the dominant oscillator to which the weighted leg will entrain? Similarly, does the addition of a weight on the trunk act to change frequency/amplitude, or are the natural frequency and amplitude still dominant?

Of course, it is not possible to state that the infants were always using a "preferred" speed or natural frequency in every trial and therefore the interpretation of the data should be seen through this light. Nevertheless, for both frequency (stride time) and amplitude (stride length), clear indications of structural stability (i.e., attractivity to natural no-weight values) were not apparent until the infants had been walking for six months (see Figure 5). For example, in terms of timing,

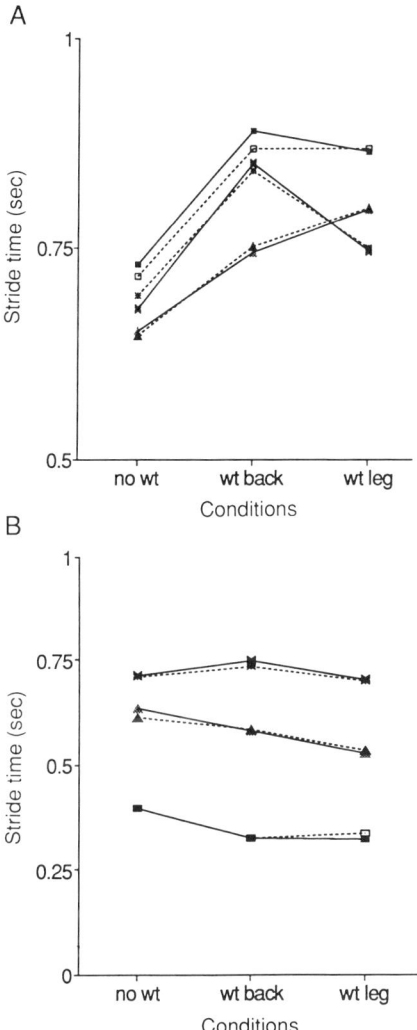

FIGURE 5 Stride times (A) and lengths (B) across age and condition.

at one month the frequency of the weighted limb decreases but the unweighted limb also slows down and 50% phasing is still seen. In other words, frequency unlocking occurs with the weighted leg appearing to entrain the unweighted leg. A similar situation occurs at three months of walking and it is not until six months that both legs remain close to the unweighted frequency. Here, then, we have a situation where the effect of changing inertial properties of the limb (by adding mass) was greater in the younger infant. That is, the dynamics of the component oscillators (legs) were changed although the coupling between them, as reflected by the 50% temporal phasing, did not change.

The amplitudes of the legs' motion were also affected by the weight, resulting in shorter strides. In this case, however, the legs show a slippage in their phase relationship, since the phasing changed from 50/50 to 40/60%. As with frequency, the amplitudes of the legs are equal. By six months of walking, the stride lengths are no different from those in the unweighted conditions. This evidence of frequency and amplitude unlocking is also reflected in the conditions with the weight on the back, indicating that it is the presence of additional weight per se that affects the developing system until six months of walking.

Overall, the results of this second experiment confirm the earlier results in that infants have a less stable coupling relationship between footfalls than adults have until they have been walking for at least three months. It also indicates that temporal and spatial coordination have slightly different rates of development and/or different coupling properties with regard to phasing stability for this particular task. It seems possible that temporal and spatial coordination are not tightly linked phenomena themselves and have independent behavioral manifestations, at least in the young infants.

II. The Role of Intralimb Coordination

The above experiments have demonstrated that the legs appear to be loosely entrained at the emergence of walking but, with experience, the interlimb coupling stabilizes. This occurs, presumably, at least partially as a result of increasing postural control. However, there are other aspects of the neuromuscular system that might well contribute to the improvement in interlimb coordination. As Newell (1985) argues, the definition of a movement pattern includes the relative motions of the body segments involved. So far, we have looked at the relationship between the legs at footfall, but we should also consider the relationships between the different segments of each leg. That is, how does the *intralimb* coordination of each leg contribute to the overall interlimb coordination between the legs? Does interlimb coordination emerge with intralimb coordination or is there a specific interlimb coupling function that operates independently of the intralimb coordination functions? In other words, is it possible to be tightly coordinated across the legs while being loosely coordinated within each leg? Alternatively, does intralimb coordination develop in advance of interlimb coordination and therefore contribute to the changes we see in interlimb coordination? To examine these questions, we assessed the intralimb coordination of the same babies used in the previous study. If the time course for obtaining adultlike stability of coordination is similar for both intra- and interlimb coordination, then this would lend support to the idea

However, if one studies Thelen's data carefully, it is clear that the measures of intralimb coordination reflect a more quantitative level of analysis than those used for interlimb coordination. Given these measurement differences, it is possible that the stability of intralimb coordination is, in fact, greater than that seen in interlimb coordination. This interpretation would support the findings in the animal literature where Bekoff (1981) suggests that, in general, intrajoint coordination (muscles around the same joint) tends to precede interjoint coordination (within a leg), which in turn precedes interlimb coordination. If humans do, indeed, develop their intralimb coordination for non-weight-bearing activities before interlimb coordination, then this sequential adaptation may be a developmental phenomenon occurring only in the early months of life, so that later coordination patterns are acquired in a more synchronous fashion. Alternatively, each specific coordination task may have differing rates of stability in intra- and interlimb components, according to the specific constraints of the task. Careful descriptions of the acquisition of coordination for other tasks are necessary to test these hypotheses.

III. The Development of Walking

In dynamical terms, we can summarize the developmental story for walking in the following manner. The alternating phasing pattern is the attractor state for this class of movements (bipedal locomotion) to which the system (infant) evolves. Given the constraints of the young bipedal system, the infant is attracted to a symmetrical coupled limit-cycle system for her upright locomotion. In fact, as demonstrated in all our studies (Clark et al., 1988; Clark & Phillips, 1993; Whitall et al., 1992), the infants quickly discover the same dynamic solution. However, at least three months of practice appear to be necessary for the phasing relationships to stabilize; this holds true for both intralimb and interlimb coordination. Put another way, the limit-cycle attractor for locomotion appears to be stronger after several months of walking and more equivalent to adults. One important subsystem that seems to underlie the interlimb and intralimb coupling stability is postural control.

After three months of walking experience, the infant appears to have a stable base of operation that affords functional adaptability and flexibility. For example, it is at three months of walking that we notice "gesturing" in the infant, that is, a nonverbal expressive quality in her movement. Clearly, by this time, the infant is quite capable in her walking of expressing her disinterest in filming one more trial. A short while later, the infant begins to explore specific adaptations of upright locomotion, namely, running and galloping. The acquisition of these that they are interdependent and emerge together. If the time c differs, then it would suggest that one may be leading the other, o another way, one may be rate limiting.

In a recent paper, we have described the development of intra coordination in three infants followed longitudinally from their walking steps through to their first year of walking (Clark & Phil 1993). Analyses included examination of the thigh and shank pl plane trajectories as well as the relative phasing relationships betw the two segments. While the newly walking infants seem to be attrac to the same dynamic solution as the adult walker, theirs is an unsta solution. It is not until the infants had been walking three months t both the phase plane trajectories and the relative phasing measu settled into a stable regime. Just as we had observed in interli coordination, intralimb coordination appears to be adultlike by the ti the infant has been walking for three months. Indeed, the results fro the intralimb investigation would suggest that intralimb and interlin coordination develop together. Examination of gait parameters, such the percentage of time spent in double support (considered an indicat of postural stability), again points to the importance of balance in th scaffolding of the coordinative relationship. Those constraints that sur round and shape the system at one level of organization (i.e., within the limb) are influential at the next level of organization (i.e., between the limbs). Therefore, for example, postural stability affects the dynamics of each component leg's oscillations as well as the coupling between them, as noted earlier.

Other studies on developing intralimb coordination in humans have shown fairly tight intralimb synergies at younger ages for newborn stepping and kicking (Thelen & Fisher, 1982, 1983). Thelen, Ridley-Johnson, & Fisher (1983) noted, also, that alternating leg movements were present in the newborn period, which they interpreted as evidence for the presence of bilateral interlimb coordination synergies. Developmental changes in both intra- and interlimb coordination followed a similar time course over the months preceding the acquisition of walking. Specifically, the interjoint couplings became less tightly coupled and then changed their relationship before combining again into more tightly coupled and larger units. This process of individuation of joints (differentiation of diffusely organized motor patterns) was accompanied by asymmetrical loss of early interlimb coordination. As the intralimb coordination reintegrated into coherent, larger functional groups, interlimb coordination reemerged in the form of simultaneous as well as alternating coordination (Thelen, 1985). The parallel time course of intra- and interlimb coordination during this time period reflects our own findings of synchronous rather than sequential development of intralimb and interlimb coordination during acquisition of bipedal locomotion.

new gait forms represent what Saltzman and Munhall (1992) call quantitative bifurcations among a set of topologically identical forms exhibited by the system. We now turn to a brief discussion of these other gait forms.

IV. Later Emerging Gait Forms: Running and Galloping

The gait of running typically emerges a few months after independent walking and has been distinguished from the latter by a flight phase in place of the double support phase found in walking (Forrester, Phillips, & Clark, 1993). The interlimb coordination at footfall, however, remains at 50% phasing as the legs alternate striking the ground. Later still, around two years or older, galloping will emerge as a viable gait form, although it does not replace walking and running as the predominant mode of locomotion. Galloping occurs when one leg is in front of the other with the leading leg executing a walking step while the rear leg executes a running (or leaping) step (see Figure 6). This gait

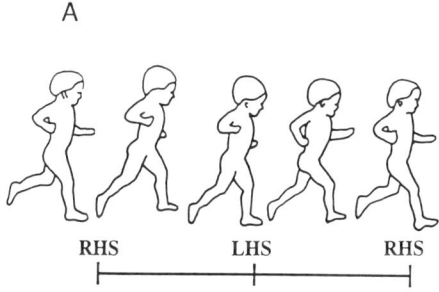

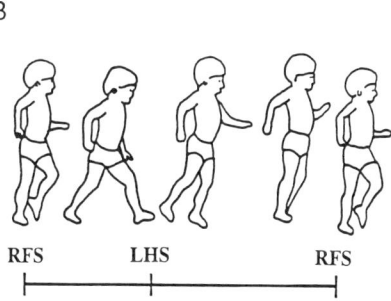

FIGURE 6 Running (A) and galloping (B) locomotor skills showing foot-strike patterns. RHS, right heel strike; LHS, left heel strike; RFS, right foot strike. (Reprinted with permission from Whitall, 1989.)

mode results in both double support and flight phases indicative of walking and running, respectively. However, detailed kinematic analysis suggests that the gait may more accurately be described as an asymmetric run (Whitall & Caldwell, 1992). In any event, the important aspect of galloping is that it is the first time that a child learns to intentionally change the coupling of the lower legs from the alternating phasing patterns and adopt an asymmetric coordination.

Our investigation of these two gaits follows two lines: one ascertaining the stability of form and the other examining transitions. In an early study, the properties of phase locking, entrainment, and structural stability were examined in preferred-speed running and galloping of 40 subjects ranging from two years to adulthood (Whitall, 1989). For the run, all ages demonstrated quite stable alternating phasing (temporal and amplitude) with low variability and no effect of unilateral leg weighting. Stability of phasing did increase slightly with age, however. Similarly, *all* ages accomplished a gallop that was only slightly less stable than the run in terms of both variability and weighting effect. The most interesting finding was that subjects did not find the same phase-locked mode of coordination for the gallop. Regardless of age, there was a tendency to cluster around two predominant temporal phase modes, namely 33/66 and 25/75% phasing (120° and 90°), with the former appearing more attractive than the latter. Curiously enough, dancers have long known about these attractive phase-locking modes, referring to them as binary and ternary gallops, respectively. Amplitude phasing was also somewhat variable between subjects, but here no attractive modes appeared; rather, the foot placement seemed to be a matter of personal style.

A further interesting feature is that the actual temporal phase mode exhibited is not idiosyncratic to the subject but related to the progression velocity, so that higher velocity moves the subject from 33 to 25% phasing (Whitall & Clark, 1986). Thus, the gallop is unique in the study of human interlimb coordination, since it illustrates an attractive asymmetric (off-phase) coupling, which is speed related and apparently requires relatively little practice (cf. Zanone & Kelso, Chapter 22, this volume). Since other lower-limb gait patterns such as crawling, walking, running, and skipping all show the stable out-of-phase pattern, these characteristics are presumably directly related to the orientation of the legs and the concomitant changes in their biomechanical characteristics. For example, in order to go faster, the system requires more muscular effort that is specific and different for each leg. The consequences of this differential input to each leg result in a change in phasing relationship. In effect, one leg (the leading) acts like a forcing oscillator on the other leg as the legs move closer to being in-phase.

The fact that, when infants are young, the lower limbs show properties of interlimb coordination similar to those of generic coupled nonlinear limit-cycle oscillators does not answer the developmental question of how these shifts in coordination are acquired. More recent work focuses on the transition from walking to running (and galloping). How does the infant manage the changing gait forms and what are the developmental (as opposed to real-time) control parameters that enable the transition to occur? In two longitudinal case studies we have quantified fast walking in the month preceding the first run, the first observed runs, and runs performed 1–2 months later. As the infants walk faster, the period of double support becomes shorter and eventually disappears. However, during the transitional period of several months, there are often cycles with flight along with cycles of no-flight even within the same trial (Forrester et al., 1993; Whitall & Getchell, 1993). Despite these inconsistencies, relative temporal phasing between the limbs remains stable (low variability) at 50% throughout. Thus, unlike the situation in the novice walker, where high variability of relative phase accompanied a gait transition, the new runner's interlimb coordination was consistent despite the change of gait mode. Rather than disrupt the coordination, the increased force production required to produce a flight phase actually maintained or increased the stability of coupling between the limbs.

If the interlimb coordination is not reorganized at the footfall level, is coordination actually reorganized at the intralimb level? Using qualitative dynamics derived from the relative phasing coordination of the knee (i.e., thigh and shank segmental coordination) for a walk and run, Forrester et al. (1993) suggest there are topological similarities between the two gaits. The only differences they noted were at foot strike, where the trajectory does not always display an inner loop in walking. Presumably this is due to less damping of the system with the lower ground reaction forces of walking. These data, along with those of adults, suggest that there is little intralimb coordination change at the onset of running (see Forrester et al., 1993, for more details of this argument). If the shift from walking to running does not occur at the level of behavioral coordination, then what prevents a child from running as soon as walking appears? Two prime candidates for rate-limiting control parameters are the ability to produce a force of two times body weight on one leg and the ability to land and balance in single support (Clark & Whitall, 1989). Recently we have hypothesized that a related factor in the shift from one gait to another is the ability to manage different energy strategies that characterize specific gait forms (Forrester et al., 1993). Changing energetics have long been implicated in real-time gait shifts but this idea has not received much attention in a developmental context. Current work includes an examination of this issue in gait development (e.g., Forrester *et al.*, 1993) as well as adult comparisons (e.g., Caldwell & Whitall, 1993).

V. Summary

Employing the conceptual framework of dynamical systems, we have sought to understand the development of bipedal interlimb coordination. This perspective has led us to examine the phasing relationships both within and between the legs of human infants and adults. In a series of experiments, we have described these coordinative relationships as infants first attempt to walk, and thereafter until they are able to change the coupling of the symmetric alternating gait form to perform the asymmetric galloping. The general model of coupled nonlinear limit-cycle oscillators has properties that we have tested in a behavioral and topological (though not yet in a mathematical) manner. Building a mathematical model for developmental transitions will not be a trivial task in the future. However, even at a metaphorical and qualitative level, the framework of dynamical systems theory appears to be potentially rich for those seeking to understand the development of bipedal locomotion as well as other developmental phenomena (cf. Fogel & Thelen, 1987; Thelen & Ulrich, 1991; Wolff, 1987). This theory not only provides an understanding about how such systems are organized, but also offers potential solutions for how transitions to new states occur. While this approach has afforded us insights into the organization of the infant's limbs in the year following those first unsteady but independent steps, it is clear there is still much to be understood about the development of bipedal interlimb coordination.

References

Bekoff, A. (1981). Embryonic development of the neural circuitry underlying motor coordination. In W. M. Cowan (Ed.), *Studies in developmental neurobiology* (pp. 134–170). New York: Oxford University Press.

Bekoff, A., & Trainer, W. (1979). The development of interlimb coordination during swimming in postnatal rats. *Journal of Experimental Biology*, **83**, 1–11.

Bradley, N. S., & Smith, J. L. (1988). Neuromuscular patterns of stereotypic hindlimb behaviors in the first two postnatal months. I. Stepping in normal kittens. *Developmental Brain Research*, **38**, 37–52.

Bril, B., & Brenière, Y. (1992). Postural requirements and progression velocity in young walkers. *Journal of Motor Behavior*, **24**, 105–116.

Caldwell, G. E., & Whitall, J. (1993). *An energetic comparison of symmetrical and asymmetrical human gait*. Manuscript submitted for publication.

Clark, J. E., & Phillips, S. J. (1993). A longitudinal study of intralimb coordination in the first year of independent walking: A dynamical systems analysis. *Child Development*, **64**, 1143–1157.

Clark, J. E., & Whitall, J. (1989). Changing patterns of locomotion: From walking to skipping. In M. Woollacott & A. Shumway-Cook (Eds.), *Development of posture and gait across the lifespan* (pp. 128–151). Columbia, SC: University of South Carolina Press.

Clark, J. E., Whitall, J., & Phillips, S. J. (1988). Human interlimb coordination: The first 6 months of independent walking. *Developmental Psychobiology,* **21,** 445–456.

Clark, J. E., Truly, T. L., & Phillips, S. J. (1990). A dynamical systems approach to understanding the development of lower limb coordination in locomotion. In H. Bloch & B. I. Bertenthal (Eds.), *Sensory-motor organizations and development in infancy and early childhood* (pp. 363–378). Dordrecht: Kluwer.

Clark, J. E., Truly, T. L., & Phillips, S. J. (in press). On the development of walking as a limit cycle system. In E. Thelen & L. Smith (Eds.), *Dynamical systems in development: Applications.* Cambridge, MA: MIT Press.

Cohen, A. H. (1988). The evolution of vertebrate central pattern generator locomotion. In A. H. Cohen, S. Rossignol, & S. Grillner (Eds.), *Neural control of rhythmic movements in vertebrates* (pp. 129–166). New York: Wiley.

Crutchfield, J. P., Farmer, J. D., Packard, N. H., & Shaw, R. S. (1987). Chaos. *Scientific American,* **254**(12), 46–57.

Fentress, J. C. (1978). *Mus musicus:* The developmental orchestration of selected movement patterns in mice. In G. M. Burghardt & M. Bekoff (Eds.), *The development of behavior: Comparative and evolutionary aspects* (pp. 321–342). New York: Garland STPM Press.

Fogel, A., & Thelen, E. (1987). The development of expressive and communicative action in the first year: Reinterpreting the evidence from a dynamic systems perspective. *Developmental Psychology,* **23,** 747–761.

Forrester, L. W., Phillips, S. J., & Clark, J. E. (1993). Locomotor coordination in infancy: The transition from walking to running. In G. J. P. Savelsbergh (Ed.), *The development of coordination in infancy* (pp. 359–393). New York: Elsevier.

Garfinkel, A., Spano, M. L., Ditto, W. L., & Weiss, J. N. (1992). Controlling cardiac chaos. *Science,* **257,** 1230–1235.

Kay, B. A., Kelso, J. A. S., & Saltzman, E. L. (1991). Steady-state and perturbed rhythmical movements: A dynamical analysis. *Journal of Experimental Psychology: Human Perception and Performance,* **17,** 183–197.

Kelso, J. A. S., & Schöner, G. (1988). Self-organization of coordinative movement patterns. *Human Movement Science,* **7,** 27–46.

Kelso, J. A. S., & Tuller, B. (1984). Converging evidence in support of common dynamical principles for speech and movement coordination. *American Journal of Physiology 246 (Regulatory, Integrative, Comparative Physiology),* **15,** R928–R935.

Kelso, J. A. S., Holt, K. G., Rubin, P., & Kugler, P. N. (1981). Patterns of human interlimb coordination emerge from the properties of nonlinear limit cycle oscillatory processes: Theory and data. *Journal of Motor Behavior,* **13,** 226–261.

Kopell, N. (1988). Toward a theory of modeling central pattern generators. In A. H. Cohen, S. Rossignol, & S. Grillner (Eds.), *Neural control of rhythmic movements in vertebrates* (pp. 369–413). New York: Wiley.

Kopell, N., & Ermentrout, G. B. (1988). Coupled oscillators and the design of central pattern generators. *Mathematical Biosciences,* **90,** 87–109.

McGraw, M. B. (1940). Neuromuscular development of the human infant as exemplified in achievement of erect locomotion. *The Journal of Pediatrics,* **17,** 741–771.

Newell, K. M. (1985). Coordination, control and skill. In D. Goodman, R. B. Wilberg, & I. M. Franks (Eds.), *Differing perspectives in motor learning, memory, and control* (pp. 295–317). Amsterdam: North-Holland.

Pavlidis, T. (1973). *Biological oscillators: Their mathematical analysis.* New York: Academic Press.

Prechtl, H. F. R. (1986). Prenatal motor development. In M. G. Wade & H. T. A. Whiting (Eds.), *Motor development in children: Aspects of coordination and control* (pp. 53–64). Dordrecht, The Netherlands: Martinus Nijhoff.

Raibert, M. H. (1986). *Legged robots that balance.* Cambridge, MA: MIT Press.

Rand, R. H., Cohen, A. H., & Holmes, P. J. (1988). Systems of coupled oscillators as models of central pattern generators. In A. V. Cohen, S. Rossignol, & S. Grillner (Eds.), *Neural control of rhythmic movements in vertebrates* (pp. 333–367). New York: Wiley.

Rosen, R. (1970). *Dynamical system theory in biology. Volume I: Stability theory and its application.* New York: Wiley.

Saltzman, E. L., & Munhall, K. G. (1992). Skill acquisition and development: The roles of state-, parameter-, and graph-dynamics. *Journal of Motor Behavior,* **24,** 49–57.

Schmidt, R. C., Treffner, P. J., Shaw, B. K., and Turvey, M. T. (1992). Dynamical aspects of learning an interlimb rhythmic movement pattern. *Journal of Motor Behavior,* **24,** 67–84.

Thelen, E. (1983). Learning to walk is still an "old" problem: A reply to Zelazo (1983). *Journal of Motor Behavior,* **15,** 139–161.

Thelen, E. (1985). Developmental origins of motor coordination: Leg movements in human infants. *Developmental Psychobiology,* **18,** 1–22.

Thelen, E. (1989). Evolving and dissolving synergies in the development of leg coordination. In S. A. Wallace (Ed.), *Perspectives on the coordination of movement* (pp. 259–281). New York: Elsevier.

Thelen, E., & Fisher, D. M. (1982). Newborn stepping: An explanation for a "disappearing reflex." *Developmental Psychology,* **18,** 760–775.

Thelen, E., & Fisher, D. M. (1983). The organization of spontaneous leg movements in newborn infants. *Journal of Motor Behavior,* **15,** 353–377.

Thelen, E., & Ulrich, B. D. (1991). Hidden skills: A dynamic systems analysis of treadmill stepping during the first year. *Monographs of the Society for Research in Child Development,* **56**(1) (Whole Serial No. 223).

Thelen, E., Ridley-Johnson, R., & Fisher, D. M. (1983). Shifting patterns of bilateral coordination and lateral dominance in the leg movements of young infants. *Developmental Psychobiology,* **16,** 29–46.

Thelen, E., Kelso, J. A. S., & Fogel, A. (1987). Self-organizing systems and infant motor development. *Developmental Review,* **7,** 39–65.

Thompson, J. M. T., & Stewart, H. B. (1986). *Nonlinear dynamics and chaos.* New York: Wiley.

von Holst, E. (1939/1973). *The behavioral physiology of animal and man.* Coral Gables, FL: University of Miami Press.

Whitall, J. (1989). A developmental study of the interlimb coordination in running and galloping. *Journal of Motor Behavior,* **21,** 409–428.

Whitall, J., & Caldwell, G. E. (1992). Coordination of symmetrical and asymmetrical human gait: Kinematic patterns. *Journal of Motor Behavior,* **24,** 339–354.

Whitall, J., & Clark, J. E. (1986). *The interlimb coordination of galloping: Theoretical predictions and data.* Paper presented at the annual meeting of the North American Society for the Psychology of Sport and Physical Activity, Scottsdale, Arizona, June.

Whitall, J., & Getchell, N. (1993). *From walking to running: Using a dynamical systems approach on the development of locomotor skills.* Manuscript submitted for publication.

Whitall, J., Block, M. B., & Clark, J. E. (1992). The development of walking: Interlimb coordination as coupled limit cycle systems. In M. Woollacott & F. Horak (Eds.), *Proceedings of XIth International Symposium on Posture and Gait: Control Mechanisms* (pp. 315–318). Eugene, OR: University of Oregon.

Williams, T. L., Sigvardt, K. A., Kopell, N., Ermentrout, G. B., & Remler, M. P. (1990). Forcing of coupled nonlinear oscillators: Studies of intersegmental coordination in the lamprey locomotor central pattern generator. *Journal of Neurophysiology,* **64,** 862–871.

Wolff, P. H. (1987). *The development of behavioral states and the expression of emotions in early infancy.* Chicago, IL: University of Chicago Press.

Woollacott, M., & Horak, F. (1992). *Posture and gait: Control mechanisms* (Vol. II) (p. 317). Eugene, OR: University of Oregon.

20

Shifting Patterns of Interlimb Coordination in Infants' Reaching:
A Case Study

Daniela Corbetta and Esther Thelen

Department of Psychology
Indiana University
Bloomington, Indiana

I. Introduction
II. Part 1: The Underlying Dynamics of Reaching
   A. The Context of the Study
   B. The Interlimb Strategy of Reaching
   C. The Temporal Organization of Spontaneous Arm Movements
   D. Defining Preferred Patterns of Interlimb Coordination
   E. Interlimb Movement Repertoire during the First Year of Life
   F. Preliminary Discussion
III. Part 2: The Transition from Bimanual to Unimanual Reaching
   A. The Dynamic Processes of Interlimb Coupling
   B. Bilateral Torques in Reaching
   C. Conclusions
IV. General Discussion
    References

I. Introduction

Since the classic studies of Halverson (1931) and Piaget (1936), developmental psychologists have sought to understand how infants learn to reach and grasp objects. Several studies have provided rich and detailed descriptions of the improvements in accuracy, speed, and smoothness of the reach trajectory, coordination of vision and reaching, and successive adaptation of the reach to the grasped object (Bushnell, 1985; Fetters & Todd, 1987; Mathew & Cook, 1990; McDonnell, 1975, 1979; Mounoud, 1983; von Hofsten, 1979, 1982, 1991; Thelen, Corbetta, Kamm, Spencer, Schneider, & Zernicke, 1993. These studies have focused primarily on how infants bring one hand to contact an object. However, casual observations of young infants reveal that the well-studied one-handed reach is only part of the story.

Indeed, a few studies have shown that the ability of infants to reach for objects with one hand is often preceded by an initial period of two-handed responses. White, Castle, and Held (1964), for example, observed that the one-handed reaches of 5- to 6-month-old infants emerged from earlier initial bilateral and symmetrical object-oriented responses. They also found that these bilateral symmetrical responses of both arms predominated in the repertoire of spontaneous movement of infants 4 to 6 weeks before the onset of reaching. Flament (1974, 1975) reported similar observations. She found that early reaches for small objects were often performed with a bilateral arm extension toward the object even if the object was then only unimanually grasped. Finally, two other studies reported that the development of reaching for small graspable objects followed a similar story (Bresson, Maury, Piéraut-Le Bonniec, & de Schonen, 1977; Ramsay & Willis, 1984). These researchers found that the ability to reach for an object on a support with one hand was preceded by a stage of two-handed collaborative reaching, where one hand approached the support while the second one reached for the object. Thus, all these studies showed that when the development of reaching is considered as a two-limb system, the ability of infants to reach with one arm emerged from an initial period of two-handedness.

This developmental picture of two-handed reaching contrasts with the ability of adults to differentiate the use of one or two arms for reaching. Indeed, adults generally lift large or heavy objects using two arms and seize small objects using one arm. Young infants, on the contrary, do not seem to show such ability when they begin to reach.

From these observations, several questions arise. First, why does a period of two-handedness precede the development of one-handed reaching? In other words, why do young infants prefer to use two hands for reaching when the object is small enough to be grasped with one hand? And second, what changes in the perceptual motor abilities of infants when they shift from two hands to one hand? The studies reviewed above did not address these questions.

This first problem related to infants' early two-handedness can be paralleled to a second puzzling aspect of the development of infant's reaching; two-handed patterns appear and disappear at different times in development. Indeed, the few studies that have investigated the development of reaching as a two-handed behavior emphasized the changing nature of interlimb coordination over the infants' first year. Here again, the development of reaching seems to be more complex than improvements in a single reaching arm. Rather, the development is marked by periods of upper-arm bilateral responses, alternating with periods of unilateral activity, with shifts between symmetrical and asymmetrical movements occurring several times during the first year (Ames, 1949; Flament, 1974, 1975; Gesell, 1939, 1946; Gesell & Ames, 1947; Ramsay, 1985). In particular, during the second half of the first year, the spatiotemporal linkage of interlimb coordination undergoes several fluctuations. Goldfield and Michel (1986a,b), for example, found that movement simultaneity and movement direction varied according to both the age of the subjects and the task conditions. These reports gave an accurate description of the developmental changes of infants' interlimb coordination but they did not explain why these seemingly mysterious shifts occur in the bilateral organization of the reach. Only Rochat (1992) suggested that postural control might influence the development of upper-arm bilateral organization. He found that when infants were able to sit alone they developed more lateralized reaches, whereas prior to independent sitting their reaches were primarily bilateral.

In summary, the study of early reaching has progressed along two parallel, but largely unrelated, tracks. On one hand, we have gained considerable understanding of infants' developing control of one arm. On the other hand, we have documented the puzzling shifts of bilateral coordination, but the reason why these shifts occur remains yet unresolved.

In the present chapter we aim to show that these developmental processes are related and that shifts in interlimb coordination can be understood through a new approach to the study of movement. Since Bernstein (1967), it has been widely recognized that coordination and control are not limited to a single limb or body part, but that every act is in a sense a whole body act. This is especially true in infants, whose movements are often generalized and diffuse. Our purpose, therefore, is to view infant reaching in a new way—from the perspective of the more general interlimb dynamics—to better understand the nature of the shifts of bilateral coordination.

The work we present here is inspired by the dynamical systems perspective. This perspective presupposes that the formation of movement patterns emerges naturally from the cooperative coupling of the ensemble of components that constitute the behavior (i.e., the collective activity of the neural, muscular, skeletal, and vascular components of the body segments involved in the movement). Several theoretical and experimental investigations have demonstrated that this emergent cooperative coupling is governed by well-defined dynamics of coordination, the "intrinsic dynamics." The intrinsic dynamics are defined by characteristic stable and unstable movement patterns that reflect the preferred or nonpreferred coordination tendencies of the system. These studies also have shown that these intrinsic coordination tendencies do not only underlie the actual performed patterns of coordination, but correspond to an entire dynamics of coordination within various energy parameters, across a normal range of physical environments, and with or without specific task demands (Jensen, Ulrich, Thelen, Schneider, & Zernicke, in press; Kelso, 1984; Kelso & Schöner, 1988; Kugler & Turvey, 1987; Schöner & Kelso, 1988; Schöner, Zanone, & Kelso, 1992; Zanone & Kelso, 1991; for a tutorial review, see Jeka & Kelso, 1989, and chapters in Part II, this volume).

In terms of infant reaching, then, we expected that the patterns of both one- and two-handed reaching are part and parcel of the same dynamics that govern nonreaching movements and that they are subject to the same principles of coordination and control. This dynamical systems perspective led us to study infant reaching in a much wider context, therefore, than the ones previously adopted by most researchers in the field. First, we did not limit our measurements to the movements performed by the reaching arm alone. Rather, we studied the movements performed by both arms during the reach itself and within the movement context from which the reach emerges—the entire intrinsic dynamic. Second, in contrast to the usual cross-sectional studies that compare groups of subjects at different ages, we narrowed our investigations to only a few subjects, and observed them at frequent intervals over an extended period of development to capture their significant behavioral changes.

In adopting such methods, our goals were twofold. First, we wanted to describe the coordination repertoire of two-handed movement or coupling dynamics that underlie the developmental course of infants' reaching. In other words, we wanted to identify if characteristic patterns of interlimb coordination predominated during certain periods in development and, if so, to track the evolution of these coordination patterns during the first year of life. Second, we wanted to capture the processes that underlie behavioral shifts in the development of reaching. In particular, we wanted to understand what changed in the coordination of the two-limb system to entail a modification of the coupling dynamics from one preferred pattern of behavior during one period of development (i.e., reaching using two hands) to another new pattern of behavior at another developmental period (i.e., reaching with only one hand).

This chapter is organized into two parts that will approach these goals respectively. The work we present here is based on new results from the developmental story of one subject, Nathan, whom we observed closely over his first year. In the first part, we introduce a new method to describe the underlying dynamics of infants' upper-limb movement repertoire. We identify the characteristic patterns of spontaneous interlimb coordination that predominate at certain periods in development and show how these characteristic patterns of interlimb coordination affected Nathan's age-related tendencies to use one or two hands for reaching. In the second part of the chapter, we focus on one period of transition, when Nathan shifted from a predominantly two- to a predominantly one-handed type of reaching. A deeper analysis of Nathan's bimanual reaches allowed us to identify the coordination processes that underlie the strong "two-handedness" of the first period and the subsequent shift to a predominant lateralized behavior that occurred a few weeks after. In our conclusions, we will discuss the implications of our findings for understanding the development of interlimb coordination in infancy and consider the generalization of our findings to other subjects.

II. Part 1: The Underlying Dynamics of Reaching

The goal of this section is to determine whether the type of reach (one- or two-handed) used during different periods in development is a reflection of changes in the underlying dynamics of the motor system. In other words, we are asking if the background of spontaneous non-object-oriented movements from which the reach emerges show related intrinsic dynamics, that is, similar tendencies to couple or decouple the movements of both arms. To address this problem we developed a method to describe continuously in time the modulations that occurred between the simultaneous movements of each arm before, during, and after the reach. This method allowed us to evaluate the behavioral changes that occurred during different periods in development and to define whether characteristic patterns of spontaneous interlimb coordination were related to one- or two-hand use during the reach. We shall briefly present the protocol of the study, and then we shall present three successive steps of movement analyses that were designed to identify periods of stability and periods of change in both Nathan's reaching and spontaneous behavior.

A. The Context of the Study

The data we present here are preliminary results from a longitudinal study of the development of reaching in four infants (Thelen *et al.*, 1993). In this chapter, we introduce the data of one subject only, Nathan, whose movement analysis has been completed. Nathan visited the laboratory every week from the age of 3 weeks until 30 weeks, and then every other week from 30 to 52 weeks. At every visit, Nathan sat in a slightly reclined chair without armrests. Nathan's torso was stabilized with a wide strap, but his arms were free to move fully in every direction. Chairs of increasing size but identical design were used during the year of testing in order to adapt to Nathan's growing size. Furthermore, at every session, we used small, attractive graspable toys that we presented at midline at eye height. We always used the same toys from the first to the last visit of Nathan.

During the experiment, the 3-dimensional positional data of the shoulder, elbow, wrist, and hand of both Nathan's arms were recorded with a WATSMART motion analysis system using a sampling rate of 150 Hz. At every session, we collected multiple 14-sec trials to capture object-directed movements as well as spontaneous movements. A few seconds after beginning data collection, we brought the toy to Nathan's reaching space and held it there until he grabbed it. If he grabbed the toy before the end of the trial, we also collected data on manipulatory behaviors.

The kinematic data that we present here are a subset of the total amount of data collected. We include here all the data (1) in which Nathan was active and alert, regardless of whether he was reaching or not, and (2) which satisfied our data quality criteria. We excluded from our data set all the trials or portions of trials in which Nathan was sleepy or passive (when sucking on a pacifier, for example). Moreover, we excluded many manipulatory behaviors that did not involve major arm activity. Within this first behavioral selection, we used the 3-dimensional data only when the markers were seen by the camera during at least 70% of the trial and only when the gaps of unseen markers were smaller than one-third of sampling frequency. The remaining segments of data were then splined linearly and filtered using a fourth-order Butterworth filter (for more details on the methods and data handling, see Thelen *et al.,* 1993). The analyses of the kinematic data presented in the following sections include only the data that contained information for both hands at the same time.

B. The Interlimb Strategy of Reaching

In this section we identify periods of predominantly one- or two-handed types of reaching in Nathan's first year. These interlimb strategies were coded from the videotapes as bimanual when both hands moved together while reaching for the toy. We coded them as unimanual when only one hand was used for reaching, while the other hand remained steady along the side of the body. These data are depicted in Figure 1 for the whole period of testing, from 5 to 52 weeks old.

Nathan's first reaches emerged at week 12, when he began to successfully contact the toy. The bars at weeks 10 and 11 represent arm extensions produced in the direction of the toy while Nathan was looking at it. They were coded as reaches even though they did not result in a successful contact with the toy. Figure 1 shows clear and distinct periods in the bilateral organization of Nathan's reaches, with periods predominated by one type of reaching strategy and abrupt transitions from one type of strategy to another. From 12 to 20 weeks, Nathan mostly used two hands to reach for the toy. From 21 to 42 weeks, he predominantly used one hand. Toward the end of the first year, he mostly used two hands again for reaching.

During the first period of bimanual hand use, Nathan did not show any hand preference; either hand could contact the toy at first. However, from week 21, when Nathan shifted into a predominantly unilateral behavior, he also started to show a preference in the hand used for reaching. During a 6-week period, from 21 to 27 weeks old, Nathan always reached for the toy using the right hand first. Then, from week 28, he returned to a more ambidextrous although predominantly unilateral behavior, using sometimes the right and sometimes the left hand for reaching.

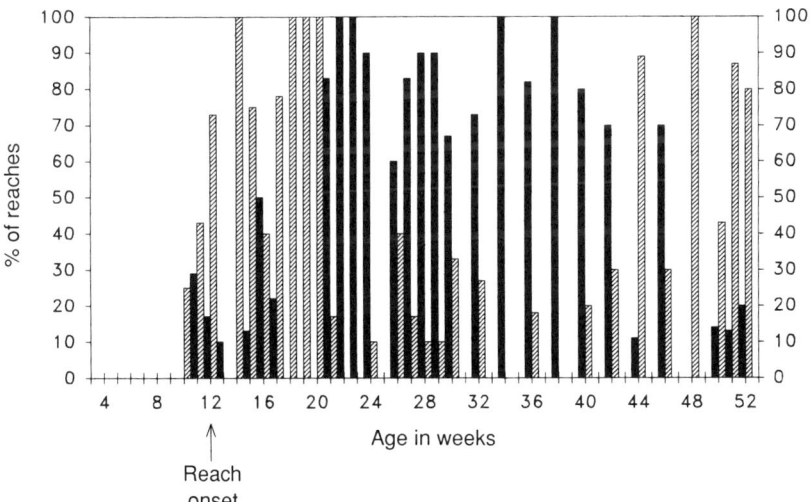

FIGURE 1 Percentage of reaches performed using one or two hands during Nathan's first year of life. These values are normalized as a function of the total number of toy presentations. Bimanual, ▨ ; unimanual, ■ .

C. The Temporal Organization of Spontaneous Arm Movements

Do we see similar developmental shifts when we look at the movements Nathan produced during his spontaneous activity? Is there a stronger synchrony in movement coupling between limbs when Nathan predominantly used two hands for reaching? To address these problems we examined the temporal organization of Nathan's arm movements performed before, during, and after the reach.[1] In other words, during the 14 seconds of data we collected for each trial, we looked at how Nathan was moving both arms together during the 1-2 seconds of the reach and also during the other 12-13 seconds of the trial when Nathan was *not* reaching. In general, each reach represented only a small portion of the total trial and was often embedded in ongoing spontaneous movements such as arm "flapping" or arm waving (see also descriptions in Thelen, Jensen, Kamm, Corbetta, Schneider, & Zernicke, 1991; Thelen *et al.*, 1993). At a later age, when Nathan was able to grasp the toy without difficulty, the spontaneous movements sometimes included seemingly directed activities such as shaking the toy or throwing the toy. We nonetheless called these nonreaching movements "spontaneous." These movements could involve the activity of one arm only or the activity of the two arms together. When moving both arms, the movements exhibited either similar or different patterns. By analyzing these free and unconstrained movements, we sought to determine if the spontaneous coordination patterns of these nonreaching movements showed similar characteristics to the coordination patterns observed during the reach. In this section, we introduce the main elements of the analysis we developed to describe the natural coordination dynamics of Nathan's spontaneous arm movements. We will use several examples to show how this method can describe different patterns of interlimb coordination and how it allows the assessment of preferred patterns of interlimb coordination in development.

To describe the temporal organization of Nathan's spontaneous interlimb movements we used a running correlation technique that we applied to the continuous resultant velocity profiles of both hands.

---

[1] We emphasize here that the technique we present to analyze the coordination dynamics of Nathan's spontaneous interlimb movements captures only whether both arms accelerate and decelerate in synchrony. This measure does not provide information about movement directions in the 3-dimensional space. Thus, it does not indicate whether movements occurred in a symmetrical or asymmetrical fashion. To facilitate our task and to emphasize that the presented technique is primarily a measure of movement synchrony and not of movement symmetry, we will refer to it as the analysis of the temporal characteristics of the movements, even though we are aware that velocity is composed of both temporal and spatial information (such as movement amplitude).

A 1-sec window size was shifted continuously frame-by-frame (every 7 msec) along the velocity profile of the total segment of data. At every shift of the window a correlation value was calculated. These successive shifts of 7 msec resulted in a continuous correlation function that fluctuated progressively between +1 and −1 values and described the different temporal patterns of interlimb coordination that Nathan produced during every segment of data. Exemplar plots of correlation functions with the corresponding velocity profiles of both hands are presented in Figures 2 and 3.

Figure 2 shows two short segments of Nathan's *related spontaneous arm movements*. In the example on the left, the bimanual velocity profiles show similar temporal patterns: both hands accelerate and decelerate at the same time; the corresponding running correlation function fluctuates around high positive values. This means that during the 4.5 sec of movement plotted on Figure 2, Nathan's right and left arms predominantly moved in synchrony. In contrast, the right side of the figure is an example of movements with reversed temporal pat-

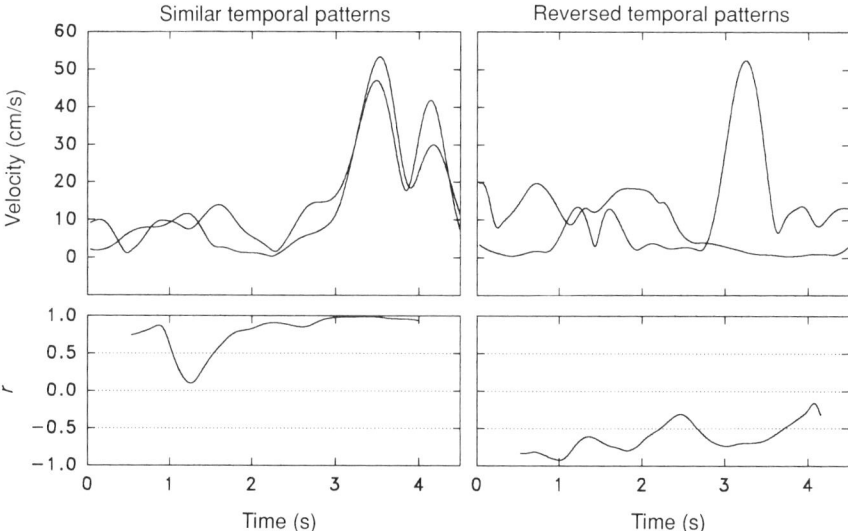

FIGURE 2 Exemplar plots of 4.5 sec of Nathan's spontaneous movements showing related arm movements. Top panels display the resultant 3-dimensional velocities for right and left hands and bottom panels show the corresponding correlation functions. Left panels show an example of similar temporal patterns: Both hands speed up and slow down at the same time; the correlation function fluctuates around high positive values. Right panels show an example of reversed temporal patterns: When one hand speeds up the other one slows down; the corresponding correlation function fluctuates around negative values.

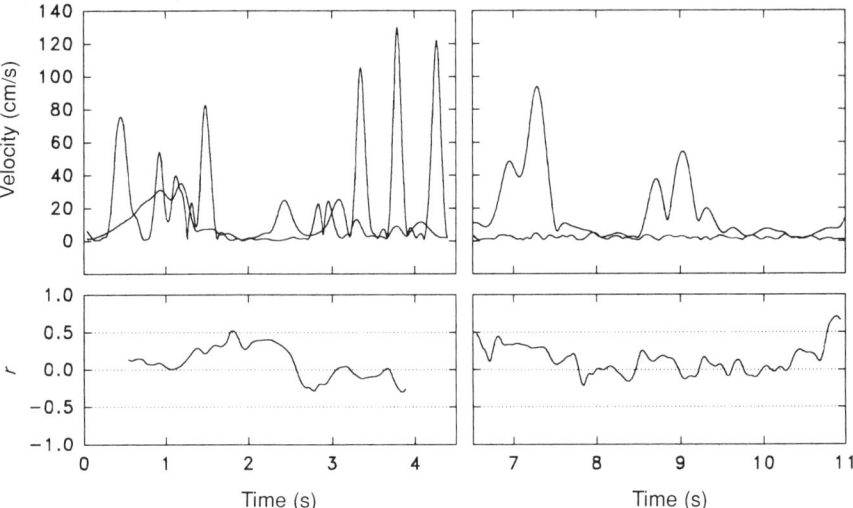

FIGURE 3 Exemplar plots of 4.5 sec of Nathan's spontaneous movements showing unrelated arm movements. Top panels display the resultant 3-dimensional velocities for right and left hands and bottom panels show the corresponding correlation functions. Both examples show that both hands performed different temporal patterns. Left panels show that while one hand moves slowly the other one performs several rapid velocity changes. Right panels show that while one hand moves the other remains steady. Both examples correspond to correlation values that fluctuate around zero.

terns. The velocity profiles reveal that when one hand was speeding up, the other hand was slowing down. This pattern of movement resulted in a negative correlation function, as shown in the corresponding graph.

Unrelated movement patterns can also be described using the same technique. The two short intervals of movement plotted on Figure 3 show that both arms produced different patterns of movement. The left side of the figure illustrates that while one hand moved quickly with rapid changes in velocity, the other hand moved slowly with a totally different temporal pattern. The right side shows that while one hand moved the other hand was still. Both examples gave correlation functions that fluctuated around zero, as shown by the two graphs below. This means that during these two segments Nathan moved both arms, producing quite different temporal patterns.

D. Defining Preferred Patterns of Interlimb Coordination

This running correlation technique allowed us to determine whether preferred patterns of interlimb coordination predominated in Nathan's repertoire of spontaneous movement at different periods in development. Preferred patterns of interlimb coordination were determined by calculating the frequency rate of each correlation value measured from the ensemble of trials of a same session. Figures 4 and 5 show two examples of correlation functions with their frequency rate for two data segments of 9 sec.

Figure 4 illustrates a portion of Nathan's spontaneous movements where a *preferred pattern* of interlimb coordination predominated. As shown in the bottom graph, the correlation function largely fluctuated around high positive correlation values. Accordingly, the frequency rate of the correlation values represented on the small histogram on the lower right shows a high peak around high positive values. This indicates that during these 9 sec of movement, Nathan had a tendency to move his right and left arm together with similar temporal patterns. There are no negative correlation values in this example, which means that even when the movements were not highly correlated they never showed reversed temporal patterns, and remained in a range of values that were closer to movement synchrony.

Figure 5 shows a different pattern of temporal organization. In this example, *no preferred pattern* of interlimb coordination predominated. The frequency histogram reveals that the frequency rate of correlation

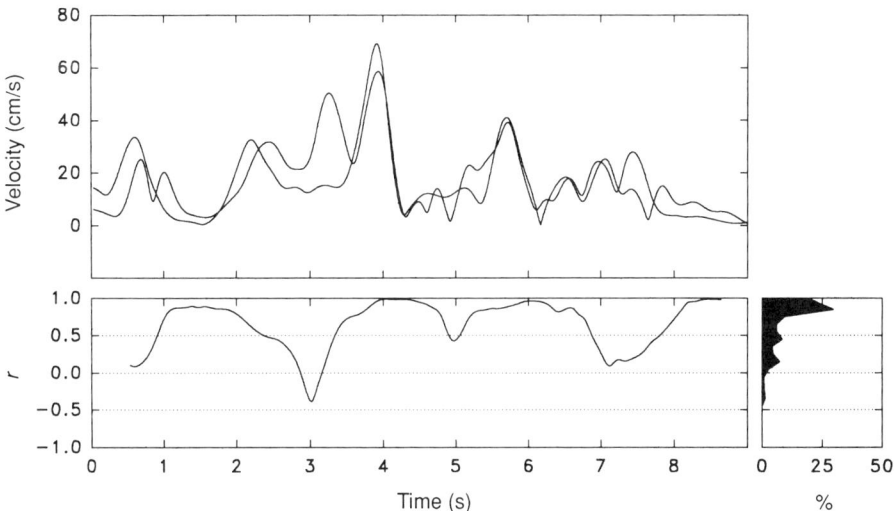

FIGURE 4 Exemplar plots of 9 sec of Nathan's spontaneous movements showing preferred pattern of interlimb coordination. Top panel displays the resultant 3-dimensional velocities for right and left hands. Bottom panels show the corresponding correlation function and frequency rate of the measured correlation values. During these 9 sec, Nathan moved both hands using predominantly similar temporal patterns. The correlation function fluctuates around positive values and the histogram in the lower right corner reveals an asymmetrical distribution of the correlation values, with a large peak around high positive values.

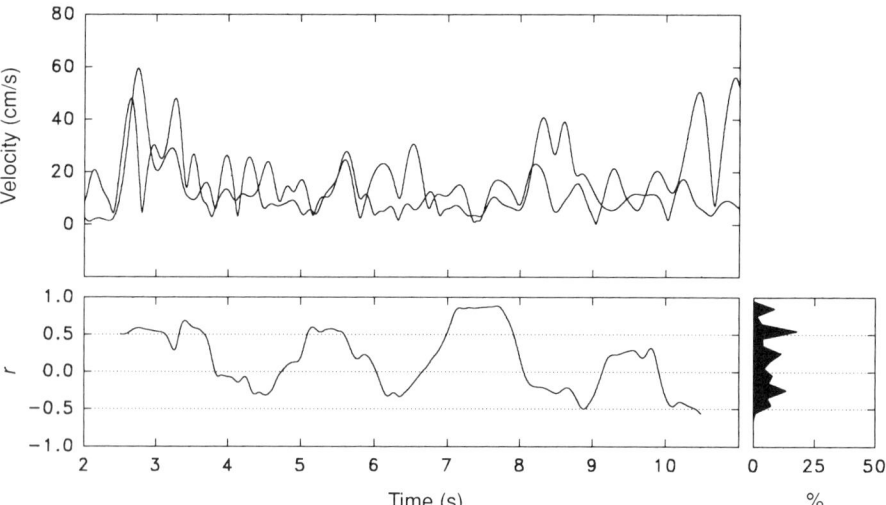

FIGURE 5 Exemplar plots of 9 sec of Nathan's spontaneous movements showing no preferred pattern of interlimb coordination. Top panel displays the resultant 3-dimensional velocities for right and left hands. Bottom panels show the corresponding correlation function and frequency rate of the measured correlation values. During these 9 sec, Nathan moved both hands, changing continuously the movement patterns and thus showing no predominant temporal pattern of interlimb coordination. The correlation function fluctuates back and forth from positive to negative values and the histogram in the lower right corner reveals a wide and almost equal distribution of correlation values in many different categories.

values is distributed almost equally among many different correlation states. Indeed, in this example, the correlation function fluctuates back and forth from positive to negative correlation values, revealing that during these 9 sec of movement Nathan continuously changed his pattern of interlimb coordination.

E. Interlimb Movement Repertoire during the First Year of Life

We wanted to know whether some preferred patterns of interlimb coordination were characteristic of certain periods in Nathan's repertoire of spontaneous interlimb movement. To address that problem, we took the data of every trial within every weekly session and performed a running correlation on every velocity profile. For each week we calculated the frequency rate of the ensemble of correlation values obtained from the different trials. The results of this analysis are represented in Figure 6 as a 3-dimensional landscape describing Nathan's underlying movement repertoire.

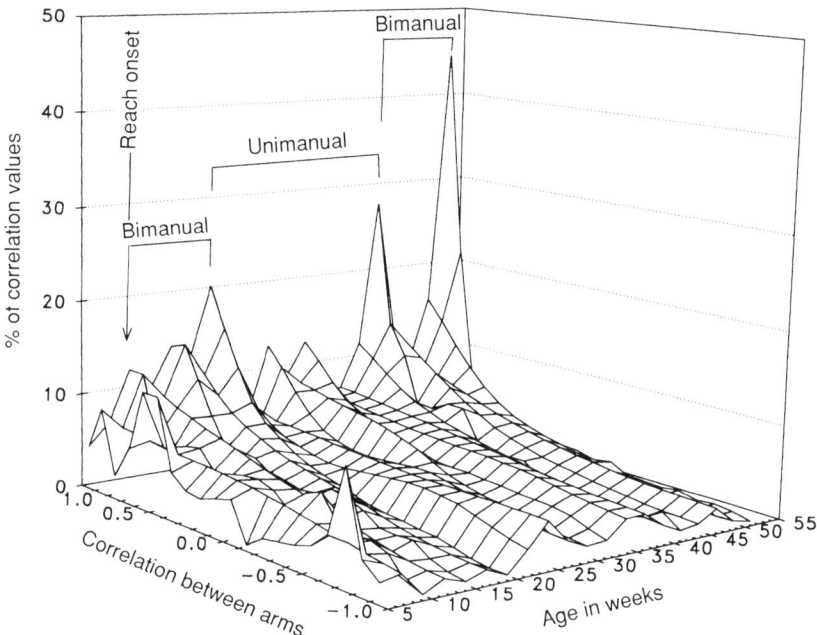

FIGURE 6 Three-dimensional landscape of Nathan's underlying interlimb movement repertoire during his first year of life. The preferred strategies used for reaching (one- or two-handed) at different times in development are displayed together with the frequency rate of correlation values used to describe the patterns of spontaneous interlimb coordination.

The horizontal axis on the right displays the weekly sessions from 5 to 52 weeks old. The horizontal axis on the left shows the different correlation values obtained by using the running correlation technique on the velocity profiles. The vertical axis shows the normalized frequency rate of the correlation value obtained for every weekly session. In other words, the frequency histograms we described above are displayed on Figure 6 for the entire data set collected over Nathan's first year of life. On this 3-dimensional landscape, the *similar temporal patterns* are represented in the back plane of the graph where positive correlation values are displayed. The *reversed temporal patterns,* indicated by negative correlation values, are shown in the front plane. The surface in the middle, from correlation values of +0.5 to −0.5, displays *changing and unrelated interlimb movement patterns.*

Similar to the interlimb strategy of the reach described above, this 3-dimensional landscape shows clear transitional periods in the bilateral organization of Nathan's spontaneous movements. The first peak of related interlimb movements appears at 12 weeks of age, when

Nathan first started to reach. Then, movements predominantly stayed positively correlated until 20 weeks. During that same period Nathan mostly used two hands to reach for the toy. From 21 to 42 weeks, this 3-dimensional landscape shows a big valley, where no preferred mode of interlimb coordination seems to predominate; moderate peaks of similar temporal movement patterns and moderate peaks of reversed temporal movement patterns are equally represented. During that same period, from 21 to 42 weeks, Nathan mostly used one hand to reach. Toward the end of the first year, the return of high peaks reveals that Nathan's arm movements were again highly positively correlated. At this time, Nathan predominantly used two hands for reaching. Thus, Figure 6 indicates that when Nathan used two hands to reach for the toy, the predominant mode of his movement repertoire also showed coupling between his two arms. And when Nathan's spontaneous arm movements were not predominantly positively correlated, Nathan primarily reached with one hand.

F. Preliminary Discussion

What do these findings tell us about infant bimanual reaching? As mentioned before, the objects used in this experiment to elicit reaching were small attractive toys that could be grasped with only one hand. Nonetheless, at certain ages Nathan preferred to use two hands for reaching even though he was capable of using only one hand (Figure 1 shows indeed that during the periods that Nathan preferred to use two hands for reaching, he also performed a small percentage of reaches using only one hand). Why did Nathan predominantly use two hands when reaching with one hand would suffice for the task? The analysis we introduced here to describe the underlying dynamics of Nathan's interlimb movement repertoire during his first year of life provided a preliminary answer to this question. We observed that Nathan's strong tendency to couple the movements of both his arms when reaching for a toy was not just a characteristic of the reach itself, but was fundamentally related to the intrinsic dynamics of his spontaneous interlimb movements. Indeed, even though Nathan's spontaneous interlimb movements were not object-oriented, they had similar coordination characteristics and similar age-related changes to those observed in the development of reaching. Bimanual coupling characterized both reaching and nonreaching movements, and bimanual decoupling was apparent in both reaching and spontaneous movements. In the next section, we analyze the underlying processes that characterize these observed regularities and changes in the development of Nathan's first period of interlimb coordination.

III. Part 2: The Transition from Bimanual to Unimanual Reaching

In the first part of this chapter, we showed how the coupling dynamics of Nathan's spontaneous interlimb coordination affected his reaching strategy. We also showed how these coupling dynamics underwent qualitative changes during Nathan's first year. Our concern in this second part of the chapter is to understand why these changes occurred. Indeed, we still may ask why, at certain ages, patterns of related interlimb coordination strongly predominated in Nathan's movement repertoire and why, at other ages, no predominant coordination pattern emerged. What processes underlie the strong synchrony characteristic of Nathan's bimanual periods and what changed in Nathan's interlimb movement repertoire when this strong interlimb synchrony disappeared to give rise to predominantly one-handed reaches?

Recent investigations on adult interlimb coordination and infant reaching suggested to us an interesting way to address these questions. We will first briefly review the findings of these studies in order to explain how they led us to develop additional analyses of Nathan's interlimb coordination. Then we will show how these new movement analyses helped us to uncover the dynamical processes that underlie (1) Nathan's first period of "two-handedness" from 12 to 20 weeks old, and (2) Nathan's first transition of interlimb coordination at 21 weeks old, when he shifted from a predominantly bimanual to a predominantly unimanual reaching behavior.

A. The Dynamic Processes of Interlimb Coupling

Research on adult interlimb coordination has shown that modulation of the dynamics of interlimb coupling arise not only from the modifications of the external task requirements, but also from modifications that occur in the internal energetic status of the limbs. For example, the abrupt transitions from walking to galloping in quadrupedal gait (see Kelso & Tuller, 1984), or shifts from asymmetrical to symmetrical finger/wrist oscillatory movements occurred when the movement speed reached a critical point (Cohen, 1971; Kelso, 1984; Kelso, Holt, Rubin, & Kugler, 1981; Kelso, Scholz, & Schöner, 1986; Kelso & Schöner, 1988). The dynamic interpretation of these phenomena is that as speed increases, one mode of coordination loses stability and becomes energetically expensive to maintain. Thus, near a critical value, the system shifts into a new and more stable mode of coordination whose energetic cost is more adapted to the new speed conditions (Kelso, 1984; Kelso & Tuller, 1984). Similarly, the coordination of interlimb patterns that are spatiotemporally asymmetrical is difficult to perform because movements tend naturally to be attracted toward symmetrical and synchronized coordination. This attraction toward movement synchronization, and the corollated difficulty to dissociate the movement patterns, increases as movement speed increases (Swinnen, Walter, Serrien, & Vandendriessche, 1992). However, when subjects reduce the acceleration peaks of their movements, they can decouple their movements and maintain the asymmetrical pattern (Swinnen, 1991; Walter & Swinnen, 1992). Accordingly, Walter and Swinnen (1990) suggested that the tendency to synchronize bimanual movements might be related to the torque (or stiffness) that accommodates the inertial load associated with the movement. The higher the torques needed (or the stiffer the system), the more bimanually synchronized the movements.

How can these findings be related to the shifting patterns of interlimb coordination observed in Nathan? As we saw in our description of Nathan's interlimb coordination, the changes that occurred in his coupling dynamics were not related to modifications of the experimental context; the posture in which Nathan was tested as well as the objects we presented to Nathan for reaching were kept identical during the entire year of testing. Thus, as suggested by these studies on adult interlimb coordination, an alternative explanation might be that the changes we observed in Nathan's interlimb coordination reflected changes in the energetic status of the limbs or in the control of the forces delivered to the limbs.

Two arguments support this interpretation. The first is related to the method we used to discover the underlying coupling dynamics of Nathan's interlimb coordination. Indeed, our movement analysis was based entirely on the velocity profiles and measured whether both hands were increasing and decreasing movement velocities together or not. It is possible that these changes in Nathan's tendency to couple or uncouple movement velocities reflect changes in his ability to control bilaterally the forces of his movements. The second argument is based on recent observations on infants' reaching. We found that the way infants were resolving the task of "reaching" was strongly related to the amount of energy they were imparting to their arm before and during the reach (Thelen et al., 1993). Additionally, we observed that highly active infants, who produced large and fast spontaneous movements, adapted the intrinsic dynamics of their reaching arm by damping down the muscle forces moving their joints, to stabilize their arm in space and to bring their hand near the target location. The more quiet infants primarily adapted the intrinsic dynamics of their reaching arm to counteract the forces of the gravitational field, to lift their arms near the toy. However, in this earlier study, the observations focused only on the arm that first contacted the toy; we did not document how the adaptation of the intrinsic dynamics of the reaching arm might have affected the contralateral arm. This question is especially important if both hands are used for reaching. Indeed, if the energy status of the limb affects the way infants will control forces to bring their hand to the toy, how do these forces affect the activity of the contralateral arm?

In light of these considerations, we analyzed Nathan's transition from bimanual to unimanual reach, from 12 to 20 weeks old, by examining the torques produced by both arms during the reach. In doing so we wanted to understand whether the modulations we observed in Nathan's coupling dynamics were reflecting changes in the process of control of the bilateral forces associated with the reaching task. These data will be presented and discussed in the next two sections.

B. Bilateral Torques in Reaching

The ensemble of requirements needed for the calculations of torques in infants' limbs reduced the amount of data we could use for this analysis compared to kinematics alone (for details, see Thelen et al., 1993, and Schneider, Zernicke, Ulrich, Jensen, & Thelen, 1990). In addition, Nathan reached little in the two weeks following the onset of reaching, for motives known only to him. For these reasons, we are unable to give a complete and quantitative account of the changes that occurred in Nathan's interlimb coordination during the few weeks that followed his onset of reaching. As a consequence, the descriptions to follow are based only on exemplar trials from weeks 12, 15, 17, and 20, the sole weeks from which kinematic and kinetic data for both hands during the reach were available. Nonetheless, we still have been able to discern regular patterns in the dynamic process that underlies bimanual coupling during the first weeks of reaching. By detecting changes in that dynamic process, we have been able to uncover the changes in Nathan's interlimb coordination dynamics when he shifted from a bimanual to a unimanual reaching behavior.

These results are presented in Figures 7 to 11. The graph at the top of each figure represents the velocity profiles of both hands before, during, and after the reach. The two graphs below the velocity profile show the corresponding time series of the shoulder torques of both arms during the reach segment. The graph entitled "Leading hand" presents the torques of the hand that contacted the object first, while "Following hand" shows the torques of the contralateral hand during the same temporal interval until it also contacted the target. In these figures, the NET torque corresponds to the total rotational force acting on the shoulder joint. This torque is composed of three different kinds of forces: the gravitational forces (GRA) that act on the center of mass of the segment; the motion-dependent forces (MDT) resulting from the forces

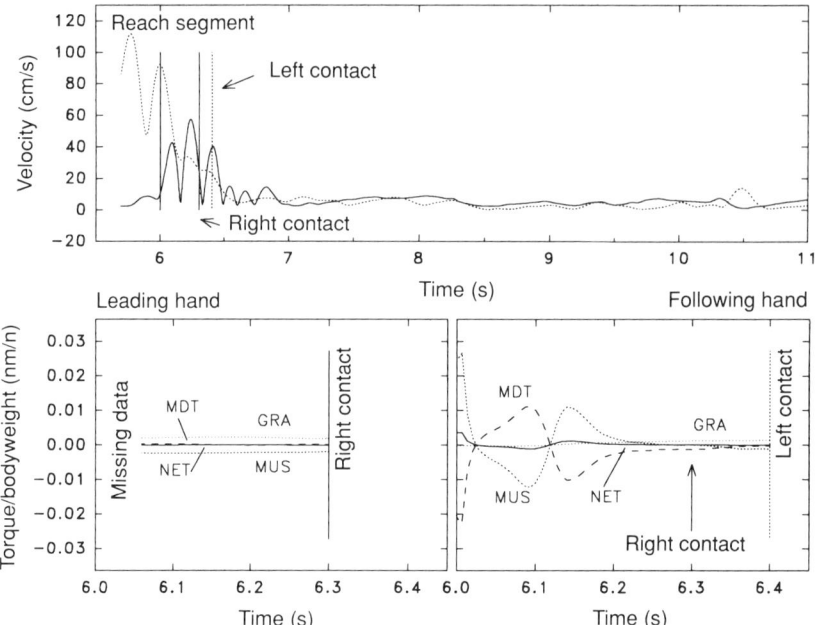

FIGURE 7 Top panel: Exemplar plot of resultant 3-dimensional velocities for right and left hands during 5.5 sec of Nathan's interlimb movements including reach and spontaneous activity at 12 weeks old. The length of the segment is determined by the visibility of the markers on the hands, thus before 5.5 sec into the trial, markers were not visible. Bottom panels: Torques at the shoulder associated with the reaching segment for the leading hand, which contacted the toy first (left panel) and the following hand, which contacted the toy second (right panel). Negative torques work to flex the joints. NET, Sum of all torques rotating the shoulder joint; GRA, torques as a result of the pull of gravity. Note that gravity is extensor at the shoulder; MDT, torques rotating the shoulder that result from the movement of the other mechanically linked segments of the arm; MUS, torques rotating the shoulder that arise from muscle contraction and tissue deformation. The low MUS of the leading hand counteracts GRA, while high MUS of the following hand is associated with rapid movements at the shoulder that generate high MDT. Nm/N, Newton meters/Newton.

acting on one segment that have been mechanically transmitted to the other segments of the limb; and the muscle torque (MUS), which corresponds to the forces arising from muscle contraction (for more details on the measurement of these variables, see Schneider *et al.*, 1990; Schneider & Zernicke, 1992).

The most striking feature that emerges is the consistency of certain movement characteristics across weeks 12, 15, and 17, and the disappearance of those characteristics at week 20, just before Nathan shifted to unimanual reaching. Figures 7, 8, and 9 all show lower velocity peaks in the hand that first contacted the target than in the hand that made contact subsequently. Moreover, in all three examples the kinetic data show asymmetrical patterns between hands, with higher MDT and MUS torques in the following hand than in the leading hand. These strong differences between arms are not present at week 20, where a more symmetrical pattern in the force distribution between hands predominates (see Figures 10 & 11). These data show that, while the leading hand presented a constant pattern of movement for reaching across weeks, the following hand did not. Thus, changes in Nathan's behavior primarily occurred in the activity of the following hand. Indeed, the high velocity peaks and motion-dependent torques present at weeks 12, 15, and 17 are scaled down at week 20 as if Nathan learned to control forces in the arm contralateral to the reaching arm.

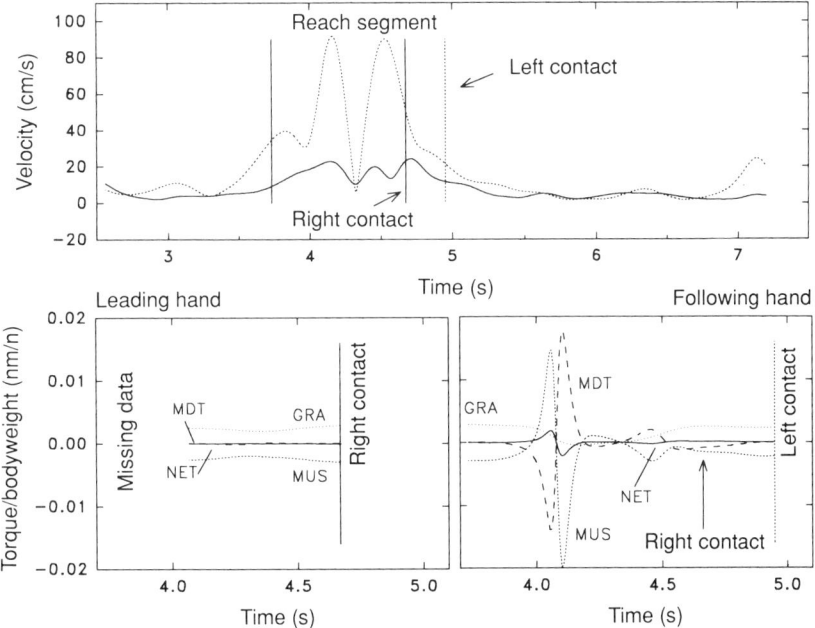

FIGURE 8 Top panel: Exemplar plot of resultant 3-dimensional velocities for right and left hands during 5 sec of Nathan's interlimb movements including reach and spontaneous activity at 15 weeks old. The length of the segment is determined by the visibility of the markers on the hands, thus before 2.5 sec into the trial, markers were not visible. Bottom panels: Torques at the shoulder associated with the reaching segment for the leading hand, which contacted the toy first (left panel) and the following hand, which contacted the toy second (right panel). Similar to week 12, the low MUS of the leading hand counteracts gravity. The high MUS of the following hand is primarily associated with high spikes of MDT. Nm/N, Newton meters/Newton.

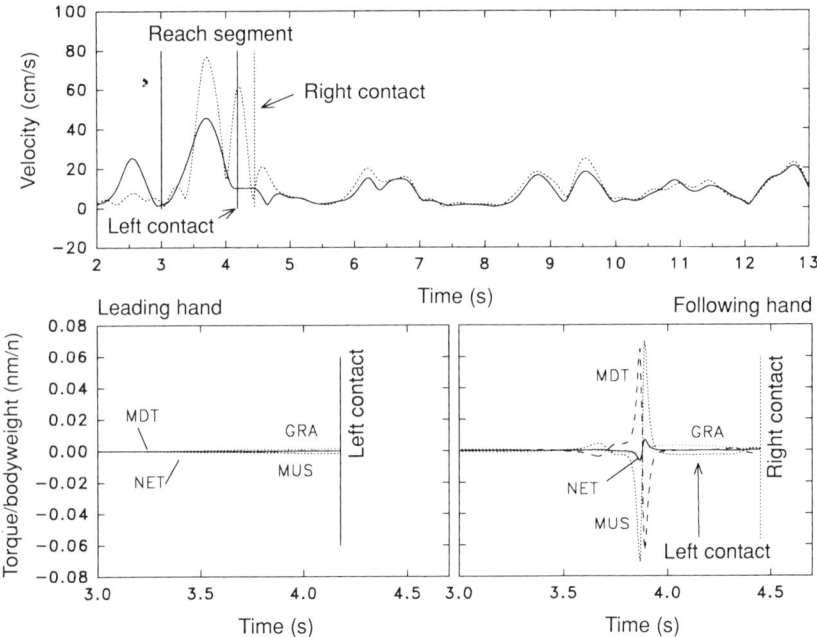

FIGURE 9 Top panel: Exemplar plot of resultant 3-dimensional velocities for right and left hands during 11 sec of Nathan's interlimb movements including reach and spontaneous activity at 17 weeks old. Bottom panels: Torques at the shoulder associated with the reaching segment for the leading hand, which contacted the toy first (left panel) and the following hand, which contacted the toy second (right panel). Similar to weeks 12 and 15, the low MUS torque of the leading hand counteracts gravity. The high MUS of the following hand is primarily associated with high spikes of MDT. Nm/N, Newton meters/Newton.

C. Conclusions

We asked if Nathan's shift from coupled to uncoupled patterns of interlimb coordination reflected modifications in the control of forces imparted to the limbs. We indeed found a change in force patterns associated with this developmental shift, although the modulation occurred not in the leading arm, but in the following arm. What does this tell us about the dynamic processes that underlie Nathan's first tendency to couple the movements and then the transition in behavior from bimanual to unimanual reaching? In the weeks that Nathan showed a strong tendency to use both hands, the shoulder of the following hand generated high MDT and high MUS only during the time that the leading hand moved toward the toy. As soon as the leading hand contacted the toy, the torques of the contralateral arm decreased even

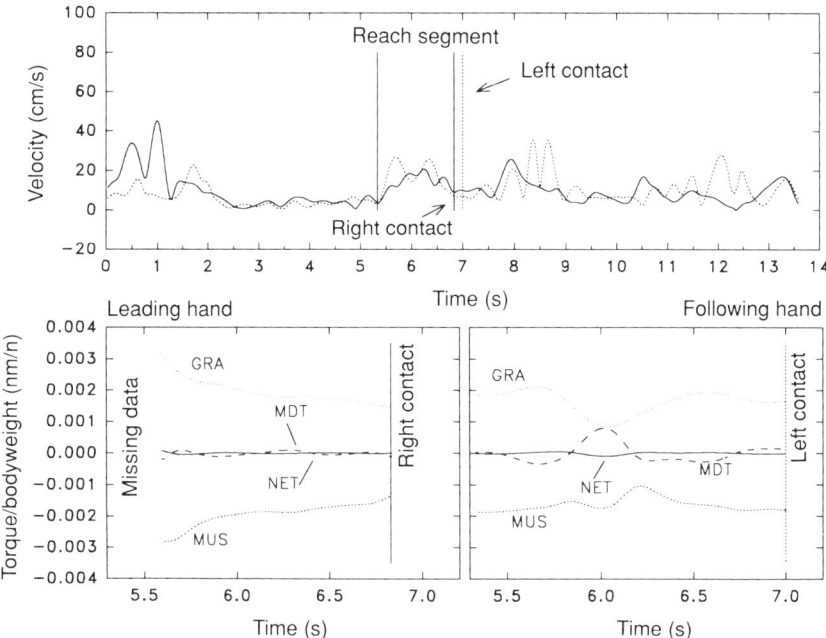

FIGURE 10 Top panel: Exemplar plot of resultant 3-dimensional velocities for right and left hands during 11 sec of Nathan's interlimb movements including reach and spontaneous activity at 20 weeks old. Bottom panels: Torques at the shoulder associated with the reaching segment for the leading hand, which contacted the toy first (left panel) and the following hand, which contacted the toy second (right panel). MDT is low at both shoulders. Low MUS counteracts either GRA (leading hand) or both GRA and MDT (following hand). Nm/N, Newton meters/Newton.

though that second hand was still actively moving toward the toy. This phenomenon, which occurred repeatedly during Nathan's period of two-handed reaches, illustrates the linked nature of interlimb coordination during that period of Nathan's life. These patterns of bilateral movement suggest that a high activity was generated in the following hand as long as the leading hand was moving. However, as soon as the leading hand stopped, the forces in the following hand damped as if both hands were strongly connected to each other at the level of forces. At week 20, both hands still moved together for reaching, however, the torques generated by the following hand were not as high as in the weeks before. Finally, at week 21, Nathan used only one hand for reaching; the following hand simply was no longer active.

These data clearly show that from 12 to 20 weeks the tendency to use two arms for reaching arose from an asymmetrical pattern of the inertial and muscle forces associated with the reaching activity. The decou-

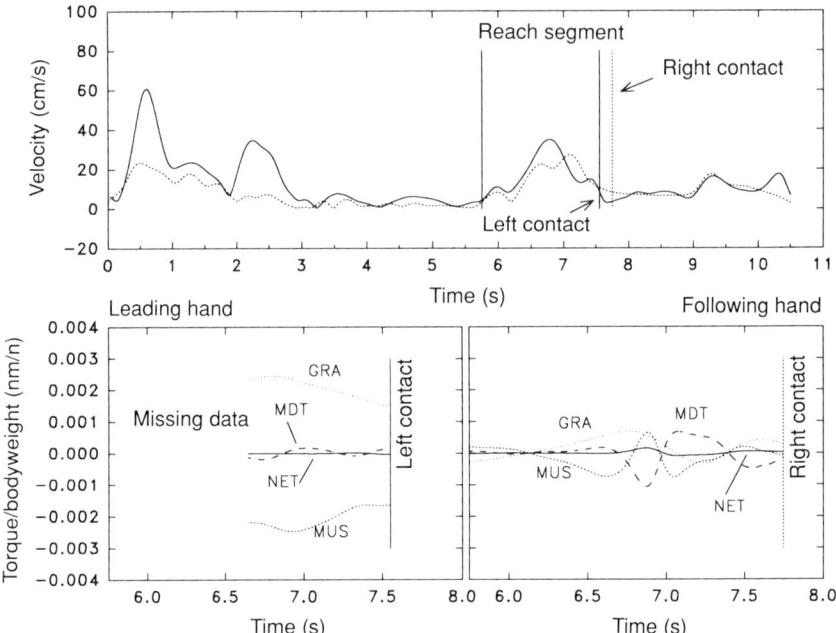

FIGURE 11 Top panel: Exemplar plot of resultant 3-dimensional velocities for right and left hands during 14 sec of Nathan's interlimb movements including reach and spontaneous activity at 20 weeks old. Bottom panels: Torques at the shoulder associated with the reaching segment for the leading hand, which contacted the toy first (left panel) and the following hand, which contacted the toy second (right panel). Low MUS at both shoulders primarily counteracts GRA. Nm/N, Newton meters/Newton.

pling that occurred at week 21 emerged from Nathan's ability to control the associated activity by controlling the forces of his contralateral arm.

IV. General Discussion

Our main concern in this chapter has been to explain the surprising tendency of infants at certain ages to use two hands for reaching and then to shift to one hand. To provide that explanation, we studied the developmental course of reaching in a larger context than the one usually adopted by traditional infant research. The context of our study included an analysis of both the reach and the spontaneous movements from which the reach emerges. These analyses allowed us to discover the underlying dynamics of reaching and to show in particular that the periods of two-handed reaching were governed by underlying coordination dynamics that predominantly tended toward interlimb coupling.

During times of one-hand use, spontaneous interlimb coordination was much more varied, with different patterns of coupling and uncoupling.

The second focus of the chapter was on the shifting patterns of interlimb coordination in development, where periods of predominantly "two-handedness" in the reach alternate with periods of predominantly "one-handedness." We hypothesized that these shifting patterns of interlimb coordination were related to changes in the energetic status of the limbs and thus to a change in the ability to control the forces imparted to the movements. The torque analyses we performed on the reaching movements revealed that Nathan's transition from bimanual to unimanual reaching between 12 to 21 weeks of age was related to a control of the forces of the contralateral arm. This new-found ability, that we observed only during the reach, probably affected the entire dynamics of coordination, since we also observed a decrease of movement coupling in the repertoire of spontaneous movement.

These findings provide new insights for studies of the development of reaching and help to build the bridge between the development of single-limb reaching, two-handedness, and shifting patterns of interlimb coordination in early infancy. Nevertheless, the analyses presented here still leave several questions unanswered. In particular, we did not address why Nathan returned to a predominant bimanual activity at the end of his first year of life. Nor did we address whether the return to two-handed reaching was also characterized by a change in the control of forces of the arm contralateral to the reaching arm, as was the case for the first transition. Additional investigations will be necessary to complete this developmental picture. However, despite the incomplete story, the data we presented here clearly demonstrated that the development of reaching does not result only from the control of the reaching arm, and does not emerge solely from the progressive coordination between visual and arm systems, as traditionally accepted (McDonnell, 1975, 1979; Piaget, 1936; von Hofsten, 1979, 1982). As we have shown here, the development of reaching also involves the control and coordination of the arm contralateral to the reaching arm, and this control progresses by mastering the forces intrinsic to the movements.

Another question that remains unanswered is whether our findings, based on only one subject, Nathan, can be verified with other subjects. Observations of other infants who participated in this and another study seem to support these findings. In these preliminary observations, we found that the transition from two-handed to one-handed reaching or from one-handed to two-handed reaching occurred at different times in development, and varied enormously from infant to infant. However, despite these individual differences, shifts in the interlimb strategy of the reach always seemed to occur concomitantly with a change in the control of the forces of the contralateral arm. Additionally, we found that when infants developed control over the forces of their contralateral arm, they also were able to adapt their one- or two-handed movements to external task demands. For example, one-handed reaches were used for only small objects and bimanual reaches were used primarily for large objects (Corbetta & Thelen, 1992a,b).

The development of reaching, like the development of walking (see Thelen & Ulrich, 1991), must arise from a complex process that involves the coordination of many components together. As shown by numerous studies, the development of reaching involves a coordination between visual and arm systems (McDonnell, 1975, 1979; Piaget, 1936; von Hofsten, 1979, 1982), between visual and tactile perception (Hatwell, 1986), and between head control and trunk control (Fontaine, 1984; Grenier, 1981; Rochat, 1992). In addition, as we have shown here, the development of reaching also involves the coordination and control of the activity of the arm contralateral to the reaching arm.

Acknowledgments

We would like to warmly thank Nathan and his parents for their valuable participation in this yearlong research project. We are also grateful to Karen E. Adolph, Deanna Berkoben, Dexter Gormley, Jody L. Jensen, Kathi Kamm, Jürgen Konczak, Michael Schoeny, John P. Spencer, and Gregory A. Smith, who provided major help during data collection and data processing. Special thanks are also addressed to Klaus Schneider and Ronald F. Zernicke, who provided the model to perform the torque analyses presented in this chapter. This research was funded by Grant HD223800 from the National Institute of Child Health and Human Development and by a Research Scientist Development Award from the National Institute of Mental Health to E. Thelen. D. Corbetta was supported by the Swiss National Science Foundation, Grant 8210-025926.

References

Ames, L. B. (1949). Bilaterality. *Journal of Genetic Psychology,* 75, 45–50.
Bernstein, N. (1967). *The coordination and regulation of movements.* Oxford: Pergamon Press.
Bresson, F., Maury, L., Piéraut-Le Bonniec, G., & de Schonen, S. (1977). Organization and lateralization of reaching in infants: An instance of asymmetric functions in hands collaboration. *Neuropsychologia,* 15, 311–320.
Bushnell, E. W. (1985). The decline of visually guided reaching during infancy. *Infant Behavior and Development,* 8, 139–155.
Cohen, L. (1971). Synchronous bimanual movements performed by homologous and nonhomologous muscles. *Perceptual and Motor Skills,* 32, 639–644.
Corbetta, D., & Thelen, E. (1992a). Bimanual reaching in 5- to 8-month-olds: Task effects and neuromotor mechanisms. *Infant Behavior and Development,* 15, 199. (Abstract of the 8th International Conference on Infants Studies, May 6–10, 1992, Miami, Florida).

Corbetta, D., & Thelen, E. (1992b). Mechanism underlying shifts in interlimb movement patterns in infancy: A perturbation study. *Society for Neuroscience Abstracts,* **18,** 516.
Fetters, L., & Todd, J. (1987). Quantitative assessment of infant reaching movements. *Journal of Motor Behavior,* **19,** 147–166.
Flament, F. (1974). Intelligence pratique et latéralité: Étude génétique de la synergie et de la prévalence manuelle chez le nourrisson. *Bulletin de Psychologie,* **27,** 681–684.
Flament, F. (1975). *Coordination et prévalence manuelle chez le nourrisson.* Paris: Editions du C.N.R.S.
Fontaine, R. (1984). Fixation manuelle de la nuque et organisation du geste d'atteinte chez le nouveau-né. *Comportements,* **1,** 119–121.
Gesell, A. (1939). Reciprocal interweaving in neuromotor development: A principle of spiral organization shown in the patterning of infant behavior. *The Journal of Comparative Neurology,* **70,** 161–180.
Gesell, A. (1946). The ontogenesis of infant behavior. In L. Carmichael (Ed.), *Manual of child psychology* (pp. 295–331). New York: Wiley.
Gesell, A., & Ames, L. B. (1947). The development of handedness. *The Journal of Genetic Psychology,* **70,** 155–175.
Goldfield, E. C., & Michel, G. F. (1986a). The ontogeny of infant bimanual reaching during the first year. *Infant Behavior and Development,* **9,** 81–89.
Goldfield, E. C., & Michel, G. F. (1986b). Spatiotemporal linkage in infant interlimb coordination. *Developmental Psychology,* **19,** 259–264.
Grenier, A. (1981). La "motricité libérée" par fixation manuelle de la nuque au cours des premières semaines de la vie. *Archives Françaises de Pédiatrie,* **38,** 557–561.
Halverson, H. M. (1931). An experimental study of prehension in infants by means of systematic cinema records. *Genetic Psychology Monographs,* **12,** 107–285.
Hatwell, Y. (1986). *Toucher l'espace: La main et la perception tactile de l'espace.* Paris: Presses Universitaires de Lille.
Jeka, J. J., & Kelso, J. A. S. (1989). The dynamic pattern approach to coordinated behavior: A tutorial review. In S. A. Wallace (Ed.), *Perspectives on the coordination of movement* (pp. 3–45). Amsterdam: North-Holland, Elsevier Science Publishers.
Jensen, J. L., Ulrich, B. D., Thelen, E., Schneider, K., & Zernicke, R. F. (in press). Adaptive dynamics of the leg movement patterns of human infants: I. The effect of posture on spontaneous kicking. *Journal of Motor Behavior.*
Kelso, J. A. S. (1984). Phase transitions and critical behavior in human bimanual coordination. *American Journal of Physiology,* **246,** R1000–R1004.
Kelso, J. A. S., & Schöner, G. (1988). Self-organization of coordinative movement patterns. *Human Movement Science,* **7,** 27–46.
Kelso, J. A. S., & Tuller, B. (1984). A dynamical basis for action systems. In M. S. Gazzaniga (Ed.), *Handbook of cognitive neuroscience* (pp. 321–356). New York: Plenum.
Kelso, J. A. S., Holt, K. G., Rubin, P., & Kugler, P. N. (1981). Patterns of human interlimb coordination emerge from the properties of non-linear limit cycle oscillatory processes: Theory and data. *Journal of Motor Behavior,* **13,** 226–261.
Kelso, J. A. S., Scholz, J. P., & Schöner, G. (1986). Non-equilibrium phase transitions in coordinated biological motion: Critical fluctuations. *Physics Letters A,* **118,** 279–284.
Kugler, P. N., & Turvey, M. (1987). *Information, natural law, and the self-assembly of rhythmic movement.* Hillsdale, NJ: Erlbaum.
Mathew, A., & Cook, M. (1990). The control of reaching movements by young infants. *Child Development,* **61,** 1238–1258.
McDonnell, P. M. (1975). The development of visually guided reaching. *Perception and Psychophysics,* **19,** 181–185.
McDonnell, P. M. (1979). Patterns of eye-hand coordination in the first year of life. *Canadian Journal of Psychology,* **33,** 253–267.

Mounoud, P. (1983). L'évolution des conduites de préhension comme illustration d'un modèle du développement. In S. de Schonen (Ed.), *Le développement dans la première année* (pp. 75–106). Paris: Presses Universitaires de France.

Piaget, J. (1936). *La naissance de l'intelligence chez l'enfant.* Neuchâtel & Paris: Delachaux & Niestlé.

Ramsay, D. S. (1985). Infants' block banging at midline: Evidence for Gesell's principle of "reciprocal interweaving" in development. *British Journal of Developmental Psychology,* **3,** 335–343.

Ramsay, D. S., & Willis, M. P. (1984). Organization and lateralization of reaching in infants: An extension of Bresson et al. *Neuropsychologia,* **22,** 639–641.

Rochat, P. (1992). Self-sitting and reaching in 5- to 8-month-old infants: The impact of posture and its development on early eye-hand coordination. *Journal of Motor Behavior,* **24,** 210–220.

Schneider, K., & Zernicke, R. F. (1992). Mass, center of mass, and moment of inertia estimates for infant leg segments. *Journal of Biomechanics,* **22,** 805–817.

Schneider, K., Zernicke, R. F., Ulrich, B. D., Jensen, J. L., & Thelen, E. (1990). Understanding movement control in infants through the analysis of limb intersegmental dynamics. *Journal of Motor Behavior,* **22,** 493–520.

Schöner, G., & Kelso, J. A. S. (1988). Dynamic pattern generation in behavioral and neural systems. *Science,* **239,** 1513–1520.

Schöner, G., Zanone, P. G., & Kelso, J. A. S. (1992). Learning as change of coordination dynamics: Theory and experiment. *Journal of Motor Behavior,* **24,** 29–48.

Swinnen, S. P. (1991). Bimanual movement control: Dissociating the metrical and structural specifications of upper-limb movements. In J. Requin & G. E. Stelmach (Eds.), *Tutorials in motor neuroscience II* (pp. 87–94). Dordrecht, Netherlands: Kluwer.

Swinnen, S. P., Walter, C. B., Serrien, D. J., & Vandendriessche, C. (1992). The effect of movement speed on upper-limb coupling strength. *Human Movement Science,* **11,** 615–636.

Thelen, E., & Ulrich, B. D. (1991). Hidden skills. *Monographs of the Society for Research in Child Development,* **56,** (Serial No. 223).

Thelen, E., Jensen, J. L., Kamm, K., Corbetta, D., Schneider, K., & Zernicke, R. F. (1991). Infant motor development: Implications for motor neuroscience. In J. Requin & G. E. Stelmach (Eds.), *Tutorials in motor neuroscience II* (pp. 43–57). Dordrecht, Netherlands: Kluwer.

Thelen, E., Corbetta, D., Kamm, K., Spencer, J. P., Schneider, K., & Zernicke, R. F. (1993). The transition to reaching: Mapping intention and intrinsic dynamics. *Child Development,* **64,** 1058–1098.

von Hofsten, C. (1979). Development of visually directed reaching: The approach phase. *Journal of Human Movement Studies,* **5,** 160–178.

von Hofsten, C. (1982). Eye-hand coordination in the newborn. *Developmental Psychology,* **18,** 450–461.

von Hofsten, C. (1991). Structuring of early reaching movements: A longitudinal study. *Journal of Motor Behavior,* **23,** 280–292.

Walter, C. B., & Swinnen, S. P. (1990). Kinetic attraction during bimanual coordination. *Journal of Motor Behavior,* **22,** 451–473.

Walter, C. B., & Swinnen, S. P. (1992). Adaptative tuning of interlimb attraction to facilitate bimanual decoupling. *Journal of Motor Behavior,* **24,** 95–104.

White, B. L., Castle, P., & Held, R. (1964). Observations on the development of visually directed reaching. *Child Development,* **35,** 349–364.

Zanone, P. G., & Kelso, J. A. S. (1991). Experimental studies of behavioral attractors and their evolution with learning. In J. Requin & G. E. Stelmach (Eds.), *Tutorials in motor neuroscience II* (pp. 121–133). Dordrecht, Netherlands: Kluwer.

21

Manual Strategies and Interlimb Coordination during Reaching, Grasping, and Manipulating throughout the First Year of Life

Jacqueline Fagard

Laboratoire de Psycho-biologie du Développement
Ecole Pratique des Hautes Etudes Centre National de la Recherche Scientifique
Unite de Recherche Associee
Paris, France

    I. Strategies at the Onset of Functional Reaching
        A. Bilateral Patterns at Birth
        B. Strategies for Reaching at 2 to 5 Months of Age: Unilateral, Bilateral, Unilateral Again
        C. Influence of Posture on Unilateral versus Bilateral Reaching
        D. Unilateral versus Bilateral Reaching as a Function of Object Characteristics and Presentation
        E. Methodological Problems, or Another Source of Variability for the Observed Reaching and Grasping Strategy
        F. Temporal Coupling of Bilateral Movements
   II. From Early Object Exploration to Hand Role Differentiation in Object Manipulation
        B. Complementary Role for Means–End Behaviors
        C. From Complementary Sequential to Complementary Simultaneous Actions
  III. Conclusion
        References

Skills, in particular manual skills, are phylogenetically and developmentally important in that they participate in problem solving between a subject and his environment. The fact that manual skills require almost systematically, and at various degrees, the cooperation between the two hands provides interest to the understanding of bimanual coordination. The goal of this chapter is to review the data concerning the emergence of bimanual coordination from the early patterns observable at birth to the bimanual cooperation involved in object manipulations typical of a one-year-old child, and to try to analyze which theoretical approaches best account for such development.

From studies on adult bimanual coordination, which received much interest during the last 15 years, the most striking, emerging picture concerns the tendency to spontaneously coordinate the two hands (Cohen, 1971; Peters, 1977; Kelso, Southard, & Goodman, 1979; Marteniuk & MacKenzie, 1980), and the difficulty in making them operate at different velocities or rhythms (Yamanishi, Kawato, & Suzuki, 1980; Kelso, 1981; Shaffer, 1982; Deutsch, 1983; Klapp, Hill, Tyler, Martin, Jagacinski, & Jones, 1985; Summers, 1987). Differentiation[1] between the outputs of the two hands cooperating to reach a unique goal is precisely what characterizes the bimanual patterns involved in most common manual skills (in contrast, the tasks most frequently studied in adults involved movements, often repetitive and sometimes rhythmical, where each hand must use a different pace to perform similar action such as tapping, pointing, or the like). Not surprisingly, most adult models for bimanual coordination were developed to explain the formation of coordinative patterns in these tasks, where the two hands do the same thing at a different pace. Some postulate the existence of a central processor responsible for coordinating the activation of the two hands, either through a central timekeeper (Wing & Kristofferson, 1973), or through a program or schemata that would commonly define some of the parameters for the two hands, but not all parameters (Schmidt, Zelaznik, Hawkins, Frank, & Quinn, 1979). From a different point of view, the dynamic system approach is particularly useful to understand the spontaneous coordination of the two hands moving together, which becomes constraining when each hand should move at a different velocity. This approach originated on the basis of Bernstein's work on one side (Bernstein, 1967), and on physical theory of nonlinear dynamics on another side, and was developed by Kelso, Holt, Kugler, and Turvey (1980). According to the dynamic system theory, bimanual coordination

---

[1] *Hand differentiation* here refers to the different patterns of movements assumed by the two hands. The difference may bear only on the spatiotemporal parameters for similar actions (*differentiation between output of the two hands*), or on the complementary roles assumed by each hand in performing a unique action (*hand role differentiation*).

does not result from *a priori* prescription centrally originated, but is the *a posteriori* consequence of low-level functional coupling between the two hands constrained to act as a unit (Kelso *et al.*, 1979). One model, however, the kinematic chain model, concerns bimanual cooperation, where each hand plays a different and complementary[2] role with a unified goal (Guiard, 1987). According to this model, each hand represents a motor, and the two hands typically cooperate as two motors assembled in series, thereby forming a kinematic chain governed by a few high-order principles.

Several studies showed evidence that bimanual cooperation develops mainly toward the end of the first year of life, but more precocious forms of bilateral coordination can be observed, for instance during reaching and grasping. In fact, reaching and grasping are the first steps toward object exploration and manipulation, which in turn require bimanual coordination. However, compared with the large body of literature on progress in eye–hand coordination during reaching and grasping, not much work has been devoted so far to the understanding of the onset of bimanual coordination. Although the different bimanual patterns used preferentially at each age have been fairly well described, understanding of the mechanisms underlying progress in bimanual coordination is still at what can be termed the "babbling" stage.

General developmental theories of motor development reflect the central/dynamic opposition. It is usually recognized that motor development corresponds to the building together of synergies, starting from a set of basic patterns inherently present at or before birth. Theoretical disagreements bear on the relationships between these inherent patterns and mature functional patterns, on the process responsible for change, and on the neural basis of the motor repertoire. From a traditional approach (maturational or constructivist), development arises from an increased control of the higher functions over the skeletomotor system. This is supposedly made possible either by maturation of the CNS allowing inhibition of the primitive responses and the development of voluntary cortical control (McGraw, 1941; Gesell, 1946), or by cognitive progresses allowing increasing representations of schemes (Piaget, 1936). In contrast to these traditional perspectives, the dynamic point of view postulates that new spatiotemporal orders emerge not from centrally prescribed programs but from the system dynamics (Thelen, Kelso, & Fogel, 1987). Concerning the development of bimanual coordination, the specific questions bear on which inherent patterns form the basis on which bimanual coordination develops, when does intentional bimanual coordination first emerge, and what best accounts

---

[2] The term *complementary* refers to the fact that each hand assumes a different role within a common and unique goal.

for the progresses in bimanual coordination observable during the first year of life.

This chapter will first review the developmental chronology of bimanual coordination. In a first section, early manifestations of interlimb coupling will be reviewed, as well as the fluctuating uni/bilateral patterns during reaching and grasping. The diversity of the patterns as a function of posture or object presentation will be described. In addition, the temporal relationship between the two hands during bilateral movements will be examined. The second section will analyze progress in object manipulation, from patterns with undifferentiated roles for each hand, or patterns with only one hand active at a time, to patterns of coordination involving differentiated hand roles. The difficulty in temporal coordination within the last patterns mentioned will be examined. Finally, a tentative conclusion on the processes underlying such bimanual development will be proposed.

I. Strategies at the Onset of Functional Reaching

A. Bilateral Patterns at Birth

Primitive form of symmetrical bilateral coordination is sometimes associated with reflexes, such as the Moro reflex, that involve symmetrical arm abduction and extension with the hands open and fingers generally flexed (Mitchell, 1960; Illingworth, 1975), or such as the first respiratory movements of the newborn (Bergeron, 1948). This might indicate the presence of symmetrical pathways linking the upper limbs. However, the precursor value for bimanual coordination of these patterns is questionable, since they concern reflexive behavior and since the precocious forms of visuomotor patterns do not involve a bilateral strategy. For instance, experimental and clinical observations have shown that when experimentally provided with postural support, newborns exhibit reaching movements toward an object (Bower, 1974; Grenier, 1981; von Hofsten, 1982); these "prereaching" movements consist of one arm being directed toward the object (Bower, 1974; see Table I). The next section will examine when and in which conditions bilateral strategies are used during intentional reaching as functional reaching first emerges, that is, in normal conditions around 2–3 months of age.

B. Strategies for Reaching at 2 to 5 Months of Age: Unilateral, Bilateral, Unilateral Again

Although mature reaching for small objects is mostly unimanual, spontaneous strategies preceding this mature stage fluctuate between unilaterality and bilaterality. There are a few studies concerning bilat-

TABLE I Main Steps in the Development of Manual Patterns in the First Year of Life

Age	At birth	3 months	4–5 months	6–7 months	8 months	9 months	12 months
Change parameters	ATNR[a]	ATNR ↓	Postural control of upper body ↑ Control force of second hand ↑ Visual guidance ↑	Second hand free Need for new motor schemes, for object exploration		Maturation of interhemispheric cooperation	
Motor patterns	Bilateral reflexive	Bilateral proximal	Unilateral distal	Unilateral	Differentiated bimanual cooperation Sequential bimanual cooperation	Differentiated bimanual cooperation Simultaneous timing−	Differentiated bimanual cooperation Simultaneous timing+
	Unilateral proximal			Undifferentiated or passive/active			
	(Moro reflex)	(Early reaching)	(Reaching)	(Grasping, mouthing, fingering, banging at midline...)	(Pulling then grasping)	(Grasping while pulling, pulling while orienting...)	
	(Prereaching)	(Clasping hands at midline)					

[a] ATNR, asymmetric tonic neck reflex.

eral coordination during the first weeks of life. The now classic White, Castle, and Held (1964) study where infants were observed in a supine position and were provided with an object at a reaching distance either at the midline or at a lateral position, shows that between 2 and 3 months of age, response to object presentation consists of unilateral hand raising, and is more likely to occur if the object is presented on the side of the commonly viewed hand, which is the hand extended in the favored tonic neck reflex position (White, et al., 1964). As the asymmetrical tonic neck reflex declines and posture becomes more symmetrical at around 3 months of age, unilateral arm approaches decrease in favor of bilateral patterns, such as hands clasping to the midline. Bilateral arm activity in response to object presentation becomes more frequent up to 4 1/2 months but as the object begins to be crudely grasped at about 4 months, unilateral responses reappear, to predominate at 5 months in what White et al. term "top level reaching" (see Table I). This consists in a rapid lifting of one hand toward the object. Top level reaching is visually controlled although visual control over the hands fades.

This U-shaped development of bimanual responses to object presentation (which rises between 2 and 3 months of age, and then decreases after 4 months of age), is also observed for mouthing a tactually presented object (Rochat, 1992). After placing an object in either the left or the right hand of 2- to 5-month-old infants, Rochat analyzed the pattern used by the infants to transport the object to the mouth. He found this pattern to be unimanual at 2 months, bimanual at 3 months (and also at 4 months for one of the two experiments reported), and to become primarily one-handed again at 5 months. Such fluctuations between unilaterality and bilaterality of reaching were first observed by Gesell and Ames (Gesell & Ames, 1947; Ames, 1949).

Reaching requires control of the upper limb (arm, hand), which is the last part of the effector chain also involving postural adjustments. Consequently, the use of a strategy (uni- vs. bimanual) must be affected by the postural adjustments the infant can or has to make. This can be observed during development or by changing the postural conditions of testing, as will be seen in the next section.

C. Influence of Posture on Unilateral versus Bilateral Reaching

Postural gains in the first few months affect the bimanual system in a variety of ways: the decline in postural asymmetry during the third month of life results in increasingly bilateral (or increasingly symmetrical) responses to object presentation. More indirectly, it is probable that greater neck and upper-body control results in a subsequent increase of unilateral responses to object presentation after 4 1/2–5 months: the more distal the segments used, the more unilateral (and contralateral) the control (Brinkman & Kuypers, 1973), and as reaching becomes more visually guided (Bushnell, 1985), the control over distal segments would then favor the use of one system only when the stimulus affords it (see Table I). The impact of posture on infants' manual strategy can also be seen in studies that vary postural testing conditions. Postural conditions during testing may even influence the grasping strategy in older infants. Six-month-olds reach bimanually in the supine position but tend to reach unimanually when seated, whereas eight-month-olds reach unimanually in both postures (Rochat & Stacy, 1989).

At an identical stage of postural control and identical experimental postural conditions, object physical properties, including shape, size, and mode of support for presentation influence manual strategy, as will be mentioned in the next section.

D. Unilateral versus Bilateral Reaching as a Function of Object Characteristics and Presentation

Some coupling between perception and action has been observed during prereaching: hand opening varies according to object size (Bower, 1972), and more generally object size affects infants' manual activity before they are capable of reaching (Bruner & Koslowski, 1972). As the reaching movement becomes less ballistic and is gradually guided rather than triggered by vision (von Hofsten, 1984), the reaching movement becomes more clearly discriminated according to object physical properties. In a study where manual strategy was investigated at the time of grasping, and where infants were tested in a seated position, manual strategy depended on the object properties of size and shape for infants as young as 4 months of age. Their strategies were mostly unimanual for small objects, and mostly bimanual for the large object, except when presented in the open mode (opening face up) (Newell, Scully, McDonald, & Baillargeon, 1989).

In the same vein, the support on which the object is presented also influences the reaching pattern. When a small object is presented on a support (palm, block, or hand), a two-handed indirect reaching (when one hand arrives on the support first and the second hand lands either on the support or directly on the object) persists longer before being replaced by a direct and presumably unilateral reaching than when the object is presented on the tips of the fingers (Bresson, Maury, Pieraut-Le Bonniec, & de Schonen, 1977). The authors term the first stage "the differentiated roles approach," meaning that it is a precursor of hand cooperation, one hand serving as spatial reference, the other hand being active, as in the active/passive bilateral patterns (one hand holds an object/the other hand acts on it, as in peeling an apple, for example). I personally believe that this stage when the two hands approach the object differently is just an indirect precursor of bimanual cooperation, only because unimanual grasping frees the other hand for manipulation (see Table I).

Although all these results converge in forming a coherent picture of the main strategies and their change as reaching and grasping progress, some problems arise when comparing the results of the several studies, as will be shown in the next section.

E. Methodological Problems, or Another Source of Variability for the Observed Reaching and Grasping Strategy

A first difficulty arises from differences in testing situations relative to postural conditions, object size, and/or object presentation, all parameters which, as we saw, influence the strategy used by the infant. More importantly, comparison of the results is hindered by the lack of uniformity of the criteria used to label a movement as bimanual versus unimanual. The criteria for coding reaching as bimanual include that the "second" hand moves at least halfway toward the object by the time the leading hand has touched or grasped the object (Lockman, Ashmead, & Bushnell, 1984), when both hands contact the object simultaneously (Humphrey & Humphrey, 1987), or within 1 sec of each other (Goldfield & Michel, 1986a). Bilaterality is sometimes coded only at the time of grasping and only for successful grasping (Newell et al., 1989). Another criterion, used for characterizing bringing the object to the mouth as bimanual, is an observed decrease in the distance between the mouth and both hands that occurs simultaneously with a decrease in the distance between hands (Rochat, 1992). Last, in some studies, the criterion for coding a movement as bimanual is not mentioned (White et al., 1964; Bresson et al., 1977).

A more rigorous definition of bimanual strategies, an analysis of all trials, including those ending with failure to grasp, and of the whole sequence, starting from initiation of the movement following object presentation until the object is properly secured, or until there is an unsuccessful attempt to grasp, would help to draw reliable conclusions on the developmental course in the use of one or two hands for reaching and grasping. In order to understand the development of interlimb coupling during bimanual movements, we need information on more than the strategy globally defined as bimanual. We need to learn about the temporal relationships within the bimanual strategy. Does the coupling within the bimanual synergy become stronger as the infants' reaching skills develop? Most studies investigating early onset of bimanual patterns have been conducted in conditions that made fine evaluation of interlimb coordination unfeasible. As we will see next, however, some recent research considers this point.

F. Temporal Coupling of Bilateral Movements

According to McDonnell, Corkum, and Wilson (1989), upper-limbs synchronization does not change significantly between 1 and 5 months of age. The authors used movement sensors that were attached to the infant's limbs and connected to a microcomputer. They defined synchronization as a cooccurrence of movements within 100 msec. However, they collapsed conditions with and without stimulus, at least for this result, which restricts possible conclusions.

In a sophisticated and intensive longitudinal study, Thelen and collaborators observed the development of reaching of four infants from the age of 3 weeks, recording 3-dimensional coordinates of the upper-limbs as well as EMG data from the muscles involved. They found that infants initiate a goal-directed movement bimanually up to 20 weeks, unimanually from 20 weeks to the end of the first year, then bimanually again. During bimanual movements, they observed a lag between the two hands. The leading hand showed more stiffening and damping than the following hand and infants shifted to a unimanual strategy as they learned to control forces in the nonreaching hand (Thelen & Corbetta, 1992; see also Corbetta & Thelen, Chapter 20, this volume).

In an experiment designed to study changes in bimanual coordination during reaching and grasping as a function of object size at 7, 9, 11, and 13 months of age, we found that at all ages some of the reaching movements were initiated bimanually, especially for the large objects, but the majority started unimanually, and the second hand was activated before, during, or right after the leading hand had touched the object (Fagard & Jacquet, submitted). The time lag between initiations of the two hands decreased as the size of the object increased. A major difference between the two youngest groups (7 and 9 months) and the two oldest groups (11 and 13 months) was the capacity of the oldest to initiate the movement of the second hand while the first hand was still on its way to the object, whereas the youngest rarely started their second hand before reaching of the first hand was completed. These findings suggest that the first modes of interlimb coordination are somewhat restricted to either synchronous[3] or sequential[4] movements (see Table I). As they grow older, infants develop different types of interlimb coordination, where the two hands can be simultaneously activated on different temporal patterns (with a delay between the two initiations, for instance).

---

[3] The term *synchronous* is used here to refer to two actions occurring at the same time after a simultaneous initiation.

[4] *Sequential* means that actions of the two hands start one after another without overlapping.

A change in the spatiotemporal linkage between the upper limbs between 7 and 11 months of age was also found by Goldfield and Michel (1986a). In their study of bimanual reaching for toys presented inside a transparent toy box (only the trials where the two hands contacted the target within 1 sec for each other were analyzed), these authors found that bimanual movements tend to be more simultaneous at 7 than at 11 months of age, and that the shape of the trajectories suggests that 7-month-olds move both hands in the same direction and 11-months-olds in mirror (the authors call it complementary) directions. In another study, Goldfield and Michel (1986b) also found that 11-month-olds (as compared with 7–10 and 12-month-olds) were less likely to have their bimanual reach perturbed by a barrier placed in the path.

From the studies on reaching and grasping, it appears that upper-limb coordinative patterns are quite flexible from the start, and that their appearance depends on factors such as posture and object physical properties. These early bilateral patterns are, however, limited to actions similar for the two hands (both hands reach or both hands grasp), with synchrony being the only mode of temporal coordination. One-handed grasping, which becomes the prevalent mode to grasp small objects as posture and visuomanual coordination progress, opens the way for manipulation with the other hand, and therefore for bimanual manipulation with each hand assuming a different role, what we call complementary bimanual coordination. We will see in the next section that this form of bimanual coordination develops during the second half of the first year of life, starting with some kind of sequential patterns, each hand being active one after the other, before temporal coordination allows for complementary actions of the two hands to occur simultaneously.

II. From Early Object Exploration to Hand Role Differentiation in Object Manipulation

A. Early Patterns of Exploratory Manipulation

As soon as the infant can grasp an object, (s)he starts to explore and thus to manipulate it. Even at 3 or 4 months of age, if the object is directly inserted into the infant's hand before (s)he is able to voluntarily grasp it, (s)he takes the object to mouth or "fingers" it (Rochat, 1989, 1992).

The second half of the first year of life has been shown to be a time of tremendous development in object manipulations of increasing complexity (Piaget, 1936; McCall, 1974; Lézine, 1978). Mouthing, fingering, shifting the object from hand to hand, rotating an object while looking at it, and banging objects at the midline are behaviors that are commonly observed between the sixth and twelfth month of age (see Table I). At 6 months of age infants evidence some difficulty in shifting objects from one hand to the other (letting go of the object is difficult at this age). Shifting the object from hand to hand becomes more common at 7 months (Lézine, 1978). Mouthing is at its peak around 6 and 7 months of age (Palmer, 1989) and decreases in frequency thereafter (McCall, 1974; Lézine, 1978; Ruff, 1984), more or less depending on object properties (Palmer, 1989). At 7 months of age the infant's manipulative repertoire also includes banging objects at the midline (Fenson, Kagan, Kearsley, & Zelazo, 1976; Ramsay, 1985), and scratching with the finger of one hand the object held in the other hand (Lézine, 1978; Ruff, 1984). Whereas mouthing decreases with age, fingering and shifting the object from one hand to the other increases until 9 months of age (Lézine, 1978; Ruff, 1984). Object rotation is another frequent behavior and changes from being one handed at 6 and 9 months of age to two handed at 12 months of age (Ruff, 1984). After 7 months of age, infants manipulate objects in increasingly long and better organized sequences (Lézine, 1978).

These motor schemes, which clearly have an exploratory function, differ from the outset as a function of object characteristics. Even though motor schemes become increasingly finer with age (with a sharp increase between 9 and 13 months; Fenson *et al.*, 1976), infants vary their action with the object physical properties as soon as they start exploring them. This involves fingering for texture changes, transferring for shape change, more banging and waving with objects that make sounds (McCall, 1974; Lézine, 1978; Ruff, 1984; Palmer, 1989).

Most of these manipulatory patterns are two handed. Even early manifestations of mouthing, although they do not necessarily involve bilateral transport of the object to the mouth (Rochat, 1992), involve mostly a bilateral support of the object during oral contact (Rochat, 1989). As opposed to the proximal coordination evident in bilateral reaching, the first manipulations involve more distal interlimb coordination. In addition, as patterns of manipulation become more diversified, increasing hand-role differentiation can be observed. For instance, whereas mouthing requires an undifferentiated role for each hand, during fingering and shifting an object from hand to hand, each hand assumes a different role (holding/fingering; grasping/letting go). This may be considered as a precursor of the latter complementary bimanual coordination. However, in fingering or shifting from hand to hand, only one hand is active at a time, which may reduce the need for temporal coordination between the two hands. As we will see in the next section, the first means-end behaviors in which each hand must actively assume a different role in order to achieve a unique goal are succeeded with a sequential strategy before being succeeded with a simultaneous control of the two hands on different temporal parameters.

B. Complementary Role for Means–End Behaviors

Piaget (1936) first showed that around the eighth month, during what he called stage 4 of sensorimotor development, schemata coordination, and the exploitation of cause–effect and spatial relationships in means–end behaviors emerge. At that age, infants acquire the ability to apply known means in order to solve new problems: for instance, they can pull a string in order to get a toy attached to it and too far away to be grasped directly. More recently, Willatts (1985) observed that intentionality to use supports to retrieve distant objects appears at 7 months of age, but that only 8-month-olds rapidly become successful at this task. Therefore, after 8 months of age infants can succeed at means–ends using a bimanual strategy where both hands play a complementary role (see Table I). They can pull a cloth to retrieve an object too far away to be grasped directly (Casati & Lézine, 1968; Willatts, 1985), or pull a box near them with one hand and reach for the toy inside with the other hand (Diamond, 1991). Although the two hands play a different role, and although this makes these manipulations a first clear instance of manual differentiation, the two parts of the task can be performed in sequence without too much temporal coordination between the two hands. In addition, a unimanual strategy can be used and is not rare at first.

Around 9–10 months of age, radically new bimanual behaviors emerge, which improve considerably during the last trimester. Infants at this age can coordinate different actions of the two hands simultaneously such as holding up a transparent lid to grasp something from underneath it (Bruner, 1970), lift a box with one hand and simultaneously reach for the toy inside with the other hand (Diamond, 1991), and orient an object to pick up another object from inside (Flament, 1975; Fagard & Jacquet, 1989; see Table I). What is new in these bimanual schemes is that the first action must be maintained for the second one to be carried out. The following section will be devoted to the analysis of the difficulties met by the infants when they first try to do this.

C. From Complementary Sequential to Complementary Simultaneous Actions

Before they are completely mastered, success on these bimanual coordinated activities takes place with poor temporal coordination. For instance, Bruner (1970) showed that in the first bimanual strategies at the "Box task" (where the lid must be opened and held open in order to retrieve a toy from inside it), infants sometimes do not hold the lid in the open position long enough for the other hand to grasp the toy easily (this imperfect bimanual strategy follows a stage in which infants try to solve the problem unimanually, with what Bruner calls a "worming technique" that is frequent among 9–11-month-olds). Within his frame of coding, Bruner reports a low incidence of this badly timed bimanual strategy.

Using a frame-by-frame analysis with a 40 msec window, I found that this difficulty in temporal coordination in fact characterizes the earliest bimanual strategies that emerge around 9 months of age in the box task, and features in other means-end tasks as well. In a study designed to investigate progress in intermanual coordination during the second half of the first year of life, I tested 24 infants of 6–7, 9–10, and 11–12 months of age (8 infants in each group) on four tasks involving two steps of action with a differing temporal relationship between the two steps. In two of these tasks, the sequential tasks ("String" and "Doll/cover"), the two actions could be made one after the other (pulling or raising *then* grasping). The two other ones, the simultaneous tasks ("Box" and "Tube/container"), required the first action to be maintained while the second one took place (grasping *while* holding up or pulling *while* holding in the correct orientation).

1. The "String"[5] task used a small multicolored cylindrical cage with a bell trapped inside, attached to the end of a 26-cm white cotton string. It was presented so that the string was reachable whereas the cage was out of direct reach. Therefore, in order to grasp the object, the string had to be pulled before the object could be within reach. The same hand could do the two steps but a quicker strategy consisted in pulling the string with one hand while reaching for the object with the other hand.

2. In the "Doll/cover" task[6] a small plastic playschool doll (5.1 × 2.4 cm) was presented under a transparent semicircular cover (6.7 cm high and 21.0 cm in circumference). To retrieve the doll, the infant had to displace the cover first. For this task, also, the same hand could do the two steps, but lifting the cover with one hand and grasping the doll with the other one was a more efficient strategy.

3. In the "Box" task, a 17 × 11 1/2 × 4 cm opaque box with a hinged lid was presented to the infant. A toy was inserted in the box as the infant watched. To get the toy, the infant had to raise the lid and keep the lid up with one hand while getting the toy out of the box with the

---

[5] This is one of the Piagetian tasks adapted by Casati and Lézine (1968) in their test.

[6] This task was first designed and used by J. Larouche (Larouche, 1991).

other hand. A unimanual strategy is almost, but not entirely, prevented by the constraints of the task.

4. The "Tube/container" task used a small plastic tube inside a wooden container (10.4 × 2.3 × 2.3 cm), with the orange cap of the tube sticking out from the container. To extract the tube, one hand had to pull on the tube while the other hand held the container. As in the preceding task, a successful unimanual strategy is almost impossible.

The tasks were presented in a random order. The infants were seated on one experimenter's lap facing the testing table. Objects were presented at a midline position on the table by a second experimenter. Each session was analyzed frame by frame. The main events concerning each hand (initiations, touches, grasping, picking up, releasing, stopping, etc.) and gaze (on the object, off the object, elsewhere, etc.) were coded as well as their timing (in msec). In addition, for each event coded for any of the three items (right hand, left hand, gaze), the two other items were described as well.

For all four tasks, the success rate increased significantly with age. The two sequential tasks were performed successfully earlier than the two simultaneous tasks. All the 6-month-olds failed the latter (see Figure 1).

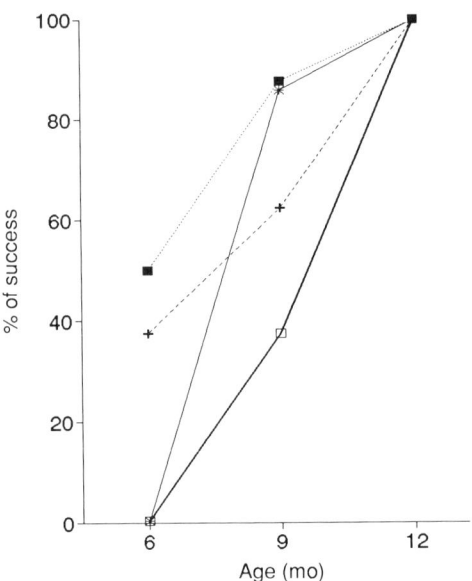

FIGURE 1 Success at sequential (String and Doll/cover) versus simultaneous (Box and Tube/container) tasks. String, ■ ; doll/cover, ┈┼┈ ; box, ✕ ; tube/container, ⊟ .

Several factors influenced success at means–end, and the difference in rate of success between the two kinds of tasks was not entirely due to the more complex bimanual coordination involved in simultaneous in comparison with sequential tasks. The two most important are: (1) The level of difficulty of the movement required in the first step. For instance, although grasping the string was not easy at first because of its small diameter, pulling it toward oneself was easy even at 6 months, since this is part of the grasping/mouthing pattern that is already acquired at this age. In contrast, pulling upward and away from the body, as in lifting up the box lid, is less usual and more difficult; (2) Sustaining interest in the second step after performing the first one depends partly on the task configuration. For instance, many of the youngest infants mouthed the cover for a length of time, or banged on the box lid, or shook the string with the object hanging at the end. Depending on the type of object used, and the latitude of its mobility with regard to these kinds of side activities, infants will be more or less "hooked" into staying at the first step and trapped in another game, different from the target one implied by the task.

Some bimanual patterns that could lead to success in the two sequential tasks were observed as of 6 months of age. At the String task, a bimanual shifting strategy between the two hands was typical of younger infants. This consisted in pulling the string with one hand, then the other, switching back and forth. This could be done without looking at the toy and obviously with the sole goal of mouthing the string. Thus, infants could inadvertently draw the toy near enough to grasp it as soon as they caught sight of it or touched it, succeeding on a "one-at-a-time" basis. This alternative pattern of bimanual coordination, as was mentioned earlier, emerges between 6 and 7 months of age. A unimanual strategy was more common in older infants because when pulling was clearly intended to get the object, the pulling movement was quite energetic, resulting in a lateral displacement of the object rather than a straight one. Therefore only the pulling hand could reach the object.

In the Doll/cover task, a differentiated bimanual pattern (one hand raising the cover and the other hand grasping the doll) was used by the majority of the 12-month-old infants (see Figure 2). None of the infants in the two younger groups succeeded in using a differentiated bimanual pattern. Most of the youngest infants tended to grasp the cover bimanually; the percentage of bimanual grasping of the cover decreased significantly from 6 to 12 months of age. Bimanual grasping usually followed a unimanual initiation, where the second hand arrived on the cover secondarily to help the first grasp it. Therefore, this strategy could not be solely attributed to a lack of planning of the second step, but rather to the difficulty encountered when grasping the cover.

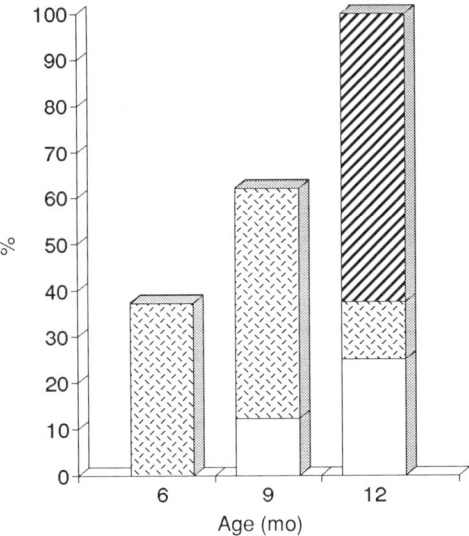

FIGURE 2 Strategies at the Doll/cover task ending with success. Bim. diff. ▨, one hand raises the cover/the other hand grasps the doll; Bim.-unim. ▦, two hands raise the cover/one hand grasps the doll; Unim. ☐, the same hand raises the cover and grasps the doll.

The bimanual coordination involved in the other two tasks was more demanding. In addition to success occurring later than on the sequential tasks, the simultaneous tasks were initially performed in an awkward manner with an evident difficulty to temporally coordinate the two hands. Almost half of the infants in the two older groups tried to perform the Box task unimanually (one hand raised the lid, wiggled into the box, and tried to grasp the object). Whereas the 12-month-olds shifted quickly to a successful bimanual strategy, two 9-month-olds stuck to the unimanual strategy and succeeded, although awkwardly, in retrieving the toy that way. The other strategies were mostly bimanual–unimanual for the 9-month-olds (two hands raise the cover/one hand lets go of the cover and goes inside the box to grasp the toy). Twelve-month-olds were equally distributed between this bimanual–unimanual strategy and the complementary strategy in which one hand raises the lid and the other hand grasps the object. These results are congruent with data from Ramsay and Weber (1986), who found that, for a similar task, by 18 months of age, all infants use a complementary strategy with complete hand-role differentiation. In addition to the greater number of complementary strategies at 12 as compared to 9 months of age, another striking difference between the two age groups was the temporal coordination during the bimanual strategies. Some infants tended to release the lid before the grasping hand was out of the box, such that the lid fell on the grasping hand, making the end of the action rather awkward. The holding hand did not let go at the same time as the hand heading for the toy, rather the holding hand let go of the lid as soon as the grasping hand contacted the toy. The percentage of such trials with lack of temporal coordination between the two hands was significantly higher at 9 than at 12 months (see Figure 3).

Similar difficulties in temporal coordination between differentiated actions of the two hands could also be observed in the Tube/container task. Almost all 12-month-olds started the movement with one hand and successfully used a complementary strategy to get the tube out of the container (one hand grasped the wooden container, orientated it and held it, while the other hand pulled out the tube). Six-month-olds could not do the task, even after demonstration or with an easier version of the task (where the tube is almost half out of the container and presented with orientation). The few 9-month-olds who succeeded in pulling the tube with one hand while holding the container with the other hand let the container drop on the table before the tube was totally out. As in the Box task, this difficulty in maintaining the part of the first hand as soon as the second hand does its active part gives the impression of difficulty in temporally coordinating the two hands. This type of difficulty disappears by the end of the first year of life.

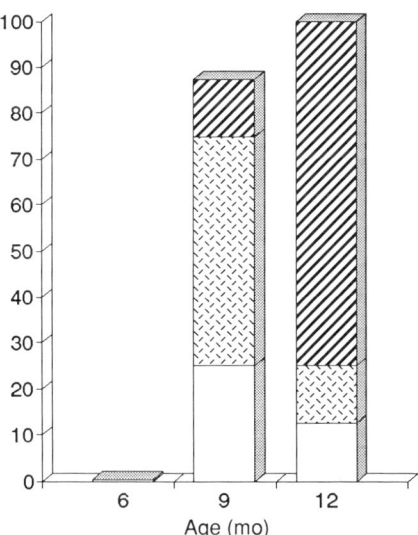

FIGURE 3 Strategies in the Box task ending with success. Unim, ☐: same hand raises the lid and grasps the object; Bim, ▨: good timing, one or two hands raise the lid/one hand grasps the object; Bim, ▩: poor timing, the hand holding the lid releases it before the other hand grasps the object.

In conclusion, 6-month-olds could coordinate several sequences of actions using the two hands, on similar or different patterns of action. It is only after 9 months of age that infants became able to coordinate simultaneous and differentiated activation of the two hands cooperating in a single action. Progress between 9 and 12 months of age concerned the skillfulness with which these cooperative patterns were mastered (see Table I). The completely differentiated strategy used (each hand performing a different part of the action) and the intermanual temporal coordination observed at 12 months gave an impression that the tasks were skillfully completed at that age.

III. Conclusion

Whereas some interlimb coordination inherently exists at birth, major developmental changes in voluntary bimanual coordination occur during the first year of life. The proximal coordination of the first reaching movements is followed by more distal coordination during early object manipulation. This distal coordination is at first relatively undifferentiated, in terms of hand role, before differentiation appears in the complementary roles assumed by each hand in more elaborated manipulations such as in means–end behavior. Along with the capacity to assume a different role emerges the capacity to differentiate the temporal patterns devoted to simultaneous bilateral movements. For instance, the strategies in two-step means-end behaviors go from being sequential before 9 months to being simultaneous at the end of the first year of life, and temporal bimanual coordination progresses particularly during this last trimester. By the end of the first year, mature forms of bimanual cooperation can be observed.

Going back to the specific questions concerning the development of bimanual coordination raised in the introduction, one can ask which inherent patterns form the basis on which bimanual coordination develops, when does intentional bimanual coordination first emerge, and what best accounts for the progress in bimanual coordination observable during the first year of life. The fact that the reflexive bilateral coordination coincides at birth with a period of unilateral prereaching makes it unlikely that these primitive coordinated patterns form *stricto sensu* a basis for manual strategy during reaching. However, the bilateral coordination observable in primitive reflexes reflects the bilateral organization of neural control over proximal segments (ipsilateral as well as contralateral), which may induce some low-level coupling in the spinal network. This may favor the use of a bilateral synergy for upper-limb activity as postural asymmetry disappears, increasing the likelihood that the two hands might meet at the midline. The tactual and visual consequences then reinforce the use of the scheme (Piaget's circular reaction). Later, the unilateral reaching begins to predominate as infants learn to control forces in the nonreaching hand (Thelen & Corbetta, 1992). This is made possible by the development of upper-body control and when control over upper limbs shifts from being proximal to distal. In my opinion, the emergence of true unimanual reaching is a result of an increase in the control over distal effectors, for which pathways are contralateral (Brinkman & Kuypers, 1973), and of the reduction of the low-level coupling, which lowers the chance of bilateral coactivation. At the same time, the increase in visual control of reaching (Bushnell, 1985) also encourages infants to use only one system, in particular the right hand. Once infants use only one hand to grasp small objects, then the other hand is free to manipulate the object held in the grasping hand. This paves the way for increasingly complex manipulations, with differentiated roles for each hand. In fact, after objects are successfully reached and grasped, there is a growing need for bimanual patterns destined for exploration.

As infants explore objects more thoroughly, they gather information about object properties but they also develop new schemes of intermanual coordination. The first possible kind of differentiated coordination between the two hands is either passive/active or sequential: one hand plays its part, stops, and then the second hand initiates its own action. Because of the spontaneous tendency to synchronize the two hands, it is possible that simultaneous coordination of differentiated actions of the two hands necessitates inhibition of symmetrical coactivation at several levels of coupling before differentiation could occur. Progress in temporal bimanual cooperation may then partly be due to the maturation of the interhemispheric cooperation through the callosal connections, in particular between the supplementary motor areas (SMAs), as suggested by Diamond (1991) (see Table I).

Therefore, the conceptual view being proposed here differs from the conceptions of motor development discussed in the introduction in the importance that it assigns to the interactions between maturation and experience across the whole development of bimanual coordination. It is suggested that changes in bimanual coordination during the first year of life are produced by the interaction between the developmental changes occurring in the underlying neural structures and the constant feedback that the infant receives from his or her actions. Specifically, the described changes have either been documented or hypothesized to be the result of gain in postural control, shift of control from proximal to distal effectors, increased opportunities to explore new schemes of manipulation, and maturation of interhemispheric cooperation. New forms of manual behaviors are then considered to emerge as the consequence of the dynamic relationships between these various factors, each playing the role of change parameters at a certain moment, and new forms of behaviors in turn opening the way for the influence of other factors.

Acknowledgments

I gratefully thank David Lewkowicz for his critical reading of the manuscript. I also thank the anonymous reviewers for their helpful comments.

References

Ames, L. B. (1949). Bilaterality. *Journal of Genetic Psychology,* **75,** 45–50.
Bergeron, M. (1948). *Les manifestations motrices spontanées chez l'enfant.* Paris: Hermann.
Bernstein, N. (1967). *The coordination and regulation of movements.* Oxford: Pergamon Press.
Bower, T. G. R. (1972). Object perception in infants. *Perception,* **1,** 15–30.
Bower, T. G. R. (1974). The development of motor behavior. In B. A. Aktinson, J. Freedman, & R. F. Thomson (Eds.), *Development in infancy,* Vol. 6, (pp. 135–187). New York: Academic Press.
Bresson, F., Maury, L., Pieraut-Le Bonniec, G., & de Schonen, S. (1977). Organization and lateralization of reaching in infants: an instance of asymmetric functions in hands collaboration. *Neuropsychologia,* **15,** 311–320.
Brinkman, J., & Kuypers, H. G. J. M. (1973). Cerebral control of controlateral and ipsilateral arm, hand and finger movements in the split-brain rhesus monkey. *Brain,* **96,** 653–673.
Bruner, J. S. (1970). The growth and structure of skill. In K. Connolly (Ed.), *Mechanisms of motor skill development* (pp. 63–94). New York: Academic Press, Inc.
Bruner, J. S., & Koslowski, B. (1972). Visually preadapted constituents of manipulatory action. *Perception,* **1,** 3–14.
Bushnell, E. W. (1985). The decline of visually guided reaching during infancy. *Infant Behavior and Development,* **8,** 139–155.
Casati, I., & Lézine, I. (1968). *Les étapes de l'intelligence sensori-motrice: épreuves adaptées de J. Piaget.* Paris: Centre de Psychologie Appliquée.
Cohen, L. (1971). Synchronous bimanual movements performed by homologous and nonhomologous muscles. *Perceptual and Motor Skills,* **32,** 639–644.
Deutsch, D. (1983). The generation of two isochronous sequences in parallel. *Perception and Psychophysics,* **34**(4), 331–337.
Diamond, A. (1991). Neuropsychological insights into the meaning of object concept development. In S. Carey & R. Gelman (Eds.), *Biology and knowledge: Structural constraints on development* (pp. 37–80). Hillsdale, NJ: Erlbaum.
Fagard, J., & Jacquet, A. Y. (1989). Onset of bimanual coordination and symmetry versus asymmetry of movement. *Infant Behavior and Development,* **12,** 229–236.
Fagard, J., & Jacquet, A. Y. (submitted for publication). Changes in reaching and grasping objects of different sizes between 7 and 13 months of age.
Fenson, L., Kagan, J., Kearsley, R. B., & Zelazo, P. R. (1976). The developmental progression of manipulative play in the first two years. *Child Development,* **47,** 232–236.
Flament, F. (1975). *Coordination et prévalence manuelle chez le nourisson.* Paris: Editions du Centre National de la Recherche Scientifique.

Gesell, A. (1946). The ontogenesis of infant behavior. In L. Carmichael (Ed.), *Manual of child psychology*. New York: Wiley.

Gesell, A., & Ames, L. B. (1947). The develoment of handedness. *Journal of Genetic Psychology,* **70,** 155–175.

Goldfield, E. C., & Michel, G. F. (1986a). Spatio-temporal linkage in infant interlimb coordination. *Development Psychobiology,* **19,** 259–364.

Goldfield, E. C., & Michel, G. F. (1986b). The ontogeny of infant bimanual reaching during the first year. *Infant Behavior and Development,* **9,** 87–95.

Grenier, A. (1981). "Motricité libérée" par fixation manuelle de la nuque au cours des premières semaines de la vie. ["Liberated motricity" by manual fixation of the neck in the course of the first weeks of life]. *Archive Française de Pédiatrie,* **38,** 557–561.

Guiard, Y. (1987). Asymmetric division of labor in human skilled bimanual action: the kinematic chain as a model. *Journal of Motor Behavior,* **19**(4), 486–517.

Humphrey, D. E., & Humphrey, G. K. (1987). Sex differences in infant reaching. Note in *Neuropsychologia,* **25**(6), 971–975.

Illingworth, R. S. (1975). *The development of the infant and young child*. Edinburgh: Churchill Livingstone.

Kelso, J. A. S. (1981). On the oscillatory basis of movement. *Bulletin of the Psychonomic Society,* **18,** 63.

Kelso, J. A. S., Southard, D. L., & Goodman, D. (1979). On the nature of human interlimb coordination. *Science,* **203,** 1029–1031.

Kelso, J A. S., Holt, K. G., Kugler, P. N., & Turvey, M. T. (1980). On the concept of coordinative structures as dissipative structures: II. Empirical lines of convergence. In G. E. Stelmach & J. Requin (Eds.), *Tutorials in motor behavior*. New York: North-Holland.

Klapp, S. T., Hill, M. D., Tyler, J. G., Martin, Z. E., Jagacinski, R. J., & Jones, M. R. (1985). On marching to two different drummers: Perceptual aspects of the difficulty. *Journal of Experimental Psychology: Human Perception and Performance,* **11,** 814–827.

Larouche, J. (1991). *The activation of bimanual coordination in infants seven and one half to sixteen months*. Memoire: Columbia University in Paris.

Lézine, I. (1978). Premières organisations des activitiés manipulatoires chez des enfants de 5 à 9 mois. *Arch. Psychol.,* **XLVI,** 177.

Lockman, J. J., Ashmead, D. H., & Bushnell, E. W. (1984). The development of anticipatory hand orientation during infancy. *Journal of Experimental Child Psychology,* **37,** 176–186.

Marteniuk, R., & MacKenzie, C. L. (1980). A preliminary theory of two hand coordinated control. In G. E. Stelmach & J. Requin (Eds.), *Tutorials in motor behavior*. Amsterdam: North-Holland: Elsevier.

McCall, R. B. (1974). Exploratory manipulation and play in the human infant. *Monographs of the Society for Research in Child Development,* **39,** (2, Serial No 155).

McDonnell, P., Corkum, V. L., & Wilson, D. L. (1989). Patterns of movement in the first 6 months of life: New directions. *Canadian Journal of Psychology,* **43**(2), 320–339.

McGraw, M. B. (1941). Development of neuromuscular mechanisms as reflected in the crawling and creeping behavior in the human infant. *Journal of Genetic Psychology,* **58,** 83–111.

Mitchell, R. G. (1960). The Moro Reflex. *Cerebral Palsy Bulletin,* **2,** 135.

Newell, K. M., Scully, D. M., McDonald, P. V., & Baillargeon, R. (1989). Task constraints and infant grip configurations. *Developmental Psychobiology,* **22**(8), 817–832.

Palmer, C. F. (1989). The discriminating nature of infants' exploratory actions. *Developmental Psychology,* **25**(6), 885–893.

Peters, M. (1977). Simultaneous performance of two motor activities, *Neuropsychologia,* **15,** 461–465.

Piaget, J. (1936). *La naissance de l'intelligence chez l'enfant*. Delachaux and Niestlé: Neuchâtel.
Ramsay, D. S. (1985). Infants' block banging at the midline: Evidence for Gesell's principle of "reciprocal interweaving" in development. *British Journal of Developmental Psychology*, **3**, 335–343.
Ramsay, D. S., & Weber, S. L. (1986). Infants' hand preference in a task involving complementary roles for the two hands. *Child Development*, **57**(2), 300–307.
Rochat, P. (1989). Object manipulation and exploration in 2- to 5-month-old infants. *Developmental Psychology*, **25**(6), 871–884.
Rochat, P. (1992). Hand-mouth coordination in the newborn: morphology, determinants, and early development of a basic act. In G. J. P. Savelsbergh (Ed.), *The development of coordination in infancy*. Amsterdam: Elsevier.
Rochat, P., & Stacy, M. (1989). *Reaching in various postures by 6 and 8 month-old infants: The development of mono-manual grasp*. Poster presented at the Biennial Meeting of the Society for Research in Child Development, Kansas, Missouri.
Ruff, H. A. (1984). Infants' manipulative exploration of objects: effects of age and object characteristics. *Developmental Psychology*, **20**(1), 9–20.
Schmidt, R. A., Zelaznik, H. W., Hawkins, B., Frank, J. S., & Quinn, J. T. (1979). Motor output variability. A theory for accuracy of rapid motor acts. *Psychological Review*, **86**, 415–451.
Shaffer, L. H. (1982). Rhythm and timing in skill. *Psychological Review*, **89**(2), 109–122.
Summers, J. J. (1987). *The production of polyrhythms*. Paper presentd at the Second Workshop on Rhythm Perception and Production, Marburg, Germany.
Thelen, E., & Corbetta, D. (1992). Dynamics of interlimb coordination in the arms and legs of infants. Presentation at the HFSP, *"The control and modulation of patterns of interlimb coordination: a multidisciplinary perspective."* Leuven, Belgium.
Thelen, E., & Ulrich, B. D. (1991). Hidden skills: a dynamic systems analysis of treadmill stepping during the first year. *Monographs of the Society for Research in Child Development*, Serial No. **223**, 56.
Thelen, E., Kelso, J. A. S., & Fogel, A. (1987). Self-organizing systems and infant motor development. *Developmental Review*, **7**, 39–65.
von Hofsten, C. (1982). Eye-hand coordination in the newborn. *Developmental Psychology*, **18**(3), 450–461.
von Hofsten, C. (1984). Developmental changes in the organization of prereaching movements. *Developmental Psychology*, **20**, 378–388.
White, B. L., Castle, P., & Held, R. (1964). Observations on the development of visually directed reaching. *Child Development*, **35**, 349–364.
Willatts, P. (1985). *Development and rapid adjustment of means–ends behavior in infants aged six to eight months*. Communication at the Eighth Biennial Meeting of the International Society for the Study of Behavioral Development. Tours, France.
Wing, A. M., & Kristofferson, A. B. (1973). The timing of interresponse intervals. *Perception and Psychophysics*, **13**, 455–460.
Yamanishi, J., Kawato, M., & Suzuki, R. (1980). Two coupled oscillators as a model for coordinated finger tapping of both hands. *Biological Cybernetics*, **37**, 121–225.

22

The Coordination Dynamics of Learning
Theoretical Structure and Experimental Agenda

Pier-Giorgio Zanone and J. A. S. Kelso

Program in Complex Systems and Brain Sciences
Center for Complex Systems
Florida Atlantic University
Boca Raton, Florida

I. Introduction
    A. What Changes with Learning?
    B. How Does Learning Occur?
    C. What Form Do Changes Due to Learning Take?
    D. What Governs the Rate of Learning?
    E. What Determines the Form That Changes Due to Learning Take?
II. Theoretical Structure
III. Initial Tests of Predictions: Transitions and Transfer
IV. Experimental Agenda
    A. Generalization of Learning
    B. Form of Learning
    C. Rate of Learning
    D. Routes to Learning
    E. Transition Paths after Learning
V. Conclusions
    References

I. Introduction

Learning, development, and evolution are separate fields that are now studied largely in isolation of each other. Yet, all deal, in one form or another, with pattern formation processes, that is, the emergence of (new, more, different) structure, organization, or order. And all deal, in one form or another, with the fundamental issues of stability and change. Thus, separation of these domains does not necessarily negate the possibility that common principles may exist that underlie pattern stability and change. The present chapter aims at identifying principles of adaptive change, seen as a pattern-formation process, through the window of human perceptual–motor learning.

Learning, though central to psychology and biology, remains a concept characterized by a variety of definitions, approaches, theories, and experimental methods. General laws of learning were sought early on, emphasizing behavioral (e.g., Humphrey, 1933; Pavlov, 1927/1960; Skinner, 1938; Thorndike, 1911) and neurophysiological (e.g., Hebb, 1949) approaches. Recently, following the seminal work by McCulloch and Pitts (1943), strictly computational models of learning (i.e., computer-simulated, artificial neural networks) constitute a lively field of research.

In this context, the originality and the pertinence of a dynamical approach to learning resides in the identification of constraints that affect the learning process at the behavioral level itself. Roughly speaking (see Section II, for further development), its basic tenet is that what is learned behaviorally as a spatiotemporal *coordination pattern* results from modification of underlying *coordination dynamics* (which, in some conceptual context, we also refer to as *intrinsic* dynamics; see below). These coordination dynamics express the organism's *preferred and stable coordination tendencies* that exist at any moment, and therefore, systematically affect its ability to learn. The conceptual framework of coordination dynamics may incorporate innate biological constraints, but also encompasses functional constraints that reflect previous experience. Of course, such experience may be specific to the task to be learned as well as related to a large ensemble of tasks that exhibits no obvious commonality with the task at hand. In spite of, or because of, such nonspecific influences, we believe that the coordination dynamics are just as important to incorporate into computational theories of learning as, say, biologically plausible properties of single neurons (Abbott, 1990). Biological models of learning could also gain from the idea of intrinsic dynamics in accounting for the already complex initial state from which only certain stimuli, and not others, are capable of evoking certain responses (Eimas & Galaburda, 1990). Consequently, an advantage of the present approach is that it treats individual differences explicitly. In psychology, the fact that individual subjects bring different backgrounds and capacities into the learning environment has often been raised as a problem for constructing a theory of learning. A traditional solution is to attempt to equate these differences by having subjects learn as arbitrary a task as possible (viz., a task that none of the subjects is likely to have previously experienced), thereby "controlling" for different initial states. In contrast, the present approach provides a means for evaluating the coordination dynamics of each individual *before* exposure to a new task. Furthermore, the nature of the learning process itself can be exposed by probing the coordination dynamics as practice proceeds. Thus, our approach may inform the older quest for laws of learning, in the sense of providing generic principles leading to adaptive change.

The benefit of knowing what the coordination dynamics are before learning is that, in principle, the following set of essential questions pertaining to learning become tractable and new issues arise, constituting an agenda for future research.

A. What Changes with Learning?

A central hypothesis of the present framework is that the entire layout of the coordination dynamics changes with learning, not simply the particular coordination pattern being learned. This chapter presents preliminary findings about transfer of learning (Zanone & Kelso, 1992a) that strongly support this idea. In expanding the topic of transfer to that of *generalization* of learning, one is led to ask the question of *what is learned* when a given behavioral pattern is acquired, and in particular, how independent such learning is from that which was actually involved in the original task.

B. How Does Learning Occur?

Learning occurs as specific modifications of the coordination dynamics in the direction of the behavioral pattern to be learned. For the most part, these modifications tend to reduce competition that may arise between the task to be learned and the existing coordination dynamics. Competition occurs when extrinsic learnign requirements do not coincide with a stable state of the current coordination dynamics. By contrast, when they do coincide, cooperative processes run the show. This chapter reports experimental evidence (Zanone & Kelso, 1992a,b; 1992c) that confirms the above theoretical predictions (Schöner, 1989; Schöner & Kelso, 1988a,b; Schöner, Zanone, & Kelso, 1992), through systematic and direct study of the ongoing modifications of the coordination dynamics with learning, rather than through the assessment of simple improvements in performance in a single task. The theoretical tenet remaining to be proved is that the required pattern indeed becomes a constitutive part of the coordination dynamics.

C. What Form Do Changes Due to Learning Take?

Theory posits that dpending on the relationship between the to-be-learned task and the existing coordination dynamics, change may be continuous or abrupt. In particular, if an externally imposed learning requirement is far away from attractive states of the coordination dynamics, the learning process may involve steep changes. If the task to be learned is close to an existing preferred state, the learning process may be smooth and gradual. One of the goals of our proposed agenda is to render this concept of distance rigorous.

D. What Governs the Rate of Learning?

Theoretically, the degree of competition between learning requirements and coordination dynamics determines the rate of learning. Suppose the individual coordination dynamics are known in advance through appropriate probes that show that some states are less stable than others. Suppose also that the learning requirement is closer to the less stable than the more stable state. Then, a more rapid learning is expected in the former (where competition is small) than in the latter case (where competition is large). Such a prediction is open to experimental verification.

E. What Determines the Form That Changes Due to Learning Take?

Competitive or cooperative mechanisms largely determine the behavior observed at any given point in time. In particular, these hypothetical mechanisms are reflected by the variability of the observed pattern and its deviation toward the system's intrinsically preferred states as practice proceeds. Suppose the coordination dynamics were not known before the learning task is introduced, and that in one subject the learned task was accomplished quickly and in another, much more slowly. Without a theory of individual differences, the responses of both subjects would typically be grouped together, and one would be none the wiser about why learning was rapid and efficient in one case and not the other. In the present approach, the identification of so-called intrinsic dynamics, which capture the spontaneous coordination tendencies of the system before introducing a learning requirement, allows us not only to understand such differences in learning, but to predict them. One important consequence of the individual- and task-specific nature of the coordination dynamics is that the individual learner is the relevant unit of analysis.

II. Theoretical Structure

The present approach to learning as a pattern-formation process in complex systems (i.e., with many degrees of freedom) stems from the concepts of self-organization in nonequilibrium systems (e.g., Haken, 1983; Nicolis & Prigogine, 1989; Yates, 1987), and uses the tools of nonlinear dynamical systems. The aim is to come up with a theoretical model of the time course of a phenomenon at a given level of description, namely, to identify its dynamics, or equations of motion. We define a coordination pattern as an observable, reproducible, and stable relationship among the components of a biological system. The nature of the system's *coordination dynamics* is revealed around *phase transitions* or *bifurcations*,[1] points at which *spontaneous* changes in the coordination pattern occur under the influence of some parameter. Such spontaneous changes signify a qualiltative alteration of the system's underlying coordination dynamics. Thus, phase transitions are important in two respects. At the methodological level, various behaviors can be distinguished from each other as different stable patterns before and after transitions. At the theoretical level, such nonlinear, qualitative change among patterns leads to the identification of so-called collective variables or *order parameters* that describe the patterns themselves. The (low-dimensional) coordination dynamics can be then reconstructed by mapping the observed, stable coordination patterns onto *attractors* (viz., asymptotically stable solutions) of the collective variable equation of motion and by studying the stability of such collective states (see Kelso & Schöner, 1988; Schöner & Kelso, 1988c,d).[2] Thus, the switching from one coordination pattern to another observed at the behavioral level can be unequivocally linked, at the theoretical level, to the vanishing of an attractor from the system's coordination dynamics, typically a qualitative change or bifurcation.

According to the theory, boundary conditions (e.g., environmental, task, energetic constraints) act as parameters on the collective variable dynamics, thereby effecting changes in the coordination pattern. If these parameters are nonspecific (i.e., they do not stipulate a particular coordination pattern), they are called *control parameters*. Control parameters lead to a phase transition when they attain a certain critical value. In contrast, if the parameters are specific, they represent behav-

---

[1] Mathematicians typically coin such phenomena as bifurcations, while physicists, for historical reasons, tend to stick to the term phase transitions. Here, we will use these expressions interchangeably.

[2] Due to this mapping, the concept of coordination dynamics may be used here synonymously with the collective variable equation of motion.

ioral requirements of any origin (e.g., environmental, perceptual, intentional, memory) influencing the observed pattern. Such requirements are captured theoretically as *behavioral information,* which acts as a force that attracts the collective variable to a specific value. That is, the required pattern specifies an attractor of the collective variable dynamics. An important point is that *information is thus expressed by a variable of the same type as the collective variable* used to characterize the coordination patterns, that is, information is specific to the coordination dynamics. Therefore, the full "coordination dynamics" represent *spontaneous* coordination tendencies alone, as in Equation (1), defining so-called *intrinsic* dynamics, as well as the influence of specific informational requirements on these tendencies, as in Equation (2) (cf. Figure 1A and B). Another important point is that information exists only to the extent that it modifies the coordination dynamics, that is, to the extent that it alters the behavioral pattern in a specific fashion. As such, there is no more "information" in behavioral information than there is in the intrinsic dynamics: these are joint theoretical constructs that separate out the respective contribution of the system's spontaneous coordination tendencies and of specific (environmental, task, energetic, etc.) constraints to the full coordination dynamics.

On the other hand, we do not wish to be interpreted as saying that information is irrelevant to the intrinsic dynamics. The collective variable is clearly a meaningful informational variable reflecting the coupling between components or between components and the environment (e.g., Kelso, Ding, & Schöner, 1992). The definition of such a collective variable, we stress again, is always task dependent. Nonetheless, we reseve the term behavioral information for those influences that extend beyond the spontaneous coordination tendencies that reflect an individual subject's behavioral repertoire at a given instant, in particular, before a learning task is introduced.

To study learning as modifications of existing coordination tendencies requires an experimental model system in which the spontaneous coordination dynamics can be identified. A strength of the present approach to learning is that a test field already exists, pertaining to previous work on bimanual coordination. Typically, rhythmic movements of symmetric fingers exhibit only two patterns of coordination that are stable at a low frequency, namely, in-phase (simultaneous activation of homologous muscles) and anti-phase (alternate activation) (Kelso, 1984). However, when frequency increases (viz., a nonspecific control parameter), a spontaneous change from the anti-phase to in-phase pattern occurs at a critical parameter value, whereas the in-phase pattern remains stable over the same range of required frequencies. Such switching between coordination patterns has been modeled as a bifurcation from bistable to monostable dynamics of the collective variable, relative phase (ϕ) between the moving components (Haken, Kelso, & Bunz, 1985; Schöner, Haken & Kelso, 1986). The model of the bimanual intrinsic dynamics (Haken *et al.*, 1985) expresses the rate of change in relative phase, $\dot{\phi}$, as a function of the derivative of its potential, $V(\phi)$:

$$\dot{\phi} = -\frac{dV(\phi)}{d\phi} + \sqrt{Q}\xi_t, \qquad (1)$$

where $V(\phi) = -a\cos(\phi) - b\cos(2\phi)$ and $\sqrt{Q}\xi_t$ is Gaussian white noise of strength Q. Noise is introduced in Equation (1) because all real systems described by low-dimensional dynamics are coupled to many subsystems at a more microscopic level (e.g., at a neuromuscular level), which act as stochastic forces on the collective variable, ϕ. One may then view noise as a continuously applied perturbation that produces deviations from the stable state. Complying with periodicity and symmetry requirements, Equation (1) captures the observed bifurcation, namely, for $0 < a < 4b$, two stable states $\phi = 0°$ and $\phi = \pm 180°$ exist, while for $a > 4b > 0$, only $\phi = 0°$ remains stable. This is not just a nice compact mathematical description of the phenomena. In addition, typical features of the theoretical model system defined in Equation (1) have been observed in experiments. In particular, enhancement of relative phase fluctuations (Kelso & Scholz, 1985; Kelso, Scholz, & Schöner, 1986) and critical slowing down, seen as an exponential increase in relaxation time of the anti-phase mode (Scholz, Kelso, & Schöner, 1987; Scholz & Kelso, 1989), have been assessed on the way to the transition. These critical phenomena attest to the nonequilibrium character of the transition. That is, the generic mechanism for pattern change is *instability*.

Figure 1A plots the potential of the bistable dynamics as defined by Equation (1).[3] The relative stability of the two attractors at 0 and 180° is indicated by the depth of each well, and their attraction by the slope at each point of the curve. Such a picture makes it clear how the system will eventually relax into one of the two attractors, depending on its initial condition. Any small perturbation causes the system to fall into the *basin of attraction* of one of the two stable modes. Such an "attractor layout" reflects which stable patterns are observed in the absence of specific requirements, such as those of a learning task. This layout

---

[3] The rate of change in the system state is given by the slope of the potential function. Hence, fixed points correspond to points with zero slope, minima being attractors and maxima, repellers.

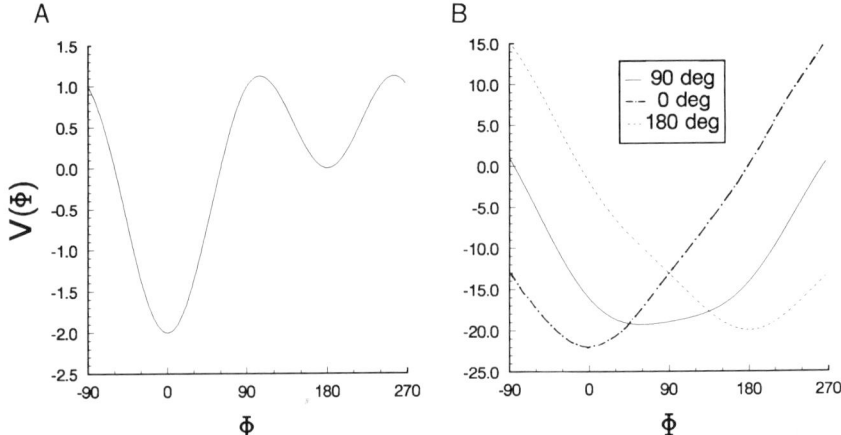

FIGURE 1 Visualization of the potential of the bimanual coordination dynamics. (A) The potential of the coordination dynamics is displayed in the bistable regime, using Equation (1) with a and $b = 1$ Hz. (B) Behavioral information is introduced attracting ϕ to $\psi = 0°$ (dotted line), $\psi = 180°$ (dashed line), and $\psi = 90°$ (solid line), using Equation (2) with a and $b = 1$ Hz, and $c = 20$ Hz.

corresponds to the intrinsic version of the coordination dynamics, namely, the *intrinsic* (coordination) dynamics.

Consider now the situation in which a required pattern is specified by the environment or from memory. The model when such so-called *behavioral information* specifying a required relative phase, ψ, is included, reads:

$$\dot{\phi} = -\frac{dV_\psi}{d\phi} + \sqrt{Q\xi_t}, \qquad (2)$$

where now, $V_\psi = V(\phi) - c \cos[(\phi - \psi)/2]$. $V(\phi)$ is the potential of the intrinsic dynamics given in Equation (1), and the second term is the simplest function[4] that attracts relative phase toward the required phasing. The parameter c represents the strength with which environmental information acts on the coordination dynamics.

Figure 1B plots the potential defined by Equation (2) for three required relative phases, $\psi = 0$, 90, and 180° (dotted, solid, and dashed curves, respectively) and exhibits two main features. On one hand, when the required relative phase coincides with one of the stable intrinsic patterns, the minimum of the potential is exactly at the required relative phase (i.e., $\psi = 0°$ or 180°), and its shape is well articulated (less so for anti-phase than for in-phase, reflecting the differential stability of

---

[4] To conform with certain periodicity requirements, the function is 4π-periodic (see Schöner & Kelso, 1988a).

these two collective states). This case exemplifies the *cooperation* between extrinsic requirements and spontaneous coordination dynamics. On the other hand, if the required relative phase does not correspond to one of the intrinsically stable patterns, for instance $\psi = 90°$ (cf. the solid curve in Figure 1B), *competition* between the two "forces" leads to a deformed potential. On one hand, the minimum is pulled away from the required relative phase, implying a larger error in performance, and, on the other hand, the minimum exhibits a wider, less articulated shape, signifying enhanced variability. Thus, the extent to which behavioral information cooperates or competes with the intrinsic dynamics determines the observed coordination patterns. The important operational consequence of this competitive or cooperative interplay is that by *scanning* a large set of required phasings, the experimenter obtains a *means to evaluate the attractor layout* systematically, that is, to probe the underlying coordination dynamics at any point in time. Using such a "scanning procedure" with phasing requirements specified either by the enviornment (Tuller & Kelso, 1989) or from memory (Yamanishi, Kawato, & Suzuki, 1980), the bistability of the bimanual dynamics at low movement frequencies has been unquestionably revealed. Similar methods of probing the coordination dynamics at different moments during the learning process constitute the backbone of the experiments presented in this chapter.

Assume now that a subject must learn a phasing pattern specified by the environment that does *not* correspond to one of his/her intrinsically stable patterns, say, $\psi = 90°$. At the point in time when this new relative phase is learned, the influence of the bistable coordination dynamics attracting the collective variable toward $\phi = 0°$ and $\phi = 180°$ theoretically tends to dwindle, due to the progressively overwhelming attraction to the pattern being learned. Thus, the process of learning a new coordination pattern is predicted to involve a qualitative change in the coordination dynamics, that is, a phase transition, with the emergence of a new attractor at $\phi = 90°$ and loss of stability of the previously stable coordination patterns (Schöner, 1989; Schöner & Kelso, 1988a,b).[5] It is important to realize that we are referring here to the evolution of the learning process in time, that is, to the changes occurring in the coordination dynamics at the moment when the task requirement (i.e., $\psi = 90°$) is eventually achieved. *A priori*, this does not have to correspond to the final form of the intrinsic dynamics, that is, to what the spontaneous bimanual coordination tendencies are after

---

[5] Theoretically, 90° is an unstable fixed point (repeller) of the coordination dynamics before learning (cf. the bump of the potential curve in Figure 1A). So, the bifurcation predicted with learning transforms such a repeller into an attractor of the coordination dynamics, leading to a so-called *pitchfork bifurcation*.

learning under nonspecific parametric influences or a large set of different specific requirements.

III. Initial Tests of Predictions: Transitions and Transfer

The prediction that changes due to learning take the form of a phase transition has been recently confirmed (Zanone & Kelso, 1992b; see Schöner et al., 1992, which emphasizes the theory–experiment link). The thrust of the study was to determine systematically the individual attractor layout *throughout* the process of learning a new relative phase, $\psi = 90°$.[6] By probing the coordination dynamics *before, during,* and *after* practice of the to-be-learned pattern, using the scanning procedure mentioned above, we aimed at tracing the predicted modifications of the attractor layout with learning.

The experimental design was as follows. Five subjects participated in the experiment. The learning procedure involved fifteen trials per day over five consecutive days, in which the 90° phasing pattern was practiced. A visual metronome displayed the phasing pattern to be performed, which the subjects had to match with the coordinated periodic motion of homologous fingers. The trajectories of both hands and the required phasing signals were recorded continuously in real time. Each day, four probes of the coordination dynamics were carried out, in which the required phasing was scanned between 0 and 180° in successive steps of 15°, namely, one before, two during, and one after the practice trials. In order to test the permanence of the learning effects, two recall trials were administered one week later. Subjects had to produce the pattern required in the learning trials from memory. Then, a probe of the pattern dynamics was performed using the same scanning procedure.[7]

The results in the learning trials show that over the five days of practice, performance tended gradually to the required 90° pattern, with a substantial decrease in within- and across-trial fluctuations and between-subject variability. Typically, the produced coordination pat-

---

[6] As a matter of convention, positive values of relative phase imply that the right event (finger motion or metronome signal that specifies the phasing pattern to be performed) leads with respect to the left event, and conversely.

[7] Two arguments can be raised against the assumption that the probes themselves may induce specific learning. First, no knowledge of the results or feedback was given during or after the scanning trials. Yet, such information is essential for learning a phasing pattern (Zanone & Kelso, 1992a). Second, if learning were to occur during a scanning trial, its influence should be equivalent for *all* the patterns practiced in the various plateaus. Hence, the comparison across plateaus and probes is still valid.

tern stabilized to the required pattern by the third day of practice. Such learning proved to be robust over the retention interval, as shown by the recall trials, which did not differ significantly from the final practice trials. None of this is any different from findings of thousands of practice/learning studies.

Our interest, however, focuses on how the coordination dynamics evolve as practice proceeds and performance improves. The individual attractor layouts in the first and the last probes (i.e., before and after practice) and that of the recall session are presented in Figure 2 for two typical subjects, TM and MS. On the top (solid) curves, the variable delta RP, namely, the difference between the required and the actually produced relative phase, is plotted as a function of the required phasing. (When the required phasing is overestimated, delta RP has a positive value, and inversely.) The bottom (dotted) curves plot the corresponding SD of delta RP.

Let us focus on the initial probe on Day 1 for subject TM (left part of Figure 2A). In the first probe, delta RP (top curves) exhibits a humped curve as a function of the required pattern, that is, the error is lowest when the required phasing is 0 to 180°. Moreover, below 180°, the produced relative phase departs from the required phasing in the direction of the 180° pattern, and such a mismatch is roughly proportional to the difference between the required pattern and 180°. Thus, the negative slope of delta RP, about 180°, signifies that this pattern attracts nearby phasing patterns. The interval spanned by such a "negative-slope effect" reflects the width of the basin of attraction of the underlying attractive state, hence its strength.[8] The bottom curves show the SD of delta RP is lowest at 0 and 180°, while it increases markedly at intermediate values. The 0 and 180° patterns appear to be much more stable than the others. Thus, negative slope and low SDs are indicative that the in-phase and anti-phase patterns constitute attractive states of the coordination dynamics.

At first blush, a discrepancy seems to arise here with respect to the theoretical model exposed in Equation (1). The probe of the initial coordination dynamics for subject TM, which shows a stronger negative slope effect for 180° than for 0°, suggests that the latter is a more stable

[8] Strictly speaking, the "negative-slope" effect represents attraction in parameter space, that is, performance in nearby phasing conditions is biased in the direction of the intrinsic attractors, and not an attraction with time in phase space (the space of the dynamic variable, ϕ), as required by the definition of an attractor of the dynamics (i.e., asymptotically stable stationary solution). Nonetheless, attraction in parameter space reflects the underlying coordination dynamics. Because the requirements are expressed in the same type of variable as the performed pattern, attraction in parameter space occurs toward those required values that correspond to attractors of the coordination dynamics in phase space. Therefore, the terms stable state, attractor, stable fixed point, and attractive state are used synonymously in this chapter.

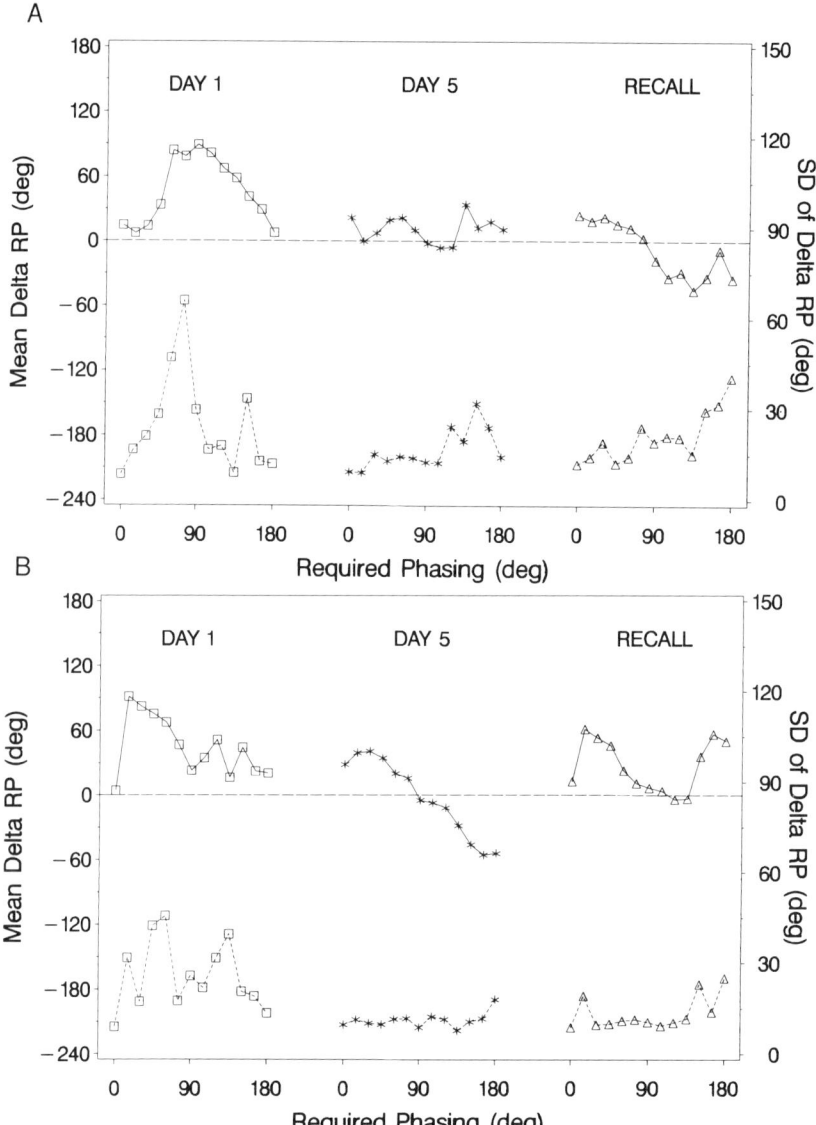

FIGURE 2 Individual attractor layouts before and after practice, and during the recall session. The upper solid graphs plot mean delta RP (produced minus required relative phase) as a function of the required phasing. The bottom dotted graphs present the corresponding SD.

attractor than the former, whereas Figure 1A indicates that the reverse is true for the intrinsic dynamics. Now, such an apparent incompatibility is easily understood in reference to the aforementioned distinction between the full coordination dynamics and spontaneous coordination tendencies (viz., intrinsic dynamics). In the scanning procedure, the passage from a requirement of 0 to 15° between the first and second plateau is not the same as, say, between 45 and 60°. The reason is that the 15° requirement represents the first change from a purely in-phase pattern, whereas later on, the learner must distinguish between different values of out-of-phase patterns. Such an implicit requirement (i.e., "do not in-phase any longer") constitutes behavioral information that contributes to the coordination dynamics and may kick the system out of the 0° intrinsic attractor. However, in spite of such information specific to the scanning task, the presence of spontaneous coordination tendencies (viz., intrinsic dynamics) is still revealed through competitive and cooperative processes. Ultimately, the essential quality of the coordination dynamics becomes obvious, namely, bistability at in- and anti-phase.

The probe of the coordination dynamics after five days of practicing the 90° pattern is presented in the middle graphs of Figure 2A for subject TM. Now, not only the 0 and 180° patterns act as attractive states for nearby phasing patterns and are stable, but the 90° pattern does as well: the collective variable dynamics appear to be tristable within the 0–180° interval. Thus, consonant with our main theoretical prediction, *learning involves a phase transition* from bistable to tristable coordination dynamics. After the transition, behavioral information defining the learning requirement and the new learned attractor then cooperate and stabilize the produced pattern at $\phi = 90°$. In other terms, the bifurcation associated with learning leads from a competitive to cooperative interplay between the spontaneous coordination tendencies and explicit behavioral information defined explicitly by the learning task. Moreover, such a qualitative modification of the attractor layout persists over one week. Indeed, the picture for the recall probe (right graphs of Figure 2A) still exhibits the typical features of the last probe (cf. middle graphs). Such persistence of the learned pattern as an attractive state attests to the stability of the underlying coordination dynamics.

For subject MS, the initial probe of the coordination dynamics (left part of Figure 2B) exhibits a negative slope of delta RP (upper solid curves) for intermediate required phasings, as well as below 180°. The pattern variability (lower dotted curves) appears to be smallest at 0 and 180°, as well as for several intermediate values around 90°. These features suggest that the intrinsic dynamics are tristable *before* any practice of the 90° pattern. After learning, the final coordination dynamics (middle part of Figure 2B) appears to be bistable at 0 and 90°. Delta RP shows a negative slope all the way across the interval between 45 and 180°, intersecting the zero axis at 90°, while the SD is low, comparable to that of the 0° pattern. The 90° pattern has become a strong attractor, which "sucks in" the 180° pattern itself. Thus, the destabilization of the 180° attractor with learning involves a bifurcation from tristability to bistability. Such an alteration of the coordination dynamics with learning again persists over a one-week interval, as shown in the right part of Figure 2B.

In sum, the foregoing findings provide strong arguments in favor of the approach to learning taken in the present chapter, answering two of the main questions asked in Section I, namely, what changes with learning and how this change occurs. Our main theoretical prediction is confirmed: learning is not just a gradual improvement with practice but also involves a phase transition process in which the required pattern becomes stabilized as an attractive state of the coordination dynamics. Changes in behavior with learning, traditionally measured as simple improvements in performance in the learning task itself, are then to be viewed as the outcome of modifications of the *entire* layer of the coordination dynamics. Since spontaneous coordination tendencies differ across individuals, the learning process can only be understood as the result of the interplay between coordination dynamics and the learning task if the individual's existing coordination tendencies are known before exposure to the learning task.

Transfer paradigms are deemed to be a powerful means of assessing learning, albeit rather indirectly. Such an assumption is at the origin of much research in the field of motor behavior (see Adams, 1987, for a review). Admittedly, learning does not only affect the specific task to be learned but may also affect behaviors other than the ones actually practiced. In our model system of bimanual coordination, a precise prediction may be drawn regarding transfer of learning, based on the fact that the intrinsic dynamics prove to be *invariant* over the transformation $+ \phi \rightarrow - \phi$. Due to the definitions of the collective variable, this means that the coordination dynamics are *symmetric* with respect to a left–right distinction, hence, independent of any lead-lag preference between the components. Such symmetry is broken as soon as behavioral information specifies a required pattern (i.e., with a definite left or right lead, say, $\psi = +90°$). Moreover, when such a pattern becomes a new attractive state of the coordination dynamics (at $\phi = +90°$) with learning, the question arises of whether the symmetry partner (i.e., $\phi = -90° = 270°$) becomes an attractive state as well. If such were the case, one would be led to three conclusions. First, this finding would provide firm evidence, according to traditional criteria, that learning has actually occurred, because of such (positive) transfer of a learned task to a nonpracticed one. Second, a central tenet of the theory, namely, that learning implies an alteration of the entire coordination dynamics, would be enhanced. Third, one could assume that what is learned is an *abstract* timing relation, independent of the order between the components that instantiate the pattern. Therefore, it is crucial to test whether learning a specific phase relationship automatically entails learning the symmetric phasing pattern. Transfer then may be said to follow a principle of *symmetry conservation* of the coordination dynamics.

This issue of learning by transfer has recently been tackled. Only two exemplary findings will be reported here (see Zanone & Kelso, 1992a, for more details). Similar methods as in the above study on learning were adopted in order to induce learning and to follow in time modifications of the coordination dynamics. A difference was, of course, that such a probe had to be carried out over the full 0–360° span, encompassing right- and left-lead phasing patterns (i.e., positive and negative values of the collective variable, ϕ, respectively), which we did not do in earlier work (cf. Figure 2). Here is the basic procedure. On the first day, a 0–360° probe of the coordination dynamics was carried out, establishing the initial attractor layout. Thus, the to-be-learned pattern (ψ) was set on an individual basis such that it did not correspond to an intrinsic stable pattern.[9] That is, $\psi = 90°$ or $\psi = 135°$, depending on whether the 90° pattern was an intrinsic attractor or not, respectively. In this case, competition between behavioral information and the coordination dynamics was expected, and learning was predicted to involve a phase transition. Still, on the first day, twenty practice trials of the selected phasing pattern were then administered. On the second day, a probe of the dynamics was performed between 0 and 180° only, in order to avoid practice of any negative-signed (left-lead) pattern. Then, thirty more practice trials were administered, followed by a complete probe of the final coordination dynamics.

Figure 3A presents the attractor layouts in the first and last probes for subject JG (i.e., before and after practice of the to-be-learned pattern), the intermediate half-probe being skipped for the sake of simplification. Before practice on Day 1, the initial layout (left two graphs) exhibits the characteristic features of bistable dynamics. Note that the picture is quite symmetric. For this subject the learning task was then set at 90° of relative phase. By the end of the learning procedure (right two graphs), the presence of a novel attractive state at 90° is obvious, reflected by the negative slope effect and low SD about that value. This

---

[9] Indeed, we know from our previous experiment on learning that some subjects may exhibit spontaneous coordination tendencies that already contain the 90° attractor prior to learning (cf. Figure 2B).

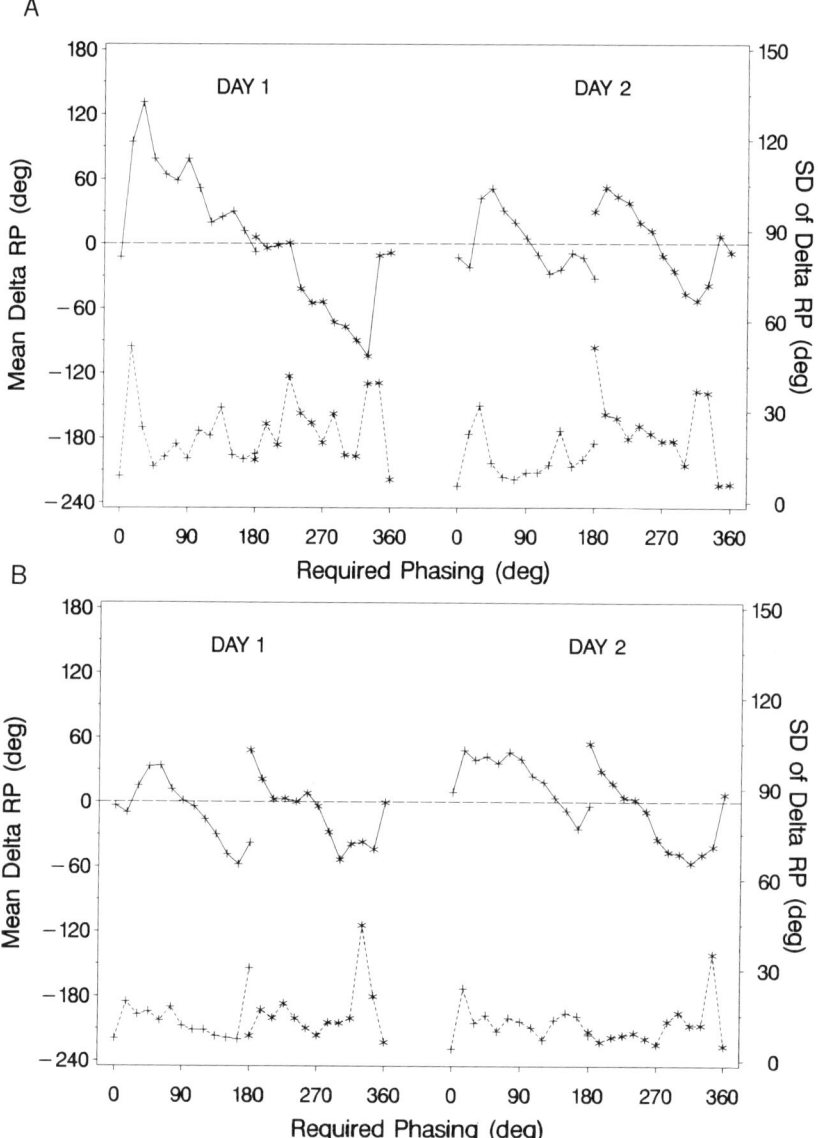

FIGURE 3 Individual attractor layouts before and after learning. (See legend for Figure 2.)

is a phase transition with learning, similar to that shown in the original learning experiment (cf. Figure 2A). Most remarkable is that a new attractor appears at 270° as well, although such a pattern has *not been practiced* at all. Let us emphasize that not only is the 270° pattern produced quite stably when required, but also it strongly attracts nearby phasing patterns. So, a phase transition is also observed in the interval 180–360°, resulting from learning the 90° pattern. This constitutes a clear instance of *transfer of learning*. Learning the 90° pattern involved a bifurcation from bistable dynamics at 0 and 180° to a multistable regime in which the 0, 90, 180, *and* 270° patterns are attractive states. Let us note that the precise extent to which the 180° intrinsic attractor loses stability with learning has to be checked out, because this determines which type of bifurcation represents the learning process. If the 180° pattern is almost completely destabilized (cf. Figure 2B), a pitchfork bifurcation prevails. If not, a more suitable model is a saddle-node bifurcation.[10]

Figure 3B displays the initial and final attractor layouts for subject BF. Initially (left two graphs), the pattern dynamics happen to be stable at about 0, 90, 180, and 270°. Note that the picture is also fairly symmetric. So, this subject practiced 135° of relative phase. By the end of Day 2 (right two graphs), no qualitative change is detected, the dynamics remaining in the same regime. Nevertheless, a substantial shift of the 90° attractor is noticeable, such that it is finally quite close to the required pattern, about 135°. Now, a concomitant and symmetric shift of the initial attractor at 270° is observed, reaching 225° (i.e., −135°) by the end of the experiment. Such evolution of an attractive state at 270° mirroring that of the 90° pattern suggests strongly that transfer of learning occurred. Nonetheless, a point of theoretical interest is that another route to learning is identified here, involving only parametric, quantitative change in the pattern dynamics: the *drift of an attractor* or stable fixed point in parameter space.

To sum up, whatever the learned pattern and the form that learning takes, the symmetry partner to the learned pattern is shown to stabilize with learning as well, although it has not been practiced. The fact that the to-be-learned pattern and its symmetry partner are both learned suggests strongly that transfer of learning occurs automatically. *Symmetry conservation* appears to be a feature of the bimanual coordination dynamics. Furthermore, this finding provides an answer to our original question (see Section I) of what is learned. Learning seems to occur at a rather abstract level, namely, independent of any lead-lag ordering between the components. It is important to emphasize again that with-

---

[10] Another clue (readily available from the data) of a saddle-node bifurcation is the drift of the stable fixed point before transition, which is not expected in a pitchfork bifurcation.

out knowledge of the individual attractor layout prior to practice, none of these findings would make any sense.

IV. Experimental Agenda

The last part of this chapter returns to questions in Section I about learning that remain to be answered, and focuses on new issues that have arisen from the discussion of the foregoing results. This section highlights how theoretical concepts and methods come to grips with the issues, proposing typical means of operationalization. Let us note from the outset that these methods and concepts, developed here for phase-locked patterns, can be expanded easily to learning frequency relations between coordinating components that are not 1 : 1 (e.g., Beek, Peper, & van Wieringen, 1992; DeGuzman & Kelso, 1991), as well as to learning specific end effector trajectories in multijoint movements. Indeed, there is nothing in principle that prevents extension of the same concepts and methods to learning in general, such as learning to speak, read, play the piano, or to use a prosthetic device (Wallace & Carlson, 1992). The caveat is, of course, that the individual coordination dynamics should be identified before learning. For the most part, the solution lies in carefully distinguishing the spontaneous coordination tendencies (viz., intrinsic dynamics) from specific, momentary, and local informational forces that may be brought into play simultaneously (see Section III).

A. Generalization of Learning

The conclusion of our transfer study was that what is learned is an abstract phase relationship between the rhythmically moving components. A straightforward extension of this conclusion is to posit that learning a phase relationship is independent of the bodily segments actually recruited in the movement. Eventually, such generalized, abstract learning sits at the origin of the individual's ability to achieve the same outcome irrespective of the effectors employed, a phenomenon traditionally embraced by the term "motor equivalence" (see, e.g., Wright, 1990, for a recent discussion). Therefore, through the study of learning a relative timing pattern and its generalization, one has at hand a convenient window to establish whether an essential mechanism underlying adaptability in biological systems pertains to the *abstract* and *dynamic* nature of what is learned.

From the current perspective, the issue of generalization of learning can be operationalized as follows. Suppose that learning a new phasing pattern is transferred between two different coordination systems, the initial coordination dynamics of which are known. Then, learning by generalization should involve alterations of both attractor layouts in the direction of the (same) to-be-learned pattern. A particularly adequate situation is to study transfer of learning between bimanual coordination and coordination between the leg and the arm on the same side of the body, an experimental model system extensively investigated by Kelso and Jeka (1992). Both systems are fundamentally bistable under specific boundary conditions, showing attraction to in-phase and antiphase. Beyond this similarity, however, a basic difference is that for the ipsilateral system, the sign of a relative phase value (e.g., $+90°$ vs. $-90°$) is no longer a matter of convention but, due to biomechanical differences, expresses a definite lead-lag relationship between the leg and the arm. Unlike the bimanual system then, the ipsilateral system is not symmetric (viz., invariant under the transformation $+\phi \rightarrow -\phi$). Thus, the two systems are not governed by identical coordination dynamics even though both possess similar features. Practically, this means that, using the convention of the Kelso and Jeka study, a relative phase with a positive value (e.g., $+90°$) implies that the leg lags with respect to the arm, and *not vice versa* as it would be in a symmetric system. Another difference is that in the bimanual system, in-phase refers to simultaneous activation of homologous muscles, leading the limbs to perform mirror movement with respect to the body symmetry axis. In contrast, in the ipsilateral system, in-phase means that both limbs move spatially in the same direction, that is, simultaneously upward or downward (see Baldissera, Cavallari, & Civaschi, 1982). We remark again that such results attest to the functional, or task-specific nature of the coordination dynamics (see also Saltzman & Kelso, 1987).

A simple procedure can be proposed to evaluate transfer between the bimanual and ipsilateral coordination systems. First, in order to check the initial bistability of both dynamics at in- and anti-phase, the coordination dynamics of bimanual coordination and those of the (right) ipsilateral coordination are probed prior to practice for each subject by scanning the required relative phase between 0 and 180°, as in the experiments presented in Section III. Second, subjects practice 90° of relative phase with the bimanual system,[11] until meeting a given criterion for learning. Finally, a probe of the coordination dynamics is carried out with the ipsilateral system, then with the bimanual system, in order to allow for comparison with the initial attractor layout.

In order to demonstrate generalization of learning conclusively, it is necessary to show that the to-be-learned pattern constitutes an attractive state of *both* coordination systems after practice, whereas such is

---

[11] A more complete design might involve learning 270° for a second group of subjects, in order to control the possible effects of the lead-lag relationshp between the fingers in the learning task.

not the case initially. This means that not only does learning involve changes in the coordination dynamics of the system that practiced the task, but also those of other coordination systems. Beside this critical assumption, two issues may be easily pursued using the experimental setup. The first issue is whether transfer of learning occurs automatically to the "other" ipsilateral system, that is, the arm and leg coordination on the left side (which, up to now, has not been involved in the experiment at all). If so, the symmetry of the coordination dynamics between the two ipsilateral systems is maintained. Such a test of symmetry conservation (see Section III) opens up the classical issue of bilateral, left–right (a)symmetry to investigation from a dynamical perspective. The second issue is whether the symmetry partner of the learned phasing (i.e., $-90°$), which we know to be learned automatically in the bimanual system, is transferred to the ipsilateral system. If so, solid proof would be provided that the degree of abstraction, hence the generalization and transferability of a learned phasing pattern, is high, and that it can "overcome" different and rather stringent constraints.

Another extension of the foregoing work is to study coordination *within* a hand. It is reasonable to assume that the coordination dynamics of coordination between two adjacent fingers of the same hand are comparable to those of bimanual coordination, namely, exhibiting attractive states at 0 and 180°, and, perhaps, in addition, at 90 and 270°. Thus, not only could all the experiments presented in the present chapter be conducted using this within-hand model system, but transfer of learning *between* hands is open to investigation. For instance, a finger pattern that does not belong to the coordination dynamics of either the right or the left hand before learning may be practiced with one hand. After learning, the existence of this very pattern as an attractive state of the coordination dynamics of the other hand would constitute strong evidence for transfer. Such a study of the learning dynamics in within- and across-hand coordination may lead to understanding skill acquisition in manual behaviors such as typing or playing an instrument. We stress again that experimental model systems are chosen as windows or tools that aid the discovery of learning principles and mechanisms. The extent to which the latter transcend the particular material components involved is an open question.

B. Form of Learning

Theory predicts that the form (abrupt vs. smooth) that the learning process exhibits depends on the relationship between the task to be learned and the intrinsic component of the coordination dynamics, namely, the distance in parameter space between the to-be-learned pattern and the closest intrinsic attractor. One hypothesis is that when the to-be-learned pattern is far from an intrinsic attractor, learning will be abrupt, whereas otherwise it will be smooth. This assumption of the role of the distance between competing extrinsic and intrinsic "forces" on the coordination dynamics can be addressed using the bimanual system. The main feature of such an experiment is that in order to assess the abruptness of the phase transitions expected with learning a novel phasing pattern, the modifications of the coordination dynamics must be traced systematically by successive probes throughout the learning process, in a similar fashion to the Zanone and Kelso experiment (1992b). Thus, how the learning process unfolds with time, hence, its form, is captured not only in terms of changes in performance but also in terms of the evolution of the entire attractor layout.

A possible operationalization of this issue is the following. After establishing the bistability of the initial coordination dynamics at $\phi = 0$ and 180° (tested by the same probing method of scanning phasing values between 0 and $+180°$, as above), different groups of subjects learn a phasing pattern of, say, 30, 45, 60, 75, or 90°. Every fifth learning trial,[12] the coordination dynamics are probed systematically. According to the foregoing hypothesis, when the to-be-learned pattern is far from the 0° attractor (e.g., a difference of 60–90°), learning will be abrupt, whereas when it is near, the learning process will be smooth. The abruptness versus smoothness of such a change in the coordination dynamics can be assessed by the number of successive probes between the moment when the first modification is detected in the initial attractor layout and the moment when the learned pattern emerges as a novel stable attractive state.

Two mechanisms are hypothesized to underlie such an effect. First, change due to learning is smooth to the extent that the initial competition between the task requirement and the coordination dynamics is weak. Second, the preliminary results on transfer reported above (see Section III) suggest that learning a pattern fairly close to an intrinsic attractor may take the form of a smooth shift of an existing attractor toward the to-be-learned pattern instead of a more abrupt phase transition. In the proposed experiment, these two hypotheses may be separated, depending on how the coordination dynamics evolve during the learning process, that is, whether a qualitative modification of the coordination dynamics (viz., a bifurcation) is observed or not.

C. Rate of Learning

The working hypothesis is that the extent to which the learning requirement competes with the coordination dynamics determines the

---

[12] Previous findings (see Zanone & Kelso, 1992b) show that dramatic modifications of the coordination dynamics may occur with only ten practice trials.

rate of learning. Theoretically, learning rate should vary inversely with the stability of the closest intrinsic attractor to the required pattern. In other words, a task requirement near a stable attractive state should be learned more slowly than a requirement close to a less stable attractor. Such a claim can be tested easily using the bimanual coordination system, since the 0° pattern is more stable than the 180° pattern. Here again, theoretical considerations imply that the rate of learning not only be assessed as improvements in performance in the learning task but also as specific alterations of the initial coordination dynamics.

A simple experiment can be devised to test this prediction specifically. Subjects exhibiting bistable dynamics before practice learn either 45 or 135° of relative phase. Thus, both task requirements are distant from the 0 or 180° intrinsic attractor by an absolute value of 45°. As previously, systematic probes of the coordination dynamics are carried our periodically throughout practice in order to trace the modifications in the attractor layout with learning.

Two results can be anticipated, following the foregoing theoretical premises. First, the 135° pattern should be learned more rapidly than the 45° pattern, due to the weaker competition from the intrinsically less stable state (i.e., $\phi = 180°$). Second, according to the predictions of Section IV,B, one can expect that such learning will not involve a smooth but rather an abrupt transition, because of the large distance between the to-be-learned pattern and the intrinsic attractors. Moreover, systematic probes of the coordination dynamics during practice may provide arguments to confirm or reject the alternative interpretation that differences in learning rate might result from different types of learning that do or do not involve a phase transition (see Section IV,D for a specific test).

D. Routes to Learning

The question raised in this section concerns the ways in which the initial attractor layout can be modified so that the learned pattern becomes a stable attractive state of the coordination dynamics. Several scenarios may be envisaged:

1. Both the old and newly learned stable patterns constitute distinct attractors of the collective variable dynamics. Thus, the number of stable patterns increases with learning. Such a process is illustrated in Figure 2A, which shows a transition from bistable to tristable coordination dynamics.

2. A previously existing intrinsic attractor destabilizes and is absorbed into the basin of attraction of the learned attractor. Such a reduction in the number of the initial attractors is exemplified in Figure

2B. Note that we have previously shown that the 180° pattern may destabilize, at least temporarily, as the learned pattern is being established (see Zanone & Kelso, 1992b). This depends, again, on the initial attractor layout defining the relative stability of coordinated states.

3. No bifurcation occurs with learning, but instead, a parametric change is observed. The number of attractive states does not change, but such stable patterns may stabilize at different parameter values. In particular, the attractor closest to the required pattern and/or the least stable coordinative state is pulled toward the to-be-learned value. Such shift of a stable fixed point is represented in Figure 3B.

4. None of the above, which means, by definition, no learning has occurred at all. The coordination pattern is either readily attracted to a nearby existing stable pattern, without any change with practice, or it remains highly variable, because of, in theory, unresolved competition between the intrinsic dynamics and behavioral information.

What factors dictate whether the system adopts one or another of these learning routes? Let us examine the available data once more (see Section III). When learning involved a bifurcation, the coordination dynamics were initially bistable at 0 and 180°, and the learning requirement was set at 90° (cf. Figure 2). In contrast (cf. Figure 3B), the drift of a stable fixed point occurred with learning when the initial dynamics were multistable (i.e., attraction to 0, 90, 180, and 270°), with the to-be-learned pattern set at 135°. Three factors seem to be relevant:

1. The *distance in parameter space* beween the task requirement and the closest intrinsic attractor(s). The idea is that when the distance in parameter space is small, a moving fixed point route is observed, whereas when it is large, a bifurcation occurs. This hypothesis has been partially addressed in Section IV,B.

2. The *stability* of attractive states closest to the task requirement. The assumption is that intrinsically less stable coordination patterns are more susceptible to shifts in parameter space than are highly stable attractive states. This hypothesis has already been invoked in Section IV,C.

3. The number of attractors in parameter space, that is, their *density*. In the case of bistable initial dynamics, there may be "room enough" to create a novel attractor in the 0–180° interval. In contrast, when three attractive states are already present prior to learning, some kind of maximal density may be reached, so that no novel attractor can be created without moving or removing an already existing stable state. To some degree, the resolution of this issue may depend on the ability of the learners to produce various required phase differences, that is, how sensitive they are to task requirements.

Since the first two hypotheses (1. and 2. above) are also testable by some of the above experiments concerning bistable bimanual coordination systems, more insight into the issue may be gained by testing specifically the "maximal density" hypothesis (point 3. above) involving multistable dynamics. One way to do this is to select subjects, following an initial probe between 0 and 180°, who exhibit attraction to 0, 90, and 180°. The task, then, is to learn a 45° phase relation, which is an unstable fixed point of the dynamics between the attractors at 0 and 90°. A final probe of the dynamics carried out after practice should reveal the expected modifications of the attractor layout with learning.

Several patterns of results may be observed. If, across subjects, both the 0 and 90° attractors move to the required pattern with learning, the distance hypothesis (point 1. above) is verified. If only the 90° stable pattern does so, an explanation in terms of relative stability (point 2. above) is more probable. As suggested by previous findings (Zanone & Kelso, 1992b), it is quite unlikely that the 0° attractor is ever completely destabilized, because of its intrinsic stability. Finally, if a phase transition is observed, leading to the emergence of the 45° pattern as a novel attractor, the density hypothesis (point 3. above) may be ruled out. Of course, such conclusions would gain in strength, were they corroborated by the results of analogous experiments with bistable dynamics (cf. section IV,B and IV,C).

E. Transition Paths after Learning

To complete this preliminary experimental agenda (with a bang as it were), the fundamental theoretical prediction that the learned pattern becomes a constitutive part of the coordination dynamics should receive a rigorous test. This implies that with practice, the learned pattern becomes an attractive state of the dynamics *in the absence* of specific task requirements. In other words, were we to examine spontaneous coordination tendencies after learning, the newly learned pattern would belong to such intrinsic dynamics.

A robust method to test this idea is the "phase transition paradigm" used by Kelso (1984) (see Section II), in which by systematically varying a control parameter (i.e., frequency), the system is induced to switch to a more stable pattern, following loss of stability of the initially established pattern. In the case of multistable coordination dynamics, as, for instance, after learning, a consistent sequence of switching among existing attractive states should be observed when frequency is gradually increased, under the caveat that the frequency at which each pattern loses stability depends on its intrinsic stability. Such transition paths also inform how stable the learned pattern is relative to those attractive states that existed before learning.

The basic design might involve subjects with bistable coordination dynamics at 0 and 180° who must learn a phasing pattern of 90°. Once the pattern is learned, a probe of the coordination dynamics is carried out by scanning between 0 and 180°. Then, the procedure of scaling-up frequency can be conducted for two different initial conditions, namely, the 90° (learned) or the 180° stable state. According to previous findings, the 0° state is likely to remain the most stable pattern after learning. Thus, we can expect that the 0° pattern will be the final pattern observed when frequency is increased. Nonetheless, en route to such a monostable regime, the sequence of switching should depend on the relative stability of the other attractive states and on the initial conditions (viz., the initially established pattern). If the pattern is 90°, the sequence should be 90° → 0° if the 90° learned pattern is more stable than the 180° initial attractor, and 90° → 180° → 0° otherwise. If the initial mode is 180°, the sequence should be 180° → 90° → 0° if the 90° learned pattern is more stable than the 180° initial attractor, and 180° → 0° otherwise.

Incidentally, the results of the foregoing experiment provide an objective comparison between both methods of examining the coordination dynamics, namely, the "phase transition" and the "scanning" procedures. The same number of attractors should be revealed through both procedures, and their relative stability, measured, for instance, by the bifurcation frequency or by the extent of the negative-slope effect, should be similar. Furthermore, other methods exist to assess specifically the relative stability of the attractive states, such as relaxation and switching time measures (see Section II). These methods can be adopted to refine our quantitative definition of the differences between learned versus initially existing attractive states.

V. Conclusions

The present chapter proposed a research agenda that could lead to the experimental verification of explicit theoretical predictions regarding the dynamics of the learning process itself. The approach is unique because it provides operational methods to probe the underlying coordination dynamics throughout the course of learning. In particular, the approach comes to grips with the issue of individual differences in learning, which can be reassessed in terms of cooperative and competitive processes between the coordination dynamics reflecting each individual's spontaneous coordination tendencies and behavioral information capturing specific task constraints. The experimental thrust is aimed at identifying some of the generic principles that govern the dynamics of learning. Somehow paradoxically, such principles emerge by studying the individual as the unit of analysis. Once these laws of learning are known, they will in turn be open to experimental confirmation or falsification, because precise predictions can be made about how the learning process unfolds depending on task requirements and the individual initial coordination dynamics.

It may be useful to emphasize again (see Section III) that if the laws and mechanisms presented here are generic, the coordination dynamics are specific to the system under scrutiny. Indeed, such dynamics encompass the system's spontaneous coordination tendencies as well as continuously acting "informational forces," which may specifically influence the system's current state. As shown above in a dramatic fashion, the mere fact that subjects wiggle homologous fingers in a frequency-locked fashion does *not* dictate that the system's coordination dynamics be bistable at in- and anti-phase. That another stable coordination pattern exists (e.g., 90°) does not pertain to the individual, nor the milieu, nor the task, but to their interaction. That is, given coordination dynamics arise when the system is set in the task. In this sense, "bimanual coordination" is a misnomer, albeit the underlying coordination dynamics may show common characteristics across individuals and/or tasks. Moreover, not only may the same "real" implementation (e.g., two homologous fingers) exhibit a wide variety of dynamical regimes, but altogether different substrates may be governed by comparable dynamics. Such dynamical flexibility may underlie the ability of biological systems for adaptation, and also provides a challenge to experimenters, who may feel they are facing an elusive object of study. This is the reason why the precise identification of the initial coordination dynamics is a mandatory step in the way to understanding the learning process.

As a first extension of this dynamic approach to learning, it is likely that similar mechanisms and principles are at the origin of "antilearning," that is, *forgetting*. Classically, one distinguishes forgetting by decay, in which the "memory trace" wanes with time, or by interference, in which a learned behavior vanishes due to the growing influence of some other acquired behavior. In dynamical terms, such bifurcations related to forgetting may bear upon competitive processes between various behavioral attractors (as discussed above). Certain findings of ours (see Zanone & Kelso, 1992b), showing that the relative stability of the attractive states may be subject to noticeable changes during the retention interval, are in keeping with such an assumption. Some insight into the mechanisms of forgetting may be given by the observation of the interplay between the attractors belonging to the initial coordination dynamics, those only recently acquired, and those just about to be learned. Thus, determinants of the learning process such as those investigated or proposed here (the distance, the relative stability between attractors, their number, etc.) are likely to govern the process of forgetting as well. These factors are then open to systematic study following the general procedures we have expounded here. The reason is that the coordination dynamics and other behavioral constraints can be expressed in the same (collective variable) space, that is, a common metric is available to nail these constraints down experimentally.

Another line of research, which pertains to the topic of memory, is the study of the joint mechanisms of *recall and recognition*. In motor behavior studies, this issue has enjoyed some interest centered around the notion of schemata (e.g., Schmidt, 1988) and has been revitalized in the totally different language of artificial neural networks, which contain content-addressable memory (e.g., Hinton & Anderson, 1989). In the present framework, a first theoretical sketch (see Schöner, 1989) has been drawn that conceives of recall as a phase transition that activates memorized coordination patterns in response to environmental requirements. Thus, a great deal of experimental work is awaiting in order to elucidate the putative mechanisms underlying recall and recognition, such as relaxation to memorized attractors, change in the strength (viz., relative stability) of memorized versus environmental behavioral information, or competition among existing attractive states, and so forth. Here again, factors such as the distance between the memorized pattern and the required pattern, and the number of activated memorized patterns are likely to play a crucial role in the dynamics of recall and recognition.

Finally, it should be kept in mind that the dynamical mechanisms and principles exposed in the present chapter in the case of learning are meant to be abstract, hence, independent of the characteristic level of description, time scale, and substrate of the phenomenon under investigation. Thus, they are likely to constitute a general framework for understanding behavioral change in biological systems at different levels of description and on various speed and time scales. For instance, we have argued (Zanone & Kelso, 1991) that in biological systems, learning at a more microscopic level bears on stabilization and elimination of neuronal connections, the dynamics of which are akin to those of the cooperative and competitive processes we have described in our (macroscopic) model system. Along developmental time, also (e.g., Zanone, Kelso, and Jeka, 1992d), selection of those coordination patterns that are best tuned or adapted to the baby's specific environment can be interpreted as the outcome of the interplay between existing coordination dynamics and environmental constraints (see Zanone & Kelso, 1991, 1992a). Therefore, the present approach, striving for a lawful description of change at the behavioral level and on the time scale of (skill) learning, promises to provide the building blocks of a general theory of adaptive change in biological systems.

Acknowledgments

This research was funded partly by NSF Grant DBS-9213995, NIMH Grant MH 42900, BRSG Grant NSS 1-SO7-RR07258-01, and U.S. ONR Contract N00014-88-J-119.

References

Abbott, L. F. (1990). Learning in neural network memories. *Network*, **1**, 105–122.
Adams, J. A. (1987). Historical review and appraisal of research on the learning, retention, and transfer of human motor skills. *Psychological Bulletin*, **101**(1), 41–74.
Baldissera, F., Cavallari, P., & Civaschi, P. (1982). Preferential coupling between voluntary movements of ipsilateral limbs. *Neuroscience Letters*, **34**, 95–100.
Beek, P. J., Peper, C. E., & van Wieringen, P. C. W. (1992). Frequency locking, frequency modulation, and bifurcation in dynamic movement systems. In G. E. Stelmach & J. Requin (Eds.), *Tutorials in Motor Behavior*, II (pp 599–622). Amsterdam: North-Holland.
DeGuzman, G. C., & Kelso, J. A. S. (1991). Multifrequency behavioral patterns and the phase attractive circle map. *Biological Cybernetics*, **64**, 485–495.
Eimas, P. D., & Galaburda, A. M. (1990). *Neurobiology of congnition*. Cambridge, MA: MIT Press.
Haken, H. (1983). *Synergetics, an introduction: Non-equilibrium phase transitions and self-organization in physics, chemistry and biology*. Berlin: Springer.
Haken, H., Kelso, J. A. S., & Bunz, H. (1985). A theoretical model of phase transitions in human hand movements. *Biological Cybernetics*, **51**, 347–356.
Hebb, D. O. (1949). *The organization of behavior*. New York: Wiley.
Hinton, G. E., & Anderson, J. A. (1989). *Parallel models of associative memory*. Hillsdale, NJ: Erlbaum.
Humphrey, R. (1933). *The nature of learning in its relation to the living system*. London: Kegan.
Kelso, J. A. S. (1984). Phase transitions and critical behavior in human bimanual coordination. *American Journal of Physiology: Regulatory, Integrative and Comparative Physiology*, **15**, R1000–R1004.
Kelso, J. A. S., & Jeka, J. J. (1992). Symmetry breaking dynamics of human multi-limb coordination. *Journal of Experimental Psychology: Human Perception and Performance*, **18**(3), 645–668.
Kelso, J. A. S., & Scholz, J. P. (1985). Cooperative phenomena in biological motion. In H. Haken (Ed.), *Complex systems: Operational approaches in neurobiology, physical systems and computers* (pp. 124–149). Berlin: Springer.
Kelso, J. A. S., & Schöner, G. S. (1988). Self-organization of coordinative movement patterns. *Human Movement Science*, **7**, 27–46.
Kelso, J. A. S., Scholz, J. P., & Schöner, G. S. (1986). Non-equilibrium phase transitions in coordinated biological motion: Critical fluctuations. *Physics Letters*, **A118**, 279–284.
Kelso, J. A. S., Ding, M., & Schöner, G. (1992). Dynamic pattern formation: A primer. In J. E. Mittenthal & A. B. Baskin (Eds.), *Principles of organization in organisms, SFI*

Studies in the sciences of complexity, Vol. XIII (pp. 397–439). Reading, MA: Addison Wesley.
McCulloch, W. S., & Pitts, W. H. (1943). A logical calculus of ideas immanent in nervous activity. *Bulletin of Mathematical Biophysics*, **5,** 115–133.
Nicolis, G., & Prigogine, I. (1989). *Exploring complexity: An introduction.* Salt Lake City, UT: Freeman.
Pavlov, I. M. (1927/1960). *Conditioned reflexes.* New York: Dover.
Saltzman, E. L., & Kelso, J. A. S. (1987). Skilled actions: A task dynamic approach. *Psychological Review*, **94,**(1), 84–106.
Schmidt, R. A. (1988). *Motor control and learning.* Champaign, IL: Human Kinetics.
Scholz, J. P., & Kelso, J. A. S. (1989). A quantitative approach to understanding the formation and change of coordinated movements patterns. *Journal of Motor Behavior*, **21**(2), 122–144.
Scholz, J. P., Kelso, J. A. S., & Schöner, G. S. (1987). Non-equilibrium phase transitions in coordinated biological motion: Critical slowing down and switching time. *Physics Letters*, **A123,** 390–394.
Schöner, G. S. (1989). Learning and recall in a dynamic theory of coordination patterns. *Biological Cybernetics*, **62,** 39–54.
Schöner, G. S., & Kelso, J. A. S. (1988a). A synergetic theory of environmentally-specified and learned patterns of movement coordination. I. Relative phase dynamics. *Biological Cybernetics*, **58,** 71–80.
Schöner, G. S., & Kelso, J. A. S. (1988b). A synergetic theory of environmentally-specified and learned patterns of movement coordination. II. Component oscillator dynamics. *Biological Cybernetics*, **58,** 81–89.
Schöner, G. S., & Kelso, J. A. S. (1988c). Dynamic pattern generation in behavioral and neural systems.. *Science*, **239,** 1513–1520.
Schöner, G. S., & Kelso, J. A. S. (1988d). A dynamic theory of behavioral change. *Journal of Theoretical Biology*, **135,** 501–524.
Schöner, G. S., Haken, H., & Kelso, J. A. S. (1986). A stochastic theory of phase transitions in human hand movement. *Biological Cybernetics*, **53,** 442–452.
Schöner, G. S., Zanone, P. G., & Kelso, J. A. S. (1992). Learning as change of coordination dynamics: Theory and experiment. *Journal of Motor Behavior*, **24**(1), 29–48.
Skinner, B. F. (1938). *The behavior of organisms.* New York: Appleton-Century-Crofts.
Thorndike, E. L. (1911). *Animal intelligence.* New York: MacMillan.
Tuller, B., & Kelso, J. A. S. (1989). Environmentally-specified patterns of movement coordination in normal and split-brain subjects. *Experimental Brain Research*, **75,** 306–316.
Wallace, S. A., & Carlson, L. E. (1992). Critical variables in the coordination of prosthetic and normal limbs. In G. E. Stelmach & J. Requin (Eds.), *Tutorials in motor behavior II* (pp. 321–341). Amsterdam: North-Holland.
Wright, C. E. (1990). Generalized motor programs: Reexamining claims of effector independence in writing. In M. Jeannerod (Ed.), *Attention and performance, XIII* (pp. 294–320). Hillsdale, NJ: Erlbaum.
Yamanishi, T., Kawato, M., & Suzuki, R. (1980). Two coupled oscillators as model for the coordinated finger tapping by both hands. *Biological Cybernetics*, 37, 219–255.
Yates, E. F. (1987). *Self-organizing systems. The emergence of order.* New York: Plenum.
Zanone, P. G., & Kelso, J. A. S. (1991). Relative timing from the perspective of dynamic pattern theory. In J. Fagard & P. Wolff (Eds.), *The development of timing control and temporal organization in coordinated action* (pp. 69–92). Amsterdam: North-Holland.
Zanone, P. G., & Kelso, J. A. S. (1992a). Learning ahd transfer as paradigms for behavioral change. In G. E. Stelmach & J. Requin (Eds.), *Tutorial in motor behavior II* (pp. 563–582). Amsterdam: North-Holland.

Zanone, P. G., & Kelso, J. A. S. (1992b). The evolution of behavioral attractors with learning: Nonequilibrium phase transitions. *Journal of Experimental Psychology: Human Perception and Performance,* **18**(2), 403–421.

Zanone, P. G., & Kelso, J. A. S. (1992c). *Elementary dynamics of learning and transfer.* Manuscript submitted for publication.

Zanone, P. G., Kelso, J. A. S., & Jeka, J. J. (1992d). Concepts and methods for a dynamical approach to behavioral coordination and change. In G. J. P. Savelsbergh (Ed.), *The development of coordination in infancy.* (pp. 89–135). Amsterdam: North-Holland.

23

The Formation and Dissolution of "Bad Habits" during the Acquisition of Coordination Skills

Charles B. Walter

Motor Control Laboratory
School of Kinesiology
University of Illinois at Chicago
Chicago, Illinois

Stephan P. Swinnen

Laboratorium Motorische Controle
Departement Kinantropologie
Katholieke Universiteit Leuven
Leuven, Belgium

I. Introduction
II. The Nature of Task-Specific Bias (Bad Habits)
    A. Behavioral Bias in "Traditional" Motor Learning Tasks
    B. Behavioral Bias in Coordination Tasks
III. Sources of the Formation of Bad Motor Habits
    A. Learning and Memory
    B. Alternative Sources
IV. Overcoming Systematic Coordinative Bias: Dissolving Bad Habits
    A. The Experimental Paradigm
    B. Augmented Information Feedback
    C. Adaptive Tuning
    D. Part–Whole Transfer of Learning
V. Task Differences and Individual Differences
    A. Task Differences
    B. Individual Differences
VI. Concluding Remarks
References

I. Introduction

The emergence of a "bad habit" is often the bane of a performer attempting to acquire a novel motor skill. A bad habit is characterized by a consistent bias toward an incorrect action. Even casual observation suggests that this frustrating behavior is a common phenomenon during the acquisition of certain motor skills. Other skills seem to elicit more random variation in performance rather than systematic bias when performed by the novice. That is, bad habits appear more frequently during the acquisition of some skills than others. Individual differences in the tendency to display a consistent bias are evident as well. The habit can be quite difficult to overcome for learners who are strongly predisposed to an incorrect behavior, which may contribute significantly to motor learning difficulties.

These observations raise a number of interesting questions that bear directly on accounts of motor learning and control. Why, during the acquisition of some skills, is the same incorrect pattern repeated time after time? Why is this tendency apparent when learning some skills but not others? Why do certain individuals display this pattern of biased motor behavior more than others? Finally, what training techniques can be used to overcome the inappropriate action in favor of the correct movement pattern?

This chapter provides a discussion of the nature of bad habits, or systematic behavioral biases, and their role in the acquisition of coordination skills. The intent is not to provide definitive answers to the questions posed above, but rather to prompt further efforts to understand the basic phenomena and to apply this understanding to skill training. This issue is especially relevant to the acquisition of tasks that require the concerted movement of different limbs, since these skills appear to be particularly prone to generate unintended behavioral bias. As will become evident, relatively little experimental attention has been devoted to this problem for complex tasks. Moreover, it appears that conceptual development in this area has been limited because of the nature of the tasks examined and the theoretical frameworks previously used. These general limitations are discussed. Finally, implications of adopting coordination skills for examining traditional issues such as useful forms of augmented feedback, the effectiveness of part–whole practice, and the notion of individual differences are explored.

II. The Nature of Task-Specific Bias (Bad Habits)

The selection of a task with which to examine issues of motor behavior is rarely arbitrary, and the implications of the selection are often profound. The nature of the task largely determines the type of theory that best accounts for the acquisition and/or control of the action. If task demands indeed drive theoretical development, it is useful to consider the nature of "traditional" motor learning tasks and the effect that task characteristics might have had both on the biases observed and on the explanations proposed to account for them. Although our primary focus is on coordination tasks, it is useful to first briefly consider some forms of bias that have been noted for more traditional, unimanual tasks. Some of these effects may be specific to single-limb actions, but others may apply to coordination tasks as well.

A. Behavioral Bias in "Traditional" Motor Learning Tasks

Laboratory tasks used to examine motor learning issues have been classified along a number of dimensions (e.g., Magill, 1989; Schmidt, 1988). A few dimensions include the speed of the movement (slow or ramp vs. rapid or ballistic), the nature of the physical goal (e.g., positioning or timing), the assumed underlying ability (rhythm, hand-eye coordination, etc.), and the definitiveness of the initiation and termination of the task (e.g., discrete aiming vs. continuous oscillations). Despite these differences, most of the tasks that have received primary attention in the last 50 years have one common element: virtually all of the actions have required the movement of a single limb. For example, of the 36 tasks examined by Fleishman in two studies of individual differences in perceptual–motor abilities (Fleishman, 1958a,b), only 5 required the concerted movement of more than one limb. Aiming, positioning, timing, and tracking tasks, which have accounted for a good deal of the basic motor learning research in this period, have all typically taken a unimanual variation.

The selection of unimanual tasks to examine principles of motor learning and control has had a substantial effect on the nature of the systematic bias observed. Response bias has most often been observed in "short term motor memory" (STMM) paradigms, where the subject attempts to reproduce a goal movement outcome (typically a specific movement time, amplitude, or force level). The bias is measured by an outcome score termed "constant error" (CE), which indicates a consistent tendency to move too slow, too far, and so forth. A number of factors influence CE under these conditions. The factors that have most consistently influenced response bias are discussed briefly here (see Laabs & Simmons, 1981, for a more complete review).

Several forms of bias observed during unimanual STMM tasks can be termed "serial effects," since the effect only appears in the context of a series of trials. A relatively weak but consistent finding is a general negative CE (i.e., undershooting) following fairly long intertial intervals (e.g., Faust-Adams, 1972). A prior movement with a substantially different parameter (faster, shorter, more forceful, etc.) can also influence the CE of the subsequent movement. The error is often in the direction of the previous action (e.g., Laabs, 1974). Since the observed outcome is essentially a combination of the previous movement and the one intended, the effect has been described as serial "assimilation." The opposite effect can occur for relatively large differences between task parameters, however. This "contrast" effect renders systematically greater differences between the tasks than intended. Most of these serial effects have been attributed to learning and memory (see Section III,A).

Another consistent bias in unimanual movements falls in the general category of a "range effect," that is, the effect is different for tasks performed in different ranges of the scale of an independent variable. Generally, CE is positive at low goal magnitudes or intensities and negative at high goal values (e.g., Faust-Adams, 1972). This is not a serial effect, but is rather dependent on the physical properties of the intended action. As such, it appears to be due more to a methodical error in movement organization than to decay or interference in the memory representation.

The serial and range effects observed for unimanual tasks have received little attention for coordination tasks. This is perhaps because different, even stronger forms of bias are noted for the latter. Additionally, whereas the effects noted above are relatively easy to overcome intentionally, the bad habits associated with coordination skills are often frustratingly difficult to eliminate. The forms of bias that emerge during the acquisition of some coordination tasks are discussed in the next section.

B. Behavioral Bias in Coordination Tasks

Examining coordination tasks rather than unimanual tasks affords the opportunity to observe qualitatively different forms of bias. Specifically, there are systematic performance errors that are evident in the *relationship* between the two movements. The interlimb biases are seen both in movement outcome scores and in limb trajectories. The general phenomena are analogous to the behavior of a "kinematic pair" in mechanics, where two objects mutually constrain relative motion. The mechanical linkage among limbs through the torso has little to do with the coordination tendencies observed for bimanual tasks, however. The general nature of these biases is briefly discussed next.

Most of the coordination tasks that have been systematically examined have required different trajectories and/or timing patterns to be produced bilaterally. The strongest general behavioral tendency for these tasks can be characterized as a bias toward interlimb symmetry. The bimanual interaction is manifested during oscillations of similar frequencies as a strong tendency to produce the movements in-phase. A somewhat weaker attraction toward alternation appears as well (Cohen, 1971; Kelso & Scholz, 1985). This relative phase attraction clearly biases movements that must depart from the preferred modes, rendering such actions difficult to perform initially (Lee, Swinnen, & Verschueren, 1993; Swinnen, DePooter, & Delrue, 1991a; Zanone & Kelso, 1992). Similar phase attraction appears for multifrequency actions, particularly for lower order frequency ratios (deGuzman & Kelso, 1991). The interlimb attraction is sufficiently strong in some subjects that identical frequencies emerge when different bilateral frequencies are attempted (Walter, Pollatou, Corcos, Swinnen, & Pan, 1993).

A similar general effect is noted for discrete bimanual movements. If the intended outcomes of concurrent limb movements differ in some respect, a tendency toward interlimb symmetry typically occurs. This constitutes a systematic bias toward similar bilateral scales of a given variable. The effect is particularly strong along the dimension of time. If movements ordinarily take different lengths of time to complete when produced in isolation, they are drawn toward a common duration when performed together (Kelso, Southard, & Goodman, 1979). This tendency toward "isochrony" is again a manifestation of assimilation, in that the movement requiring the greater magnitude tends to be reduced and the one with the lower magnitude increases. The same trend is somewhat evident for movement distance (Sherwood, 1990). The preferred outcome in both cases is one of scalar or "metrical" symmetry. This assimilation differs from that discussed for unimanual tasks in that it derives from concurrent movements rather than serial ones, suggesting different sources for the effects (see section III,B).

If bimanual movements are fairly similar, but differ in some topological respect, each limb tends to incorporate characteristics of the other's pattern in another form of assimilation (e.g., Franz, Zelaznik, & McCabe, 1991; Swinnen, Walter, & Shapiro, 1988). The preferred, symmetrical coordination pattern can be described as a tendency toward a linear relative motion trajectory (Walter & Swinnen, 1990a). Other preferred patterns will perhaps be discovered as tasks with more complex, asymmetrical demands are examined. Just as unimanual tasks have limited the forms of motor bias previously observed, the relatively simple laboratory coordination tasks examined thus far perhaps limit the forms of preferred relative motion exhibited (Walter, in press).

The biases described above occur on the first attempt at the task. They are thus qualitatively different from the serial biases noted for unimanual tasks. Preferred coordinated patterns are also evident for unimanual tasks that involve more than one joint (e.g., Soechting & Lacquaniti, 1981). The rigid links among limb segments, however, make it much more difficult to separate coordination patterns that are primarily determined by mechanical interactions from those that emerge from other sources. More generally, the focus on single-limb (often uniarticular) actions has limited the investigation of preferred relative motion patterns. As a result, bad habits that are manifested as consistently inappropriate patterns of coordination have received much less attention than the various forms of outcome bias (CE).

III. Sources of the Formation of Bad Motor Habits

As noted above, a habitual behavior is characterized by its relatively high frequency. A habit is generally defined as a "customary practice," a "dominant or regular disposition or tendency," or an "acquired behavior pattern regularly followed until it has become almost involuntary" (Webster's Encyclopedic Unabridged Dictionary, 1989). The last definition is consistent with accounts of "habit" found in most psychology texts. According to traditional learning theory, a "habit" is an acquired behavior with a high probability of being elicited in a given situation, and "habit strength" is indicated by the frequency of behavior actually observed. A bad habit from this view is an inappropriate or incorrect behavior that is due to *learning*. This potential source of bias is discussed first. Other possible sources, most of which are especially relevant to coordination tasks, are presented subsequently.

A. Learning and Memory

Three general sources of the effect of learning and memory on motor bias have been proposed. The first is simply based on forgetting the memory trace for the goal action. The notion is that the trace "shrinks" over time, causing the systematic undershooting following long intertrial intervals for unimanual tasks (Pepper & Herman, 1970). This assumes, of course, that forgetting occurs in a given direction. That is, the representation of the proper action changes such that its scale decreases. An alternative view of forgetting is that the proper representation simply degrades, in which case systematic error could not be attributed to this factor.

A second potential learning source of motor bias falls in the category of "negative transfer of learning." Transfer of motor learning refers to the effect of a previously learned motor skill on the acquisition of a new skill. Negative transfer of learning implies that the previous task detrimentally influences the new task. The bias due to negative transfer of learning has traditionally taken one of two forms. It is evident either as the tendency to consistently select a specific action that is incorrect, or as the tendency to perform the correct general action in an incorrect (but consistent) manner. The first form is more common. Such errors are often observed in situations in which a new motor response must be paired with a stimulus that was previously associated with a different response (e.g., Lewis, McAllister, & Adams, 1951). For example, it is now legal to turn right at a red light in most areas of the United States. This response is fairly automatic for experienced drivers if no traffic is present. Since this practice is prohibited in Europe, however, Americans must consciously suppress this response to avoid a potentially dangerous situation. This inappropriate behavior is particularly prone to emerge under conditions of reduced vigilance. When driving lanes were changed from the left side to the right side of the road in Sweden, old driving patterns (and thus accidents) were particularly evident when the driver was tired or intoxicated (M. Peters, personal communication).

It is important to note that the negative transfer described above influences the selection of an action, not its technique. As Fitts and Posner (1967) note, "it is not the learning of an opposite response per se which leads to negative transfer, but it is the necessity of making an opposite response to the same or similar stimulus cues" (p. 20). The other form of negative transfer, perhaps more relevant to the theme of the present chapter, influences the movement pattern itself. The correct general response is attempted, but some aspect of the movement pattern or scale (speed, amplitude, etc.) is systematically incorrect. A commonly cited example is the negative effect of having learned a racquet sport that requires a specific technique (e.g., the relatively stiff wrist of a tennis forehand) on learning a new racquet sport requiring a slightly different technique (the wrist "snap" of a forehand for racquetball). Even if the correct action is selected (e.g., a forehand stroke), the old movement pattern may consistently interfere with critical characteristics of the new one. When applied to discrete, unimanual tasks, the effect may be noted as the assimilation of disparate movement scales for primary and interpolated activities, which was discussed in Section II,A (e.g., Laabs, 1973).

The possibility of the latter form of negative transfer of learning has been discussed for some time. Few experimental examples of it have been provided beyond STMM paradigms, however. Moreover, negative transfer is often fairly easy to overcome when it does appear (Schmidt, 1988). Perhaps the most stubborn cases involve the inability to ignore a previously learned timing pattern (Summers, 1975; Shapiro, 1977). It has thus been hypothesized that negative transfer of movement patterns is most likely to occur when two tasks have a similar sequence of movements but a different timing structure (Schmidt & Young, 1987). Finally, the very definition of transfer of learning implies that it only applies to serial conditions. Considering the obstinance of some incorrect movement patterns that can appear spontaneously, it seems that negative transfer cannot be considered as the only source of bad habits.

A third learning source of systematic bias is the notion that the acquisition of the task at hand has simply progressed incorrectly. The improper movement representation, in other words, has been encoded because the same incorrect movement pattern has been repeated during practice. This could occur, for example, in a situation where an instructor is not present and the task does not inherently provide adequate information feedback for the learner to correct errors, or if the learner is consistently too fatigued to perform the skill correctly. Improper learning seems to be the consensus view among various skill instructors for the emergence of bad habits, yet is has received little empirical attention. This account ignores, however, the question of why a *particular,* incorrect movement pattern may be displayed across a number of different learners.

To summarize, it appears that previous experience can have a relatively strong influence on the tendency to select inappropriate actions in some situations. But it is much more difficult for negative transfer and incorrect initial learning to account for systematic variations in movement patterns themselves, particularly if the same bias occurs across different learners. Learning factors may well be causally related to the serial biases observed for unimanual tasks, but it is unclear how these sources could derive the specific bimanual biases toward bilateral structural and metrical symmetry that were reviewed in section II,B. Moreover, the tendency to produce symmetric movements typically appears on the first trial, precluding a serial effect. Additional factors that may contribute to the formation of bad habits must therefore be considered, particularly for coordination skills.

B. Alternative Sources

A number of alternative sources of preferred movement patterns have been proposed, many of which apply directly to coordination tasks. A brief survey of some of the potential mechanisms follows. Since these factors influence behavior in a systematic, predictable manner, they can be viewed as sources of constraint for coordination skills. It is useful to recall that, to the extent that the preferred pattern differs from the goal pattern, these factors may contribute to the formation of bad habits.

1. Informational Sources
a. Attentional Limitations One possible source of the tendency to perform similar movements with different limbs is a limitation in processing the information used to control the action. Coordination skills are essentially viewed as dual (or multiple) tasks from this per-

spective, with each limb requiring distinct control of its movement. Limitations in processing feedback from different limbs (Cohen, 1971) and in organizing disparate "motor programs" concurrently (Schmidt, 1988) have both been proposed as factors affecting coordination skills. It is assumed in both cases that the processing demands imposed by the task exceed the capability to handle the information. Only one of the two channels is processed, and the same sensory preception and/or motor organization is ascribed to both limbs as a result.

Recent information processing views of dual task interference have also focused on the related concepts of "outcome conflict" and "microsimilarity." Outcome conflict occurs when two tasks require processing that is critically different, leading to confusion or delay (Navon & Miller, 1987). According to Wickens (1989), the degree of confusion or crosstalk is based on microsimilarity between the tasks, that is, "similarity that can be expressed in a quantitative scale such as the closeness in space, similarity of color, or similarity of meaning" (p. 72). Such similarity appears to facilitate motor tasks that are bilaterally symmetric, while tasks that are asymmetric in some manner may suffer (e.g., Peters, 1977).

b. Information Flow in the Perception–Action Cycle A somewhat different interpretation of interlimb constraint is based on a coupling of movement and sensation in a continuous perception–action cycle (e.g., Turvey, 1990; Turvey & Carello, 1986). The tendency for interlimb entrainment is attributed to the dynamics of information flow fields generated during movement from this perspective rather than to limitations in the ability to process the information. Information content is the critical factor, as similar coupling phenomena may be observed for different modes of transmission. For example, the bimanual coupling noted within a subject exhibits a number of characteristics common with the coupling observed between subjects, despite the former being mediated neurally and the latter visually (Schmidt, Carello, & Turvey, 1990).

2. Neural Sources

a. Neural Connectivity The structure of neural connectivity could also influence systematic bias toward similar limb movements. Relevant neuroanatomical features include bilateral connections between sensory and motor centers at the supraspinal level and bilateral distribution of some descending motor pathways (Brinkman & Kuypers, 1973; Preilowski, 1987), particularly for more proximal limb musculature. These connections provide a hard-wired explanation of interlimb assimilation and support an important role for inhibition in the acquisition of coordination skills (Marteniuk, MacKenzie, & Baba, 1984; Swinnen & Walter, 1988). It should be noted that the interference arising from these structures could occur during a number of information-processing functions (sensory processing, movement organization, response execution, etc.). Neural connectivity can thus be viewed either as an alternative to processing accounts of crosstalk, or as the physical mechanism through which cognitive mediation and/or direct perception effects occur.

b. Reflexes The influence of neural connections on preferred patterns of coordination is also manifested through the effect of innate reflexes (e.g., Easton, 1972; Fukuda, 1961). Dorsiflexion of the head, for example, elicits the symmetric neck reflex. The reflex tends to facilitate bilateral extension of the arms. This natural synergy is useful for a gymnast performing a handstand, where the supporting function of the arms is potentiated as the head is tipped back (Swinnen, 1990). It may render other skills more difficult, however. The same gymnast performing a back flip needs to flex the arms to hold the knees in a "tuck" position, while dorsiflexing the head to spot the ground for landing. These movements are incompatible with the coordination pattern elicited by the reflex, requiring the inhibition of its effect.

c. Central Pattern Generators Central pattern generators (CPGs), particularly in the form of coupled neural oscillators, provide another potential source of coordinative constraint (Delcomyn, 1975; Grillner, 1985). A CPG is a group of neurons connected such that they provide a patterned, typically rhythmical output (oscillation) when stimulated. Coupled CPGs can exert a mutual influence on each other for each of the three factors that characterize an oscillation: amplitude, frequency, and phase. Plasticity in the coupling of "unit CPGs" has been proposed as one potential source of learning complex, coordinated skills (Grillner, 1985).

d. Lesions Neural damage can also cause inappropriate behavior to consistently appear in a given situation. Frontal lesions in monkeys, for example, can result in the repetition of a previously learned behavior in an inappropriate setting, termed "perseveration" (e.g., Mishkin, Prockop, & Rosvold, 1962). The tendency toward such sterotypic behavior suggests that damage to certain centers may inhibit the ability to reorganize behavior in new situations. This apparently facilitates negative transfer from previous experience, perhaps contributing to the formation of a bad habit in the novel setting.

3. Optimization

Another factor that should be considered as a possible determinant of preferred coordinated patterns is the fact that movement trajectories are often consistent with the optimization of some variable (e.g., Hasan, 1986; Hogan, 1984). This general principle has primarily been examined for relatively simple tasks, but it might be especially critical for more complex skills—particularly for gross, potentially fatigue-inducing skills (e.g., Sparrow & Irizarry-Lopez, 1987). If the desired movement pattern does not coincide with the most efficient way to execute the skill, an incorrect action pattern may emerge.

4. Intrinsic Dynamics

Others have proposed that relatively global, dynamical descriptions are useful for examining coordinative constraints (e.g., Kelso & Schöner, 1988). "Intrinsic dynamics" are inferred from the observation of "collective variables," whose behavior is the result of a number of different sources. The combined effect of the various sources is a vector field that draws coordination toward preferred modes (e.g., Schöner, Zanone, & Kelso, 1992). Complex, preferred behaviors can emerge from simple changes in some control parameters. Frequency scaling for a system of nonlinearly coupled nonlinear oscillators (limit-cycle oscillators), for example, can account for transitions in the relative phasing of limb movements (e.g., Haken, Kelso, & Bunz, 1985). This type of mechanism is potentially critical for locomotion in some species, as the behavior that it produces is qualitatively similar to gait transitions in quadrupeds. The notion that nonlinearities can facilitate the emergence of other forms of behavior has been proposed as a general mechanism that may aid the acquisition of novel skills (Kelso & Schöner, 1988).

5. Physical Fitness

A final consideration for the emergence of a particular action pattern is the physical fitness of the individual. For present purposes, the strength and flexibility of the performer might be critical, especially for the acquisition of gross skills. If either of these factors is insufficient to produce the desired movement pattern, they may work in concert with the above influences to constrain the action to a specific, incorrect topology. This problem is observed quite frequently during the acquisition of many gymnastics skills.

In summary, a variety of factors potentially influence the tendency to generate preferred patterns of coordinated movement. If the preferred patterns differ from the desired movement and persist over time, they are identified as bad habits. A number of the factors described above are absent for many of the more traditional, single-limb (often uniarticular) laboratory tasks. Perhaps this is why the implications of systematic bias in movement production have been virtually ignored in the development of learning theories and training techniques. The latter issue is discussed next.

IV. Overcoming Systematic Coordinative Bias: Dissolving Bad Habits

We have thus far attempted to demonstrate that many of the laboratory tasks previously used for studies of motor learning elicit limited forms of bias. Attempts to learn many "real world" skills, however, often result in a much greater tendency to form bad habits. Laboratory tasks that require coordination among different limbs exhibit more bias than single-limb tasks, suggesting that some of its sources are unique to multilimb (or multisegment) behavior. It is clear that a rigorous examination of the formation and dissolution of bad habits requires a paradigm that reliably elicits systematic bias when the task is initially attempted.

We embarked about 8 years ago on a series of studies examining a laboratory coordination task that is particularly useful for observing systematically biased motor behavior. Although individual differences are evident in the degree of initial bias for the task, its qualitative nature is quite consistent. The paradigm thus provides a useful tool for observing the phenomenon of the bad habit as we have operationally defined it here. The task has allowed us to examine a number of factors that affect the strength of the habit or bias. We have also examined several training manipulations to aid subjects to overcome the incorrect behavior. In this section we review several studies that are representative of this line of work in an attempt to elaborate alternative approaches to some traditional motor learning issues.

A. The Experimental Paradigm

All of the studies in this section refer to the same basic task, so it will be described here in some detail. The general goal of the subject is to produce different arm movements simultaneously. Specifically, the task requires that one arm generate a unidirectional movement concurrent with a contralateral, sequential movement (Figure 1A). The subjects sit midway and slightly behind two horizontal levers that are affixed to a table. They rest their forearms on the levers during the experiment to minimize fatigue and to control shoulder position. An analog potentiometer or a digital shaft encoder is located at the base of the axle supporting each lever to detect its angular position. The unidirectional movement is typically a horizontal elbow flexion to a target located between 60 and 90° from the starting position. The sequential movement is composed of a flexion–extension–flexion movement to the same target location. Movement duration (and thus speed) is varied from quite slow velocities to rapid velocities, and is controlled through feedback concerning movement time. Subjects are instructed to initiate

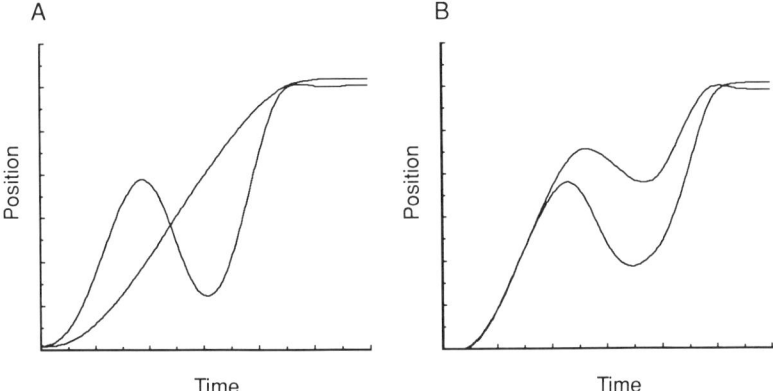

FIGURE 1 Position–time traces for (A) the intended unidirectional and reversal movements and (B) a trial exhibiting strong interlimb interference in the form of spatiotemporal assimilation.

and terminate the two movements together, and to perform the two movement patterns as smoothly as possible. Interference between the limbs is often evident initially (Figure 1B). The typical behavior is for the limb attempting to perform the sequential movement to produce a reversal of slightly lower magnitude than that intended, while the unidirectional limb unintentionally pauses or reverses direction concommitantly (e.g., Swinnen et al., 1988).

Figure 2 contains angle–angle plots for the desired pattern and for an initial trial exhibiting interlimb interference. The preferred but unsuccessful relative motion pattern for subjects approaches linearity (Figure 2B), indicating that both limbs tend to generate similar spatiotemporal trajectories. It should be noted that this does not suggest that each individual trajectory is linear. The tendency is toward both metrical symmetry and topological or "structural" symmetry, with the latter effect being somewhat greater. The more complex, reversal pattern yields less than the simpler, unidirectional pattern (Swinnen & Walter, 1991), and the degree of interlimb attraction is less when the reversal is performed with the preferred limb than with the nonpreferred limb in right-handers (Walter & Swinnen, 1990b).

These tendencies all represent the systematic biases (i.e., the bad habits) that must be overcome to successfully perform the task. The fact that disparate patterns such as these generally derive greater variability than bilaterally symmetric actions (Walter, Swinnen, & Franz, 1993) further renders the task quite difficult to perform. Simply repeating the task does not substantially decrease interlimb symmetry, (e.g., Swinnen et al., 1988), and in some cases actually serves to increase

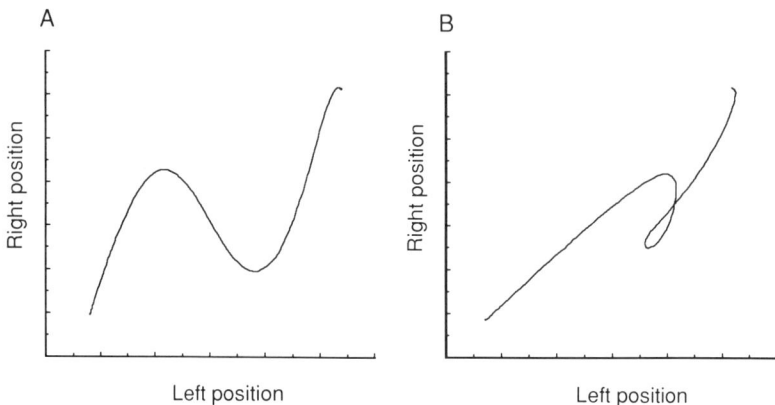

FIGURE 2 Position–position trajectories for (A) a successful trial and (B) an unsuccessful, coupled trial. The trials are the same as those plotted in Figure 1.

the degree of coupling (Swinnen, Walter, Pauwels, Meugens, & Beirinckx, 1990). The task is thus quite useful for examining various training manipulations.

B. Augmented Information Feedback

Perhaps the most often studied variable in the acquisition of motor skills is augmented information feedback (see Newell, Morris, & Scully, 1985, and Salmoni, Schmidt, & Walter, 1984, for reviews). Information concerning the discrete outcome of an action is termed knowledge of results (KR) and information about the movement pattern itself is termed knowledge of performance (KP), or kinematic feedback. KR is particularly useful for learning motor tasks of the type often examined in the laboratory. As Newell (1985) has pointed out, these tasks usually involve scaling a given movement parameter (amplitude, speed, duration, etc.) to coincide with a specific goal, rather than acquiring a novel movement pattern. It has been suggested that KP or kinematic feedback may be required for more complex tasks with the latter requirements (Fowler & Turvey, 1978; Newell & Walter, 1981).

Previous work has demonstrated that kinematic feedback provided as position–time traces facilitates performance for the bimanual task described above (Swinnen & Walter, 1991). We recently compared kinematic feedback with KR for acquiring this task (Swinnen, Walter, Lee, & Serrien, in press). KR took the form of within-trial correlations between the acceleration-time traces for each limb. This provides a sensitive measure of the degree of bilateral coupling, or kinematic symmetry, as a single value. Subjects either received no feedback throughout training, KR, KP, or a combination of KR and KP (Experiment 2). Retention tests without feedback were given 10 minutes and five months after the completion of 100 practice trials, which were performed over two sessions. Measures of interlimb coupling included the interlimb acceleration correlation coefficients and a coefficient of work dissociation. The latter coefficient is based on the ratio between the net work generated by each limb, and provides a measure of metrical dissociation (see Swinnen, Walter, Beirinckx, & Meugens, 1991b, for details).

All three feedback groups significantly improved with practice, that is, the interlimb correlations decreased (Figure 3) and the work dissociation improved. No differences among the feedback groups were noted in either retention test for the correlations. The combination of KR and KP appeared to yield the greatest dissociation in net work. The findings suggest that KR may indeed facilitate the acquisition of multilimb movements if it contains information that is relevant to achieving the goal of the task. It is probably particularly useful when the subjects have a "good idea" of the task, which diminishes the need for information concerning the criterion pattern of a movement (Newell, Carlton, & Antoniou, 1990). Additional information may be required for movement patterns that are more difficult to comprehend, however, as bimanual tasks of this type are quite difficult to perform (Franz, Zelaznik, Walter, & Swinnen, 1993).

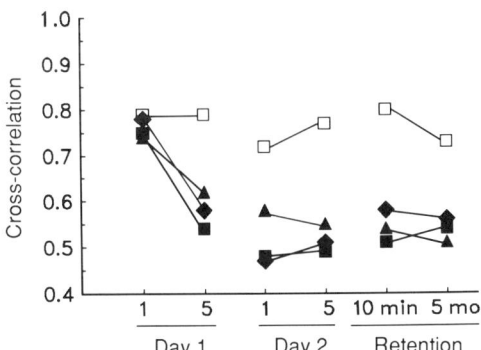

FIGURE 3 Cross-correlations between right and left acceleration-time traces for trial blocks 1 and 5 for training Day 1, for blocks 1 and 5 for training Day 2, and for retention tests given 10 minutes and 5 months later. □, No feedback group; ■, group with kinematic feedback; ◆, group with goal feedback; ▲, group with a combination of kinematic and goal feedback. (Adapted from Swinnen *et al.*, in press.)

C. Adaptive Tuning

A somewhat different strategy for overcoming a bad habit is to attempt to identify the source of the habit in order to diminish its effect. Speed (Swinnen, Walter, Serrien, & Vandendriessche, 1992) and torque (Walter & Swinnen, 1990a,b) are directly related to the degree of attraction toward spatiotemporal symmetry for our bimanual task. This tendency can be represented by a three-dimensional gradient potential with a linear minimium in the angle–angle plane (Figure 4A). The nature of the valley minimum suggests that bimanual actions that fall within this basin are drawn toward similar movement patterns. Decreasing the speed or torque required by the task diminishes the strength of attraction toward the preferred pattern (Figure 4B). This may aid subjects in overcoming the emergent "habit" of assimilating the two movement patterns.

A recent investigation of this "adaptive tuning" of interlimb attraction suggests that the technique may be useful for decoupling limb movements (Walter & Swinnen, 1992). One group of subjects practiced at a relatively rapid criterion speed for 80 trials (the "constant" group). The other group (the "adaptive" group) practiced at a reduced speed for 20 trials. This was followed by 20 trials at each of three progressively greater speeds, the last of which was the criterion rate. All subjects received feedback concerning their position-time patterns following every fifth trial during acquisition. A no-feedback retention test was administered at the end of the training session. The subjects with the adaptive tuning clearly demonstrated greater learning than those in the constant group. Their interlimb correlations dropped significantly more from a pretest to a no-feedback retention test than for the group with constant practice. Moreover, the adaptive group was nearly 100% successful in producing trials without unintentional reversals in the unidirectional limb by the end of practice (Figure 5). This form of

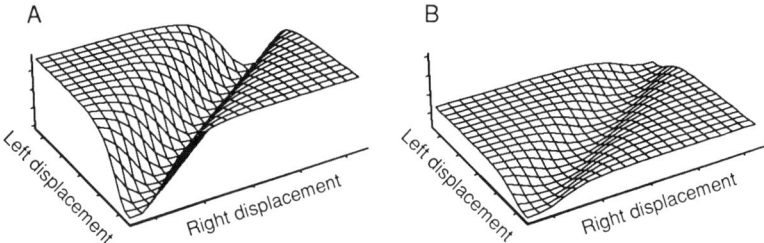

FIGURE 4 Schematic gradient representing the concept of adaptive tuning. Strong attraction toward bilateral symmetry is demonstrated in (A). The attraction has been "tuned down" in (B) to facilitate departure from the preferred coordination pattern.

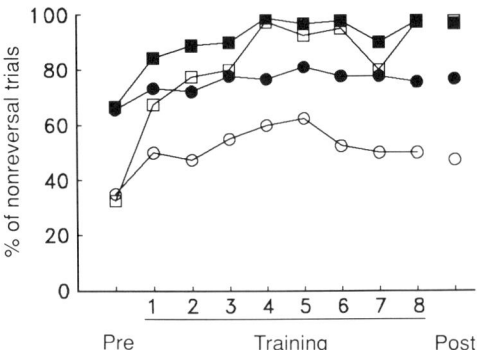

FIGURE 5 Percentage of trials where the unidirectional limb successfully maintained movement continuity for groups with constant practice at the criterion speed (●, ○) and with adaptive tuning of speed (■, □). Subjects have been stratified into those demonstrating initial success rates of less than 50% (○, □) and greater than 50% (●, ■).

training appeared to be especially effective for subjects initially displaying the greatest tendency toward similar bilateral spatiotemporal patterns (Figure 5). The usefulness of the adaptive tuning manipulation is consistent with the effectiveness of speed progression strategies that are often adopted for the acquisition of musical skills.

Other forms of adaptive training (i.e., systematic progression of task difficulty) have had mixed results for tracking tasks (Lintern & Gopher, 1978). It should be noted, however, that the performance errors for those tasks do not ordinarily exhibit the same kind of strong, systematic bias as the present task. Subjects are not overcoming bad habits in tracking tasks so much as they are reducing variability. Initially reducing, then increasing the speed of movement over practice trials has not proven to be a consistent method for simply improving movement accuracy (Fulton, 1942; Solley, 1952,; Sturt, 1921). Adaptive training in the form of "tuning" the bias may thus be more useful for tasks that systematically elicit bad habits than for those that exhibit more random variability.

D. Part–Whole Transfer of Learning

A number of issues fall under the general heading of transfer of learning. The potential effect of negative transfer on systematic movement bias was discussed above. Another transfer issue that is particularly relevant for coordination tasks is that of "part–whole" training. This refers to the techinque of decomposing a skill into logical components or "parts" that are initially practiced individually. Once a reasonable degree of success has been reached in performing the components, they are reassembled into the "whole" skill. Theoretically, this technique may facilitate the acquisition of some multicomponent skills by simplifying their performance.

Although we have not performed a systematic study of part–whole transfer using this task, a general indication of its efficacy can be gained by comparing studies with different protocols. Most of our experiments have required subjects either to initially practice with the same movement in both arms to gain a feel for the correct duration, or to perform the goal, disparate bimanual task from the outset. Two studies, however, have required subjects to perform each arm movement individually for a series of trials before practicing the action bimanually (Swinnen et al., 1990; Swinnen, Young, Walter, & Serrien, 1991c). It is admittedly unwise to place too much faith in a direct comparison of data among studies. But it is interesting to note that the average interlimb correlations at the beginning of bimanual practice for the latter studies (.5 to .6) are slightly lower than those for studies that have not provided initial practice with each arm individually, which have ranged from .5 to .9 (most falling between .6 and .8; Swinnen & Walter, 1991; Swinnen et al., in press; Walter & Swinnen, 1990a, 1992). Practicing each limb individually may thus improve performance, to a modest degree, as compared to no practice at all. Subjects generally fail to approach the levels attained by subjects with a similar amount of bimanual practice, however, if augmented feedback is provided.

V. Task Differences and Individual Differences

Two questions that were posed in the Introduction remain to be addressed. These concern differences among tasks in their tendencies to evoke bad habits, and differences among individual performers in the development of the biases. Although these are important issues, little experimental attention has been focused on them in relation to other issues.

A. Task Differences

One principled basis for predicting whether a given coordination task will tend to produce a bad habit is based on the notion of intrinsic dynamics. As noted above, intrinsic dynamics emerge from a combination of sources, and determine a basin of attraction with the preferred coordination pattern or mode at the minimum (e.g., Figure 4). Once the topology of the basin is established, the direction and extent of systematic bias can be predicted for a novel task by mapping the desired relative motion pattern onto the basin. The examination of more complex, asymmetrical actions may yield new forms of basins, that is, the naturally preferred pattern may depend on the intended action. The position of a novel task in relation to the toplogy of its related basin may prove a useful means for helping to determine task difficulty as well (Walter & Swinnen, 1992), another traditional concept that has proven quite difficult to operationalize.

B. Individual Differences

A number of factors that may be particularly relevant sources of bias for coordination skills were discussed in section III,B. It is obvious that many of these factors may differ among individuals. It should come as no surprise, therefore, that rather large individual differences are evident in the degree to which performers display the systematic bias described above when attempting to perform our bimanual task (e.g., Figure 5). Additional differences were described in our earliest paper (Swinnen et al., 1988), and concern the nature of the preferred relative motion pattern exhibited. Individual differences have been noted in virtually all of our investigations, as well as in related studies (e.g., Kelso, Putnam, & Goodman, 1983).

The point that we wish to make concerning this issue is that some performers may have particular difficulty in dissolving bad habits that emerge early in learning. These individuals may especially benefit from training strategies that are specifically designed to help them depart from their preferred movement pattern (specific forms of augmented feedback, adaptive tuning, etc.). This approach to the effect of individual differences on appropriate training techniques has not been rigorously examined to our knowledge. It not only holds potential for the acquisition of novel skills in healthy individuals, but for relearning skills in patients with neuromuscular pathologies. The notion is that a particular pathology may exert a systematic effect on critical dynamics of the CNS, which collectively determine the emergence of inappropriate movement patterns.

VI. Concluding Remarks

We have suggested here that the types of tasks examined in many motor learning studies have, until relatively recently, limited the forms of bias observed during acquisition. Traditional tasks have most often required the use of a single limb. This has precluded the study of preferred movement patterns, which result from an interaction between limbs, in favor of outcome effects. Task selection has largely influenced theoretical interpretations of bias as well. Single-limb effects are primarily serial in nature, suggesting a systematic alteration in memory structure as their primary source. Conversely, preferred relative motion patterns that appear when learning coordination tasks are often evident immediately, and are likely influenced by the ensemble effect of a number of disparate mechanisms. Finally, single-limb effects can be overcome fairly easily with the aid of augmented feedback by intentionally scaling up or down a given movement parameter. The "bad habits" that appear when attempting a novel coordination task are typically quite persistant.

The notion that natural coordination tendencies or modes significantly influence skill acquisition (e.g., Schöner et al., 1992; Walter & Swinnen, 1992; Zanone & Kelso, 1992) is a relatively new and promising approach to motor learning. It has important implications for such traditional issues as effective training techniques, task difficulty, and individual differences. Many problems concerning these research topics have been left unresolved and are currently somewhat stagnant. It is our feeling that this framework has the capacity to inject new life into their investigation.

References

Brinkman, J., & Kuypers, H. G. J. M. (1973). Cerebral control of contralateral and ipsilateral arm, hand, and finger movements in the split-brain Rhesus monkey. *Brain,* **96,** 653–674.
Cohen, L. (1971). Synchronous bimanual movements performed by homologous and non-homologous muscles. *Perceptual and Motor Skills,* **32,** 639–644.
deGuzman, G. C., & Kelso, J. A. S. (1991). Multifrequency behavioral patterns and the phase attractive circle map. *Biological Cybernetics,* **64,** 485–495.
Delcomyn, F. (1975). Neural basis of rhythmic behavior in animals. *Science,* **210,** 492–498.
Easton, T. A. (1972). On the normal use of reflexes. *American Scientist,* **60,** 591–599.
Faust-Adams, A. S. (1972). Interference in short-term retention of discrete movements. *Journal of Experimental Psychology,* **96,** 400–406.
Fitts, P. M., & Posner, M. I. (1967). *Human performance.* Belmont, CA: Brooks/Cole.
Fleishman, E. A. (1958a). An analysis of positioning movements and static reactions. *Journal of Experimental Psychology,* **55,** 13–24.
Fleishman, E. A. (1958b). Dimensional analysis of movement reactions. *Journal of Experimental Psychology,* **55,** 438–453.
Fowler, C. A., & Turvey, M. T. (1978). Skill acquisition: An event approach with special reference to searching for the optimum of a function of several variables. In G. E. Stelmach (Ed.), *Information processing in motor control and learning* (pp. 1–40). New York: Academic Press, Inc.
Franz, E. A., Zelaznik, H. N., & McCabe, G. (1991). Spatial topological constraints in a bimanual task. *Acta Psychologica,* **77,** 137–151.
Franz, E. A., Zelaznik, H. N., Walter, C. B., & Swinnen, S. P. (1992). *Levels of constraint in a bimanual task.* Submitted manuscript.

Fukuda, T. (1961). Studies on human dynamic postures from the viewpoint of postural reflexes. *Acta Oto-Laryngologica,* **161,** 1–52.
Fulton, R. E. (1942). Speed and accuracy in learning a ballistic movement. *Research Quarterly,* **13,** 30–36.
Grillner, S. (1985). Neurobiological bases of rhythmic motor acts in vertebrates. *Science,* **228,** 143–149.
Haken, H., Kelso, J. A. S., & Bunz, H. (1985). A theoretical model of phase transitions in human hand movements. *Biological Cybernetics,* **51,** 347–356.
Hasan, Z. (1986). Optimized movement trajectories and joint stiffness in unperturbed, inertially loaded movements. *Biological Cybernetics,* **53,** 373–382.
Hogan, N. (1984). An organising principle for a class of voluntary movements. *Journal of Neuroscience,* **4,** 2745–2754.
Kelso, J. A. S., & Scholz, J. (1985). Cooperative phenomena in biological motion. In H. Haken (Ed.), *Complex systems: Operational approaches in neurobiology, physics, and computers* (pp. 124–149). New York: Springer-Verlag.
Kelso, J. A. S., & Schöner, G. (1988). Self-organization of coordinative movement patterns. *Human Movement Science,* **7,** 27–46.
Kelso, J. A. S., Southard, D. L., & Goodman, D. (1979). On the coordination of two handed movements. *Journal of Experimental Psychology: Human Perception and Performance,* **5,** 229–238.
Kelso, J. A. S., Putnam, C. A., & Goodman, D. (1983). On the space-time structure of human interlimb coordination. *Quarterly Journal of Experimental Psychology,* **35A,** 347–375.
Laabs, G. J. (1973). Retention characteristics of different reproduction cues in motor short-term memory. *Journal of Experimental Psycology,* **100,** 168–177.
Laabs, G. J. (1974). The effect of interpolated motor activity on short-term retention of movement distance and end-location. *Journal of Motor Behavior,* **6,** 279–288.
Laabs, G. J., & Simmons, R. W. (1981). Motor memory. In D. H. Holding (Ed.), *Human skills* (pp. 119–151). Chichester: Wiley.
Lee, T. D., Swinnen, S. P., & Verschueren, S. (1993). *Increased resistance to phase wandering as a function of learning a bimanual pattern.* Manuscript in preparation.
Lewis, D., McAllister, D. E., & Adams, J. A. (1951). Facilitation and interference in performance on the modified Mashburn apparatus: I. The effects of varying the amount of original learning. *Journal of Experimental Psychology,* **41,** 247–260.
Lintern, G., & Gopher, D. (1978). Adaptive training of perceptual-motor skills: Issues, results, and future directions. *International Journal of Man-Machine Studies,* **10,** 521–551.
Magill, R. A. (1989). *Motor learning: Concepts and applications* (3rd Ed.). Dubuque, IA: Wm. C. Brown.
Marteniuk, R. G., MacKenzie, C. L., & Baba, D. M. (1984). Bimanual movement control: Information processing and interaction effects. *Quarterly Journal of Experimental Psychology,* **36A,** 335–365.
Mishkin, M., Prockop, E. S., & Rosvold, H. E. (1962). One-trial object-discrimination learning in monkeys with frontal lesions. *Journal of Comparative and Physiological Psychology,* **55,** 178–181.
Navon, D., & Miller, J. (1987). The role of outcome conflict in dual-task interference. *Journal of Experimental Psychology: Human Perception and Performance,* **13,** 435–448.
Newell, K. M. (1985). Coordination, control and skill. In D. Goodman, R. B. Wilberg, & I. M. Franks (Eds.), *Differing perspectives in motor learning, memory, and control* (pp. 295–317). Amsterdam: Elsevier.
Newell, K. M., & Walter, C. B. (1981). Kinematic and kinetic parameters as information feedback in motor skill acquisition. *Journal of Human Movement Studies,* **7,** 235–254.

Newell, K. M., Morris, L. R., & Scully, D. M. (1985). Augmented information and the acquisition of skill in physical activity. In R. L. Terjung (Ed.), *Exercise and sport sciences reviews* (pp. 235–261). Lexington, MA: Collamore Press.

Newell, K. M., Carlton, M. J., & Antoniou, A. (1990). The interaction of criterion and feedback information in learning a drawing task. *Journal of Motor Behavior, 22,* 536–552.

Pepper, R. L., & Herman, L. M. (1970). Decay and interference effects in short-term retention of a discrete motor act. *Journal of Experimental Psychology Monograph, 83,* 2.

Peters, M. (1977). Simultaneous performance of two motor activities: The factor of timing. *Neuropsychologia, 15,* 461–465.

Preilowski, B. (1987). The role of corollary motor discharges, the corpus callosum, and the supplementary motor cortices in bimanual coordination. *Behavioral and Brain Sciences* (commentary), *10,* 322–323.

Salmoni, A. W., Schmidt, R. A., & Walter, C. B. (1984). Knowledge of results and motor learning: A review and critical reappraisal. *Psychological Bulletin, 95,* 355–386.

Schmidt, R. A. (1988). *Motor control and learning* (2nd Ed.). Champaign, IL: Human Kinetics.

Schmidt, R. A., & Young, D. E. (1987). Transfer of movement control in motor skill learning. In S. M. Cormier & J. D. Hagman (Eds.), *Transfer of learning* (pp. 47–79). Orlando, FL: Academic Press, Inc.

Schmidt, R. C., Carello, C., & Turvey, M. T. (1990). Phase transitions and critical fluctuations in the visual coordination of rhythmic movements between people. *Journal of Experimental Psychology: Human Perception and Performance, 16,* 227–247.

Schöner, G., Zanone, P. G., & Kelso, J. A. S. (1992). Learning as change of coordination dynamics: Theory and experiment. *Journal of Motor Behavior, 24,* 29–48.

Shapiro, D. C. (1977). A preliminary attempt to determine the duration of a motor program. In D. M. Landers & R. W. Christina (Eds.), *Psychology of motor behavior and sport*. Champaign, IL: Human Kinetics.

Sherwood, D. E. (1990). Practice and assimilation effects in a multilimb aiming task. *Journal of Motor Behavior, 22,* 267–291.

Soechting, J. F., & Lacquaniti, F. (1981). Invariant characteristics of a pointing movement in man. *Journal of Neuroscience, 1,* 710–720.

Solley, W. H. (1952). The effects of verbal instruction of speed and accuracy upon the learning of a motor skill. *Research Quarterly, 23,* 231–240.

Sparrow, W. A., & Irizarry-Lopez, V. M. (1987). Mechanical efficiency and metabolic cost as measures of learning a gross motor skill. *Journal of Motor Behavior,19,* 240–264.

Sturt, M. (1921). A comparison of speed with accuracy in the learning process. *British Journal of Psychology, 12,* 289–300.

Summers, J. J. (1975). The role of timing in motor program representation. *Journal of Motor Behavior, 7,* 229–241.

Swinnen, S. P. (1990). Motor control. In R. Delbucco (Ed.), *Encyclopedia of human biology* (Vol. 5) (pp. 105–120). San Diego: Academic Press, Inc.

Swinnen, S. P., & Walter, C. B. (1988). Constraints, in coordinating movements. In A. M. Colley & J. R. Beech (Eds.), *Cognition and action in skilled behavior* (127–143). Amsterdam: North-Holland.

Swinnen, S. P., & Walter, C. B. (1991). Toward a movement dynamics perspective on dual-task performance. *Human Factors, 33,* 367–387.

Swinnen, S. P., Walter, C. B., & Shapiro, D. C. (1988). The coordination of limb movements with different kinematic patterns. *Brain and Cognition, 8,* 326–347.

Swinnen, S. P., Walter, C. B., Pauwels, J. M., Meugens, P. F., & Beirinckx, M. B. (1990). The dissociation of interlimb constraints. *Human Performance, 3,* 187–215.

Swinnen, S. P., De Pooter, A., & Delrue, S. (1991a). Moving away from the in-phase attractor during bimanual oscillations. In P. J. Beek, R. J. Bootsma, & P. C. W. van Wieringen (Eds.), *Studies in perception and action. Proceedings of the VIth international conference on perception and action* (pp. 315–319). Amsterdam: Rodopi.

Swinnen, S. P., Walter, C. B., Beirinckx, M. B., & Meugens, P. F. (1991b). Dissociating the structural and metrical specifications of bimanual movement. *Journal of Motor Behavior*, **23**, 263–279.

Swinnen, S. P., Young, D. E., Walter, C. B., & Serrien, D. J. (1991c). Control of asymmetrical bimanual movements. *Experimental Brain Research*, **85**, 163–173.

Swinnen, S. P., Walter, C. B., Lee, T. D., & Serrien, D. J. (in press). Acquiring bimanual skills: Contrasting forms of information feedback for interlimb decoupling. *Journal of Experimental Psychology: Learning, Memory, and Cognition*.

Swinnen, S. P., Walter, C. B, Serrien, D. J., & Vandendriessche, C. (1992). The effect of movement speed on upper-limb coupling strength. *Human Movement Science*, **11**, 615–636.

Turvey, M. T. (1990). Coordination. *American Psychologist*, **45**, 938–953.

Turvey, M. T., & Carello, C. (1986). The ecological approach to perceiving-acting: A pictorial essay. *Acta Psychologica*, **63**, 133–155.

Walter, C. B. (in press). The acquisition of repulsive skills: A case for convergent kinesiology. In D. W. Edington (Ed.), *Capstone knowledge in kinesiology*. Ann Arbor: University of Michigan Press.

Walter, C. B., & Swinnen, S. P. (1990a). Kinetic attraction during bimanual coordination. *Journal of Motor Behavior*, **22**, 451–473.

Walter, C. B., & Swinnen, S. P. (1990b). Asymmetric interlimb interference during the performance of a dynamic bimanual task. *Brain and Cognition*, **14**, 185–200.

Walter, C. B., & Swinnen, S. P. (1992). Adaptive tuning of interlimb attraction to facilitate bimanual decoupling. *Journal of Motor Behavior*, **24**, 95–104.

Walter, C. B., Pollatou, E., Corcos, D. M., Swinnen, S. P., & Pan, H. (1993). *Interlimb interference during bimanual oscillations of "simple" harmonic ratios.* Manuscript in preparation.

Walter, C. B., Swinnen, S. P., & Franz, E. (1993). Stability of symmetric and asymmetric discrete bimanual actions. In K. M. Newell & D. M. Corcos (Eds.), *Variability and motor control.* (pp. 359–380) Champaign, IL: Human Kinetics.

Wickens, C. D. (1989). Attention and skilled performance. In D. H. Holding (Ed.), *Human skills* (pp. 71–105). Chichester: Wiley.

Zanone, P. G., & Kelso, J. A. S. (1992). Evolution of behavioral attractors with learning: Nonequilibrium phase transitions. *Journal of Experimental Psychology: Human Perception and Performance*, **18**, 403–421.

24
Learning to Coordinate Redundant Biomechanical Degrees of Freedom

Karl M. Newell

College of Health and Human Development
The Pennsylvania State University
State College, Pennsylvania

P. Vernon McDonald

Krug Life Sciences
Houston, Texas

I. Introduction
II. Degrees of Freedom and Changes in Coordination Mode
III. Order to the Changes in the Coordination Mode
IV. Coordination Mode and the Perceptual–Motor Workspace
V. Search Strategies, the Perceptual–Motor Workspace, and the Acquisition of Coordination
    A. Changes in Perceptual Invariants with Practice
    B. The Dual Control Problem
VI. Concluding Comments
    References

I. Introduction

Prolonged practice at a physical activity can lead to quite remarkable changes in the organization of the sensorimotor system that supports the action. One manifestation of this change in organization is that the movement *form* of the skilled performer is usually quite different from that of the beginner, a finding that is prevalent in a range of phylogenetic and ontogenetic activities. The changes in movement organization that accompany practice are reflective of the performer learning to effectively and efficiently coordinate redundant biomechanical degrees of freedom (Bernstein, 1967).

In this paper we discuss one key issue that is central to the development of a general theory of the acquisition of coordination, namely, that of the coordination and control of *redundant* biomechanical degrees of freedom. By redundant degrees of freedom we are referring to the situation where there are more degrees of freedom available than are required to solve the task, so that several potential joint configurations can realize a solution to the task demands. This issue of redundant degrees of freedom is central to changes in coordination modes that arise from practice, and encompasses both the domains of motor learning and motor development (Newell, 1986; Newell & van Emmerik, 1990). This is not to suggest that there are not differences of conceptual or empirical emphasis in motor learning and development, but rather to suggest that the issue of *change* in coordination is a general one no matter what the age of the performer and the nature of the task at hand.

Our investigation of change in motor learning and development is driven by an ecological approach to perception and action (Kugler & Turvey, 1987; Turvey, 1990; Turvey & Kugler, 1984). In this emerging framework there are a number of interrelated constructs that pertain to the question of change in movement organization (Newell, 1991; Newell, Kugler, van Emmerik, & McDonald, 1989a). These constructs are not mutually exclusive and include (1) the constraints to action (Newell, 1986); (2) the nature of the perceptual–motor workspace (Kugler & Turvey, 1987; Newell, McDonald, & Kugler, 1991) or intrinsic dynamics (Schöner, 1989; Schöner & Kelso, 1988; Zanone & Kelso, 1992) supporting the emerging coordination modes; (3) the search strategies used to explore the dynamical perceptual–motor workspace supporting action (Fowler & Turvey, 1978; Newell *et al.*, 1989a; Newell & McDonald, 1992); and (4) the role of augmented information in changing coordination modes (Newell, 1991, 1992).

In this chapter we focus on the issue of change in the organization of the redundant biomechanical degrees of freedom as a function of practice. The particular emphasis is the relation between the organization of the degrees of freedom in the coordination mode and the degrees of freedom of the dynamical layout in the perceptual–motor workspace. Thus, Bernstein's (1967) intuitions about the mastery of redundant degrees of freedom are examined both at the behavioral level in terms of the organization of the coordination mode and as seen through the mapping of information and dynamics of the perceptual–motor workspace (Shaw & Alley, 1985).

II. Degrees of Freedom and Changes in Coordination Mode

The degrees of freedom of the human movement system may be considered at a variety of levels of analysis. As one considers increasingly microlevels of analysis (joints, muscles, motor units, etc.) there is a concomitant increase in the number of degrees of freedom that require coordination. In this context, degrees of freedom are viewed to be the number of independent coordinates required to uniquely specify the configuration of the system without violating the geometry of the individual components. The general problem of developing a theory of coordination requires a consideration of the degrees-of-freedom problem both *within* and *between* levels of analysis. Here, we consider the problem of change in the coordination mode at only the macro level of the dynamics of joint relations in the conduct of action.

Before proceeding it is worth emphasizing that the issue of identifying the contributing degrees of freedom, even at a relatively macroscopic biomechanic level, is not quite as straightforward as one might imagine. The interaction of contralateral and ipsilateral joints, segments, and/or muscles can play a crucial role in the organization of certain movement patterns. Even more difficult is the determination of the degrees of freedom at the level of the perceptual–motor workspace. While we may discuss these workspace degrees of freedom at a theoretical level, their identification remains exceptionally challenging because it is these degrees of freedom that emerge from the confluence of the informational and action aspects of the task at hand.

Bernstein (1967) viewed motor skill learning as essentially a reflection of the mastery of the redundant degrees of freedom. He proposed a two-stage learning process in the changing coordination and control of the biomechanical degrees of freedom. This two-stage process to coordination changes emerged from the learner initially freezing the limb and torso segments in movement execution, thus reducing the number of biomechanical degrees of freedom at the periphery to a minimum. Thus with the learner initially having frozen the degrees of freedom, the first stage of learning Bernstein identified corresponds to the lifting of all restrictions, that is, to the incorporation of all possible degrees of freedom into the coordination mode. The second stage corresponds to the utilization of the reactive phenomena (forces) that arise from the freeing of the degrees of freedom. This second stage of utilizing and exploiting, rather than reacting to, the reactive forces is essential to the efficient and effective production of movement. There have been very few, if any, demonstrations of the complete set of coordination changes that accompany practice and learning, in part because the tasks and skill level of the performers used for study have tended to produce constraints that emphasize only certain phases of the learning process.

There have been a number of isolated demonstrations of each of these two stages of coordination changes in a range of tasks and subject types. The initial freezing and subsequent freeing of the biomechanical degrees of freedom has been demonstrated in both the development of the fundamental phylogenetic movement skills (e.g., Bernstein, 1967; Gesell, 1929; Shirley, 1931; crawling, Sparrow & Irizarry-Lopez, 1987) and more recently in the acquisition of a number of ontogenetic tasks (e.g., dart throwing, McDonald, van Emmerik, & Newell, 1989; handwriting, Newell & van Emmerik, 1989; pistol shooting, Arutyunyan, Gurfinkel, & Mirski, 1968; Arutyunyan, Gurfinkel, & Mirski, 1969; simulated skiing, Vereijken, van Emmerik, Whiting, & Newell, 1992). The accommodation of the reactive forces arising from the freeing of the degrees of freedom has also been demonstrated as a function of practice (e.g., Bernstein, 1967; Schneider & Zernicke, 1989; Schneider, Zernicke, Schmidt, & Hart, 1989).

In spite of these developments, there is still not a comprehensive set of studies of the initial stages of learning that involve the release of the frozen degrees of freedom or of the highly skilled performer that may still, for example, be seeking higher order perceptual invariants as a support for further shifts of organizational control. In short, much of motor learning has studied the middle phase of learning in which the structural organization of the movement output was essentially determined either by the demands of the task constraints or by the established practice level of the performer. We have previously dubbed this empirical phase of motor learning as that of nonoptimal control (Newell, 1985, 1992).

III. Order to the Changes in the Coordination Mode

One of the most striking types of change in coordination that accompanies learning is seen in the directional trends to the organization of the torso and limb degrees of freedom. These directional trends were initially reported in regard to the changes in patterns of coordination that occur in infancy with the development of the fundamental movement patterns (Gesell, 1929, 1946). However, directional trends to coordination mode changes are also evident in adult learning and relearning protocols. That is, there are directional trends to the way in which the redundant degrees of freedom are brought on-line into the organization of the coordination mode, irrespective of performer age and task category. In this section we discuss the generality of these directional trends to the behavioral strategies evident in the change of coordination as a prelude to considering why such systematic changes to the form of movement should occur with practice.

Gesell (1929, 1946) observed that infants tended to change the organization of the movement coordination pattern of the fundamental movement skills in a cephalo–caudal, proximal–distal, or ulnar–radial direction. This postulate of directional trends to the development of coordination patterns was one of Gesell's set of developmental principles. These same directional trends have also been observed with adults in learning or relearning protocols that require the change in organization of the biomechanical degrees of freedom to effectively and efficiently satisfy the task constraints. The generality of the directional trend in coordination mode changes, no matter what the age of the performer or the task type, has been masked by the prevalence of adult studies examining primarily single-degree-of-freedom tasks. To illustrate the directional principles of bringing on-line redundant biomechanical degrees of freedom as a function of practice we provide two rather distinct examples.

In Figure 1 is shown the developmental stages of prone progression as observed by Shirley (1931) in her study of the development of 21

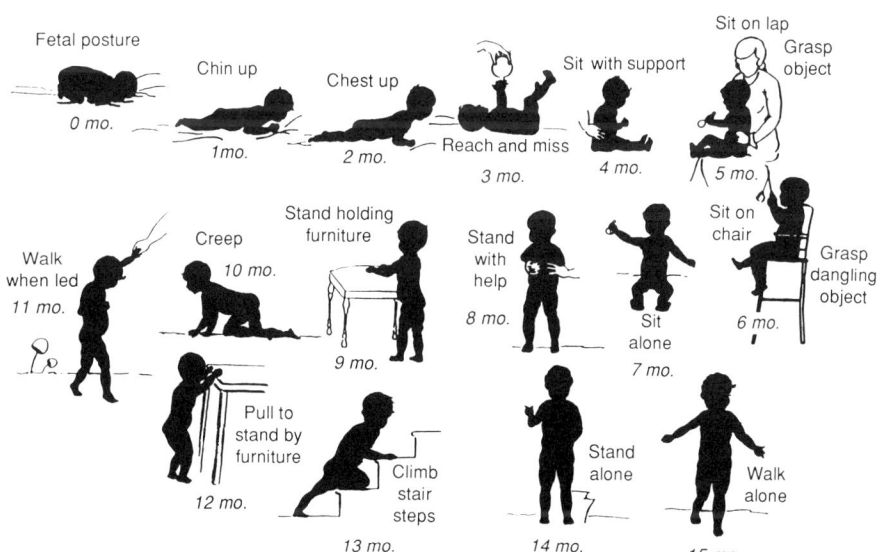

FIGURE 1 The developmental stages of prone progression as observed by Shirley (1931).

infants. This static depiction of the dynamic postural changes occurring through the initial year of life clearly shows the cephalo–caudal directional trend to the emerging postural control in infancy. These cephalo–caudal trends in postural development, together with the accompanying proximal–distal and ulnar–radial changes, were seen by Gesell (1929) as existing in all infants due to the genetic influence of maturation.

In Figure 2 is shown the organization of the arm links in handwriting as a function of dominant/nondominant arm (Newell & van Emmerik, 1989). Figure 2A shows the configurational relations of the highly practiced dominant limb in a single subject writing his signature at a normal checkbook-size height. Figure 2B shows the same configuration relations when the same subject is using the unpracticed nondominant limb to write the signature in the same way. It is apparent that the subject freezes the wrist and elbow joints in writing with the nondominant limb and essentially produces the signature from the shoulder joint. The nondominant signature appears more structured and orderly in motor output but this is a consequence of what is, in effect, overconstraint on the organization of the motor system.

These two examples are reflective of general trends that appear in coordination mode changes irrespective of task or age of the performer (see Newell *et al.*, 1989a; Newell & van Emmerik, 1990, for more extensive reviews). While these behavioral strategies leading to change in coordination modes have been observed in a variety of situations, there are several associated issues that are not well established empirically. One issue is that although there are emerging trends *within* each of the cephalo–caudal, proximal–distal, and ulnar–radial directions, the relation *between* these three directional trends has not been emphasized. It seems reasonable to propose that the relation between directions is even more task dependent than the order to the changes within a given directional dimension. In addition, observation suggests that learners do not always approach a new task by reducing the degrees of freedom in the coordination mode to a *minimum,* as was postulated by Bernstein (1967). The number of degrees of freedom frozen and the initial organization of these degrees of freedom in the coordination mode is strongly *task* and *performer* dependent. These points suggest that we need further examination of the form of both the freezing and the form of the freeing of the degrees of freedom in the coordination mode in the examination of both developmental and more mature movement learning. The general perspective is that the phenomena of freeing and freezing biomechanical degrees of freedom suggest additional principles in the change of coordination modes beyond those outlined by Bernstein's (1967) original perspective.

In particular, the form of the freezing of degrees of freedom can be considered as a behavioral indication of the initial conditions of each individual learner. One might expect the learner's initial conditions, defined as the form of the freezing of the degrees of freedom, to determine (in part) the form of the subsequent learning dynamic. Even though the learners may eventually gravitate to a relatively small number of stable states (as many dissipative and biological systems do) the process of reaching these states need not be tightly determined. Thus the behavioral sequence exhibited during the acquisition process may well be individually specific. This postulate would have significant implications for the type of instruction and information provided to the learner to facilitate the acquisition of skill. While these claims currently lack empirical support, they are consistent with a dynamical characterization of the process of skill acquisition. Moreover, the clear identification of the initial conditions may permit a more robust determination of the novelty of the new skill to be learned, relative to each learner.

The prevalence of these directional principles, however, suggests that there are general trends to the way in which changes in coordination patterns occur as a function of practice. A more significant implication is that a single theoretical orientation may be able to accommodate these changes without the need for age- or task-specific theories of change in movement coordination or control. The time scales to the attractor dynamics[1] arising from the various sources of structural and functional constraint to coordination (particularly organismic constraints) may vary as a function of age but the dynamical principles supporting action may be the same in both phylogenetic and ontogenetic activities. It is often assumed that the time scale is longer for the coordination mode changes in the acquisition of phylogenetic as opposed to ontogenetic activities. Certainly, the observations of the time span required for the development of independent postural support in infancy give indirect evidence for this assumption. However, there are so few systematic observations of adults learning a motor task that induces change in the coordination mode over practice conditions, that direct evidence is not available on the relative time scale issue for motor learning and development. In a recent study on the acquisition of handwriting with the nondominant limb in adults (Newell & van Emmerik, 1989), we found that there was little change in the introduction of more distal control of the limb segments, in spite of 10,000 trials of practice over a 3-month duration. This finding suggests that the time scale issue in the change of coordination mode is highly task dependent

---

[1] A discussion relevant to this issue can be found in Saltzman and Munhall (1992), who address the problem with reference to parameter and graph dynamics.

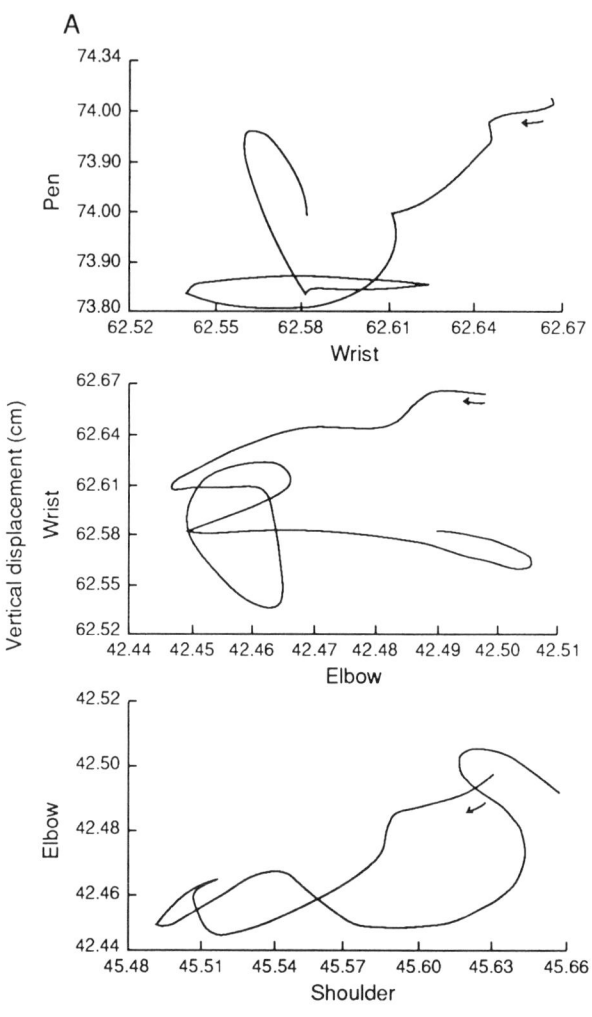

FIGURE 2 Position–position plots of pairs of pen and joint movements for signatures written with the dominant (A) and nondominant (B) hands. (Adapted from Newell & van Emmerik, 1989).

and, furthermore, that the assumption about developmental differences in the time scale for coordination changes is an empirical question.

It should be noted in closing this section that we are not proposing here that these directional trends *must* emerge from practice-induced

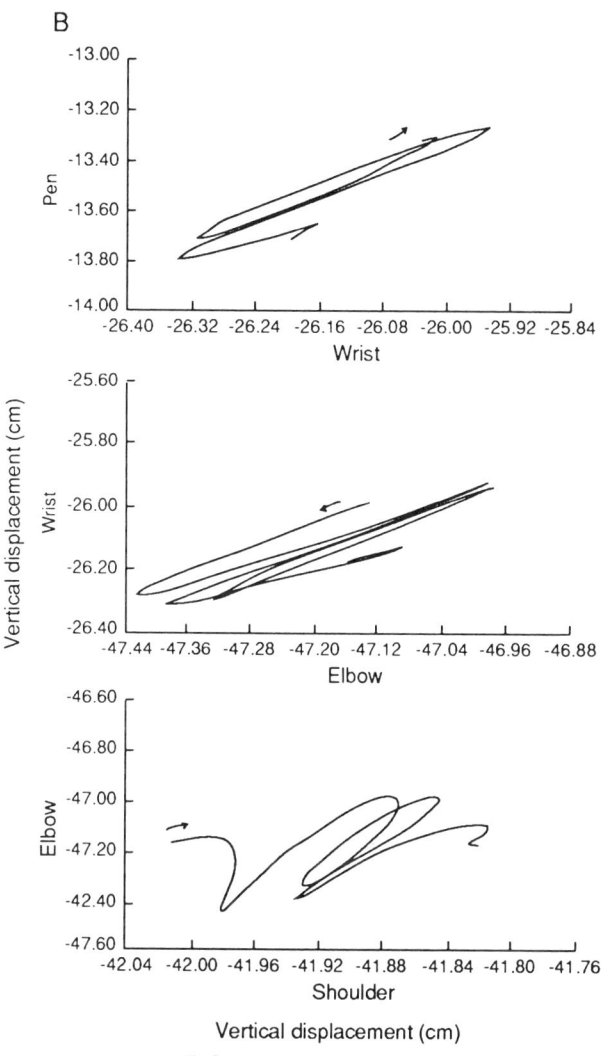

FIGURE 2 Continued

changes in coordination patterns. Rather, we propose that there is a strong tendency for these directional trends to be pervasive in motor learning and development. Given the principle that coordination modes are softly assembled (Kugler & Turvey, 1987) and the fact that a range of task constraints have yet to be examined (Newell & van Emmerik, 1990), it is possible that a different strategy to the change in coordination mode could occur with a different set of task constraints. Thus, the directional trends to practice induced coordination mode changes dominate in motor learning and development but they may not be reflective of exclusive behavioral strategic principles.

Why should these directional changes of the coordination mode occur as a function of practice? Bernstein (1967) proposed two possible hypotheses for the directional trends. One hypothesis was due to physiological factors, while the other hypothesis was mechanical in origin. Gesell (1929) also elaborated from anatomical structural accounts of cephalo–caudal, proximal–distal, ulnar–radial development to functional accounts of the directional trends in the maturation of the developing nervous system. The physiological, or more generally, the biological hypothesis postulates that the nerve dynamics of the distal musculature develop at a slower rate than those of the more proximal muscles. This developmental physiological account has been used to explain, for example, the proximal–distal trend to the bringing in of the leg musculature in the development of walking in the infant. In contrast, the mechanical hypothesis holds that because the muscle mass of the more proximal muscles is larger than that of the distal muscles, they can overcome the resistance of the upper limbs more easily than those of the foot because the relative velocities are as a rule higher for the distal as opposed to the proximal segments. Bernstein (1967) emphasized the mechanical hypothesis for the directional trends apparent in the change in coordination modes (including the development of walking) because he felt it was unlikely that the development of the distal nerve dynamics would be so strikingly different from those of the proximal muscles over the span of the time period that encompassed the observed developmental differences in the walking pattern.

There have been no direct tests of these hypotheses, either in motor learning or motor development, regarding why the change in coordination modes occurs (although see the work of Thelen and colleagues on infant locomotion, Thelen, 1986). Certainly the mechanical hypothesis is general to movement coordination irrespective of developmental stage and category of physical activity. On the other hand, anatomical and physiological development are clearly sources of constraint to the observed changes in the fundamental movement skills, particularly at significant transitional segments of the life span (Newell, 1984). Thus, both structural and functional sources of constraint to action play a role in determining practice-induced changes to coordination modes. However, as we will subsequently discuss, the mechanical hypothesis holds particular relevance to an examination of the search strategies used to explore the dynamical properties of the perceptual–motor workspace.

IV. Coordination Mode and the Perceptual–Motor Workspace

Bernstein's (1967) account of movement coordination gave emphasis to the degrees-of-freedom problem in motor control and skill acquisition (see also Greene, 1969; Turvey, Shaw, & Mace, 1978). This account primarily addressed the degrees of freedom at the behavioral level, that is, those degrees of freedom that are being organized independently in the coordination mode. Even at this relatively macroscopic level the number of potential configurations of the system is quite large because of the many joint configurations that are possible and the redundancy that is generally available in the seeking of solutions to various task demands. Indeed, it is the redundancy in the biomechanical degrees of freedom that affords flexibility to the satisfying of task solutions in human action.

In spite of the prevalence of this biomechanical redundancy, there is a tendency for individuals to gravitate to use a small set of coordination modes and often even a *single* preferred mode for realizing a given task goal. This trend is evident both within an individual and between individuals when an appropriate frame of reference is constructed to consider the organism–environment interaction in relation to a given task goal. It is also the case that this trend often prevails where the task demands do not explicitly require the subject to minimize or maximize a given system or task parameter.

A striking illustration of this proposition has emerged in recent studies of infant and adult grip configurations, as a function of object and task constraints, which showed that subjects (even 4-month-old infants) tend to use a single preferred grip configuration for a given set of constraints (see Newell, Scully, Tenenbaum, & Hardiman, 1989b; Newell, Scully, McDonald, & Baillargeon, 1989c). The objects were light, small, and in conjunction with the task constraints, afforded considerable redundancy with respect to possible grip configurations for the prehensile act. For example, in considering the number of digits utilized to pick up objects, there are 1023 different combinations of digits possible, and yet infant through adult subjects tended to use a single preferred set of digits for a given object. The actual grip configuration utilized was also shown to be strongly related to body scale and independent of the sensory mode (vision or vision plus haptics) used to specify prospectively the grip configuration for action. These illustrations of the hand in action, with its many degrees of freedom in terms of finger joint configurations, are powerful examples of the significance of preferred coordination modes to a given set of constraints, in spite of the high degree of biomechanical redundancy afforded to the subjects.

In the ecological account of perception and action (Kugler & Turvey, 1987; Turvey & Kugler, 1984), the coordination modes are emergent properties of the constraints to action as a result of the mapping of perception and action at a hypothetical dynamical interface called the perceptual–motor workspace. From this perspective the coordination mode is defined over both the informational field arising from perception and the kinetic field arising from action. Learning is reflected in the increased coordination of the mapping of the perception and action fields (Shaw & Alley, 1985).

The attractor organization of this dynamical interface is viewed as providing the building blocks for the emergent relational properties of the coordination mode. This theoretical formulation seeks to describe the mapping of perception and action as a low-dimensional system within the perceptual–motor workspace. Shaw and Alley (1985) proposed that the functions for perception and action are duals as defined in the mathematical sense. Briefly, such a formulation permits the establishment of an isomorphic (one-to-one) correspondence between perception and action.

One of the attractive features of this theoretical formulation is that potentially it holds a principled rationale to accommodate the degrees of freedom that are organized at the level of the coordination mode. In other words, the degrees of freedom that are required to define the attractor layout in the perceptual–motor workspace are usually lower than those evident at the behavioral level in the coordination mode. In this view, the degrees of freedom controlled are defined by an abstract perceptual–motor workspace and not the observable joint space of the coordination mode. Figure 3 provides a schematic of the proposed relations between the different frames of reference to considering the degrees-of-freedom construct.

The emphasis of this theoretical orientation is to understand the degrees of freedom of the attractor layout in the perceptual–motor

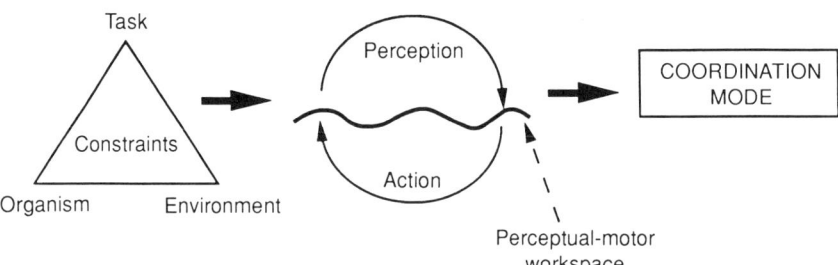

FIGURE 3 Schematic representing the different frames of reference for considering the degrees-of-freedom construct.

workspace and how these degrees of freedom in the workspace map to the degrees of freedom evident in the coordination mode. Schöner and Kelso (1988) have proposed an abstract attractor construct for action similar to that of the perceptual–motor workspace (Kugler & Turvey, 1987) and characterize the physical layout of this attractor layout as the *intrinsic dynamics* supporting action. The properties of the attractors of the dynamical interface of the peceptual–motor workspace can be described formally by principles of dynamical systems and nonlinear dynamics. To date, the most systematic work characterizing the perceptual–motor workspace or intrinsic dynamics has been limited to the rhythmical motion of two-degree-of-freedom tasks (e.g., Haken *et al.*, 1985; Schmidt, Beek, Treffner, & Turvey, 1992), where no redundancy existed in the joint space degrees of freedom. While there is redundancy at levels of analysis other than the joint space in these movement tasks, it was the latter frame of reference that constituted the focus of the empirical observations. Beek (1989) has provided an in-depth dynamic analysis of the task of juggling, but this work is confined to the interface of the hand–ball contact, without consideration of the within-limb dynamics. Thus, empirical work is still required that demonstrates the formal link between the coordination mode and the perceptual–motor workspace, where redundant biomechanical degrees of freedom are available.

An important property of the attractor layout of the perceptual–motor workspace is that it is nonstationary, that is, the nature of the workspace changes over time because of a variety of influences, including ongoing interactions of the performer with the environment. Thus, one way in which coordination models may change over time is due to the changing nature of the attractor layout supporting action (see Zanone & Kelso, 1992, for an example). The changing attractor layout provides a formal basis to reconsider many traditional problems in the learning, retention, and transfer of motor skills. In other words, the problems of motor learning should be considered at the level of the perceptual–motor workspace rather than the focus being limited to the observable behavioral level of so-called task characteristics.

V. Search Strategies, the Perceptual–Motor Workspace, and the Acquisition of Coordination

The concept of search strategies refers to the way in which the performer explores the perceptual–motor workspace to solve the movement problem (Fowler & Turvey, 1978; Newell & McDonald, 1992; Newell *et al.*, 1989a). The search strategy identifies the choice of path through a multidimensional state space that maps perception and action consistent with the constraints of the task goal. Thus, search strategies are defined by some qualitative feature of the pathway taken through the perceptual–motor workspace. There are potentially a range of search strategies that could be used in the regulation of biological motion (Newell & McDonald, 1992; Newell et al., 1989a).

Motion about the workspace both creates and annihilates informative features of the dynamical interface of the perceptual–motor workspace. Consequently the perception–action interface necessitates exploratory behavior in order to facilitate learning. Following Gibson (1966, 1979), it is assumed that the invariant properties of the dynamic interface provide information about the evolution and layout of the space. The identification of the perceptual invariants arising from the workspace is a major element in developing a theory of how performers search through the space.

There is a strong link between the nature of the workspace and the search strategies used to explore the space. In this context of considering the change in coordination modes, it is also postulated that there is a relation between the search strategies of the critical and gradient regions of the perceptual–motor workspace and the change in coordination modes observed at the behavioral level (Newell et al., 1989a). In other words, there is a relation between the search strategies observed in the perceptual–motor workspace and the trends reviewed earlier to the directional changes in coordination mode. The formal establishment of this link is a central component to the development of a theory of the acquisition of coordination.

In the remainder of this chapter we discuss two aspects of the role of search strategies in the change of coordination mode that build on our previous theorizing on this issue (Newell & McDonald, 1991; Newell et al., 1989a). First is the issue of perceptual invariants and perceptual degrees of freedom and their relation to the degrees of freedom in the coordination mode. Second is the issue of priorities in searching through the perceptual–motor workspace that we identify under what has been called the dual control problem. These two issues combine to emphasize the significant role that information plays in the directional changes in coordination modes outlined earlier, and are consistent with the proposition that the change of coordination mode issue cannot be confined to mechanical considerations.

A. Changes in Perceptual Invariants with Practice

The search through the perceptual–motor workspace is guided by the utilization of perceptual invariants that arise from the layout of the space. This space is formed from information garnered via all the sensory receptors, although following Gibson (1979), it may be more intuitive to consider the concepts within the single sensory mode of vision. There has been a growing empirical effort to understand the invariants of the visual-flow field that may guide action (Koenderink, 1990), with the most strongly oriented action-related work focusing on the prospective control of timing actions with the environment (Lee, 1980). The majority of this work has not considered the influence of learning of the pickup of perceptual invariants, in spite of the long-held postulate within the ecological approach to perception, that perceptual learning is essentially a matter of enhanced differentiation and the education of attention (Gibson, 1969).

There is a small literature in perceptual–motor skills that provides some evidence for the influence of practice on the changing perception of movement-related information cues, although the theoretical orientation of this literature is distinct from the ecological approach to perception and action. In the tracking literature, Fuchs (1962), following the work of Paul Fitts, suggested that with practice, learners increasingly utilize higher order derivatives of the error signal. That is, as the skill level of the performer increases, he or she switches from controlling the positional properties of the tracking task to the higher order kinematic derivatives of velocity and then acceleration. Fuchs also showed that skilled subjects under stress tend to switch to lower level derivatives of the error signal. This dual strategy of changing the nature of the informational cues of the error signal utilized in performance as a function of practice and task conditions was called the progression–regression hypothesis by Fuchs (1962). More recent tracking studies have confirmed this switch in informational control that accompanies practice at a task (Franks & Wilberg, 1982; Marteniuk & Romanow, 1983), although there are a number of related studies that have not shown such effects. Fowler and Turvey (1978) also proposed that one consequence of practice at a perceptual–motor search task is a switch to higher order properties of the error signal provided to the subject.

The nature of perceptual learning in motor learning and development is not well articulated. In the theoretical framework pursued here the mapping of the perception to action is an integral element of the way change is induced in coordination modes with practice. Indeed, perceptual constraints to action may be as much a factor to changing the coordination mode as the physical constraints that arise in the conduct of action. Revealing the nature of the informational cues used by the learner as a function of the task and stage of practice is a key element to formalizing the search strategy approach to the acquisition of motor skills (Newell *et al.*, 1989a).

In general, the issue can be considered as one of assimilating actions and information within an appropriate frame of reference. The crucial point is that actions generate information and in turn information guides action. Consequently, it follows that if one is unable to reliably generate certain actions, one will preclude the opportunity to be exposed to certain flows of information. An excellent example arises in the study of wielding by Turvey and colleagues (Solomon, Turvey, & Burton, 1989a; Solomon, Turvey, & Burton 1989b). In order for the wielder to determine the properties of the object wielded, a certain set of transformations must be undergone to reveal the appropriate informational invariant. These transformations are generated by a particular set of actions. Thus the link between action and information is an intimate and circular one. With respect to the bodies of evidence discussed earlier in this section, perhaps a greater understanding of the changing focus of perception in learning would arise from a methodology that is sensitive to the complementary nature of action and information.

B. The Dual Control Problem

Skill acquisition can be considered as the discovery of the dynamical laws that organize information and action. In searching for solutions to the task goal the performer is in essence discovering the dynamic characteristics of the system, where the system is defined over the organism, environment, and task. In the adaptive control literature the problem of discovering the dynamic characteristics of the system, while at the same time trying to control the system, has been called the dual control problem (Feld'baum, 1965). This issue of dual control has a number of implications for behavior in the perceptual–motor system. The significance of this dual control problem varies with the nature of the task constraints but it would seem particularly important in whole-body tasks with many, often redundant, biomechanical degrees of freedom. Such situations may necessitate the use of conflicting control requirements in order to simultaneously maintain exploration of the system dynamics *and* system integrity through stability. Hence, the interaction of these multiple control demands on the system may act to change the movement form and search strategies of the observed behavior.

The dual control problem is particularly relevant to perceptual–motor skill acquisition, given that in searching the perceptual–motor workspace to create new patterns of coordination, there is an increased likelihood of approaching the unstable regions of the dynamical layout. However, these instabilities may be more or less important according to the task demands and the relation of the organism to the environment in satisfying the task constraints. For example, in laboratory tasks that have subjects seated performing a finger-coordination task, the general stability of the whole body is accommodated by the chair and the seated posture of the subject. Instabilities in the relative phase of the two fingers probably has little bearing on the overall system integrity and the problem of postural stability. Thus, in these laboratory fine-motor tasks the instabilities of the whole body system are probably only marginally influenced (if at all) by the direct nature of the finger motion. On the other hand, one may speculate that the coordination stability of the finger motions may be specific to the fact that the subject may be seated. In general then, one problem of dual control, namely that of preserving some overall body stability while searching for a particular coordination solution to a given task constraint, is often minimized in laboratory coordination tasks.

Contrast this situation with the whole-body actions of learning to walk, learning to skate, learning to ride a bicycle, and even vehicular control, such as driving a car or flying an airplane. In these tasks there is considerably more salience to the constraint of preserving system integrity, body posture, or the alignment of the vehicle. The constraint of self preservation (e.g., not crashing the plane) is a significant factor in searching the stable and unstable regions of the perceptual–motor workspace in the acquisition of movement sequences that directly engage the whole body and/or vehicle. In other words, the search to realize new task demands is tempered by the costs and payoffs of the loss of overall stability. The dual control problem is, therefore, a significant factor in constraining the search through the perceptual–motor workspace. To our knowledge there has been no research on the role of this dual control problem in the acquisition of new coordination modes.

There are a number of well-known examples in which an artificial environment is created in whole, or in part, to reduce the cost of realizing the instabilities of the system during the search for new patterns of coordination. An established example is a flight simulator. An empirical question is whether the differences in the costs to instabilities in the natural and artificial situation lead to some influence on the form of the search behavior, as well as the fidelity of the whole system. This is a particular problem when examining postural control with respect to vehicular motion. In most simulators pilots will not experience the effects of increased or decreased g-force associated with many flight maneuvers, and consequently the behavioral responses to these tasks will be modulated. Coordination of postural control and vehicular control should be explicitly considered in the evaluation of experiential and action fidelity in a simulator (Riccio, in press).

Sometimes artificial aids are constructed to partially support the performer and reduce the system integrity costs to approaching the postural instabilities. For example, children learning to ride a bicycle often have a preparatory phase in which training wheels are attached to the two-wheel bicycle. The addition of these training wheels adds considerably to the stability of the bicycle but it also changes considerably the layout of the perceptual–motor workspace supporting the action of bicycling. Indeed, while the training wheels undoubtedly enable young children to transport themselves through the environment when they otherwise would not be able to do so with the regular bicycle, there is no evidence that this artificial support is an appropriate or optimal training aid to facilitate learning to ride a bicycle. Thus, the provision of environmental supports to facilitate system stability during the solving of the dual control problem leads to changes in the layout of the perceptual–motor workspace, and the possible neutral or negative transfer to the natural task conditions that ultimately are required to be learned. This is a relevant example of the earlier claim that the learning, transfer, and retention of motor skills need to be approached at the level of the perceptual–motor workspace and not merely the behavioral task characteristics. It should also be noted that there is a strong relation between instructional strategies and the dual control problem that is not well understood.

The dual control problem also has implications for learning to coordinate redundant degrees of freedom in movement tasks. In many of these movement situations the preservation of posture or the relation of the body to the environment is fundamental. Thus, in searching for solutions to task demands, the dual control problem may impose a constraint that leads to priorities in realizing particular coordination solutions. Consequently, the change of coordination modes with practice may be organized around the perception–action relation that preserves the postural integrity of the system. The complementary changes in the other (often more peripheral) degrees of freedom may be linked to changes in this important postural dynamic.

A recent study by Vereijken (1991) on the acquisition of a ski-simulator task is consistent with this proposition. With practice there was an increase in the number of degrees of freedom brought into the coordination mode that accompanied increments of movement amplitude and velocity of the ski board (Vereijken *et al.*, 1992). However, it appeared that the organizing dynamical relation of the task was the relative phase of the mass of the body with the moving board. Vereijken (1991) showed that this dynamic appeared to change with practice through behavior modeled as a balancing pendulum, to a hanging pendulum, and finally to a buckling compound-pendulum model. The change in this postural relation between the performer and the board presumably facilitated the accompanying change in coordination of the more peripheral limb degrees of freedom.

Thus, the mass relations of the body parts, together with the preservation of the system integrity, contribute to channeling the search through the perceptual–motor workspace and the accompanying change in the coordination mode. The order to the changing redundant degrees of freedom may be found, therefore, in a simple relation defined over the organism, environment, and task, even when several redundant biomechanical degrees of freedom are available. The search for low-dimensional accounts of the coordination mode and its changes with practice is a major focus of a dynamical account of motor skill acquisition.

VI. Concluding Comments

There is considerable order and regularity to the changes in coordination modes or movement form that accompany motor learning and development. The predominant behavioral strategies to the bringing on-line of biomechanical degrees of freedom into the emerging coordination mode are the directional trends of cephalo–caudal, proximal–distal, and ulnar–radial. These trends are evident in the domains and experimental protocols of motor learning, motor development, and rehabilitation. The directional trends to the change of coordination modes are, therefore, relatively independent of performer age and task constraints. However, these directional trends are probably not behavioral strategies exclusive to the potential set of strategies that could characterize change of coordination modes. Further empirical work is required to fully characterize the freezing and freeing of the biomechanical degrees of freedom in movement coordination as a function of the confluence of constraints to action.

Our theoretical strategy is to map the practice-induced behavioral changes observed in the coordination mode to the search strategies revealed in the exploration of the dynamical interface of the perceptual–motor workspace (Newell *et al.*, 1989a). At this point there is little empirical evidence that has examined this theoretical perspective, in part, because few skill-oriented practice studies have afforded redundancy in the available biomechanical degrees of freedom. The general challenge of this orientation is to characterize the mapping of perception and action that accompanies practice (Shaw & Alley, 1985). The expectation is that relatively low dimensional solutions exist to this mapping if appropriate dimensions are used to describe the information and dynamics of event perception and action.

Acknowledgments

We would like to thank Stephan Swinnen, Chuck Walter, and an anonymous reviewer for their helpful comments on an earlier version of this manuscript.

References

Arutyunyan, G. H., Gurfinkel, V. S., & Mirskii, M. L. (1968). Investigation of aiming at a target. *Biophysics,* **13,** 536–538.
Arutyunyan, G. H., Gurfinkel, V. S., & Mirskii, M. L. (1969). Organization of movements on execution by man of an exact postural task. *Biophysics,* **14,** 1162–1167.
Beek, P. J. (1989). *Juggling dynamics.* Amsterdam: Free University Press.
Bernstein, N. (1967). *The co-ordination and regulation of movements.* New York: Pergamon.
Feld'baum, A. A. (1965). *Optimal control systems.* New York: Academic Press.
Fowler, C. A., & Turvey, M. T. (1978). Skill acquisition: An event approach with special references to searching for the optimum of a function of several variables. In G. E. Stelmach (Ed.), *Information processing in motor control and learning* (pp. 1–40). New York: Academic Press, Inc.
Franks, I. M., & Wilberg, R. B. (1982). The generation of movement patterns during the acquisition of a pursuit tracking task. *Human Movement Science,* **1,** 251–272.
Fuchs, A. (1962). The progression–regression hypothesis in perceptual–motor skill learning. *Journal of Experimental Psychology,* **63,** 177–192.
Gesell, A. (1929). Maturation and infant behavior pattern. *Psychological Review,* **36,** 307–319.
Gesell, A. (1946). The ontogenesis of infant behavior. In L. Carmichael (Ed.), *Manual of child psychology* (pp. 295–331). New York: Wiley.
Gibson, E. J. (1969). *Principles of perceptual learning and development.* New York: Appleton Century and Crofts.
Gibson, J. J. (1966). *The senses considered as perceptual systems.* Boston: Houghton Mifflin.
Gibson, J. J. (1979). *The ecological approach to visual perception.* Boston: Houghton Mifflin.
Greene, P. H. (1969). Seeking mathematical models for skilled actions. In D. Bootzin & H. C. Muffley (Eds.), *Biomechanics (Proceedings of the first Rock Island Arsenal biomechanics symposium).* New York: Plenum Press.
Haken, H., Kelso, J. A. S., & Bunz, H. (1985). A theoretical model of phase transitions in human hand movements. *Biological Cybernetics.* **51,** 347–356.
Koenderink, J. J. (1990). *Solid shape.* Cambridge: MIT Press.
Kugler, P. N., & Turvey, M. T. (1987). *Information, natural law, and self-assembly of rhythmic movement: Theoretical and experimental investigations.* Hillsdale, NJ: Erlbaum.
Lee, D. N. (1980). Visuo-motor coordination in space–time. In G. E. Stelmach & J. Requin (Eds.), *Tutorials in motor behavior* (pp. 281–296). Amsterdam: North-Holland.
Marteniuk, R. G., & Romanow, S. K. E. (1983). Human movement organization and learning as revealed by variability of movement, use of kinematic information, and fourier analysis. In R. A. Magill (Ed.), *Memory and control of action* (pp. 167–197). Amsterdam: North-Holland.
McDonald, P. V., van Emmerik, R. E. A., & Newell, K. M. (1989). The effects of practice on limb kinematics in a throwing task. *Journal of Motor Behavior,* **21,** 245–264.
Newell, K. M. (1984). Physical constraints to the development of motor skills. In J. R. Thomas (Ed.), *Motor development during childhood and adolescence* (pp. 105–120). Minneapolis: Burgess.
Newell, K. M. (1985). Coordination, control and skill. In D. Goodman, R. B. Wilberg, & I. M. Franks (Eds.), *Differing perspectives in motor learning, memory, and control* (pp. 295–317). Amsterdam: North-Holland.
Newell, K. M. (1986). Constraints on the development of coordination. In M. G. Wade &

H. T. A. Whiting (Eds.), *Motor development in children: Aspects of coordination and control* (pp. 341–360). Boston: Martinus Nijhoff.
Newell, K. M. (1991). Motor skill acquisition. *Annual Review of Psychology, 42,* 213–237.
Newell, K. M. (1992). Augmented information and motor skill acquisition. In R. Daugs & K. Bliscke (Eds.), *Motor learning and training.* Shormmdorf: Hofmann.
Newell, K. M., & van Emmerik, R. E. A. (1989). The acquisition of coordination: Preliminary analysis of learning to write. *Human Movement Science, 8,* 17–32.
Newell, K. M., & van Emmerik, R. E. A. (1990). Are Gesell's developmental principles general principles for the acquisition of coordination? In J. E. Clark & J. H. Humphrey (Eds.), *Advances in motor development research* (Vol. 3) (pp. 143–164). New York: AMS Press.
Newell, K. M., & McDonald, P. V. (1992). Searching for solutions to the coordination function: Learning as exploratory behavior. In G. E. Stelmach & J. Requin (Eds.), *Tutorials in motor behavior II* (pp. 517–532). Amsterdam: North-Holland.
Newell, K. M., Kugler, P. N., van Emmerik, R. E. .A, & McDonald, P. V. (1989a). Search strategies and the acquisition of coordination. In S. A. Wallace (Ed.), *Perspectives on the coordination of movement* (pp. 85–112). Amsterdam: North-Holland.
Newell, K. M., Scully, D. M., McDonald, P. V., & Baillargeon, R. (1989b). Task constraints and infant prehensile grip configurations. *Developmental Psychobiology, 22,* 817–832.
Newell, K. M., Scully, D. M., Tenenbaum, F., & Hardiman, S. (1989c). Body scale and the development of prehension. *Developmental Psychobiology* **22,** 1–13.
Newell, K. M., McDonald, P. V., & Kugler, P. N. (1991). The perceptual–motor workspace and the acquisition of skill. In J. Requin & G. E. Stelmach (Eds.), *Tutorials in motor neuroscience* (pp. 95–108). Dordrecht: Kluwer.
Riccio, G. E. (in press). Coordination of postural control and vehicular control: Implications for multimodal perception and simulation of self motion. In J. Flach, P. Hancock, J. Caird, & K. Vicente (Eds), *The ecology of human-machine systems.* Hillsdale, NJ: Erlbaum.
Saltzman, E. L., & Munhall, K. G. (1992). Skill acquisition and development: The roles of state-, parameter-, and graph-dynamics. *Journal of Motor Behavior, 24,* 49–57.
Schmidt, R. C., Beek, P. J., Treffner, P. J., & Turvey, M. T. (1992). Dynamical substructure of coordinated rhythmic movements. *Journal of Experimental Psychology: Human Perception and Performance, 17,* 635–651.
Schneider, K., & Zernicke, R. F. (1989). Jerk-cost modulations during the practice of rapid arm movements. *Biological Cybernetics,* **60,** 221–230.
Schneider, K., Zernicke, R. A., Schmidt, R. A., & Hart, T. J. (1989). Changes in limb dynamics during practice of rapid arm movements. *Journal of Biomechanics,* **22,** 805–817.
Schöner, G. (1989). Learning and recall in a dynamic theory of coordination patterns. *Biological Cybernetics,* **62,** 39–54.
Schöner, G., & Kelso, J. A. S. (1988). A synergetic theory of environmentally-specified and learned patterns of movement coordination. *Biological Cybernetics,* **58,** 71–80.
Shaw, R. E., & Alley, J. R. (1985). How to draw learning curves: Their use and justification. In T. D. Johnston & A. T. Pietrewicz (Eds.), *Issues in the ecological study of learning* (pp. 275–304). Hillsdale, NJ: Erlbaum.
Shirley, M. M. (1931). *The first two years: A study of twenty-five babies: Vol. 1. Postural and locomotor development.* Minneapolis: University of Minnesota Press.
Solomon, H. Y., Turvey, M. T., & Burton, G. (1989a). Gravitational and muscular variables in perceiving extent by wielding. *Ecological Psychology,* **1,** 265–300.
Solomon, H. Y., Turvey, M. T., & Burton, G. (1989b). Perceiving rod extents by wielding: Haptic diagonalization and decomposition of the inertia tensor. *Journal of Experimental Psychology: Human Perception and Performance,* **15,** 58–68.
Sparrow, W. A., Irizarry-Lopez, V. M. (1987). Mechanical efficiency and metabolic cost as

measures of learning a novel gross motor skill. *Journal of Motor Behavior,* **19,** 240–264.
Thelen, E. (1986). Development of coordinated movement: Implications for early human development. In M. G. Wade & H. T. A. Whiting (Eds.), *Motor development in children: Aspects of coordination and control* (pp. 107–124. Dordrecht: Martinus Hijhoff.
Turvey, M. T. (1990). Coordination. *American Psychologist,* **45,** 938–953.
Turvey, M. T., & Kugler, P. N. (1984). An ecological approach to perception and action. In H. T. A. Whiting (Ed.), *Human motor actions: Bernstein reassessed* (pp. 373–412). Amsterdam: North-Holland.
Turvey, M. T., Shaw, R. E., & Mace, W. (1978). Issues in a theory of action: Degrees of freedom, coordinative structures and coalitions. In J. Requin (Ed.), *Attention and performance VII* (pp. 557–595). Hillsdale, NJ: Erlbaum.
Vereijken, B. (1991). *The dynamics of skill acquisition.* Ph.D. Thesis, Free University, Amsterdam.
Vereijken, B., van Emmerik, R. E. A., Whiting, H. T. A., & Newell, K. M. (1992). Free(z)ing degrees of freedom in skill acquisition. *Journal of Motor Behavior,* **24,** 133–142.
Zanone, P. G., & Kelso, J. A. S. (1992). Evolution of behavioral attractors with learning: Nonequilibrium phase transitions. *Journal of Experimental Psychology: Human Perception and Performance,* **18,** 403–421.

25

Constraints and Coordination in Whole-Body Actions

Wynne Ashley Lee

*Programs in Physical Therapy
and The Northwestern Institute for Neuroscience
Northwestern University Medical School
Chicago, Illinois*

Aileen Marie Russo

*Programs in Physical Therapy
Northwestern University Medical School
Chicago, Illinois*

I. Introduction
II. Modeling Coordination in Whole-Body Actions
    A. Local Constraint Models
    B. Global Constraint Models
    C. Summary and Desiderata for Future Models
III. Coordination during Pulls Made while Standing
    A. The Stand-and-Pull Task: Paradigm, Issues, Results
    B. Global and Local Coordination during Pulling
    C. Summary
IV. Conclusions
    Appendix
    References

I. Introduction

Coordination can be defined empirically as the patterns of mechanical and neuromuscular events that characterize tasks performed skillfully by a system with many degrees of freedom (DFs). Currently, most scientists who study coordination in biological systems accept Bernstein's premise that biological systems have too many redundant DFs to be easily controlled (Bernstein, 1967). To perform specific actions, the system's DFs must somehow be organized to generate the necessary patterns of perceptual, neural, and biomechanical activity (Bernstein, 1967; Easton, 1972; Gelfand, Gurfinkel, Tsetlin, & Shik, 1971; Turvey, Shaw & Mace, 1978). Consequently, contemporary research on coordination in biological systems is dominated by efforts to understand the principles that underlie the organization of DFs. The overarching question is: How are effective coordinated patterns of whole-body action constrained so that the performer can achieve task goals?

The inherent complexity of this problem has historically made it difficult to address either analytically or empirically. Over the past decade, however, two distinct methods have emerged that are beginning to make the problem of coordination in whole-body actions more tractable. Some scientists have devloped analytical and computer simulation models of actions performed by standing humans. Those models enable researchers to ask how specific constraints determine coordinated action. Other scientists have adopted an empirical approach in which they seek invariant movement or neuromuscular patterns within classes of actions (e.g., locomotion, postural reactions, reaches or pulls made while standing). The observed invariants are used as a basis for making inferences about underlying constraints on the system's DFs. Few research programs on whole-body coordination have merged the analytical and empirical approaches.

The purpose of this chapter is to briefly and selectively review some of the research on whole-body coordination that is based on these two approaches, and to describe some insights these methods are yielding into how the CNS may coordinate actions involving the whole body. It is suggested that combining the analytical and empirical methods will be necessary in future work on this problem. The review focuses on analytical and empirical studies that attempt to identify the local or global biomechanical constraints (defined below) that may underlie organization of whole-body actions performed by standing humans. Comparable research is currently being done on neurophysiological mechanisms by which the CNS achieves coordination (Massion, 1992), but that work is beyond the scope of the present review.

The chapter is divided into two parts. The first part briefly introduces several analytical models that explore how different constraints, or sets of constraints, can account for coordination and goal achievement in whole-body actions performed by standing subjects. These models are important because they enable multidimensional coordinative patterns to be quantitatively analyzed. Some problems with purely analytical approaches are discussed, and desiderata are proposed for future constraint-based models of coordination in whole-body actions. The second part reviews recent research on a task in which freely standing human subjects learn to accurately produce pulling forces against a handle. This research which combines analytical and empirical methods, has begun to elucidate how global and local coordination may be controlled and learned in this act. An analytical model is described that operationally defines global coordination in the task; empirical studies provide at least partial support for that model. Preliminary data will be reported on practice-related changes in analytically defined local (joint) coordination. These data argue against the hypothesis that joint kinematic DFs increase as skill on a multijoint action improves.

II. Modeling Coordination in Whole-Body Actions

There are at least two distinct ways to model how the system's DFs may be constrained to achieve particular goals or coordinative patterns in whole-body actions. These methods differ in their emphasis on what might be termed "local" versus "global" constraints (cf. Droulez & Berthoz, 1986; Massion, 1992). Local constraints relate to joint kinematic and kinetic variables or muscle variables (e.g., stiffness, joint torque, onset order of muscle activations). Global constraints relate to abstract goals of the system, such as maintaining or accelerating the center of mass (CM) to a particular location over the support base, reaching to a point in space, or producing a specified force on an external object. Both approaches share the premise that coordinated kinematic, kinetic, or neuromuscular patterns are not explicitly planned, stored, or the result of direct guidance via some regulatory process. Rather, patterns emerge as the net effect of all constraints that are currently operating on the system's DFs. Adherents of both approaches generally have avoided the inverse dynamics method that has characterized many models of reaching and other upper-limb actions (Flash & Hogan, 1985; Flash & Henis, 1991). Perhaps the whole-body system is so complex that it renders implausible the idea that coordination is achieved by the computation of inverse dynamics.

A. Local Constraint Models

Raibert and his colleagues have developed a particularly ingenious scheme of scheduled local constraints to control locomotion in robotic systems (for review, see Raibert, 1990). Their robots achieve dynamically balanced locomotion by an algorithm that sequentially controls the vertical thrust of the legs during stance, foot placement, and the angle of the upper-body mass to the legs. This work shows elegantly how a "bottom up" system of controlling *local* variables can yield the *global* behavior of maintaining balance during locomotion without explicit reference to or constraint on global variables. This method worked even for different gait patterns and during external perturbations.

Other engineering-based approaches have adopted impedance-based models for controlling joint motion and equilibrium during standing. For example, Barin and Stockwell (1985) postulated that the small perturbations associated with natural, quiet standing might be controlled indirectly through joint stiffness. They estimated joint stiffness matrices for the ankle, knee, hip, and neck joint angle and torque data measured from healthy standing humans, using multivariate regression analyses. The estimated stiffness matrices allowed them to simulate joint motion and overall equilibrium during standing. The simulated results compared quite well with observed motions.

However, none of the local constraint control schemes described above has been extended much beyond the initial conditions under which they were developed. Consequently the value of such models has not yet been demonstrated over a range of actions such as gait initiation, bending over, or reaching forward to pick something up.

Before discussing global constraint-based models of coordination, it is worth noting that neuroscientists have long advanced the idea that the nervous system has "evolved in" low-level (especially spinal) structures that mediate intra- and interlimb coordination patterns. For example, simple and complex reflexes, central pattern generating circuits, and parallel branching pathways from supraspinal structures are believed to govern the organization of electromyographic (EMG) activity in leg, trunk, and neck muscles that control posture, equilibrium, and locomotion in vertebrate and invertebrate species (for reviews, see Easton, 1972; Gallistel, 1980; Grillner, 1981). Such neural constraints are believed to contribute to the generation and modulation of automatic and voluntary actions that are performed while standing. Probably the best-known example of this idea from postural research is the hypothesis that spinally mediated circuitry governs the organization of distinct leg and trunk muscle synergies evoked by perturbing the support surface on which subjects stand (Nashner & Woollacott, 1979). While discussing analytical models of neural constraints is beyond the scope of this chapter, recent studies have demonstrated the plausibility of formally modeling the effects of such constraints in whole-body actions. For example, Ramos and Stark developed a model for stance control that has reciprocal spinal-level inhibitory relationships between agonist and antagonist muscles (Ramos & Stark, 1990a; Ramos & Stark, 1990b). Simulations with this model shed light on the contributions of passive mechanical, feedforward, and stretch reflex feedback processes to maintaining equilibrium during voluntary arm flexion and trunk bending motions. Connectionist network models also can model constraints on multisegmental systems (Bullock & Grossberg, 1988; Jordan, 1990). Such models have not yet been used to control the kinds of whole-body motions that are the focus of this review.

B. Global Constraint Models

The second modeling approach has explored how global constraints contribute to whole-body actions that have goals such as maintaining upright stance, reaching to a target, or a combination of goals. The models described below illustrate some different ways that global constraints might determine effective coordinated movements made by standing subjects.

The first example is from work by Gordon, Zajac, and their colleagues, who have examined how global constraints influence coordination during the maintenance of upright standing in humans (Gordon, 1990; Gordon, Zajac, Khang, & Loan, 1988; Kuo & Zajac, 1992). This group has developed a 4-segment model of stance in the sagittal plane (foot, shank, thigh, head–arms–trunk) with 14 mathematically orthogonal musculotendonous actuators. Without any constraints, this system could generate essentially unlimited combinations of hip, ankle, and knee accelerations. But empirical evidence suggests that standing subjects show restricted patterns of joint motion and EMG activity when balance is perturbed (for review, see Nashner & McCollum, 1985). Rather than directly specifying joint angle motion patterns, Zajac and his colleagues proposed that the system has several global goals, such as maintaining balance. These goals can be mathematically defined as constraints on system movement (e.g., keep the center of pressure within the base of support; maximize the acceleration of the CM toward a point over the axis of ankle rotation). When thus constrained, the physics of the multisegmental system yields three-dimensional "feasible acceleration sets" for ankle, knee, and hip angular accelerations, which are a subset of all the acceleration patterns that the unconstrained system might generate. Specifically, for initial body positions like those observed during normal sway and postural perturbation studies, the simulation predicts a $3:1$ ratio of hip to ankle acceleration when knee acceleration is zero. Although experimental analysis has yet to confirm or refute this predicted coordinative pattern between hip and ankle accelerations, the example shows how a relatively small number of global constraints applied to the modeled musculoskeletal system can synthesize emergent patterns of coordinated joint motion, as well as CM motion.

The second example of a global constraints approach to coordination is represented by Hinton's 2-dimensional simulation of sagittal-plane reaching by a 7-segment, open kinematic chain (1984). His model demonstrated how multiple global constraints could generate trajectories of whole-body motion. The model had one fixed end (the "foot"); the tip of the seventh segment (the "hand") was free to move. The goal of the simulation was to have the system start from any physically realistic initial configuration and have the hand reach toward targets in the 2-dimensional space. An iterative algorithm was adopted in which two global constraints were applied independently to each joint at each discrete time step, yielding a net torque that in turn determined simulated joint motions. The change at each joint's motion on each iteration was a linear function of (1) a torque proportional to the current distance of the hand to the target, and (2) a torque proportional to the current anterior–posterior distance of the CM to the midpoint of the foot. (Local constraints, e.g., diminished torques as each joint reached its physiological limits, also were applied.) Computer simulations showed that this algorithm enabled the tip to reach the desired point, while the CM remained above the base of support. No kinematic or kinetic trajectory plan was required to achieve these goals.

Because Hinton was interested in showing that multiple, parallel constraints applied independently could solve the dual reaching and balance problems, he did not explore interjoint coordination per se. He did, however, find that adding further constraints on relative joint motions of adjacent segments was useful. In particular, the simulation got "stuck" in certain regions of the postural configuration-target space, where a change in one joint angle based on the constraint rules would be cancelled out by an independently determined change in an adjacent joint. This problem was eliminated by postulating another independent set of "synergy" constraints that changed two adjacent joint motions by reciprocally related amounts. The synergy constraints reduced the number of iterations needed to complete the reach. However, because each joint angle change was still determined by the sum of multiple independent constraints, it is unclear if this parallel constraint model would generate highly coordinated joint motions.

Dynamic global constraints are important in the third example. Saltzman and Kelso (1987) sought a general conceptual framework for the organization of skilled actions in systems with multiple DFs. They postulated that end-effector motions in many skilled actions can be characterized in terms of attractor dynamics. For example, trajectories of target-directed actions (reaching to a target; moving the CM to a desired location) often exhibit at least some properties of a point attractor (i.e., if perturbed, the CM moves toward a point in space), while rhythmic actions have trajectories that can be characterized by simple periodic or limit-cycle attractors (i.e., trajectories are cyclic and, for a limit cycle, return to the cycle after perturbations). Those resemblances led Saltzman and Kelso to argue that organizing the system's DFs is tantamount to establishing the appropriate dynamic equations of constraint at the task (global) level. Simulations showed that correctly selected and appropriately parameterized equations can, indeed, generate endeffector trajectories similar to those that have been observed empirically. (They noted, however, that selecting and subsequently parameterizing the dynamic equations are nontrivial problems.)

The salience of Saltzman and Kelso's model for joint coordination lies in the fact that the global, task-level dynamic equations and parameters constrain the "articulator" equations that govern joint motion. Like Hinton, Saltzman and Kelso focused on developing a principle that would enable the segmental DFs to be constrained to achieve global goals. Consequently, they did not explore whether the model generated joint kinematic or kinetic patterns that resemble those observed empirically. Such comparisons are needed, because motor equivalence is a hallmark of redundant systems. One cannot assume that simulated interjoint movement patterns will be biologically realistic simply because the trajectory of the endeffector of the system is modeled with reasonable accuracy.

Saltzman and Kelso's (1987) approach differs from those of Zajak, Gordon and colleagues (Gordon, 1990; Gordon et al., 1988; Kuo & Zajac, 1992) or Hinton (1984) who did not require their equations of constraint to yield trajectories with specific dynamic properties. Nonetheless, all three approaches share the premise that global constraints that are directly related to task goals provide a basis for coordination in whole-body actions. Finally, it should be noted that the hypothesis that flexibly imposed global constraints may underlie different patterns of movement coordination has its parallel in the neuroscience literature on postural control (see Massion, 1992, for review).

C. Summary and Desiderata for Future Models

Local and global constraint-based models provide ways to probe how different kinds of constraints might lead to the emergence of coordinated whole-body actions. These two approaches are not, of course, mutually exclusive. Indeed, the models by Hinton (1984) and by Gordon et al. (1988) included local as well as global constraints, although the above simplified introduction to the models emphasized the latter. Combining local and global constraints is consistent with neurophysiological data, which emphasize that coordination depends on both lower and higher level neural processes. Moreover, recently developed connectionist models of multisegmental actions also are hybrids with multiple levels of parallel constraints (e.g., Jordan, 1990). As more is learned about whole-body control and coordination in a range of tasks, it seems probable that hybrid models will be used with increasing frequency.

As promising as these constraint-based models are, it is unclear whether they will support the development of principled models that yield testable predictions about whole-body coordination under a range of conditions. This concern is not trivial, since optimization approaches to modeling of systems with redundant neuromuscular DFs have revealed that many optimization criteria (which act as system constraints) can yield comparable patterns of muscle activation. Similar problems could plague constraint-based models of whole-body coordination. Perhaps, then, it would be useful to lay out some desiderata for a reasonably comprehensive constraint-based model of whole-body action.

First, it would be useful to have a model that, when constrained in different ways, could account for global and local coordination in the four uniplanar "postural" tasks that have been studied extensively. These are tasks in which standing subjects (1) maintain balance after support-surface perturbations; (2) point or reach to targets with an upper or lower extremity; (3) adopt a new base of support by standing up from a chair; or (4) exert forces on objects in the environment, for example, pull on a handle. Second, it would be useful for the model to predict how specific aspects of coordination (e.g., spatial, timing, and amplitude relationships among system DFs) would vary with the direction, force, and speed of the perturbation or voluntary act. Finally, it would be advantageous to develop a model that could simulate how coordination might change with development, skill acquisition, or selected neural or musculoskeletal pathologies.

The call for a general model of whole-body coordination might seem foolhardy, given how many actions humans can perform (even assuming uniplanar motion). However, such a goal can be justified in two ways. First, if the nervous system coordinates multijoint actions via constraints, as speculated, then it should be theoretically possible to identify sets of constraints that define different tasks and then analytically impose those constraints on a fuller model of the musculoskeletal system. Variations in coordination would then "fall out" of the DFs being subject to different task constraints. For example, a reasonably full model for sagittal-plane pulling motions (e.g., the stand-and-pull task described in Section III,A) would have at least five segments (forearm, upper arm, trunk, thigh, shank) and permit contact forces at the wrist as well as the ankle. That model could be adapted to other sagittal-plane tasks. Consider, for example, the well-studied task of maintaining stance against perturbations induced by support surface displacements. In that task, subjects are told to keep their arms crossed, which means that the arms' motions relative to the trunk must equal zero. Moreover, no contact forces occur at the wrist. Under these two additional constraints, the "full" model is reduced to a form specific for the perturbation task. This approach is tacitly adopted in biomechanical models of stance that treat the head, arms, and trunk as one segment—except that the full model does not exist! The limitation to the constraints approach thus may lie *not* in the fact that the system behaves differently in various tasks, but rather in the problem of developing tractable but relatively realistic models of the musculoskeletal system and task constraints.

The second justification is more philosophical, namely that science often seeks general, not condition-specific, explanations of phenomena. Determining which phenomena and analyses are essential to modeling coordination in whole-body motions is admittedly a serious problem, given the current state of scientific knowledge. Nonetheless, it does not seem unreasonable to suggest that a legitimate, if ambitious, *long-term* goal of research in this area is the development of a general model of whole-body coordination.

A practical hindrance to developing a general model of whole-body coordination has been the lack of parallel empirical research that tests model predications. Researchers who develop analytical models inevitably must make arbitrary assumptions and simplifications. Without empirical testing, models remain merely interesting speculations. Similarly, empiricists can now record more biomechanical data than can be readily interpreted; and some questions (e.g., the relative contributions of passive mechanical and active neural processes to control) are difficult, if not impossible, to address purely through empirical studies, without some guiding analytical or theoretical framework.

Although the advantages of combining analytical and empirical studies are fairly evident, there are still relatively few instances where a combined approach has been used to understand whole-body coordination. An example is Zajac *et al.*'s (1988) model, which was designed to explore questions about "postural synergies" that are elicited by sudden motion of the support surface. The data on human postural reactions reported by Nashner and colleagues (for review, see Nashner & McCollum, 1985) provide some *post hoc* tests of the model.

Section III of this chapter reviews another such research effort, the goal of which has been to understand coordination in a task in which freely standing subjects learn to produce voluntary pulling forces against a handle. That work, although in its initial stages, has provided insights into what the relevant global DFs and coordination are for this task, and has permitted preliminary tests of hypotheses about how local coordination changes with practice.

III. Coordination during Pulls Made while Standing

Over the past few years, we have begun to study global and local coordination during a whole-body task that involves motion of the arms, trunk, and lower extremities (Lee, 1992; Lee, Michaels, & Pai, 1990a; Lee, Michaels, & Pai, 1990b; Lee & Rogers, 1987; Michaels & Lee, 1993; Michaels, Lee, & Pai, 1993). Specifically, we have examined how freely standing adult humans produce, and learn to produce, accurate impulse-like peak forces on a handle without losing their balance (the "stand-and-pull" task; Figure 1A,B).

Section III,A briefly describes the basic paradigm and results of some empirical studies that sought to determine how coordinative patterns change with task conditions. Section III,B summarizes modeling and empirical results that show that subjects learn a global coordinative function that links one of the body's kinematic DFs, CM_{AP}, to pulling force. That function, which is qualitatively invariant across subjects, characterizes whole-body motions well for all but very low-force pulls. Section III,C reports preliminary analyses that suggest that local coordination among lower extremity joints is much more variable than global coordination. The results also provide preliminary evidence against the hypothesis that the learning of multijoint actions entails an increase in the number of joint kinematic DFs (Bernstein, 1967).

A. The Stand-and-Pull Task: Paradigm, Issues, Results

1. Experimental Paradigm

The basic experimental paradigm is as follows. Subjects are asked to make single, brief, bilateral pulls (mean time to peak force <200 msec) in the posterior direction when presented with visual targets that represent forces from 5 to 95% of their maximum (Lee et al., 1990a) or estimated maximum pulling force (Lee, 1992). The pulls are bilaterally symmetrical and so can be modeled in the sagittal plane. Subjects initiate the pulls under self-paced conditions, after meeting two criteria: (1) starting in the same quasi-static posture with the center of mass (estimated by center of pressure, recorded by a force plate) in a specific position; and (2) exerting no more than 1.00 N on the handle. The target force is always predictable. EMG activity in the leg muscles, ground reaction forces, pulling force, and coordinates of joint motion are recorded. Joint-angle displacements and center of mass coordinates in the anterior–posterior and vertical dimensions (CM_{AP} and CM_V are computed using the joint-motion coordinates that are recorded from a WATSMART™ system. Data were typically recorded from 5–10 trials in each condition. Figure 2 shows representative trajectories for CM and leg joint angle variables from one well-practiced subject performing a 55% pull.

2. Are Patterns Invariant?

The initial question underlying these studies was, How do patterns of leg-muscle activity, ankle torque, and joint motion change with varia-

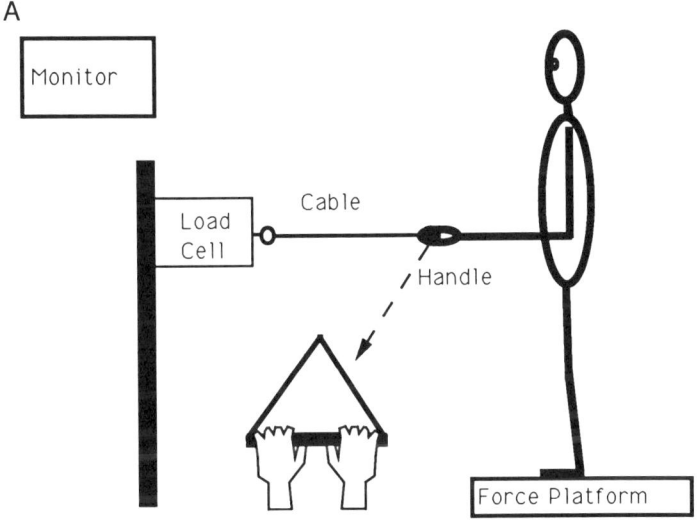

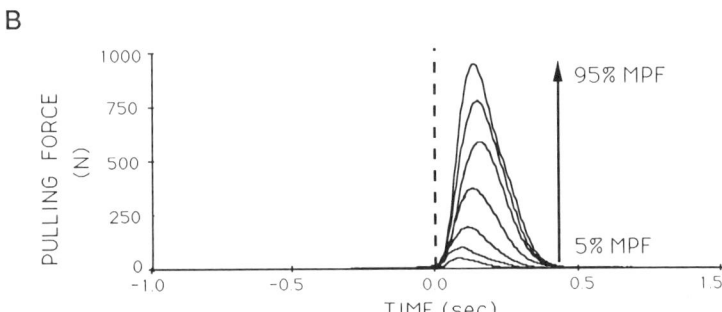

FIGURE 1 (A) The initial configuration of a person in the stand-and-pull task. An inelastic cable connects the handle to a load cell, which is fixed to an immobile support structure. Prior to, during, and after the pull, the subject can move her or his body, providing the feet stay flat on the floor. The arms are supposed to be held parallel to the floor. (Adapted from Michaels *et al.*, 1993.) (B) Representative 5–95% pulling forces, overlaid and time locked to force onset, for one subject. (Adapted from Lee *et al.*, 1990a.)

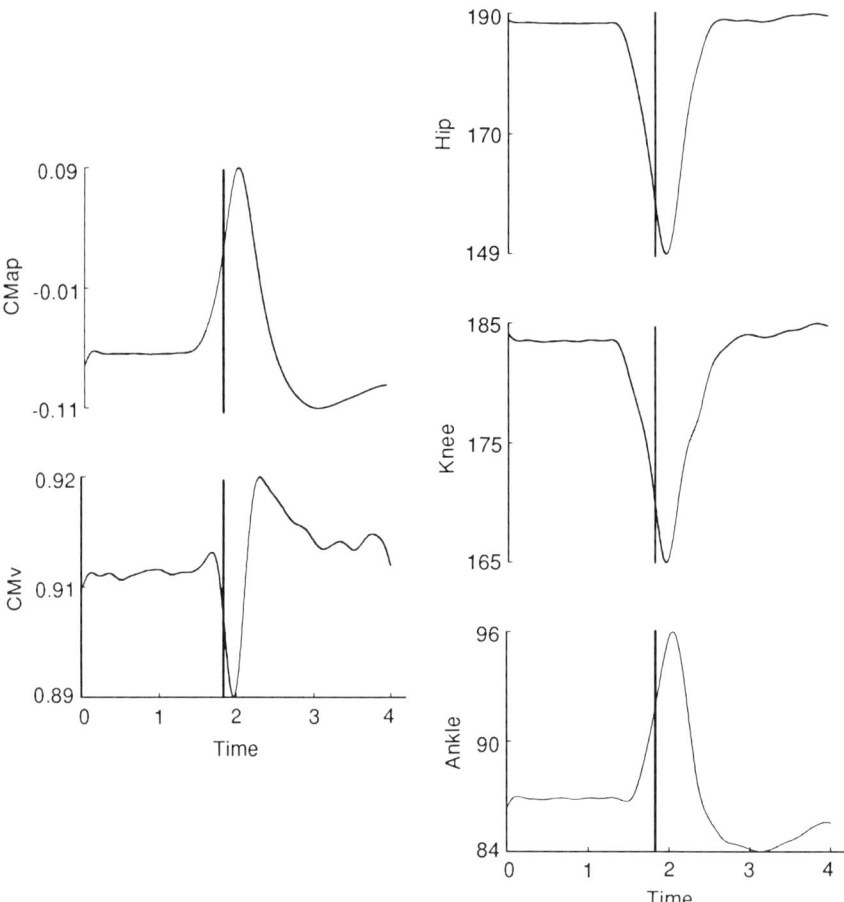

FIGURE 2 Anterior–posterior and vertical CM (CM_{AP}, CM_V, in meters), and hip, knee, and ankle joint angle (in degrees) time histories from one subject making a pull to 55% of maximum. Positive CM values represent posterior and upward motion. More positive joint-angle values represent joint extension. The vertical lines indicate the onset of pulling force.

tions in task parameters such as peak pulling force or the length of the base of support? Each study tested two alternative hypotheses. One hypothesis was that subjects would use the same qualitative patterns (e.g., the order of EMG onset times, ratios of torque impulses, or slopes of angle–angle plots) despite differences in force or base length. Such invariance could reflect an underlying constraint on the systems DFs. The predictions were based on the idea that movements of a given class share the same form of coordination and need only have their parameters scaled to produce actions with different durations, magnitudes, or directions (Bernstein, 1967; Schmidt, 1975).

The alternative hypothesis was that coordinative patterns might change qualitatively (in shape or structure) as a function of peak pulling force or the length of the support base. This hypothesis was motivated by observations that biomechanical or EMG patterns change with some scalar property of other actions performed by standing subjects. For example, altered coordination has been observed during trunk-bending movements made at slow and rapid speeds (Crenna, Frigo, Massion & Pedotti, 1987). Similarly, postural reactions that are elicited in subjects by support-surface motion have different forms of coordination when the length of the surface is varied (Horak & Nashner, 1986). Finally, many systems have essential nonlinearities that lead to different stable states for different input or "control parameter" domains (Kugler, Kelso, & Turvey, 1980; Scholz & Kelso, 1989; Schöner & Kelso, 1988). If pulling force were such a control parameter, then coordination might change qualitatively when force is increased beyond some critical value.

3. Empirical Results

In one study, subjects made voluntary maximal pulls on a handle while standing on support surfaces that ranged from 20 to 100% of foot length (Lee & Rogers, 1987). These subjects only quantitatively altered their patterns of hip-angle motion and muscle activation when the support base length was changed (Figure 3). Similarly, monotonic relationships were observed between pulling force and the onset times and amplitudes of EMG activity of leg muscles for pulls with peak forces from 5 to 95% of maximum (Lee et al., 1990b).

Although these results might suggest that a single constraint governed the organization of EMG during pulling, qualitative shifts *did* occur in the ankle torque impulses (Lee et al., 1990a) and in relative motions of the ankle, knee, and hip joints for 5–95% pulls (Lee et al., 1990b). Subjects maintained approximately static postures while producing low forces (5–10% of maximum), used "inverted pendulum" rotations about the ankle for 20–40% forces, and used patterns with simultaneous hip, ankle, and often knee-joint motions for forces above 40% (Figure 4). At the higher forces, two of six subjects used a "jackknife" pattern that was roughly comparable to the hip strategy pattern noted by Horak and Nashner during postural reactions (1986). The other four subjects adopted different joint-angle patterns. The switch from the pendulum pattern to movements with considerable hip flexion resembles the switch in joint-angle patterns reported by Horak and Nashner when the support surface was varied (1986). In contrast to

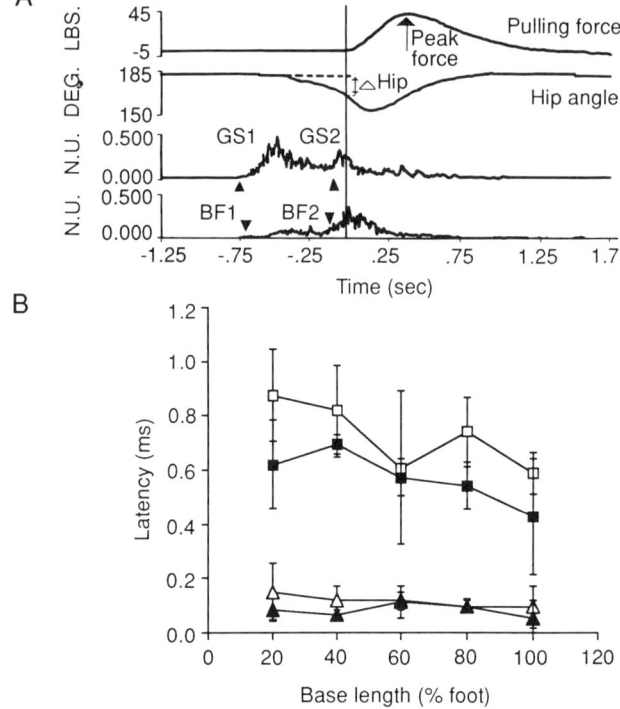

FIGURE 3 (A) A single trial of pulling force, hip-flexion angle, and EMG activity in the medial gastrocnemius (GS) and biceps femoris (BF) muscles for a maximal pull made while standing on a support base equal to foot length. Amplitudes are normalized to maximal values for each muscle (N.U., normalized units). Two distinct bursts of GS and BF EMG activity precede the pull. This two-burst pattern was typical of most subjects. (B) Group means and standard deviations ($N = 9$) of onsets of the earlier (GS1, BF1) and later (GS2, BF2) EMG bursts versus the length of the support base (100% = foot length). ■, ▲, GS; □, △, BF; ■, □, first EMG burst; ▲, △, later EMG burst. Time zero indicates the onset of the pull.

those qualitative variations in joint coordination, CM_{AP} phase portraits (displacement-velocity plots) had the *same* shape as pulling force increased for all subjects (Figure 5).

4. Summary

The above results suggest that the answer to the seemingly simple question, "Do qualitatively invariant coordinative patterns characterize whole-body motions?" depended greatly on which variables were analyzed. Similar results have been observed in human locomotion, where individual joint motions vary considerably across steps (even within subjects), while net support moments are much less variable (Winter, 1984). Such observations led us to analyze the stand-and-pull task and its kinematic DFs more carefully and to postulate a simple dynamic model of the constraining relationship between the global variable, CM_{AP}, and the primary outcome variable, peak pulling force. The results of those analyses are reviewed below (for details, see Michaels et. al., 1993).

B. Global and Local Coordination during Pulling

The body-segment system during the stand-and-pull task has fewer kinematic DFs than it might initially seem, given that the subject's hand, forearm, upper arm, trunk, thigh, shank, and foot segments all *might* move. When all physiological and experiment-specific task constraints are considered, it becomes clear that the system is a closed, 4-link kinematic chain with only two kinematic DFs. We chose the coordinates of CM (CM_{AP}, CM_V) to represent those DFs.[1]

This choice was governed by our interest in *explicitly* linking the task's two potentially conflicting global goals, the production of pulls with particular peak forces and the maintenance of equilibrium before and after the pull. Most whole-body tasks performed by standing subjects require that the CM in the horizontal plane be either regulated, that is, held constant (Bouisset & Zattara, 1987; Crenna et al., 1987; Friedli et al., 1988; Pedotti, Crenna, Deat, Frigo, & Massion, 1989) or controlled, that is, shifted deliberately from one quasi-static state to another (Mouchino, Aurenty, Massion & Pedotti, 1992; Pai & Rogers, 1990; Rogers & Pai, 1990). In general, both static and dynamic equilibrium in the sagittal plane are related directly to the anterior–posterior trajectory of the CM. Choice of CM coordinate space rather than joint-angle space to represent the system's two kinematic DFs thus ensured that the model of global coordination between body motion and pulling force would include a variable related to equilibrium control. Such a model is described below.

1. Global Coordination

Equations of motion show that CM_{AP} motion contributes directly to producing horizontal pulling force, with CM_V motion contributing much less. The equations demonstrate that the CM_{AP} and pulling force components of the stand-and-pull task are mutually dependent. With the exception of low pulling forces that could be developed isometrically, movement of the CM_{AP} is *essential* to the development of pulling force, that is, pulling force is not merely a perturbation of some constant reference value of CM_{AP} that must be corrected by "postural adjustments."

---

[1] Two joint angles also could have been used, because joint configuration maps uniquely to CM coordinates in this task.

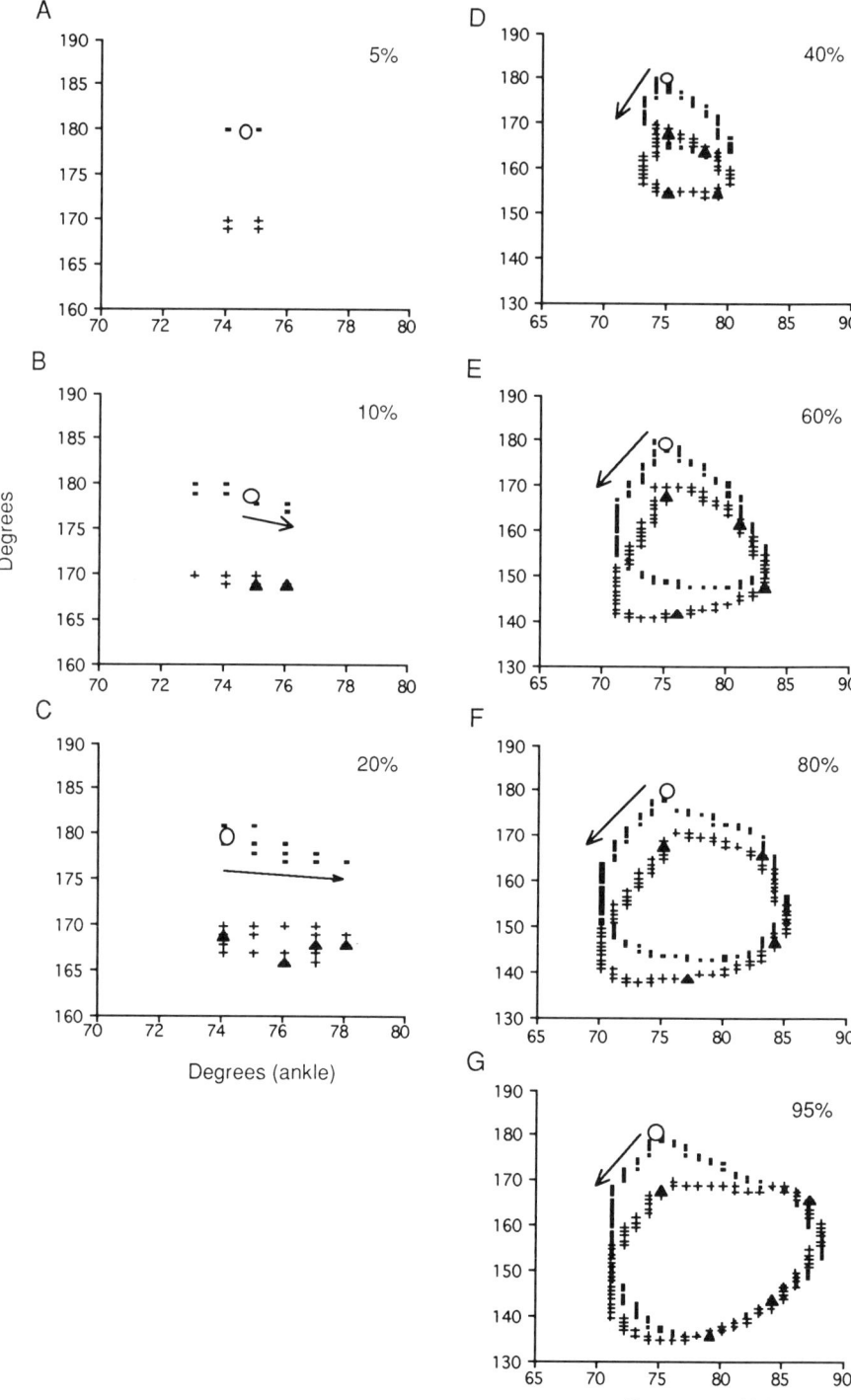

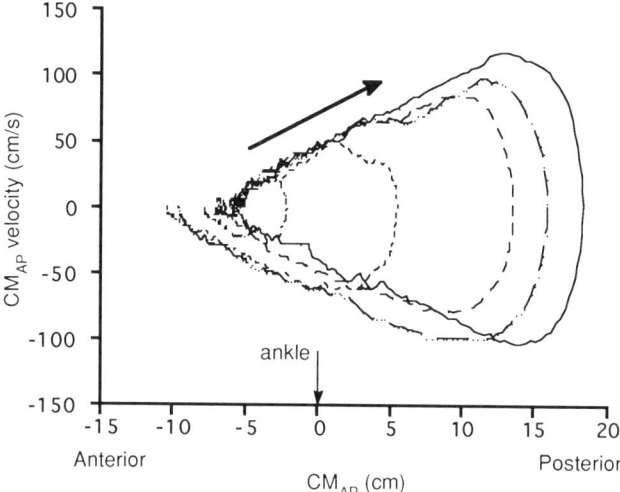

FIGURE 5 Overlaid CM_{AP} phase planes for 20–95% pulls for a representative subject. Phase plot size increases consistently from the smallest (20%) to largest (95%) pulling forces. Positive values indicate posterior displacements and velocities. CM_{AP} motion begins at approximately −6 cm for all traces and proceeds clockwise, in the direction of the arrow. Zero on the x-axis represents the midpoint of the lateral malleolus; the most posterior edge of the foot is about 2 cm posterior to that point. Note that CM_{AP} displacement on pulls greater than 40% moves behind the back edge of the foot. (Adapted from Lee et al., 1990b.)

While the equations of motion define the basic kinetics of the task, they do not elucidate the particular dynamic structure of the CM_{AP} trajectories that precede, accompany, and follow the pull. The lenticular shapes of the CM_{AP} phase planes (Figure 5) and their approximate symmetry about the zero-velocity axis suggested that the relationship between CM_{AP} motion and pulling force might be modeled as a simple dynamic system (Michaels et al., 1993). The model, which was developed in the spirit of Saltzman and Kelso's task dynamics framework (1987), also has much in common with other lumped models of whole-body actions such as postural reactions, running, hopping, and simulated ski-slalom motions (Greene & McMahon, 1979; Johansson, Mag-

---

FIGURE 4 Hip–ankle (■) and knee–ankle (+) angle–angle plots for 5, 10, 20, 40, 60, 80, and 95% pulls for one subject. Motions start at the large open circle and progress counterclockwise, in the direction of the arrow. Triangles on knee-angle plots indicate the times of movement onset, onset of pulling force, peak pulling force, and force offset. Note the different x- and y-axis scales for smaller (left column) and larger (right column) pulls. (Adapted from Lee et al., 1990b.)

nusson, & Akesson, 1988; McMahon & Cheng, 1990; McMahon & Greene, 1979; Vereijken, Emmerik, Whiting, & Newell, 1992).

a. Dynamic Model We modeled the relationship between pulling force and CM_{AP} motion as an inverted pendulum,[2] with a point mass at the top and the bottom end attached to the axis of a pulley (Figure 6). An initially slack elastic cord (rather like a bungie cord) is attached between the point mass and a fixed object in the environment, at the height of the mass. Motion of the pendulum is determined by gravitational force, a torque associated with a weight applied through the moment arm of the pulley, and an elastic force applied to the mass whenever the slack in the cord has been taken out. We assumed that the weight applied to the pulley would always equal the person's weight, so pulley torque depends solely upon its moment arm, that is, the pulley radius. The elastic force of the cord depends on its stiffness and the distance that the CM_{AP} must move backward to remove the slack. The slack term represents a threshold nonlinearity in the equations: the force due to the cord is applied to the pendulum mass only when CM_{AP} exceeds the slack term. The CM_{AP} trajectory is entirely determined once the pulley radius, cord stiffness, and cord slack are set.

The behavior of the model can be described by a second-order differential equation, with a threshold nonlinearity due to the slack term. That equation represents a global constraint, or *coordinative function*[3] between pulling force and CM_{AP} motion. If the hypothesized coordinative function provides a good estimate of the global constraint, then it should produce simulated CM_{AP} trajectories that match those that are observed empirically. As noted above, this coordinative function implies that control of the "postural" (CM_{AP} motion) and "focal" (pulling force) components of the task are not mechanically independent. This model differs from Hinton's approach, in which goals related to postural and focal terms were independent constraints on joint motions. The absence of any optimized cost function on the CM_{AP} term distinguishes this dynamic model from that used by Gordon and co-workers (1988, 1990).

b. Fit of Model to Data We have asked three sets of questions about the adequacy and utility of this model of global coordination. The first and most crucial question was: How well does the model fit the data for

---

[2] This is a *virtual* pendulum; we are explicitly not suggesting that the body as a whole be modeled as an inverted pendulum.

[3] We use the term *coordinative function* to differentiate it from *coordinative structure*, a term that Easton (1972) used to designate the neural circuitry that mediates spinal reflex patterns, and that Turvey (1977) and others have used to describe muscles that are constrained to act as a unit.

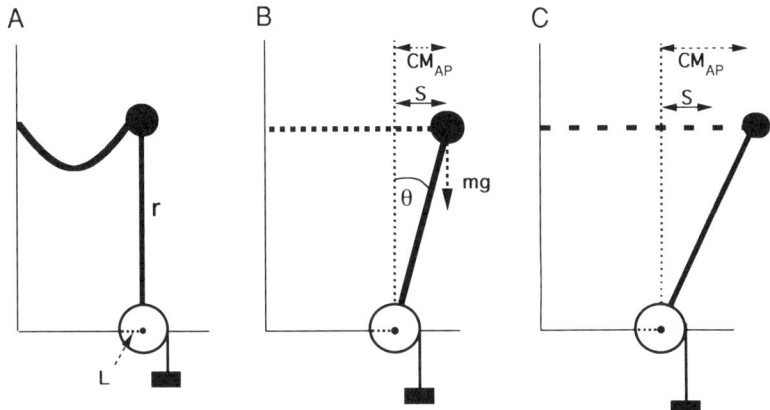

FIGURE 6 Inverted pendulum model of the relationship between horizontal pulling force and CM_{AP} motion. The cord elasticity (K), pulley radius (L) and slack in the cord (S) plus gravity determine the force on the pendulum mass. The cord exerts force on the pendulum only when $CM_{AP} \geq S$. The K, L, and S parameters are assumed to be set at the start of a trial; their values determine both the dynamic CM_{AP} trajectory and peak pulling force. (Adapted from Michaels et al., 1993.)

pulls of different peak force in well-practiced subjects? The second set of questions was: Which parameters change, and how do they change, with pulling force? The third set of questions relates to learning of the pulling task, specifically: Do subjects show evidence of learning the hypothesized relationship between pulling force and CM_{AP} motion, or, do they simply learn to parameterize the equation more appropriately and consistently after practice?

The answer to the first question is that for well-practiced subjects, the model accounts reasonably well for CM_{AP} trajectories during pulls to force targets $\geq 20\%$ of maximum. The least-squares fits of the model to the CM_{AP} trajectory data were high for near maximal pulls (coefficient of determination, $R^2 > .98$; Michaels et al., 1993). Fits of the model for moderate pulls (20–80%) were somewhat lower ($R^2 \geq .92$; median $R^2 = .98$), and fits often were poor for pulls less than 20% of maximum (Michaels & Lee, 1993). In general, as peak pulling force increased, the slack term increased linearly for all subjects, changes in stiffness were small but inconsistent, and the pulley radius term changed little. Computer simulations suggest that low forces could be generated just by varying pulley radius, which would represent a special case of the model. Such results are consistent with reports that subjects can support static loads up to about 15% of their body weight without any change in posture (Lee, Deming and Sahgal, 1988). They also suggest that independent regulation of CM_{AP} position (equivalent to the slack term in the model) and pulling force is feasible for low force pulls.[4]

A second study evaluated how CM_{AP} trajectories change when subjects learn to pull accurately to 25, 40, and 55% of their estimated maximum (Lee, 1992). CM_{AP} trajectories and force accuracy were evaluated for the first and last ten attempts at each force target (total of 90 trials per target, spread over 3 days). The fit of the model to the CM_{AP} data improved significantly from the beginning to the end of practice, from a median $R^2 = .81$ on Day 1 to a median $R^2 = .96$ on Day 3. The improvements were most dramatic for the 25% pulls, probably because of a ceiling effect on model fit for the other two forces: even on Day 1, fits between the predicted and observed CM_{AP} accelerations were high ($R^2 \geq .85$) for the 40 and 55% pulls. Stiffness decreased significantly with practice. Variability in the slack parameter also decreased with practice, in parallel to decreased variability in peak-force error. Slack increased linearly with pulling force, as had been seen in the earlier study (Michaels & Lee, 1993). Practice-related changes were not, however, observed in the pulley radius parameter. It remains to be determined whether the improvements in the fit of the model and model parameters can be sustained over time, thereby reflecting genuine learning. However, studies of trunk bending and leg lifts have shown that subjects with and without gymnastics or dance training differ in their abilities to perform those tasks, suggesting that learning does occur in whole-body tasks that require the coordination of equilibrium and focal control processes (Pedotti *et al.*, 1989; Mouchino *et al.*, 1992).

We interpret these data as follows. First, subjects showed evidence of learning the coordinative function between pulling force and CM_{AP} motion, as reflected in the improved fit of the model to the data on Day 3. This is consistent with the hypothesis that global coordinative functions are developed during learning. Second, the decreased stiffness is consistent with Vereijken *et al.*'s report (1992) that subjects' simulated ski-slalom motions became more compliant with practice, supporting the hypothesis that subjects become less stiff as they learn motor tasks (Bernstein, 1967). The lack of practice-related changes in pulley radius might be due to subjects maximizing that variable, either voluntarily or because the magnitude of the pulley radius is constrained physically by the length of the feet.

c. Limitations of the Dynamic Model The model captures only the coordinative function that governs CM_{AP} motion. It does not fully de-

---

[4] Low forces also could be generated by pulling with the arms, with little or no motion of the CM_{AP}. The movement analysis system lacked the resolution necessary to detect such small movements. Current studies with a more accurate system will be able to resolve this question.

scribe whole-body movement patterns or how they change with practice. The coordinative function(s) that may govern CM_V motion remain to be determined. Preliminary examination of CM_V trajectories and the relationship between CM_{AP} and CM_V motions suggests that some of the most interesting practice-related changes may occur in motions of the CM_V, the less task-constrained DF during pulling. As noted above, CM_V phase planes are much more variable than CM_{AP} phase planes, so it may be harder to develop one coordinative function (or set of functions) for CM_V, especially if individuals impose different implicit constraints on CM_V motion.

Finally, joint configuration is codetermined by CM_{AP} and CM_V. Local joint coordination patterns thus could simply emerge as a consequence of coupled coordinative functions over CM_{AP} and CM_V. However, more direct approaches also can be taken to describing joint coordination and how it changes with task conditions, which do not require that global coordinative functions be defined. One such approach is described below. We used that approach to address questions about how joint coordination changed with practice.

2. Local Coordination in the Lower Extremity

a. Conceptual Issues We seek to answer three sets of questions about local (interjoint) coordination in the stand-and-pull task. The first set of questions relates to system order (number of joint kinematic DFs). Specifically, how many joint motions are functionally related, and what are the structure and the fit of those functions? Are the order, structure, and fit of the functions constant for all subjects, practice levels, and pulling force? The questions about changes in system order are of particular theoretical importance, because of the long-standing hypothesis that novices "freeze" most DFs of the system, whereas skilled performers use all, or at least more, DFs (Bernstein, 1967; Fowler & Turvey, 1978). The implication is that the dimensionality of the system *increases* with practice. This hypothesis has received empirical support for several multisegmental tasks (e.g., Newell & van Emmerik, 1989; Newell, van Emmerik, & McDonald, 1989; Vereijken et al., 1992). The converse hypothesis that new constraints over DFs emerge with practice, resulting in *decreased* system dimensionality, also has been proposed and has received some empirical support (Arutyunyan, Gurfinkel, & Mirskii, 1968; Arutyunyan, Gurfinkel, & Mirskii, 1969; Mouchino et al., 1992). A third alternative, namely that the *structure* but not the *number* of constraints may change with practice, has received little attention. As will be seen below, this last possibility characterized some subjects' practice-related changes in joint coordination during the stand-and-pull task.

The second set of questions concerns how the parameters of joint coordinative functions change with conditions. Some changes in joint coordination may be due simply to altered parameters of the function that links joint motions. Indeed, some parametric changes (e.g., in the sign of the slope of a linear relationship between two joint motions) might yield distinct motion patterns that could be interpreted incorrectly as qualitative alterations in the structure of the coordinative function.

Third, empirically determining system order provides a cross-check on the claim that subjects' sagittal-plane body movements in the stand-and-pull task have only two kinematic DFs (Michaels et al., 1993). That claim was based on several assumptions, including the one that subjects adhere to all "soft" instructional constraints. That might not be true, especially when subjects are initially learning the task. If the claim is correct, then motions of at least one pair of leg joint angles (e.g., hip–ankle, hip–knee, ankle–hip) should be linearly related.

The following sections present preliminary results on lower extremity joint coordination, which address these three questions for the stand-and-pull task.

b. Methods Time series analysis[5] provides one way to evaluate how strongly pairs of signals during discrete movements are linearly related, making minimal *a priori* assumptions about the form of the signals (Box & Jenkins, 1976; Chatfield, 1989). This approach, combined with frequency analysis, has recently been used in charting changes in lower extremity joint coordination during sit-to-stand movements made by stroke patients during the course of their rehabilitation (Ada, O'Dwyer, & Neilson, 1993). Below, we present initial results of a time series analysis of the relationships between pairs of lower extremity joints, for a single pulling force (55% of maximum) made on the first and third day of practice in the experiment described above (Lee, 1992). The primary purpose of presenting these data is to illustrate how this method can help answer questions about interjoint coordinative functions, rather than to fully describe joint coordination during pulling. The example will be restricted to coordination between pairs of ankle, knee, and hip angular displacements.

Briefly, the method consists of estimating the input–ouput relationship between one (or more) input and one (or more) output from sets of empirical time series data. The analysis depends on the inherent form

---

[5] Several important assumptions must be met for time series analysis to be valid, especially with data from discrete actions that typically have trajectories that are not stationary. For an overview of these cautions, see Chatfield (1989); for more extensive treatment, see Box and Jenkins (1976).

of the data, rather than on particular assumptions about its frequency content. Time series methods generally are used to evaluate causal relationships among signals, but we have used them in a purely descriptive manner, since there is no reason to assume that motion at one joint causes motion at another. The method is used to obtain the best fitting and simplest autoregressive linear model between input and output signals. The end result is an equation that relates the input(s) to the output(s), with a defined system lag, linear parameters on model terms, and a measure of the fit of the model to the data. That measure is called the variance accounted for (VAF), which is broadly comparable to the coefficient of determination (R^2) in standard linear regression. We estimated models of the pairwise input–output relationships between hip and knee, ankle and knee, and hip and ankle angular displacements for 55% pulls performed by each subject on the first and third days of practice (for details, see Appendix).

We selected a minimal mean VAF of .90 as the criterion for a sufficiently strong linear relationship between a pair of joints to argue that they were acting as a single DF. Unlike standard regression models, guidelines for judging the statistical likelihood of autoregressive time series models are not clear-cut or wholly agreed upon. Since in standard regression analysis, fits lower than .90 are often statistically significant and considered "high," our choice of a minimal VAF criterion of .90 needs some justification. Our choice was governed by comparing predicted and actual angular trajectories for the "output" of the joint system. Figure 7 presents sample comparisons of three trials with VAFs of .80, .90, and .99. As can be seen, even VAFs of .80 yield only fair to poor agreement between predicted and observed values (Figure 7A); .90 VAFs yield better results (Figure 7B). Figure 7C shows the match for a trial in which the VAF equaled .99.

c. Results VAFs were used to answer questions about the number of DFs of the system and about the functions that linked joint motions on Day 1 and Day 3 (Table I). Four subjects showed one or two joint DFs (2 or 1 tightly coupled joint motions) on both days. This result is consistent with the claim that the system maximally has two DFs. One subject had no tightly coupled joints on Day 1 (i.e., the movement had three DFs), suggesting that she may not have adhered to all instructional constraints on Day 1. Three subjects (DL03, EK04, EB05) did not change their number of DFs with practice; on both days, the VAFs reveal one pair of coupled joints. Subject MC01 *increased* the number of coupled joints from none to one, for a decrease from three to two DFs. The remaining subject (BM02) *decreased* the number of coupled joints from two to one, for an increase in the DFs. While these findings must be regarded as preliminary given the small sample size and subjects'

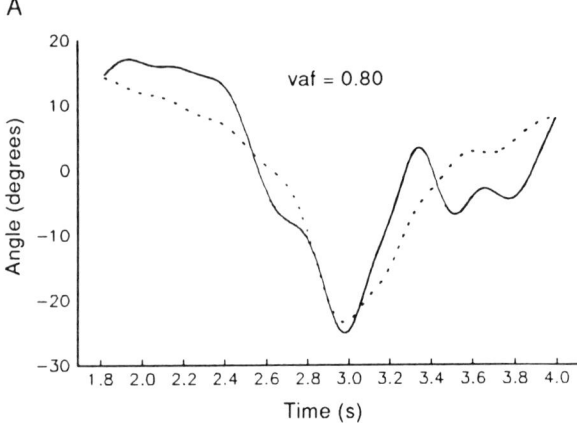
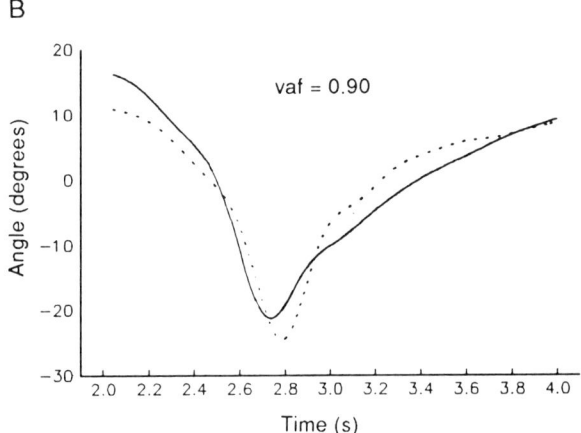
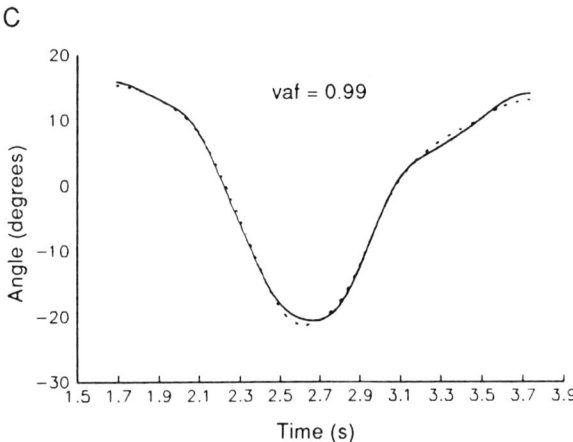

TABLE I VAFs for Each Joint Pair on Day 1 and Day 3

Subject		Hip–Knee		Ankle–Knee		Hip–Ankle	
		Day 1	Day 3	Day 1	Day 3	Day 1	Day 3
MC01	M[a]	0.77	0.93[b]	0.73	0.93[b]	0.82	0.66
	SD	0.19	0.02	0.17	0.02	0.07	0.12
	B1		0.07		170.21		
	B2		1.08		−167.78		
	Lag		65.00		10.00		
BM02	M	0.99[b]	0.98[b]	0.75	0.34	0.90[b]	0.58
	SD	0.01	0.01	0.15	0.18	0.07	0.21
	B1	2.59	2.32			−6.97	
	B2	−2.04	−1.59			6.77	
	Lag	20.00	0.00			120.00	
DL03	M	0.69	0.42	0.71	0.46	0.92[b]	0.93[b]
	SD	0.15	0.12	0.14	0.15	0.10	0.03
	B1					−5.67	−2.81
	B2					4.53	1.79
	Lag					2.50	0.00
EK04	M	0.96[b]	0.84	0.47	0.77	0.68	0.94[b]
	SD	0.07	0.15	0.33	0.18	0.35	0.06
	B1	2.67					−3.01
	B2	−2.05					2.58
	Lag	2.50					20.00
EB05	M	0.73	0.75	0.66	0.72	0.97[b]	0.97[b]
	SD	0.22	0.14	0.24	0.12	0.02	0.02
	B1					−11.56	−5.41
	B2					10.85	4.44
	Lag					0.00	7.50

[a] M, Mean; SD, standard deviation; B1 and B2, model parameters for joint pairs with VAF ≥ .90 (means are in dimensionless units); Lag, median lag (in msec) for joint pairs with VAF ≥ .90.
[b] Joint pair systems with mean VAF ≥ .90.

relatively limited amount of practice, it is interesting to note that most subjects showed *no* change in the number of DFs over three days of practice, in contrast to the common predictions that system order should either increase or decrease with practice.

FIGURE 7 Comparison of empirically measured (solid line) and simulated output (dotted line) joint angular trajectories for trials with time series models having unsatisfactory (.80), minimally acceptable (.90), and excellent (.99) VAFs. The simulated angles are derived from the estimated time series model between one angle (which served as the "input") and a second angle (which served as the "output").

The VAFs along with the estimated model parameters (B1 and B2)[6] reveal the *structure* of the joint coordinative functions observed in subjects on different days. Two subjects (DL03, EB05) used the same constraint on both days. That constraint on ankle and hip angular motion resulted in a "jackknife" pattern in which the ankle extended (plantarflexed) while the hip flexed (cf. Figure 8A EB05; see also Michaels et al., 1993). Subject EK04 also demonstrated one constraint, but the constraint and hence the joint motion pattern changed with practice: on Day 1, she used a "sitting down" pattern with the hip and knee flexing or extending together; on Day 3, she adopted the jackknife pattern (see Figure 8B). This case is interesting because it shows that movement patterns can change qualitatively even if the *number* of constraints remains constant. The case of Subject BM02 is interesting for a different reason. She retained the stronger of the two constraints observed on Day 1 (which yielded the "sitting down" pattern); the other constraint (between ankle and hip) weakened rapidly to a low VAF by the beginning of Day 3.

Questions about practice-related changes in model parameters can be validly answered only when the same coordinative function is used on both days. In this data set, only three subjects met that criteria: BM02 (hip–knee), DL03 (hip–ankle), and EB05 (hip–ankle). Although quantitative changes in the magnitudes of the parameters were observed, the dominant parameter signs remained constant. Hence, the basic directional form of the coupling was similar on Day 1 and Day 3.

Finally, model lags provide insight into the phase relationships between those tightly coupled joints.[7] As shown in Table I, median lags were 20 msec or less for nine of the eleven cases where pairs of joints had high VAFs. The median lag for all eleven cases was 2.5 msec (interquartile range = 0.0 to 20 msec). This means that most pairs of linearly coupled joint motions were in-phase on the first and third days of practice.

d. Summary and Limitations: Local Coordination These initial quantitative analyses of joint coordinative functions make two contributions to understanding interjoint coordination. First, they provide a way to directly test Bernstein's hypothesis (1967) about practice-related changes in joint coordination. Although the small number of subjects and practice days preclude overgeneralization of the findings,

---

[6] In general, the sign of the relative motion between the two joints was determined by the sign of the parameter (B1 or B2) with the higher absolute value. Positive values for the higher valued parameter indicated that the joints moved in the same way (e.g., hip flexion and ankle flexion).

[7] Two pairs of joints can be tightly coupled while the third pair is not if the lags for joint-system pairs differ, for example, subject MC01, where hip–knee and ankle–knee pairs were tightly coupled, but not hip–ankle (Table I).

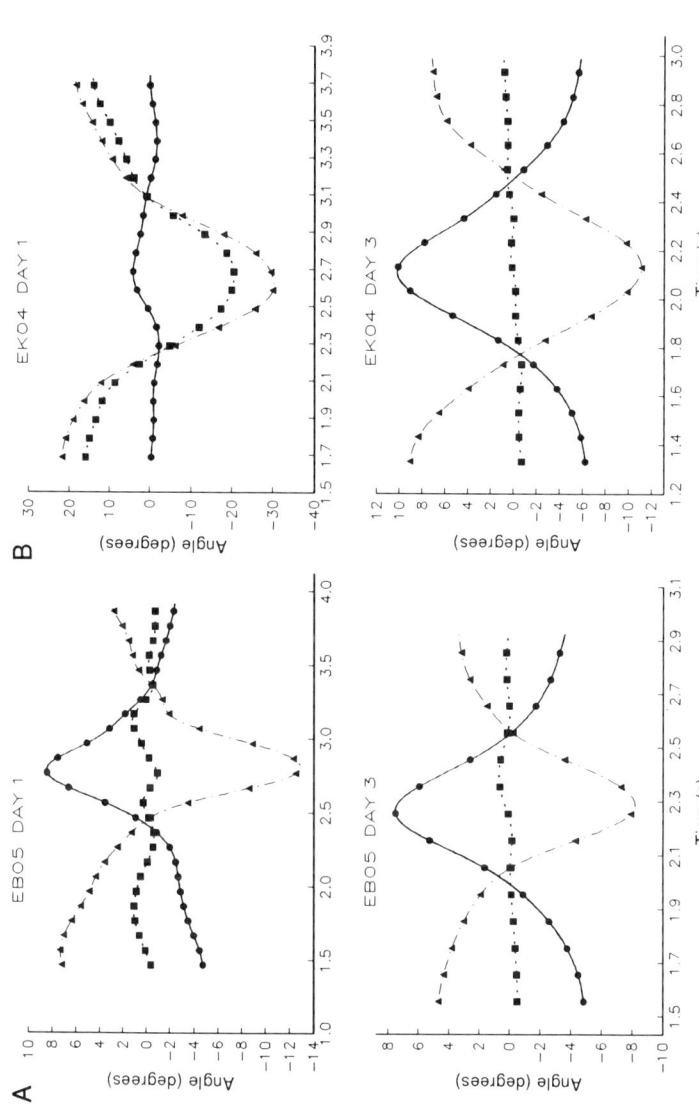

FIGURE 8 Ankle-, knee-, and hip-angle time histories for two subjects on their first and third days of practice. Both subjects had only one significant pair of coupled joints (VAF ≥ .90) on both days. (A) Subject EB05 used the same pattern of ankle-hip coupling on Day 1 and Day 3. (B) Subject EK04 changed the hip-knee joint coupling pattern on Day 1 to the ankle-hip coupling on Day 3. y-axes are angular changes, in degrees, from trial mean values. Negative angles represent joint flexion and positive angles represent joint extension.

the data nonetheless suggest that classical hypotheses about the learning of multijoint coordination may be incorrect, and that new hypotheses—for example, that subjects may switch coordinative patterns, or even simply alter parameters of coordinative functions with practice—need further investigation. Second, the fact that individuals use different patterns of joint coordination to produce comparable body-scaled forces underscores the need for further studies of whole-body coordination actions to determine *why* this is the case.

Thus, time series analysis provides a way to analytically model interjoint coordination in a way that affords insights into the number of joint DFs used by subjects and how those DFs change with practice. However, while such analyses can describe dynamic systems with nonzero time lags, they cannot capture many other kinds of nonlinear relationships. Moreover, the specific models and parameter values should not be interpreted too strongly, because other types of autoregressive models could be selected that might yield equally good fits but have different parameters. And different models might reveal tighter linear coupling between joint pairs than is evident for the models described above and in the Appendix. (Of course, these are general criticisms which apply to any "curve fitting" analysis of experimental data; they are not unique to time series analyses.) Despite these limitations, the time series models provide a way to rigorously evaluate changes in coordination across days and force levels when systems are fit well by the same model structure. However, principle-based theories still need to be developed to guide our understanding and analysis of the coordinative functions that govern interjoint movement patterns in whole-body actions.

C. Summary

The invariance in global CM_{AP} coordination paired with the variability in CM_V and lower extremity joint coordination suggest that coordination should be examined at both levels. CM_{AP} invariance presumably "falls out of" the physics that relate pulling force to horizontal motion of the CM. The variability in CM_V and joint motion remains puzzling, but at least three explanations are possible. First, perhaps only some subjects discovered joint coordinative functions that were optimal. With additional practice and feedback, all subjects might eventually converge on similar movement patterns. Alternatively, individual anthropomorphic variations (segment lengths, weight, muscle strength, etc.) might dictate varied "optimal" movement patterns for different subjects. Finally, the observed variations in CM_V and joint coordination may be genuinely equivalent with respect to producing pulling force. The fact that global coordinative functions between CM_{AP} and pulling force were similar even when subjects used different joint coordination patterns supports this alternative. Individuals may simply impose different implicit (and yet-to-be-discovered) constraints on CM_V motion, which is relatively *un*constrained by the requirement to produce horizontal pulling force. This possibility could be explored by simulation studies like those described in Section II.

IV. Conclusions

The principles and mechanisms that govern coordination in whole-body actions are not well understood. While scientists generally agree that *some* kinds of constraints mediate coordinated action in systems with many DFs, less agreement exists about the precise nature, level, and mechanisms by which constraints emerge or are imposed. Nevertheless, advances have been made over the past decade in understanding how the multiple DFs in whole-body actions are coordinated. This chapter has reviewed some advances that have emerged from two methodological approaches. First, analytical studies have begun to provide mathematically rigorous models that can be used to simulate how various constraints determine coordination in a particular system. Such simulations are important because the constraints that underlie coordinated actions in biological systems cannot be directly measured. Second, recent empirical studies have begun to yield information about the structure and acquisition of global and local coordination during pulling actions performed by standing humans. Empirical studies also provide the means to test predictions that are derived from analytical models about how constraints govern coordination.

However, new puzzles (e.g., the high variability in joint coordination during the stand-and-pull task) can arise from empirical studies. These puzzles may be difficult or impossible to resolve purely by empirical methods. It seems likely, therefore, that future investigations into the mechanisms of coordination during whole-body actions will increasingly combine analytical and empirical methodologies. It is anticipated that such studies ultimately will yield a general and principled model of coordination and its acquisition in whole-body actions.

Acknowledgments

We greatly appreciate Joseph Given's guidance on the time series analysis, Fang Gao's programming assistance, and Jeff Springer's help in figure preparation. The work was supported by NSF Grant BNS9021487 to W.A. Lee.

Appendix

One general time series description for a linear system, called the autoregressive or ARX model (Box & Jenkins, 1976; Chatfield, 1989; Ljung, 1987) is described by the equation:

$$y(t) + A1y(t-1) + \ldots + A_{na}y(t-na) =$$
$$B1u(t-nk) + B2u(t-nk-1) + \ldots + B_{nb}u(t-nk-nb+1) \quad (1)$$

where nk is the number of delays from input to output. An ARX process containing na autoregressive terms, nb extra terms, and nk lag terms is said to be an ARX process of order (na nb nk), abbreviated ARX(na nb nk). Three single input–single output systems were arbitrarily defined (causal relationships are not assumed):

> System 1: $u(t)$ = hip angle; $y(t)$ = knee angle
> System 2: $u(t)$ = ankle angle; $y(t)$ = knee angle
> System 3: $u(t)$ = hip angle; $y(t)$ = ankle angle

Using this model, system identification was carried out for each trial on Day 1 and Day 3 for each subject, using a two-stage process. The MATLAB™ System Identification Toolbox was used for all procedures. In the first stage, the impulse response estimate was used to select the delay nk for each system. The cross-correlation function between the (prewhitened) input and output is directly proportional to the impulse response function. The cross-correlation function of the filtered input and filtered output was plotted. A significant peak in the cross-correlation function at lag d indicates that the input is related to the output when delayed by time d. This corresponds to the delay nk of the system.

After estimating nk in this manner, we examined a set of six first-order and second-order candidate models: ARX(01nk), (02nk), (11nk), (12nk), (21nk), and (22nk). Our procedure was to test the six models with nk selected from the impulse response estimate and then to use the model that gave the best overall fit for a second round of modeling. The "best" model structure was chosen by comparing a criteria of fit obtained for each of these six model structures. For a criterion of fit, the variance accounted for (VAF) = $1 - \text{variance}_{\text{residuals}}/\text{variance}_{\text{output}}$ was determined for each model. The ARX(02nk) yielded the highest overall VAFs.

In the second stage of modeling, the nk that gave the best fit for the ARX(02nk) was selected and model parameters estimated. The medians, means, and standard deviations of the VAFs, lags, and B1 and B2 parameters subsequently were computed for each subject from each block of trials.

References

Ada, L., O'Dwyer, N., & Neilson, P. (1993). Recovery following stroke as reflected in joint kinematics and inter-joint coordination. *Human Movement Science,* **12,** 137–154.
Arutyunyan, G. H., Gurfinkel, V. S., & Mirskii, M. L. (1968). Investigation of aiming at a target. *Biophysics,* **13,** 536–538.
Arutyunyan, G. H., Gurfinkel, V. S., & Mirskii, M. L. (1969). Organization of movements on execution by man of an exact postural task. *Biophysics,* **14,** 1162–1167.
Barin, K., & Stockwell, C. W. (1985). A mathematical model of human postural control in the sagittal plane. In M. Igarashi & O. Black (Eds.), *Vestibular and visual control of postural and locomotor equilibrium* (pp. 29–33). Basel: Karger.
Bernstein, N. (1967). *The coordination and regulation of movement.* New York: Pergamon.
Bouisset, S., & Zattara, M. (1987). Biomechanical study of the programming of anticipatory postural adjustments associated with voluntary movement. *Journal of Biomechanics,* **20,** 735–742.
Box, G. E. P., & Jenkins, G. M. (1976). *Time series analysis: Forecasting and control* (rev. ed.). San Francisco: Hodlen-Day.
Bullock, D., & Grossberg, S. (1988). Neural dynamics of planned arm movements: Emergent invariants and speed-accuracy properties during trajectory formation. *Psychological Review,* **95,** 49–90.
Chatfield, C. (1989). *The analysis of time series* (4th Ed.). New York: Chapman and Hall.
Crenna, P., Frigo, C., Massion, J., & Pedotti, A. (1987). Forward and backward axial synergies in man. *Experimental Brain Research,* **65,** 538–548.
Droulez J., & Berthoz, A. (1986). Servo-controlled (conservative) versus topological (projective) model of sensory motor control. In W. Bles & T. Brandt (Eds.), *Disorders of posture and gait* (pp. 83–97). New York: Elsevier.
Easton, T. A. (1972). On the normal use of reflexes. *American Scientist,* **60,** 591–599.
Flash, T., & Henis, E. (1991). Arm trajectory modifications during reaching towards visual targets. *Journal of Cognitive Neuroscience,* **3,** 220–230.
Flash, T., & Hogan, N. (1985). The coordination of arm movements: An experimentally confirmed mathematical model. *Journal of Neuroscience,* **5,** 1688–1703.
Fowler, C. A., & Turvey, M. T. (1978). Skill acquisition: An event approach with special reference to searching for the optimum of a function of several variables. In G. E. Stelmach (Ed.), *Information processing in motor control and learning* (pp. 1–40). New York: Academic Press, Inc.
Friedli, W. G., Cohen, L., Hallet, M., Stanhope, S., & Simon, S. R. (1988). Postural adjustments associated with rapid voluntary arm movements. II. Biomechanical analysis. *Journal of Neurology, Neurosurgery, and Psychiatry,* **51,** 232–243.
Gallistel, C. R. (1980). *The organization of action.* Hillsdale, NJ: Erlbaum.
Gelfand, I. M., Gurfinkel, V. S., Tsetlin, M. L., & Shik, M. L. (1971). Some problems in the analysis of movements. In I. M. Gelfand, V. S., Gurfinkel, S. V. Fomin, & M. L. Tsetlin (Eds.), *Modes of the structural-functional organization of certain biological systems.* (pp. 160–171). Cambridge, MA: MIT Press.
Gordon, M. E. (1990). *An analysis of the biomechanics and muscular synergies of human standing.* Ph.D. Dissertation, Stanford University.
Gordon, M. E., Zajac, F. E., Khang, G., & Loan, J. P. (1988). Intersegmental and mass center accelerations induced by lower extremity muscles: Theory and methodology with emphasis on quasi-vertical standing postures. In R. L. Spilker & B. R. Simon (Eds.), *Ninth winter annual meeting of The American Society of Mechanical Engineers* (pp. 481–492). Chicago, IL: American Society of Mechanical Engineers.

Greene, P. R., & McMahon, T. A. (1979). Reflex stiffness of man's anti-gravity muscles during kneebends while carrying extra weights. *Journal of Biomechanics,* **12,** 881–891.
Grillner, S. (1981). Control of locomotion in bipeds, tetrapods, and fish. In J. M. Brookhart & J. M. Mountcastle (Eds.), *Handbook of physiology* (pp. 1179–1236). Washington, D.C.: American Physiological Society.
Hinton, G. (1984). Parallel computations for controlling an arm. *Journal of Motor Behavior,* **16,** 171–194.
Horak, F. B., & Nashner, L. M. (1986). Central programming of postural movements: adaptation to altered support-surface configurations. *Journal of Neurophysiology,* **55,** 1369–1381.
Johansson, R., Magnusson, M., & Akesson, M. (1988). Identification of postural dynamics. *IEEE Transactions on Biomedical Engineering,* **35,** 858–869.
Jordan, M. I. (1990). Motor learning and the degrees of freedom problem. In M. Jeannerod (Ed.), *Attention and performance XIII,* Hillsdale, NJ: Erlbaum.
Kugler, P., Kelso, J. A. S., & Turvey, M. (1980). On the concept of coordinative structures as dissipative structures: I. Theoretical lines of convergence. In G. E. Stelmach & J. Requin (Eds.), *Tutorials in motor behavior* (pp. 1–47). Amsterdam: North-Holland.
Kuo, A. D., & Zajac, F. E. (1992). An analysis of biomechanical constraints on the coordination of standing posture. In M. Woollacott & F. Horak (Eds.), *Posture and gait: Control mechanisms* (pp. 344–347). Eugene, OR: University of Oregon Press.
Lee, W. A. (1992). Learning to control body center of mass during arm pulls made while standing. In M. Woollacott & F. Horak (Eds.), *Posture and gait: Control mechanisms* (pp. 364–367). Eugene, OR: University of Oregon Press.
Lee, W. A., & Rogers, M. W. (1987). Mechanical and cognitive constraints on standing influence postural adjustments and maximal forces during pulling. *Neuroscience Abstracts,* **13,** 347.
Lee, W. A., Deming, L., & Sahgal, V. (1988). Quantitative and clinical measures of static standing balance in hemiparetic and normal subjects. *Physical Therapy,* **68,** 970–976.
Lee, W. A., Michaels, C. F., & Pai, Y.-C. (1990a). The organization of torque and EMG activity during bilateral handle pulls by standing humans. *Experimental Brain Research,* **82,** 304–314.
Lee, W. A., Michaels, C. F., & Pai, Y. C. (1990b). Variability and invariance of postural adjustments during abrupt pulls of different force made by standing humans. In T. Brandt, W. Paulus, W. Bles M. Dietrerich, S. Krafczyk, & A. Staube (Eds.), *Disorders of posture and gait* (pp. 98–102). Prien, F. G. R.: Thieme Press.
Ljung, L. (1987). *System identification: Theory for the user.* Hillsdale, NJ: Prentice-Hall.
Massion, J. (1992). Movement, posture and equilibrium: Interaction and coordination. *Progress in Neurobiology,* **38,** 35–56.
McMahon, T. A., & Cheng, G. (1990). The mechanics of running: How does stiffness couple with speed. *Journal of Biomechanics,* **23**(Suppl. 1), 65–78.
McMahon, T. A., & Greene, P. R. (1979). The influence of track compliance on running. *Journal of Biomechanics,* **12,** 893–904.
Michaels, C. F., & Lee, W. A. (1993). The organization of multisegmental pulls: II. Submaximal pulls. (Submitted).
Michaels, C. F., Lee, W. A., & Pai, Y.-C. (1993). The organization of multisegmental pulls made by standing humans: I. Near-maximal pulls. *Journal of Motor Behavior,* **25,** 107–124.
Mouchino, L., Aurenty, R., Massion, J., & Pedotti, A. (1992). Coordination between equilibrium and head–trunk orientation during leg movement: A new strategy built up by training. *Journal of Neurophysiology,* **67,** 1587–1598.

Nashner, L. M., & McCollum, G. (1985). The organization of human postural movements: A formal basis and experimental synthesis. *Behavioral and Brain Science,* **8,** 135–172.

Nashner, L. M., & Woollacott, M. (1979). The organization of rapid postural adjustments in standing humans: An experimental-conceptual model. In R. E. Talbot & D. R. Humphrey (Eds.), *Posture and movement* (pp. 243–257). New York: Raven Press.

Newell, K. M., & van Emmerik, R. E. A. (1989). The acquisition of coordination: A preliminary analysis of learning to write. *Human Movement Science,* **8,** 17–32.

Newell, K. M., Kugler, P. N., van Emmerik, R. E. A., & McDonald, P. V. (1989). Search strategies and the acquisition of coordination. In S. A. Wallace (Ed.), *Perspectives in the coordination of movement* (pp. 86–122). Amsterdam: Elsevier.

Pai, Y.-C., & Rogers, M. W. (1990). Control of body mass transfer as a function of speed of ascent in sit-to-stand. *Medicine and Science in Sports and Exercise,* **22,** 378–384.

Pedotti, A., Crenna, P., Deat, A., Frigo, C., & Massion, J. (1989). Postural synergies in axial movements: short and long-term adaptation. *Experimental Brain Research,* **74,** 3–10.

Raibert, M. H. (1990). Trotting, pacing and bounding by a quadruped robot. *Journal of Biomechanics,* **23,** 79–98.

Ramos, C. F., & Stark, L. W. (1990a). Postural maintenance during fast forward bending: a model simulation experiment determines the "reduced trajectory." *Experimental Brain Research,* **82,** 651–657.

Ramos, C. F., & Stark, L. W. (1990b). Postural maintenance during movement: Simulations of a two joint model. *Biological Cybernetics,* **63,** 363–375.

Rogers, M. W., & Pai, Y.-C. (1990). Dynamic transitions in stance support accompanying leg flexion movements in man. *Experimental Brain Research,* **81,** 398–402.

Saltzman, E., & Kelso, J. A. S. (1987). Skilled actions: A task-dynamic approach. *Psychological Review,* **94,** 84–106.

Schmidt, R. A. (1975). A schema theory of discrete motor skill learning. *Psychological Review,* **82,** 225–261.

Scholz, J. P., & Kelso, J. A. S. (1989). A quantitative approach to understanding the formation and change of coordinated movement patterns. *Journal of Motor Behavior,* **21,** 122–144.

Schöner, G., & Kelso, J. A. S. (1988). Dynamic pattern generation in behavioral and neural systems. *Science,* **239,** 1513–1520.

Turvey, M. T. (1977). Preliminaries to a theory of action with reference to vision. In R. E. Shaw & J. Bransford (Eds.), *Perceiving, acting and knowing* (pp. 211–265). Hillsdale, NJ: Erlbaum.

Turvey, M. T., Shaw, R. E., & Mace, W. (1978). Issues in the theory of action: Degrees of freedom, coordinative structures and coalitions. In J. Requin (Ed.), *Attention and performance VII* (pp. 557–595). Hillsdale, NJ: Erlbaum.

Vereijken, B., Emmerik, R. E. A., Whiting, H. T. A., & Newell, K. M. (1992). Free(z)ing degrees of freedom in skill acquisition. *Journal of Motor Behavior,* **24,** 133–142.

Winter, D. A. (1984). Kinematic and kinetic patterns in human gait: variability and compensating effects. *Human Movement Science,* **3,** 51–76.

26
Coordinating the Two Hands in Polyrhythmic Tapping

Jeffery J. Summers

Department of Psychology
University of Southern Queensland
Toowoomba, Queensland, Australia

Jeff Pressing

Department of Music
La Trobe University
Bundoora, Victoria, Australia

I. Introduction
II. Constraints on Bimanual Performance
III. The Production of Polyrhythms
    A. Studies of Polyrhythmic Tapping
    B. Integrated versus Parallel Motor Organizations
    C. Coordinating the Hands
IV. A Model of Polyrhythmic Tapping
V. Conclusion
    References

I. Introduction

Most people experience great difficulty when asked to perform two simple motor tasks, one with each hand, at the same time. The party trick of rubbing one's stomach with one hand in a circular motion and tapping the head with the other at an independent tempo is a classic example of this limitation on bimanual performance. Typically in this dual-task situation, one of the activities takes precedence over the other, and people either tap or perform rubbing movements with both hands. At the other extreme, some people, such as highly skilled musicians, appear to have achieved independent control of the two hands. They can perform different rhythms at the same time and vary the tempo in each hand separately.

Any theory of bimanual coordination, therefore, must account not only for the limitations on bimanual performance but also for how these limitations can be overcome. In this chapter we shall focus on the latter issue and examine the way people, both skilled (i.e., musically trained) and unskilled, coordinate the hands when two differently timed motor activities are performed concurrently.

II. Constraints on Bimanual Performance

A great deal of evidence has now accumulated from laboratory studies that there are severe temporal constraints on the performance of simultaneous motor actions by the two hands (see Kelso, Chapter 5, and Peters, Chapter 27, this volume, for more comprehensive coverage of this literature). These constraints are particularly evident when each hand must conform to a different rigid temporal structure. For example, Peters (1977) asked subjects to tap as fast as possible with one hand while tapping a 1-2-3 and 4 rhythm with the other hand. Only 15 out of the 150 subjects tested were able to perform the dual task. That the difficulty in concurrently performing differently timed activities is not restricted to the two hands was demonstrated in the next experiment in which subjects recited the nursery rhyme "Humpty Dumpty" while tapping the 1-2-3 and 4 rhythm with the hand. None of the 100 subjects could perform the task! When the concurrent motor activities share a common time base (e.g., performing two taps with one hand against one with the other), however, little difficulty is experienced.

Temporal constraints on bimanual coordination have also been demonstrated in two-handed aiming movements (e.g., Kelso, Southard & Goodman, 1979) and in simple cyclical movements of the two hands (e.g., Kelso, Holt, Rubin, & Kugler, 1981). The general finding that has emerged from this work is that there is a disposition toward simple timing relations in the coordination of the two hands.

Although there is general agreement on the existence of mutual interference in the bimanual performance of temporally constrained actions, the locus and cause of these effects has generated considerable interest among researchers in motor control. One reason for this is that the topic of bimanual coordination provides a fertile testing ground for two competing models of motor action: the cognitive, where action is considered to originate in central mechanisms (mental "software"), and the increasingly accepted "dynamical," where coordination is considered to emerge from the physical properties of the neuromotor system by the setting of appropriate boundary conditions and system parameters (the system is typically modeled as a set of coupled oscillators).

One "cognitive model," for example, places the limitation on bimanual coordination in a central attentional system. Peters (1990), has argued that the interlimb and manual/vocal interference effects observed lend strong support to "the concept of a central unitary scheduler, the unitary acting self, that assigns priorities and determines which of two concurrent activities is in the foreground of attention" (p. 564). Furthermore, performance asymmetries evident in a variety of bimanual tapping tasks suggest that there is a natural dispostion of attentional focus toward the preferred hand (see Peters, Chapter 27, this volume). According to this view, when two differently timed actions are produced concurrently, the scheduler gives priority to one action sequence and the other activity is integrated into the dominant rhythm. If the two activities cannot be interlaced into a harmonic structure the tasks cannot be performed concurrently (Peters, 1990).

An apparently radically different account of bimanual interactions is offered by proponents of the dynamical approach to motor behavior. Rhythm is seen as an emergent property of the dynamical behavior of the neuromotor system itself, and the temporal constraints on coordinating the two hands are seen as reflecting the entrainment properties of coupled nonlinear oscillators. Yamanishi, Kawato and Suzuki (1980), for example, trained subjects to perform a bimanual tapping task in which the phase relation between the hands was varied in ten steps from 0.0 (synchronous tapping) through 0.5 (alternation) to 0.9. For both skilled (musically trained) and unskilled subjects accurate and stable performance was evident when the relationship between the hands was in-phase (0.0) or anti-phase (0.5). Furthermore, intermediate phases produced unstable performance and a tendency to shift to the nearest stable phase. Similar tendencies have been shown in split-brain subjects (Tuller & Kelso, 1989). In the dynamical orientation to bimanual coordination, in-phase and anti-phase relations represent intrinsic phase lockings between the hands. The anti-phase relation, however, is less stable as spontaneous shifts from an anti- to in-phase relation occur as the cycling rate of the hands is increased (Kelso & Schöner, 1988; Scholz & Kelso, 1989). Kelso and DeGuzman (1988) argue that the acquisition of new temporal coordinations necessitates the creation of an attractor corresponding to the required task (e.g., a 5:3 ratio). The process involves cooperative and competitive interactions between the intrinsic patterns (i.e., 1:1 ratio) and the intended ratio. The influence of the intrinsic phase lockings gradually disappears with practice and the required relative phasing emerges.

The well-documented performance and training techniques of musical performers, and the requirements of many musical scores, however, suggest that the in-phase and anti-phase modes are not the only readily accessed phase relations. Specifically, subdivision of beats into 2, 3, 4 or more parts is a common musical procedure. For example, if fingers 1, 2, and 3 (thumb, index, middle) of a single hand are depressed rapidly and sequentially, as is common in some piano scale passages (123123123. . .), phase relations of 120° are easily achieved. This corresponds naturally to time subdivision into 3 parts, or triplet phasing. In this regard the step sizes used by Yamanishi *et al.* (1980) were possibly not optimal, as they disregard musicians' predilections for such time subdivision into a *small* number of equal parts. It seems likely that this practice corresponds to a path of optimal cognitive simplicity for both trained and untrained subjects.

In sum, two quite different explanations have been offered for the bias toward harmonic timing between the two hands when they are functioning together. One emphasizes the central control of timing and limitations/biases in a central attentional mechanism, the other stresses interaction between coupled nonlinear oscillators operating at a functionally low level in the motor system. The advantage of the oscillator model is that it minimizes the need for "cognitive" processes in interlimb coordination. It should be noted, however, that until recently the model has been applied mainly to simple cyclical movements of the fingers and hands. The extent to which it can be extended to more complex coordination situations, without recourse to cognitive processes, remains to be seen (for a recent attempt at modeling intentionally within a dynamical framework, see Shaw, Kadar, Sim, & Repperger, 1992). The case of musical coordination raises this point with perhaps particular acuteness. For example, in romantic piano music, the presence of consistent slight note-by-note accents and deviations from metronomic temporal positioning in only one hand (found, for example, in the performance recommendations of Frederic Chopin and Leopold Mozart and in computer microrecordings of some expert performers) clearly suggests some sort of hierarchical parametrically controllable program. Another example is the musical grace note, which is a small note attached to a more important one, and which is found in many musical styles. The executed duration of a grace note can depend on local context and tempo at execution time, which implies either a cognitive decision-making structure or a complex set of nonlinear conditions.

Whichever perspective of explanation is adopted, it must also be granted that temporal interference is quite limited in some situations: usually those characterized by a high level of overlearning, or where muscle systems are highly differentiated. For example, the rhythms of walking and talking do not normally interfere. Even the previously mentioned "impossible" task of rubbing one's stomach with one hand and tapping the head with the other is in fact performable to a considerable degree by experienced percussionists (Pressing, 1984; Summers & Pressing, 1993). In the system dynamic model, this may be handled by the supposition that the oscillators describing the two system components have a weak or zero coupling. In the case of the attentional model, overlearning is interpreted as reducing the conscious attentional load of highly overlearned behaviors to nearly zero.

III. The Production of Polyrhythms

To gain a better understanding of how the hands interact in complex coordination tasks, a number of researchers have examined the production of polyrhythms. Polyrhythms require the simultaneous production of two conflicting but isochronous motor sequences. For example, in a three versus five polyrhythm one hand beats three times to five beats of the other hand. Thus, they are repeating rhythmic patterns with the beats of each hand coinciding once per cycle.

Polyrhythms possess a number of features that make them ideally suited to an examination of the possible role of cognitive processes in bimanual coordination. First, in a polyrhythmic sequence, each hand makes an independent contribution to a common goal and, as such, closely resembles many everyday tasks. Second, the two hands move at different frequencies, a situation that has been shown to produce strong tendencies toward phase and frequency synchronization of the hands (e.g., Kelso et al., 1981; Kelso & Jeka, 1992). Third, unlike the simple cyclical movements previously studied, polyrhythms involve complex, coherent rhythmic structures, and accurate performance requires the precise phasing of the two hands. Finally, polyrhythms are extremely difficult to perform, especially for the musically naive. However, with extensive practice, people such as skilled musicians can accurately perform even the most complex polyrhythms.

A. Studies of Polyrhythmic Tapping

Typically, in studies of polyrhythm performance, subjects are required to tap out repetitively both simple rhythms and polyrhythms on two keys using the index fingers of each hand. The rhythms are presented to subjects as two parallel tone trains either through headphones or speakers, one tone train for each hand sequence. Two different response tasks have been used to study polyrhythm performance: synchronization and continuation. In the synchronization task subjects attempt to synchronize their right hand taps with tones presented through the right headphone/speaker and their left hand taps with the tones presented through the left headphone/speaker. In the continuation paradigm, after an initial period of synchronization, the presented tones stop and the subject attempts to continue tapping out the rhythm from memory at the same tempo. Across studies (given the obvious limitations of such comparisons) the continuation paradigm appears to produce better overall performance than the synchronization task. It may be that the cognitive load of matching responses to two clearly distinguished (i.e., by ear) externally imposed rhythms is the cause of the poorer performance in the synchronization task. In the continuation paradigm subjects are free to "self-organize" their responses.[1]

Across the studies performance of the following polyrhythms has been examined: 3-against-2, 5-against-2, 4-against-3, 5-against-3, and 5-against-4. A graphic description of these five polyrhythms is shown in Figure 1. One cycle of each polyrhythm is presented and the intervals between adjacent taps with either hand are labeled. Each cycle of a polyrhythm is initiated by a simultaneous left- and right-hand response.

B. Integrated versus Parallel Motor Organizations

It is clear that accurate performance of a complex polyrhythm requires the overcoming of the inherent temporal constraints on bimanual coordination. A question of considerable theoretical importance, then, is how these temporal constraints are overcome.

There are a number of possible ways through which crosstalk between the hands may be eliminated or reduced. One possibility is that with practice relative independence between the limbs is achieved, perhaps through the development of some form of insulation against neural crosstalk (Kinsbourne & Hicks, 1978; Marteniuk, MacKenzie, & Baba, 1984). Shaffer (1981) has argued, for example, that the rhythmic interplay evident between the hands of highly skilled pianists during a performance can be achieved only with independent timing mechanisms for the two hands. Yamanishi et al. (1980) also noted that the entrainment between the hands in bimanual tapping was much stronger for unskilled than skilled (musically trained) subjects.[2] However, it is unclear whether the weakening of the interaction be-

---

[1] The authors would like to thank an anonymous reviewer for pointing out this distinction between synchronization and continuation tasks.

[2] This difference between musicians and nonmusicians, however, was not replicated in a study by Tuller and Kelso (1989).

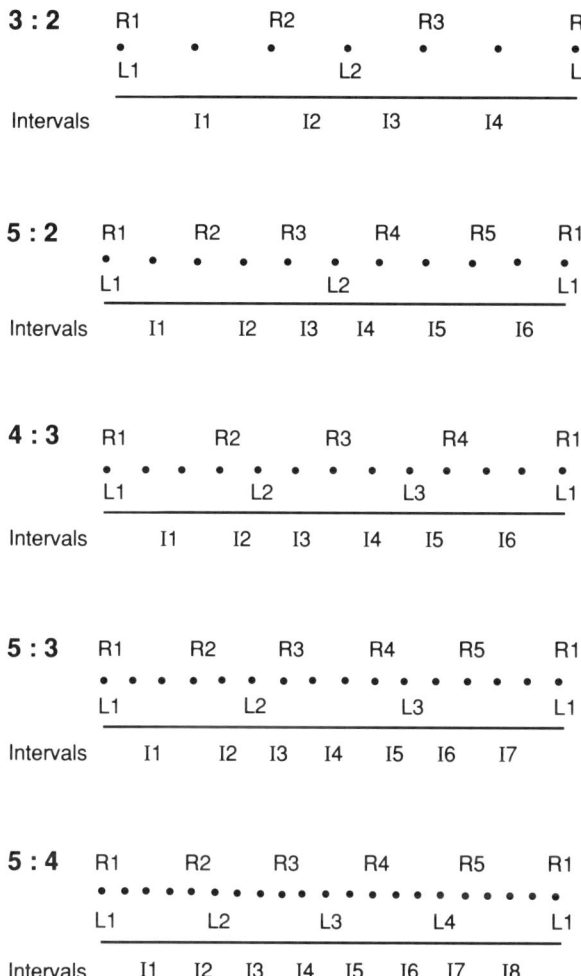

FIGURE 1 Schematic of the temporal relationships between the two hands required in the performance of a 3:2, 5:2, 4:3, 5:3, and 5:4 polyrhythm (one cycle of each repeating pattern is shown). In these examples the right hand (R) takes the faster beat and the left hand (L) the slower beat. R1 to Rn are taps with the right hand, L1 to Ln taps with the left hand. Each cycle of a polyrhythm is initiated by a simultaneous R and L response. I1 to In are within-hand and between-hand intertap intervals. (From Summers et al., 1993b. Reprinted by permission.)

tween the hands with learning reflects the increasing influence of high-level control centers (Craske & Craske, 1986; Yamanishi et al., 1980), or a reduction in the influence of a higher level process that is constrained to handle single inputs from independent timing mechanisms (Keele & Ivry, 1987).

Alternatively, it may be that the constraints on the concurrent performance of differently timed activities are overcome by combining the initially separate activities into a common time base. In this way the task becomes a single rhythm performed with two hands, rather than two autonomous rhythms performed by distinct hands. Deutsch (1983), for example, has argued that the ability to generate two isochronous sequences concurrently depends on the degree to which an internal representation of the patterns as an integrated whole can be developed.

A central issue in the polyrhythm work, therefore, has been to determine whether in such tasks subjects interweave the timing of the two hands (an integrated organization) or independently time the movements of the hands (a parallel organization). In an integrated organization the two hands operate on a common time base, whereas in a parallel organization the hands are controlled separately. However, as successful polyrhythm performance requires that the hands are synchronized at the start of each cycle (see Figure 1), it is unlikely that complete independence between the hands can be achieved.

Deutsch (1983) compared the synchronization performance of musically trained subjects in both simple rhythms (e.g., 1-against-1, 2-against-1) and polyrhythms (e.g., 3-against-2, 4-against-3). She argued that if a parallel organization was used, no difference in performance should be observed between simple rhythms and polyrhythms. However, tapping variability was found to increase dramatically for the polyrhythmic sequences compared to the simple patterns, suggesting that subjects were using an integrated organization. Furthermore, for the polyrhythmic sequences, accuracy of performance was inversely related to the complexity of the associated integrated representation.

Jagacinski, Marshburn, Klapp, and Jones (1988) provided a more direct test of the use of integrated versus parallel organizations in a 3-against-2 polyrhythm by examining the pattern of variances and covariances among intertap intervals. Three integrated models of motor organization in which the two hands are based on a common timekeeper were contrasted with three parallel models assuming separate timekeepers for each hand (see Jagacinski et al., 1988, for details). The analysis of the tapping data was based on the Wing and Kristofferson (1973) model of the timing of repetitive movements and its extension to the bimanual situation of Vorberg and Hambuch (1978, 1984). The basic model assumes two processes operating in the timing of intertap intervals: a central timer that generates a series of internal events, each of which initiates a motor response; and an implementation process that introduces variable delays in the execution of the response. Each process is assumed to operate as a series of independent, randomly varying intervals. The actual observable intertap intervals, therefore, are assumed to reflect both timekeeper and motor delay processes.

In this study, a synchronization task was used and the perceptual organization of the polyrhythm was manipulated by varying the pitch difference between tone trains. For four musically trained subjects the pitch difference was small (262 Hz and 349 Hz, respectively), leading to the perception of a serially integrated pattern. For the other four musically trained subjects the pitch difference was large (262 Hz and 2794 Hz), leading to a streamed percept (Bregman & Campbell, 1971), in which two independent parallel streams of tones are perceived. Although performance with the integrated tones was superior to performance with the streamed tones, the different perceptual organizations did not induce different motor organizations. All eight subjects in this experiment exhibited an integrated motor organization.

Similar findings have been reported in a recent study by Summers, Todd, and Kim (1993c). Again, a 3-against-2 polyrhythm was the test pattern and perceptual organization was manipulated by varying the pitch difference between tone trains. In addition, the study examined whether some independence between the hands may be achieved by delaying the initiation of the movement sequence for one hand relative to the other hand. There is some evidence to suggest that independent operation of the limbs may be easier to achieve if the initiation of one movement is delayed with reference to the other (Swinnen, Walter, & Shapiro, 1988).

In this experiment, performance on a normal 3-against-2 polyrhythm (Figure 2, Simultaneous) was compared with performance on a pattern (Figure 2, Shifted), in which the three taps of the faster (right) hand were shifted 100 msec later in the cycle (cycle duration 1300 msec). Thus, shifting the right-hand pattern relative to the left-hand pattern changed the time relations between hands but not within a hand. Two groups of 24 subjects (12 musicians, 12 nonmusicians) practiced extensively either the simultaneous or shifted version of the 3-against-2 polyrhythm before transferring to the other version.

Although streaming the tones and/or shifting the motor pattern for one hand relative to the other significantly reduced performance accuracy in a continuation paradigm, these factors had no obvious effect on the motor organization adopted by skilled and unskilled subjects. Once again, all subjects in this experiment appeared to use, with varying degrees of success, an integrated motor organization.

An interesting finding was that more nonmusicians were able to accurately reproduce the required relative phasing for the shifted pattern than musically trained subjects. A number of the musicians appeared to distort the nonmetric time pattern to a metric one by reducing the interval between the initial left- and right-hand taps (see Figure 3A). Some nonmusicians, in contrast, showed a tendency to increase the interval between the initial left- and right-hand taps, shorten the following right hand interval (R1-R2), and then lengthen the between-

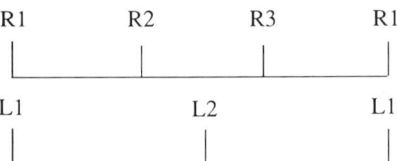

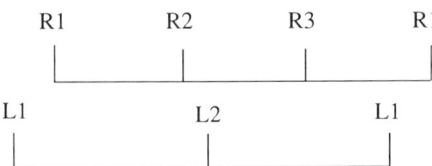

FIGURE 2 Schematic of the temporal relationships between the two hands required in the performance of the simultaneous and shifted 3-against-2 patterns (one cycle of each repeating pattern is shown). Each vertical line represents a tap with either the right (R) or left (L) hand.

hand interval (R2-L2). The aim of these adjustments seemed to be to place the left, slow hand response (L2) close to the middle of the fast hand interval (R2-R3), so that the R2-L2-R3 triplet involved a simple alternation of the hands (see Figure 3B). The complexity of the shifted pattern, therefore, brought forth two rather different cognitive strategies. The strategy depicted in Figure 3(1) was to keep the structure of each hand intact but shift them toward synchronization by approximate alignment of starting times. The strategy shown in Figure 3(2) was to distort the time structure of each hand to facilitate their integration into a pattern of only 5 beats duration.

Other studies using more complex polyrhythms also suggest that using separate timing mechanisms for the two hands may not be possible in polyrhythmic tapping. Summers, Ford, and Todd (1993a), for example, gave two musicians (GN and RB) and two nonmusicians (PA and TM) eight days of practice on a 5-against-3 polyrhythm (see Figure 1), with the right hand taking the faster beat. Over the course of training, each subject performed 4224 cycles of the polyrhythm (2112 cycles of synchronization and 2112 cycles in a continuation paradigm). Of particular interest in this experiment was whether a shift from an integrated to a parallel motor organization would be observed over the

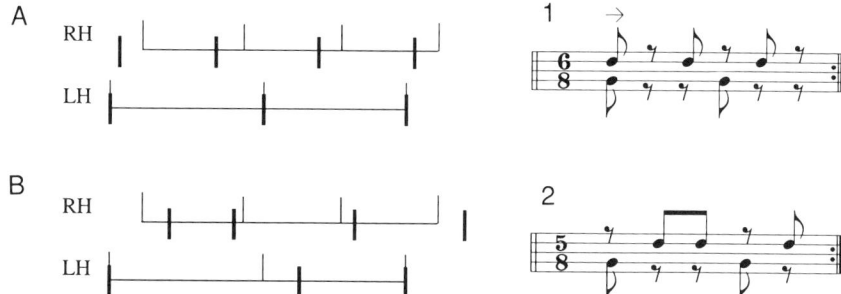

FIGURE 3 Schematic of the distortions in the reproduction of the 3-against-2 shifted pattern (one cycle of each repeating pattern is shown). Thin vertical lines represent the required target positions of taps with either the right (RH) or left (LH) hand. Thick vertical lines indicate actual tap positions. (A) The distortions from the target positions produced by a musically trained subject. (B) The distortions produced by a nonmusician. (1) and (2) Closely approximate musical notation for (A) and (B), respectively.

course of training. Following training, subjects were asked to perform the 5:3 polyrhythm under a variety of conditions, including reversing the hand assignment and tapping the rhythm as fast as possible. The purpose of the transfer trials was to examine the issue of specificity of learning.

The pattern of variances and covariances among intertap intervals indicated that both musically trained subjects used an integrated form of organization throughout the training period. Mean absolute timing error for these subjects, defined as the difference between the corresponding intervals in the response sequence and the target sequence, steadily declined over training from 33.9 msec on Day 1 to 17.9 msec on Day 8 for subject GN, and from 25.7 msec on Day 1 to 11.9 msec on Day 8 for subject RB.

The nonmusician TM experienced great difficulty in performing the polyrhythm and, in fact, was never able to produce the correct timing (see later discussion). He was able, however, to show a large reduction in timing error over training days, from 323.6 msec on Day 1 to 94.3 msec on Day 8. Furthermore, the covariance analyses suggested the development of an integrated organization with practice. Mean timing error for the other nonmusician (PA) decreased from 102.2 msec on the first day of training to 37.6 msec on the final day. The covariance analyses for this subject, however, showed no clear pattern with regard to integrated versus parallel organizations.

The general finding that emerged from this study was that none of the subjects tested showed evidence for the development of a parallel organization with practice. Rather, all the subjects appeared to adopt, to a greater or lesser extent, some form of integrated motor organization.

Finally, two studies have examined whether learning the motor pattern for each hand separately before combining them would encourage the use of a parallel organization (Klapp, Martin, McMillan, & Brock, 1987; Summers & Kennedy, 1992). In the Summers and Kennedy study, the target polyrhythm was again 5-against-3 with the right hand taking the faster beat and a cycle duration of 1500 msec. One group of subjects (6 musicians and 6 nonmusicians) received extensive practice at tapping out the rhythm for the right hand (one tap every 300 msec) and the left hand (one tap every 500 msec) in isolation. Following training, subjects were asked to produce the two rhythms at the same time but try to keep the two streams separate, synchronizing the two hands only at the beginning of each cycle. None of the nonmusicians was able to perform the bimanual task.

Four of the musically trained subjects produced patterns resembling a 5-against-3 polyrhythm, but in all cases the left (slow) hand intertap intervals (ITIs) exhibited high variability relative to right (fast) hand ITIs. The response profile for these subjects is shown in Figure 4A and suggests an integrated strategy in which movements of the slow hand are interlaced with the movements of the fast hand. The remaining two musically trained subjects showed quite different response profiles (Figure 4B and 4C). Both subjects exhibited a strong tendency toward synchronizing taps with the slow and fast hand. For one subject (B), this was achieved by changing the fast-hand stream to produce a 2:1 pattern, whereas for the other subject (C), slow-hand responses were "attracted" toward the fast-hand taps.

It is useful to compare musical notations for the played patterns, and these are given in Figure 4 (1)–(4). All musically trained subjects were unable to produce a close approximation to the 5:3 polyrhythm. This may reflect the experience level of the performers and the fact that the subjects were pianists rather than percussionists. In percussion training, complex rhythms are slowed down, worked out precisely on the basis of subdivision of pulse, and gradually brought up to correct tempo. This system is used for simpler polyrhythms by virtually all professional pianists, but more complex polyrhythms are often executed by only coordinating endpoints and note order within the pattern, letting intuition or other musical factors like melodic design determine the details of timing. In the cases here, where there is no melodic material, this approach might well lead to the simplifications observed, which may be described as follows.

The standard musical notation for the 5:3 rhythm, Figure 4(1), shows that the repeating time-cycle would typically be conceptualized as being divided into 5 fast pulses (5/4 meter). Each fast pulse is then conceptually divided into 3 parts, to allow the slower hand taps to be slotted in the appropriate subdivision locations. Hence the cycle is

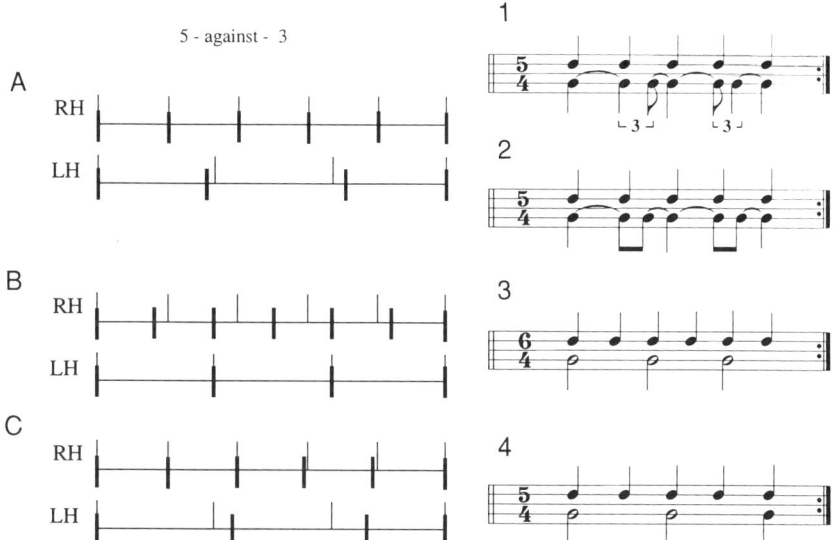

FIGURE 4 Schematic of the distortions in the reproduction of the 5-against-3 polyrhythm by musically trained subjects (one cycle of each repeating pattern is shown). Thin vertical lines represent the required positions of taps with either the right (RH) or the left (LH) hand. Thick vertical lines indicate actual tap positions in the reproductions of subjects. (A) The pattern produced by four subjects. (B) and (C) The patterns produced by the two remaining musically trained subjects. Corresponding musical notation is also given: (1) notates the target 5:3 polyrhythm; (2), (3), and (4), respectively, show close musical approximation to the actual performances of (A), (B), and (C).

divided overall into 15 parts. In strategy 4(2), used by four of the six subjects, the meter remains with 5 pulses per cycle (5/4) but the pulse subdivision is into only 2 parts and so the entire time cycle is now divided into only 10 parts. In strategy 4(3), the meter has now 6 pulses per cycle (6/4) with no further subdivisions. Finally, strategy 4(4) divides the cycle into only 5 pulses, with slow-hand pulses shifted toward neighboring fast-hand pulses.

In summary, training each hand independently did not lead to the use of a parallel organization when the two simple rhythms had to be produced concurrently. A similar finding was obtained by Klapp et al. (1987) for a 3-against-2 polyrhythm. The musicians in the Summers and Kennedy (1992) study found polyrhythmic 5:3 patterns difficult to produce and simplified them to patterns of roughly comparable time proportions whose level of hierarchical complexity was more manageable. Cycle subdivision was reduced from 15 to 10, 6, and even 5 parts. In all cases, played tap relations were in-phase or alternate-phase. The musicians preserved the number and order of taps in nearly all cases.

These simplifications do not follow known bifurcation routes in mathematical and natural systems, represented by the Farey tree (see Peper, Beek, & Van Wieringen, 1991), in any explicit fashion, which might be considered possible according to dynamic models. They are explainable, however, in dynamic terms to a considerable extent on the basis of in-phase and anti-phase entrainment. In cognitive terms they are explainable on the basis of reduction of cognitive load with minimal cognitive "dissonance."

Manipulation of a number of factors related to polyrhythm performance, therefore, has failed to induce the adoption of a parallel organization in either skilled or unskilled subjects. Before concluding that the operation of independent timing mechanisms for the two hands is not possible in concurrent tasks, it must be recognized that the tapping tasks studied in the laboratory studies are quite different from the piano performances studied by Shaffer (1981). It may also be that independence between the hands can be achieved, as is the case with highly skilled musicians, only after years of intense training.

C. Coordinating the Hands

Studies of polyrhythmic tapping indicate that when people have to execute two differently timed action sequences at the same time, the temporal constraints on bimanual coordination are overcome by interweaving the timing of the two hands. There are, however, a number of ways in which the two hands can be integrated in the performance of temporally dissonant patterns. In all cases, however, the assumption is that the hands operate on a common time base. One form of integrated organization involves the linear chaining of intervals within a cycle with the timing of a response being cued by the immediately preceding response, irrespective of hand. Alternatively, a hierarchical form of integrated organization may be employed involving higher order time intervals.

To determine the form of integrated organization adopted by subjects in studies of polyrhythmic tapping, various aspects of the tapping data can be examined. A hierarchical organization, for example, in which the movements of one hand are subordinate to movements of the other hand, may be inferred from an examination of the precision with which fast-hand and slow-hand intervals are reproduced over a cycle of a polyrhythm. The pattern of correlations between adjacent intertap intervals within a cycle has also been used to identify coordination strategies. For example, a hierarchical integrated organization in which movements of the slow hand are subordinate to the movements of the fast hand would predict strong negative correlations for adjacent between-hand intervals (Peters & Schwartz, 1989). In the 5:3

polyrhythm shown in Figure 1 the relevant correlations are between intervals I2,I3 and intervals I5,I6. Finally, regression analyses have been employed to distinguish between specific models of integrated motor organization (Jagacinski et al., 1988).

The general conclusion that emerges from these analyses is that subjects adopt a variety of strategies for integrating the timing of the two hands (e.g., see Figures 3 and 4). In the Summers et al. (1993c) study, for example, a chaining structure was the most frequently used organization for the simultaneous version of the 3-against-2 polyrhythm (see Figure 2), whereas for the shifted 3:2 pattern, a hierarchical structure was adopted by the majority of subjects.

In the performance of the more complex polyrhythms (i.e., 4:3, 5:3, 5:4), however, support has been consistently found for the use of a hierarchical organization in which the fast-hand beats form the time base into which movements of the slow hand are interlaced. First, in these patterns slow-hand responses usually exhibit greater variability than fast-hand responses. This trend is shown in Figure 5, where coef-

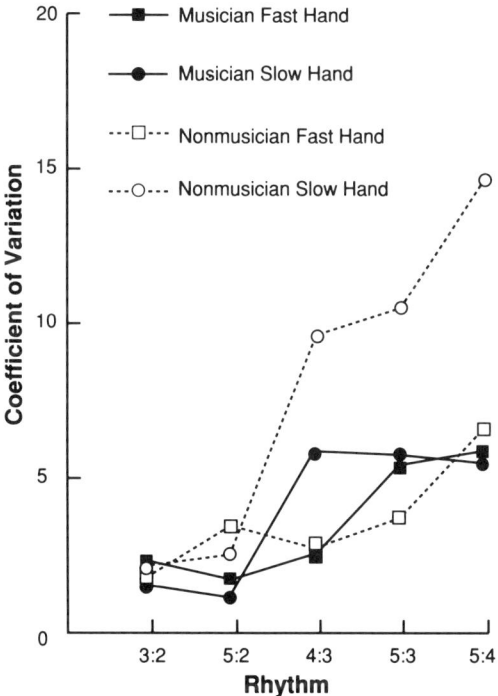

FIGURE 5 Coefficients of variation for fast-hand intervals and slow-hand intervals in the reproductions of five polyrhythms by musicians and nonmusicians. (From Summers et al., 1993b. Reprinted by permission.)

ficients of variation (i.e., standard deviations relative to the mean expressed as percentages) were plotted for fast-hand intervals and slow-hand intervals for five polyrhythms varying in complexity (Summers, Rosenbaum, Burns, & Ford, 1993b). As can be seen, the main difference in the performance of musicians and nonmusicians was in the precision with which slow-hand responses were made. Strong negative correlations for adjacent between-hand intervals have also been consistently obtained in the performance of the more complex polyrhythms (see Table I).

Further analysis of the intertap profiles produced in a continuation paradigm by nonmusicians in the Summers et al. (1993b) study showed that, across all the polyrhythms tested, these subjects maintained a relatively regular series of taps with the fast hand. Accurate reproduction of slow-hand intervals was also evident for the 3:2 and 5:2 pat-

TABLE I Expected and Obtained ITIs and Correlations between Adjacent ITIs within a Cycle for Subjects RB and TM[a]

Dy8	I1	I2	I3	I4	I5	I6	I7
			Subject RB				
EXP	426	284	142	426	142	284	426
OBS	430	275	142	418	135	295	430
SD	15.5	21.4	21.4	17.4	29.4	30.7	18.8
COR	−.38	−.70	−.18	−.29	−.89	−.11	
CV	Fast (right) hand = 1.9%			Slow (left) hand = 2.8%			
REV (Day 9)							
EXP	438	292	146	438	146	292	438
OBS	439	290	145	425	142	314	440
SD	16.7	35.6	22.1	19.6	24.7	32.1	16.9
COR	−.26	−.76	.07	−.34	−.73	−.13	
CV	Fast (right) hand = 2.74%			Slow (left) hand = 3.52%			
			Subject TM				
EXP	405	270	135	405	135	270	405
OBS	343	325	324	190	183	305	348
SD	18.8	24.7	23.6	16.1	25.4	26.4	20.2
COR	−.20	−.36	−.08	.01	−.31	−.03	
CV	Fast (right) hand = 42.6%			Slow (left) hand = 2.9%			
REV (Day 9)							
EXP	597	398	199	597	199	398	597
OBS	496	520	469	238	305	482	474
SD	80.3	81.7	45.0	19.1	65.7	58.9	94.3
COR	.46	.17	.11	.63	−.11	.20	
CV	Fast (right) hand = 48.3%			Slow (left) hand = 2.72%			

[a] ITI values in milliseconds; Dy8, Day 8; EXP, expected; OBS, observed; REV, reverse hands; SD, standard deviation; COR, correlation between adjacent ITIs; CV, coefficient of variation.

terns. In the performance of the 4:3, 5:3, and 5:4 polyrhythms, however, these subjects showed a systematic tendency to place slow-hand taps close to the middle of the relevant fast-hand beat. This tendency increased with speed of performance and was evident regardless of which hand (left or right) took the faster beat. These trends are illustrated in Figure 6, which shows the deviations from the expected proportions in the reproduction of the 4:3, 5:3, and 5:4 polyrhythms (cycle duration = 1500 msec) by an unskilled subject. Musical notations for the performed patterns are also given, and they make explicit the clear tendency to replace the ideal triplet subdivision with a division of pulse into two parts. This rather robust trend is easily explained by both models of motor behavior. In the emergent systems dynamic approach this is seen as a preference toward simpler attractors (in-phase or alternate-phase) when the system becomes less stable. In the cognitive approach, the use of simpler rhythmic subdivisions reduces cognitive (attentional) load.

Although musically trained subjects are more accurate than nonmusicians in the performance of polyrhythms (especially the slow-hand responses), they also appear to adopt a hierarchical timing system. Mean ITIs and correlations between adjacent ITIs within a cycle of the 5:3 polyrhythm (cycle duration = 2100 msec) produced by a musician (RB) in the Summers *et al.* (1993a) study are shown in Table I. The

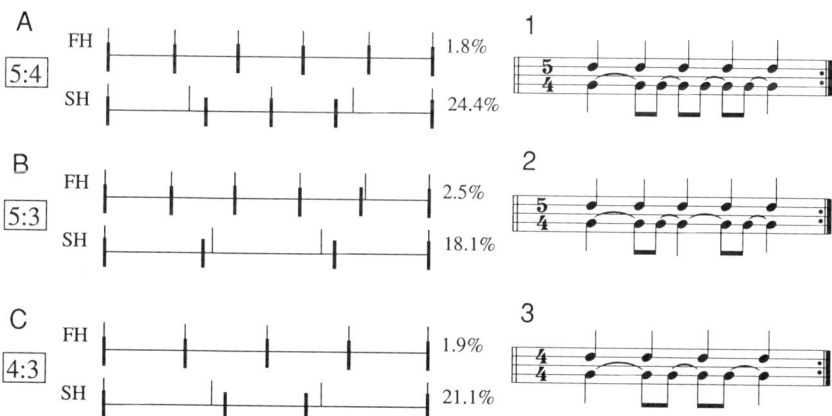

FIGURE 6 Schematic of the distortions in the reproduction of a 4:3, 5:3, and 5:4 polyrhythm by a nonmusician (one cycle of each repeating pattern is shown). Thin vertical lines represent the required positions of taps with either the fast (FH) or slow (SH) hand. Thick vertical lines indicate actual tap positions in the reproductions of the subject. Coefficients of variation (in percent) for fast-hand and slow-hand responses are also shown.

data shown are for the last day (Day 8) of training and the transfer condition (REV) on Day 9, in which the subject was asked to perform the polyrhythm with the reversed hand arrangement (i.e., fast hand = left, slow hand = right). To examine the extent to which subjects followed the metric structure of the task we divided the overall average cycle duration by 15 (i.e., 5 × 3, the smallest common denominator for the task), and then compared the expected and observed values for the intervals I1 to I7 (see Figure 1). For subject RB, for example, the shortest interval separating two taps should be 142 msec (2130/15) on Day 8 and 146 msec (2190/15) in the REV condition. As can be seen in Table I, RB adhered closely to the metric structure of the task. The ITI pattern, however, does not provide a clear indication as to the form of integrated organization (i.e., nonhierarchical vs hierarchical) adopted by this subject. The coefficients of variation for fast-hand (1.9%) and slow-hand (2.8%) responses and the pattern of correlations, in contrast, suggest a hierarchical timing strategy in which the fast-hand beats formed the time base into which movements of the slow hand were interlaced. Furthermore, an almost identical pattern of results was obtained when the subject performed the polyrhythm with the reverse hand arrangement.

Although using the faster beat as the time base has been the most commonly observed strategy in our studies, not all subjects used this form of hierarchical organization. For example, the ITI pattern produced by TM, a nonmusician, is also shown in Table I. Clearly, this subject was unable to use the metric structure inherent in the task. Rather, the subject appears to have adopted a hierarchical organization, but one in which slow hand beats were used as the time base for integrating the two hands. On Day 8, for example, fast-hand intervals varied greatly in duration (343, 649, 190, 488, 348 msec), whereas slow-hand beats were reproduced relatively accurately (668, 697, 660 msec). It appears that TM attempted to solve the concurrent task problem by subdividing each slow-hand beat into equal intervals. This strategy was used throughout the training period and in the reverse-hands condition. Although this strategy allowed TM to produce the correct sequence of key taps, he was never able to reproduce the correct timing pattern for the 5 : 3 polyrhythm.

The use of a hierarchical integrated organization is consistent with the attentional model of bimanual coordination (Peters, 1990). Subjects, when faced with the task of executing temporally dissonant motor sequences, give priority, through the allocation of focal attention, to one rhythm and interlace the other activity into the dominant rhythm. In the production of polyrhythmic sequences there appears to be a prefered allocation of attentional focus toward the faster of the two movement streams, regardless of hand arrangement (Peters & Schwartz, 1989).

Nevertheless, attention can be selectively given to the slower movement stream by some performers. Such attention appears to correspond to the placement of the fundamental pulse of the musical meter. A standard training procedure for percussionists, for example, is to learn a given polyrhythm (e.g., 2:3) with maximal security of execution and redundancy of memory coding by learning it three separate ways: with 2 as the fundamental pulse, 3 as the fundamental pulse, and as a linear composite. In a forthcoming paper (Pressing, Summers, Magill, 1993) we present evidence that experienced performers can adopt several explicit and distinct cognitive models for executing the same physical pattern, and that measured patterns of correlation readily distinguish between these cognitive models. The cases studied are 3:4 polyrhythms (executable with 2 distinct cognitive strategies) and a displaced 1:2 polyrhythm that can be executed using 5 different cognitive strategies. The different strategies are based on different cognitive choices made by the performer about what is to act as the fundamental pulse stream into which the taps played by the hands are to be inserted. This appears to indicate an irreducible cognitive component in timed behavior in some situations, even rather simple ones. It might be feasible, however, to interpret these cognitive models in the oscillator framework by specifying particular combinations of oscillator couplings and driving frequencies (e.g., deGuzman & Kelso, 1991), although we do not do so here. However (1) this may not offer any conceptual simplification, and (2) this points to a sense of equivalence between the language of oscillators and cognition rather than an incompatibility.

IV. A Model of Polyrhythmic Tapping

In this final section we shall briefly outline a model that has been proposed for how, using a counting strategy, subjects could perform various polyrhythms (see Summers et al., 1993b) for a full account of the model). The two basic assumptions of the model are: (1) that subjects adopt a hierarchical timing system in which movements of the slow hand are normally subordinate to movements of the fast hand; and (2) that a single mechanism (a counter) controls the serial ordering and timing of responses.

In a hierarchical timing system, two kinds of information would seem necessary to correctly perform slow-hand responses in polyrhythms. The first involves determining into which fast-hand beat(s) slow-hand responses are to be inserted. The second involves determining when within the relevant beat(s) a slow-hand response should be executed. The requirements for correctly placing slow-hand responses in the correct beats is illustrated in Figure 7 for a 4-against-3 polyrhythm with a 1500-msec cycle duration (i.e., fast-hand taps every

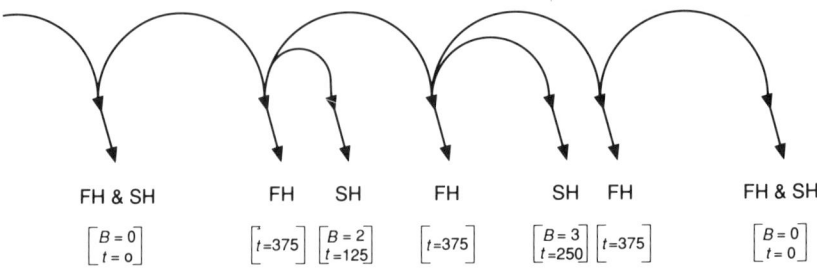

FIGURE 7 A model for a 4-against-3 polyrhythm performed at a 1500-msec cycle duration. (From Summers *et al.*, 1993b. Reprinted by permission).

375 msec and the slow hand every 500 msec). Slow-hand responses should occur at the beginning of each cycle (synchronous with a fast-hand tap), 125 msec after the second fast-hand response, and 250 msec after the third fast-hand response. These requirements suggest that one operation that must be performed is the counting of beats within a cycle. Furthermore, when all the beats in a response cycle have been completed, the fast-hand and the slow-hand response can be performed simultaneously and the beat count reset to zero.

Another operation is determining the timing of fast-hand responses. According to the model, this decision is made both by incrementing a timer that is reset after each beat is completed, and by waiting for the timer to reach a target value before the next fast-hand response is made. Slow-hand responses are assumed to be initiated when requisite delays have transpired after the triggering of preceding fast-hand responses. To trigger slow-hand responses at the correct time, however, two conditions must be met: The number of beats (B) completed in the cycle must have the desired value, and the delay (t) within the current beat must also have the desired value (see Figure 7). In this model, therefore, the timing of slow-hand responses is dependent on the timing of fast-hand responses, but the timing of fast-hand responses is independent of slow-hand response timing. This general model provided a good fit to the performance data obtained from both skilled and unskilled subjects and across a variety of polyrhythms (Summers *et al.*, 1993b).

V. Conclusion

In this chapter we have presented a review of recent work on the production of polyrhythmic sequences. The general finding that has emerged is that subjects attempt to overcome the temporal constraints on bimanual coordination in these tasks by interweaving the timing of the two hands. The most common integration strategy observed was to use the fast hand as the time base into which slow-hand responses are inserted.

We have also highlighted the differing and often equally viable explicative credibility of dynamical and cognitive approaches to the organization of such patterns. One further example of this difference may be given by looking at the acquisition of new patterns. In the dynamic/action approach, a new pattern comes from a decoupling process, and by defining or stabilizing new attractors in the phase space of the problem. In the cognitive approach, a programmable memory is explicitly invoked. For simple patterns both methods seem to have ready applicability. For more complex patterns, or sophisticated timed motor skills such as music performance, it is hard to see how the equations of motion involving the relevant neuromotor components can encode the specificity of action required, or indeed what methodology would allow the derivation of suitable equations, to say nothing of their solution. The boundary conditions would be, for example, highly complex and time varying. It should also be remembered that software systems used in robot control use an explicitly cognitive (motor program) design, and this must also cast doubt on the practical explicative power of a purely dynamical approach, even as the constraints of oscillator theory are incontrovertibly elegant and well founded, and readily observable in simpler types of action.

Acknowledgment

The studies of bimanual coordination from J. J. Summers's laboratory that are reported in this chapter were supported by Australian Research Council Grant A78831548.

References

Bregman, A. S., & Campbell, J. (1971). Primary auditory segregation and perception of order in rapid sequences of tones. *Journal of Experimental Psychology,* **89,** 244–249.
Craske, B., & Craske, J. D. (1986). Oscillator mechanisms in the human motor system: Investigating their properties using the aftercontraction effect. *Journal of Motor Behavior,* **18,** 117–145.
deGuzman, G. C., & Kelso, J. A. S. (1991). Multifrequency behavioral patterns and the phase attractive circle map. *Biological Cybernetics,* **64,** 485–495.
Deutsch, D. (1983). The generation of two isochronous sequences in parallel. *Perception & Psychophysics,* **34,** 331–337.
Jagacinski, R. J., Marshburn, E., Klapp, S. T., & Jones, M. R. (1988). Tests of parallel versus integrated structure in polyrhythmic tapping. *Journal of Motor Behavior,* **20,** 416–442.

Keele, S. W., & Ivry, R. I. (1987). Modular analysis of timing in motor skill. In G. Bower (Ed.), *The psychology of learning and motivation* (pp. 183–228). New York: Academic Press, Inc.

Kelso, J. A. S., & deGuzman, G. C. (1988). Order in time: How the cooperation between the hands informs the design of the brain. In H. Haken (Ed.), *Neural and synergetic computers* (pp. 180–196). Berlin: Springer-Verlag.

Kelso, J. A. S., and Jeka, J. J. (1992), Symmetry breaking dynamics of human multilimb coordination. *Journal of Experimental Psychology: Human Perception and Performance*, **18,** 645–668.

Kelso, J. A. S., & Schöner, G. (1988). Self-organization of coordinative movement patterns. *Human Movement Science*, **7,** 27–46.

Kelso, J. A. S., Southard, D. L., & Goodman, D. (1979). On the coordination of two-handed movements. *Journal of Experimental Psychology: Human Perception and Performance*, **5,** 229–238.

Kelso, J. A. S., Holt, K. G., Rubin, P., & Kugler, P. N. (1981). Patterns of human interlimb coordination emerge from the properties of non-linear, limit-cycle oscillatory processes: Theory and data. *Journal of Motor Behavior*, **13,** 226–261.

Kinsbourne, M., & Hicks, R. E. (1978). Functional cerebral space: A model for overflow, transfer and interference effects in human performance. In J. Requin (Ed.), *Attention and performance VII* (pp. 345–362). Hillsdale, NJ: Erlbaum.

Klapp, S. T., Martin, Z. E., McMillan, G. C., & Brock, D. T. (1987). Whole-task and part-task training in dual motor tasks. In L. S. Mark, J. S. Warm, & R. L. Hutton (Eds.), *Ergonomics and human factors*. New York: Springer-Verlag.

Marteniuk, R. G., MacKenzie, C. L., & Baba, D. M. (1984). Bimanual movement control: Information processing and interaction effects. *Quarterly Journal of Experimental Psychology: Human Experimental Psychology*, **36,** 335–365.

Peper, C. E., Beek, P. J., & Van Wieringen, P. C. W., (1991). Bifurcations in bimanual tapping: In search of Farey principles. In J. Requin & G. E. Stelmach (Eds.), *Tutorials in motor neuroscience* (pp. 413–431). N.A.T.O. Advanced Study Institute. Dordrecht: Kluwer.

Peters, M. (1977). Simultaneous performance of two motor activities: The factor of timing. *Neuropsychologia*, **15,** 461–464.

Peters, M. (1990). Interaction of vocal and manual movements. In G. E. Hammond (Ed.), *Cerebral control of speech and limb movements* (pp. 535–574). Amsterdam: North-Holland.

Peters, M., & Schwartz, S. (1989). Coordination of the two hands and effects of attentional manipulation in the production of a bimanual 2:3 polyrhythm. *Australian Journal of Psychology*, **41,** 215–224.

Pressing, J. (1984). Cognitive processes in improvisation. In R. Crozier & A. Chapman (Eds.), *Cognitive processes in the perception of art* (pp. 345–363). Amsterdam: North-Holland.

Pressing, J., Summers, J. J., & Magill, J. (1993). *Cognitive multiplicity in motor coordination*. Manuscript in preparation.

Scholz, J. P., & Kelso, J. A. S. (1989). A quantitative approach to understanding the formation and change of coordinated movement patterns. *Journal of Motor Behavior*, **21,** 122–144.

Shaffer, L. H. (1981). Performances of Chopin, Bach, and Bartok: Studies in motor programming. *Cognitive Psychology*, **13,** 326–376.

Shaw, R. E., Kadar, E., Sim, M., & Repperger, D. W. (1992). The intentional spring: A strategy for modeling systems that learn to perform intentional acts. *Journal of Motor Behavior*, **24,** 3–28.

Summers, J. J., & Kennedy, T. (1992). Strategies in the production of a 5:3 polyrhythm. *Human Movement Science,* **11,** 101–112.

Summers, J. J., & Pressing, J. (1993). *Doing impossible motor coordination tasks.* Manuscript in preparation.

Summers, J. J., Ford, S., & Todd, J. A. (1993a). Practice effects on the coordination of the two hands in a bimanual tapping task. *Human Movement Science,* **12,** 111–188.

Summers, J. J., Rosenbaum, D. A., Burns, B. D., & Ford, S. K. (1993b). Production of polyrhythms. *Journal of Experimental Psychology: Human Perception and Performance,* **19,** 416–428.

Summers, J. J., Todd, J. A., & Kim, Y. H. (1993c). Influence of perceptual and motor factors on bimanual coordination in a polyrhythmic tapping task. *Psychological Research,* **55,** 107–115.

Swinnen, S., Walter, C. B., & Shapiro, D. C. (1988). The coordination of limb movements with different kinematic patterns. *Brain and Cognition,* **8,** 326–347.

Tuller, B., & Kelso, J. A. S. (1989). Environmentally-specified patterns of movement coordination in normal and split-brain subjects. *Experimental Brain Research,* **75,** 306–316.

Turvey, M. T. (1990). Coordination. *American Psychologist,* **45,** 938–953.

Vorberg, D., & Hambuch, R. (1978). On the temporal control of rhythmic performance. In J. Requin (Ed.), *Attention and performance VII* (pp. 535–555). Hillsdale, NJ: Erlbaum.

Vorberg, D., & Hambuch, R. (1984). Timing of two-handed rhythmic performance. In J. Gibbon & L. Allan (Eds.), *Timing and time perception* (pp. 390–406). New York: New York Academy of Sciences.

Wing, A. M., & Kristofferson, A. B. (1973). Response delays and the timing of discrete motor responses. *Perception & Psychophysics,* **14,** 5–12.

Yamanishi, J., Kawato, M., & Suzuki, R. (1980). Two coupled oscillators as a model for the coordinated finger tapping by both hands. *Biological Cybernetics,* **37,** 219–225.

27

Does Handedness Play a Role in the Coordination of Bimanual Movement?

Michael Peters

Department of Psychology
University of Guelph
Guelph, Ontario, Canada

I. Introduction
II. The Role of Handedness As Inferred through Observations of Naturally Occurring Bimanual Behaviors
    A. Hand Roles
    B. Handedness Classification
III. The Role of Handedness in Experiments That Examine Bimanual Coordination
    A. Right-Handers
    B. Left-Handers
IV. Summary and Conclusions
    References

I. Introduction

In order to put coordinated bimanual movement in humans into proper perspective, a brief quasi-evolutionary comment on interlimb coordination is appropriate. During locomotion, there is an obligatory coordination between the limbs in the sense that reciprocal innervation ensures the smooth collaboration of extensors and flexors in the opposite upper and lower limbs. In some vertebrate classes (amphibians, birds) bilaterally symmetrical movement of limbs in locomotion is common. In mammals, bilaterally symmetrical movement is widespread only among marsupials, although several eutherian orders show bilaterally symmetrical hopping (e.g., some rodents and lagomorphs), and bilaterally symmetrical movements are seen during flying in bats and swimming in pinnipeds. The most common form of interlimb coordination in mammals, however, consists of alternating movements of the limbs. In the higher vertebrates, the coordinations that reach across the midline can be readily interrupted in the case of both alternating and symmetrical movements, when directional maneuvres are required. The ability to interrupt the flow of bilateral coordinations, essential to locomotion in a world that offers obstacles and requires sudden changes in direction, forms the evolutionary basis for skilled bimanual coordination in humans. Supraspinal mechanisms have to be able to disrupt the spinal oscillatory network that underlies locomotion, so that asymmetry in movement can be achieved (much like a slip differential in a car will allow one or the other rear wheel to revolve more quickly). It is probable that skilled collaborative coordination of the hands in bimanual movement builds on the ability to disengage oscillatory networks from their fixed relationship with motor neurons. This is not to say that disengagement means independence from oscillators. Indeed, the ability to rejoin a common collaborative mode of movement after momentary independence presupposes some common time base that is used to integrate the limbs.

The unusual aspect of bimanual skilled movement in humans is that performance is intermittent and goal oriented. While it is abundantly clear that the separate movements of the two hands have to take place relative to some common shared timing reference (Peters, 1985, 1990a), it is true that everyday life bimanual activities are characterized by the assignment of qualitatively different tasks to the two hands. It is also true that bimanual coordination involves the differential allocation of attention to the two hands so that the awareness of what the two hands are doing is not equivalent. As an example, when peeling a potato, the foreground of the attention is on the hand that wields the peeler while the positioning movements (invariably carried out by the nonpreferred hand) are not directly attended to. This is just another way of saying that handedness is an important aspect of bimanual coordination and that the skilled coordinations in bimanual movement in humans are really quite different from the more obviously rhythmic activities during locomotion. By and large, the literature dealing with bimanual coordination does not address the question of handedness. With few exceptions (e.g., Peters, 1985; Walter & Swinnen, 1990a), the role of the two hands is considered equivalent and hand task assignments are not counterbalanced.

One of the reasons for the neglect of handedness lies in the nature of the experimental tasks that are most commonly used to study interlimb coordination. Our description of the fundamentals of interlimb coordination has stressed the occurrence of coupled movements across the midline, in which some quite marked adherence to a common oscillator network is important. In many of the experimental approaches to interlimb coordination, notably in kinesiology, the two hands perform qualitatively similar trajectories, and rhythmical repetition is frequently a feature of such experiments. In a way, subjects are asked to behave *as if* their movements are driven by oscillators, and the paradigms show strong resemblances to naturally occurring locomotory behaviors. In contrast, skilled bimanual movements in performance is, as has been noted, intermittent and goal directed.

It is reasonable to assume that theories of interlimb coordination that are based on the "oscillator" tasks will differ from theories or models concerned with tasks that more directly reflect real-life bimanual activities. In the former, factors like differential skill and attentional asymmetries are not highly important and are therefore not emphasized. This allows considerable elegance in the design and analysis of experiments. When aspects such as handedness and attentional asymmetries are introduced, vague hypotheses replace elegant theories and experimental control becomes difficult. Nevertheless, there are some heuristic advantages to a broadened perspective on bimanual coordination. A simple illustration can be given of how attentional processes can make a significant impact on bimanual coordination. Imagine holding the forearms horizontally in front of the body, so that the hands are in the same plane, and almost meet at the midline. Then, attempt to perform counterclockwise circular movements with the forearms, moving one forward as the other moves backward. This is an exceedingly hard task. However, the task becomes manageable when the preferred hand is made to rotate more quickly relative to the nonpreferred hand, and by deliberately attending to the preferred hand. While the outcome of this simple manipulation is not difficult to interpret within an attentional model of bimanual coordination, it is rather more difficult to explain in terms of oscillator networks and coordination linkages.

This particular chapter invites the reader to consider bimanual coordination in a broader context. The question asked here is whether the observed preference asymmetries that underlie handedness play a significant role in bimanual coordination. Does the specific assignment of hand roles matter in terms of the quality of coordination, and how would evidence that this is so affect models of manual coordination? Asymmetries in bimanual coordination can arise from asymmetries in the motor performance characteristics of the two hands, and these, in turn, can arise from or interact with differential motor experience in the two hands (e.g., van Emmerik, 1992). Because of such interactions, the influences of underlying lateral asymmetries and differential motor experience cannot be controlled for in a comprehensive way. Why, then, pursue the question? There are two reasons. At worst, experimental evidence will provide practical information about human/machine design in terms of what activities should be assigned to the preferred hand and what activities should be assigned to the nonpreferred hand in order to ensure efficient and safe coordination. At best, study at the neuropsychological level can point out the range of behavioral phenomena that have to be accounted for by those who work at the more basic neurophysiological level, or for those who construct models of coordinative movement.

II. The Role of Handedness as Inferred through Observations of Naturally Occurring Bimanual Behaviors

A. Hand Roles

One of the most obvious approaches to the question of bimanual asymmetries is to see how humans allocate hand/task roles in everyday behavior. In right-handers the situation is quite clear, as some selected examples will show. In shaping a stone tool, one of the earliest cases in which the handedness of the toolmaker can be ascertained, the preferred right hand wields the flaking tool while the nonpreferred left hand positions the artifact that is shaped. In writing, the right hand writes, and the left hand positions the writing substrate so that the position of the writing hand is relatively constant (Guiard & Millerat, 1984). In peeling an apple, the right hand performs the peeling action while the left hand supports and positions the apple so that the stroke of the right hand is in the correct location. In shooting a rifle, the right hand determines when exactly the rifle fires while the left hand positions and stabilizes the rifle. In playing the violin, the right hand produces the sound while the left hand prepares the string so that the bow can act on it. On a more complex level, in playing keyboard music, the right hand tends to provide the melodic line while the left hand provides the rhythmic structure. Naturally, there are many exceptions to this last statement in more difficult piano pieces but in terms of both the history of keyboard music as well as in the ontogenetic progression of the keyboard player the statement is valid (Peters, 1986). In all of these activities, the complementarity of the hand roles is obvious. It is also clear that the coordination of manual activity involves the sequencing of the separate acts of the hands in time. In the majority of bimanual activities, the nonpreferred hand will perform its action *before* the preferred hand does its work. This is even the case for the earliest manifestation of bimanual coordination that has been studied properly, the reaching for objects by infants. Here, deSchonen (1977) reports that the left hand is initially extended before the right hand, to provide a spatial reference for the right hand that reaches out for the object. In experimental tasks, where there is no complementary aspect to the two component movements, the order of hand movements may be quite different (e.g., Kay, Saltzman, Kelso, & Schöner, 1987).

Several aspects of hand-role allocation in these examples of naturally occurring bimanual activities in right-handers are noteworthy:

1. In all cases, the primary goal of the coordinated movement is more directly realized through the preferred hand, and the nonpreferred hand acts in support.

2. In all of these tasks, whether simple or complex, coordination involves the precise timing of the onset of right-hand movement relative to the preparatory action of the left hand. Because the left nonpreferred hand acts in preparation to activities of the right hand, its movements will in many cases precede the movements of the preferred hand. In exceptional cases, as in keyboard playing, the activities of the right and left hand may be concurrent, or intercalated. In this particular case, a common time base is likely used in determining the onset and offset of movement in the two hands.

3. In general, the right hand will perform movements that tend to require continuous precise adjustment of speed and force. The left hand will tend to move intermittently, assuming hand postures and positions.

4. When attention is allocated to the two hands in turn, the requirements for attention to the activities of the left hand tend to be brief and intermittent, while attention to the preferred right hand has more of an "on-line," continuous quality. The preferred hand normally receives focused attention (figure) while the nonpreferred hand is not directly attended to (ground), or receives only subsidiary attention (Peters, 1990a).

In general, the above principles are reflected in the operation of machines, musical instruments, and tools and toys in general. In the design of these, the hand-role preference of right-handers is the predominant influence in deciding hand-role allocation. While the designers may not necessarily have always been right handed, the designs do not take the preferences of left-handers into account. This even applies to foot laterality. For instance, in cars, the accelerator is operated by the right rather than the left foot and this arrangement remains the same regardless of whether the vehicle is steered with a right- or a left-handed arrangement. When the movements of hands and feet have to be coordinated, hand movements are given precedence in general, and the right foot will tend to be the foot that is chosen as the foot that fulfills a function that is most directly related to that of the right hand (Peters, 1988). This is exemplified by the controls of a motorbike. Here, the right hand operates the accelerator while the right foot operates the brake. Both the right hand and right foot are directly involved in changing the rate of speed of the vehicle. The left hand (operates the gear shift) and the left foot (operates the clutch) act in a supportive and preparatory way for the actions of the right hand. In contrast to the design of motor cars, where the position of the steering wheel dictates the hand that operates the gear shift, the design of controls of motorbikes reflects the universal preference patterns of right-handers, and is not influenced by the side of the road that is used for driving in a particular country. In general, preferences in multilimb coordination are such that preference is given to the right body half over the left body half, and feet are given a lower priority than hands (Forster & Webster, 1991; Peters, 1988).

In conclusion, observation of naturally occurring tasks shows that asymmetries in multilimb coordination are ubiquitous. The fact that asymmetries of this kind are common does not allow a conclusion about the origins of these asymmetries. One approach to distinguish between compelling or merely habit-formed sources of the asymmetries would be to see how well individuals adapt to role reversals. However, naturally occurring bimanual activities are characterized by a double dissociation; it is not only the case that the preferred hand is specialized for certain activities, with the nonpreferred hand taking the default role, but the nonpreferred hand is specialized as well, both in terms of its movement role and its relation to conscious intent. Any thorough experimental manipulation of role reversal would involve prohibitively long training sessions. This is so for right-handers, but also for left-handers. However, left-handers, far more often than right-handers, encounter hand/role reversal as a consequence of living in a world designed by and for right-handers. Whether or not the ability of left-handers to adapt to role reversals better than right-handers is due to a different history of motor experience or due to fundamental (brain organization) differences between right-handers and left-handers can be answered only with reference to a brief discussion of a current understanding of handedness.

B. Handedness Classification

The classification of handedness has proven to be an unexpectedly difficult task, and, considering how many investigators have looked at this question, there is remarkably little consensus. Briefly, handedness can be defined on the basis of hand preference for certain tasks or hand performance. In terms of preference patterns, cultural influences interact with inherent predispositions so that the prevalence, for example, of left-handed writing may differ by less than 2% in some cultures, to more than 12% in North America. In the following, North American prevalence figures will be used because it is assumed that in this culture the pressure against use of the left hand for certain activities, while not totally absent, is relatively small. Recent figures, based on more than 1 million U.S. respondents (Gilbert & Wisocki, 1992), suggest that approximately 14% of males and 12% of females in the current generation are left-handed writers. This is likely an underestimate of genotypic left-handers but represents a figure that is closer to the "real prevalence" of left-handers than other, lower, estimates.

An immediate concern in terms of comparisons between left-handers and right-handers is whether the two handedness groups are equivalent or not. Some investigators maintain that right-handedness is the normal handedness status of all individuals and that left-handedness is a result of pathology, however defined (Bakan, 1990; Coren & Halpern, 1991). If so, little can be learned from left-handers that would be of use for right-handers. However, careful evaluation (Harris & Carlson, 1988; Harris, in preparation) makes it clear that the claims for left-handed pathology can, at most, apply to a small proportion of left-handers. Judging by motor performance, there is no evidence that there is any pathology involved in left-handedness in the population most often studied in experimental work on handedness, university students. Another concern involves the subclassification of left-handers. By definition, left-handers prefer to use the left hand for activities involving fine manual skill. However, almost 50% of left-handed writers prefer the right limb for throwing (Gilbert & Wisocki, 1992); they throw better with the right arm and the right arm is stronger than the left arm (Peters, 1990b; Peters & Servos, 1989). Because of the dissociation between strength and fine manual skill in left handers, studies in which left-handers are to be compared to right-handers must involve a subdivision of left-handers into at least two subgroups. Whether or not further subgrouping is advisable remains uncertain; cluster analysis of a large number of individuals shows that there may be as many as five handedness subgroups that are recognized as distinct by a discriminant function analysis (Peters & Murphy, 1992).

How does handedness relate to bimanual coordination? The question is best approached through Liepmann's (1905) model of apraxia. Apraxia denotes a disturbance in higher order motor organization that cannot be reduced to difficulties in movement execution but concerns the ability to carry out sequences of movements that serve a particular goal. Liepmann noted apraxic disturbances in the *left* limb after damage to the left cerebral cortex. In contrast, apraxia in the *right* limb after right hemisphere damage was rarely observed. From this, Liepmann drew the conclusion that the left hemisphere was specialized for the planning and sequencing of voluntary motor activities. An immediate implication of this hypothesis is that in bimanual motor control, the commands to the executing motor machinery in the right and left hemisphere are issued by the left hemisphere only. Such commands would reach the executing motor machinery (i.e., primary motor cortex) of the left hemisphere, and therefore the right hand, directly. Access to the right hemisphere executing machinery for the left hand would be indirect. In 1905, Liepmann expressed this in a somewhat extreme way: "Righteousness implies that the right hand can do many things which the left hand cannot do. Our results suggest that even those things which can be done by the left hand cannot be attributed to right hemisphere function, but are due to left hemisphere activity" (p. 2375, loosely translated by author). Much of the evidence of mirror movements in the left limb after loss of the right preferred limb (Schott, 1980) is in harmony with this postulate. It should be noted here that because the preferred hand and arm are used for most highly skilled manipulative activities in which hand choices can be made, there is no equivalent situation of mirror movements in the preferred right arm after the left has been lost. This, in itself, is a rather strong reflection of asymmetries in arm- and hand-movement roles. Although Liepmann acknowledged the possibility that there would be exceptions to this arrangement, and particularly so in left-handers, he felt that left hemisphere dominance for "eupraxis" could be considered established for most right-handers. Even early sceptics of Liepmann's model admitted that there was, overall, an asymmetry favoring the left hemisphere of right-handers (Brun, 1921). Faglioni and Basso (1984), in a more recent review, suggest that in left-handers, the converse arrangement holds. That is, even in spite of the fact that left-handers as a group show a clear left hemisphere specialization for motor speech, they will tend to show apraxic disturbances predominantly after *right* hemisphere damage.

In summary, Liepmann's (1905) model suggests that in bimanual coordination, one brain half is primarily involved in the planning and sequencing of skilled motor behavior. For right-handers, this is the left hemisphere, and for left-handers it is (with less certainty) the right hemisphere. However, it is necessary to intercalate an additional level between planning and general sequencing of movement and the final implementation. The sequencing of movement components, especially when learned movements are concerned, requires the ordering of movement elements in the same sense that syntactic rules order sentence elements. In movement plans, such ordering must draw on information from both hemispheres and the element of laterality enters only when movement commands are issued for the two hands. Because both hands may be required to commence or terminate movement sequences at precisely the same time, there is an advantage to an arrangement in which the initiation and termination of the movement trajectories in the two hands are issued by a unilateral source (Peters, 1990a). In right-handers, then, the commands for the initiation and termination of movements in both hands originate from the left hemisphere. This means that in right-handers the left hemisphere is specialized for the planning and sequencing of skilled movements, the initiation and termination of movement trajectories in both hands (the essential aspect of coordination), and the actual implementation and guidance of movement in the preferred right hand. In addition, and in logical support of such an arrangement, an argument has also been made that there is an attentional bias toward right-hand activities that harmonizes with the idea of a specialization of the left hemisphere in generative action (Corballis, 1989). This allows the prediction that even in the situation where simple movements of the two hands are used, bimanual coordination will be most successful if attention is focused on the preferred hand, and when the preferred hand performs the activity that is most demanding in attentional terms.

The preceding consideration of handedness indicates that left-handers cannot be considered to be simple mirror images of right-handers if only for the reason that fine manual skill in left-handers is executed via the right hemisphere, while the important motor speech specialization lies in the left hemisphere. Our current understanding of handedness does not allow a prediction of how exactly left-handers will differ from right-handers in managing bimanual coordinated movements, and in the ability to adapt to hand/role reversals. Some empirical evidence on this question is available from relatively simple experimental motor tasks. Such tasks tend to have the advantage that they are not contaminated with specific practice effects, although the attentional biases described before can be expected to influence coordination strategies. The available evidence will be discussed in the following section. For purposes of greater clarity, the results for right-handers and left-handers will be treated separately.

III. The Role of Handedness in Experiments That Examine Bimanual Coordination

A. Right-Handers

The reader might object that the distinction between normally occurring and laboratory tasks is forced. No claim is made here that laboratory motor tasks differ categorically from normally occurring tasks, but it is true that laboratory tasks tend to involve simple movements, and a great effort is made to reduce the individual components to such an extent that they can be measured with some degree of accuracy. In addition, the tasks are chosen to suit the relatively limited models that are meant to account for their performance. In the following review, it will be seen that, in many cases, the tasks that are studied with the greatest degree of accuracy, and for which quite exhaustive analytical models are offered, do not involve a control for hand and hand roles when bimanual activities are involved. Those studies in which hand roles and handedness were incorporated into the experimental design also happen to involve more loosely defined tasks and incompletely developed formal analysis. Nevertheless, there is sufficient evidence on the role of handedness in bimanual coordination to suggest that this dimension should be given some consideration in dynamic approaches to questions of motor skill.

One final point needs to be considered before looking at the literature. When questions about coordination are asked, there are two basic approaches that can be taken. First, one can carefully look at the complementary behavior of the two hands in tasks that require collaboration. Second, one can challenge the ability of the two hands (i.e., the brain) to guide different motor tasks, with the intent to see what the limitations in performing different concurrent tasks can tell us about basic coordinative mechanisms.

One of the earliest reports on bimanual concurrent motor activities was published by Welch (1898), who asked subjects to produce a stable tracking force with one hand while changing force rhythmically with the other hand. Welch examined the effects of the rhythmic force changes on the steadiness of tracking. He found that tracking performance of the right hand was more adversely affected by concurrent rhythmic changes in force in the left hand than in the converse arrangement. This is one of the first documentations of an asymmetric interference effect. Corresponding asymmetries have been found in other, quite different tasks. Perhaps the most directly comparable task is that used by Walter and Swinnen (1990a), in which subjects were required to perform a single unidirectional movement with one arm and a more complex movement with the other arm in which the direction of movement was changed. Subjects had to perform the movements within a given time spain. There was an interference asymmetry in the sense that performance was worse when the left arm performed the more demanding reversal movement and the right arm performed the unidirectional movement than in the converse condition. A task that was similar in principle was used by Schmidt, Treffner, Shawn, and Turvey (1992), but in this case no systematic hand-role reversal was explored.

However, in their case, 3 out of the 4 subjects spontaneously chose the hand/role combination that is natural for right-handers, with the right hand taking the faster of the two movements. Jeeves, Silver and Jacobson (1988) used a tracking task that yielded asymmetries comparable to those reported by Walter and Swinnen (1990a). Subjects had to track a diagonal line by moving a pen with two cranks; one moved the pen in the vertical direction and the other moved it in the horizontal direction. The steepness of the diagonals could be controlled by the ratio of turns in the two cranks. Jeeves *et al.* found that performance was best when the right hand turned the crank more quickly than the left hand. These findings were made with acallosal patients, and Preilowski (1972), who had made similar observations, interpreted the asymmetry in terms of the role of the forebrain commissure in bimanual coordination. There is little doubt that the corpus callosum does play an important role in bimanual coordination, but the same kind of asymmetry is, of course, seen in intact individuals. A different kind of asymmetry was reported by Marteniuk, MacKenzie and Baba (1984). They used a Fitts-type paradigm in which movement times of the two hands were shown to be differentially affected when the other hand was or was not burdened with an additional weight. In this work, the movement time of the left hand was linked to the weight condition in the right hand, but the converse was not true. In these studies, the movements required are still somewhat complex, so that asymmetries in movement execution could have a bearing on the results.

However, such asymmetries are even observed when the movements themselves are exceedingly simple. When subjects are required to tap rhythmically with one hand while performing slow or fast tapping movements with the other hand, asymmetries emerge that cannot easily be accounted for in terms of asymmetries of the executing machinery. Peters (1981) asked subjects to tap in synchrony with a metronome with one hand while tapping as quickly as possible with the other hand. Here, too, asymmetries emerged. Subjects did far better when the left hand tapped in synchrony with the relatively slow beat of the metronome, and the right hand tapped as quickly as possible, than with the converse condition. In a somewhat similar experiment, Ibbotson and Morton (1981) also showed a performance asymmetry. Subsequent studies (Peters, 1985) confirmed that in tapping tasks of this kind, performance in right-handers was better when the right hand took the activity that was more demanding of attention than the activity for the left hand. In general, the condition that was most demanding of attention was when the hand was to tap as quickly as possible. The general principle that performance in right-handers is better when the right hand performs the faster pace, even when the faster pace is considerably slower than the maximal rate that can be achieved, is tacitly acknowledged by the experimental design of studies in which there is no counterbalancing of hand roles (e.g., Farnsworth & Poynter, 1931; McLeod, 1975; Kelso, Holt, Rubin & Kugler, 1981; Klapp, 1979; Deutsch, 1983; Swinnen, Walter, & Shapiro, 1988; Walter & Swinnen, 1990b). It is of some interest to note that hand asymmetries are not generally observed when the hands perform polyrhythms that require both hands to follow an external pacing source (Summers, 1989). While in this particular case subjects will prefer to perform the faster of the two temporal chains with the preferred hand, if given a choice, they will perform equally well with either combination if both are tested. However, the general principle that performance is better when attention is paid to the "faster" hand still holds, even in the case of polyrhythms, when the direction of attention focused on the two hands is experimentally manipulated (Peters & Schwartz, 1988). This latter point is important. When the bimanual organization essentially assigns an equal status to the movements of the two hands, and when these movements are counted off from a single temporal grid the resolution of which is determined by the common denominator of the temporal chains in the two hands, an asymmetry in the movements of the two hands is not expected. An asymmetry is expected only when there is a differential focusing of attention and an unequal allocation of "effort" (Kahneman, 1973) to the two hands. The dimension of effort is not trivial. For instance, if one hand taps a rhythm and the other taps as quickly as possible, an intuitive guess might be that attention is focused on the hand that taps the rhythm. After all, the generation of a repetitive rhythm would be considered by many to be a more complex task than simple speeded tapping, and would intuitively be assumed to be more demanding of attention. However, the production of the rhythm can proceed in an automated, or semiautomated way (Kahneman & Treisman, 1983) while the speeded tapping demands continuous "on-line" attention.

The previous selection of studies shows that asymmetries in the coordination of the two hands can be observed in a variety of experimental conditions. The question is: can such asymmetries also be seen in left-handers? It has previously been stated that apraxic disturbances in left-handers tend to occur after right, rather than left, hemisphere damage (Faglioni & Basso, 1984), and this would lead to the expectation that left-handers would tend to show asymmetries that are the reverse of those seen in right-handers. However, there is also some legitimate question as to whether such an expectation is justified because left-handers do have the speech motor machinery in the left hemisphere, and numerous writers (e.g., Kimura, 1979) have pointed out the fact that motor sequencing requirements for speech and manual movements are quite similar.

B. Left-Handers

That the situation with regard to left-handers is likely to be confusing can already be anticipated on the basis of the studies that examine the interference effects of concurrent skilled hand movement and speech. In right-handers, interference effects tend to be seen most often with concurrent speech and right-hand movement. This is explained on the basis of an "interference in cerebral space" model that stresses the common left hemisphere origin of preferred-hand motor control and motor speech control in right-handers (Kinsbourne & Hicks, 1978). In left-handers, the evidence is less consistent, but a number of researchers have reported greater left-hand performance decrement with concurrent speaking (Orsini, Satz, Soper, & Light, 1985; Simon & Sussman, 1987; van Strien & Bouma, 1988). This is not easily explained with the same explanation that is used for the selective right-hand interference in right-handers, because left-handers presumably also have motor speech control in the left hemisphere.

Little work is available on hand performance asymmetries in left-handers for concurrent manual tasks. However, our own recently completed work allows some preliminary conclusions. We have used the 2:1 task, in which subjects have to tap once for every two taps in the other hand. Subjects were asked to tap as quickly as possible without compromising the rhythm. No external pacing was used. The task has the advantage that it is simple in terms of motor requirements, can be performed well with minimal practice by most subjects, and allows measurement of the salient aspects of the hand movements. We tested 63 right-handed subjects and 122 left-handed subjects. Any work with left-handers that concerns manual skill requires at the very least a subgrouping of left-handers into those who show consistent left-hand preferences ($N = 73$) and those who prefer the left hand for fine manual activities, such as writing, and the right hand for strength activities, such as throwing and holding a racquet ($N = 46$), because these groups are not equivalent in terms of hand preference and performance (Peters & Servos, 1989). Right-handers showed the expected performance asymmetry: they performed better when the right hand tapped twice for every single tap in the left hand than in the converse condition ($F(1,61) = 15.2, p < .00001$). This was the case even though the rate of tapping in the faster right hand (mean = 2.6 taps/sec) was about half the single-hand speeded tapping rate measured for this group. This asymmetry has been replicated independently in other studies (Ferron, 1992; Forster & Webster, 1991; Webster, 1990). The variability of inter-tap intervals was measured as well. It was seen that in this preferred condition, *both* hands performed more regularly than was the case for the converse arrangement. In other words, not only did the right hand perform more regularly than the left hand when the "2 tap" performance was compared, but it performed less regularly than the left hand when it performed a single tap for every two taps in the left hand. The variability measure, therefore, shows a double effect for the bimanual task, with better performance for *both hands* in their preferred condition, which was "two right/one left" (R2L1).

No significant asymmetries were observed for the consistent left-handers; subjects performed as quickly in the "R2L1" condition as in the "L2R1" condition. In addition, this group did not show an asymmetry with regard to the variability of intertap intervals. Finally, the inconsistent left-handers tapped more quickly in the R2L1 condition than in the L2R1 condition ($F(1,44) = 6.2, p < .02$), and the inconsistent left-handers also performed more regularly with the R2L1 condition than with the L2R1 condition ($F(1,45) = 4.1, p < .05$. Thus, the inconsistent left-handers showed the same pattern of asymmetry as the right-handers, with regard to both speed and regularity of performance. Finally, it should be pointed out that there were no significant overall differences in performance between the groups. The results for the inconsistent left-handers are labile; in several replication studies a significant difference was seen in some but not in other replications. In all cases, the means favored the R2L1 combination.

In a different bimanual task, in which subjects had to tap as quickly as possible with one hand while tapping slowly and regularly with the other, an asymmetry favoring one particular hand combination (right fast/left slow) was seen in right-handers only. The two groups of left-handers did not show any significant performance asymmetry. Overall, the results suggest that the asymmetry patterns in bimanual tasks are different for left-handers than for right-handers (Peters & Murphy, in preparation).

In summary, right-handers tend to be consistent in terms of the direction of bimanual asymmetries. Their performance conforms with the four basic principles of bimanual coordination described for naturally occurring tasks. The congruency of bimanual performance and single-hand preference in right-handers is so clear that one would be tempted to seek an answer in the asymmetry in bimanual coordination in more basic unilateral asymmetries. However, consideration of the bimanual performance of left-handers indicates that the picture is more complex than anticipated. Left-handers with consistent left-hand preference choices appear to be just as lateralized in the left hand as right-handers are in the right hand when unimanual performance and preference is compared. If unilateral asymmetries are a determining factor in the directionality of bimanual asymmetries, these consistent left-handers should show the opposite bimanual asymmetries from right-handers. This is not the case. Instead, the consistent left-handers do not show any significant group asymmetries favoring either the left or the right hand in the bimanual tasks. In contrast, left-handers who dissociate the hand used for fine manual skill (left hand) and the hand used for strength and ballistic activities (right hand), show an altogether different pattern. These individuals perform better with the left hand on fine manual skill activities, with smaller between-hand differences than either the consistent left-handers or the right-handers. However, when performing bimanual tasks, the inconsistent left-handers will show asymmetries in the direction of those seen in *right-handers*. In this particular case, the asymmetries of unimanual and bimanual specialization go in the opposite direction. The results suggest that, at least for left-handers, there is no congruence of lateral asymmetries for unimanual and bimanual activities. The fact that no directionally consistent asymmetries are observed in bimanual activities in left-handers indicates that the *direction* of asymmetry in bimanual performance cannot be explained in terms of simple asymmetries in unimanual performance. From this it follows that directionally consistent asymmetries in bimanual performance are not a corollary of hand preference, as based on unimanual performance.

A preliminary interpretation of these results emphasizes differences in the way in which individuals allocate themselves to the two hands during concurrent task performance. The maintenance of adequate performance requires an asymmetrical allocation of attention but the direction of this allocation shows a strong directionally consistent bias only in right-handers. Alternative explanations involving the location of a unilateral source that initiates and terminates action sequences in the two hands await further, more comprehensive work.

IV. Summary and Conclusions

The objective of this chapter was to evaluate the scant evidence on the role of handedness in bimanual coordination. In naturally occurring tasks, there is a clear interaction between hand preference and which hand will perform what role in bimanual activities. In "stripped-down" experimental tasks that involve simple movement trajectories, but where the timing of the onset and offset of component movements in the two hands is crucial, hand preference also interacts with performance of bimanual tasks. For right-handers, there tends to be a double dissociation: if the more attention-demanding task of the two concurrent tasks is labeled as "A," and the less attention-demanding task is labeled as "B," the preferred hand performs "A" better than the nonpreferred hand, *and* the nonpreferred hand performs "B" better than the preferred hand. The fact that the nonpreferred hand is predisposed to perform supportive actions in naturally occurring bimanual activities is therefore also reflected in the simplified laboratory tasks as long as these require or permit a differential allocation of attention. For left-handers, the situation is unclear and remains to be explored fully. It is possible that left-handers, because they have extensive practice in focusing attention to either hand, are not subject to the same constraints in bimanual coordination as right-handers. Alternatively, the way in which attention comes to bear on movement initiation and termination is possibly quite different in the two handedness groups. If sequencing and higher order motor control of movement in general are linked to motor speech specialization, the motor organization of left-handers can be assumed to be different from right-handers in a fundamental way because left-handers share with right-handers a left hemisphere specialization for motor speech but they differ from right-handers in terms of hand preference for fine, manual, skilled activities.

The source of asymmetries in bimanual activities could, of course, be sought in structural and physiological asymmetries that have been documented at various levels of the motor system (Glick & Shapiro, 1984; Kertesz & Geschwind, 1971; Nathan, Smith & Deacon, 1990; Tan, 1989). However, the differential results for right-handers and left-handers cause us to question whether asymmetries in bimanual performance can be reduced to such simple structural asymmetries. Finally, a comment about practical applications of work on handedness and bimanual coordination is appropriate. Where machine controls, tools, and musical instruments require bimanual coordination, the task assignment and attentional allocation is such that the preferences of right-handers are favored. To the extent that left-handers appear to be less biased in terms of a directional preference for bimanual tasks, this arrangement appears quite acceptable. However, it is also true that various claims have been made that left-handers are involved in accidents as a result of being left-handers in a right-handed world (Coren & Halpern, 1991). To what extent this claim can be supported remains to be seen.

When irreconcilable attentional demands are made during concurrent motor tasks during experimental tasks in the laboratory, subjects will often completely cease to attend to one of the tasks. When this happens, the subject may not, in fact, be aware that only one of the motor processes is guided as intended. The subject focuses on one hand and neglects the performance of the other hand. In laboratory tasks this can be done. In real-life activities, the results can be catastrophic. One factor that also needs to be considered in practical terms is age. There is some evidence that bimanual coordination becomes disproportionately difficult in aging (Ferron, 1992; Stelmach, Amrhein, & Goggin, 1988). Of particular interest is Ferron's (1992) study because she evaluated the performance of the "2:1" task as a function of age. Two aspects of her findings are noteworthy. First, even though there was no significant difference in terms of the speed of single-handed finger tapping between the young control subjects and the aged subjects (two groups: from 65 to 74 years of age, and 75 and older), both groups of older subjects showed a striking loss in the ability to perform the 2:1 task. Second, some of the older subjects showed marked difficulties in lifting the left finger from the tapping key when the right hand performed the faster beat; in the converse condition, the right finger was able to lift off the key when the left hand performed the faster beat. The results are of particular interest because the 2:1 task is remarkably simple compared to many other tasks and mental activities these adults can perform, and the question of why they suffer such dramatic losses in bimanual task performance deserves detailed evaluation. Specifically, models of multilimb coordination will have to address the fact that in these individuals, normal, multilimb coordination in everyday activities, in terms of complex, naturally occurring tasks, bimanual tasks, and locomotion, is relatively intact, while this quite simple artificial task shows such remarkable losses.

Another practical aspect of an understanding of multilimb coordination lies in the design of robots with flexible manipulative skills. When machines are to produce coordinated manipulations, some control processes that seem complex can, in fact, be managed quite simply. For instance, in programming a computer to play a Bach fugue, where the two "hands" play concurrent voices, it is possible to simply program the notation for each "hand" and then to link the two processes on a common time scale. It would not be necessary, as is likely the case for a human musician, to assign different priorities and representations for the voices. For this application, asymmetries in programming of the control processes are neither required nor desirable (although, in order for the fugue rendition to sound musical, the programming would likely have to simulate the inevitable deviations from the precise metric that are characteristic for human pianists). However, it is conceivable that when robots are asked to generate a flexible action plan where responses are not strictly a function of predetermined response sequences or specified external inputs, programming would involve the formulation of a principal goal of action, with subsidiary processes serving a supportive function, as is the case in normally occurring human activities. Of course, robots are not limited to two hands or arms, but it is clear that if robots were to be provided with more than two manipulating appendages, the necessity for an overall plan of action, into which the separate appendages feed their subsidiary and complementary contributions, would be even greater. The actual way in which subsidiary processes are coordinated toward an intended goal is determined by a learning process that in turn involves processes such as intent, direction of attention, correction, and storage in memory of the coordination patterns. At this level, control of movement and "thinking" become rather indistinguishable.

This brings us back to the distinction that was made in the beginning of the chapter between "oscillator-type" tasks that are rhythmically controlled and that have no other goal than the performance of the movement itself, and tasks that are closer to real-life, bimanual activities. The very nature of the former tasks fosters symmetries and, if not eliminates, drastically reduces the importance of attentional asymmetries. It is not surprising, therefore, that models concerned with such tasks stress self-organization (Turvey, 1990) and find little room for attention as a variable that exists outside the execution of movement itself. The factor of handedness would, at most, play a role when minute differences in timing between the preferred and nonpreferred hand, attributed to differential experience, would introduce minor aberrations to otherwise quite perfectly symmetrical expressions of interacting oscillators. However, the argument presented in this chapter stresses that handedness in skilled behavior is an expression of asymmetrically directed attention and that it reveals a fundamental aspect of how movement intent is expressed through the hands. The choice of tasks that would not permit the expression of handedness will narrow the generality of conclusions that can be reached about the nature of skilled behavior, and will not allow meaningful statements about higher level voluntary movement that involves bimanual coordination.

Acknowledgments

Support by NSERC Grant No. A-7054 and the Human Frontier Science Program Organization (HFSPO) is gratefully acknowledged.

References

Bakan, P. (1990). Non-righthandedness and the continuum of reproductive casualty. In S. Coren (Ed.), *Left-handedness* (pp. 33–74). Amsterdam: North Holland.

Brun, R. (1921). Klinische und anatomische Studien über Apraxie. *Schweizer Archiv für Neurologie und Psychiatrie,* **9,** 29–64.

Corballis, M. C. (1989). Laterality and human evolution. *Psychological Review,* **96,** 492–505.

Coren, S., & Halpern, D. F. (1991). Left-handedness: a marker for decreased survival fitness? *Psychological Bulletin,* **109,** 90–106.

DeSchonen, S. (1977). Functional asymmetries in the development of bimanual coordination in human infants. *Journal of Human Movement Studies,* **3,** 144–156.

Deutsch, D. (1983). The generation of two isochronous sequences in parallel. *Perception & Psychophysics*, **34**, 331–337.
Faglioni, P., & Basso, A. (1984). Historical perspectives on neuroanatomical correlates. In E. A. Roy (Ed.), *Neuropsychological studies of apraxia and related disorders* (pp. 3–44). Amsterdam: North-Holland.
Farnsworth, P. R., & Poynter, W. F. (1931). A case of unusual ability in simultaneous tapping in two different times. *American Journal of Psychology*, **43**, 633.
Ferron, D. (1992). *Changes with aging in right hemisphere activation as reflected in bimanual and dihaptic task performance.* Unpublished doctoral dissertation, Carleton University, Ottawa, Canada.
Forster, D. C., & Webster, W. G. (1991). Concurrent task interference in stutterers: dissociating hemispheric specialization and activation. *Canadian Journal of Psychology*, **45**, 321–335.
Gilbert, A. N., & Wisocki, C. J. (1992). Hand preference and age in the United States. *Neuropsychologia*, **30**, 601–608.
Glick, S. D., & Shapiro, R. M. (1984). Functional and neurochemical asymmetries. In N. Geschwind & A. M. Galaburda (Eds.), *Cerebral dominance: the biological foundations*, (pp. 147–166). Cambridge: Harvard University Press.
Guiard, Y. (1987). Asymmetric division of labour in human skilled bimanual action: the cinematic chain as a model. *Journal of Motor Behaviour*, **19**, 486–517.
Guiard, Y., & Millerat, F. (1984). Writing postures in left-handers: inverters are handcrossers. *Neuropsychologia*, **22**, 535–538.
Harris, L. J. (1993). Do left-handers die sooner than righthanders? A commentary on Coren and Halpern's "Left-handedness: a marker for decreased survival fitness". *Psychological Bulletin*.
Harris, L. J., & Carlson, D. F. (1988). Pathological left-handedness: An analysis of theories and evidence. In D. Molfese and S. J. Segalowitz (Eds.), *Brain lateralization in children* (pp. 289–372). New York: Guilford Press.
Ibbotson, N. R., & Morton, J. (1981). Rhythm and dominance. *Cognition*, **9**, 125–135.
Jeeves, M. A., Silver, P. H., & Jacobson, I. (1988). Bimanual coordination in callosal agenesis and partial commissurotomy. *Neuropsychologia*, **26**, 833–850.
Kahneman, D. (1973). *Attention and effort.* Englewood Cliffs, NJ: Prentice-Hall.
Kahneman, D., & Treisman, A. (1983). Changing view of attention and automaticity. In R. Parasurman, R. Davies, & J. Beatty (Eds.), *Varieties of attention* (pp. 29–61). New York: Academic Press, Inc.
Kay, B. A., Saltzman, E. L., Kelso, J. A. S., & Schöner, G. (1987). Space-time behaviour of single and bimanual rhythmical movements: Data and limit cycle model. *Journal of Experimental Psychology*, **13**, 178–192.
Kelso, J. A. S., Holt, K. G., Rubin, P., & Kugler, P. N. (1981). Patterns of human interlimb coordination emerge from the properties of nonlinear limit cycle oscillatory processes. *Journal of Motor Behaviour*, **13**, 226–261.
Kertesz, A., & Geschwind, N. (1971). Patterns of pyramidal decussation and their relationship to handedness. *Archives of Neurology*, **24**, 326–332.
Kimura, D. (1979). Neuromotor mechanisms in the evolution of human communication. In H. D. Steklis & M. J. Raleigh (Eds.), *Neurobiology of social communication in primates* (pp. 197–219). New York: Academic Press, Inc.
Kinsbourne, M., & Hicks, R. E. (1978). Functional cerebral space: a model for overflow, transfer and interference effects in human performance: a tutorial review. In J. Requin (Ed.), *Attention and performance VII* (pp. 345–401). New Jersey: Erlbaum.
Klapp, S. T. (1979). Doing two things at once: the role of temporal compatibility. *Memory and Cognition*, **7**, 375–381.
Liepmann, H. (1905). Die linke Hemisphäre und das Handeln. *Münchener Medizinische Wochenschrift*, Nov. 28, 2322–2326, 2375–2378.

Marteniuk, R. G., MacKenzie, C. L., & Baba, D. M. (1984). Bimanual movement control: Information processing and interaction effects. *The Quarterly Journal of Experimental Psychology*, **36A,** 335–365.

McLeod, P. D. (1975). *Response interference in dual task performance*. Doctoral Dissertation, Cambridge University, Cambridge.

Nathan, P. W., Smith, M. C., & Deacon, P. (1990). The spinocortical tracts in man. *Brain*, **113,** 303–324.

Orsini, D. L., Satz, P., Soper, H. V., & Light, R. K. (1985). The role of familial sinistrality in cerebral organization. *Neuropsychologia*, **23,** 223–232.

Peters, M. (1981). Attentional asymmetries during concurrent bimanual performance. *Quarterly Journal of Experimental Psychology*, **33,** 95–103.

Peters, M. (1983). Differentiation and lateral specialization in motor development. In G. Young, C. Corter, S. J. Segalowitz, & S. Trehub (Eds.), *Manual specialization and the developing brain: longitudinal studies* (pp. 141–159). New York: Academic Press, Inc.

Peters, M. (1985). Constraints in the coordination of bimanual movements and their expression in skilled and unskilled subjects. *Quarterly Journal of Experimental Psychology*, **37A,** 171–196.

Peters, M. (1986). Hand roles and handedness in music. *Psychomusicology*, **6,** 29–34.

Peters, M. (1988). Footedness: Asymmetries in foot preference and skill and neuropsychological assessment of foot movement. *Psychological Bulletin*, **103,** 179–192.

Peters, M. (1990a). Interaction of vocal and manual movements. In G. Hammond (Ed.), *Motor cerebral control of speech and limb movement* (pp. 535–574). Amsterdam: Elsevier.

Peters, M. (1990b). Subclassification of lefthanders poses problems for theories of handedness. *Neuropsychologia*, **28,** 279–289.

Peters, M., & Murphy, K. (1992). Cluster analysis reveals at least three, and possibly five distinct handedness groups. *Neuropsychologia*, **30,** 373–380.

Peters, M., & Murphy, K. (1993). Interference between manual performance and speech in righthanders and subgroups of lefthanders.

Peters, M., & Schwartz, S. (1988). Coordination of the two hands and effects of attentional manipulation in the production of a bimanual 2:3 polyrhythm. *Australian Journal of Psychology*, **41,** 215–224.

Peters, M., & Servos, P. (1989). Performance of subgroups of lefthanders, and righthanders. *Canadian Journal of Psychology*, **43,** 341–358.

Preilowski, B. F. B. (1972). Possible contribution of the anterior forebrain commissures to bilateral motor coordination. *Neuropsychologia*, **10,** 267–277.

Schmidt, R. C., Treffner, P. J., Shaw, B. K., & Turvey, M. T. (1992). Dynamical aspects of learning an interlimb rhythmic movement pattern. *Journal of Motor Behaviour*, **24,** 67–83.

Schott, G. D. (1980). Mirror movements of the left arm following peripheral damage to the preferred right arm. *Journal of Neuropsychology, Neurosurgery, and Psychiatry*, **43,** 768–773.

Simon, T. J., & Sussman, H. M. (1987). The dual task paradigm: Speech dominance or manual dominance? *Neuropsychologia*, **25,** 559–569.

Stelmach, G. E., Amrhein, P. C., & Goggin, N. L. (1988). Age differences in bimanual coordination. *Journal of Gerontology: Psychological Sciences*, **43,** P18–23.

Summers, J. (1989). Temporal constraints in the performance of bimanual tasks. In D. Vickers & P. L. Smith (Eds.), *Human information processing: measures, mechanisms and models* (pp. 155–168). Amsterdam: North-Holland.

Swinnen, S., Walter, C. B., & Shapiro, D. C. (1988). The coordination of limb movements with different kinematic patterns. *Brain and Cognition*, **8,** 326–347.

Tan, U. (1989). The H-reflex recovery curve from the wrist flexors: lateralization of motoneuronal excitability in relation to handedness in normal subjects. *International Journal of Neuroscience,* **48,** 271–284.

Turvey, M. T. (1990). Coordination. *American Psychologist,* **45,** 285–325.

van Emmerik, R. E. A. (1992). Kinematic adaptations to perturbations as a function of practice in rhythmic drawing movements. *Journal of Motor Behaviour,* **24,** 117–131.

van Strien, J. W., & Bouma, A. (1988). Cerebral organization of verbal and motor functions in left-handed and right-handed adults: effects of concurrent verbal tasks on unimanual tapping performance. *Journal of Clinical and Experimental Neuropsychology,* **10,** 139–156.

Walter, C. B., & Swinnen, S. P. (1990a). Asymmetric interlimb interference during the performance of a dynamic bimanual task. *Brain and Cognition,* **14,** 185–200.

Walter, C. B., & Swinnen, S. P. (1990b). Kinetic attraction during bimanual coordination. *Journal of Motor Behaviour,* **22,** 451–473.

Walter, C. B., & Swinnen, S. P. (1992). Adaptive tuning of interlimb attraction to facilitate bimanual decoupling. *Journal of Motor Behaviour,* **24,** 95–104.

Webster, W. G. (1990). Evidence in bimanual finger tapping of an attentional component to stuttering. *Behavioral Brain Research,* **37,** 93–100.

Welch, J. C. (1898). On the measurement of mental activity through muscular activity and the determination of a constant of attention. *American Journal of Physiology,* **1,** 283–306.

28

Temporal Organization of the Prehension Components in a Bimanual Task

Umberto Castiello* and George E. Stelmach

† Department of Exercise Science and Physical Education
Arizona State University
Tempe, Arizona

I. Introduction
   A. Experiment A
   B. Experiment B
II. Discussion
   References

This chapter assesses the organization of the components of a reach-and-grasp movement performed bimanually. Two pilot experiments examine the control mechanisms subserving bimanual tasks which require the simultaneous performance of more than one movement component. In Experiment 1 two different grasping actions were activated simultaneously: a precision grip with one hand and whole hand prehension with the other. In Experiment 2 the same grasping action was performed by both limbs. Movement duration was the same for both limbs in both experiments despite unique grasp-related kinematic profiles for the nonhomologous tasks. Thus for the limb performing a precision grip, a shortening of the early transport component phase allowed for a prolonged approach time. Movement duration is thus an

* *Current address:* Dipartimento di Psicologia, Università di Bologna, 40126 Bologna, Italy

important parameter for the temporal coupling of complex movements involving more than one component.

I. Introduction

Various theories have been proposed to explain the central mechanisms which underlie the control of bimanual movements. Since movement duration and other temporal aspects of bimanual nonhomologous pointing tasks are similar for both hands, Kelso et al. (1979) proposed that limbs are constrained to act as a single unit. As dictated by the end goal, muscles temporarily group to act as functional units (Turvey, 1977) or "coordinative structures" (Easton, 1972). This theory implies that the central control can flexibly alter synergic interactions according to the required motor output (Bernstein, 1967; Turvey, 1977). Marteniuk et al. (1984) found that movement duration is not similar for the two limbs when subjects performed a nonhomologous bimanual task with styli. Given, however, that the movement duration of one limb is influenced by that of the other, Marteniuk et al. (1984) proposed that the commands for the left and right limbs are delivered via separate channels. With transmission of the output signals, the two channels engage in "neural cross-talk" at various levels of the central nervous system (CNS). The bilateral projection of cortical neurons to the motoneurones of proximal arm musculature could thus be a means by which the neural impulses to one arm could influence those of the other (Brinkman & Kuypers, 1973; Lawrence & Kuypers, 1968a,b). Another concept explaining bilateral motor control uses the theory of motor programs (Keele, 1981; Welford, 1968). For example, Schmidt et al. (1979) suggested that the framework for a bimanual movement is determined by a central motor program. In the execution of this program some controls are common to both limbs (e.g., movement duration; cf. "global" aspects of Heuer, 1985) while others are individual to each limb (e.g., movement distance; cf. "local" aspects of Heuer, 1985).

Previous experiments of bimanual tasks have largely focused on single component movements such as transport of the limb to a target. For example, Kelso et al. (1979) found that for the performance of nonhomologous Fitts' aiming tasks, the longer movement duration of the limb with the higher index of difficulty was also evident for the limb with the lower index. Swinnen et al. (1988, 1990) compared the performance of a unidirectional elbow movement with the simultaneous performance of a bidirectional elbow movement by the contralateral limb and found that when subjects first attempted this task, the pattern of the bidirectional movement was imposed upon the limb that performed a unilateral movement. Similarly, Franz *et al.* (1991) reported that when subjects drew a circle with one hand and a straight line with the other, the circle resembled a straight line and the line resembled a circle.

However, in some respects these tasks do not examine the natural level of intercoordination between limbs. This may only be possible when the CNS is not overloaded with the task of coordinating "unusual" actions, such as the previously mentioned component movements. Activities of daily living rarely manifest breakdown into movement units. Most commonly, each upper limb performs a complex series of actions that do not resemble those of the contralateral limb.

Typically, the limbs work in a coordinated fashion. Thus it is common to stabilize with one hand and manipulate with the other. For example, when opening a bottle, both limbs reach for the object. One hand then firmly grasps the bottle while the other opens the top. These bimanual activities are complex from two perspectives. First, each limb must couple two distinct movement components: the reach and the grasp (Jeannerod, 1981, 1984). The performance of a reach-to-grasp movement with *one* limb has been well characterized. Jeannerod (1981, 1984) described two main components of this movement. One is the transport whereby the hand is brought to the target by the reaching arm. The other is the manipulation whereby the hand prepares for and then grasps the target. These components are thought to be subserved by independent neural channels (Brinkman and Kuypers, 1975; Muir & Lemon, 1983; Rizzolatti *et al.*, 1988) which are activated in parallel. However, it has been demonstrated that the kinematic parameterization of transport is not completely independent from that of manipulation. For example, studies have shown that the organization of not only the manipulation component but also of the transport component is different when a subject performs a reach with a precision grip then when a reach with whole hand prehension is performed (Castiello *et al.*, 1992; Gentilucci *et al.*, 1991). Thus, bimanual prehension requires the neural organization for both the proximal and the distal components of the movement. From a second perspective, complexity arises with the need to combine the performance of one type of grasp (precision grip) by one limb with the performance of another type of grasp (whole hand prehension) by the other limb. The neural channels that subserve each grasp are thought to be different (Muir and Lemon, 1983; Rizzolatti *et al.*, 1988).

This chapter assesses the kinematic coordination between the upper limbs during the performance of complex movements that contain two distinct components: reaching (transport) and grasping (manipulation) (Jeannerod, 1981, 1984).

A. Experiment 1

Experiment 1 investigates the bimanual and simultaneous activation of two different distal motor tasks.

1. Methods

a. Procedure Six right-handed subjects were instructed to reach and then grasp the target, a cylindrical object, similar to an empty container, with a pull tab on its top surface (Figures 1A and 1B). It was placed 35 cm from the hands along the subject's midline. The diameter of the cylinder was 7.5 cm (height 20 cm) and the diameter of the pull tab was 2 cm. In 10 practice trials, the subject was required to grasp the cylinder with the right or left hand and open the lid using the pull tab with the other. After practicing, all subjects adopted two clear patterns of grasp according to the diameter of the targets. The pull tab was grasped and lifted with a precision grip (PG) consisting of opposition between the index finger and thumb (Napier, 1956). The cylinder was grasped with a whole hand prehension (WHP) characterized by flexion of all the fingers around the cylinder (Figures 1A and 1B).

Movements were recorded by an Optotrak 3D system. The cameras monitored the displacements of active markers (infrared-emitting diodes, IREDS) which were attached to the skin overlying the following areas on the dorsal surface of the right and left arms: (1) distal styloid process of the radius (wrist IRED), (2) medial to the lower ulnar corner

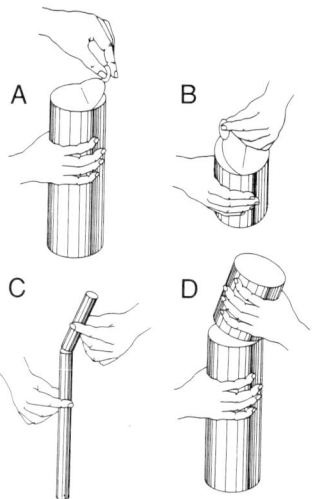

FIGURE 1 The apparatus used for Experiment 1, as viewed from the side (A) and from above (B). (C and D) The apparatus used for Experiment 2.

of the thumb nail (thumb IRED), and (3) lateral to the lower radial corner of the index finger nail (finger IRED). The wrist IRED was used to measure the displacement, velocity, and acceleration of the arm. The finger and thumb IREDS were used to measure the displacements of the finger and thumb and the size of the grip aperture (finger–thumb distance). A spatial error of 0.3 mm was determined by dynamic accuracy tests. Position of the IREDS was sampled at 250 Hz and stored on an IBM 386 computer. Subjects were instructed to commence each trial following an acoustic signal.

The unimanual control trials consisted of the four following blocks: left hand grasping the tab or cylinder; right hand grasping the tab or cylinder. The bimanual tasks consisted of the two following blocks: right hand grasping the cylinder while the left hand simultaneously grasps the tab; left hand grasping the cylinder while the right hand simultaneously grasps the tab.

b. Data Processing. The X, Y, and Z trajectories of each IRED and the velocity of the wrist IRED were computed following filtering (Butterworth dual-pass filter; cutoff frequency 10 Hz). Acceleration data were derived by differentiating the velocity data. Movement time for each limb was measured as the time from the onset of the wrist IRED movement to contact of the fingers with the object. This latter contact was established by analyzing the grip size profile to determine the point of constant aperture.

For the transport component the following parameters were determined: (1) the time from movement onset to the first peak of acceleration; (2) the time from movement onset to the peak of velocity; (3) the time from movement onset to the maximum trough of the acceleration profile, i.e., time to peak deceleration; and (4) the time from the velocity peak (zero crossing of the acceleration curve) to the end of the movement, i.e., deceleration time. The following parameters were computed for the manipulation component of each trial: (1) the time from movement onset to maximum grip aperture; and (2) the amplitude of the maximum grip aperture.

Note that all temporal kinematic parameters were computed from movement onset and that Figures 2 and 3 were plotted from movement onset.

2. Results

Movements performed to grasp the cylinder will be referred to as whole-hand prehension (WHP) trials; movements performed to grasp the pull tab will be referred to as precision grip (PG) trials. Movement time and the kinematic parameters of the transport and manipulation components are shown in Table I (each value represents the average of all subjects).

TABLE I Kinematic Parameters of the Transport and Manipulation Components (Experiment 1)

	Unimanual				Bimanual			
	PG[a]		WHP		PG		WHP	
	LH	RH	LH	RH	LH	RH	LH	RH
MT	829	800	775	777	813	810	810	800
(msec)	(65)	(39)	(38)	(43)	(87)	(84)	(77)	(56)
TPA	209	217	188	217	180	194	220	229
(msec)	(16)	(26)	(15)	(16)	(21)	(14)	(19)	(32)
TPV	355	375	350	352	357	360	375	384
(msec)	(33)	(36)	(24)	(30)	(39)	(32)	(38)	(36)
TPD	505	543	504	507	516	560	555	564
(msec)	(58)	(20)	(41)	(35)	(41)	(58)	(58)	(43)
DT	474	425	432	415	475	450	435	426
(msec)	(47)	(27)	(43)	(44)	(51)	(43)	(45)	(30)
TGA	453	458	485	476	437	481	514	531
(msec)	(43)	(22)	(51)	(35)	(56)	(58)	(49)	(51)
AGA	59	58	120	113	51	56	119	115
(mm)	(7)	(5)	(22)	(10)	(5)	(3)	(12)	(63)
%	54	57	62	61	53	72	63	63

[a] PG, precision grip; WHP, whole-hand prehension; LH, left hand; RH, right hand; MT, movement time; TPA, time to peak of acceleration; TPV, time to peak velocity; TPD, time to peak deceleration; DT, deceleration time; TGA, time of maximum grip aperture; AGA, amplitude of maximum grip aperture; %, time to maximum grip aperture expressed as a percentage of MT. Standard deviations (SD) in parentheses.

a. Movement Time For the unimanual trials, movement time varied significantly according to the type of grasp. The average time for a PG movement was longer than that for the WHP. Movement time also varied according to the hand used. For PG movements performed with the left hand, movement time was significantly longer than for those movements performed with the right hand. No differences were found in the bimanual condition: the movement time of the hand adopting the PG was similar to that of the hand adopting WHP. Similarly, the left and right hands showed no differences in the bimanual tasks.

b. Transport Component The wrist IREDS (right and left arm) were used for kinematic analysis of the transport component. For both hands and with the unimanual and bimanual conditions the wrist movements displayed a typical single-peak velocity profile (Figure 2).

Differences in the transport component between the unimanual and bimanual conditions were largely revealed during the first or acceleration phase of the movement. In the unimanual task, the time to peak velocity for PG movements was no different than that for WHP movements. This contrasted to the findings for this parameter under bi-

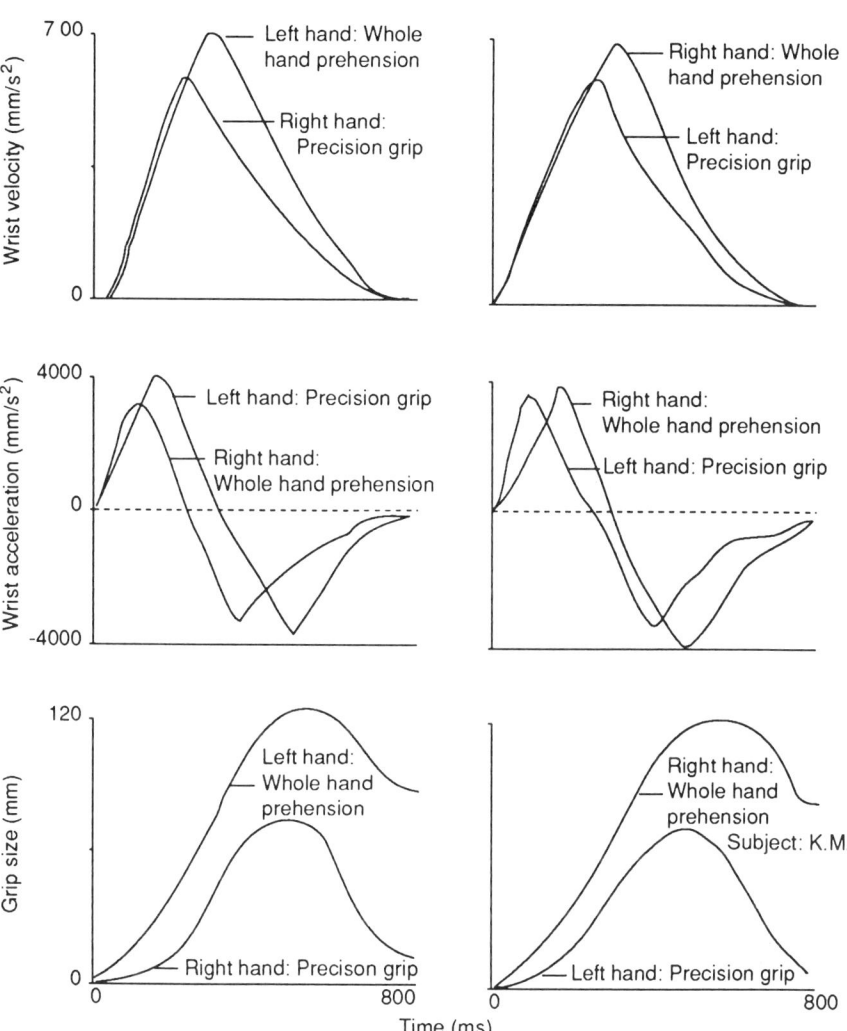

FIGURE 2 Representative examples of the kinematics of prehension movements in the nonhomologous bimanual condition (Experiment 1). (*Left*) Whole-hand prehension (left hand) and precision grip (right hand). (*Right*) Precision grip (left hand) and whole-hand prehension (right hand).

manual conditions: peak velocity occurred significantly earlier for PG than for WHP movements. Peak acceleration also occurred at an earlier time for the former than for the latter under the bimanual conditions. The shortening of the acceleration phase when a precision grip was used by one hand during the bimanual task seemed necessary to allow for an adequate amount of time for the deceleration phase.

In contrast to the acceleration phase, the deceleration phase of the transport component showed similar patterns under both unimanual and bimanual conditions. Differences arose when comparing the two hands. Peak deceleration for PG movements performed with the left hand under both unimanual and bimanual conditions was reached significantly earlier than when PG movements were performed with the right hand (Figure 2). The deceleration time also varied according to hand. Thus the time from peak deceleration to the end of the movement was longer for the left than for the right hand. This parameter was also greater for PG than for WHP movements (both for unimanual and bimanual conditions). Overall, the deceleration time for PG movements performed by the left hand under the unimanual condition was the longest.

c. Manipulation Component During arm transport, grip aperture increased to a maximum before closing around the object (Figure 2). As expected, the amplitude of the maximum grip aperture was related to the type of grasp adopted: it was smaller for PG movements. The time of this maximum grip aperture was also related to the type of prehension. Under both unimanual and bimanual conditions, the peak for WHP was later than that for PG (Figure 2). For the left hand, maximum grip aperture for PG movements occurred at an earlier time than under any other condition. These results were further confirmed when the time to maximum grip aperture was expressed as a percentage of the total movement time. Time to maximum grip aperture was relatively earlier for PG than for WHP movements both under unimanual and bimanual conditions. It was earliest for PG movements performed with the left hand.

B. Experiment 2

Results from the first experiment indicate that the kinematics of the transport and manipulation components differ according to whether or not the different types of grasps were performed individually (by one limb) or simultaneously (by both limbs). Experiment 2 was conducted in order to further verify if these differences were related to the unique features of each type of prehension or to a more general plan adopted for the coordination of the two arms. In this experiment the bimanual tasks required the same distal action (PG or WHP) for both hands.

1. Methods

a. Subjects Six naive subjects were selected according to the criteria of Experiment 1.

b. Apparatus and Procedure For the first session of this experiment (Session A) the large cylinder was replaced by two cylinders set on top of each other. Subjects were instructed to grasp the top cylinder (height 8 cm; diameter 7.5 cm) with the left hand and the bottom cylinder (height 12 cm; diameter 7.5 cm) with the right hand and vice versa (see Figure 1D). The procedure was identical to the bimanual condition of Experiment 1. In 10 practice trials the subjects were required to grasp the bottom cylinder while grasping and lifting the top cylinder. All subjects adopted whole-hand prehension (as described previously) for the grasp of both cylinders (see Figure 1D).

In Session B two small cylinders were used. Again one cylinder was set on top of the other (the top cylinder was 8 cm high, and the bottom cylinder was 12 cm high; both had a diameter of 2 cm). Subjects were instructed to grasp the bottom cylinder while grasping and lifting the top cylinder. For 10 practice trials subjects used a precision grip in order to grasp both the top and the bottom cylinders (Figure 1C). The order of the two sessions (A and B) was counterbalanced across subjects in order to avoid practice effects.

2. Results

For each subject, the mean values of each dependent measure were calculated for the Session (A, WHP or B, PG) and Hand (right or left). The kinematic parameters of the transport and manipulation components are shown in Table II where each value represents the mean value across subjects. Since the difference between the two sessions is in the type of grasp adopted, for sake of clarity Session A is hence referred to as WHP and Session B as PG.

a. Movement Time As was found for the unilateral condition of Experiment 1, movement varied according to the type of grasp: for PG movements transport time was longer than that for WHP movements.

b. Transport Component No differences were found between PG and WHP movements during the first or acceleration phase of the movement (Figure 3). However, and as was found for both unimanual and bimanual conditions of Experiment 1, the deceleration time was longer for PG than for WHP movements. The times of peak velocity, time to peak acceleration, and peak deceleration were the same for both left and right hands.

c. Grasp Component The pattern of the manipulation component in a bimanual homologous task showed no differences from that of a unimanual or of a bimanual nonhomologous task. The mean value of the maximal hand aperture was wider for WHP than for PG movements. The time to maximal aperture between the digits was once again related to the type of grasp. For WHP movements this maximal aperture occurred later than for PG movements (Figure 3). This result was confirmed when expressing the time of maximum grip aperture as a percentage of the total movement time.

TABLE II Kinematic Parameters of the Transport and Manipulation Components (Experiment 2)

	Bimanual WHP[a]		Bimanual PG	
	LH	RH	LH	RH
MT (msec)	835 (88)	823 (100)	882 (88)	886 (73)
TPA (msec)	229 (36)	234 (21)	233 (22)	243 (24)
TPV (msec)	371 (29)	369 (31)	354 (32)	361 (45)
TPD (msec)	534 (55)	538 (59)	521 (52)	534 (53)
DT (msec)	476 (61)	454 (42)	528 (52)	525 (50)
TGA (msec)	518 (51)	525 (63)	505 (58)	504 (49)
AGA (mm)	54 (9)	58 (8)	110 (11)	117 (10)
%	62	63	57	56

[a] Refer to Table I for an explanation of the abbreviations.

Since the kinematic profiles were similar for both limbs, temporal landmarks of the transport and manipulation components were compared in order to further verify the observed synchrony of performing the same grasping action with both limbs. A series of correlation coefficients was calculated. The Fisher Z transformation of data was used for homogeneity of variance and to counteract any nonnormal distributions. The significance of each correlation was assessed with the Student's t test. The results indicated that the time to peak acceleration of one limb was significantly correlated to that of the other limb. Such significant bimanual correlations were also found for times of peak velocity, of peak deceleration and of maximum grip aperture.

II. Discussion

The present study uses the reach and grasp movement for the study of bimanual tasks. Its purpose is to assess the coordination of the proximal (transport) and the distal (manipulation or grip) components of this well-defined movement when two different grasping actions

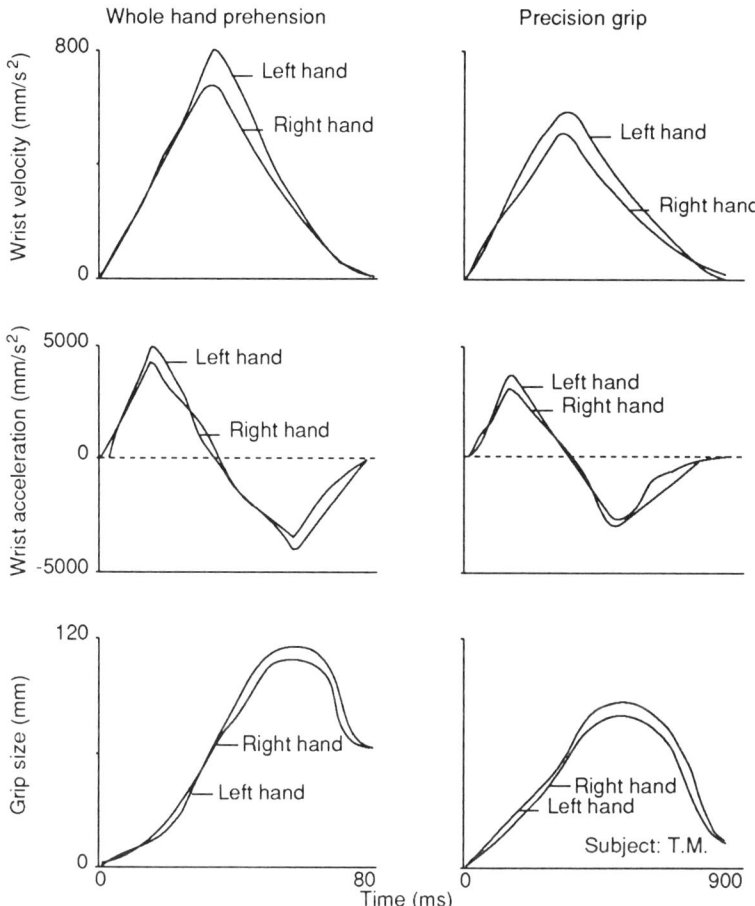

FIGURE 3 Representative examples of the kinematics of prehension movements in the homologous bimanual condition (Experiment 2).

(whole-hand prehension and precision grip) are simultaneously activated.

During the unimanual condition, movement time is greater when subjects use a precision grip. In particular, this is more pronounced for the left or nondominant hand. This finding is in agreement with previous results that demonstrated greater influences upon the nondominant hand with decreases in the size of the target to be reached (Flowers, 1975; Sheridan, 1973).

Kinematic analysis reveals that the increase in movement time is primarily due to a prolongation of the deceleration phase. This effect is more pronounced for left-hand precision grip movements. The extension of the later phase of the transport component with the more precise movement is in accordance with previous studies of prehension movements: the time from peak velocity to the end of the movement is greater when reaching to grasp more fragile (Marteniuk et al., 1987) or smaller (Gentilucci et al., 1991) objects.

The finding that a prolonged deceleration phase is present with the movement requiring greater accuracy and precision rather than a more gross flexion of all the digits may indicate the importance of allowing sufficient time for effective error correction. By extending the period of approach the subject is given additional time to adjust for the requirements of the more "precise" object and to perhaps allow for the independent use of the finger and thumb from the more ulnar digits. The more pronounced findings for the nondominant hand may further support the idea of the need for greater correction time (Todor & Smiley, 1985). Marteniuk et al. (1987) suggested that the duration of this deceleration phase could even be used as a reliable index of the precision needed for the performance of a prehension task. An alternative view is that this time could be employed to recruit more complex or higher cortical centers needed for the greater levels of calculation afforded by the opposition grip. For example, the programming of corticomotoneuronal cells, which discharge during the precision grip (Muir & Lemon, 1983; Muir, 1985), may take a longer time than the programming for the more gross movement (whole-hand prehension), which could employ different and possibly lower neural centers.

The current study divides the transport component into two phases: an acceleration phase from movement onset to the peak of velocity (the point at which the acceleration profile recrossed the zero line) and a deceleration phase from the velocity peak to the end of the movement. As was found for previous studies, kinematic parameters measured during the acceleration phase are not affected by the type of stimulus or the grasp adopted during a unimanual task. It is only by adding the condition that a precision grip be executed at the same time as a whole-hand prehension that the acceleration or earliest phase of the transport component is altered: the peaks of acceleration and velocity occur earlier for the limb that performs the precision grip than for the limb that performs the whole-hand prehension. Again this effect is more pronounced for the left hand. The simultaneous execution of two grasping actions and thus the restriction of movement duration for the limb performing the more accurate task reveals that the early phase of the transport component can clearly be affected by changes in the distal manipulation component. First, this provides quite direct evidence for the modification of the transport component in relation to different distal grasping actions (see Castiello et al., 1992). Second, it indicates that such modification can occur during the initial phases of movement.

When studying parameters of the manipulation component, object processing and effector selection by the distal channel unquestionably occur in the early phases of reaching: the hand initiates its opening with the beginning of the reaching or transport movement and reaches a unique grasping shape well in anticipation of its contact with the stimulus target. It is thus not entirely unexpected that modification of the proximal component could occur at a very early stage. Given that the precision grip, and particularly one performed by the left hand, "needs" a longer approach and thus deceleration time, it is indeed sensible to arrange for a shortening of the acceleration phase to compress all requirements within the time allotted.

With each limb executing the same distal grasping action, the kinematic parameters do not show differences between the two hands, i.e., the acceleration and deceleration phases show the same profiles according to the grasp adopted and temporal parameters measured from both limbs are tightly correlated. For these homologous tasks, the proximal and distal components require the same coupling of kinematic parameters and the activation of similar sets of muscles for each limb. Similarly, and as was found by Kelso et al. (1979) with bilateral pointing studies, the kinematic profiles of each limb are very similar. Kelso et al. (1979) proposed that this "fixed and reproducible" (p. 236) interlimb coordination reflected a coordinative structure (Turvey, 1977): control signals act to group the muscles of both limbs as a single functional unit for the purpose of attaining the bimanual goal. The high degree of interlimb kinematic coordination does not favor separate programming of each limb.

Consider, however, the rarity of performing exactly the same task simultaneously with both hands. What purpose is served by gross synergic groupings that largely ignore differences between two limbs? The results from our bimanual experiments, whereby a different grasp is required by each hand, do not support the idea of both limbs acting as a single unit. Despite the activation of corresponding muscle groups for the transport component and consequently the recruitment of the same but contralateral neural pathways, the kinematic organization differs according to the grasping action adopted. Thus the parameter "movement duration" may present an optimized time frame within which the nervous system builds and temporally scales the coordination between individual motor programs. Accordingly, this parameter may be adopted by the system for the coordination of bimanual movements requiring the simultaneous execution of different actions (Franz et al., 1991).

This study supports the well-documented findings of a covariation of grip aperture with object size (Gentilucci et al., 1991), Jeannerod, 1981; 1984; Martenuik et al., 1990; Wallace & Weeks, 1988; Wing et al., 1986).

For both unimanual and bimanual tasks the time to maximum grip aperture is less when a precision grip is performed. The hand thus demonstrates an earlier anticipation of the object's characteristics when a more precise grip is required. The object to be grasped and thus the grasp to be adopted govern computational requirements of both proximal and distal components of a reach to grasp movement. Command signals ensure that the former component anticipates the more accurate task by executing a prolonged deceleration phase as if steadying the limb and that the latter component has attained a maximum aperture well in advance of its contact with the object.

When a subject performs a reaching task that requires the distal programming of a gross grip with one hand and a precision grip with the other hand, central processing normalizes the bimanual task to a common movement duration. It thus may be possible that this parameter is then used for the determination of the relative timing of the kinematic parameters for each limb. It has been found, for example, that primates with unilateral supplementary motor area (SMA) lesions showed inappropriate mirror-symmetric movements in tasks that required independent use of the limbs (Brinkman, 1981, 1982). With subsequent section of the corpus callosum, the lesioned animals regained the ability to perform these bimanual tasks (Brinkman, 1981). On the basis of these results and those obtained from neurophysiological and regional cerebral blood flow studies (Orgogozo and Larsen, 1979; Roland et al., 1980), Goldberg (1985) proposed that both SMAs are active for nonhomologous bimanual movements under normal circumstances. Each functions not only to influence the ipsilateral primary motor cortex and thus contralateral motor control, but also, via callosal connections, to suppress the influence of the contralateral SMA. When one SMA is damaged the activity of the undamaged SMA is unchecked and, via its ipsilateral and contralateral connections to the motor cortex (Pandya et al., 1969), can promote the passage of one output response to both sides of the body. With lesion of the corpus callosum, the influence on the contralateral motor cortex, and thus of ipsilateral motor control, is largely forfeited. In the undamaged section, Goldberg (1985) proposes that the neural influence of SMA activity from one hemisphere on that of the other may function to establish "an overall temporal structure for the task."

Acknowledgments

This research was supported by grants from APA and from NIH (NS17421) to G. E. Stelmach. The authors thank Carl Waterman for implementing the computer programs for data analysis and Dr. K. M. S. Bennett for reviewing the manuscript.

References

Brinkman, J. (1981). *Neurosci. Lett.*, **27,** 267.
Brinkman, J. (1982). *Soc. Neurosci. Abstr.* **8,** 734.
Brinkman, J., & Kuypers, H. G. J. M. (1973). *Brain* **96,** 653.
Castiello, U., Bennett, K. M. B., & Paulignan, Y. (1992). *Behav. Brain Res.*, **50,** 7.
Easton, T. A. (1972). *Am. Sci.*, **60,** 591.
Flowers, K. (1975). *Brit. J. Psychol.*, **66,** 39.
Franz, E. A., Zelaznik, H. N., & McCabe, G. (1991). *Acta Psychol.*, **77,** 137.
Gentilucci, M., Castiello, U., Scarpa, M., Umiltá, C., & Rizzolatti, G. (1991). *Neuropsychologia*, **29,** 361.
Goldberg, G. (1985). *Behav. Brain Sci.* **8,** 567.
Heuer, H. (1985). *J. Mot. Behav.*, **17,** 335.
Jeannerod, M. (1981). *In* J. Long and A. Baddeley (eds.), *Attention and performance IX* (p. 153). Hillsdale NJ: Erlbaum.
Jeannerod, M. (1984). *J. Mot. Behav.*, **162,** 35.
Keele, S. W. (1981) *In* V. B. Brooks (ed.), *Handbook of Physiology* p. 1391. Bethesda MD.
Kelso, J. A. S., Southard, D. L., & Goodman, D. (1979). *J. Exp. Psychol.: Hum. Percept. Perf.*, **5,** 229.
Lawrence, D. G., & Kuypers, H. G. J. M. (1968a). *Brain,* **91,** 1.
Lawrence, D. G., & Kuypers, H. G. J. M. (1968b). *Brain,* **91,** 15.
Lemon, R. N., Mantel, G. W. H., & Muir, R. B. (1986). *J. Physiol.* **381,** 497.
Marteniuk, R. G., Leavitt, J. L., MacKenzie, C. L., & Athenes, S. (1990). *H. Mov. Sci.*, **9,** 149.
Marteniuk, R. G., MacKenzie, C. L., & Baba, D. M. (1984). *Q. J. Exp. Psychol.*, **36A,** 335.
Marteniuk, R. G., MacKenzie, C. L., Jeannerod, M., Athenes, S., & Dugas, C. (1987). *Can. J. Psychol.*, **41,** 365.
Muir, R. B. (1985). *Exp. Brain Res. Suppl.* **10,** 155.
Muir, R. B., & Lemon, R. N. (1983). *Brain Res.*, **261,** 312.
Napier, J. R. (1956). *J. Bone Jt. Surg.*, **38B,** 902.
Orgogozo, J. M., & Larsen, B. (1979). *Science,* **206,** 847.
Rizzolatti, G., Camarda, R., Fogassi, L., Gentilucci, M., Luppino, G., & Matelli, M. (1988). *Exp. Brain Res.*, **71,** 491.
Schmidt, R. A., Zelaznik, H. W., Hawkins, B., Frank, J. S., & Quinn, J. T. (1979). *Psychol. Rev.* **86,** 415.
Sheridan, M. R. (1973). *J. Mot. Behav.*, **5,** 199.
Swinnen, S. P., Walter, C. B., & Shapiro, D. C. (1988). *Brain Cogn.*, **8,** 326.
Swinnen, S. P., Walter, C. B., Beirinckx, M. B., & Meugens, P. F. (1990). *Hum. Perform.* **3,** 187.
Todor, J. I., & Smiley, A. L. (1985). *In* E. A. Roy (ed.) *Neuropsychological studies of apraxia and related disorders* (p. 108). Amsterdam: North-Holland.
Turvey, M. T. (1977). *In* R. Shaw and J. Bransford, (eds.), *Perceiving, acting, and knowing* (p. 52). Hillsdale, NJ: Erlbaum.
Wallace, S. A., & Weeks, D. L. (1988). *J. Mot. Behav.* **20,** 81.
Welford, A. T. (1968). *Fundamentals of skill.* Methuen, London.
Wing, A. M., Turton, A., & Fraser, C. (1986). *J. Mot. Behav.*, **18,** 245.

Index

Ability, 493
Absolute coordination, *see* Coordination
Acallosal, *see* Corpus callosum, Callosal agenesis, Split-brain
Acquisition, *see* Learning
Adaptive tuning, 506–507
Afferent, 374
    flexor reflex, 99
    group I, 99–100, 129, 135, 139
    group II, 99, 129, 135
    group III, 129, 135
    group IV, 129
Alien hand syndrome, 185, 187
Alpha motoneuron, *see* Neuron
Anti-phase coordination, *see* Coordination
Aphasia, 181
Apraxia, 180, 182, 601
Articulatory movements, 346
Assimilation effect, 191, 494, 506
Association cortex, 180, 202
    frontal, 187
    parietal, 186–187, 191–192,
    temporal, 202
Attention, 269, 573, 612
Attractor, 16, 22, 297, 308, 315, 327, 343, 355, 394, 404, 465–467, 471–474, 477–484, 486, 526–527, 542, 573
Axon
    corticospinal tract, 36–37
    rubrospinal tract, 37–39
    vestibulospinal tract, 39–44
Automatic gain control, 101

Beharrungstendenz, *see* Maintenance tendency
Bereitschaftspotential, 183–184
Bifurcation, 301, 303, 312–313, 316, 465, 469, 474, 485, *see also*
    codimension-1, 313–314
    codimension-2, 313
    Hopf, 313–314
    pitchfork, 305, 312–314
    saddle-node, 305, 312–314
    transcritical, 313
Bilateral deficit, 259–272
Bimanual coordination, *see* Coordination
Bipedal locomotion, *see* Locomotion
Brainstem, 35, 93, 106, 140–141, 190, 201

Callosal agenesis, 212, 215–216, 220, 222
Center of mass, 539, 541–542, 550–557, 563–565
Central pattern generator, 4, 50, 55–59, 101, 110, 118, 176, 500
Cephalo-caudal development, *see* Development

Cerebral blood flow, 630
Cerebral hemispheres,
    left-right, 238, 602–603, 606, 610
Cerebellum, 36, 169, 171, 185, 192, 201
Collective variable, see Parameter
Control parameter, see Parameter
Coordination
    absolute-relative, 2, 3, 17–18, 50, 282, 286, 301, 304, 315–316, 399, 584
    between-person, 305–307
    bimanual, 11, 12–14, 21, 413–438, 439–458, 461–488, 491–510, 571–591, 595–612, 617–630
    homolateral, 14, 229–241
    heterolateral, 229–242
    in-phase, anti-phase, 14–18, 20, 231–241, 281–298, 320–333, 345, 391–408, 467–487, 495, 573–574, see also Relative phase
    intralimb, 402–404
Coordinative structure, 12, 181, 188, 201, 340, 555, 618, 629
Corpus callosum, 185–187, 190, 193, 202, 209–227, 271, 605, 630, see also Acallosal, Callosal agenesis
Corticospinal tract, see lateral descending motor tract group
Coupled oscillators, 277–300, 301–317, 319–337, 391–411
Crawfish, 51–71
Critical slowing down, 312
Crossed reflex, see Reflex
Crossed-uncrossed differences (CUD), 209–224
Crosstalk, 191, 220, 499–500, 576, 618

Deafference, see Deafferentation
Deafferentation, 13, 110, 210, 243–258, 348
Decerebration, 78, 109

Degrees of freedom, 2, 12, 301, 313–314, 346, 516–533, 537–538, 546, 551, 557–558, 561, 563, 565
Desynchronization, 305, 315, 491–513
Development, 19–20, 371–387, 391–408, 413–438, 439–460
    cephalo-caudal, 373, 386, 519
    proximo-distal, 373, 457, 519
Double hemisected spinal cord, see Spinal cord
Dual task, 498, 571–572
Dynamical systems perspective, 14–19, 277–300, 301–318, 319–321, 339–368, 392, 415, 440, 461–489
Dyspraxia, 180
Dynamics, 14–19

Eigenfrequency, 295, 297, 335, 397, 399
Electroencephalography, 128
Energy landscape, 281
Entrainment, 314–315, 397–398, 406, 584, see also Phase entrainment
Equilibrium point hypothesis, 342
Exitatory postsynaptic potential, see Postsynaptic potential
Extrapyramidal neuron, see Neuron

Fastigiospinal tract, see Medial descending motor tract group
Feedback, see Information feedback
Fitts' law, 344–345
Flexor reflex afferent, see Afferent
Floor reaction forces, 77
Force-coding receptors, 64–68
Forgetting, 486
Frequency locking, 281, 287–293

Gait patterns, 344, see also Locomotion
Galloping, 405–408, 427, see also Locomotion

Golgi tendon organ, 8, 172, 176
Grasping, 413–438, 439–460, 617–630
Gravitational forces, 429–432
Group I-II-III-IV afferent, *see* Afferent

Habit, 496
Handedness, 596–612, *see also* Left-hander, Right-hander
Hemifield, 211–212, 219
Hemiplegia, left-right, 236–238
Hemisection, *see* Spinal cord
Hemispheres, *see* Cerebral hemispheres
Heterolateral coordination, *see* Coordination
Hierarchical organization, 584–591
Homolateral coordination, *see* Coordination
H-reflex, *see* Reflex
H-wave, *see* Wave
Hysteresis, 305, 312

Ideomotor apraxia, 186
Individual differences, 508–509
Information feedback, 504
  knowledge of performance, 504–505
  knowledge of results, 504–505
Inhibition
  reciprocal, 270–271
  interhemispheric, 271–272
Inhibitory postsynaptic potential, *see* Postsynaptic potential
In-phase coordination, *see* Coordination
Intralimb coordination, *see* Coordination
Interference, 503, 573–574, 607, *see also* Crosstalk
Interhemispheric transfer, *see* Transfer
Interlimb dissociation–decoupling, 13, 23
Interlimb reflex, *see* Reflex

Intermanual conflict, 185, 187, 200
Intermittent dynamics, 315
Interstitiospinal tract, *see* Medial descending motor tract group
Ipsilateral reflex reversal, *see* Reflex

Kinesthetic afference, 232–241

Labyrinth, 149
Labyrinthine reflex, *see* Reflex
Lamprey, 76
Lateral descending motor tract group, 12, 31–47, 189, 220–221, 445, 457, 499
Laterality, 599, *see also* Handedness
Learning, 21–24, 189, 461–490, 491–510, 515–533
Left-handers, 599–603, 607–611
Limit cycle, 356, 362, 394–395, 397, 501, 542
Locomotion, 4–7, 49–73, 97–126, 391–408
  fictive, 106–107
Longitudinal myelotomy, *see* Spinal cord

Magnet effect, 3, 282–283, 289, 315
Maintenance tendency, 3, 282–283, 289
Manipulation, 448–456
Mass-spring, 349, 353
Medial descending motor tract group, 12, 31–47, 220–221, 445, 457, 499, 618
Medial frontal cortex, 185, 187, 202
Mirror movement, 185
Motion-dependent forces, 429–432
Motor cortex, 34, 36, 79, 179–207, 192–193
Motor equivalence, 181, 347, 478, 543
Motor program, 104, 181, 201, 499, 618
Motor time, *see* Reaction time
Moving room experiment, 383

Multifunctionality, 305, 309
Multiple axon collateral, 12
Multistability, 305, 309
Muscle stiffness, 385
Muscle strength,
  grip, 261–262
  lower limb, 262–264
  upper-limb, 260–261
M wave, *see* Wave

Neural crosstalk, *see* Crosstalk
Neuron
  alpha motoneuron, 130
  extrapyramidal, 141
  propriospinal, 141
  pyramidal, 141, 192
  reticulospinal, 171

Oligosegmental pathway, 134, 138
Optic flow, 157, 375, 377, 529
Order parameter, *see* Parameter
Otolith, 150, 162

P1–P2 response, 101–104, 113–114, 117
Parameter
  order, 15, 279–280, 301, 303, 307, 309, 320–321, 339, 342–343, 465–466, 487, 501
  control, 15, 279, 286, 309, 320, 465, 484, 501
Parietal lobe, 140
Pattern formation, 302–303, 462, 465
Perceptual invariants, 528–530
Perceptual-motor abilities, *see* Ability
Perceptual-motor workspace, 515–533
Phase, *see also* Relative phase
  attractive-circle map, 282
  dependent response, 6, 50, 65
  dependent reflex gating, 106
  diagram, 356
  entrainment, 288–289

  locking, 288–289, 406, 478
  slipping, 305
  space, 284
  transitions, 16, 18–19, 22, 303, 320, 331, 465, 469, 473, 481–482, 484–485, 501
  wandering, 305
Polyrhythm, *see* Rhythm
Ponto-cerebellar-cortical pathway, 189
Postsynaptic potential
  excitatory, 106–107
  inhibitory, 106–107
Postural control, 7, 167–178, 371–387, 398, 444, 537–567
Postural reflex, *see* Reflex
Postural sway, 383
Potential, 280, 393
Power spectrum, 293, *see also* Spectral power
Prehension, *see* Grasping
Premotor cortex, 182, 185, 187, 190–192, 202
Premotor time, *see* Reaction time
Primary motor cortex, 185, 190–192, 202
Progression-regression hypothesis, 529
Proprioception, 13, 59–68, 110, 210, 232–236, 241, 243–258, 348
Proprioceptive reflex, *see* Reflex
Propriospinal neuron, *see* Neuron
Propriospinal pathway, 85, 108, 201
Proximo-distal development, *see* Development
Putamen, 192
Pyramidal neuron, *see* neuron

Range effect, 494
Reaching, 413–438, 439–460
Reaction time, 210–224, 241, 266–267
  premotor time, 214–215
  motor time, 214
Reciprocal inhibition, 270, *see also* Inhibition

Reference frame
  egocentric, body-centered, 148
  allocentric, space-centered, 148
  geocentric, 148
Reflex, 4, 500
  crossed, 98–99, 104, 112, 270
  exteroceptive, 105–108
  H, 6, 128–146
  interlimb, 64–68, 75–96, 97–126
  labyrinth, 9
  postural, 172
  proprioceptive, 168, 174, 176
  short-latency, 99
  stretch, 142, 167, 241, 540
  tonic neck, 9, 444, 500
  vestibulospinal, 174
Regional cerebral blood flow, 630
Relative coordination, see Coordination
Relative phase, 15, 86–92, 277–299, 301–316, 321, 340, 363, 395–403, 407, 465–488, 530–531, 579, see also Phase
Relaxation time, 208, 328, 343
Retention, 527, 532
Reticulospinal neuron, see Neuron
Reticulospinal tract, see Medial descending motor tract group
Rhythm
  simple-poly, 23, 575–591
Right-hander, 601–606, 608–611
Rubrospinal tract, see Lateral descending motor tract group
Running, 405–408, see also Locomotion

Sacculus, 150
Self-organization, 302–303, 316, 392, 465
Sensorimotor map, 385
Sensory feedback, 59–68
Short-latency reflex, see Reflex
Short-term motor memory, 493
Sign language, 181
Spectral power, 292–294
Spinal cord, 31–47, 75–96, see also
  lateral hemisection, 5, 79–84

spinal transsection, 5, 84–85
longitudinal myelotomy, 86–88
Spinal-bulbo-spinal pathway, 111
Spinal transsection, see Spinal cord
Split-brain, 11, 185–187, 189, 209–227, 271, 573
Spontaneous arm movement, 420–422
Stimulus-response compatibility, 212
Stretch reflex, see Reflex
Supplementary motor area, 183–186, 190, 192–200, 457, 630, see also Medial frontal cortex
Symmetry breaking, 305, 311–312, 330
Synergy, 45, 383, 387, 629, see also Coordinative structure

Tectospinal tract, see Medial descending motor tract group
Thalamus, 79
Time scales, 327–329
Tonic neck reflex, see Reflex
Torque, 429–432, 506, 548
Transfer
  interhemispheric, 220, 222
  of learning, 463, 474–481, 496–498, 507–508, 527, 532, 581
Transsection, see Spinal Cord
Two-thirds law, 363–364

Utriculus, 150

Ventrolateral nucleus, see Thalamus
Vestibular system, 140, 149, 172, 375
Vestibulospinal reflex, see Reflex
Vestibulospinal tract, see Medial descending motor tract group

Wave,
  M, 131
  H, 131